Mushrooms with Therapeutic Potentials

Dinesh Chandra Agrawal •
Muralikrishnan Dhanasekaran
Editors

Mushrooms with Therapeutic Potentials

Recent Advances in Research and Development

Springer

Editors
Dinesh Chandra Agrawal
Department of Applied Chemistry
Chaoyang University of Technology
Taichung, Taiwan

Muralikrishnan Dhanasekaran
Department of Drug Discovery
and Development
Auburn University, Harrison School
of Pharmacy
Auburn, AL, USA

ISBN 978-981-19-9549-1 ISBN 978-981-19-9550-7 (eBook)
https://doi.org/10.1007/978-981-19-9550-7

This Springer imprint is published by the registered company Springer Nature Singapore Pte Ltd.
The registered company address is: 152 Beach Road, #21-01/04 Gateway East, Singapore 189721, Singapore

With profound gratitude, the editors dedicate this book to their beloved parents for their unwavering support and for creating a fulfilling life

Preface

Consumption of mushrooms has significantly increased worldwide due to their immense nutritional and therapeutic advantages. Now, mycotherapy (using mushroom-based products, extracts, and bioactive compounds to treat different illnesses) is a global consolidative healthcare need. The mushroom-based nutraceuticals possess potent pharmacodynamic properties that can block various toxicological pathways. Scientific studies over the last decades have validated evidence of their efficacy in multiple diseases. Extracts and bioactive compounds obtained from different mushrooms have been used for their prophylactic and therapeutic potentials. There are ongoing research efforts on various aspects of medicinal mushrooms in different parts of the world. In this volume, chapters (mostly review articles) on medicinal mushrooms have been included considering their importance in human health. Earnest effort has been devoted to presenting new and novel perceptions about ailments alleviating effects of medicinal mushrooms, including their curative role in various peripheral and central diseases. The book contains chapters contributed by eminent researchers working with different disciplines of medicinal mushrooms in different countries across the globe. The chapters in the book have been dedicated to providing the therapeutic efficacy of mushrooms against respiratory diseases, gut microbiota, COVID-19 infection, dementia, epilepsy, cancers, neurodegenerative and kidney-related diseases, and other common pathologies. This book not only extends our knowledge about medicinal mushrooms and confirms their great potential for developing new drugs but hopefully also inspires the readers to get involved in research on mushrooms.

The editors hope that this compendium of review articles will be helpful as a reference book for advanced students, researchers, academics, business houses, and all individuals concerned with medicinal mushrooms.

Taichung, Taiwan
Auburn, AL, USA
24 October 2022

Dinesh Chandra Agrawal
Muralikrishnan Dhanasekaran

Acknowledgments

The editors thank all the invited authors to this book for preparing their valuable manuscripts. Without their contributions, this book would not have been possible.

The co-editor, Professor Dhanasekaran, wishes to place special appreciation and thanks to Professor Agrawal for his untiring efforts in handling the entire correspondence with the Springer Nature team and authors and dealing with the major editing and revision process of manuscripts and managing them from start to finish.

Editor Professor Agrawal thanks Professor Tao-Ming Cheng, President of the Chaoyang University of Technology (CYUT); Professor Wen-Goang Yang, Vice-President, CYUT; Distinguished Professor Hsi-Hsien Yang, Dean, College of Science and Engineering, CYUT; Professor Wei-Jyun Chien, Chairperson, Department of Applied Chemistry, CYUT, Taiwan for their constant support and encouragement during the progress of the book.

Editor Professor Dhanasekaran thanks the administrators, faculty, and staff at Harrison School of Pharmacy, Auburn University, and all his beloved students for their relentless dedication and inspiration.

The editors sincerely thank the entire Springer Nature Singapore Pte. Ltd. team concerned with the publication of this book.

Editors thank and appreciate their respective families (Manju, Somya, Neha, and Mihir—Family Agrawal; Madhu and Rishi—Family Dhanasekaran) for the encouragement and wholehearted support during the progress of the book.

The editors express profound gratitude towards God for providing guidance, strength, and skill to accomplish the task of handling this book.

Contents

About the Editors

Dinesh Chandra Agrawal is working as a professor in the Department of Applied Chemistry, Chaoyang University of Technology (CYUT), Taiwan. Professor Agrawal has 40 years of research experience in plant biotechnology of diverse species, including medicinal plants and medicinal mushrooms. After serving for more than 31 years, in 2013, he superannuated as a chief scientist and professor of biological sciences at the CSIR-National Chemical Laboratory, Pune, the top-ranking institute in chemical sciences under the umbrella of the Council of Scientific and Industrial Research (CSIR), Ministry of Science and Technology, Govt. of India. While in CSIR-NCL, Prof. Agrawal was a coordinator and project leader of several research projects funded by the Govt. of India. He has more than 200 publications, including eight books (six by Springer Nature) to his credit on different aspects of plant biotechnology, including medicinal plants and medicinal mushrooms. About 35 M.Tech./M.Sc. and 7 Ph.D. students have completed their thesis work under his guidance.

Professor Agrawal has been bestowed several prestigious awards and fellowships, such as the Alexander von Humboldt Fellowship (Germany), DBT Overseas Associateship (USA), British Council Scholar (UK), European Research Fellow (UK), and INSA Visiting Scientist (India). During these fellowships, he had opportunities to work in the USA, Germany, and the UK. He had a research collaboration with UMR Vigne et Vins, INRA, Centre de Recherché Colmar, France.

Professor Agrawal has reviewed many research articles for several SCI journals on plant biotechnology and served as a member of the editorial board of *Medicinal and Aromatic Plant Abstracts*, NISCAIR, Govt. of India. He is on the editorial board of the *International Journal of Applied Science and Engineering* (Scopus), serving as associate editor-in-chief of the journal.

Muralikrishnan Dhanasekaran completed his Bachelor of Pharmacy from Annamalai University and Master of Pharmacy from Jadavpur University, West Bengal, India. He obtained his Ph.D. degree from the Indian Institute of Chemical Biology, Kolkata, India, under the guidance of Dr. K.P. Mohanakumar. He then received post-doctoral training from renowned scientist Dr. Manuchair Ebadi (Prof. University of North Dakota, Grand Forks, ND, and Dr. Bala Manyam (Scott & White Clinic / Texas A & M, Temple, TX). Dr. Dhanasekaran joined Auburn University in the year 2005 and currently working as a full Professor at Harrison School of Pharmacy, Auburn University, USA. Dr. Dhanasekaran's research and interest focus on pharmacology, neuroscience, toxicology, and dietary and natural bioactives. Dr. Dhanasekaran completed the New Investigator Research Grant from Alzheimer's Association, several Auburn University grants, and several other research projects from pharmaceutical industries. Currently, 4 graduate students, 24 undergraduate students, and 7 Pharm.D students are doing research under his supervision. He has trained 16 graduate students as a mentor and 40 graduate students by serving as committee members and mentored more than 60 undergraduate students in his lab. Dr. Dhanasekaran has received several teaching awards from foreign Universities and Auburn University for teaching Pharm D., graduate, and undergraduate students. He has published more than 450 scientific abstracts, 115 peer-reviewed publications, 4 Springer Nature books, and 51 book chapters.

Contributors

Dinesh Chandra Agrawal Department of Applied Chemistry, Chaoyang University of Technology, Taichung, Taiwan

Fada Alghenaim Department of Drug Discovery and Development, Harrison College of Pharmacy, Auburn, University, Auburn, AL, USA

Amza Ali Psyence Group, Toronto, ON, Canada

Susanna M. Badalyan Laboratory of Fungal Biology and Biotechnology, Institute of Biology, Yerevan State University, Yerevan, Armenia

Alexandru Stefan Barcan Faculty of Biotechnology, University of Agronomic Science and Veterinary Medicine, Bucharest, Romania

Anush Barkhudaryan Laboratory of Fungal Biology and Biotechnology, Institute of Biology, Yerevan State University, Yerevan, Armenia
Department of Cardiology, Clinic of General and Invasive Cardiology, University Clinical Hospital № 1, Yerevan State Medical University, Yerevan, Armenia

Han Ni Booi School of Pharmacy, University of Nottingham, Malaysia Campus, Semenyih, Selangor Darul Ehsan, Malaysia

Malvi Choudhary Fungal Biotechnology Laboratory, School of Biotechnology, University of Jammu, Jammu, Jammu and Kashmir, India

Jasmina Ćilerdžić Faculty of Biology, University of Belgrade, Belgrade, Serbia

Randall Clark Department of Drug Discovery and Development, Harrison College of Pharmacy, Auburn University, Auburn, AL, USA

Jack DeRuiter Department of Drug Discovery and Development, Harrison School of Pharmacy, Auburn University, Auburn, AL, USA

Muralikrishnan Dhanasekaran Department of Drug Discovery and Development, Harrison School of Pharmacy, Auburn University, Auburn, AL, USA

Manoj K. Dhar Genome Research Laboratory, School of Biotechnology, University of Jammu, Jammu, Jammu and Kashmir, India

Jahnavi Dhingra Department of Drug Discovery and Development, Harrison College of Pharmacy, Auburn University, Auburn, AL, USA

Laura Dorina Dinu Faculty of Biotechnology, University of Agronomic Science and Veterinary Medicine, Bucharest, Romania

Shin-Yee Fung Faculty of Medicine, Department of Molecular Medicine, Medicinal Mushroom Research Group (MMRG), University of Malaya, Kuala Lumpur, Malaysia

Milica Galić Faculty of Biology, University of Belgrade, Belgrade, Serbia

Mary-Elizabeth Gifford Psyence Group, Toronto, ON, Canada
Global Wellness Institute, Miami, FL, USA

Lorenzo Gordon Psyence Group, Toronto, ON, Canada
Caribbean School of Medical Sciences, Kingston, Jamaica

Justin Grant Psyence Group, Toronto, ON, Canada
Department of Pharmacology and Toxicology, Temerty Faculty of Medicine, University of Toronto, Toronto, ON, Canada

Seren Gündoğdu Faculty of Pharmacy, Department of Pharmacognosy, Hacettepe University, Sıhhiye, Ankara, Turkey

Boris Jakopovic Dr. Myko San–Health from Mushrooms Co., Zagreb, Croatia

Ivan Jakopovich Dr. Myko San–Health from Mushrooms Co., Zagreb, Croatia

Surekha Kadannagari Department of Drug Discovery and Development, Harrison College of Pharmacy, Auburn University, Auburn, AL, USA

Sanjana Kaul Fungal Biotechnology Laboratory, School of Biotechnology, University of Jammu, Jammu, Jammu and Kashmir, India

Shannon Kim Department of Drug Discovery and Development, Harrison School of Pharmacy, Auburn University, Auburn, AL, USA

Ayşe Kuruüzüm-Uz Department of Pharmacognosy, Faculty of Pharmacy, Hacettepe University, Ankara, Turkey

Mei Kee Lee School of Pharmacy, University of Nottingham, Semenyih, Selangor Darul Ehsan, Malaysia

Kuan-Hon Lim School of Pharmacy, University of Nottingham Malaysia, Jalan Broga, Selangor, Malaysia

Keyi Liu Department of Drug Discovery and Development, Harrison School of Pharmacy, Auburn University, Auburn, AL, USA

Henry Lowe Psyence Group, Toronto, ON, Canada
Biotech R & D Institute, University of the West Indies, Kingston, Jamaica

Graham Lynn Department of Drug Discovery and Development, Harrison College of Pharmacy, Auburn, University, Auburn, AL, USA

Nick Martin Department of Drug Discovery and Development, Harrison School of Pharmacy, Auburn University, Auburn, AL, USA

Phillip McCain Department of Drug Discovery and Development, Harrison School of Pharmacy, Auburn University, Auburn, AL, USA

Timothy Moore Department of Drug Discovery and Development, Harrison School of Pharmacy, Auburn University, Auburn, AL, USA

Sylvie Morel Laboratory of Botany, Phytochemistry and Mycology, Faculty of Pharmacy, CEFE, CNRS, Univ Montpellier, EPHE, IRD, Montpellier Cedex 5, France

Rishi M. Nadar Department of Drug Discovery and Development, Harrison School of Pharmacy, Auburn University, Auburn, AL, USA

Nada Oršolić Division of Animal Physiology, Faculty of Science, University of Zagreb, Zagreb, Croatia

Nadire Özenver Faculty of Pharmacy, Department of Pharmacognosy, Hacettepe University, Sıhhiye, Ankara, Turkey

Li-Yin Pang School of Pharmacy, University of Nottingham Malaysia, Jalan Broga, Selangor, Malaysia

Rachel Parise Department of Drug Discovery and Development, Harrison School of Pharmacy, Auburn University, Auburn, AL, USA

Suhrud Pathak Department of Drug Discovery and Development, Harrison School of Pharmacy, Auburn University, Auburn, AL, USA

Sandra Kraljević Pavelić Faculty of Health Studies, University of Rijeka, Rijeka, Croatia

Diana-Roxana Pelinescu Department of Genetics, Faculty of Biology, University of Bucharest, Bucharest, Romania

Satyanarayana Pondugula Department of Anatomy, Physiology and Pharmacology, Auburn University, Auburn, AL, USA

Sindhu Ramesh Department of Drug Discovery and Development, Harrison School of Pharmacy, Auburn University, Auburn, AL, USA

Sylvie Rapior Laboratory of Botany, Phytochemistry and Mycology, Faculty of Pharmacy, CEFE, CNRS, Univ Montpellier, EPHE, IRD, Montpellier Cedex 5, France

Itika Sharma Fungal Biotechnology Laboratory, School of Biotechnology, University of Jammu, Jammu, Jammu and Kashmir, India

Didem Şöhretoğlu Department of Pharmacognosy, Faculty of Pharmacy, Hacettepe University, Ankara, Turkey

Mirjana Stajić Faculty of Biology, University of Belgrade, Belgrade, Serbia

Ileana Stoica Department of Genetics, Faculty of Biology, University of Bucharest, Bucharest, Romania

Nuriye Hilal Taştekin Department of Pharmacognosy, Faculty of Pharmacy, Hacettepe University, Sıhhiye, Ankara, Turkey

Sue-Mian Then School of Pharmacy, University of Nottingham Malaysia, Jalan Broga, Selangor, Malaysia

Kang-Nee Ting School of Pharmacy, University of Nottingham, Malaysia Campus, Semenyih, Selangor Darul Ehsan, Malaysia

Vlad Vălu Department of Natural Sciences, Faculty of Science, University of Pitesti, Pitesti, Romania

Emanuel Vamanu Faculty of Biotechnology, University of Agronomic Science and Veterinary Medicine, Bucharest, Romania

Mushrooms as Promising Therapeutic Resources: Review and Future Perspectives

Susanna M. Badalyan, Sylvie Morel, Anush Barkhudaryan, and Sylvie Rapior

Contents

Abstract Several human diseases, such as diabetes, cancer, cardiovascular and neurodegenerative disorders, increasingly affect the adult population worldwide. Therefore, scientists have tried to discover new natural sources of medicines,

S. M. Badalyan (✉) · A. Barkhudaryan
Laboratory of Fungal Biology and Biotechnology, Institute of Biology, Yerevan State University, Yerevan, Armenia
e-mail: s.badalyan@ysu.am; anush.barkhudaryan@ysu.am

S. Morel · S. Rapior
Laboratory of Botany, Phytochemistry and Mycology, Faculty of Pharmacy, CEFE, CNRS, Univ Montpellier, EPHE, IRD, Montpellier, France
e-mail: sylvie.morel@umontpellier.fr; sylvie.rapior@umontpellier.fr

D. C. Agrawal, M. Dhanasekaran (eds.), *Mushrooms with Therapeutic Potentials*,
https://doi.org/10.1007/978-981-19-9550-7_1

especially from mushrooms, to prevent and treat these diseases. Wild mushrooms and mushrooms growing on solid media or in submerged cultures belong to a large number of genera (*Agaricus*, *Auricularia*, *Ganoderma*, *Grifola*, *Hericium*, *Lentinula*, *Schizophyllum*) and may be used to produce biologically active compounds (lectins, polysaccharides, phenolics, terpenoids, and steroid derivatives) as anti-inflammatory, antimicrobial, antioxidant, antitumor, antiviral, hepatoprotective, hypocholesterolemic, hypoglycemic, immunomodulatory, and neuroprotective agents. Metabolomics and genomic studies of the unexplored biotechnological potential of mushrooms may also assist in the production of mushroom-derived biotech products. High-quality, long-term, randomized, double-blind, placebo-controlled clinical studies have been described as necessary to prove the efficacy of mushroom extracts or isolated compounds. The present review discusses the current state of knowledge and the main findings of previous studies on mycotherapeutics and healthy mycofood. This chapter is an update contribution to modern mycopharmacology and biomedicine.

Keywords Biomedicine · Clinical trials · Mushrooms · Mycopharmaceuticals · Mycotherapy · Nutraceuticals

Abbreviations

AAM	*Auricularia auricula* melanin
ABM	*Agaricus blazei* Murill (ABM)
ABTS	2,2'-Azino-bis(3-ethylbenzothiazoline-6-sulfonic acid
AE	Aqueous extract
AgNPs	Silver Nanoparticles
ALL	Acute lymphoblastic leukemia
AOA	Antioxidant activity
BDNF	Brain-derived neurotrophic factor
BGE	β-D-glucan-enriched
BW	Body weight
CGC	Chitin-glucan complex
CHS	Contact hypersensitivity
cPLA2	Phospholipase A2
CREB	cAMP response element-binding protein
DC	Dendritic cell
DMBA	7,12-Dimethylbenz(a)anthracene
DNFB	Dinitrofluorobenzene
DPPH	2,2-Diphenyl-1-picrylhydrazyl
DW	Dry weight
EAAs	Essential amino acids
EC_{50}	Half maximal effective concentration
EPS	Exopolysaccharide

FABPs	Fatty acid-binding proteins
FDA	Food and Drug Administration
GAL-Am	α-D-Galactan from *Amanita muscaria*
GAs	Ganoderic acids
GATA4	GATA binding protein 4
GC-MS	Gas chromatography - mass spectroscopy
GDH	Glutamate dehydrogenase
GLC-Am	β-D-Glucan from *Amanita muscaria*
GLF	*Ganoderma lucidum* fruiting body
GLS	*Ganoderma lucidum* spores
GLSF	*Ganoderma lucidum* spores and fruiting bodies
βgPp	β-Glucan-rich *Pleurotus pulmonarius*
HCC	Hepatocellular carcinoma
HDL	High density lipoprotein
HFD	High-fat diet
HMM	High molecular mass
HUVEC	Human umbilical vein endothelial cells
IC_{50}	Half-maximal inhibitory concentration
IL-2	Interleukin-2
INF-γ	Interferon-γ
IOP	*Inonotus obliquus* polysaccharides
IPS	Intracellular polysaccharide
LC-MS	Liquid chromatography-mass spectrometry
LDL	Low-density lipoprotein
LDLR	Low-density lipoprotein receptor
LP-7A	Latcripin-7A
LPS	Lipopolysaccharide
LX2	Human hepatic stellate cells
MDR	Multidrug-resistant
ME	Methanolic extract
MIC	Minimum inhibitory concentration
MMP	*Macrolepiota procera* mycelium polysaccharides
M^{pro}	Main protease
MS	Multiple sclerosis
NAFLD	Non-alcoholic fatty liver disease
NASH	Non-alcoholic steatohepatitis
NDD	Neurodegenerative diseases
NLRP3	Nod-like receptor protein 3
NO	Nitric oxide
NP	Nonylphenol
NPE	Neuroprotective effect
OPG	Osteoprotegerin
OSi	Oxidative stress index
oxLDL	Oxidized low density lipoprotein

PC3	Prostate cancer cell line
PF-L	Lectin purified from *Pleurotus flabellatus*
PoPE	*Pleurotus ostreatus* polar extract
RANkL	Receptor activator of nuclear factor kappa B ligand
ROS	Reactive oxygen species
S1PR1	Sphingosine-1-phosphate receptor 1
SARS-CoV-2	Severe Acute Respiratory Syndrome Coronavirus 2
SCFA	Short-chain fatty acids
TAS	Total antioxidant status
TCM	Traditional Chinese medicine
Thel	Thelephoric acid
TLR4	Toll-like receptor 4
TNFα	Tumor necrosis factor-α
TOS	Total oxidant status
TrkB	Tropomyosin receptor kinase B
TST	Tail suspension test
WHO	World Health Organization

1 Introduction

Global food production faces many challenges, including environmental impacts, climate change, water crisis, land degradation, and cultural and socioeconomic changes (Thornton et al. 2018; Pérez-Moreno et al. 2021; Vogt-Schilb et al. 2022). These challenges require research into non-traditional natural sources of human food and medicine, including plants and mushrooms collectively referred to as macrofungi (Chang and Wasser 2012; Badalyan et al. 2019; Elkhateeb et al. 2021b; Khatua and Acharya 2021; Li et al. 2021a).

Mushrooms taxonomically belong to phyla Basidiomycota (class Agaricomycetes) and Ascomycota (class Pezizomycetes) of the subkingdom Dikarya. Of the estimated number of fungal species, 0.5-(1.5)-(5.1) million, about 140,000–160,000 species are mushrooms, from which around 10% (14,000–16,000) have been taxonomically identified (Hawskworth 2012; Hibbet and Taylor 2013; Peay et al. 2016). More than 54,000 basidiomycete species will be discovered by 2030, showing a huge gap between the described and unknown mushroom diversity (He et al. 2022).

Out of about 7000 known species of mushrooms possessing varying degrees of edibility, more than 3000 species in 231 genera have been considered first-order edible fungi. About 3% of known species (at least 170, currently about 500) are poisonous, while about 700 species of 2000 known as safe mushrooms have medicinal properties (Wasser 2011, 2014; Chang and Wasser 2012; Badalyan et al. 2019; Niego et al. 2021; Gopal et al. 2022).

Edible mushrooms belonging to different ecological groups possess nutraceutical potential due to their low-fat content, unsaturated fatty acids, and high amount of dietary fiber. They are rich in proteins, vitamins, and minerals and are considered promising next-generation healthy food due to their nutritional value, attractive gourmet taste, and aroma (Chang and Wasser 2012; Badalyan and Zambonelli 2019, 2023; Włodarczyk et al. 2022).

The contents of vitamins A, C, D_2, and E (α-tocopherol), macroelements, and trace elements of hymenochaetoid wild mushrooms *Fuscoporia torulosa*, *Inonotus pachyphloeus*, *Phellinus allardii*, *Ph. fastuosus*, *Ph. gilvus*, and *Ph. sanfordii* collected from India have been recently reported (Azeem et al. 2022). This study showed that *Ph. gilvus* contained the highest number of the chemical elements. The mushrooms were rich in microelements, including Ca (80–2610 mg/kg), Cl (39.63–240 mg/kg), K (246.7–2620 mg/kg), Mg (96.6–500 mg/kg), Na (9.56–56 mg/kg), P (39.5–126.7 mg/kg), and S (69.37–170 mg/kg) in terms of dry weight (DW). Many trace elements (Co, Cr, Cu, Fe, Mn, Mo, Ni, Si, V, and Zn) and some non-essential elements (Al, Ba, Br, Rb, Sr, Ti, and Zr) were also detected in the tested species. The vitamins C (9.32 mg/100 g DW) and D_2 (1.55 mg/100 g DW) were mostly found in *F. torulosa*, while the lowest amounts were detected in *Ph. fastuosus* and *Ph. allardii*. These results will be beneficial in the formulation of mushroom-derived nutraceutical and pharmaceutical products (Azeem et al. 2022).

Medicinal mushrooms synthesize different bioactive compounds (alkaloids, lactones, polysaccharides, polyphenolics, sterols, triterpenes, terpenoids, eritadenine, chitosan, fatty acids, etc.), which are beneficial for human health (Badalyan 2012, 2016; Badalyan et al. 2019; Badalyan and Rapior 2021a, b). Fungal polysaccharides, *β*-glucans, and polysaccharide-protein complexes possess therapeutic properties, such as antiviral, antioxidant, antitumor, anti-obesity, hypocholesterolemic, hepatoprotective, immunomodulatory, anti-aging, and others. The bioactive ingredients may be extracted from edible and medicinal mushrooms and incorporated into health-enhancing functional food products and supplements to prevent or treat several human diseases (Badalyan and Zambonelli 2019, 2023; Badalyan et al. 2019; Badalyan and Rapior 2021a, b; Haq et al. 2022).

Since ancient times, several edible agaricomycete and ascomycete mushrooms, belonging to the genera *Agaricus*, *Boletus*, *Flammulina*, *Lyophyllum*, *Lentinula*, *Morchella*, *Pleurotus*, *Tuber*, and others with worldwide distribution, have been reported as a valuable food with healing properties (Badalyan 2012; Badalyan et al. 2019; Kües and Badalyan 2017; Badalyan and Rapior 2021a, b; Li et al. 2021a).

Ganoderma species have been used in traditional Chinese medicine (TCM) for more than six millennia for strengthening the immune system, treatment of hypertension, arthritis, bronchial asthma, anorexia, gastritis, hemorrhoids, hepatitis, vascular problems, cancer, and other disorders (Wasser 2017; Badalyan et al. 2019; Xu et al. 2021). Polysaccharides and triterpenes are the main bioactive ingredients isolated from *Ganoderma* species, particularly from *Ganoderma lucidum* (Kües and Badalyan 2017; Badalyan et al. 2019; Badalyan and Rapior 2020). These compounds have shown positive synergetic antitumor effect when used in combination with radiotherapy and chemotherapy. The analysis of complex molecules derived from *G. lucidum* for future cancer therapy has been recently reported (Xu et al. 2021).

The well-known medicinal polyporoid mushroom *Fomitopsis officinalis* has been used by humans as medicine for over 5000 years. The chemical structure of therapeutically active compounds (triterpenoids, polysaccharides, organic acids, coumarins, and phenolics) with anti-inflammatory, cytotoxic, and antimicrobial effects from fruiting bodies of this polypore (Muszyńska et al. 2020). Agaricoid species *Auricularia auricula-judae* possess anti-inflammatory, antioxidant, and other health-enhancing effects due to substantial amounts of phenolics, flavonoids, carbohydrates, β-glucans, melanin derivatives, 5'-nucleotides, vitamin D_2, ergosterol, and ergothioneine. The available data may provide a basis for formulation of different mushroom-derived healthy biotech products from this fungus (Islam et al. 2022).

The Ascomycetes species from the genus *Cordyceps* (Ophiocordycipitaceae, Hypocreales) has been used in Asian countries for longevity and healthy life. The fruiting bodies of these entomophagous fungi develop and erupt from the head of larva and adult stages of other insect species. Numerous *Cordyceps* species were described in *Materia Medica* and used in TCM for more than 2000 years. Among these species, *Cordyceps sinensis* (syn. *Ophiocordyceps sinensis*), with various hosts, including *Hepialus armoricanus,* is the most widely used fungus. *Cordyceps militaris,* the orange caterpillar fungus, has a similar chemical composition and biological activities to *C. sinensis.* Although many *in vitro* and *in vivo* studies of *Cordyceps* species have been performed, it is still debatable whether they are healthy food supplements or therapeutics. Nonetheless, the *Cordyceps* industry has significantly advanced and offers thousands of products commonly available worldwide (Elkhateeb and Daba 2022b; Mapook et al. 2022). Nowadays, the major *Cordyceps*-based companies are ALOHA MEDICINALS (USA, https://www.alohamedicinals.com); DOCTORS BEST (USA, https://www.drbvitamins.com); HOSTDEFENSE MUSHROOMS (USA https://hostdefense.com); HERBSENS (China, https://www.herbsens.com); RealHerbs (USA http://www.irealherbs.com); TEREZIA (Czech Republic, https://www.terezia.eu); ZEINPHARMA (Germany, https://www.zeinpharma.com); THE REALLY HEALTHY (UK, https://www.healthy.co.uk) and others.

Thus, mushrooms possess unexplored potential to develop different mycoproducts, such as pharmaceuticals, nutraceuticals, nutriceuticals, and cosmeceuticals (Badalyan et al. 2019, 2022; Badalyan and Rapior 2021a, b; Elkhateeb and Daba 2022a). The sustainable approach for using nutritional values and health benefits of bioactive metabolites of edible and medicinal mushrooms will assist in their further biomedical and biotechnological applications (Badalyan 2012; Kües and Badalyan 2017; Badalyan et al. 2019; Badalyan and Rapior 2021a, b; El-Ramady et al. 2022; Badalyan et al. 2022). However, extensive pre-clinical and clinical studies are warranted to explore the economic sustainability of different mushroom species and evaluate their potential applications in biomedicine. New nano-technological approaches, such as nano-biofortification, nano-emulsion techniques, and usage of silver nanoparticles (AgNPs), are needed to supply edible

Fig. 1 (**a**–**f**) The fruiting bodies of wild-growing medicinal mushrooms used as therapeutic agents: (**a**) *Ophiocordyceps (=Cordyceps) sinensis* (Courtesy of Rioult JP), (**b**) *Ganoderma lucidum* (Courtesy of Rioult JP), (**c**) *Trametes versicolor* (Courtesy of Nespoulous D), (**d**) *Lentinula edodes* (Courtesy of Nebel RE), (**e**) *Pleurotus ostreatus* (Courtesy of Nespoulous D), (**f**) *Hericium erinaceus* (Courtesy of Bessière JM)

mushrooms with essential nutrients and increase their bioactive ingredients (Badalyan et al. 2019; Kaplan et al. 2021; Chandrawanshi et al. 2022; El-Ramady et al. 2022; Mishra and Kaladhar 2022). Further metabolomic and genomic studies of unexplored biotechnological potential of mushrooms may assist in the production of novel mushroom-derived biotech products for human welfare (Chaturvedi et al. 2018, 2019; Badalyan and Zambonelli 2019, 2023; Bolaniran et al. 2021; Kaplan et al. 2022).

The advancements in the study of mushrooms as promising therapeutic resources and perspectives for their biotechnological and biomedical application are discussed in this Chapter. Selected wild-growing fruiting bodies of medicinal mushrooms used as therapeutic agents are shown in Fig. 1a–f.

2 Mushroom-Derived Biomolecules and Mycopharmacology

The mushrooms-derived biomolecules mainly belong to polysaccharides, polyphenolics, sterols, terpenoids, proteins, and alkaloids which can be a source for developing novel health-enhancing formulations used in mycotherapy (Da Silva de Souza et al. 2017; Badalyan et al. 2019; Badalyan and Rapior 2021a, b; Girometta 2019; Elsayed et al. 2021; Das et al. 2022b; Elkhateeb and Daba 2022a; Kuo et al. 2022; López-Hortas et al. 2022; Meade et al. 2022) (Table 1). Identification of bioactive molecules is necessary to assess their pharmaceutical potential, safety, and efficacy.

Table 1 Therapeutic potential of mushroom-derived bioactive molecules

Species	Bioactive compound(s)	Therapeutic effect(s)	References
Amanita muscaria	α-D-Galactan (GAL-Am), *β*-D-Glucan (GLC-Am)	Antitumor	Zavadinack et al. (2021)
Amylosporus cf. *graminicola* and *Amylosporus* cf. *campbelii*	Colletochlorin B -amylosporanes derivatives	Antibacterial, cytotoxic	Kemkuignou et al. (2022); Hamad et al. (2022)
Astraeus hygrometricus	Astrakurkurone	Cytotoxic	Dasgupta et al. (2019)
Auricularia auricula	Melanin	Antioxidant, hypolipidemic, hepatoprotective	Chen et al. (2014); Hou et al. (2021)
Calocybe gambosa	α-Tocopherol, organic and fatty acids	Antioxidant with moderate antimicrobial potential	Petrović et al. (2022)
Cordyceps sinensis	Cordymin, Cordycepin, adenosine, *β*-(1-3)-D-glucan, ergosterol, mannitol	Hypoglycemic, antitumor, anti-inflammatory, antioxidant, antimicrobial	Elkhateeb and Daba (2022b)
Cordyceps militaris	Cordycepin, mannitol	Anti-inflammatory, hypolipidemic, antitumor	Chen et al. (2014); Phull et al. (2022); Elkhateeb and Daba (2022b)
Cortinarius sp.	Polyketides rufoolivacin E viocristin B	Antioxidant	Song et al. (2022a)
Fomitopsis (=*Laricifomes*) *officinalis*	Chlorinated coumarins from mycelia and lanostane triterpenoids from basidiomes	Antiviral, antibacterial, trypanocidal	Girometta (2019)
Fomitopsis palustris	Polyporenic acid B, palustrisolides A, C, and G	Cytotoxic	Zhao et al. (2018)
Fomitopsis pinicola	Lanostane triterpenoids, pinicopsic acid F, 16-α-hydroxy-3-oxolanosta-7,9(11),24-trien-21-oic acid	Anti-inflammatory	Liu et al. (2022)
Favolaschia calocera	Ergostane triterpenoid favolon, favolon C, norergostane laschiatrion	Antifungal	Palasarn et al. (2022)
Ganoderma lucidum	Triterpenes and aromatic meroterpenoids	Antioxidant, neuroprotective	Badalyan et al. (2019); Badalyan and Rapior (2021a, b); Wang et al. (2019)
Ganoderma resinaceum	Branched mannogalactan [O-2-*β*-D-mannosyl-(1→6)-*α*-D-galactan] and highly branched *β*-D-glucan (1→3)(1→4)(1→6)-*β*-D-glucan	Antitumor, immunomodulatory	Badalyan et al. (2019); Badalyan and Rapior (2021b); Bleha et al. (2022);

(continued)

Table 1 (continued)

Species	Bioactive compound(s)	Therapeutic effect(s)	References
Ganoderma tsugae	Tsugaric acid F, palmitamide	Cytotoxic	Lin et al. (2016)
Hericium erinaceus	Terpenoid erinacine A	Neuroprotective, antioxidant, antitumor, anti-aging, hypolipidemic, gastroprotective	Badalyan and Rapior (2021b); Kuo et al. (2022)
Inonotus nidus-pici	Citropremide, 3,4-dihydroxybenzalacetone, lanosterol, ergost-6,8,22-trien-3β-ol, and ergosterol peroxide	Antimicrobial, antioxidant, anti-proliferative	Garádi et al. (2021)
Inonotus obliquus	Inotodiol, trametenolic acids, polysaccharides IOP	Cytotoxic, hepatoprotective	Zhang et al. (2018b); Li et al. (2019); Khoroshutin et al. (2021); Lee et al. (2021)
Isaria sinclairii	Amino acid myriocin	Immuno-suppressive	Ayzenberg et al. (2016); Chiba (2020); Cuello-Oderiz and McGraw (2022)
Lentinula edodes	Peptide latcripin 7A (LP-7A)	Antitumor, proapoptic	Din et al. (2020)
Neonothopanus nimbi	Phenolic scaffolds neonambiterphenyls-A	Antiviral, anti-COVID-19, anti SARS-CoV-2	Sen et al. (2022)
Ophiocordyceps gracilis (anamorph *Paraisaria dubia*)	Polysaccharides	Antioxidant	Tong et al. (2022)
Phellinus orientoasiaticus	Cyclohumulanoids of illudane-, sterpurane-, and tremulane-type scaffolds	Anti-inflammatory	Pham et al. (2022)
Piptoporus betulinus	Piptolinic acid A	Cytotoxic	Tohtahon et al. (2017)
Pleurotus flabellatus	Lectin PFL-L	Antibacterial	Murugesan and Gunasagaran (2021)
Polyozellus multiplex	Phenolics and flavonoids	Antiviral (anti-COVID-19)	Sen et al. (2022)
Polyozellus multiplex	Terphenylquinone pigment thelephoric acid (Thel)	Enhance alkaline phosphatase activity and bone matrix mineralization in pre-osteoblasts, cell adhesion, and migration in osteoblast differentiation	Park et al. (2022)

(continued)

Table 1 (continued)

Species	Bioactive compound(s)	Therapeutic effect(s)	References
Taiwanofungus camphoratus	Triterpenoids, phytosterols, antcin A, B, C, H, and K, dehydroeburicoic and dehydrosulphurenic acids	Hepatoprotective inducing hepatocellular carcinoma apoptosis, antitumor	Hsieh et al. (2011); Huang et al. (2012); Lu et al. (2020); Du et al. (2012); Wang et al. (2022a)
Trametes versicolor	Protein musarin	Anti-proliferative, cytotoxic	He et al. (2021b)
Tricholoma pardinum	Pardinols A–H, saponaceol B	Anti-inflammatory, cytotoxic	Zhang et al. (2018a, b)
Wolfiporia cocos (=*Poria cocos*)	Poricoic acid GM, lanostane triterpenoids	Anti-inflammatory, antioxidant, immunomodulatory	Bao et al. (2022)
Xylaria sp.	Eremophilanes, xylariones A1–B2 (2,5-diarylcyclopentenones), xylaripyone A-G (*α*-pyrone derivatives), xylaripyone H (γ-pyrone, derivative), diketopiperazine cyclo-(L-Leu-N-ethyl-L-Glu)	Antitumor, antioxidant anti-proliferative, anti-inflammatory antibacterial, anti-obesity, antiviral	Yuyama et al. (2017); Song et al. (2022b)
Xylodon flaviporus	Drimane-type sesquiterpene	–	Pham et al. (2022)

Evaluation of bioactive metabolites is also required to avoid the risk of post-market withdrawal of different bioproducts, including mycoproducts.

2.1 *Polysaccharides*

Despite advancements in cancer treatment, including chemotherapy, radiotherapy, hormonal therapy, and surgery, researchers are pursuing novel bioactive compounds to treat malignancy to avoid adverse side effects. Fungal polysaccharides have shown a huge therapeutic potential in this field of clinical research (Jin et al. 2012; Aras et al. 2018; Dasgupta et al. 2019; Daba et al. 2020; Din et al. 2020; He et al. 2021a, b; Xu et al. 2021; Chandrawanshi et al. 2022; Mishra and Kaladhar 2022).

Polysaccharides are classified according to their structure (linear and branched), sugar composition (homo- and hetero-polysaccharides), and type of bonds between monomers, such as β-(1→3), β-(1→6), or α-(1→3), etc. Fungal polysaccharides mainly belong to β-glucans with a backbone of glucose residues linked by β-(1→3)-glycosidic bonds and an attached β-(1→6) branch. The antitumor activities of fungal polysaccharides mainly depend on their molecular mass, degree of branching, conformation, and structure modification. The polysaccharides isolated particularly from medicinal mushrooms *G. lucidum* (Reishi or Lingzhi), *Grifola frondosa* (Maitake), *Trametes versicolor,* and *Schizophyllum commune* exert their antitumor effect through their immunomodulatory activity (Ferreira et al. 2015; Wasser 2017; Badalyan et al. 2019). Higher molecular weight and low degree of branching of polysaccharides increase immunomodulatory effect of these molecules. The role of primary structures (sidechain length, branching) and conformations (single helix, triple helix, or random coil) is important for this activity and remain to be clarified (Han et al. 2020).

Polysaccharides isolated from the aqueous extracts (AE) of *Ganoderma resinaceum* contain two main polysaccharides: mannogalactan, a branched *O*-2-*β*-D-mannosyl-(1→6)-*α*-D-galactan and a highly branched (1→3)(1→4)(1→6)-*β*-D-glucan. Mannogalactan predominated in cold-water extract, whereas *β*-D-glucan was the main product of hot-water extract. Three sub-fractions from the hot-water soluble fraction contained branched *β*-D-glucans. The alkaline extract contained a linear (1→3)-*α*-D-glucan and a weakly branched (1→3)-*β*-D-glucan having terminal *β*-D-glucosyl residues attached to *O*-6 of the backbone. Following all extractions, the insoluble part of extract was identified as a polysaccharide complex containing chitin and *β*-D-glucans (Bleha et al. 2022).

The antitumor potential of *α*-D-galactan (GAL-Am) and *β*-D-glucan (GLC-Am) chemically characterized and purified from fruiting bodies of *Amanita muscaria* have been studied using B16-F10 murine melanoma and non-tumorigenic fibroblast BALB/3T3 clone A31 cell line. Both polysaccharides showed a selective reduction in proliferation against melanoma cells and did not affect non-tumorigenic fibroblast cells. GAL-Am or GLC-Am did not modulate the adhesion of B16-F10 cells. They selectively reduced melanoma cell viability and proliferation, suggesting further investigation to evaluate their anti-melanoma properties (Zavadinack et al. 2021).

The administration of polysaccharides extracted from edible medicinal oyster mushroom *Pleurotus ostreatus* showed anti-cancer and anti-ascitic effects in *in vivo* and *in vitro* experiments. They significantly decreased the tumor cell metastasis and increased survival in mouse models of H22 malignant ascites. The downregulation of regenerative genes Foxp3 and Stat3 and secretion of immunological factors IL-2, TNF-α, and INF-γ were observed after treatment with the extracted and partially purified polysaccharide derived from *P. ostreatus*. Consistent with tumor suppression hypothesis *in vivo*, polysaccharides decrease invasion and migration capabilities and induce the chain of gene regulation processes leading to apoptosis in the hepatocellular carcinoma (HCC) cell line (Khinsar et al. 2021).

It has been shown that sulfated *P. ostreatus*-derived polysaccharides revealed a stronger anticoagulant activity *in vitro* than in other conditions in the intrinsic pathway, which indicate that they significantly improved the plasma clot formation inhibitory activity by intrinsic and extrinsic pathways compared to native polysaccharides. Moreover, the cytotoxicity of polysaccharides against two normal cell lines was reduced. These findings suggest that *P. ostreatus*-derived polysaccharides may be an alternative to anticoagulant therapy (Rizkyana et al. 2022).

The method for enhancing intracellular polysaccharide (IPS) production by the anamorph (*Paraisaria dubia*) of *Ophiocordyceps gracilis* has been developed. The highest IPS and exopolysaccharides (EPS) yield in a 5 L bioreactor reached 83.23 ± 1.4 mg/mL and 518.50 ± 4.1 mg/L, respectively. Both IPS and EPS showed high antioxidant activity (AOA) (Tong et al. 2022).

The effects of polysaccharides isolated from medicinal mushrooms, *Lentinus edodes* and *Agaricus blazei* Murill (ABM), on the gelation process of the Pluronic® F127 copolymer have been evaluated. Their structural characterization revealed a *β*-glucan from *L. edodes* and a proteoglycan complex from ABM. Both possess rheological properties to develop pharmaceutical formulations with AOA and low cytotoxicity against human neutrophils (Menezes et al. 2022).

The dietary supplements obtained from mycelial extracts of *Lentinus* (*Pleurotus*) *sajor-caju* may improve their beneficial properties by increasing the amount of proteins or polysaccharides about 3.5- and 4.5-fold, respectively. *In vitro* antioxidant and anticancer activities against human normal colon epithelial cells (CCD 841 CoTr) and colorectal cancer cells (HT-29, SW948, and LS 180) have been reported. The viability of cancer cells was reduced along with pro-apoptotic and NO (nitric oxide)-secreting effects of their extracts. However, the mechanisms for this activity remain to be elucidated. This study suggests that polysaccharides have multiple nutritional and anticancer properties and may be used as a source of therapeutic biomolecules or functional foods (Zając et al. 2021).

S. commune is a source of chitin-glucan complex (CGC) with biodegradable, biocompatible, antioxidant, and antibacterial properties. It is also considered a powerful agent in wound-healing therapy. The optimization of CGC nanofibers by electrospinning showed that CGC/PVA/gelatin nanofibers inhibited the growth of *Escherichia coli* and *Staphylococcus aureus* by 25% and 78% after 24 h, respectively. These nanofibers are non-toxic to fibroblast cells and improve their proliferation and adhesion. *S. commune* CGC nanofibers showed 86% wound-healing effect in Wistar rats (Zeynali et al. 2022).

In vitro antioxidant and immunomodulatory potential of a crude polysaccharide obtained from *Termitomyces medius* has been previously reported using cell proliferation, stimulation of phagocytosis, NO release capacity, and reactive oxygen species (ROS) production in murine macrophage cell line RAW 264.7. However, *in vivo* trials are required for the development of pharmaceutical applications and to study medicinal properties of these polysaccharides (Mitra et al. 2021).

2.2 Terpenoids, Steroids, Sterols, and Lipids

Since ancient times, humans have been interested in natural products derived from plants and fungi. Mushrooms are considered sources of different terpenoids, steroids, sterols, and lipids with antioxidant, antibacterial, antimitotic, neuroprotective, anti-inflammatory, cytotoxic, immunomodulatory, and other pharmacological effects (Corrěa et al. 2017; Wasser 2017; Morel et al. 2018, 2021; Badalyan et al.

2019; Badalyan and Rapior 2021a, b; Wang et al. 2019; Akiba et al. 2020; Diallo et al. 2020, 2021).

However, mushrooms, as a source of bioactive terpenoids, have not been sufficiently investigated; mushroom-derived terpenoids and their therapeutic properties have been used in pre- and post-clinical trials (Badalyan et al. 2019; Dasgupta and Acharya 2019).

Ganoderma species possess two main types of biomolecules: polysaccharides with antioxidant and immunomodulatory activities and lanostane-type triterpenoids, such as ganoderic acids (GAs) with antimicrobial, anticancer, antiviral, and immunomodulatory effects. Triterpenoids constitute more than 0.5% of *Ganoderma* fruiting bodies on a DW basis. At least 561 triterpenoids have been isolated from *Ganoderma* mushrooms, particularly from *G. lucidum, G. lingzhi, G. sinense*, and *G. leucocontextum*. The *Ganoderma* triterpenoids are structurally different. Among these GAs, ganoderiols, lucidones, lucidenic acids, and ganolucidic acids are considered the most important for their bioactivity (Kües and Badalyan 2017; Angulo-Sanchez et al. 2022).

GAs are highly oxygenated triterpenoids with different functional groups attached to the lanostane skeleton, allowing new drugs to be developed to treat multiple illnesses, including cancer. Cytotoxic effects of GAs have been associated with inhibitory activity of specific targets, such as STAT3, to induce apoptosis and increase NK-cell activity. Due to the bioactivity of *Ganoderma* terpenoids, novel strategies are developing for their synthesis. The current knowledge on *Ganoderma* triterpenoids and their production, biosynthesis, pharmacological properties, gene expression in liquid culture, and their pharmacological potential has been reported (Angulo-Sanchez et al. 2022). According to the authors, the investigation of mechanisms of action of GAs is still warranted despite extensive information about their bioactivities, particularly in cancer, as well as their pharmacokinetic and toxicological properties to advance their application in clinical models. Moreover, it is necessary to characterize different phases of the life cycle (primary and secondary mycelia and fruiting bodies) of strains since biotic and abiotic ecological factors, combined with genetic intraspecies variability, play a significant role in the synthesis of bioactive compounds.

Around 431 secondary metabolites, 380 terpenoids, 30 steroids, and other bioactive compounds have been isolated from 22 *Ganoderma* species (Baby et al. 2015). The structure-activity relationship of these molecules has been revealed (Castellano and Torrens 2015).

Cancer chemopreventive agents, such as a new lanostanoid named tsugaric acid F and a new palmitamide with weak antioxidant and cytotoxic activities against PC3 cells (human prostate cancer cell line) were isolated and characterized from *Ganoderma tsugae* (Lin et al. 2016).

Fourteen lanostane triterpenoids, nine *Ganoderma* acids, and five *Ganoderma* alcohols were isolated from *Ganoderma hainanense* fruiting bodies. Reviewing *G. hainanense*, a species similar to *G. lucidum* and *Ganoderma sinense*, containing lanostane triterpenoids, the former might also have a broad spectrum of activities, especially as an anticancer agent against HL-60, SMMC-7721, A-549 and MCF-7

(Peng et al. 2015). Three undescribed lanostane triterpenoids, ganoellipsic acids A-C, and seven known *Ganoderma* lanostanoids have been isolated from cultivated fruiting bodies of *Ganoderma ellipsoideum*. The chemical structures of these compounds have been elucidated (Sappan et al. 2022).

Seven secondary metabolites, including a new lanostane triterpene, two known aromatic meroterpenoids, and four known triterpenes, were isolated from the fruiting bodies of *G. lucidum*. They showed *in vitro* AOA and neuroprotective effect (NPE) against H_2O_2 and aged Aβ-induced cell death in SH-SY5Y cells (thrice-subcloned cell line derived from the SK-N-SH neuroblastoma cell line). Therefore, *G. lucidum*-derived meroterpenoids may be suggested as potential antioxidants and neuroprotective functional food ingredients to prevent the development of neurodegenerative diseases (NDD) (Wang et al. 2019).

Terpenoid erinacine A extracted from the mycelial biomass of russuloid mushroom *Hericium erinaceus* was intravenously and *per os* administered to Sprague-Dawley rats. The bioavailability of erinacine A in rats after *per os* administration at 2.381 g/kg body weight (BW), *H. erinaceus* mycelium extract (equivalent to 50 mg/kg BW of erinacine A) was 24.39%. These preclinical studies initially showed that erinacine A could pass the blood-brain barrier of rats by passive diffusion. Further studies of metabolomic profiles of erinacine A to develop *H. erinaceus* mycelium-based neuroprotective drugs are warranted (Tsai et al. 2021).

Thirteen undescribed lanostane triterpenoids, including six C25–C27 nor-lanostane derivatives and nine known derivatives, were isolated from fruiting bodies of *Fomitopsis pinicola*. Anti-inflammatory assays indicated that pinicopsic acid F and 16α-hydroxy-3-oxolanosta-7,9(11),24-trien-21-oic acid showed moderate inhibitory effects against lipopolysaccharide (LPS)-induced NO production in RAW 264.7 cells, with IC_{50} values of 24.5 and 25.7 μM, respectively (Liu et al. 2022).

Fifteen undescribed lanostane-like C31-triterpenoid derivatives and five known derivatives (palustrisoic and polyporenic acids) were isolated from the hydro-alcoholic extract of cultivated *Fomitopsis palustris* fruiting bodies. Polyporenic acid B demonstrated strong cytotoxicity against cell lines HCT116, A549, and HepG2, while weak cytotoxicity of palustrisolids A, C, and G was shown (Zhao et al. 2018).

The chemical study of cultivated edible medicinal mushroom *Wolfiporia cocos* (= *Poria cocos*) revealed 46 lanostane triterpenoids containing 17 new compounds. Evaluation of anti-inflammatory activities of these compounds showed that poricoic acid GM significantly inhibited NO production in LPS-induced RAW264.7 murine macrophages ($IC_{50} = 9.73$ μM). Moreover, poricoic acid GM induced HO-1 protein expression, inhibited iNOS and COX2 protein expression, and released PGE2, IL-1β, IL-6, TNF-α, and ROS in murine macrophages. The poricoic acid GM suppressed IκBα protein phosphorylation to regulate antioxidant genes. The lanostane triterpenoids proved to be the key compounds responsible for anti-inflammatory properties of *W. cocos* (Bao et al. 2022).

Isolation and structure elucidation of three new meroterpenoids and seven known compounds from chloroform extracts of agaricomycete fungus *Albatrellus yasudae* and their Aβ-aggregation inhibitory activity have been reported. Apart from known grifolin, grifolic acid, neogrifolin, confluentin, 2-hydroxyneogrifolin, daurichromenic acid, and a cerebroside derivative, three novel compounds were identified (Akiba et al. 2020).

The steroids and polysaccharides isolated from the sclerotia of *Polyporus umbellatus* play an important role in diuresis and nephroprotection. The metabolic pathways indicated that steroids, fatty acids, and carbohydrates were actively produced when *P. umbellatus* sclerotia was infected by *Armillaria mellea*. The contents of ergosterol, polyporusterone A, and B are increased by 32.2%, 75.0%, and 20.0%, respectively. The comprehensive metabolomic and transcriptomic information will contribute to studying the mechanisms of *P. umbellatus* sclerotial formation infected by *A. mellea* (Xing et al. 2021).

Eight undescribed lanostane triterpenoids, pardinols A-H, and a previously reported lanostane triterpenoid saponaceol B were isolated from *Tricholoma pardinum* fruiting bodies. Pardinols B and E-H demonstrated inhibitory effects on NO production with an IC_{50} value ranging from 5.3 to 14.7 mM. The cytotoxicity against five human cancer cell lines with IC_{50} values less than 40 mM was also shown (Zhang et al. 2018a).

According to Tohtahon et al. (2017), a lanostane triterpenoid piptolinic acid A isolated from methanolic extract (ME) of *Piptoporus betulinus* exhibited both cytotoxic activity against human promyelocytic leukemia cell line HL-60 and human acute monocytic leukemia cell line THP-1 (IC_{50} = 1.77 mM and IC_{50} = 8.21 mM, respectively). Lanostane and ergostane triterpenoids have also been identified in the medicinal mushroom *Antrodia cinnamomea* (Qiao et al. 2015).

Bioactive eremophilane-type sesquiterpenes with a broad spectrum of bioactivity, including antibacterial, anti-inflammatory, anti-obesity, antiviral and cytotoxic, have been observed in several macrofungi, especially from the ascomycete genus *Xylaria* (Yuyama et al. 2017).

A bis-epoxide ergostane triterpenoid favolon, its undescribed derivative favolon C, and a biogenetically related norergostane laschiatrion were isolated from the cultures of agaricoid species *Favolaschia calocera*. The study initially described the stereochemistry of favolon and its derivatives (Palasarn et al. 2022).

Further chemical screening of mushrooms allows the discovery of new pharmacologically promising terpenoids, steroids, and sterols with therapeutic effects.

2.3 Phenolics

Phenolics represent a diverse group of bioactive compounds, including flavonoids, phenolic acids, quinones, tocopherols, tannins, etc. Mushroom-derived phenolics are known for their various therapeutic effects, including anti-inflammatory, antioxidant, analgesic, and neuroprotective (Badalyan et al. 2019; Dhakal et al. 2019; Badalyan and Rapior 2021a, b). Phenolic compounds, flavonoids, ascorbic acid, β-carotene, and lycopene have been detected in fruiting body extracts of

G. frondosa, S. commune, Volvariella volvacea (Acharya et al. 2015, 2016; Yao et al. 2016; Butkhup et al. 2018).

The study of total phenolic content, as well as antioxidant, antimicrobial, and inhibitory effects against cholinesterase, tyrosinase, α-amylase, and α-glucosidase activities of ME and AE obtained from *Ganoderma applanatum* suggest that they may be considered a source of new food supplements and represent a model for the development of new drug formulations (Zengin et al. 2015).

Studies of AOA of cultivated edible mushrooms *A. bisporus* and *P. ostreatus,* as well as several wild-growing edible species, such as *Boletus edulis, Cantharellus cibarius, Russula alutacea,* and *Trametes* species, identified them as sources of phenolics and flavonoids (Buruleanu et al. 2018). The total phenolic contents (coumarins, flavanols, flavonols, isoflavonoid derivatives, phenolic acids) and AOA of *T. versicolor* and *T. gibbosa* were evaluated using ME and AE. The highest AOA was observed in ME, whereas the highest polyphenol and flavonoid contents were detected in AE (Pop et al. 2018).

Recently, Psurtseva et al. (2022) analyzed three *Sparassis crispa* strains on various agar and liquid media for growth and production of phenolic compounds, such as sparassol (methyl-2-hydroxy-4-methoxy-6-methylbenzoate), methyl ester of sparassol, and methyl ester of orsellinic acid. The strain LE-BIN 2902 of *S. crispa* was considered promising for the production of sparassol.

2.4 Amino Acids and Proteins

A natural immunosuppressive product, myriocin—a complex amino acid, was isolated from culture broths of ascomycete fungus *Isaria sinclairii.* After extensive chemical modification and pharmacological evaluation, a highly potent immunosuppressor, fingolimod, was developed. Fingolimod is a sphingosine-1-phosphate receptor 1 (S1PR1) modulator used as a therapeutic agent for treatment of multiple sclerosis (MS) and autoimmune diseases of the central nervous system. The analysis of mechanism of action of fingolimod revealed its molecular target as S1PR1, which plays an essential role in lymphocyte circulation. Phosphorylated fingolimod acted as an antagonist of S1PR1, modulates lymphocyte circulation, and shows potent immunosuppressive activity (Chiba 2020). Previous studies have showed that fingolimod significantly reduced the relapse rate of MS and has been approved as a new medicinal product in many countries (Kappos et al. 2015; Derfuss et al. 2016).

The unphosphorylated fingolimod reduces the ability of cytotoxic CD8 T-cells to kill their target cells which can increase susceptibility to viral infections and enhance therapeutic efficacy in MS. Fingolimod modulates the proliferation of anti-inflammatory M2 phenotype macrophages, their morphology, and ability to release cytokines. It has been reported to be a cannabinoid receptor antagonist, a phospholipase A2 (cPLA2) and a ceramide synthase inhibitor, it also stimulated the repair processes of glial cells and glial precursor cells after injury. Preclinical and clinical data have revealed NPE of fingolimod, which may open a perspective for future therapeutic approaches and diagnosis of NDD, such as stroke, Alzheimer's, and other diseases (Ayzenberg et al. 2016; Chiba 2020; Cuello-Oderiz and McGraw 2022).

The development of effective chemotherapeutic drugs with few side effects is still continuing. The peptide latcripin-7A (LP-7A), extracted from *L. edodes*, causes apoptosis and autophagy of MCF-7 and MDA-MB-231 breast cancer cells by inducing their growth arrest at G_0/G_1 phase and decreasing mitochondrial membrane potential without adverse effects on MCF-10A normal breast cells (Din et al. 2020).

The amino acids, protein, mineral, and fatty acid composition and nutritional values of uncooked and pressure-cooked gasteroid fungus *Astraeus hygrometricus* have been studied. The results have shown that all samples possess nutritional values with sufficient protein, high carbohydrate and fiber contents, and low-fat amount with antioxidant properties (Pavithra et al. 2018).

2.5 *Lectins and Other Compounds*

Lectins, also known as hemagglutinins, are proteins that specifically and reversibly bind to certain carbohydrates and are involved in various biological processes at the level of recognition between cells.

Lectins are largely distributed in mushrooms and account for their potential medicinal properties, including antitumor, mitogenic/antimitogenic, nematocidal, entomocidal, immunomodulatory, antiviral, and other effects (Sabotič et al. 2016; Kües and Badalyan 2017; Singh et al. 2017; Badalyan et al. 2019; Badalyan and Rapior 2021a, b). The antibacterial effect of 18 kDa lectin purified from *Pleurotus flabellatus* (PF-L) has been reported against Gram-positive (*Bacillus subtilis, and S. aureus*) and Gram-negative (*Pseudomonas aeruginosa, E. coli, and Klebsiella pneumoniae*) human pathogens. PF-L also possesses a potent antioxidant effect (Murugesan and Gunasagaran 2021).

The mushroom-derived alkaloids and related compounds have recently been described by Zorrilla and Evidente (2022). The authors classified five subgroups as β-carboline alkaloids, pyrroloquinoline alkaloids, pyrrole alkaloids, indole alkaloids, and miscellaneous alkaloids. The isolation, structure, biological activities, and pharmacological potential of fungal alkaloids to develop drugs and agrochemicals have been reported. Among bioactive indole alkaloids, psilocybin and psilocin are two the most studied hallucinogenic compounds (Dinis-Oliveira 2017). Moreover, psilocin-containing mushrooms, as viable chemotherapeutic agents against Severe Acute Respiratory Syndrome Coronavirus-2 (SARS-CoV-2), have recently been proposed (Khan et al. 2022) (**see Part 3.5**).

3 Mushrooms as Therapeutics

The literature data, case studies, and information derived from traditional medicine and ethno-mycological data show that mushrooms possess promising pharmacological potential (anti-allergic, antibacterial, anti-depressant, antifungal, anti-inflammatory, antioxidant, antiviral, cardio-, hepato- and neuroprotective, cytotoxic, hypotensive, immunomodulatory, etc.). Therefore, they can be considered promising sources for developing new mycopharmaceuticals and functional food with health-enhancing effects (Wasser 2011, 2014; Özcan and Ertan 2018; Sharma et al. 2018, 2022a, b; Badalyan and Zambonelli 2019; Badalyan et al. 2019; Badalyan and Rapior 2021a, b).

The traditional usage of mushrooms and a growing scientific interest due to their valuable chemical composition, nutritional value, and therapeutic properties allow considering them as natural source for the development of mushroom-derived bioproducts, including healthy food, pharmaceuticals, nutraceuticals, and cosmeceuticals (Chang and Wasser 2012; Badalyan and Zambonelli 2019, 2023; Badalyan et al. 2019, 2022; Badalyan 2020; Badalyan and Gharibyan 2020; Badalyan and Rapior 2021a, b; Bolaniran et al. 2021; Cateni et al. 2021; Elkhateeb et al. 2021b; Elkhateeb and Daba 2022a; El-Ramady et al. 2022) (Table 1).

Isolation, purification, identification, and characterization of bioactive compounds from edible mushrooms, the study of their chemical and biological properties, extraction techniques, and food applications have recently been updated (López-Hortas et al. 2022).

The chemical composition (13 minerals, 23 polyphenols, 11 organic acids, 22 carbohydrates) and medicinal properties (antimicrobial, antioxidant, hypoglycemic, neuroprotective, and cytotoxic) of methanolic and acetone extracts of edible giant polypore *Meripilus giganteus* have been reported (Petrović et al. 2022). According to authors, the antimicrobial activity of tested extracts ranged from 0.002 to 20 mg/mL. The AOA was detected by DPPH assay as a half maximal inhibitory concentration (IC_{50}) was 673.42 and 712.31 μg/mL for methanolic and aceton extracts, respectively, and by reducing power assay from 0.042 to 0.099 mg/mL. The total phenolic content was determined as 4.86 mg gallic acid equivalents (GAE)/g for the ME and 5.06 mg GAE/g for the aceton extract. The hypoglycemic effects of tested extracts ranged from 30.66% to 38.67% and 11.06% to 17.08%, respectively. The NPE of *M. giganteus* tested by acetylcholinesterase inhibition assay ranged from 4.54% to 9.31%, while the cytotoxic activity revealed by microtetrazolium assay—196.24 μg/mL to 322.8 μg/mL. Based on obtained data, *M. giganteus* extracts were considered natural sources of therapeutic compounds without side effects (Petrović et al. 2022).

The edible medicinal mushroom *H. erinaceus* is known for its nutraceutical and pharmaceutical properties, particularly, antioxidant, antitumor, anti-aging, hypolipidemic, and anti-inflammatory. Erinacine A has been reported as the main bioactive compound in the mycelium of *H. erinaceus* as a potential source of functional food widely used to treatment of neurological diseases. The consumption of erinacine A-enriched mycelium has shown significant nutraceutical effects in Alzheimer's and Parkinson's diseases, ischemic stroke, and other NDDs (Badalyan and Rapior 2021a; Kuo et al. 2022).

For the first time, the metabolic process of erinacine A and identification of its metabolites from the rat and human liver S9 fraction were described. The results have shown that 75.44% of erinacin A was metabolized within 60 min in rats and 32.34% within 120 min in humans. Five major metabolites of erinacin A were detected and identified. A better understanding of metabolic process of erinacin A and creation of a database on its metabolic profile may facilitate further nutraceutical application and discovery of related biomarkers of this unique bioactive molecule (Kuo et al. 2022).

The exotic edible fungus *Dictyophora indusiata* (= *Phallus indusiatus*) is known for its numerous medicinal properties. Comprehensive reviews of chemistry, pharmacology and potential therapeutic applications of extracts and compounds obtained from *D. indusiata* have been published (Habtemariam 2019). The chemistry of polysaccharides, as major bioactive compounds (β-(1→3)-D-glucan with side branches of β-(1→6)-glucosyl units) are discussed, whereas low molecular weight compounds include terpenoids and alkaloids. The biochemical and cellular mechanisms of different therapeutic actions, such as antioxidant and anti-inflammatory, were reported along with potential applications in cancer therapy, immunotherapy, and NDDs (Habtemariam 2019).

The medicinal mushroom *Leucocalocybe mongolica* possesses high nutritional, pharmaceutical, and therapeutic values. It has been used in a wide range of chronic diseases in TCM. Around 100 compounds (polysaccharides, sterols, lectins, laccase, amino acids, and volatiles) with significant pharmacological effects (antitumor, antiproliferative, hypoglycemic, hepatoprotective, and hypotensive) have been identified in this fungus. Since polysaccharides can increase NO production, they may also have an strong hypotensive effect. However, further studies are needed to assess the pharmacological potential of *L. mongolica* (Zaki et al. 2022).

Bioactive molecules extracted from edible and inedible bracket mushrooms show favorable biological potential and can be used to develop various health care bioproducts (Barros et al. 2008; Heleno et al. 2015a, b; Reis et al. 2017; Badalyan et al. 2019; Badalyan and Zambonelli 2023).

Chen et al. (2010) reported induction of ovulation in women with polycystic ovary syndrome by Maitake (*G. frondosa*) extract. In addition, amino acids, aromatic acids, flavones, polysaccharides, triterpenes, and other bioactive molecules isolated from the inedible bracket *Phelllinus linteus* exhibited anticancer, anti-inflammatory, antioxidant and hypoglycemic properties and contributed to the regulation of the immune system (Chen et al. 2016).

3.1 Antitumor and Immunomodulatory

Cancer is a leading cause of morbidity and mortality worldwide. Chemotherapy has been extensively used to treat certain types of cancer; however, side effects and drug resistance are the drawbacks of these agents. Therefore, the development of new strategies with minimal adverse effects, including natural compound therapy, is required.

Cancer cells reprogram their metabolism to meet the demands of uncontrolled proliferation and survival. This re-programming of lipid metabolism supports tumor growth, cancer metastasis, and therapy resistance. Therefore, targeting this process is regarded as a potential therapeutic strategy.

Bioactive molecules from plants and mushrooms have not been used as first-line therapy in oncology. The main anticancer and cytotoxic substances of fungal origin are polysaccharides, particularly immunomodulatory β-D-glucans and triterpenoids, such as ganoderic acids, the antagonists of growth factor receptors and inhibitors of cyclin-dependent kinase. However, the anticancer effects of mushrooms were typically demonstrated *in vitro* and *in vivo* models, and clinical studies conducted in humans are limited. Evaluation of the efficacy of adjuvant therapy with mushroom-derived molecules in combination with evidence-based medicine, as well as precisely designed experimental and clinical studies, could be an effective therapeutic approach (Joseph et al. 2018; Zmitrovich et al. 2022).

The anticancer activities of mushroom extracts tested *in vitro*, *in vivo,* and *in silico* experimental models were recently prospected using scientific electronic databases (Nowakowski et al. 2021a, b). The extracts obtained from 92 species, using 12 different solvents, reduced the viability of 38 various cancers (breast, cervical, colorectal, gastric, lung, ovarian, prostate etc.), and 61 extracts showed cytotoxicity against breast cancer in a broad spectrum of experimental models. Various anticancer mechanisms of action of mushroom extracts have been reported, as well (Nowakowski et al. 2021a).

Different signaling pathways, particularly WNT, SHH, TGF-β/Smad, and JAK/STAT, have been shown to modulate cancer development and progression. There is a growing evidence that genetic/epigenetic mutations and loss of apoptosis also require a "multi-molecular" perspective for development of cancer therapy. An overview of mushrooms' regulation of different signaling pathways and their bioactive compounds were reported (Aras et al. 2018). The regulation of WNT and JAK-STAT pathways by mushrooms has been extensively reviewed. However, information on the regulation of TGF-β /Smad, Notch, and TRAIL-induced signaling pathways is lacking due to superficial data. The mechanisms of modulation of oncogenic and tumor suppressor microRNAs by mushrooms in different cancers have not been sufficiently investigated, yet. Therefore, a detailed insight related to targeting multiple pathways by mushroom extracts or their bioactive compounds will be useful to fill the knowledge gap and transform medically valuable bioactive molecules into clinically effective therapy (Aras et al. 2018).

Mushroom-derived compounds also impact cancer cells, terminating the cell cycle and inhibiting proliferative signaling pathways PI3K/AKT, Wnt-CTNNB1, and NF-κB. Therefore, mushroom-derived bioactive molecules and extracts as therapeutic agents may be used in studies of multiple pathways (Joseph et al. 2018).

Gastric cancer (GC) is the fourth most commonly diagnosed cancer and the second leading cause of death worldwide. The medicinal fungus *Phellinus igniarius* is used in traditional medicine as an anticancer agent. The ethanolic extract of *Ph. igniarius* was studied against five human tumor cell lines HepG-2, AGS, SGC-7901, Hela, and A-549. The extract was the most cytotoxic against SGC-7901 cells *in vitro* and strongly inhibited tumor growth in xenografted nude mice *in vivo*. After treatment, typical morphological changes due to cell apoptosis, including chromatin condensation and nuclear fragmentation with the formation of apoptotic bodies, were observed. Fungal extract blocked the SGC-7901 cell cycle at G0/G1 phase and induced apoptosis by downregulating cyclin D1 expression. It caused a remarkable decrease in mitochondrial membrane potential and induced mitochondrial-dependent apoptosis by triggering the activation of caspases 9 and 3 and cleavage of PARP in SGC-7901 cells. Furthermore, *Ph. igniarius* extract increased Bax/Bcl-2 ratio in SGC-7901 cells *in vitro* and *in vivo*. Thus, *Ph. igniarius* may be a promising therapeutic agent for prevention and treatment of GC, as it may induce apoptosis of cancer cells through a mitochondria-dependent pathway (Wang et al. 2018).

Medicinal bracket fungus *Sanghuangprous vaninii* (= *Phellinus vaninii*) is used to treat vaginal bleeding, leucorrhea, and abdominal pain in patients with different gynecological tumors. The antitumor potential of a hydro-ethanolic extract derived from *S. vaninii* was evaluated *in vitro* on three human cancer cell lines: human gastric cancer cells (SGC7901), human ovarian cancer cells (SK-OV-3), and human cervical cancer cells (SiHa), as well as on a 4-week-old BALB/c *in vivo* female mice. The results have shown that the extract possesses anticancer activity by inducing apoptosis of SiHa cells. According to the results, *S. vaninii* was suggested as a potential antitumor agent against cervical cancer (He et al. 2021a).

Recent literature on the medicinal properties of the Chaga mushroom (*Inonotus obliquus)* and its usage as a supplementary medicine for cancer therapy have been updated (Khoroshutin et al. 2021).

Breast cancer is among the most common cancers causing the death of women worldwide. The cytotoxic effect of ethanolic extract of *I. obliquus* was observed after *per os* administration in 4T1 tumor-bearing *BALB/c* mice. The lanostane triterpenoid inotodiol and trametenolic acids have been identified as the main cytotoxic constituents of *I. obliquus* (Lee et al. 2021). The effect of inotodiol on breast cancer was also reported in streptozotocin (STZ)-induced diabetic rats. The results have shown that inotodiol lowered blood glucose levels, reduced plasma levels of cholesterol, triglyceride, and high-density lipoprotein (HDL), and induced apoptosis via downregulation of β-catenin signaling in rats (Zhang et al. 2018b).

Thus, *I. obliquus* as an anticancer agent may be recommended as a preventive medicine for treatment of breast cancer, particularly in diabetic patients (Zhang et al. 2018b; Lee et al. 2021).

The mycochemical study of the methanol extract of *Inonotus nidus-pici*, a relative species of *I. obliquus,* revealed five compounds (citropremide, 3,4-dihydroxybenzalacetone, lanosterol, ergost-6,8,22-trien-3*β*-ol, and ergosterol peroxide) which exhibited moderate antimicrobial activities against Gram-positive (*B. subtilis* subsp. *spizizenii*, and *Rhodococcus fascians)* and Gram-negative (*Pseudomonas syringae* pv. *maculicola*, and *Aliivibrio fischeri*) bacteria, as well as an AOA in DPPH assay, an anti-proliferative potential on A431 human skin-derived and epidermoid carcinoma cell lines (Garádi et al. 2021).

A novel 12-kDa protein musarin, purified from *T. versicolor* hot-water extract has shown a strong anti-proliferative potential against human colorectal cancer stem cell-like CD24+CD44+HT29 and demonstrated tyrosine kinase-inhibitory activity *in vitro*. As a promising new drug, musarin may be used in colorectal cancer therapy (He et al. 2021b).

Novel anticancer triterpene GL22 isolated from *Ganoderma leucocontextum* significantly inhibited the growth of liver cancer cell line Huh7.5 *in vitro* and Huh7.5-derived tumor xenografts *in vivo*, inducing mitochondrial dysfunction and cell death (Liu et al. 2018).

The cytotoxic triterpene astrakurkurone, isolated from a wild edible mushroom *A. hygrometricus* at low doses, was shown to be active against HCC cell lines (Hep 3B and Hep G2) (Nandi et al. 2019). The astrakurkurone acts by inducing apoptosis of cells, disrupting mitochondrial membrane potential, and inducing the expression of Bcl-2 family proteins Bax, caspases 3 and 9. Previous molecular study predicted a direct interaction of the drug with anti-apoptotic proteins, Bcl-2 and Bcl-xL. Thus, astrakurkurone can be suggested as a promising therapeutic agent for cancer treatment (Dasgupta et al. 2019).

The selenium (Se)-enriched EPS fraction, isolated from *L. edodes,* mainly consists of a highly branched 1-6-α-mannoprotein (molecular weight 4.5×10^6 Da). It significantly improves cell viability when incubated with human umbilical vein endothelial cells (HUVEC) absent from human cervical HeLa cells. The Se-EPS fraction also possesses antioxidant and immunosuppressive activities. At concentrations of 10–100 μg/mL, it inhibited mitogen-induced T-cell proliferation without revealing a significant effect on B cells (Górska-Jakubowska et al. 2021).

The aqueous, 70% and 95% ethanolic extracts derived from *C. cibarius, Coprinus comatus, Lactarius deliciosus,* and *Lycoperdon perlatum* were investigated for their anticancer effect on U87MG, LN-18 glioblastoma, and SVGp12 normal human astroglial cell lines. *C. comatus* and *L. deliciosus* demonstrated the greatest anticancer effects. The activities of ethanolic extracts were higher than those of aqueous extracts. The anti-glioma mechanism of *C. comatus,* based on the inhibition of cancer cell proliferation and induction of apoptosis, was associated with the termination of subG1 or G2/M phase of the cell cycle and inhibition of metalloproteinase activity (Nowakowski et al. 2021b).

Fifteen compounds from submerged mycelial cultures of *Amylosporus* cf. *graminicola* and *Amylosporus* cf. *campbelii* were isolated. Seven novel amylosporanes' derivatives, five known metabolites, and colletochlorin B have been identified (Kemkuignou et al. 2022). Colletochlorin B showed a cytotoxic activity against *B. subtilis* (MIC = 2 μg/mL), stronger than oxytetracycline, and cytotoxicity against squamous cancer A431 cells with an IC_{50} value of 4.6 μM (Hamad et al. 2022).

The cold-water extract obtained from freeze-dried *Lignosus rhinocerus* sclerotial powder TM02® was tested on a panel of human oral cancer cell lines. MTT (3-(4, 5-dimethylthiazolyl-2)-2, 5-diphenyltetrazolium bromide) proliferation assay indicated that oral cancer cells ORL-48 (derived from gingiva), ORL-188 (derived from the tongue), and ORL-204 (derived from buccal mucosa) were inhibited by *L. rhinocerus*. A high molecular mass fraction from the cold-water extract induced apoptosis on ORL-204 and arrest G0/G1-phase cell cycle through caspase-3/7 cleavage. These results support the usage of *L. rhinocerus* as a potential dietary compound for cancer prevention and treatment (Yap et al. 2022).

The methanolic and ethyl acetate extracts of agaricomycetous mushrooms, *A. hygrometricus*, *Lentinus* sp., *Serpula* sp., *Tricholoma* sp., and *Phallus* sp. were investigated for their anti-proliferative activities on Jurkat leukemic cell line. A high and more efficient anti-proliferative activity of ME ($IC_{50} = 22.7 \pm 0.23$ μg/mL) was detected compared to ethyl acetate extract of *A. hygrometricus* ($IC_{50} = 68.9 \pm 0.33$ μg/mL) (Pal et al. 2021).

Reishi mushroom (*G. lucidum)* has traditionally been used to treat immunosuppression and cancer. Skin cancer-preventive effects of commercial products containing spore and fruiting bodies of *G. lucidum* have been tested on JB6 cells. The products prevented the development of skin cancer, probably via attenuating UV-induced immunosuppression (Shahid et al. 2022). Silver and titanium dioxide nanoparticles prepared with an AE of medicinal bracket fungus *Fomes fomentarius* revealed a strong cytotoxic effect on HCT-116 human colorectal carcinoma cells (Rehman et al. 2020). Polyphenol-rich and β-glucan-containing aqueous alkali extract obtained from *F. fomentarius* fruiting bodies (FFE) were studied on human colorectal adenocarcinoma (Caco-2) and cutaneous melanoma (COLO-818) cells. Caco-2 cells did not react on FFE, which may indirectly suggest its safety for human intestinal epithelium. The melanoma cells were dose-dependent at 0.01-0.05 mg/mL concentrations of FFE (Storsberg et al. 2022).

Fourteen undescribed compounds including five 2,5-diarylcyclopentenone (xylariaones A1-B2), seven α-pyrone derivatives (xylaripyones A-G), one γ-pyrone derivative xylaripyone H, one diketopiperazine cyclo-(L-Leu-N-ethyl-L-Glu), and two known diketopiperazines, have been isolated from cultures of an endophytic *Xylaria* sp. (Song et al. 2022c). These compounds were also investigated for their potential anti-proliferative effects against PC3 and A549 human tumor cell lines. The authors reported that xylaripyone D exhibited a moderate inhibitory activity against the proliferation of PC3 cell lines with an IC_{50} value of 14.75 μM. In addition, xylariaone A3 and xylaripyone F showed weak inhibitory effects on NO production in RAW 264.7 murine macrophages (IC_{50} of 49.76 and 69.68 μM, respectively).

The AE derived from three ascomycetous species of edible ectomycorrhizal desert truffles (*Terfezia claveryi, T. boudieri,* and *T. olbiensis*) were investigated for their cytotoxic and apoptosis-inducing effects on pancreatic cancer cell line

(PANC-1). The results showed a strong, dose-dependent inhibition of PANC-1 cell growth by inducing apoptosis via upregulation of pro-apoptotic genes BAX, CDKN1A, and TP53 and downregulation of the anti-apoptotic gene *BCL2*. These results suggest that *Terfezia* truffles may be used as a functional food with anticancer effects (Saleh et al. 2022).

Doxorubicin (DOX) is a widely used chemotherapy drug with a cardiotoxic effect. The effect of *Morchella esculenta* extract in attenuating DOX-induced cardiotoxicity *in vitro* by MTT assay using H9c2 cardiomyoblast cells was evaluated. The mushroom extract reduced cytotoxicity at concentrations of 150 and 200 μg ($p < 0.05$ and $p < 0.01$, respectively) and restored near-normal levels of endogenous antioxidants, such as SOD, GPx, and GSH, depleted by DOX administration (Das et al. 2022b).

3.2 Antioxidant

Many chronic diseases, such as cancer and inflammation, develop due to the adverse effect of free radicals (hydroxyl, peroxy-nitrite, nitrite, DPPH, superoxide anion, hydrogen peroxide). As active producers of phenolic compounds and flavonoids, mushrooms are considered a natural source of dietary antioxidants (Huang et al. 2022). Various methods, such as ABTS, DPPH, hydrogen peroxide, and lipid peroxidation, were used to assess the antioxidant potential of edible and medicinal mushrooms. The potential of mushrooms as a source of natural antioxidants and their application in mycotherapy of different diseases have been assessed (Mwangi et al. 2022).

The ethanolic extract of *Trametes hirsuta* was studied for its total antioxidant status (TAS) and total oxidant status (TOS). The oxidative stress index (OSi; ratio of TOS to TAS values) was outlined. The TAS and TOS values of *T. hirsuta* were 3.466 ± 0.148 mmol/L and 13.482 ± 0.234 μmol/L, respectively. The OSi value was estimated to be 0.390 ± 0.018. The most effective antimicrobial activity of ethanolic extract of *T. hirsuta* was detected at concentrations of 100, 200, and 400 μg/mL (Akgul et al. 2021).

The chemical composition of three extracts of *Ganoderma australe*, and the role of IPS in the induction of maturation and activation of bone marrow-derived dendritic cells (DCs), have been studied. Glucose, mannose, and galactose prevailed in all extracts. The IPS extract obtained from fungal mycelia showed a significant AOA, possibly related to high amounts of phenolic compounds. The effect of AE of mycelial polysaccharides derived from *G. australe* on the maturation and activation of mouse DCs and prevention of oxidative processes was originally described in this study (Gallo et al. 2022).

The screening of AOA of decoction, infusion, and hydro-methanolic extracts isolated from edible medicinal *Flammulina velutipes* in *in vitro* ABTS and DPPH assays showed that the decoction extract was capable of better scavenging of radicals followed by the hydro-methanolic extract; in contrast, the infusion extract exhibited a promising effect on metal ion chelating ability (Sharma et al. 2022a). The amount of phenols in the decoction was the highest (14.85 μg gallic acid equivalent/mg extract), followed by the infusion (12.02 μg gallic acid equivalent/mg extract). Additionally, lycopene (1.4 μg/mg of extract) and ascorbic acid (1.56 μg/mg of extract) were detected in greater quantity in the hydro-methanolic fraction. At 100 μg/mL and 500 μg/mL concentrations, both hydro-methanolic extract and infusion fraction inhibited around 25% and 77% radicals, respectively. The decoction fraction exhibited the best potential by inhibiting more than 32% and 90% radicals, respectively. According to obtained data, *F. velutipes* was considered a source of mycopharmaceuticals with AOA (Sharma et al. 2022a).

A moderate antimicrobial potential and AOA of *Calocybe gambosa* have recently been reported. Based on the chemical analyses of macronutrients, α-tocopherol, free sugars, organic and fatty acids of *C. gambosa*-enriched oatmeal cookies, this mushroom was suggested as a potential functional food with antimicrobial and antioxidant effects (Petrović et al. 2022).

Two new polyketides, rufoolivacin E and viocristin B, a new natural product of 1-hydroxy-3,6,8-trimethoxyanthraquinone, 13 known compounds with inhibitory effects against glutamate dehydrogenase (GDH) and AOA have been isolated and identified from wild growing *Cortinarius* sp. Four compounds exhibited significant AOA with IC_{50} values of 7.0 ± 0.3, 8.6 ± 0.1, 7.5 ± 0.1, and 2.8 ± 0.2 μg mL^{-1}, respectively. Thus, *Cortinarius* sp. may be used in food and drug industries as a natural source of polyketides (Song et al. 2022a, b)

The chemical study of a hydro-ethanolic extract of *Russula pseudocyanoxantha* and its functional secondary metabolites with AOA was performed. The extract was highlighted to be enriched with various phenolic compounds, such as *p*-coumaric acid, cinnamic acid, and pyrogallol, as well as ascorbic acid and carotenoids. A significant *in vitro* radical scavenging effect (hydroxyl, DPPH, and ABTS), ferrous ion chelating capacity, and potency reduction were recorded at EC50 values of 15–2674 μg/mL. Incorporation of *R. pseudocyanoxantha* into daily diet, as a novel mycofood, may provide health benefits. It may be further used in formulating nutraceutical, cosmeceutical, and pharmaceutical bioproducts (Badalyan et al. 2022; Khatua and Acharya 2022).

The chemical screening of therapeutic potential of lignicolous fungus *Lentinus squarrosulus* revealed that AE, hydro-ethanolic, and ethanolic extracts of *L. squarrosulus* were almost free of tannins, poor in the total flavonoids and moderately rich in reducing sugars. The aqueous and ethanolic extracts were rich in total polyphenols, whereas aqueous and hydro-ethanolic extracts—in alkaloids. The AE was rich in saponosides and hydro-ethanolic extract in coumarin derivatives. These results have shown that tested extracts possess low to moderate AOA. The highest activity has demonstrated the ethanolic extract of *L. squarrosulus,* which could be used as an antioxidant agent (Ndong et al. 2021). Evaluation of the nutritional profile and digestion-stimulating effect of *L. squarrosulus* suggested that this fungus promotes the growth of selected probiotic bacteria, especially *Bifidobacterium* strains, and may be used as a functional food to improve gut microflora (Ayimbila et al. 2022). The antioxidant effects of water and ethyl acetate extracts from the ascomycete mushroom *Morchella steppicola* were also reported (Sarikurkcu et al. 2022).

3.3 Anti-Inflammatory and Antimicrobial

The numbers of human pathogens resistant to antibiotics have increased worldwide. The World Health Organization (WHO) considers resistant antimicrobial infections to be a threat to global health, accounting for a high percentage of annual deaths (https://www.ox.ac.uk/news/2022-01-20-estimated-12-million-people-died-2019-antibiotic-resistant-bacterial-infections) (Prestinaci et al. 2015; Saha and Sarkar 2021; Murray et al. 2022). Particularly, multidrug-resistant ESKAPE pathogens (*Enterococcus faecium*, *S. aureus*, *K. pneumoniae*, *Acinetobacter baumannii*, *P. aeruginosa*, and *Enterobacter* spp.) have become the universal origin of skin and soft-tissue infections in the WHO priority pathogens list (https://www.who.int/news-room/fact-sheets/detail/antimicrobial-resistance).

Mushrooms are a source of different antimicrobial compounds. The cytokines IL-1β, IL-8, and TGF-β involved in signaling inflammatory processes, which may be produced by HT-29 epithelial cell line, were selected to study *in vitro* anti-inflammatory activity of crude EPS isolated from *T. versicolor*. IL-1β and IL-8 mediate inflammatory processes, whereas TGF-β is the signaling peptide related to the anti-inflammatory processes. The medium, containing 2 mg/mL of crude EPS, significantly reduced the concentration of pro-inflammatory IL-8 and IL-1β cytokines. In contrast, a 2.5-fold increase in TGF-β concentration was related to the anti-inflammatory properties of *T. versicolor* EPS (Angelova et al. 2022).

The antimicrobial potential of different extracts of hymenochaetoid bracket fungi *Phellinus tuberculosus* and *Fuscoporia ferruginosa* against Gram-positive (*S. aureus* ATCC25923, *B. subtilis* ATCC6633, *Streptococcus mutans* ATCC 35,665) and Gram-negative (*P. aeruginosa* ATCC9027, *A. baumannii* BAA-744, and *E. coli* ATCC8739) bacteria, as well as of *Candida albicans* ATCC10231 strains showed that Gram-positive bacteria were more sensitive to fungal extracts compared to Gram-negative bacteria. No activity of fungal extracts against *C. albicans* yeast was found. The activities of methanol and ethanol extracts were similar, while the antibacterial effect of AE was weak. The extracts of *P. tuberculosus* showed the highest antibacterial activity. The methanolic and ethanolic extracts of *F. ferruginosa* were highly resistant to *P. aeruginosa*, whereas *S. mutans* showed the highest sensitivity to these extracts (Dokhaharani et al. 2021). The ethanolic extract of polypore mushroom *Trametes hirsuta* at concentrations of 100, 200, and 400 μg/mL was effective against six bacterial strains (*A. baumannii, Enterococcus faecalis, E. coli, P. aeruginosa, S. aureus*, and *S. aureus* MRSA), as well as *Candida* species (*C. albicans, C. glabrata,* and *C. krusei*) (Akgul et al. 2021).

The extracts from culture liquid and mycelial biomass of *Lentinus arcularius* were tested against bacteria (*E. coli*, *P. aeruginosa,* and *S. aureus*) and fungi (*C.

albicans, Saccharomyces cerevisiae, and *Aspergillus niger*). The best antimicrobial effect revealed ethyl acetate and n-butanol fractions (Yen et al. 2022).

The antimicrobial and anti-biofilm properties of aqueous and methanolic extracts obtained from *Boletus edulis* and *Neoboletus luridiformis* against multidrug-resistant bacteria were revealed. The most relevant antibacterial effect was detected in aqueous and methanolic extracts of *B. edulis* against Gram-positive *S. aureus* and Gram-negative *P. aeruginosa.* At the same time, the highest bacterial biomass-removing ability has shown the AE of *B. edulis* in *S. aureus* and *E. coli* (Garcia et al. 2021).

P. ostreatus polar extract (PoPE) obtained from cultivated fruiting bodies was evaluated for antimicrobial and cytotoxic effects. The extract was characterized by its phenolic and flavonoid contents, which were 6.94 and 0.15 mg/g, respectively. PoPE showed the potent antimicrobial activity against fungal and bacterial pathogens (*C. albicans, S. aureus, Micrococcus luteus,* and *E. coli*). PoPE was found to inhibit phytopathogenic fungi *Fusarium oxysporum* (47%), *F. solani* (28%), and *Rhizoctonia solani* (21%). The extract was 13 times more selective and toxic to MCF-7 cells than Vero normal cells at an IC_{50} value of 4.5 μg/mL, promoting cell cycle arrest in the sub-G1 stage and cell apoptosis. It significantly increased TNF-α production, while decreasing IL-6 levels. The total AOA, lipid peroxide, and glutathione-reductase activity were recorded at 0.14 ± 0.02 mmol/L, 15.60 ± 0.015 nmol/mL, and 9.50 ± 1.30 U/L, respectively (Hamad et al. 2022)

The antibacterial, antifungal, and antioxidant activities were detected in the ethanolic extract derived from the poisonous fungus *Entoloma sinuatum*. The extract was effective against bacteria at concentration 200 and 400 μg/mL and fungi at 50 μg/mL. This fungus was identified as a natural antioxidant and antimicrobial agent (Bal et al. 2022).

The co-cultivation of microbes and fungi is used as a strategy to induce the biosynthesis of desirable bioactive metabolites. Randomized co-culturing of *Phellinus orientoasiaticus* (Hymenochaetaceae) and *Xylodon flaviporus* (Schizoporaceae) showed no antagonistic growth. Three new sesquiterpenes and five known analogues have been isolated and characterized. The LC-MS analysis suggested that cyclohumulanoids of illudane-, sterpurane-, and tremulane-type scaffolds were produced by *P. orientoasiaticus*, whereas a drimane-type sesquiterpene by *X. flaviporus*. None of the isolates exhibited antifungal activity or cytotoxicity. The compounds produced by *P. orientoasiaticus* promoted NO production of LPS-treated RAW276.4 cells ranging from 15.9% to 38.0% at 100 μM (Pham et al. 2022).

Fomitopsis officinalis, also known as *Laricifomes officinalis*, is a medicinal polypore used for millennia (Agarikon) to treat several, particularly pulmonary diseases. There is rich ethno-mycological information; however, isolation and chemical characterization of single compounds and *in vitro* bioactivity tests have only recently been performed (Girometta 2019). According to several reports, a broad-spectrum of antibacterial and antiviral activities of *F. officinalis* against *Mycobacterium tuberculosis, Yersinia pseudotuberculosis, S. aureus,* and *Ortopox* virus are available. The chlorinated coumarins derived from mycelia and lanostane triterpenoids obtained from basidiomes are directly responsible for antiviral, antibacterial, and trypanocidal activities, respectively (Girometta 2019).

The antibacterial activities of 28 extracts from seven Basidiomycota and Ascomycota mushrooms against Gram-positive and Gram-negative bacteria have been evaluated (Morel et al. 2021). The cyclohexane extract of *Rubroboletus lupinus* was active on a methicillin-resistant *S. aureus* (MRSA) (MIC = 125 μg/mL) and mildly active on methicillin-sensitive *S. aureus* (MSSA) and *B. subtilis* (MIC = 250 μg/mL). Cyclohexane extract from *Neoboletus luridiformis* was active against MSSA with MIC=125 μg/mL; while *Gyroporus castaneus* (cyclohexane and chloroform) extracts against MSSA and MRSA (MIC = 125 μg/mL). Among the tested extracts, *Gyromitra esculenta* cyclohexane extract has shown the strongest antibacterial activity with MIC = 31 μg/mL against MSSA, MSSA, and *B. subtilis* strains (Morel et al. 2021).

In the context of antibiotic resistance, the antibacterial potential of 70 extracts from 31 French mushrooms against wild-type and multidrug-resistant (MDR) bacteria *E. faecalis*, *E. coli*, *P. aeruginosa*, *Staphylococcus epidermidis*, and *S. aureus*, were screened. The results have shown that 95% of extracts contained antibacterial compounds. The ethyl acetate extracts exhibited more active compounds than ME. All extracts were mainly active against Gram-positive bacterial strains. The most promising mushroom extracts were tested against various MDR-strains of *S. aureus* and *E. coli*. The activity was weaker on MDR strains. However, *F. pinicola* and *Scleroderma citrinum* contained several compounds which inhibited the growth of MDR pathogenic bacteria. The stearic acid was identified as an ubiquitous compound contributing to the antibacterial defense of mushrooms. The results revealed the antibacterial potential of mushrooms. Nevertheless, further assays are needed to consider fungal compounds as therapeutic agents to counter antibiotic resistance (Huguet et al. 2022).

Evaluation of antimicrobial activities of aqueous and hydro-alcoholic extracts from the mycelia of *Ph. linteus* and *Pleurotus albidus* against *Bacillus cereus*, *B. subtilis*, *P. aeruginosa*, and *S. aureus* showed that *P. albidus* extracts showed a stronger activity against *Bacillus* strains. In contrast, *Ph. linteus* extract was effective against *S. aureus* and *P. aeruginosa*. The AE of *P. albidus* and 30% hydro-alcoholic extraction for *Ph. linteus* were the best for obtaining bioactive compounds. MS analyses allowed the identification of main chemical compounds, including glutathione oxidase, leucovorin, and riboflavin extracted from the mycelial biomass of these mushrooms. Considering these findings, *P. albidus* and *Ph. linteus* may be used as promising sources of bioactive molecules to develop novel health-enhancing biotech products (Contato et al. 2022).

Two *Morchella* species have been reported to possess a broad spectrum of biological activities against bacterial infections. The antibacterial potential of extracts obtained from *M. esculenta* and *M. conica* were evaluated against *S. aureus*, MRSA, and *Streptococcus pyogenes*. The extracts of *M. conica* inhibited the growth of *S. aureus* with an inhibitory zone ranging from 10.66 ± 0.3 to 21.00 ± 1.5 mm. The MIC of bacterial growth ranged from 3.33 ± 0.6 to 16.0 ± 0 mg/mL. *Morchella* extracts prevent the growth of *S. aureus* at 8-16 mg/mL showing a bactericidal effect and were more effective against MRSA than currently available antibiotics. Therefore, *Morchella* extracts may be suggested as potential antibacterial agents (Haq et al. 2022).

Identification of bioactive constituents of medicinal mushroom *C. militaris*, including cordycepin, essential amino acids, sterols, and polysaccharides with preventive and therapeutic effects in chronic diseases, including cancer, diabetes, and allergies, have been reported (Sharma et al. 2022b). The molecular mechanisms contributing to the anti-inflammatory properties of *C. militaris* have recently been updated (Phull et al. 2022).

3.4 Antiviral

Severe Acute Respiratory Syndrome Coronavirus-2 (SARS-CoV-2) is a highly pathogenic virus that caused a global pandemic which has promoted researchers worldwide to develop of a novel vaccines or small molecular therapeutics for SARS-CoV-2. Although several vaccines have already been discovered, no medication has been previously approved by FDA for the treatment of COVID-19.

Studies have identified promising hits against the main protease (M^{pro}) of SARS-CoV-2 from edible mushrooms. A structure-based virtual screening of 2433 compounds derived from mushrooms was performed with M^{pro} protein 6LU7. New data showed that Kynapcin-12 (M_78), Kynapcin-28 (M_82), and Kynapcin-24 (M_83) are potent inhibitors of M^{pro} protein 6LU7 available in the edible mushroom *Polyozellus multiplex*. These novel phenolic scaffolds may be developed to potential SARS-CoV-2 inhibitors. The identified molecules could be further explored as antiviral therapeutic agent against this virus. Another promising compound, neonambiterphenyls A (M_366), is recently detected in the poisonous mushroom *Neonothopanus nimbi* (Sen et al. 2022).

3.5 Hypoglycemic and Hypocholesterolemic

The polysaccharide (β-D-glucan) and vitamin D contents of mushrooms determine medicinal components and prevention of diabetes mellitus. Within the selected 23 macrofungi, only 11 demonstrated *in vivo* or *in vitro* hypoglycemic activities. Although species of genera *Pleurotus*, *Grifola,* and *Ganoderma* were the most studied, *in vivo* and *in vitro* investigations and clinical tests should be performed due to unknown mechanisms of therapeutic action (Das et al. 2022a).

In vivo (inhibition of α-amylase and α-glucosidase) and *in vitro* (oral starch tolerance and glucose tolerance tests) hypoglycemic activities of 70% ethanolic extracts of bracket fungi *Inonotus pachyphloeus* and five *Phellinus* species (*Ph. allardii*, *Ph. fastuosus*, *Ph. gilvus*, *Ph. sanfordii*, and *Ph. torulosus)* have been reported. The extracts of *Ph. fastuosus* ($IC_{50} = 27.33 \pm 1.45$ mg/mL) and *Ph.*

sanfordii (IC_{50} = 30.33 ± 0.88 mg/mL) strongly suppressed of α-amylase and α-glucosidase activities. Their most active extracts (200 and 400 mg/kg BW, respectively) significantly reduced *in vivo* postprandial hyperglycemic peaks in rats. Further pharmacological studies are needed to develop *Ph. fastuosus* and *Ph. sanfordii*-derived and clinically applicable therapeutic agents for the menagement of diabetes (Azeem et al. 2021).

The extracts from cultivated Agaricomycetes mushrooms *S. commune* (IC_{50} = 15.3 mg/mL; hexane extract) and *L. edodes* (IC_{50} = 12.9 mg/mL; ethanolic extract), as well as wild mushrooms *Phaeogyroporus portentosus* (IC_{50} = 15.7 mg/mL; ethanolic extract) and *Craterellus aureus* (IC_{50} = 25.9 mg/mL; ethanolic extract) have shown potent anti-amylase and anti-glucosidase activities. Both inhibitory enzyme activities are associated with diabetes and obesity. Furthermore, 80% of ethanolic extracts (v/v) had an anti-lipase activity as for *L. edodes* (IC_{50} = 8.85 mg/mL) and *P. portentosus* (IC_{50} = 33.6 mg/mL). These findings promote mushroom consumption in maintaining good health, as well as controlling diabetes and obesity by enzyme inhibitions (Wunjuntuk et al. 2022).

Cholesterol absorption inhibitors, statins, and fibrates are the main therapeutics to reduce a high level of blood lipids. In TCM, herbs and mushrooms as healthy food (i.e., Pu-erh tea, *Crataegus, A. auricula-judae*) are used to treat hyperlipidemia. Cordycepin obtained from *C. militaris* revealed a weak regulatory activity on the level of blood plasma lipids but more significant hepatoprotective effect (Chen et al. 2014; Badalyan et al. 2019).

A randomized, controlled, double-blind clinical trial was performed for eight weeks on volunteers from 18 to 65 years old ($n = 52$) with untreated mild hypercholesterolemia (Morales et al. 2021). The subjects consumed a *β*-D-glucan-enriched (BGE) mixture (10.4 g/day) obtained from shiitake mushroom (*L. edodes*) equivalent to 3.5 g/day of fungal *β*-D-glucans or a placebo incorporated in three various commercial creams. No inflammatory or immunomodulatory effects were shown, and no changes in IL-1β, IL-6, TNF-α, or oxLDL levels were detected. Although the BGE mixture showed a hypocholesterolemic effect *in vitro* in animal studies, it did not significantly reduce the cholesterol level compared to placebo. The BGE mixture modulated the gut microbiota compared to placebo. Further clinical trials are warranted to reveal an association between BGE administration and cholesterol metabolism or microbiota (Morales et al. 2021).

3.6 Neuroprotective

Depression is a serious neuropsychiatric disorder that affects more than 260 million people worldwide. There are many recognized mental health conditions or mental illnesses. Mushrooms with NPE may be used in anti-depressive therapy (Badalyan and Rapior 2021a, b). An animal model has shown an anti-depressant-like effects of AE of *H. erinaceus* injected intraperitoneally at 10 and 25 mg/kg for 4 weeks against chronic restraint stress (CRS). The potential mechanisms of neurogenesis were elucidated. The results revealed the anti-depressant-like effects by promoting hippocampal neurogenesis and reducing neuro-inflammation by enhancing the BDNF-TrkB-CREB signaling pathway (Chong et al. 2021).

Several mushrooms (*Amanita muscaria*, *Claviceps purpurea*, *H. erinaceus*, *Pleurotus cornucopiae*, and *Psilocybe* species) are considered producers of bioactive compounds with neurotrophic effects to treat mental health disorders (addiction, anxiety, anorexia, mood, and psychosis) and pain comorbidities (Badalyan and Rapior 2021a, b). These compounds have chemical structures similar to neurotransmitters and can function as agonists of receptors implicated in psychiatric disorders (Meade et al. 2022).

In a previous study, mice were subjected to a tail suspension test (TST) to simulate sleep disruption. The results have shown that, at a dose of 150 mg/kg, *H. erinaceus* mycelium, containing erinacin A (7.20 mg/g) and erinacin C (3.35 mg/g), had reversed TST-induced sleep disturbances. This *in vivo* study suggests that *H. erinaceus* mycelium has a potential dual role in relieving anxiety and improving sleep. Future clinical trials should address these effects of *H. erinaceus* mycelium with randomized placebo-controlled trials (Li et al. 2021b).

Obesity may cause changes in behavior, while maternal obesity can contribute to metabolic disorders or metabolic syndrome in children. The effect of β-glucan-rich *Pleurotus pulmonarius* (βgPp) on the neurological behavior of female ICR mice, hippocampus and development of its offspring's hippocampus showed that βgPp significantly enhanced short-term object recognition memory in high-fat diet (HFD)-fed mice. βgPp also ameliorated histological alterations, neuronal loss, and increased ionized calcium-binding adaptor molecule-1 (Iba-1)-positive microglia in the hippocampus regions of HFD-fed mice and their male offspring. The rich extract of *P. pulmonarius* improves object recognition memory and hippocampus morphology in mice fed with an HFD. These findings demonstrated that βgPp supplementation attenuated the effects of HFD on object recognition memory, as well as alterations in hippocampal regions of maternal mice and their male offspring (Tarmizi et al. 2022).

3.7 Hepatoprotective

The hepatoprotective effect of medicinal mushrooms has been reported by many studies (Chen et al. 2014; Badalyan et al. 2019; Li et al. 2019; Hou et al. 2021).

Liver fibrosis, among other origins, is caused by non-alcoholic steatohepatitis. Non-alcoholic fatty liver disease is defined as the accumulation of fat in hepatocytes in the absence of excessive alcohol consumption. It is a common chronic liver disease that affects more than 30% of western world population and is associated with many metabolic disorders, such as obesity. Se-enriched AE obtained from *A. bisporus* was evaluated for its potential inhibitory effects on the progression of fibrosis using an *in vitro* model like LX2 cell line (human hepatic stellate cells) and *in vivo* female low-density lipoprotein (LDL) receptor knockout mice. The treatment of LX2 cells with the extract reduced fibrotic and oxidative markers levels and increased GATA4 gene expression. In LDLR-/- mice with HFD-induced liver fibrosis and inflammation, the progression of fibrosis, oxidative stress, inflammation, and apoptosis were prevented by treatment with the extract. A potential anti-atherogenic effect was also observed in the mouse model. The results have shown expression of toll-like receptor 4 (TLR4) and reduced inflammasome activation of nod-like receptor protein 3 (NLRP3) which suggests that mushroom extract exerts protective effects by alleviating inflammation and oxidative stress during the progression of liver fibrosis (Gallego et al. 2021).

The pre-treatment of human liver L02 cells with ethanol followed by *A. auricula* melanin (AAM) showed a therapeutic effect on alcohol-induced liver injury *in vitro* and *in vivo* based on the elevation of cell viability, amelioration of cell morphology, and reduction of ROS. The effect of AAM on ethanol-induced hepatocyte injury can be used as to develop hepatoprotective drug to treat alcohol liver disease (Hou et al. 2021).

Chaga mushroom (*I. obliquus*) has been traditionally used in Asian folk medicine. *I. obliquus* polysaccharides (IOP) have been shown to reduce tacrine-induced apoptosis in HepG2 cells. Inhibition of tacrine-induced ROS generation, 8-OHdG formation in mitochondrial DNA, and loss of mitochondrial transmembrane potential by IOP were also detected. In addition, IOP decreased the release of tacrine-induced cytochrome C and caspase-3 activation. The consumption of IOP may be an approach to prevent tacrine-induced hepatotoxicity (Li et al. 2019).

The isolation of 21 known compounds and structural elucidation of ten undescribed lanostane triterpenoids from static cultured mycelium of *G. lucidum* have been reported (Zhang et al. 2022). These compounds have been investigated for their hepatoprotective activity on H_2O_2-induced HepG2 cells. The obtained results demonstrated that 12 ganoderic acid derivatives possessed significant hepatoprotective activities based on suppression of ALT, AST, and LDH enzymes and increased levels of GSH in H_2O_2-injured HepG2 cells.

The chemical structure and activity of triterpenoids, including phytosterols, antcins A, B, C, H, and K, as well as dehydroeburicoic and dehydrosulphurenic acids extracted from medicinal fungus *Taiwanofungus camphoratus,* have been intensively studied to assess their hepatoprotective activity (Hsieh et al. 2011; Du et al. 2012; Lu et al. 2020; Wang et al. 2022a). It was reported that *T. camphoratus* contains more triterpenoids than *G. lucidum* and possesses a hepatoprotective effect without genotoxicity. The commercial cultivation of *T. camphoratus* is limited due to the limited resources of the host tree, *Cinnamomum camphora*. As a substrate for *T. camphoratus* cultivation, apple wood was used to study the contents of triterpenoids and the hepatoprotective activity of this fungus. The results showed that apple-wood cultivation is feasible for the commercial cultivation of *T. camphoratus* (Wang et al. 2022a). It should be noted that antcin B causes apoptosis of hepatocellular carcinoma cells by inducing oxidative stress involving reduced nicotinamide adenine dinucleotide phosphate oxidase (Hsieh et al. 2011).

3.8 Other

Various herbal medicines, including mushrooms, are considered useful for the treatment of cardio-metabolic diseases and their risk factors (hypoglycemia, diabetes, dyslipidemia, arterial hypertension, and inflammation) (Badalyan et al. 2021). *G. lucidum* has been used to prevent and treat cardiovascular and cardio-metabolic diseases by targeting their risk factors. It is also a source of antioxidant, hypotensive, hypoglycemic, hypolipidemic, and anti-inflammatory compounds.

Further clinical studies are needed to elucidate the potential health benefits of standardized preparations of *G. lucidum* in the prevention and treatment of cardio-metabolic diseases (Chan et al. 2021).

A terphenylquinone pigment thelephoric acid (Thel) with osteogenic effects on pre-osteoblasts was isolated from wild mushroom *Polyozellus multiplex*. Thel enhanced alkaline phosphatase activity, bone matrix mineralization in pre-osteoblasts, as well as cell adhesion and migration in osteoblast differentiation. These findings indicated that Thel could be used to prevent and treat bone disorders (Park et al. 2022).

Recent data has highlighted the role of gut microbiota and its various metabolites in maintaining bone health, and the usage of prebiotics in fight against degenerative bone diseases. The prebiotic potential of medicinal mushrooms *G. lucidum* and *P. ostreatus* has been studied in healthy and osteopenic women to analyze the impact of mushroom fermentation products on human osteoblasts. The results have shown that treatment with *P. ostreatus* mushroom powder and *G. lucidum* extract had positive effects on gut microbiota and SCFA (*short-chain fatty acid* analyses) production. It decreases the levels of RANkL (receptor activator of nuclear factor-κB ligand) and enhances osteoblastic activity. Thus, *G. lucidum* and *P. ostreatus* products may exert beneficial *in vitro* effects on bone physiology by altering gut microbiota and/or SCFA production (Kerezoudi et al. 2021).

The protective effects of *Macrolepiota procera* mycelium polysaccharides (MMP) on nonylphenol (NP)-induced male reproductive dysfunction have been studied. After nonylphenol (NP) administration, declined sperm count and testis index, increased deformation rate of sperms, aberrant hormone secretion, and testicular pathological injury led to a decrease in reproductive capacity. The MMP therapy reversed the changes in NP-treated mice, decreased oxidative stress, apoptosis, autophagy, inflammatory responses, and suppressed Akt/mTOR signaling pathway in testicular tissues. The results have shown MMP to be a promising therapeutic agent against the biotoxicity of NP (Wang et al. 2022b).

Uric acid is the end product of metabolism of purine compounds. Elevated concentrations of serum uric acid may cause hyperuricemia and lead to the development of gouty arthritis, arterial hypertension, cardiovascular and renal diseases. The association between mushroom consumption and hyperuricemia has been reported. The study of several edible medicinal mushrooms, *L. edodes, A. auricula-judae, P. ostreatus, F. velutipes,* and *A. bisporus* with hyperuricemia has shown that xanthine oxidase inhibitors extracted from mushrooms may alleviate hyperuricemia. The population-based prospective cohort study has demonstrated that a mushroom-rich diet is associated with a lower incidence of hyperuricemia in adult patient cohorts (Zhang et al. 2021).

4 The Advancements in Biomedical Usage of Mushrooms

4.1 Mushroom-Derived Biotech Products

Since ancient times, mushrooms have been considered an expensive gourmet and valuable food due to their high nutritional and medicinal values (Badalyan 2012; Badalyan and Zambonelli 2019, 2023). Medicinal properties (analgesic, antibacterial, antifungal, antioxidant, antitumor, antiviral, cardioprotective, immunomodulatory, neuroprotective, etc.) have been described in several genera and species of macrofungi, such as *A. bisporus, A. blazei, A. cinnamomea, G. lucidum, G. frondosa, H. erinaceus, L. edodes, P. cornucopiae, P. eryngii, P. ostreatus* and *T. versicolor.* They have also contributed to the development of pharmaceutically active ingredients, healthy and functional foods, and cosmeceutical products (Badalyan and Zambonelli 2019, 2023; Badalyan et al. 2019; 2021; 2022).

Studies on the beneficial effects of wild and cultivated medicinal mushrooms, particularly from the genus *Pleurotus*, on the nutrition and health of humans and animals have been reported (Cateni et al. 2021). An overview of the structure and composition of mycochemicals (phenolic compounds, triterpenoids and sterols, fatty acids and lipids, polysaccharides, proteins, peptides, and lectins), as well as the potential of mushrooms in the production of biotech products, including drugs, nutraceuticals, and functional food, have been discussed (Cateni et al. 2021).

The dietary consumption of biotech products obtained from *A. bisporus, G. frondosa, G. lucidum, F. velutipes,* and *L. edodes* in the form of fruiting bodies or extracts may prevent or delay the development of gastric and breast cancers (Wasser 2017).

In vitro evaluation of bioactivities of n-hexane extracts from *D. indusiata, H. erinaceus,* and *Metacordyceps neogunnii* revealed their therapeutic significance. The extract of *D. indusiata* showed the highest AOA (87.8 ± 1.2%), followed by *H. erinaceus* (84.9 ± 1.6%) and *M. neogunnii* (77.3 ± 1.3%). The extract of *M. neogunnii* revealed cytotoxicity (68.6 ± 3.6%) against HCT116 human colon cancer cell lines at a concentration of 100 μg/mL, whereas *H. erinaceus* and *D. indusiata* extracts showed weaker cytotoxic effects at the same concentration (18.3 ± 1.7% and 19.3 ± 3.2%, respectively). The extracts also showed potent *in vitro* hypocholesterolemic effects (100 ± 0%). The GC-MS analyses revealed 22 compounds in *D. indusiata*, 29 in *M. neogunnii*, and 33 in *H. erinaceus* extracts. Most of these compounds were esters of fatty acids. These results suggest that the tested mushrooms can be used as functional foods. However, further *in vivo* studies are needed to evaluate their usage as pharmaceuticals (Daba et al. 2020; Elkhateeb et al. 2021a).

The study of therapeutic potential of orally suitable preparations for infusion and decoction from wild mushrooms *Tremella fuciformis* and *Termitomyces heimii* showed that they contained a noticeable amount of bioactive metabolites, such as phenolics, flavonoids, and carotenoids. Both infusion and decoction possess AOA and have shown anti-inflammatory activities via prevention of protein denaturation. The formulations of *T. fuciformis* and *T. heimii* may be sources of antioxidant-rich healthy beverages (Ghosh et al. 2022).

Truffles are symbiotrophic hypogeous ascomycete mushrooms that have gained scientific attention over the past years for their application in medicine (Badalyan and Zambonelli 2019, 2023). They produce bioactive molecules (phenolics, terpenoids, fatty acids, lectins, ergosteroids, and polysaccharides) with antimicrobial, antiviral, antioxidant, anti-inflammatory, anti-depressant, aphrodisiac, and anti-carcinogenic effects (Badalyan 2012). The distribution, biochemical analysis, nutritional values, and health benefits of truffles have been explored, and recent information on the potential therapeutic usage have been reported (Elkhateeb et al. 2021b; Elsayed et al. 2021; Badalyan and Zambonelli 2023).

4.2 *Pre-Clinical and Clinical Studies*

Recent advances in research of medicinal mushrooms have approved the traditional knowledge about therapeutic properties of mushroom-derived bioactive compounds and extracts, as well as their usage in mycotherapy (Hapuarachchi et al. 2017; Kües and Badalyan 2017; Badalyan et al. 2019; Badalyan and Rapior 2021a). Clinical trials in European and Asian countries indicate that biotech products derived from mushrooms are used in different preventive and therapeutic strategies and other therapies (Wasser 2011, 2014, 2017; Chaiyasut and Sivamaruthi 2017). Additional data from *in vitro* and *in vivo* studies using experimental animal models have been reported. Recent reviews focused on clinical studies of several macrofungi, such as *G. lucidum, G. frondosa, H. erinaceus,* and *T. versicolor*, in the treatment of oncological, immunological diseases and diabetes mellitus (Wasser 2017; Badalyan et al. 2019). Therefore, systematic clinical trials are required to update comprehensive knowledge on mushrooms' mycopharmacological and mycotherapeutical potential.

Contemporary clinical practices rely on mushroom-derived drugs and dietary supplements to increase the tolerance of patients to cancer therapy (Steimbach et al. 2021; Twardowski et al. 2015; Tanaka et al. 2016; Tsai et al. 2016, 2021; Wasser 2017).

The edible and inedible wild and cultivated mushrooms (*Coriolus versicolor, F. fomentarius, F. officinalis, G. lucidum, G. frondosa, L. edodes, H. erinaceus, I. obliquus, O. sinensis, Ph. linteus, P. betulinus,* and *S. commune*) with pharmacological potential are undergoing clinical studies for the treatment of gastrointestinal disorders, cancer, bronchial asthma, and other diseases (Chen et al. 2016; Wasser 2017; Badalyan et al. 2019).

The bioactive compounds along with their effects and mechanisms *in vitro* and *in vivo* preclinical studies of selected medicinal mushrooms, including *A. bisporus*, *A. blazei*, *A. cinnamomea*, *C.* (=*Trametes*) *versicolor*, *G. lucidum*, *G. frondosa*, *H. erinaceus*, *L. edodes*, and *Pleurotus* spp. have been analyzed (Venturella et al. 2021).

The authors also discussed the pharmacological activities of mushrooms in clinical trials, including cancer and neuronal health, diabetes, hyperglycemia, hyperlipidemia, and cardiovascular diseases. The increasing concern in mycotherapy requires a guarantee from the scientific community to expand clinical trials and provide supplements of safe fungal origin and proven genetic purity (Venturella et al. 2021).

Overall, more than 130 medicinal properties of mushrooms (anticancer, antioxidant, antimicrobial, immunomodulating, cardiovascular, etc.) have been described (Panda and Luyten 2022). The clinical studies were conducted on medicinal mushrooms or their mushroom-based preparations but not on isolated bioactive compounds. All clinical trials, including scientific and English names of mushroom species, the type of clinical study, disease or indication, mushroom part used, dosage and duration of treatment, study outcome, and a database identifier number, have been summarized. Overall 91 published clinical trials of 22 species were analyzed. The majority of clinical studies have been carried out with seven species: *L. edodes* (19%), *A. bisporus* (15%), *A. blazei* (14%), *G. lucidum* (9%), *P. ostreatus* (7%), *G. frondosa* (5%) and *Agaricus sylvaticus* (5%). The clinical trials were mainly conducted in humans, six on chickens, one in calves, and one in dogs. Mushrooms have mainly been studied for the treatment of cancer (16%) or for their immunomodulatory effects (14%). Around 11% of trials assessed possible fitness and health benefits, such as antioxidant properties, nutritional value, vitamin content, and food intake. Many of the clinical benefits of mushrooms may be attributed to their effect on immune system. In general, a few phase III clinical trials have been performed, and in many cases, randomized, double-blind trials include a relatively small number of patients (Panda and Luyten 2022).

Medicinal properties of raw mushroom extracts and isolated bioactive molecules should be studied using bioassay-guide purification *in vitro* tests, on different cancer cell lines, or even *in vivo* animal models to test their activities on different cancers (Panda et al. 2022). Most *in vitro* cell studies use breast (43.9%), lung (14%), and colorectal (13.1%) cancer cell lines. In contrast, *in vivo* animal studies are mainly performed in mouse tumor models (58.7%). Preclinical studies may be considered for isolated bioactive molecules, as well as clinical studies, including phases from I to IV. Although around 30 mushroom species are promising for the treatment of cancer, only 11 species appear to have been clinically tested. Several mushrooms have been tested in phases I or II clinical trials, primarily for the treatment of breast (18.6%), colorectal (14%), and prostate (11.6%) cancers. Most clinical studies were performed with five species: *L. edodes* (22.2%), *C. versicolor* (13.9%), *G. lucidum* (13.9%), *A. bisporus* (11.1%) and *G. frondosa* (11.1%), involved a small number of patients and were limited to phases III or IV. Despite the availability of preclinical and clinical data, more convincing results are expected to approve the therapeutic values of mushrooms in oncology (Wasser 2017; Panda et al. 2022).

In many clinical studies, mushrooms were used as an adjuvant therapy with conventional chemo- or radiotherapy for treatment of different types of cancer (breast, stomach, liver, lung, prostate, ovary, etc.) to reduce side effects (e.g., hair loss, nausea and loss of appetite). However, no evidence has revealed appropriate dosages and duration of treatment for many species, sometimes due to poor study design, non-standardized mushroom preparations, inappropriate statistical methods, etc. (Wasser 2017; Badalyan et al. 2019).

Mycelial and fruiting body extracts, mycelial biomass, β-glucans (e.g., lentinan, Maitake D-fraction, schizophyllan, PSK, or Krestin) have been tested in clinical studies to treat oncological diseases (Ina et al. 2016; Wasser 2017; Badalyan et al. 2019). Future randomized placebo-controlled cross-over studies will elucidate the efficacy of mycotherapy, the effects of mushroom-derived anticancer compounds on long-term survival, tumor response, host immune functions, inflammation, and quality of life of cancer patients (Badalyan et al. 2019).

The oral administration of *G. lucidum*-derived β-glucans has shown a potent immunomodulatory effect *in vitro* and *in vivo*. A randomized, double-blinded, placebo-controlled clinical study was performed on asymptomatic three to five-year-old children (Henao et al. 2018). The results have shown that *G. lucidum*-derived β-glucans increased the number of immune cells in the peripheral blood, which are critical in defence against pediatric infectious diseases. However, these findings warrant clinical trials to prevent infections in healthy children and define their potential to increase lymphoid cell count during various lymphoid immune deficiencies (Henao et al. 2018).

Six preclinical and three clinical studies of the therapeutic effects of *H. erinaceus* in Alzheimer's disease have been reviewed. Preclinical trials conducted from 2011 to 2021 successfully demonstrated that extracts and bioactive compounds obtained from *H. erinaceus* have beneficial effects in improving cognitive function and behavioral deficit in animal models (mice and rats) (Yanshree et al. 2022). Three double-blind placebo-controlled clinical studies have shown similar results to preclinical studies (Mori et al. 2009, 2011; Saitsu et al. 2019; Li et al. 2020). Nevertheless, future research on *H. erinaceus* should focus on elucidating specific neuroprotective mechanisms and target sites in Alzheimer's disease.

In a previous study, Vetvicka and Vetvickova (2015) showed that the immunomodulatory effects of β-glucans were used to prevent viral infections; feeding mice infected with the influenza virus for two weeks with a mixture of glucans extracted from fruiting bodies of *G. frondosa* and mycelium of *Agaricus brasiliensis*, *I. obliquus,* and *L. edodes* improved the clinical symptoms of the disease. These results suggested that consumption of dietary glucans could be useful as a complementary or alternative approach for treatment of influenza infection. Further application of β-glucans and study of structure-activity relationship to develop functional food products fortified with β-glucans should be performed to prove this effect.

A randomized, double-blind, placebo-controlled study in pediatric patients with regular respiratory infections has shown that treatment with pleuran, a β-glucan from

P. ostreatus, reduced infection-related symptoms of atopy, associated with increased immune responses to common allergens (Jesenak et al. 2014).

Hapuarachchi et al. (2017) reviewed the medicinal properties of 30 species of *Ganoderma* (except *G. lucidum*) and their secondary metabolites; according to the authors, there is no evidence to support the use of *Ganoderma* species as potential food supplements for treatment of cancer or other diseases in humans, because no preclinical test has been carried out to date.

Preclinical trials have suggested that *G. lucidum* possesses promising anticancer and immunomodulatory properties. Over 250 clinical studies on *G. lucidum* and other *Ganoderma* species have been published (Jin et al. 2012). In this study patients with *G. lucidum* extract in their cancer regimen were 1.27 times more likely to respond to chemotherapy or radiotherapy than those who did not.

The results of a double-blind, randomized, placebo-controlled trial found no significant effect of the administration of 3 g/day *G. lucidum* or a combination of *G. lucidum* with *C. sinensis* for 16 weeks on HbA1c and fasting plasma glucose levels of a small number of patients with type 2 diabetes mellitus (Klupp et al. 2016). *G. lucidum* is also widely used to treat cardiovascular diseases (Badalyan et al. 2021).

According to Tangen et al. (2015), 40 patients with multiple myeloma scheduled to undergo high-dose chemotherapy with autologous stem cell support were randomized to receive adjuvant therapy with Andosan^{TM} mushroom extract containing 82% ABM (19 patients) or placebo (21 patients). Increased quantities of Treg cells and plasmacytoid dendritic cells were found in leukapheresis products harvested after stem cell mobilization from patients receiving Andosan^{TM}. The genome microarray demonstrated increased expression of immunoglobulin genes, killer immunoglobulin receptor genes, and HLA genes in the *Agaricus* group. This study was registered in Clinicaltrials.gov NCT00970021 (Tangen et al. 2015).

Ex vivo and *in vivo* investigations have shown anticancer, anti-inflammatory, antimicrobial, antimutagenic, antioxidant, anti-parasitic, hepatoprotective, hypodlycemic, and immunomodulatory activities in *A. blazei* Murrill sensu Heinemann (syn. *A. subrufescens*). At the same time, only 17 clinical studies and two case reports on ABM were found (Therkelsen et al. 2016). A study reported the nutritional and therapeutic properties of ABM with emphasis on its chemical composition (Da Silva de Souza et al. 2017). However, further clinical trials using reliable statistical methods and standardized preparations are needed to evaluate the efficacy of ABM as a therapeutic agent.

The medicinal properties (antitumor, immunomodulatory, anti-coagulatory, anti-inflammatory, antimicrobial, antioxidant, antiviral, neuroprotective, etc.) of *L. rhinocerotis* have been reviewed by Nallathamby et al. (2018). *In vitro* investigations have revealed that fruiting bodies and sclerotia of *L. rhinocerotis* may be considered alternative therapeutic resources in the management of non-communicable diseases. Therefore, further studies, including *in vivo* clinical trials, are needed to scientifically validate the application of chemicals derived from *L. rhinocerotis* as pharmaceuticals and medicines.

A prospective randomized study of 25 patients comparing the efficacy of *H. erinaceus* versus essential oils against *Helicobacter pylori* infection revealed that patients who received *H. erinaceus* had negative Pyloritop® test results in 89.5% of cases versus 33.3% in patients who started with essential oils (Donatini 2014). It has been suggested that *H. erinaceus* could be an alternative to antibiotic therapy for *H. pylori*-associated pathology. Additional randomized clinical trials versus reference therapy should be performed to focus on treatment without adverse effects (Donatini 2014).

A prospective phase II, randomized, simple, double-blind, placebo-controlled trial was conducted using cereal bars made from *L. edodes* (Spim et al. 2021a). Men and women aged 20-65 with at least one biochemical marker of borderline hypercholesterolemia were recruited. Sixty-eight people were randomly assigned to group I (placebo; $n = 32$) or group II (intervention; $n = 36$). Sweetened cereal bars (25 g) containing dry shiitake (3.5 g) have been consumed. A daily concentration of shiitake of 100 mg/kg is safe in a study conducted with ingestion of different concentrations of *L. edodes*. Participants in the intervention group showed a 10% reduction in triglycerides after 66 days of consuming *L. edodes* cereal bars. Moreover, after 33 days of consumption, the AOA on reduced lipid peroxidation was observed. However, the consumption of *L. edodes*-based bars caused dermatitis in 10% of individuals (Spim et al. 2021b). Therefore, this study aimed to analyze the effects of shiitake bars on borderline cholesterol and oxidative stress in humans through a phase II double-blind, randomized trial. The authors of a recent study have shown that the sweet bar 1 (SwB1) had better sensory analysis due to its stability, low production cost, and good acceptance, as well as the flexibility to add other beneficial ingredients, such as shiitake (Spim et al. 2021a).

Clinical studies of a novel bioactive substance, derived from *I. obliquus,* for cancer patients depending on the topography of tumor, disease stage, and patient age, as well as the pharmacokinetics and pharmacodynamics of the compound, were performed. The mechanisms of effects of fungal extracts obtained by different extraction methods were revealed. The recommendations for using *I. obliquus* to improve clinical outcomes in cancer patients were discussed (Khoroshutin et al. 2021).

Recently, Chan et al. (2022) investigated erinacine A-enriched *H. erinaceus* (HE) on hearing degeneration through a double-blind, randomized, placebo-controlled clinical trial: 80 hearing-impaired patients aged 50–79 were randomly divided into two groups. The results demonstrated that treatment with HE mycelium could improve hearing function, especially for high frequencies and speech recognition. This effect was observed in hearing-impaired patients of age ≧65 years.

The triterpenes of *Pleurotus tuber-regium* sclerotium have recently been investigated for their hypolipidemic effects (Wang et al. 2022c). In *in vivo* experiments, the authors showed that zebrafish tolerated total triterpenes of mushroom nuclei at a maximum concentration of 500 μg/mL. In addition, the results demonstrated that total triterpenes derived from mushroom nuclei reduced a dose effect lipid accumulation in zebrafish induced by a high-fat diet.

Based on the results of clinical studies involving medicinal mushrooms, high quality, long-term, randomized, double-blind, placebo-controlled clinical trials should be further pursued.

5 Conclusion and Future Prospects

Mushroom-derived bioactive compounds possess various medicinal properties, including immunomodulatory, anti-cancer, antiviral, antioxidant, anti-inflammatory, etc. In addition, the treatment with conventional anticancer drugs poses tremendous challenges and limitations, such as poor solubility, narrow therapeutic window, and cytotoxicity, which may cause side-effects in cancer patients.

Mushrooms are becoming applicable in nanomedicine; mushroom-derived nano-emulsion has recently been suggested for cancer therapy (Chandrawanshi et al. 2022). The therapeutic potential of AgNPs coupled with natural mushroom extracts has been extensively evaluated in various diseases. The synthesis of AgNPs using *C. versicolor* (CV-AgNPs) and *Boletus edulis* (BE-AgNPs) crude extracts was carried out (Kaplan et al. 2021). Strong antimicrobial activity of AgNPs obtained from both species was reported against Gram-negative (*P. aeruginosa* and *K. pneumoniae*) and Gram-positive (*S. aureus* and *E. faecalis*) bacteria rather than against *Candida* strains (*C. albicans* and *C. utilis*). The anti-proliferative activity of AgNPs has been detected, as well. Both species-derived AgNPs decrease proliferation of MCF-7 (breast adenocarcinoma), HT-29 (colorectal adenocarcinoma) and HUH-7 (hepatocellular carcinoma) cell lines. The synthesized nanoparticles have also shown a wound-healing effect at low concentrations on L929 cells. The obtained data may lead to an efficient nanoparticle system for drug design and delivery in the treatment of infectious diseases and cancer (Kaplan et al. 2021).

The synthesized AgNPs using *Tricholoma ustale* and *Agaricus arvensis* extracts exhibited antibacterial activity against *P. aeruginosa*, *K. pneumoniae*, *S. aureus*, *E. faecalis*, and antifungal activity against *C. albicans* and *C. utilis* strains (Kaplan et al. 2022). The AgNPs showed an anti-proliferative effect on human breast cancer (MCF-7), lung cancer (A549), osteosarcoma (Saos-2), and colon cancer (HT-29) cell lines, as well. The AgNPs stimulated intrinsic apoptotic signaling pathways via up-regulation of Bax/Bcl-2 and decreased pro-Casp9 expression in MCF-7, Saos-2, and HT-29 cells. Thus, the mushroom-derived AgNPs may be a potential metal-based nanoparticle system to treat infectious diseases and cancer (Kaplan et al. 2021, 2022).

Several human diseases, such as diabetes, cancer, cardiovascular and neurodegenerative disorders, affect adult population worldwide. Therefore, scientists have tried to discover novel natural sources of medicines, especially from plants and mushrooms, to prevent and treat these pathological conditions. This review addressed the current state of knowledge and findings of recent studies on mushroom-derived healthy food and therapeutics. Although the list of studied medicinal mushrooms and clinical approval for the application of their bioactive compounds as therapeutic agents is incomplete, this work represents an updated contribution to modern mycopharmacology, mycotherapy, and biomedicine.

Future interdisciplinary research, involving physicians, biologists, chemists, pharmacologists, and mycologists is warranted to create a basis of scientific and traditional knowledge with biomedical data for further usage of therapeutic potential of mushrooms for human welfare.

Acknowledgments This review arises from a long-standing collaboration between Yerevan State University (Armenia) and the University of Montpellier, CEFE (France) on research aimed at the identification of bioactive compounds based on the biodiversity of medicinal mushrooms to develop their pharmacological and nutritional properties and consider their biotechnological potential for human health. The work was supported by the Science Committee of Republic of Armenia, in the frames of research project n. 21T-1F228.

We are grateful to our colleagues Jean-Marie Bessière (University of Montpellier, France), RE Nebel (Kyoto Prefecture, Japan), Denis Nespoulous (University of Montpellier, France), and Jean-Philippe Rioult (EREM, Caen, France), for kindly providing photos of medicinal mushrooms (Fig. 1). We also thank Samantha C. Karunarathna (Qujing Normal University, Yunnan, China) for helping us find some photos of mushrooms (Fig. 1).

The authors have no conflict of interest that would raise questions about their independence.

References

Acharya K, Bera I, Khatua S et al (2015) Pharmacognostic standardization of *Grifola frondosa*: a well-studied medicinal mushroom. Der Pharm Lett 7:72–78

Acharya K, Ghosh S, Kundu I (2016) Pharmacognostic standardization of a well-known edible mushroom, *Volvariella volvacea*. J Appl Pharm Sci 6(11):185–190. https://doi.org/10.7324/JAPS.2016.601129

Akgul H, Aslan A, Akata I et al (2021) Phenolic content and biological activities of *Trametes hirsuta*. Fresenius Environ Bull 30(04A):4130–4135

Akiba M, Kinoshita K, Kino Y et al (2020) Isolation of three new meroterpenoids and seven known compounds from *Albatrellus yasudae* and their Aβ-aggregation inhibitory activity. Biorg Med Chem Lett 30(2):126808. https://doi.org/10.1016/j.bmcl.2019.126808

Angelova G, Brazkova M, Mihaylova D et al (2022) Bioactivity of biomass and crude exopolysaccharides obtained by controlled submerged cultivation of medicinal mushroom *Trametes versicolor*. J Fungi 8(7):738. https://doi.org/10.3390/jof8070738

Angulo-Sanchez LT, López-Peña D, Torres-Moreno H et al (2022) Biosynthesis, gene expression, and pharmacological properties of triterpenoids of *Ganoderma* species (Agaricomycetes): a review. Int J Med Mushrooms 24(6):1–17. https://doi.org/10.1615/IntJMedMushrooms.2022044016

Aras A, Khalid S, Jabeen S et al (2018) Regulation of cancer cell signaling pathways by mushrooms and their bioactive molecules: overview of the journey from benchtop to clinical trials. Food Chem Toxicol 119:206–214. https://doi.org/10.1016/j.fct.2018.04.038

Ayimbila F, Siriwong S, Nakphaichit M et al (2022) *In vitro* gastrointestinal digestion of *Lentinus squarrosulus* powder and impact on human fecal microbiota. Sci Rep 12(1):2655. https://doi.org/10.1038/s41598-022-06648-z

Ayzenberg I, Hoepner R, Kleiter I (2016) Fingolimod for multiple sclerosis and emerging indications: appropriate patient selection, safety precautions, and special considerations. Ther Clin Risk Manag 12:261–272. https://doi.org/10.2147/TCRM.S65558

Azeem U, Shri R, Dhingra GS (2021) *In vitro* and *in vivo* antihyperglycemic activities of medicinal mushrooms (Agaricomycetes) from India. Int J Med Mushrooms 23(2):29–41. https://doi.org/10.1615/IntJMedMushrooms.2021037630

Azeem U, Shri R, Dhingra GS (2022) Mineral elements and vitamins from wild wood inhabiting basidiocarps of some medicinal mushrooms (Agaricomycetes) from India. Int J Med Mushrooms 24(4):53–62. https://doi.org/10.1615/IntJMedMushrooms.2022043411

Baby S, Johnson AJ, Govindan B (2015) Secondary metabolites from *Ganoderma*. Phytochemistry 114:66–101. https://doi.org/10.1016/j.phytochem.2015.03.010

Badalyan SM (2012) Medicinal aspects of edible ectomycorrhizal mushrooms. In: Zambonelli A, Bonito G (eds) Edible ectomycorrhizal mushrooms, current knowledge and future prospects, vol 34. Springer-Verlag, Berlin/Heidelberg, pp 317–334

Badalyan SM (2016) Fatty acid composition of different collections of coprinoid mushrooms (Agaricomycetes) and their nutritional and medicinal values. Int J Med Mushrooms 18(10): 883–893. https://doi.org/10.1615/IntJMedMushrooms.v18.i10.40

Badalyan SM (2020) Medicinal coprinoid mushrooms (Agaricomycetes) distributed in Armenia (Review). Int J Med Mushrooms 22(3):257–267. https://doi.org/10.1615/IntJMedMushrooms.2020033981

Badalyan SM, Barkhudaryan A, Rapior S (2019) Recent progress in research on the pharmacological potential of mushrooms and prospects for their clinical application. In: Agrawal D, Dhanasekaran M (eds) Medicinal mushrooms—Recent progress in research and development. Springer Nature, Singapore, pp 1–70. https://doi.org/10.1007/978-981-13-6382-5_1

Badalyan SM, Barkhudaryan A, Rapior S (2021) The cardioprotective properties of Agaricomycetes mushrooms growing in the territory of Armenia (Review). Int J Med Mushrooms 23(5):21–31. https://doi.org/10.1615/IntJMedMushrooms.2021038280

Badalyan SM, Barkhudaryan A, Rapior S (2022) Medicinal macrofungi as cosmeceuticals: a review. Int J Med Mushrooms 24(4):1–13. https://doi.org/10.1615/IntJMedMushrooms.2022043124

Badalyan SM, Gharibyan NG (2020) Pharmacological properties and resource value of hymenochaetoid fungi (Agaricomycetes) distributed in Armenia: review. Int J Med Mushrooms 22(12):1135–1146. https://doi.org/10.1615/IntJMedMushrooms.2020037092

Badalyan SM, Rapior S (2020) Perspectives of biomedical application of macrofungi. Curr Trends Biomed Eng Biosci 19(5):ID.556024. https://doi.org/10.19080/CTBEB.2020.19.556024

Badalyan SM, Rapior S (2021a) Agaricomycetes mushrooms (Basidiomycota) as potential neuroprotectants. Ital J Mycol 50:30–43. https://doi.org/10.6092/issn.2531-7342/12542

Badalyan SM, Rapior S (2021b) The neuroprotective potential of macrofungi. In: Agrawal D, Dhanasekaran M (eds) Medicinal herbs and fungi—Neurotoxicity vs. neuroprotection. Springer Nature, Singapore, pp 37–77. https://doi.org/10.1007/978-981-33-4141-8_2

Badalyan SM, Zambonelli A (2019) Biotechnological exploitation of macrofungi for the production of food, pharmaceuticals and cosmeceuticals. In: Sridhar KR, Deshmukh SK (eds) Advances in macrofungi: Diversity, ecology and biotechnology, 1st edn. CRC Press, Boca Raton, pp 199–230

Badalyan SM, Zambonelli A (2023) The potential of mushrooms to develop healthy food and biotech products. In: Satyanarayana T, Deshmukh SK (eds) Fungi and fungal products in human welfare and biotechnology. Springer Nature, Singapore (in press)

Bal C, Baba H, Akata I et al (2022) Biological activities of wild poisonous mushroom *Entoloma sinuatum* (Boletales). KSU J Agric Nat 25(1):83–87. https://doi.org/10.18016/ksutarimdoga.vi.880151

Bao TR, Long GQ, Wang Y et al (2022) New lanostane-type triterpenes with anti-inflammatory activity from the epidermis of *Wolfiporia cocos*. J Agric Food Chem 70:4418–4433. https://doi.org/10.1021/acs.jafc.2c00823

Barros L, Cruz T, Baptista P (2008) Wild and commercial mushrooms as source of nutrients and nutraceuticals. Food Chem Toxicol 46:2742–2747. https://doi.org/10.1016/j.fct.2008.04.030

Bleha R, Třešňáková L, Sushytskyi L et al (2022) Polysaccharides from basidiocarps of the polypore fungus *Ganoderma resinaceum*: isolation and structure. Polymers 14(2):255. https://doi.org/10.3390/polym14020255

Bolaniran T, Jamiu AT, Garuba T et al (2021) An appraisal of the metabolites, pharmacological and biotechnological significance of edible mushrooms. Trans R Soc South Africa 75(3):257–272. https://doi.org/10.1080/0035919X.2021.1947921

Buruleanu LC, Radulescu C, Georgescu AA et al (2018) Statistical characterization of the phytochemical characteristics of edible mushroom extracts. Anal Lett 51(7):1039–1059. https://doi.org/10.1080/00032719.2017.1366499

Butkhup L, Samappito W, Jorjong S (2018) Evaluation of bioactivities and phenolic contents of wild edible mushrooms from northeastern Thailand. Food Sci Biotechnol 27(1):193–202. https://doi.org/10.1007/s10068-017-0237-5

Castellano G, Torrens F (2015) Information entropy-based classification of triterpenoids and steroids from *Ganoderma*. Phytochemistry 116:305–313. https://doi.org/10.1016/j.phytochem.2015.05.008

Cateni F, Gargano ML, Procida G et al (2021) Mycochemicals in wild and cultivated mushrooms: nutrition and health. Phytochem Rev 21:339–383. https://doi.org/10.1007/s11101-021-09748-2

Chaiyasut C, Sivamaruthi BC (2017) Anti-hyperglycemic property of *Hericium erinaceus*—A mini review. Asian Pac J Trop Biomed 7:11. https://doi.org/10.1016/j.apjtb.2017.09.024

Chan SW, Tomlinson B, Chan P et al (2021) The beneficial effects of *Ganoderma lucidum* on cardiovascular and metabolic disease risk. Pharm Biol 59(1):1159–1169. https://doi.org/10.1080/13880209.2021.1969413

Chan YC, Lin TC, Chen CC et al (2022) Effects of erinacine A-enriched *Hericium erinaceus* on elderly hearing-impaired patients: A double-blind, randomized, placebo-controlled clinical trial. J Funct Foods 97(4):105220. https://doi.org/10.1016/j.jff.2022.105220

Chandrawanshi NK, Koreti D, Kosre A et al. (2022) Mushroom-derived bioactive-based nanoemulsion: current status and challenges for cancer therapy. In: Handbook of research on nanoemulsion applications in agriculture, food, health, and biomedical sciences. IGI Global. pp. 354–376. https://doi.org/10.4018/978-1-7998-8378-4.ch016

Chang ST, Wasser SP (2012) The role of culinary-medicinal mushrooms on human welfare with a pyramid model for human health. Int J Med Mushrooms 14(2):95–134. https://doi.org/10.1615/IntJMedMushr.v14.i2.10

Chaturvedi VK, Agarwal S, Gupta KK et al (2018) Medicinal mushroom: boon for therapeutic applications. 3 Biotech 8(8):334. https://doi.org/10.1007/s13205-018-1358-0

Chaturvedi VK, Dubey SK, Singh MP (2019) Phytochemicals from medicinal plants. Antidiabetic potential of medicinal mushrooms. In: Phytochemicals from medicinal plants scope, applications, and potential health claims. AAP & CRC Press, New Jersey, USA, pp. 137–157. doi: https://doi.org/10.1201/9780429203220-7

Chen H, Miao H, Feng YL et al (2014) Metabolomics in dyslipidemia. Adv Clin Chem 66:101–119. https://doi.org/10.1016/B978-0-12-801401-1.00004-9

Chen H, Tian T, Miao H et al (2016) Traditional uses, fermentation, phytochemistry and pharmacology of *Phellinus linteus*: a review. Fitoterapia 113:6–26. https://doi.org/10.1016/j.fitote.2016.06.009

Chen JT, Tominaga K, Sato Y et al (2010) Maitake mushroom (*Grifola frondosa*) extract induces ovulation in patients with polycystic ovary syndrome: a possible monotherapy and a combination therapy after failure with first-line clomiphene citrate. J Altern Complement Med 16(12): 1295–1299. https://doi.org/10.1089/acm.2009.0696

Chiba K (2020) Discovery of fingolimod based on the chemical modification of a natural product from the fungus, *Isaria sinclairii*. J Antibiot 73(10):666–678. https://doi.org/10.1038/s41429-020-0351-0

Chong PS, Poon CH, Roy J et al (2021) Neurogenesis-dependent antidepressant-like activity of *Hericium erinaceus* in an animal model of depression. Chin Med 16:132. https://doi.org/10.1186/s13020-021-00546-8

Contato AG, Vaz de Araújo CA, Zanzarin DM et al (2022) Biological characterization and antimicrobial bioactives of mycelium extracts from medicinal mushrooms *Phellinus linteus* and *Pleurotus albidus* (Agaricomycetes). Int J Med Mushrooms 24(6):47–55. https://doi.org/10.1615/IntJMedMushrooms.2022043852

Corrêa RCG, Peralta RM, Bracht A et al (2017) The emerging use of mycosterols in food industry along with the current trend of extended use of bioactive phytosterols. Trends Food Sci Technol 67:19–35. https://doi.org/10.1016/j.tifs.2017.06.012

Cuello-Oderiz C, McGraw C (2022) Diagnostic challenges of fingolimod-associated PML-IRIS. Neuroimmunol Rep 2:100067. https://doi.org/10.1016/j.nerep.2022.100067

Da Silva de Souza AC, Corrêa VG, Goncalves GA et al (2017) *Agaricus blazei* bioactive compounds and their effects on human health: benefits and controversies. Curr Pharm Des 23(19):2807–2834. https://doi.org/10.2174/1381612823666170119093719

Daba GM, Elkhateeb W, El-Dien AN et al (2020) Therapeutic potentials of n-hexane extracts of the three medicinal mushrooms regarding their anti-colon cancer, antioxidant, and hypocholesterolemic capabilities. Biodivers J Biol Divers 21(6):2437–2445. https://doi.org/10.13057/biodiv/d210615

Das A, Chen CM, Mu SC et al (2022a) Medicinal components in edible mushrooms on diabetes mellitus treatment. Pharmaceutics 14(2):436. https://doi.org/10.3390/pharmaceutics14020436

Das S, Janardhanan KK, Thampi BSH (2022b) Bioactive extract of morel mushroom, *Morchella esculenta* (Ascomycota) attenuates doxorubicin-induced oxidative stress leading to myocardial injury. Int J Med Mushrooms 24(8):31–44. https://doi.org/10.1615/IntJMedMushrooms.2022044516

Dasgupta A, Acharya K (2019) Mushrooms: an emerging resource for therapeutic terpenoids. 3 Biotech 9:369. https://doi.org/10.1007/s13205-019-1906-2

Dasgupta A, Dey D, Ghosh D et al (2019) Astrakurkurone, a sesquiterpenoid from wild edible mushroom, targets liver cancer cells by modulating bcl-2 family proteins. IUBMB Life 71(7):99–1002. https://doi.org/10.1002/iub.2047

Derfuss T, Ontaneda D, Nicholas J et al (2016) Relapse rates in patients with multiple sclerosis treated with fingolimod: subgroup analyses of pooled data from three phase 3 trials. Mult Scler Relat Disord 8:124–130. https://doi.org/10.1016/j.msard.2016.05.015

Dhakal S, Kushairi N, Phan CW et al (2019) Dietary polyphenols: a multifactorial strategy to target Alzheimer's disease. Int J Mol Sci 20(20):E5090. https://doi.org/10.3390/ijms20205090

Diallo I, Boudard F, Morel S et al (2020) Antioxidant and anti-inflammatory potential of Shiitake culinary-medicinal mushroom, *Lentinus edodes* (Agaricomycetes), sporophores from various culture conditions. Int J Med Mushrooms 22(6):535–546. https://doi.org/10.1615/IntJMedMushrooms.2020034864

Diallo I, Morel S, Vitou M et al (2021) Ergosterol and amino acids contents of culinary-medicinal Shiitake from various culture conditions. Proceedings 70(1):78. https://doi.org/10.3390/foods_2020-07702

Din SR, Zhong M, Nisar MA et al (2020) Latcripin-7A, derivative of *Lentinula edodes* C_{91-3}, reduces migration and induces apoptosis, autophagy, and cell cycle arrest at G_1 phase in breast cancer cells. Appl Microbiol Biotechnol 104(23):10165–10179. https://doi.org/10.1007/s00253-020-10918-z

Dinis-Oliveira RJ (2017) Metabolism of psilocybin and psilocin: clinical and forensic toxicological relevance. Drug Metab Rev 49(1):84–91. https://doi.org/10.1080/03602532.2016.1278228

Dokhaharani SC, Ghobad-Nejhad M, Moghimi H et al (2021) Biological activities of two polypore macrofungi (Basidiomycota) and characterization of their compounds using HPLC–DAD and LC–ESI–MS/MS. Folia Microbiol (Praha) 66(5):775–786. https://doi.org/10.1007/s12223-021-00884-y

Donatini B (2014) Control of oral human papillomavirus (HPV) by medicinal mushrooms, *Trametes versicolor* and *Ganoderma lucidum*: a preliminary clinical trial. Int J Med Mushrooms 16(5):497–498. https://doi.org/10.1615/IntJMedMushrooms.v16.i5.80

Du YC, Wu TY, Chang FR et al (2012) Chemical profiling of the cytotoxic triterpenoid-concentrating fraction and characterization of ergostane stereo-isomer ingredients from *Antrodia camphorata*. J Pharm Biomed Anal 25(58):182–192. https://doi.org/10.1016/j.jpba.2011.09.007

Elkhateeb W, Elnahas M, Daba G (2021a) Infrequent current and potential applications of mushrooms. In: Sridhar KR, Deshmukh SK (eds) Advances in macrofungi: industrial avenues and prospects. CRC Press, pp 70–81. https://doi.org/10.1201/9781003096818-7

Elkhateeb W, Thomas P, Elnahas M et al (2021b) Hypogeous and epigeous mushrooms in human health. In: Sridhar KR, Deshmukh SK (eds) Advances in macrofungi: pharmaceuticals and cosmeceuticals. CRC Press, pp 7–19. https://doi.org/10.1201/9781003191278-2

Elkhateeb WA, Daba GM (2022a) Bioactive potential of some fascinating edible mushrooms *Flammulina, Lyophyllum, Agaricus, Boletus, Lentinula, Pleurotus* as a treasure of multipurpose therapeutic natural product. Pharm Res 6(1):000263. https://doi.org/10.23880/oajpr-16000263

Elkhateeb WA, Daba GM (2022b) *Cordyceps* more than edible mushroom—A rich source of diverse bioactive metabolites with huge medicinal benefits. J Biomed Res Environ Sci 3(5): 566–574. https://doi.org/10.37871/jbres1481

El-Ramady H, Abdalla N, Badgar K et al (2022) Edible mushrooms for sustainable and healthy Human food: nutritional and medicinal attributes. Sustainability 14(9):4941. https://doi.org/10.3390/su14094941

Elsayed EA, Alsaheb RA, Ho T et al (2021) Truffles a potential source of bioactive compounds. In: Sridhar KR, Deshmukh SK (eds) Advances in macrofungi: industrial avenues and prospects. CRC Press, pp 163–174. https://doi.org/10.1201/9781003096818-12

Ferreira ICFR, Heleno SA, Reis FS et al (2015) Chemical features of *Ganoderma* polysaccharides with antioxidant, antitumor and antimicrobial activities. Phytochemistry 114:38–55. https://doi.org/10.1016/j.phytochem.2014.10.011

Gallego P, Luque-Sierra A, Falcon G et al (2021) White button mushroom extracts modulate hepatic fibrosis progression, inflammation, and oxidative stress *in vitro* and in *LDLR*-/-mice. Foods 10(8):1788. https://doi.org/10.3390/foods10081788

Gallo AL, Soler F, Pellizas C et al (2022) Polysaccharide extracts from mycelia of *Ganoderma australe*: effect on dendritic cell immunomodulation and antioxidant activity. J Food Meas Charact 16:3251–3262. https://doi.org/10.1007/s11694-022-01444-9

Garádi Z, Dékány M, Móricz ÁM et al (2021) Antimicrobial, antioxidant and antiproliferative secondary metabolites from *Inonotus nidus-pici*. Molecules 26(18):5453. https://doi.org/10.3390/molecules26185453

Garcia J, Rodrigues F, Castro F et al (2021) Antimicrobial, antibiofilm, and antioxidant properties of *Boletus edulis* and *Neoboletus luridiformis* against multidrug-resistant ESKAPE pathogens. Front Nutr 8:773346. https://doi.org/10.3389/fnut.2021.773346

Ghosh S, Dewanjee D, Acharya K (2022) Exploring therapeutic efficacy of infusion and decoction of two wild edible mushrooms from West Bengal, India. Kavaka 58(1):11–14. https://doi.org/10.36460/Kavaka/58/1/2022/11-14

Girometta C (2019) Antimicrobial properties of *Fomitopsis officinalis* in the light of its bioactive metabolites: a review. Mycology 10(1):32–39. https://doi.org/10.1080/21501203.2018.1536680

Gopal J, Sivanesan I, Muthu M et al (2022) Scrutinizing the nutritional aspects of Asian mushrooms, its commercialization and scope for value-added products. Nutrients 14:3700. https://doi.org/10.3390/nu14183700

Górska-Jakubowska S, Klimaszewska M, Podsadni P et al (2021) Selenium-containing exopolysaccharides isolated from the culture medium of *Lentinula edodes*: structure and biological activity. Int J Mol Sci 22(23):13039. https://doi.org/10.3390/ijms222313039

Habtemariam S (2019) The chemistry, pharmacology and therapeutic potential of the edible mushroom *Dictyophora indusiata* (Vent ex. Pers.) Fischer (Syn. *Phallus indusiatus*). Biomedicines 7(4):98. https://doi.org/10.3390/biomedicines7040098

Hamad D, El-Sayed H, Ahmed W et al (2022) GC-MS analysis of potentially volatile compounds of *Pleurotus ostreatus* polar extract: *in vitro* antimicrobial, cytotoxic, immunomodulatory, and antioxidant activities. Front Microbiol 13:834525. https://doi.org/10.3389/fmicb.2022.834525
Han B, Baruah K, Cox E et al (2020) Structure-functional activity relationship of β-glucans from the perspective of immunomodulation: a mini-review. Front Immunol 11:658. https://doi.org/10.3389/fimmu.2020.00658
Hapuarachchi KK, Cheng CR, Wen TC et al (2017) Mycosphere Essays 20: Therapeutic potential of *Ganoderma* species: insights into its use as traditional medicine. Mycosphere 8(10): 1653–1694. https://doi.org/10.5943/mycosphere/8/10/5
Haq FU, Imran M, Saleem S et al (2022) Antibacterial activity of different extracts of ascomata of *Morchella conica* and *M. esculenta* (Ascomycota) against *Salmonella* species. Int J Med Mushrooms 24(9):85–95. https://doi.org/10.1615/IntJMedMushrooms.2022044572
Hawskworth DL (2012) Global species numbers of fungi: are tropical studies and molecular approaches contributing to a more robust estimate? Biodivers Conserv 21:2425–2433. https://doi.org/10.1007/s1531-012-0335-x
He MQ, Zhao RL, Liu DM et al (2022) Species diversity of basidiomycota. Fungal Divers 114:281–325. https://doi.org/10.1007/s13225-021-00497-3
He PY, Hou YH, Yang Y et al (2021a) The anticancer effect of extract of medicinal mushroom *Sanghuangprous vaninii* against human cervical cancer cell via endoplasmic reticulum stress-mitochondrial apoptotic pathway. J Ethnopharmacol 279:114345. https://doi.org/10.1016/j.jep.2021.114345
He YY, Liu S, Newburg DS (2021b) Musarin, a novel protein with tyrosine kinase inhibitory activity from *Trametes versicolor*, inhibits colorectal cancer stem cell growth. Biomed Pharmacother 144:112339. https://doi.org/10.1016/j.biopha.2021.112339
Heleno SA, Barros L, Martins A et al (2015b) Nutritional value, bioactive compounds, antimicrobial activity and bioaccessibility studies with wild edible mushrooms. LWT – Food Sci Technol 63(2):799–806. https://doi.org/10.1016/j.lwt.2015.04.028
Heleno SA, Ferreira RC, Antonio AL et al (2015a) Nutritional value, bioactive compounds and antioxidant properties of three edible mushrooms from Poland. Food Biosci 11:48–55. https://doi.org/10.1016/j.fbio.2015.04.006
Henao SL, Urrego SA, Cano AM et al (2018) Randomized clinical trial for the evaluation of immune modulation by yogurt enriched with β-glucans from Lingzhi or Reishi medicinal mushroom, *Ganoderma lucidum* (Agaricomycetes), in children from Medellin, Colombia. Int J Med Mushrooms 20(8):705–716. https://doi.org/10.1615/IntJMedMushrooms.2018026986
Hibbet DS, Taylor JW (2013) Fungal systematics: is a new age of enlightenment at hand? Nat Rev Microbiol 11:129–133. https://doi.org/10.1038/nrmicro2963
Hou R, Liu X, Wu X et al (2021) Therapeutic effect of natural melanin from edible fungus *Auricularia auricula* on alcohol-induced liver damage *in vitro* and *in vivo*. Food Sci Human Wellness 10(4):514–522. https://doi.org/10.1016/j.fshw.2021.04.014
Hsieh YC, Rao YK, Whang-Peng J et al (2011) Antcin B and its ester derivative from *Antrodia camphorata* induce apoptosis in hepatocellular carcinoma cells involves enhancing oxidative stress coincident with activation of intrinsic and extrinsic apoptotic pathway. J Agric Food Chem 59(20):10943–10954. https://doi.org/10.1021/jf202771d
Huang C-C, Hsu M-C, Huang W-C, Yang H-R, Hou C-C (2012) Triterpenoid-rich extract from Antrodia camphorate improves physical fatigue and exercise performance in mice. Evid Based Complement Alternat Med 2012:364741. https://doi.org/10.1155/2012/364741
Huang CW, Hung YC, Chen LY et al (2022) Antioxidant activities of hot water extracts from mycelial biomass of different combinations of medicinal Agaricomycetes mushrooms. Int J Med Mushrooms 24(8):21–30. https://doi.org/10.1615/IntJMedMushrooms.2022044221
Huguet C, Bourjot M, Bellanger J-M et al (2022) Screening for antibacterial activity of French mushrooms against pathogenic and multidrug resistant bacteria. Appl Sci 12(10):5229. https://doi.org/10.3390/app12105229

Ina K, Furuta R, Kataoka T et al (2016) Chemo-immunotherapy using lentinan for the treatment of gastric cancer with liver metastases. Med Sci 4(2):8. https://doi.org/10.3390/medsci4020008

Islam T, Yao F, Kang W et al (2022) A systematic study on mycochemical profiles, antioxidant, and anti-inflammatory activities of 30 varieties of Jew's ear (*Auricularia auricula-judae*). Food Sci Hum Wellness 11(4):781–794. https://doi.org/10.1016/j.fshw.2022.03.005

Jesenak M, Hrubisko M, Majtan J et al (2014) Anti-allergic effect of pleuran (β-glucan from *Pleurotus ostreatus*) in children with recurrent respiratory tract infections. Phytother Res 28: 471–474. https://doi.org/10.1002/ptr.5020

Jin X, Ruiz Beguerie J, Sze DMY et al (2012) *Ganoderma lucidum* (Reishi mushroom) for cancer treatment. Cochrane Database Syst Rev 6:CD007731. https://doi.org/10.1002/14651858.CD007731.pub2

Joseph TP, Chanda W, Padhiar AA et al (2018) A preclinical evaluation of the antitumor activities of edible and medicinal mushrooms: a molecular insight. Integr Cancer Ther 7(2):200–209. https://doi.org/10.1177/1534735417736861

Kaplan Ö, Tosun NG, İmamoğlu R et al (2022) Biosynthesis and characterization of silver nanoparticles from *Tricholoma ustale* and *Agaricus arvensis* extracts and investigation of their antimicrobial, cytotoxic, and apoptotic potentials. J Drug Deliv Sci Technol 69:103178. https://doi.org/10.1016/j.jddst.2022.103178

Kaplan Ö, Tosun NG, Özgür A et al (2021) Microwave-assisted green synthesis of silver nanoparticles using crude extracts of *Boletus edulis* and *Coriolus versicolor*: characterization, anticancer, antimicrobial and wound healing activities. J Drug Deliv Sci Technol 64:102641. https://doi.org/10.1016/j.jddst.2021.102641

Kappos L, O'Connor P, Radue EW et al (2015) Long-term effects of fingolimod in multiple sclerosis: the randomized FREEDOMS extension trial. Neurology 84(15):1582–1591. https://doi.org/10.1212/WNL.0000000000001462

Kemkuignou BM, Moussa AY, Decock C et al (2022) Terpenoids and meroterpenoids from cultures of two grass-associated species of *Amylosporus* (Basidiomycota). J Nat Prod 85(4): 846–856. https://doi.org/10.1021/acs.jnatprod.1c00975

Kerezoudi E, Mitsou EK, Gioti K et al (2021) Fermentation of *Pleurotus ostreatus* and *Ganoderma lucidum* mushrooms and their extracts by gut microbiota of healthy and osteopenic women: potential prebiotic effect and impact on human osteoblasts. Food Funct 12:1529–1546. https://doi.org/10.1039/D0FO02581J

Khan FI, Hassan F, Lai D (2022) *In silico* studies on psilocybin drug derivatives against SARS-CoV-2 and cytokine storm of human interleukin-6 receptor. Front Immunol 12:794780. https://doi.org/10.3389/fimmu.2021.794780

Khatua S, Acharya K (2021) Exploration of macrofungal wealth of West Bengal in the 21st Century. J Mycopathol Res 59(3):207–224

Khatua S, Acharya K (2022) Investigation on a novel mushroom, *Russula pseudocyanoxantha*: chemical composition and antioxidant potency. Res J Biotech 17(6):158–164

Khinsar KH, Abdul S, Hussain A et al (2021) Anti-tumor effect of polysaccharide from *Pleurotus ostreatus* on H22 mouse hepatoma ascites in-vivo and hepatocellular carcinoma in-vitro model. AMB Express 11(1):160. https://doi.org/10.1186/s13568-021-01314-51-5

Khoroshutin P, Reva G, Yamamoto T et al (2021) Pharmacokinetics and pharmacodynamics of Chaga birch mushroom components (*Inonotus obliquus*). Archiv Euromedica 11(2):31–38. https://doi.org/10.35630/2199-885X/2021/11/2/9

Klupp NL, Kiat H, Bensoussan A et al (2016) A double-blind, randomized, placebo-controlled trial of *Ganoderma lucidum* for the treatment of cardiovascular risk factors of metabolic syndrome. Sci Rep 6:29540. https://doi.org/10.1038/srep29540

Kües U, Badalyan SM (2017) Making use of genomic information to explore the biotechnological potential of medicinal mushrooms. In: Agrawal DC, Tsay HS, Shyur LF, Wu YC, Wang SY (eds) Medicinal plants and fungi: recent advances in research and development, Edition: Medicinal and aromatic plants of the world, vol 4. Springer Nature, pp 397–458. https://doi.org/10.1007/978-981-10-5978-0_13

Kuo YH, Lin TW, Lin JY et al (2022) Identification of common liver metabolites of the natural bioactive compound erinacine A, purified from *Hericium erinaceus* mycelium. Appl Sci 12: 1201. https://doi.org/10.3390/app12031201

Lee MG, Kwon YS, Nam KS et al (2021) Chaga mushroom extract induces autophagy via the AMPK-mTOR signaling pathway in breast cancer cells. J Ethnopharmacol 274:114081. https://doi.org/10.1016/j.jep.2021.114081

Li H, Tian Y, Menolli N Jr et al (2021a) Reviewing the world's edible mushroom species: A new evidence-based classification system. Compr Rev Food Sci Food Saf 20:1982–2014. https://doi.org/10.1111/1541-4337.12708

Li IC, Chang HH, Lin CH et al (2020) Prevention of early Alzheimer's disease by erinacine A-enriched *Hericium erinaceus* mycelia pilot double-blind placebo-controlled study. Front Aging Neurosci 12:155. https://doi.org/10.3389/fnagi.2020.00155

Li TJ, Lee TY, Lo Y et al (2021b) *Hericium erinaceus* mycelium ameliorate anxiety induced by continuous sleep disturbance *in vivo*. BMC Complement Med Ther 21(1):295. https://doi.org/10.1186/s12906-021-03463-3

Li Z, Mei J, Jiang L et al (2019) Chaga medicinal mushroom, *Inonotus obliquus* (Agaricomycetes) polysaccharides suppress tacrine-induced apoptosis by ROS-scavenging and mitochondrial pathway in HepG2 cells. Int J Med Mushrooms 21(6):583–593. https://doi.org/10.1615/IntJMedMushrooms.2019030857

Lin KW, Maitraie D, Huang AM et al (2016) Triterpenoids and an alkamide from *Ganoderma tsugae*. Fitoterapia 108:73–80. https://doi.org/10.1016/j.fitote.2015.11.003

Liu G, Wang K, Kuang S et al (2018) The natural compound GL22, isolated from *Ganoderma* mushrooms, suppresses tumor growth by altering lipid metabolism and triggering cell death. Cell Death Dis 9(6):689. https://doi.org/10.1038/s41419-018-0731-6

Liu Y, Liu W, Li M et al (2022) Lanostane triterpenoids from the fruiting bodies of *Fomitopsis pinicola* and their anti-inflammatory activities. Phytochemistry 193:112985. https://doi.org/10.1016/j.phytochem.2021.112985

López-Hortas L, Flórez-Fernández N, Torres MD et al (2022) Update on potential of edible mushrooms: high-value compounds, extraction strategies and bioactive properties. Int J Food Sci Technol 57(3):1378–1385. https://doi.org/10.1111/ijfs.15544

Lu KH, Pan YC, Sheen LY (2020) Combination of cut-log cultivated fruiting body and solid-state cultured mycelia of *Taiwanofungus camphoratus* ameliorates CCl_4-induced liver injury in rats. J Tradit Complement Med 10(2):166–174. https://doi.org/10.1016/j.jtcme.2019.04.008

Mapook A, Hyde KD, Hassan K et al (2022) Ten decadal advances in fungal biology leading towards human well-being. Fungal Divers 116:547–614. https://doi.org/10.1007/s13225-022-00510-3

Meade E, Hehir S, Rowan N et al (2022) Mycotherapy: potential of fungal bioactives for the treatment of mental health disorders and morbidities of chronic pain. J Fungi 8(3):290. https://doi.org/10.3390/jof8030290

Menezes TM, da Silva CM, Lima AB et al (2022) Effects of polysaccharides isolated from mushrooms (*Lentinus edodes* Berk or *Agaricus blazei* Murill) on the gelation of Pluronic® F127. Colloids Surf A: Physicochem Eng Asp 642:128684. https://doi.org/10.1016/j.colsurfa.2022.128684

Mishra ST, Kaladhar SVGK (2022) Medicinal properties and economic importance of AgNP from *Ganoderma lucidum*: a review. GJRA 11(5):95–100. https://doi.org/10.36106/gjra/5907843

Mitra P, Khatua S, Mandal NC et al (2021) Beneficial properties of crude polysaccharides from *Termitomyces medius* of West Bengal to scavenge free radicals as well as boost immunity: a new report. Res J Pharm Technol 14(2):1073–1078. https://doi.org/10.5958/0974-360X.2021.00193.1

Morales D, Shetty SA, López-Plaza B et al (2021) Modulation of human intestinal microbiota in a clinical trial by consumption of a β-D-glucan-enriched extract obtained from *Lentinula edodes*. Eur J Nutr 60(6):3249–3265. https://doi.org/10.1007/s00394-021-02504-4

Morel S, Arnould S, Vitou M et al (2018) Antiproliferative and antioxidant activities of wild Boletales from France. Int J Med Mushrooms 20(1):13–29. https://doi.org/10.1615/IntJMedMushrooms.2018025329

Morel S, Vitou M, Masnou A et al (2021) Antibacterial activity of wild mushrooms from France. Int J Med Mushrooms 23(1):79–89. https://doi.org/10.1615/IntJMedMushrooms.2020037443

Mori K, Inatomi S, Ouchi K et al (2009) Improving effects of the mushroom Yamabushitake (*Hericium erinaceus*) on mild cognitive impairment: a double-blind placebo-controlled clinical trial. Phytother Res 23(3):367–372. https://doi.org/10.1002/ptr.2634

Mori K, Obara Y, Moriya T et al (2011) Effects of *Hericium erinaceus* on amyloid β(25-35) peptide-induced learning and memory deficits in mice. Biomed Res 32(1):67–72. https://doi.org/10.2220/biomedres.32.67

Murray CJL, Ikuta KS, Sharara F et al (2022) Global burden of bacterial antimicrobial resistance in 2019: a systematic analysis. The Lancet 399(10325):629–655. https://doi.org/10.1016/S0140-6736(21)02724-0

Murugesan AK, Gunasagaran KS (2021) Purification and characterization of a synergistic bioactive lectin from *Pleurotus flabellatus* (PFL-L) with potent antibacterial and in-vitro radical scavenging activity. Anal Biochem 635:114450. https://doi.org/10.1016/j.ab.2021.114450

Muszyńska B, Fijałkowska A, Sułkowska-Ziaja K et al (2020) *Fomitopsis officinalis*: a species of arboreal mushroom with promising biological and medicinal properties. Chem Biodivers 17(6): e2000213. https://doi.org/10.1002/cbdv.202000213

Mwangi RW, Macharia JM, Wagara IN et al (2022) The antioxidant potential of different edible and medicinal mushrooms. Biomed Pharmacother 147:112621. https://doi.org/10.1016/j.biopha.2022.112621

Nallathamby N, Phan CW, Seow SL et al (2018) A status review of the bioactive activities of tiger milk mushroom *Lignosus rhinocerotis* (Cooke) Ryvarden. Front Pharmacol 8:998. https://doi.org/10.3389/fphar.2017.00998

Nandi S, Chandra S, Sikder R et al (2019) Characterization and inception of a triterpenoid astrakurkurol, as a cytotoxic molecule on human hepatocellular carcinoma cells, Hep3B. J Agric Food Chem 67(27):7660–7673. https://doi.org/10.1021/acs.jafc.9b01203

Ndong HC, Orango-Bourdette JO, Iwangou G et al (2021) A study of the therapeutic potential of *Lentinus squarrosulus* (Agaricomycetes) from Gabon. Int J Med Mushrooms 23(4):39–45. https://doi.org/10.1615/IntJMedMushrooms.2021038022

Niego AGT, Rapior S, Thongklang N et al (2021) Macrofungi as nutraceutical source: promising bioactive compounds and market value. J Fungi 7(5):397. https://doi.org/10.3390/jof7050397

Nowakowski P, Markiewicz-Żukowska R, Bielecka J et al (2021a) Treasures from the forest: evaluation of mushroom extracts as anti-cancer agents. Biomed Pharmacother 143:112106. https://doi.org/10.1016/j.biopha.2021.112106

Nowakowski P, Markiewicz-Żukowska R, Gromkowska-Kępka K et al (2021b) Mushrooms as potential therapeutic agents in the treatment of cancer: evaluation of anti-glioma effects of *Coprinus comatus, Cantharellus cibarius, Lycoperdon perlatum* and *Lactarius deliciosus* extracts. Biomed Pharmacother 133:111090. https://doi.org/10.1016/j.biopha.2020.111090

Özcan Ö, Ertan F (2018) Beta-glucan content, antioxidant and antimicrobial activities of some edible mushroom species. Food Sci Technol 6(2):47–55. https://doi.org/10.13189/fst.2018.060201

Pal A, Ray R, Acharya K et al (2021) Assessment of the anti-leukemic and antioxidant potential of the methanol extract of a wild, edible, and novel mushroom, *Astraeus hygrometricus*, and unraveling its metabolomic profile. J Adv Biotechnol Exp Ther 4(3):388–404. https://doi.org/10.5455/jabet.2021.d138

Palasarn S, Intereya K, Boonpratuang T et al (2022) Ergostane triterpenoids from the cultures of basidiomycete *Favolaschia calocera* BCC 36684 and stereochemical elucidation of favolon. Phytochem Lett 47:168–173. https://doi.org/10.1016/j.phytol.2022.01.002

Panda SK, Luyten W (2022) Medicinal mushrooms: clinical perspective and challenges. Drug Discov Today 27(2):636–651. https://doi.org/10.1016/j.drudis.2021.11.017

Panda SK, Sahoo G, Swain SS et al (2022) Anticancer activities of mushrooms: a neglected source for drug discovery. Pharmaceuticals (Basel) 15(2):176. https://doi.org/10.3390/ph15020176
Park KR, Kwon YJ, Jeong YH et al (2022) Thelephoric acid, *p*-terphenyl, induces bone-forming activities in pre-osteoblasts. J Funct Foods 91:105036. https://doi.org/10.1016/j.jff.2022.105036
Pavithra M, Sridhar KR, Greeshma AA (2018) Nutritional quality attributes of edible gasteroid wild mushroom *Astraeus hygrometricus*. In: Gehlot P, Singh J (eds) Fungi and their role in sustainable development: current perspectives. Springer, Singapore, pp 367–382. https://doi.org/10.1007/978-981-13-0393-7_21
Peay KG, Kennedy PG, Talbot JM (2016) Dimension of biodiversity in the earth mycobiome. Nat Rev Microbiol 14:434–447. https://doi.org/10.1038/nrmicro.2016.59
Peng XR, Liu JQ, Xia JJ et al (2015) Lanostane triterpenoids from *Ganoderma hainanense* J.D. Zhao. Phytochemistry 114:137–145. https://doi.org/10.1016/j.phytochem.2014.10.009
Pérez-Moreno J, Mortimer PE, Xu J et al (2021) Global perspectives on the ecological, cultural, and socioeconomic relevance of wild edible fungi. Stud Fungi 6(1):408–424. https://doi.org/10.5943/sif/6/1/31
Petrović J, Fernandes Â, Stojković D et al (2022) A step forward towards exploring nutritional and biological potential of mushrooms: a case study of *Calocybe gambosa* (Fr.) Donk wild growing in Serbia. Pol J Food Nutr Sci 72(1):17–26. https://doi.org/10.31883/pjfns/144836
Pham HT, Doan TP, Kim HW et al (2022) Cyclohumulanoid sesquiterpenes induced by the noncompetitive coculture of *Phellinus orientoasiaticus* and *Xylodon flaviporus*. J Nat Prod 85(3):511–518. https://doi.org/10.1021/acs.jnatprod.1c01022
Phull AR, Ahmed M, Park HJ (2022) *Cordyceps militaris* as a bio functional food source: pharmacological potential, anti-inflammatory actions and related molecular mechanisms. Microorganisms 10(2):405. https://doi.org/10.3390/microorganisms10020405
Pop RM, Puia IC, Puia A et al (2018) Characterization of *Trametes versicolor*: medicinal mushroom with important health benefits. Not Bot Horti Agrobo 46(2):343–349. https://doi.org/10.15835/nbha46211132
Prestinaci F, Pezzotti P, Pantosti A (2015) Antimicrobial resistance: a global multifaceted phenomenon. Pathog Glob Health 109(7):309–318. https://doi.org/10.1179/2047773215Y.0000000030
Psurtseva NV, Sazanova KV, Kiyashko AA et al (2022) Contribution to studies on rare medicinal mushroom *Sparassis crispa* (Agaricomycetes) in culture and its production of phenol compounds under various cultivation conditions. Int J Med Mushrooms 24(8):71–79. https://doi.org/10.1615/IntJMedMushrooms.2022044434
Qiao X, Wang Q, Ji S et al (2015) Metabolites identification and multi-component pharmacokinetics of ergostane and lanostane triterpenoids in the anticancer mushroom *Antrodia cinnamomea*. J Pharm Biomed Anal 111:266–276. https://doi.org/10.1016/j.jpba.2015.04.010
Rehman S, Farooq R, Jermy R et al (2020) A wild *Fomes fomentarius* for biomediation of one pot synthesis of titanium oxide and silver nanoparticles for antibacterial and anticancer application. Biomolecules 10(4):622. https://doi.org/10.3390/biom10040622
Reis FS, Martins A, Vasconcelos MH et al (2017) Functional foods based on extracts or compounds derived from mushrooms. Trends Food Sci Technol 66:48–62. https://doi.org/10.1016/j.tifs.2017.05.010
Rizkyana AD, Ho TC, Roy VC et al (2022) Sulfation and characterization of polysaccharides from Oyster mushroom (*Pleurotus ostreatus*) extracted using subcritical water. J Supercrit Fluids 179:105412. https://doi.org/10.1016/j.supflu.2021.105412
Sabotič et al (2016) Entomotoxic and nematotoxic lectins and protease inhibitors from fungal fruiting bodies. Appl Microbiol Biotechnol 100(1):91–111. https://doi.org/10.1007/s00253-015-7075-2
Saha M, Sarkar A (2021) Review on multiple facets of drug resistance: A rising challenge in the 21st century. J Xenobiot 11:197–214. https://doi.org/10.3390/jox11040014
Saitsu Y, Nishide A, Kikushima K et al (2019) Improvement of cognitive functions by oral intake of *Hericium erinaceus*. Biomed Res 40(4):125–131. https://doi.org/10.2220/biomedres.40.125

Saleh KK, Kirbag S, Dalkılıç S (2022) Inhibitory effects of *Terfezia* (Ascomycota) desert truffles on PANC-1 cell growth via upregulation of proapoptotic genes TP53, CDKN1A, and BAX and downregulation of antiapoptotic gene BCL2. Int J Med Mushrooms 24(8):61–70. https://doi.org/10.1615/IntJMedMushrooms.2022044383

Sappan M, Rachtaweea P, Srichomthonga K et al (2022) Ganoellipsic acids A–C, lanostane triterpenoids from artificially cultivated fruiting bodies of *Ganoderma ellipsoideum*. Phytochem Lett 49:27–31. https://doi.org/10.1016/j.phytol.2022.03.001

Sarikurkcu C, Solak MH, Tarkowski P et al (2022) Minerals, phenolics, and biological activity of wild edible mushroom, *Morchella steppicola* Zerova. Nat Prod Res 6(23):6101–6105. https://doi.org/10.1080/14786419.2022.2050910

Sen D, Debnath B, Debnath P et al (2022) Identification of potential edible mushroom as SARS-CoV-2 main protease inhibitor using rational drug designing approach. Sci Rep 12(1):1503. https://doi.org/10.1038/s41598-022-05349-x

Shahid A, Huang M, Liu M et al (2022) The medicinal mushroom *Ganoderma lucidum* attenuates UV-induced skin carcinogenesis and immunosuppression. PLoS One 17(3):e0265615. https://doi.org/10.1371/journal.pone.0265615

Sharma D, Singh VP, Singh NK (2018) A review on phytochemistry and pharmacology of medicinal as well as poisonous mushrooms. Mini Rev Med Chem 18(13):1095–1109. https://doi.org/10.2174/1389557517666170927144119

Sharma R, Khatua S, Acharya K et al (2022a) Mycochemical composition and antioxidant activity of *Flammulina velutipes*: a comparative study on hydromethanol, decoction and infusion extracts. Vegetos 35:607–613. https://doi.org/10.1007/s42535-022-00362-x

Sharma S, Madaan K, Kaur R (2022b) Cordycepin as a metabolite with pharmacological potential: A review. Int J Med Mushrooms 24(8):1–20. https://doi.org/10.1615/IntJMedMushrooms.2022044442

Singh V, Bedi GK, Shri R (2017) *In vitro* and *in vivo* antidiabetic evaluation of selected culinary-medicinal mushrooms (Agaricomycetes). Int J Med Mushrooms 19(1):17–25. https://doi.org/10.1615/IntJMedMushrooms.v19.i1.20

Song C, Wu M, Zhang Y et al (2022a) Bioactive monomer and polymer polyketides from edible mushroom *Cortinarius caerulescens* as glutamate dehydrogenase inhibitors and antioxidants. J Agric Food Chem 70(3):804–814. https://doi.org/10.1021/acs.jafc.1c07119

Song C, Wu M, Zhang Y et al (2022b) Correction to bioactive monomer and polymer polyketides from edible mushroom *Cortinarius caerulescens* as glutamate dehydrogenase inhibitors and antioxidants. J Agric Food Chem 70(5):1747. https://doi.org/10.1021/acs.jafc.1c07119

Song J, Xu K, Liu C et al (2022c) Bioactive specialized metabolites from the endophytic fungus *Xylaria sp.* of *Cudrania tricuspidata*. Phytochemistry 196:113079. https://doi.org/10.1016/j.phytochem.2021.113079

Spim SRV, Castanho NRCM, Pistila AMH et al (2021a) *Lentinula edodes* mushroom as an ingredient to enhance the nutritional and functional properties of cereal bars. J Food Sci Technol 58(4):1349–1357. https://doi.org/10.1007/s13197-020-04646-5

Spim SRV, Pistila AMH, Pickler TB et al (2021b) Effects of shiitake culinary-medicinal mushroom, *Lentinus edodes* (Agaricomycetes), bars on lipid and antioxidant profiles in individuals with borderline high cholesterol: a double-blind randomized clinical trial. Int J Med Mushrooms 23(7):1–12. https://doi.org/10.1615/IntJMedMushrooms.2021038773

Steimbach L, Borgmann AV, Giandotti Gomar G et al (2021) Fungal beta-glucans as adjuvants for treating cancer patients—A systematic review of clinical trials. Clin Nutr 40(5):3104–3113. https://doi.org/10.1016/j.clnu.2020.11.029

Storsberg J, Krüger-Genge A, Kalitukha L (2022) *In vitro* cytotoxic activity of an aqueous alkali *Fomes fomentarius* (Agaricomycetes) extract on murine fibroblasts, human colorectal adenocarcinoma and cutaneous melanoma cells. Int J Med Mushrooms 24(9):1–13. https://doi.org/10.1615/IntJMedMushrooms.2022044657

Tanaka A, Nishimura M, Sato Y et al (2016) Enhancement of the Th1-phenotype immune system by the intake of Oyster mushroom (Tamogitake) extract in a double-blind, placebo-controlled study. J Tradit Complement Med 6(4):424–430. https://doi.org/10.1016/j.jtcme.2015.11.004

Tangen JM, Tierens A, Caers J et al (2015) Immunomodulatory effects of the *Agaricus blazei* Murrill-based mushroom extract AndoSan in patients with multiple myeloma undergoing high dose chemotherapy and autologous stem cell transplantation: a randomized, double blinded clinical study. Biomed Res Int 2015:718539. https://doi.org/10.1155/2015/718539

Tarmizi NAKA, Kushairi N, Phan CW et al (2022) β-Glucan-rich extract of gray oyster mushroom, *Pleurotus pulmonarius*, improves object recognition memory and hippocampus morphology in mice fed a high-fat diet. J Med Food 25(3):230–238. https://doi.org/10.1089/jmf.2021.K.0121

Therkelsen SP, Hetland G, Lyberg T et al (2016) Effect of the medicinal *Agaricus blazei* Murill-based mushroom extract, AndoSan^{TM}, on symptoms, fatigue and quality of life in patients with Crohn's disease in a randomized single-blinded placebo controlled study. PLoS One 11(7): e0159288. https://doi.org/10.1371/journal.pone.0159288

Thornton PK, Kristjanson P, Förch W et al (2018) Is agricultural adaptation to global change in lower-income countries on track to meet the future food production challenge? Glob Environ Change 52:37–48. https://doi.org/10.1016/j.gloenvcha.2018.06.003

Tohtahon Z, Xue J, Han J et al (2017) Cytotoxic lanostane triterpenoids from the fruiting bodies of *Piptoporus betulinus*. Phytochemistry 143:98–103. https://doi.org/10.1016/j.phytochem.2017.07.013

Tong LL, Wang Y, Yuan L et al (2022) Enhancement of polysaccharides production using microparticle enhanced technology by *Paraisaria dubia*. Microb Cell Fact 21(1):12. https://doi.org/10.1186/s12934-021-01733-w

Tsai MY, Hung YC, Chen YH et al (2016) A preliminary randomized controlled study of short-term *Antrodia cinnamomea* treatment combined with chemotherapy for patients with advanced cancer. BMC Complement Altern Med 16(1):322. https://doi.org/10.1186/s12906-016-1312-9

Tsai PC, Wu YK, Hu JH et al (2021) Preclinical bioavailability, tissue distribution, and protein binding studies of erinacine A, a bioactive compound from *Hericium erinaceus* mycelia using validated LC-MS/MS method. Molecules 26(15):4510. https://doi.org/10.3390/molecules26154510

Twardowski P, Kanaya N, Frankel P et al (2015) A phase 1 trial of mushroom powder in patients with biochemically recurrent prostate cancer: roles of cytokines and myeloid-derived suppressor cells for *Agaricus bisporus*-induced prostate-specific antigen responses. Cancer 121:2949–2950. https://doi.org/10.1002/cncr.29421

Venturella G, Ferraro V, Cirlincione F et al (2021) Medicinal mushrooms: bioactive compounds, use, and clinical trials. Int J Mol Sci 22(2):634. https://doi.org/10.3390/ijms22020634

Vetvicka V, Vetvickova J (2015) Glucan supplementation enhances the immune response against an influenza challenge in mice. Ann Transl Med 3(2):22. https://doi.org/10.3978/j.issn.2305-5839.2015.01.08

Vogt-Schilb H, Richard F, Malaval JC et al (2022) Climate-induced long-term changes in Mediterranean fungal diversity. Fungal Ecol 60:101166. https://doi.org/10.1016/j.funeco.2022.101166

Wang C, Liu X, Lian C et al (2019) Triterpenes and aromatic meroterpenoids with antioxidant activity and neuroprotective effects from *Ganoderma lucidum*. Molecules 24(23):4353. https://doi.org/10.3390/molecules24234353

Wang C, Liu Y, Lan Y et al (2022c) Extraction of a triterpene solution and evaluation of the hypolipidemic efficacy of the *Pleurotus tuber-regium* (Fr.) Sing. sclerotium. Foods 11(18):2881. https://doi.org/10.3390/foods11182881

Wang FF, Shi C, Yang Y et al (2018) Medicinal mushroom *Phellinus igniarius* induced cell apoptosis in gastric cancer SGC-7901 through a mitochondria-dependent pathway. Biomed Pharmacother 102:18–25. https://doi.org/10.1016/j.biopha.2018.03.038

Wang L, Li WH, Zhang R et al (2022a) Study on characteristics of triterpenoids and hepatoprotective effects of fruit body of stout camphor mushroom, *Taiwanofungus camphoratus* (Agaricomycetes), cultivated with apple-wood. Int J Med Mushrooms 24(7): 53–65. https://doi.org/10.1615/IntJMedMushrooms.2022044118

Wang W, Li X, Zhang Y et al (2022b) Mycelium polysaccharides of *Macrolepiota procera* alleviate reproductive impairments induced by nonylphenol. Food Funct 13:5794–5806. https://doi.org/10.1039/d2fo00680d

Wasser SP (2011) Current findings, future trends, and unsolved problems in studies of medicinal mushrooms. Appl Microbiol Biotechnol 89:1323–1332. https://doi.org/10.1007/s00253-010-3067-4

Wasser SP (2014) Medicinal mushroom science: current perspectives, advances, evidences, and challenges. Biomed J 37(6):345–356. https://doi.org/10.4103/2319-4170.138318

Wasser SP (2017) Medicinal mushrooms in human clinical studies. Part I. Anticancer, oncoimmunological, and immunomodulatory activities: a review. Int J Med Mushrooms 19(4):279–317. https://doi.org/10.1615/IntJMedMushrooms.v19.i4.10

Włodarczyk A, Krakowska A, Lazur J et al (2022) Differences in health-promoting properties in civilization diseases of *Agaricus bisporus* fruiting bodies harvested from three flushes. Folia Hortic 34(1):17–25. https://doi.org/10.2478/fhort-2022-0002

Wunjuntuk K, Ahmad M, Techakriengkrai T et al (2022) Proximate composition, dietary fibre, beta-glucan content, and inhibition of key enzymes linked to diabetes and obesity in cultivated and wild mushrooms. J Food Compos Anal 105:104226. https://doi.org/10.1016/j.jfca.2021.104226

Xing YM, Li B, Liu L et al (2021) *Armillaria mellea* symbiosis drives metabolomic and transcriptomic changes in *Polyporus umbellatus* sclerotia. Front Microbiol 12:792530. https://doi.org/10.3389/fmicb.2021.792530

Xu J, Chen F, Wang G et al (2021) The versatile functions of *G. lucidum* polysaccharides and *G. lucidum* triterpenes in cancer radiotherapy and chemotherapy. Cancer Manag Res 13:6507–6516. https://doi.org/10.2147/CMAR.S319732

Yanshree YWS, Fung ML et al (2022) The monkey head mushroom and memory enhancement in Alzheimer's disease. Cells 11:2284. https://doi.org/10.3390/cells11152284

Yao HM, Wang G, Liu YP et al (2016) Phenolic acids isolated from the fungus *Schizophyllum commune* exerts analgesic activity by inhibiting voltage-gated sodium channels. Chin J Nat Med 14(9):661–670. https://doi.org/10.1016/S1875-5364(16)30078-4

Yap HY, Kong BH, Yap CS et al (2022) Immunomodulatory effect and an intervention of TNF signaling leading to apoptotic and cell cycle arrest on ORL-204 oral cancer cells by tiger milk mushroom, *Lignosus rhinocerus*. Food Technol Biotechnol 60(1):80–88. https://doi.org/10.17113/ftb.60.01.22.7296

Yen LTH, Thanh TH, Anh DTH et al (2022) Antimicrobial and antioxidant activity of the polypore mushroom *Lentinus arcularius* (Agaricomycetes) isolated in Vietnam. Int J Med Mushrooms 24(3):15–23. https://doi.org/10.1615/IntJMedMushrooms.2022042702

Yuyama KT, Fortkamp D, Abraham WR (2017) Eremophilane-type sesquiterpenes from fungi and their medicinal potential. Biol Chem 399(1):13–28. https://doi.org/10.1515/hsz-2017-0171

Zając A, Pięt M, Stefaniuk D et al (2021) Pro-health and anti-cancer activity of fungal fractions isolated from milk-supplemented cultures of *Lentinus (Pleurotus) sajor-caju*. Biomolecules 11(8):1089. https://doi.org/10.3390/biom11081089

Zaki AH, Zahid MT, Haiyinga B (2022) Bioactive compounds of the culinary-medicinal mushroom *Leucocalocybe mongolica* (Agaricomycetes): pharmacological and therapeutic applications - a review. Int J Med Mushrooms 24(6):19–33. https://doi.org/10.1615/IntJMedMushrooms.2022043854

Zavadinack M, de Lima Bellan D, da Rocha Bertage JL et al (2021) An α-D-galactan and a β-D-glucan from the mushroom *Amanita muscaria*: structural characterization and antitumor activity against melanoma. Carbohydr Polym 274:118647. https://doi.org/10.1016/j.carbpol.2021.118647

Zengin G, Sarikurkcu C, Gunes E et al (2015) Two *Ganoderma* species: profiling of phenolic compounds by HPLC–DAD, antioxidant, antimicrobial and inhibitory activities on key enzymes linked to diabetes mellitus, Alzheimer's disease and skin disorders. Food Funct 6(8): 2794–2802. https://doi.org/10.1039/c5fo00665a

Zeynali M, Alvandi H, Zarmi AH et al (2022) *Schizophyllum commune*-derived chitin glucan complex wound dressing: antibacterial activity and wound healing properties in a second degree burn animal model. J Nat Fibers 19(16):12870–12882. https://doi.org/10.1080/15440478.2022.2078921

Zhang SB, Li ZH, Stadler M et al (2018a) Lanostane triterpenoids from *Tricholoma pardinum* with NO production inhibitory and cytotoxic activities. Phytochemistry 152:105–112. https://doi.org/10.1016/j.phytochem.2018.05.002

Zhang T, Rayamajhi S, Meng G et al (2021) Edible mushroom consumption and incident hyperuricemia: results from the TCLSIH cohort study. Food Funct 12(19):9178–9187. https://doi.org/10.1039/d1fo00650a

Zhang X, Bao C, Long G et al (2022) Lanostane-type triterpenoids from the mycelial mat of *Ganoderma lucidum* and their hepatoprotective activities. Phytochemistry 198:113131. https://doi.org/10.1016/j.phytochem.2022.113131

Zhang X, Bao C, Zhang J (2018b) Inotodiol suppresses proliferation of breast cancer in rat model of type 2 diabetes mellitus via downregulation of β-catenin signaling. Biomed Pharmacother 99: 142–150. https://doi.org/10.1016/j.biopha.2017.12.084

Zhao J, Yang Y, Yu M et al (2018) Lanostane-type triterpenoid derivatives from the fruiting bodies of cultivated *Fomitopsis palustris*. Phytochemistry 152:10–21. https://doi.org/10.1016/j.phytochem.2018.04.012

Zmitrovich IV, Bondartseva MA, Arefyev SP et al (2022) Professor Solomon P. Wasser and medicinal mushroom science, with special attention to the problems of mycotherapy in oncology. Int J Med Mushrooms 24(1):13–26. https://doi.org/10.1615/IntJMedMushrooms.2021041831

Zorrilla JG, Evidente A (2022) Structures and biological activities of alkaloids produced by mushrooms, a fungal subgroup. Biomolecules 12:1025. https://doi.org/10.3390/biom12081025

Medicinal Mushrooms for Respiratory Health

Han Ni Booi, Mei Kee Lee, Kang Nee Ting, and Shin Yee Fung

Contents

H. N. Booi · M. K. Lee · K. N. Ting (✉)
School of Pharmacy, University of Nottingham Malaysia, Semenyih, Selangor Darul Ehsan, Malaysia
e-mail: payhnb@nottingham.edu.my; MeiKee.Lee@nottingham.edu.my; Kang-Nee.Ting@nottingham.edu.my

S. Y. Fung (✉)
Medicinal Mushroom Research Group (MMRG), Department of Molecular Medicine, Faculty of Medicine, Universiti Malaya, Kuala Lumpur, Malaysia
e-mail: syfung@um.edu.my

D. C. Agrawal, M. Dhanasekaran (eds.), *Mushrooms with Therapeutic Potentials*,
https://doi.org/10.1007/978-981-19-9550-7_2

Abstract Chronic respiratory disorders (CRDs) have been a key threat globally to the public health system. An estimated 545 million cases worldwide could cause long-term disability, multi-morbidity, and premature mortality. However, limited research funding has hampered the opportunity to proliferate research in respiratory diseases. Despite the recent advancement in respiratory treatments, the use of medicinal mushrooms observes an increasing interest. It spurs the effort to bridge the unaddressed pathways of action within the treatment algorithms. In this book chapter, we provide a comprehensive, evidence-based discussion of a collection of medicinal mushrooms that are beneficial in promoting respiratory health and potentially reducing COVID-19 symptoms in newly diagnosed patients and those who have recovered. Apart from tackling the CRDs through the immunomodulatory pathways, the discussion in this book chapter focuses on the potential bronchodilation effects of the medicinal mushrooms that are crucial in improving respiratory functions. These bioactive components in medicinal mushrooms, predominantly beta-glucans, are further reviewed to understand their roles in alleviating respiratory symptoms. Medicinal mushrooms are functional food requiring further quality, safety, and efficacy assessments. The requirements for these assessments are also highlighted to promote the future development and application of these medicinal mushrooms for better respiratory health.

Keywords Bronchodilation · COVID-19 infection · Medicinal mushroom · Respiratory disorder · Respiratory health

Abbreviations

3'UTRs	3' Untranslated Regions
ACE-2	Angiotensin-Converting Enzyme 2
ADCC	Antibody-Dependent Cellular Cytotoxicity
AHR	Airway Hyperresponsiveness
ARDS	Acute Respiratory Distress Syndrome
ASHMI	Anti-asthmatic Herbal Medicine Intervention
ASM	Airway Smooth Muscle
CAE	Cryptoporic Acid E
CAM	Complementary and Alternative Medicine
CD	Cluster of Differentiation
CFS	Chronic Fatigue Syndrome
COPD	Chronic Obstructive Pulmonary Disease
COVID-19	Coronavirus
CRD	Chronic Respiratory Disorder

EFSA	European Food Safety Authority
EMA	European Medicines Agency
FDA	Food and Drug Administration
ICS	Inhaled Corticosteroids
Ig	Immunoglobulin
IL	Interleukin
IND	Investigational New Drug
INF-γ	Interferon-gamma
LABA	Long-acting β2-Agonist
LAMA	Long-acting Muscarinic Agonist
LPS	Lipopolysaccharide
LTRA	Leukotriene Receptor Antagonist
MAPK	Mitogen-Activated Protein Kinase
MART	Maintenance and Reliever Therapy
MERS	Middle East Respiratory Syndrome Coronavirus
miR	MicroRNAs
MMP-9	Matrix Metallopeptidase 9
mRNA	Messenger Ribonucleic Acid
Mw	Molecular Weight
NF-kB	Nuclear Factor kappa B
NK	Natural Killer
PAMP	Pathogen-Associated Molecular Pattern
PKC	Protein Kinase C
PLC	Phospholipase C
PSK	Polysaccharide K
PSP	Polysaccharopeptide
SARS	Severe Acute Respiratory Syndrome
SARS-CoV-2	Severe Acute Respiratory Syndrome Coronavirus 2
STAT3	Signal Transducer and Activator of Transcription 3
Th-1	T helper 1
Th-2	T helper 2
TLR-2	Toll-like Receptor 2
TMPRSS2	Type 2 Transmembrane Serine Protease
TNF-α	Tumor Necrosis Factor-alpha
WHO	World Health Organization

1 Introduction

Respiratory disorders are pulmonary diseases that affect the airways and other structures of the lungs. These include asthma, chronic obstructive pulmonary disease (COPD), pneumonia, pulmonary fibrosis, lung cancer, and the novel coronavirus

(COVID-19) infection. It can be caused by allergens, pollutants, cigarette smoking, and bacterial or viral infections (Navarro-Torné et al. 2015).

Asthma and COPD are the two most prevalent chronic conditions affecting the respiratory system. Both conditions are inflammatory, leading to airflow limitation and hypersecretion of mucus. However, the prevalence of these diseases differs in terms of the onset of the diseases, the reversibility of the conditions, and the nature of inflammation within the airways (Cukic et al. 2012). For instance, asthma usually occurs during childhood, mainly due to genetic predisposition, while COPD is usually caused by long-term exposure to lung irritants such as tobacco and dust particles (Casas and Nemery 2014). Clinically, asthma is characterized as a reversible airway obstruction driven by eosinophils and CD4 cells. On the other hand, airway obstructions in COPD patients are irreversible, progressive, and primarily neutrophilic and CD8-driven (Cukic et al. 2012).

Conversely, influenza is an acute viral infection that primarily affects the respiratory system early. During the later stages, the infection may progress to afflict other organs, such as the heart and the brain. The influenza strains are highly contagious. These viruses can spread rapidly over a large geographical area in populations with weak immunity against them, causing an epidemic outbreak that may lead to an influenza pandemic (Moghadami 2017).

While viruses cause both influenza and the novel COVID-19 infection, the infective viral strains for both illnesses differ. Type A and type B influenza viruses are responsible for seasonal influenza, especially in the winter. In contrast, patients infected with the type C influenza strain often experience less severe flu symptoms (Krammer et al. 2018). For instance, influenza and the COVID-19 infection are contagious and lead to similar respiratory symptoms such as cough, fever, runny nose, and sore throat. However, the COVID-19 infection differs from the occurrence of influenza as a newly emerged virus causes it, generally known as the coronavirus, which is different from the conventional influenza virus (Wu et al. 2020). Compared to influenza, COVID-19 infection can spread more readily and rapidly from person to person due to spike proteins anchoring the surface membrane of the viruses. This will mediate the entry of these coronaviruses into the host cells for replication (Huang et al. 2020). Since various aspects of the pathology of the COVID-19 infection remains obscure, the application of medicinal mushrooms as part of the complementary treatment for COVID-19 infection has gained increasing interest from the public lately, owing to their well-known bioactive components.

The novel COVID-19 viral infection affects the respiratory tract acutely, where patients have been reported to experience mild to moderate respiratory symptoms such as cough and shortness of breath (Leung et al. 2020). However, if the disease is not intervened timely with proper care and treatments, the patient may be at risk of developing life-threatening complications that may include acute respiratory distress syndrome (ARDS), pneumonia, and multiple organ failures that require intensive medical interventions (Cascella et al. 2022). On top of that, despite being tested negative microbiologically, patients recovering from COVID-19 infection may still experience post-COVID syndrome, more commonly known as the long-COVID, beyond the second week of the viral clearance phase (Mahmud et al. 2021). This phenomenon associates these patients with the possibility of experiencing long-term aftereffects of COVID-19, which may affect their general well-being negatively (Booi et al. 2022).

1.1 Prevalence of Chronic and Seasonal Respiratory Diseases

Among all the respiratory diseases, asthma, COPD, and influenza were the top three leading respiratory disorders directly linked to severe illnesses and death before the COVID-19 pandemic (World Health Organization 2017).

The high prevalence of asthma and COPD remains a public health issue, with an increasing incidence rate domestically and globally. According to the estimation by WHO, in the year 2020, more than 334 million people will have asthma, and 384 million people will live with COPD worldwide (World Health Organization 2017; Global Initiative for Chronic Obstructive Lung Disease 2019).

Seasonal influenza typically takes place during the fall and winter seasons. An estimated 1 billion symptomatic cases worldwide are reported annually, with 3–five million severe cases requiring medical intervention in the hospital. The influenza-related mortality rate per 1000 is roughly 0.65 globally (World Health Organization 2019).

The COVID-19 outbreak began towards the end of 2019, causing the implementation of drastic public health measures and lockdowns globally. As of March 18, 2022, WHO reported approximately 465 million confirmed COVID-19 cases and more than six million COVID-19 related deaths worldwide. With the number of cases and deaths observing an alarming increase daily, the COVID-19 related mortality rate for every 1000 patients was evaluated to be about 12.9 individuals, a figure marking a nearly 20-fold higher fatality rate than the seasonal influenza (World Health Organization 2021).

Besides being subjected to the high mortality rate of the COVID-19 infection, patients are also prone to develop long term respiratory morbidities such as fatigue, shortness of breath, dyspnoea and persistent coughing. The occurrence rate of these complications among COVID-19 survivors, which make up the majority of reported post-acute COVID-19 symptoms, stood at over 73% (Nasserie et al. 2021). These complications cause a decline in the general well-being of the patients, their family members and the community (Booi et al. 2022).

1.2 Pathophysiology

1.2.1 Asthma

During exposure to allergens such as pollens and environmental pollutants, an allergic response mediated by immunoglobulin E (IgE) will trigger the release of inflammatory mediators such as histamine, leukotrienes, and prostaglandins upon binding to the surface receptors on the mast cells (Sinyor and Concepcion Perez 2021). These inflammatory mediators cause direct contraction of the airway smooth muscles, triggering an asthma attack. Untreated asthma attacks will result in the extensive migration of eosinophils, mast cells, and T-helper cells (CD4) to the airways, causing excessive stimulation of mucus production, airway tone, and airway responsiveness (Bush 2019). Hypertrophy of the bronchial smooth muscles and the interstitial collagen deposition occur due to the chronic inflammation occurring within the airways. These airway remodelling processes generally occur in the late phase of an asthma attack, depicting a similar persistent airflow obstruction found in patients with COPD (Doeing and Solway 2013).

1.2.2 COPD

COPD can be subcategorized into two main conditions, namely emphysema and chronic bronchitis. Emphysema is a respiratory impairment associated with dyspnoea due to the loss of alveolar structure (Goldklang and Stockley 2016). Chronic bronchitis can be termed chronic inflammation within the bronchial smooth muscles, resulting in excessive mucus secretion (Kim and Criner 2013). The pathophysiology of COPD can be characterized as the intensification of neutrophils, macrophages, and T-killer cells (CD8). Upon exposure to stimulants such as cigarette smoke, the increase in inflammatory mediators will cause a higher degree of inflammation within the airways and stimulate alveoli structural change through smooth muscle hyperplasia and declining in airway elasticity (Bourdin et al. 2009). Therefore, airflow efficiency in COPD patients is reduced, impairing the ventilation-perfusion process, which contributes to respiratory failure in the long term (Hikichi et al. 2019).

1.2.3 COVID-19 Infection

The severe acute respiratory syndrome coronavirus 2 (SARS-CoV-2) is the pathogen responsible for the COVID-19 infection which principally affects the human respiratory system. The spike protein located on the surface of the SARS-CoV-2 binds to the pulmonary alveolar epithelial cells through the angiotensin-converting enzyme 2 receptor (ACE-2) (Wiersinga et al. 2020). The protease found in the host cell, namely the type 2 transmembrane serine protease (TMPRSS2), mediates the SARS-CoV-2 uptake into the host cell through ACE-2 cleavage, which will, in turn, activate the SARS-CoV-2 protein (Hoffmann et al. 2020). At the early stages of the COVID-19 infection, the infected pulmonary alveolar epithelial cells induce a cytokine storm causing the overproduction and hyperactivation of the immune cells (Fajgenbaum and June 2020). Despite the potential elevation of the inflammatory response caused by the COVID-19 infection, 80% of the infected patients may be asymptomatic or exhibiting mild upper respiratory symptoms such as dry cough and sore throat due to a significantly lower viral load and a higher rate of viral clearance compared to those of the patients experiencing severe symptoms (Mason 2020; Kim et al. 2021). However, as the cytokine storm continues to propagate, the rate of viral replication may overtake the viral clearance rate, compromising the integrity of the tissue barriers between endothelial and epithelial cells. Consequently, the remodelling of the alveolar interstitial of the lung, impaired gaseous exchange capacity, pulmonary fibrosis, and oedema may develop as the early signs of ARDS. These conditions have been recognized as the clinical consequence of COVID-19 infection caused by the stimulation of the immune system (Wiersinga et al. 2020; Booi et al. 2022).

1.3 The Relationship between Airway Smooth Muscle (ASM) and Respiratory Diseases

The airway smooth muscle (ASM) is the structural tissue that lines the entire respiratory tract, from the trachea to the smallest airway passage, such as the bronchioles (Gu and Lee 2022). It plays an important role in managing respiratory diseases such as asthma and COPD by modulating airway responsiveness through bronchoconstriction or bronchodilation (Zuyderduyn et al. 2008). There is no scientific evidence that asthmatic patients are at an increased risk of COVID-19 infection (Hughes-Visentin and Paul 2020). However, patients with pre-existing respiratory diseases, particularly asthma, COPD, and interstitial lung disease, are at a higher risk of experiencing life-threatening complications due to COVID-19 infections (Aveyard et al. 2021). The airway luminal diameter may decrease as the ASM contracts or vice versa as a feature to maintain the airway patency (Lauzon and Martin 2016). However, excessive airway narrowing may occur in response to inordinate contractile stimuli such as chemical mediators (virulence factors, histamines, prostaglandins, leukotrienes, or allergens), suffocation due to physical activities, and climatic influence, resulting in excessive coughing and breathing difficulty (Doeing and Solway 2013). Therefore, the regulation of ASM is known to be closely related to the pathogenesis of respiratory diseases, namely hyperresponsiveness, hypercontractility, and hyperplasia of the ASM.

Airway hyperresponsiveness (AHR) is one of the clinical characteristics in respiratory diseases that indicates the potential occurrence of airway inflammation and remodeling of airways (Black et al. 2012). In response to airway inflammation, the mast cells found on the lamina propria of the respiratory tract will then migrate to the ASM to produce proinflammatory cytokines and growth factors, infiltrating the ASM with various immune mediators (Bradding 2007). However, the symptoms of AHR will persist even when the airway inflammation is well-controlled, suggesting that the cytokines may be the underlying factor that initiates irreversible hyperresponsiveness in the ASM, resulting in an independent pathogenesis pathway to airway inflammation (An et al. 2007; Black et al. 2012). This can be further substantiated by a clinical trial from Leckie et al. (2000), where anti-inflammatory therapies such as anti-immunoglobulin E and anti-interleukin (IL)-5 could not reverse the AHR symptoms in mild asthmatic patients. On the other hand, administering tumor necrosis factor-alpha (TNF-α) inhibitors to asthmatic patients has been proven to attenuate AHR symptoms with no alterations in the airway inflammation markers (Berry et al. 2006).

Prevalent AHR symptoms such as recurrent cough and wheezing may persist for several weeks after either an influenza or COVID-19 infection. These symptoms have resulted in reduced exercise tolerance reported in one-fifth of the COVID-19 survivors (Bellan et al. 2021). There has been an innumerable amount of research work focusing on discovering anti-inflammatories in treating respiratory diseases and their phenotypes based on their nature of inflammation in the last few decades. However, only a minimal amount of work has been carried out to address the extent of airway contractility using complementary and alternative medicines (CAMs), allowing a potential treatment target to be devised.

1.4 Treatment Algorithm

As of the latest development in the medicinal field, both chronic respiratory conditions such as asthma and COPD remain incurable. However, the treatment regimens available are highly effective in controlling the symptoms experienced by the patients. Patients with mild to moderate asthma may achieve a virtual cure if proper management is provided by healthcare professionals and is, in turn, adhered to by the patients (Upham and Chung 2018). However, every unique patient may respond differently to their respective treatment plans due to both conditions' multifaceted pathophysiological factors. On top of that, compliance issues arise as hand-held inhalers are usually the recommended first-line treatment for both conditions. Poor inhaler techniques may contribute significantly to the ineffective management of asthma and COPD (Ocakli et al. 2018).

Corticosteroids play a crucial role in suppressing airway inflammation and relieving signs and symptoms such as breathlessness and chronic cough in asthma and COPD. Although corticosteroids are highly effective in managing both conditions, especially during exacerbations, this medication causes various adverse effects on human physiology, ranging from local oral thrush to systemic adverse effects such as pneumonia and retarded growth in children (Heffler et al. 2018).

In addition, the chronic and repeated use of β_2 agonists for both conditions will result in the overstimulation of β_2 adrenoreceptors, resulting in the desensitization of the treatment approach (Hsu and Bajaj 2021). Although the adverse effects caused by the muscarinic antagonists, such as the dry mouth and blurred vision, may be mild and common, these should not be disregarded as the patients may not be able to fully benefit from the recommended treatment besides being at risk of adversely impacted by poor compliance and poor disease control (Naji and Gatling 2021).

Different pathophysiological factors are involved in both conditions, and the recommended treatment algorithms also vary. The treatment algorithm for both conditions can be found in Table 2 below. The overall improvement in managing the symptoms of the conditions is used as an indicator for treatment titration (National Institute for Health and Care Excellence 2021).

The administration methods for asthma and COPD therapies can be divided into three major routes: inhaled, oral, and intravenous. The five major treatment classes for asthma and COPD are listed in Table 1.

According to the information provided the WHO in 2020, the COVID-19 infection has an incubation period of up to 14 days. However, the severity of the COVID-19 infection varies with the symptoms experienced by the patients. Asymptomatic patients are often silent carriers of the COVID-19 virus. They caused more than 59% of all transmission cases based on the data obtained across 8 meta-analyses in China (Johansson et al. 2021). Therefore, despite experiencing light or no symptoms, patients who have tested positive for COVID-19 are advised to practise self-isolation to curb the spread of the virus (Booi et al. 2022). Since the influenza infection has been commonly discussed by various resources, the treatment regimen for the COVID-19 infection will be the focal point in this chapter.

Table 1 The most commonly used drugs for asthma and COPD patients are being grouped into five different classes (National Institute for Health and Care Excellence 2021)

Class	Mechanism of action	Examples	
β_2 adrenoreceptor agonist	Activation of $\beta 2$ adrenoreceptors in the airway smooth muscles, resulting in direct bronchodilation effects	Short-acting	Salbutamol, terbutaline
		Long-acting	Salmeterol, Formoterol, Vilanterol, Bambuterol
Muscarinic antagonist	Acts on M_3 receptors on airway smooth muscles, causing the inhibition of cholinergic tone and the secretion of mucus	Short-acting	Ipratropium
		Long-acting	Tiotropium Glycopyrronium, Aclidinium
Corticosteroids	Suppression of the T-cell activation and the immune system to decrease inflammation	Beclomethasone, fluticasone, prednisolone, budesonide	
Leukotriene receptor antagonist	Prevents the binding of leukotrienes and elicit blockage of leukotriene-mediated bronchoconstriction and mucus secretion	Montelukast, Zafirlukast	
Xanthine	Inhibition of phosphodiesterases in the lung tissues, causing relaxation of bronchial smooth muscles and reduction of airway responsiveness	Theophylline, aminophylline	
Immunomodulator	Known as the monoclonal antibodies, prevents the binding of IgE to the high-affinity receptors on basophils and mast cells.	Omalizumab, Benralizumab	

Table 2 A comparison among the top three respiratory disorders based on respiratory symptoms and treatment algorithms (National Institute for Health and Care Excellence 2022)

COVID-19 infection	Asthma	COPD
Respiratory symptoms		
• Dry cough • Dyspnoea • Shortness of breath • Pneumonia	• Wheezing, coughing, and shortness of breath that is persistent and severe • Blue lips • Rapid breathing	• Recurrent coughing and wheezing • Excessive mucus/sputum production • Frequent respiratory infections • Dyspnoea
Treatment guidelines		
Symptoms oriented	Stepwise approach	Based on spirometry test
Treatment algorithm (reliever and maintenance)		
For acute cough Consumption of pasteurized honey in patients ≥1-year-old ↓ Codeine or low doses of morphine in patients ≥18 years old **For dyspnoea** Breathing techniques ↓ Trial of oxygen therapy ↓ Transfer to secondary care for further evaluation ↓ Offer steroids (prednisolone/ dexamethasone) for up to 10 days **For COVID-19 patients with high risk of disease progression** Offer viral-dependent RNA polymerase inhibitor (e.g., nirmatrelvir and ritonavir (Paxlovid), remdesivir, and molnupiravir) ↓ Offer interleukin-6 inhibitors for adults with COVID-19 in hospital (e.g., tocilizumab, sarilumab)	Short-acting bronchodilators as reliever ↓ Low dose ICS ↓ Low dose ICS + LTRA ↓ Low dose ICS + LABA +/− LTRA ↓ Low dose ICS + LABA within a MART regimen +/− LTRA ↓ Increase the dose of ICS within the MART regimen from moderate to high, with fixed-dose LABA, +/− LTRA ↓ Moderate dose of ICS within the MART regimen with the trial of new drugs (e.g., LAMA/theophylline/ Immunomodulators)	Short-acting bronchodilators as reliever ↓ With asthmatic features: LABA + ICS OR Without asthmatic features: LABA + LAMA ↓ LABA + LAMA + ICS

Abbreviations: *ICS* inhaled corticosteroids, *LABA* long-acting β2-agonist, *LAMA* long-acting muscarinic agonist, *LTRA* leukotriene receptor antagonist, *MART* Maintenance and reliever therapy

The suggested treatment for COVID-19 infection varies with the symptoms and the severity of the illness. Patients with mild respiratory symptoms such as cough, dyspnoea and shortness of breath should be observed closely for any signs of disease progression following the treatment algorithm shown in Table 2.

Corticosteroids, including dexamethasone, hydrocortisone, or prednisolone, prescribed for patients with asthma and COPD exacerbations, are also recommended for COVID-19 patients under the hypoxic condition to reduce further the overall mortality rate (National Institute for Health and Care Excellence 2022).

Previous studies have elucidated that the administration of corticosteroids in the management of viral influenza that closely resembles the COVID-19 infection, such as severe acute respiratory syndrome (SARS) and Middle East respiratory syndrome coronavirus (MERS-CoV), is not ideal due to the potential inhibition of the T-lymphocyte immunity, resulting in the possibility of unceasing viral replication and delayed viral clearance (Spagnuolo et al. 2020; Romanou et al. 2021). However, using corticosteroids in COVID-19 patients will generally reduce the patients' dependency on respiratory support such as supplemental oxygen and mechanical ventilation through the dampening of cytokine storms in response to SARS-COV-2 viruses (Annane 2021; Kim et al. 2021). However, due to inconsistent efficacies and viral clearance effects, the role of corticosteroids in the treatment of COVID-19 infection remains controversial (Li et al. 2020; Lu et al. 2020; Ma et al. 2020; Hu et al. 2020). Therefore, corticosteroid treatment should only be administered to COVID-19 patients who need respiratory support. The safety of corticosteroids in treating COVID-19 infection should be consolidated with more evidence-based studies targeting the practice of precise dosage and timing control (Booi et al. 2022).

Similarly, immunomodulators such as the interleukin-6 (IL-6) inhibitors: sarilumab and tocilizumab are also proposed to treat critically ill patients with systemic inflammation to reduce the mortality rate further (World Health Organization 2017; National Institute for Health and Care Excellence 2022). Apart from the employment of corticosteroids and immunomodulators, other asthma medications have neither been known to possess relevant roles in treating COVD-19 infection nor affect the clinical outcome of the asthmatic treatment in patients contracting the viral infection (Choi et al. 2021).

The use of antivirals such as nirmatrelvir and ritonavir (Paxlovid), remdesivir, and molnupiravir is only reserved for COVID-19 patients who are at higher risk of disease progression. A five-day course of nirmatrelvir and ritonavir (Paxlovid) is recommended to be prescribed within five days of the symptom onset. For instance, adults with COVID-19 infection who do not require supplemental oxygen but at a higher risk of disease exacerbation, are the targeted population for this treatment (National Institute for Health and Care Excellence 2022).

Due to the potential drawbacks of the existing treatment strategies, there is a need to devise a new treatment methodology to optimize the treatment approach for asthma and COPD. There is growing interest in employing CAM, such as natural products, in asthma and COPD patients as part of their conventional treatment regimens (Land and Wang 2018). The discovery of natural products such as medicinal fungi provides pharmaceutical companies valuable insights into developing therapeutically significant lead molecules. These molecules may also possess unique mechanisms of action that can treat or alleviate respiratory disorders with the potential to minimize undesirable effects from current therapies. However, the use of evidence-based natural products should be encouraged to elevate the confidence level of the public in using medicinal mushrooms as part of their treatment regimens. This book chapter summarizes the knowledge about medicinal mushroom use based on reported studies on respiratory disorders.

2 Medicinal Mushrooms: Descriptions and Scientific Findings

Our ancestors discovered the use of medicinal mushrooms for their therapeutic values around the globe over a thousand years ago. They can be characterized as macroscopic fungi, appearing as a distinctive fruiting body growing on dead and decaying matter. The enzymes produced by these mushrooms have high pharmaceutical values in terms of inventing novel bioactive compounds with enormous health benefits (Ogidi et al. 2020). However, these ancient fungi remedies remain obscure to the general public due to their scarcity and great species diversity. Based on the recent finding from Hawksworth and Lücking (2017), approximately 2.2 to 3.8 million species of fungi have been identified, with a significant number of varieties remaining unidentified. However, not every fungal species is edible, as some are toxic and deadly. Therefore, contemporary approaches to cultivating and isolating the active constituents found in medicinal mushrooms are crucial to explore further and justify their potential medicinal values. The current research directions have validated the use of medicinal mushrooms as health supplements. These can be further substantiated as the mushrooms are the source of polysaccharides and polysaccharide-protein complexes which will be discussed later. Those isolated biopolymers are known to possess anticancer, anti-asthmatic, antiviral, and immunomodulatory properties, which are potentially valuable for mitigating the current pandemic outbreak besides being essential sources of medications for respiratory disorders that are yet to be discovered (Valverde et al. 2015).

A collective of 14 species of medicinal mushrooms exhibiting promising beneficial effects on respiratory health have been identified based on their histological evidence, popularity among the indigenous communities, as well as the findings from current in vivo and in vitro scientific studies and clinical trials involving human subjects.

Among all these medicinal mushrooms: [*Ganoderma lucidum*, *Ophiocordyceps sinensis*, *Lignosus rhinocerus*, *Agaricus blazei*, *Lentinula edodes*, *Pleurotus ostreatus*, *Cryptoporus volvatus*, *Phellinus linteus*, *Trametes versicolor*, and *Fomes officinalis*], despite being proven to be effective in supplementing the current treatment regimens for chronic respiratory disorders, have the potential to benefit patients with COVID-19 infection or those with the post-COVID syndrome (Slomski 2021).

Medicinal mushrooms are also an excellent source of antioxidants, with an excellent ability to scavenge free radicals that may lead to an overall reduction in oxidative stress (Mau et al. 2002). For example, these medicinal mushrooms:

[*Ganoderma lucidum, Ophiocordyceps sinensis, Agaricus blazei, Lentinula edodes, Pleurotus ostreatus, Phellinus linteus, Trametes versicolor, Lactarius deliciosus, Tremella fuciformis, Fomes fomentarius,* and *Boletus edulis*], possess a high amount of L-ergothioneine and nucleoside analogues, that have demonstrated potent anti-oxidant activities (Borodina et al. 2020). These properties play an important role in promoting respiratory health, especially in relieving respiratory symptoms such as breathlessness and chronic cough (Borodina et al. 2020).

2.1 *Ganoderma Lucidum*

Ganoderma lucidum (Fig. 1a-f) is a polypore fungus that grows on fallen wood in subtropical and temperate climate zones such as China, Korea, and Japan. Similar species can also be found in European countries such as Sweden, Denmark, and Poland (Nguyen et al. 2021). Two thousand years ago, *G. lucidum* gained popularity as a valuable traditional Chinese medicine in Asia. It is known as "Ling Zhi" in China or "Reishi" in Japan, often described as a "magic mushroom" that promotes longevity of life. Due to its woody fruiting body, the *G. lucidum* is often processed into powder form for easier consumption either directly or as a hot concoction

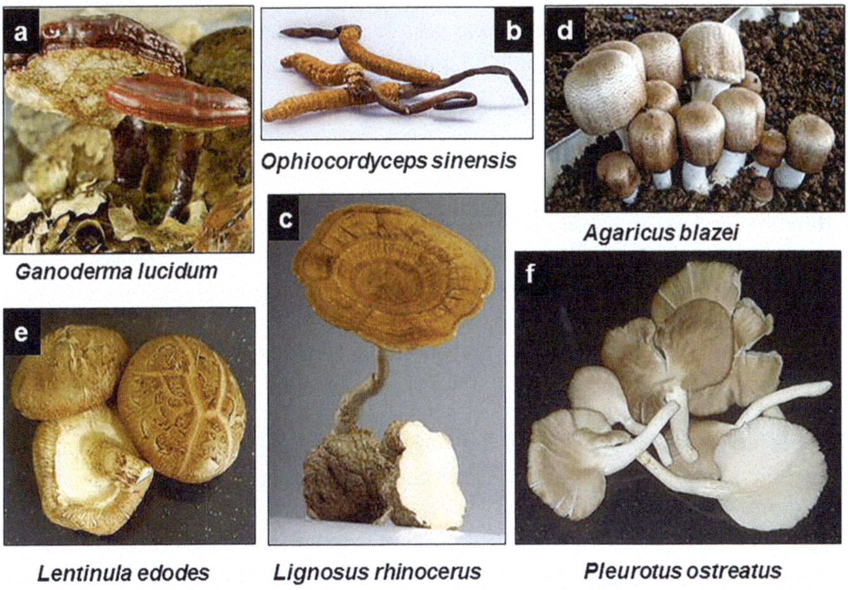

Fig. 1 (**a-f**) (**a**) *Ganoderma lucidum* in wild (Reprinted under Creative Commons CC BY 4.0 (González et al. 2020); (**b**) *Ophiocordyceps sinensis;* (**c**) The fruiting body and mycelium of *Lignosus rhinoceros* (Authors); (**d**) *Agaricus blazei* Murrill (Reprinted under Creative Commons CC BY 4.0 (Rózsa et al. 2019); (**e**) The fruiting bodies of *Lentinula edodes* (Authors); (**f**) The fruiting bodies of *Pleurotus ostreatus* (Authors)

(Wachtel-Galor et al. 2011). *G. lucidum* exists in nature in six distinctive colours of its fruiting bodies (red, black, blue, white, yellow, and purple), each capable of different health enhancing effects. Of all the subspecies of *G. lucidum*, the white *G. lucidum* species is documented to be able to boost lung functions (Nguyen et al. 2021).

The traditional use of *G. lucidum* in respiratory diseases has been studied extensively, and it is well-known for its high antioxidizing and anti-inflammatory properties. This can be further substantiated by animal studies carried out. The polysaccharide fraction of *G. lucidum* produced an anti-inflammatory effect comparable to 10 mg per kg of diclofenac. In contrast, the high levels of terpenoids (Ganoderic acid) found in *G. lucidum* could suppress the TNF-α signaling pathway in animal and cell culture studies without causing any toxicities (Liu et al. 2015). The *G. lucidum* was then combined with two other types of Chinese herbal medicines, namely Ku Shen (*Sophora flavescens*) and Gan Cao (*Glycyrrhiza uralensis*), to form a novel herbal formula known as the anti-asthmatic herbal medicine intervention (ASHMI). This formulation is the only anti-asthmatic herbal therapy that received FDA investigational new drug (IND) approval and entered clinical trials in the United States (Zhou et al. 2019). It was also proven that the anti-asthmatic effect was more persistent than dexamethasone for up to eight weeks post-exposure to allergens in the murine model (Li 2011). On a side note, the bronchodilation effects observed from the ASHMI are proposed to be associated with trifolirhizin extracted from *Sophora flavescens* through the inhibition of the acetylcholine-induced ASM contraction (Yang et al. 2013). To the best of our knowledge, it is confirmed that *G. lucidum* possesses immunomodulating and anti-inflammatory properties and there has been no research output indicating its ability to cause muscle relaxation effects yet (Lull et al. 2005).

The L-fructose-containing polysaccharide fraction from *G. lucidum* (RF3) is potentially capable of tackling COVID-19 infection. Its antiviral potency has been highlighted in a study reporting that 2 μg/ml of the RF3 extracts were molecularly active after 1280 folds of dilution in a cell-based anti–SARS-CoV-2 assay (Jan et al. 2021). The efficacy of the RF3 extracts was further proven by preliminary results obtained from the Vero E6 cell-based studies. It decreased the viral load within the COVID-19 infected golden Syrian hamster (Jan et al. 2021). The antiviral properties of this mushroom were also demonstrated through its inhibition activity against one of the SARS coronavirus replicative enzymes, RNA-dependent RNA polymerase (Fung et al. 2011). Specifically, *G. lucidum* was able to inhibit the RNA- dependent RNA polymerase activities in a concentration-dependent method with an IC_{50} of 41.9 μg/ml, blocking the synthesis process for both positive and negative RNA strands that are crucial for the viral replication (Fung et al. 2011).

Apart from its ability to hinder the rate of viral growth, the aqueous extract of *G. lucidum* benefits patients who are experiencing chronic fatigue syndrome (CFS) (Soksawatmakhin and Boonyahotra 2013). In the 12th week after consuming the aqueous extract, the level of the serum cortisol was 33% higher in the treatment group than that in the placebo group (Soksawatmakhin and Boonyahotra 2013). The increased cortisol level will result in further activation of the glucogenesis pathway, in which more blood glucose is available as the energy source for daily metabolic activities (Thau et al. 2022). This energy-boosting effect may benefit patients with CFS as part of their post-COVID-19 syndromes. These effects are also evaluated using the 12-Item Short-Form Health Survey and the Visual Analog Scale, in which a significant increase in the health-related quality of life was observed (Soksawatmakhin and Boonyahotra 2013).

These results conform to the hypothesis proposed by Wei et al. (2010), that the anti-fatigue activities of *G. lucidum* help accelerate lactic acid clearance in the blood circulation and reduce glycogen consumption. These benefits promote an overall increase in the availability of energy for improved general well-being (Booi et al. 2022).

2.2 *Ophiocordyceps Sinensis*

Ophiocordyceps sinensis (*O. sinensis*) (Fig. 1b) is a fungal species that has been widely used as a traditional medicine in Asia for its high medicinal values (Panda and Swain 2011). It is an endemic species to the alpine meadows of the Tibetan Plateau and adjoining Himalayas (Sigdel et al. 2017). The existence of *O. sinensis* is seasonal, involving the fusion between a fungus and an insect larva. The fungus will first parasitize an underground insect larva and grow within it. The insect larva will be gradually malnourished as the fungus feeds off its nutrient. Eventually, when the larva dies, the fungal fruiting body will emerge out of the ground during the harvesting season between the month of April and August annually (Panda and Swain 2011). This fruiting body is also known as the fungal gold as it is hard to harvest and is scarce.

The *O. sinensis* has shown promising beneficial effects on human physiology, ranging from the heart, lungs, kidneys, and urinary bladder (Lo et al. 2013; Pang et al. 2020). One of the many recognized potentials of the *O. sinensis*, being a potent smooth muscle relaxant, is its ability to help alleviate the symptoms of various respiratory diseases such as bronchitis, asthma, and chronic obstructive pulmonary disease (COPD) through oral administration of natural cordyceps (Wang et al. 2016). Although the source of cordyceps used in the in vivo studies has not been clearly defined, most findings that have employed water-soluble, polysaccharide fractions or alcohol extracts are associated with the relaxation of the smooth muscles in the trachea (Panda and Swain 2011).

The nucleosides within the *O. sinensis*, mainly the cordycepin, are the key bioactive molecules that possess profound antioxidant and anti-inflammatory properties. As portrayed through the employment of a cigarette smoke extract induced model for both in vitro and in vivo studies, these inhibitory effects are beneficial in asthma and COPD patients (Sun et al. 2018). For instance, it was proposed that those nucleosides have the potential to regulate the excitability of airway sensory nerves by targeting the central and peripheral synapses, relieving respiratory symptoms that may contribute to the worsening of both asthma and COPD conditions (Canning and Spina 2009; Das et al. 2021).

Based on the latest study by Li et al. (2021), where an in vitro human cell line model was used, the chemically derived fatty acids from the *O. sinensis* was able to subdue various proinflammatory cytokines that are initiated by H1N1 viral infection. These fatty acids can be incorporated into the treatment regimens for influenza viral infections or can be used as a replacement for corticosteroids to eliminate any steroid-related adverse effects. The *O. sinensis* has demonstrated its ability to slow down the cell senescence process in the airway epithelial cells by suppressing the P16 and P21 protein expressions (Ma et al. 2018). Overexpression of these proteins causes the pathogenesis of COPD manifestations, resulting in irreversible tissue damage within the airways (Wu et al. 2013). This potential of the *O. sinensis* in preventing the occurrence of lung fibrosis becomes significantly prominent with the current outbreak of the COVID-19 infection. An in vivo study using bleomycin-treated mice as the pulmonary interstitial fibrosis model reported a reduction in the number of inflammatory cells and fibroproliferative foci after the oral administration of *O. sinensis* (Wang et al. 2007; Cukic et al. 2012; Kaymakci and Guller 2020). Therefore, the use of *O. sinensis* as a supplement may help relieve post disease complications in patients afflicted with tissue damage in their lungs (Booi et al. 2022).

The cordycepin (3′ deoxyadenosine) is one of the most commonly known bioactive extracts derived from the *Cordyceps* fungi family due to its potent anti-metabolites, antioxidant and anti-inflammatory properties (Tuli et al. 2014). Its ability to lengthen the 3′ untranslated regions (3' UTRs) on the messenger RNA (mRNA) in yeast cells is found to be potentially capable of improving the stability as well as the functionality of the mRNA vaccines (Turner et al. 2021). The mRNA vaccines are one of the most widely used vaccine technologies available to manage the COVID-19 infection. With its potential to lengthen the 3'UTRs on the mRNA, it may delay the degradation process of the mRNA vaccines over time, prolonging the immunity against the COVID-19 infection and hence reducing the need for booster doses (Wadhwa et al. 2020; Barda et al. 2021; Goldberg et al. 2021).

2.3 *Lignosus Rhinocerus*

Lignosus rhinocerus (Fig. 1c), the tiger milk mushroom, is a polypore mushroom that predominantly grows in tropical forests in Southeast Asia (Nallathamby et al. 2018). Due to its scarcity and high medicinal values, it is known as the national treasure mushroom in Malaysia. The indigenous communities in Malaysia often use it as a remedy to complement the current treatment regimens in treating various conditions such as asthma, bronchitis, cough, and solid tumors, for which a complete cure is not available in the current stream of modern medicines and treatment methods (Lee et al. 2018).

The bronchodilation of *L. rhinocerus* has been established by Lee et al. (2018). The cold-water extract of the *L. rhinocerus* can exert a direct relaxation effect on the pre-contracted airway smooth muscles extracted from male Sprague–Dawley rats. The efficacy attained by *L. rhinocerus* was comparably higher than that of the current mainstream bronchodilators such as ipratropium and salmeterol. This can be further substantiated by the mechanism of action for *L. rhinocerus,* which involves the mediation of the airway relaxation effect by the calcium signaling pathway through the Gαq-coupled protein receptors in the airway smooth muscles (Lee et al. 2018). On top of that, the ability of *L. rhinocerus* to suppress airway inflammation was further proven in the airway inflammation murine model (Johnathan et al. 2021). The expression of the central genes responsible for the regulation of asthma was also downregulated after the ingestion of *L. rhinocerus* using the polymerase chain reaction array analysis (Johnathan et al. 2021).

Additionally, using *L. rhinocerus* can be potentially applied to prevent COVID-19 infections. According to the scientific report by Tan et al. (2021), the level of immunoglobulin (Ig A) within the body can be enhanced with the supplementation of 300 mg of *L. rhinocerus* fine powder. The dimeric Ig A is often found on the salivary mucosal layer as the first line of defense and is documented to be more potent than the monomeric immunoglobulin G (Ig G). This attribute is the key in neutralizing the SARS-CoV-2 viral factors through the disruption of the viral cell-attachment process (Wang et al. 2021; Sterlin et al. 2021; Booi et al. 2022).

The cold water extract and its Rhinoprolycan fraction obtained from *L. rhinocerus* were found to be capable of strengthening the immune system through the immunomodulation of the expression of IL-5, IL-6, MIP-2 and TIMP-1 on the monocyte/macrophage-like cell (Sum et al. 2020). These immunomodulatory properties, which are also found in corticosteroids, have benefited in terms of reducing the mortality rate in COVID-19 patients who require respiratory support (Feuillet et al. 2021; Sengupta et al. 2021).

2.4 *Agaricus Blazei*

Agaricus blazei (Fig. 1d) is a comestible mushroom that is endemic to Brazil. However, it is being extensively cultivated in Japan due to its culinary and medicinal properties (Firenzuoli et al. 2008). High levels of immunomodulating components such as beta-glucans and proteoglycans are discovered within *A. blazei*, making it an appealing food supplement against allergies, infections, and respiratory diseases (Hetland et al. 2011).

Hypothetically, the occurrence of asthma is closely related to the skewed ratio of T helper 1 (Th-1) and T helper 2 (Th-2) cells. The extracts obtained from *A. blazei* can modulate cytokine productions from Th-1 and Th-2 that are reciprocally inhibitory to each other (Mahmood et al. 2019). On top of that, the beneficial effects of *A. blazei* in preventing allergic asthma were further proven based on its ability to promote the balance between Th-1 and Th-2 cells, using mice as the experimental asthma model (Takimoto et al. 2008).

There were also preclinical findings in mice indicating that the daily ingestion of *A. blazei* helps alleviate allergies and airway hyperresponsiveness by suppressing IgE-mediated allergy (Ellertsen and Hetland 2009). Along with evident medicinal outcomes obtained in the animal model, the use of *A. blazei* as a supplement was further investigated in a randomized clinical trial (Mahmood et al. 2019). The oral supplementation with *A. blazei* extracts was deemed helpful in preventing pollen-induced allergies and allergies-related asthma in a group of sixty participants. However, the period of the treatment effect is unknown, while the tolerance against these medicinal extracts has not been established (Mahmood et al. 2019). On top of that, Andosan supplementation, mainly composed of water extracts from *A. blazei*, exhibits prophylactic and therapeutic effects against pneumococcal infection. This infection may worsen the COVID-19 infection (Hetland et al. 2021).

2.5 *Lentinula Edodes*

Lentinula edodes (*L. edodes*) (Fig. 1e) is commonly recognized as the Shitake mushroom in Japan or the "fragrant mushroom" in China. It is an important ingredient in Asian cuisine as it grows predominantly in Southeast Asia (Wasser 2004).

The *L. edodes* relieved airway inflammatory and AHR symptoms, which could be observed through the histological examination of the murine lung tissues by suppressing the production of proinflammatory and Th-2 cytokines (Yan and Choi 2014a). Its anti-asthmatic properties were further explored using the Western blot analysis, suggesting that the *L. edodes* suppresses the Mitogen-activated protein kinases (MAPKs) cellular signaling pathway, resulting in an overall reduction in asthmatic symptoms (Yan and Choi 2014a).

Like other medicinal mushrooms, the *L. edodes* is rich in water-soluble beta-glucans (Lentinan), which reduce inflammation in human lung epithelial cells. In other words, lentinan has the potential to be formulated to tackle the recovery treatment in lung injury, especially in COVID-19 patients with pneumonia (Murphy et al. 2020). Lentinan is also a potential therapeutic compound to relieve the ARDS symptoms in COVID-19 patients through the immunomodulation of the cytokine storm, which helps to impede the elevation of IL-6 and eventually lead to the decrease in the mortality of the disease (Murphy et al. 2020). For post-COVID recovery, lentinan has demonstrated its efficacy through the enhancement of the pO2 level in the lungs where an improvement of pulmonary compliance may be observed in COVID-19 patients with lung fibrosis (Masterson et al. 2019; Booi et al. 2022).

2.6 *Pleurotus Ostreatus*

Pleurotus ostreatus (Fig. 1f) is commonly present in the shape of an oyster shell, with large fruiting bodies and seamless stems. It can be found throughout temperate and subtropical forests worldwide. As the cultivation process for *P. ostreatus* is simple and economical, higher profitability can be achieved within a shorter period compared to that of the production of other medicinal mushrooms (Sánchez 2010).

Pleuran (beta-glucans), one of the bioactive extracts obtained from the cell wall of *the P. ostreatus*, has poor aqueous solubility compared to Lentinan found in *L. edodes* (Rop et al. 2009). Hence, its efficacy as an immunomodulator may be reduced as poor aqueous solubility causes decreased absorption and bioavailability in the body (Savjani et al. 2012).

In addition, an open-label trial with patients from similar demographic backgrounds indicated that the use of pleuran as a daily supplement was able to reduce the severity of COPD exacerbations over three months by activating both the innate and adaptive immunity (Minov et al. 2017). These findings can be further justified by a double-blind, placebo-controlled trial conducted in a group of 175 children, using pleuran supplementation (10 mg/kg) in the prevention of respiratory infections by increasing the number of the natural killer cells to further improve the immune system (Jesenak et al. 2013). The natural killer cells help maintain cellular haemostasis through the CD16-mediated antibody-dependent cellular cytotoxicity (ADCC) pathway (van Eeden et al. 2020). These properties could be advantageous in expediting the removal of COVID-19 viral infected cells from the body, which in turn lessens the disease severity as preliminary findings have shown that COVID-19 patients under critical conditions are often observed to have decreased number and function of the natural killer cells (Ahmed et al. 2020; Björkström et al. 2022; Booi et al. 2022).

2.7 *Cryptoporus Volvatus*

Cryptoporus volvatus (Fig. 2a-h) is a polypore fungus formed due to being parasitized by pine bark beetle (Park et al. 2014). It appears mainly in the wild in North America and East Asia. As the fruiting body of *C. volvatus* is woody, extraction is therefore required before the fungus can be consumed. Its bioactive polysaccharides have been widely used to treat asthma and bronchitis since the fifteenth century AD, according to the Compendium of Chinese Materia Medica (Ma et al. 2013).

These medicinal claims are then being scientifically confirmed through the in vitro and in vivo studies, using the airway tissues in asthma induced model for both murine and porcine, whereby *C. volvatus* can impede chemotaxis of eosinophils and mast cell degranulation stimulated by allergens (Tang et al. 2003; Zhao et al. 2004). With its potential to tackle several inflammatory signaling pathways, it was found that *C. volvatus* can attenuate lesions from acute lung injury ensued after

Fig. 2 (**a-h**) (**a**) *Cryptoporus volvatus* in wild (Reprinted from with permission from Elsevier (Yao et al. 2011); (**b**) The fruiting bodies of *Phellinus linteus* (with permission from Elsevier (Chen et al. 2016); (**c**) The fruiting body of *Trametes versicolor* (Reprinted under Creative Commons CC BY 4.0 (Habtemariam 2020); (**d**) *Lactarius deliciosus* with pale orange mottled cap (Reprinted under Creative Commons CC BY 2.0 (Sutton 2010); (**e**) The woody fruiting body of *Fomitopsis officinalis* (Reprinted under Creative Commons CC BY-SA 3.0 (Quarma 2003); (**f**) *Tremella fuciformis* (Authors); (**g**) *Fomes fomentarius* in wild (Reprinted under Creative Commons CC BY 4.0 (Müller et al. 2021); (**h**) *Boletus edulis* in the wild (Reprinted under Creative Commons CC BY 4.0 (Rózsa et al. 2019)

inflammation through its inhibitory effect on the lipopolysaccharide (LPS)-induced pro-inflammatory pathway (Gao et al. 2014). The ability to inhibit Schultz-Dale constriction on the airway smooth muscles has been portrayed by the low molecular weight polysaccharides of *C. volvatus*, whereby the occurrence of anaphylactic-like hypersensitivity can be prevented in the event of asthma exacerbation (Xie et al. 2006).

Lately, the aqueous extracts obtained from the *C. volvatus*, namely the $C_{M\text{-}H\text{-}L\text{-}5}$ component and Cryptoporic acid E (CAE), was proven to possess a broad spectrum of antiviral activities. The $C_{M\text{-}H\text{-}L\text{-}5}$ component was found to have similar structural properties resembling those of the current antiviral nucleoside analogues. Therefore, it has garnered interest as a potential molecule to combat the H1N1 influenza virus (swine flu) (Ma et al. 2013). Aqueous extracts from *C. volvatus* have further exhibited their ability to suppress influenza viral replication through in vivo and in vitro studies in mice (Gao et al. 2014). The discovery of CAE molecules based on the latest findings by Gao et al. (2017) suggests that CAE is key in exerting direct inhibitory effect on the viral RNA polymerase activity. This leads to a potential reduction in viral infectivity, especially the pandemic strain (H1N1/09) found in 2009 (Booi et al. 2022).

2.8 *Phellinus Linteus*

Phellinus linteus (Fig. 2b) is a fungal species that exists as a rigid, perennial fruiting body based on its ability to develop a new surface layer beyond the preceding layer annually a potential lifetime of more than 100 years (Zapora et al. 2016). Therefore, it is inedible and prone to grow on living mulberry trees in moderate climates for constant nutritional support.

Its potential as a complementary therapy for asthma was first reviewed through its inhibitory mechanisms towards mast cell activation, suppressing possible anaphylactic reactions in asthma and allergic rhinitis, utilizing both in vivo and in vitro murine models (Choi et al. 2006). The therapeutic potential of *P. linteus* as an anti-inflammatory agent is consistent with several findings by Yan and Choi (2014b), Hu et al. (2018), and Zhang et al. (2019). Generally, the anti-inflammatory effects were displayed through the hampering of MAPK, NF-κB, and PPAR inflammatory signaling pathways. A recent publication shows that the aqueous methanol extract from *P. linteus*, namely the inotilone has the highest antiviral potency compared to other isolated components obtained, including the current antiviral approach such as oseltamivir (Hwang et al. 2014). Yet, its potency is comparatively inferior to that of zanamivir, that is frequently used in resistant influenza strains (Lampejo 2020).

2.9 *Trametes Versicolor*

Trametes versicolor (Fig. 2c) is well known for the multi-colored appearance of its mushroom cap. This unique characteristic allows *T. versicolor* to be identified easily in the wild, notably on dead logs (Habtemariam 2020). It is often consumed as an herbal tea for general health-boosting effect.

According to the earliest scientific investigation dated since 1977, the polysaccharopeptide (PSP) isolated from *T. versicolor* has been utilized widely in Japan and China to promote endurance and longevity (Chang et al. 2017). The PSP is a protein-bound polysaccharide used in the treatment of chronic bronchitis to enhance the innate immunity of the patients. This medicinal claim can be further supported by two clinical trials, indicating 89.3% of overall effectiveness in treating chronic bronchitis (Teplow 2019). The polysaccharide K (PSK) is another form of protein-bound glycan that has shown a promising outcome in increasing chemotherapy or radiotherapy effectiveness for patients with lung cancer by mediating the immune effector cells (Fritz et al. 2015). Both PSP and PSK can increase the amount of serum IgG and IgM, leucocytes, NK cells, and neutrophils circulating in the body, counteracting the adverse effect of routine anticancer treatment that could eventually weaken the immune system (Zhong et al. 2019).

T. versicolor has also been selected as one of the fungi potentially capable of treating COVID-19 patients experiencing mild to moderate symptoms in a randomized, double-blinded, and placebo-controlled clinical trial (NCT04667247)

(National Institute of Health 2021). In this ongoing clinical trial, 66 human subjects were evaluated for their clinical response to evaluate the efficacy and feasibility of the treatment in COVID-19 outpatients. The outcome of the trial will be available when the trial ends in the third quarter of the year 2022 (National Institute of Health 2021; Booi et al. 2022).

2.10 *Lactarius Deliciosus*

Lactarius deliciosus (Fig. 2d), also known as the Saffron Milkcap, is widely distributed in Europe during autumn. It is popular for its culinary use due to its fruity smell and nutty flavor in nature (Mihailović et al. 2015). The high beta carotene content, an antioxidant found within *L. deliciosus,* gives the mushroom its distinctive bright auburn color. Traditionally, the *L. deliciosus* is commonly consumed by the Russians as part of their culinary heritage. Also, it is used as a folk medicine in Russia to alleviate respiratory symptoms such as asthma and chronic cough. However, no existing scientific studies support its medicinal use (Silva-Filho et al. 2020).

These medicinal benefits could be further elucidated based on its high antioxidant properties, notably the high level of nitric oxide and hydrogen peroxide scavenging activities, which could be deemed helpful in reducing oxidative stress associated with the pathophysiology of asthma (Zhu et al. 2017; Bozdogan et al. 2018). However, antioxidants are not recommended as the first-line therapy for asthma as the clinical results obtained are not consistent.

2.11 *Fomitopsis Officinalis*

Fomitopsis officinalis (Fig. 2e) is a wood-decaying perennial fungus that has been known to be able to live up to 50 years. It is predominantly distributed in old-growth forests in temperate zones, serving as a bioactive reservoir for nature. However, it is prone to become extinct due to recent extensive deforestation (Girometta 2019). Although the fruiting body is rigid and bitter, it has been used as a traditional herbal therapy to relieve pulmonary diseases, including asthma, cough, pneumonia, and tuberculosis, in North America and Mongolia (Grienke et al. 2014).

One of the methanolic extracts from *F. officinalis*, namely the lanostane triterpenes, achieved similar inhibitory effects on nitric oxide production with a smaller concentration required than dexamethasone's use in an LPS-induced cell study (Han et al. 2016). Like *T. versicolor*, the *F. officinalis* is another fungal candidate that was studied in the clinical trial (NCT04667247) mentioned earlier due to its ability to trigger differential cytokine responses (National Institute of Health 2021). The fungal extracts will then be orally administered to eligible COVID-19 patients in the form of capsules for up to two weeks. The efficacy of the treatment will be assessed based on patient-oriented parameters and cytological factors using the laboratory data from the clinical trial. The study will be completed in December 2022 (National Institute of Health 2021; Booi et al. 2022).

2.12 *Tremella Fuciformis*

Tremella fuciformis (Fig. 2f), also known as the silver ear mushroom, is a type of fungus that grows in tropical climates and is extensively cultivated in China for its culinary versatility (Shahrajabian et al. 2020). It has a jelly-like consistency, allowing it to be used as a tonic to promote general well-being and enhance one's appearance aesthetically in ancient times.

Its medicinal use is widespread among the tribal communities in China, particularly for treating chronic bronchitis and asthma (Tsai et al. 2010). These medicinal claims can be further substantiated by the recent publication from Xiao et al. (2021), in which *T. fuciformis* has shown to exhibit potent anti-inflammatory, antioxidant and immunomodulatory properties. These properties are closely associated with inhibiting the miR-155 expression and NFκB activation in the LPS-induced macrophages, reducing the overall oxidative stress responsible for airway inflammation (Ruan et al. 2018; Wu et al. 2019).

2.13 *Fomes Fomentarius*

Formes fomentarius (Fig. 2g) is a hoof-shaped fungus with a tough outer shell, mainly distributed in Europe, Asia, Africa, and North America (Peintner et al. 2019). It has been used as a traditional remedy for respiratory diseases by the central Europeans for centuries (Grienke et al. 2014). The polyphenol compound found within *F. fomentarius* is well-recognized for its free radical scavenging abilities, depicting the highest antiradical efficiency ($1/IC_{50}$) among other 31 fungus species studied (Nowacka et al. 2015). Its valuable application in respiratory diseases can be further explained through its potential to diminish the LPS-induced inflammatory response using the cell culture and the murine disease model (Choe et al. 2015).

2.14 *Boletus Edulis*

Boletus edulis (Fig. 2h) is widely recognized as the porcini mushroom in the culinary industry due to its high nutritional and medicinal values.

The bioactive polysaccharides from *B. edulis* are known to produce similar anti-inflammatory effects as dexamethasone, which can mitigate the severity of asthma through the reduction of interleukin-4 interferon-gamma in lung tissues (Wu et al.

2016). Its anti-asthmatic mechanisms could be hypothesized due to the significant selenium of up to 70 μg/g present within the dried mushroom cap. Selenium is a powerful antioxidant that could decrease airway resistance through its high antiradical activities (Vetter and Lelley 2004; Falandysz 2008; Kieliszek and Błażejak 2016).

As most culinary mushrooms contain selenium levels of less than 1 μg/g, this makes *B.edulis* an outstanding antioxidant candidate that could provide a natural means for vegan people with respiratory diseases to reduce further the risk of exacerbations of the respiratory symptoms (Falandysz 2008). However, the bioavailability of the selenium extract in *B.edulis* is inferior compared to the extraction obtained from the animal source (Mutanen 1986; Spolar et al. 1999). Hence, supplementation with adequate vitamin E is crucial as vitamin E appears to act as a synergist to the antioxidizing properties of selenium (Shreenath et al. 2021).

3 Outlook and Future Prospects

3.1 Medicinal Mushrooms and their Anti-Inflammatory Properties

Airway inflammation acts as a second line of defense of the body through activating innate immunity to eliminate noxious stimuli caused by various precipitation factors and promote tissue restoration. However, excessive chronic inflammation may lead to bronchospasm and airway remodeling, and the hallmark features present in most respiratory diseases, including the coronavirus respiratory infection (Silva-Filho et al. 2020). Hence, the use of anti-inflammatory agents is highly regarded as crucial as part of the treatment algorithm to cease the progression of the disease.

To date, corticosteroids are known as the universal anti-inflammatory agent. They have been widely used in respiratory diseases due to their ability to modulate chemical mediators responsible for bronchospasm and inflammation, including histamines, bradykinin, leukotrienes, and prostaglandins (Hall and Agrawal 2014). However, prolonged corticosteroid treatment is not recommended, owing to its diversity in the mechanism of action that may contribute to a wide variety of adverse effects. For instance, corticosteroids are potent growth factor inhibitors and bone resorption enhancers that may lead to growth retardation in children and osteoporosis in the elderly (Philip 2014).

The corticosteroids may impair the blood glucose hemostasis, besides exhibiting mineralocorticoids and immunosuppressive effects, resulting in diabetes and visual impairments such as cataracts and glaucoma (Yasir et al. 2021). Although the administration of inhaled corticosteroids is associated with fewer adverse effects, unpleasant local side effects such as the sore mouth and hoarse voice are inevitable, and systemic side effects could happen after prolonged administration (Pandya et al. 2014). Additionally, corticosteroid withdrawal syndrome may ensue after long-term exposure to corticosteroids, after which abrupt discontinuation of corticosteroid treatment could result in adrenal insufficiency, depicting hypotension, hypoglycaemia and high potassium level that may be life threatening (Alves et al. 2008).

Of all the similarities, most medicinal mushrooms studied in this chapter possess anti-inflammatory properties comparable to the effect of the standard corticosteroids such as dexamethasone. The potential bioactive extracts of these mushrooms are listed in Table 3.

3.2 Beta Glucans and their Application in Health Aspects

As of the anti-inflammatory properties mentioned earlier, the beta-glucans, also known as the fungal cell wall polysaccharides, are the primary bioactive constituents responsible for their wide variety of therapeutic benefits. They are polysaccharides consisting of glucose monomers with $1 \rightarrow 6$ glycosidic side branches, of which their degree of branching, molecular weight, and water solubility can vary greatly among the fungal species (Du et al. 2015). For instance, most fungal species with an average branching ratio between 0.2 and 0.33 shows their most remarkable ability to modulate the immune system. However, several studies have shown that the reduction of side chains in lentinan has led to an increment in their immunostimulant activities, concluding that the extent of branching and their immunomodulatory properties may be species-specific (Ren et al. 2012; Han et al. 2020).

On the other hand, beta-glucans with higher molecular weight are known to exhibit better immunomodulatory activities, mainly due to their structure stabilities, ability to be recognized by the receptors present on the immune cells as well as the reduced excretion rate from the body (Brown and Gordon 2003; Sletmoen and Stokke 2008). Polysaccharide K (PSK), a protein-bound polysaccharide obtained from *Trametes versicolor,* exhibits better immunostimulating activity at higher molecular weight fractions of greater than 200 kDa (Kim et al. 1990). The aqueous solubility of the beta-glucans is another factor that should be carefully considered during extraction. Highly water-soluble glucans tend to produce more pronounced biological effects on the immune system (Rop et al. 2009). To enhance the specific biological activities of the beta-glucans, chemical modifications can be employed to produce structural diversity to increase binding affinity to the target receptors and accommodate further their desired medical applications (Synytsya and Novák 2013).

The mechanisms of action of beta-glucans are closely related to its pathogen-associated molecular pattern (PAMP) in nature. They are the key molecules to activate innate immune responses upon detecting inflammation and infections (Camilli et al. 2018). They can perform surface ligand binding to various Dectin-1 receptors expressed on the immune cells, such as dendritic cells, macrophages, monocytes, and natural killer cells. This leads to the immunomodulation of the mucosal immune response through the regulation of cytokines such as interferon-gamma (INF)-γ, interleukin, nitric oxide, and TNF-α upon oral administration

Table 3 The bioactive extracts in medicinal mushrooms that possess anti-inflammatory activities and their proposed mechanisms of action

Species	Bioactive extracts	Proposed mechanisms of action	Reference
Ganoderma lucidum	Terpenoids (Ganoderic acid)	• Inhibition of TNF-α production • Inhibition of NF-κB signaling pathway	Liu et al. (2015)
Ophiocordyceps sinensis	Cordycepin	• Stimulation of intracellular PLC/PKC and MAPK signaling pathways to induce the production of steroid hormones	Liu et al. (2015); Yang et al. (2015)
Lignosus rhinocerus	Proteins or polysaccharide-protein complex found in the high mw and medium mw fractions	• Mediation of the calcium signaling pathway through the Gαq-coupled protein receptors found within the airway smooth muscles	Lee et al. (2018)
Agaricus blazei	Beta-glucans	• Modulation of cytokine productions from Th-1 and Th-2 that are reciprocally inhibitory to each other • Promotes the balance between Th-1 and Th-2 cells	Hetland et al. (2020)
Lentinula edodes	Lentinan	• Inactivation of OVA-induced of NF-κB and p38 MAPK in lung tissues	Yan and Choi (2014a); Murphy et al. (2020)
Pleurotus ostreatus	Pleuran	• Binds to Dectin-1 receptors to stimulate innate immune responses • Releases pro-inflammatory cytokines to improve resistance toward invading pathogens	Rop et al. (2009)
Cryptoporus volvatus	*Cryptoporus polysaccharides*	• Attenuates the LPS-induced expression of pro-inflammatory factors in human alveolar epithelial cells through the TLR2 signaling pathway	Gao et al. (2014)
Phellinus liteus	Inotilone	• Suppresses the expression of LPS-induced MMP-9, NF-κB, and MAPK activation portrayed through in vitro and in vivo studies	Huang et al. (2012)

(continued)

Table 3 (continued)

Species	Bioactive extracts	Proposed mechanisms of action	Reference
Trametes versicolor	Polysaccharide K (PSK)	• Enhances the amount of serum IgG and IgM, leucocytes, and neutrophils circulating in the body • Counteracts the adverse effect that could eventually compromise the immune system	Zhang et al. (2019)
Tremella fuciformis	Hot water extracts	• Inhibition of miR-155 expression and NFκB activation in the LPS-induced macrophages • Reduces the overall oxidative stress in airway inflammation	Li et al. (2014); Ruan et al. (2018)
Fomes fomentarius	Methanolic extracts	• Inhibits inflammatory responses in LPS-induced macrophages by suppressing STAT3 activation	Choe et al. (2015)

Abbreviations: *LPS* lipopolysaccharide, *MAPK* Mitogen-activated protein kinases, *MMP-9* Matrix Metallopeptidase 9, *Mw* Molecular weight, *NF-κB* Nuclear Factor kappa B, *PLC* Phospholipase C, *PKC* Protein Kinase C, *TLR2* Toll-like Receptor 2, *TNF- α* Tumor Necrosis Factor-alpha, *STAT3* Signal Transducer and Activator of Transcription 3

(Yadav and Schorey 2006). These novel immunomodulating properties in beta-glucans could serve as an alternative approach to dampen the inflammatory pathways and increase the formation rate of antioxidant molecules (Kofuji et al. 2012). Hence, the development of asthma through the regulation of the pathogenic immune response can be prevented. Apart from that, the beta-glucans can act as a regulator for the T-helper cells by supporting the formation of Th-1 in response to a predomination by Th-2 that is associated with the development of allergies or asthma responses (Inoue et al. 2002).

The beta-glucans have shown their medical importance at different levels of cellular organization, ranging from in vitro to in vivo studies in animals and humans based on the discussions earlier. However, the existing studies are mainly based on the crude extract obtained from fungal species rather than the purified beta-glucans. Therefore, the beneficial effects observed from different fungal species may be varied due to the potential co-existing of other chemical components present within the crude extracts. Further investigation of the chemical structures of the beta-glucans among the fungal strains is crucial to quantify the biological effectiveness in different models of interest compared to the purified beta-glucans.

3.3 *Medicinal Mushrooms and their Relevance to Respiratory Diseases*

In this chapter, we have discussed different species of medicinal mushrooms that have shown vast potential in alleviating both acute and chronic COVID-19 complications. The immunomodulatory and antioxidant properties of medicinal mushrooms have been recognized widely. Therefore, medicinal mushrooms are suitable remedial candidates that can be incorporated into the prophylactic and therapeutic management of COVID-19 to prevent excessive inflammation and tissue damage closely associated with the COVID-19 infection.

L-ergothioneine is a naturally occurring amino acid that demonstrates higher stability against oxidative stress than ascorbic acid under physiological pH (Cheung et al. 2016). It is mostly found in high concentrations between 0.14 and 7.27 mg/g of dry weight in medicinal mushrooms (Lam-Sidun et al. 2021) (Table 4). Since human beings are not able to synthesize L-ergothioneine naturally, the existence of this amino acid is greatly valued (Kalač 2016). However, the standard reference to identify the optimal amount of L ergothioneine that is both safe for ingestion and efficacious has not been established. As a reference, the average intake of L-ergothioneine by human adults of weight between 70 kg to 80 kg is estimated between 0.051 and 0.255 mg/kg bw/d based on several intake assessments conducted in several European countries and the United States (Marone et al. 2016; Ramirez-Martinez et al. 2016).

Numerous clinical trials (NCT 04323514/ NCT04264533/NCT04710329) have been conducted worldwide to study the effects of the use of ascorbic acid in hospitalized patients with COVID-19 patients (National Institute of Health 2021). The antioxidant properties of ascorbic acid have been proven to reduce the likelihood of patients requiring mechanical ventilation. Patients with acute respiratory infections will also benefit from an overall 7% to 8% shorter duration of hospital stay (Hunt et al. 1994; Hemilä and Chalker 2019). However, it is suggested that the plasma concentration of ascorbic acid in the patient may encounter ceiling effects of 12–15 mg/L even if the patient has received a large dose of ascorbic acid

Table 4 The amount of L-ergothioneine found in each medicinal mushroom species

Species	Ergothioneine (mg/g in dry weight)	Reference
Ganoderma lucidum	0.56	Kalaras et al. (2017)
Ophiocordyceps sinensis	0.142	Chen et al. (2012)
Agaricus blazei	0.0796	Chen et al. (2012)
Lentinula edodes	0.92	Kalaras et al. (2017)
Pleurotus ostreatus	1.21	Kalaras et al. (2017)
Phellinus linteus	0.1818	Chen et al. (2012)
Trametes versicolor	0.013	Chen et al. (2012)
Tramella fuciformis	0.58	Halliwell et al. (2018)
Boletus edulis	7.27	Kalaras et al. (2017)

intravenously (Blanchard et al. 1997). This nonlinear pharmacokinetics relationship could be attributed to the saturable gastrointestinal absorption and elimination capacity (Wagner 1973; Blanchard et al. 1997). Hence, repeated administration of small doses of ascorbic acids is necessary for improved cellular functions (Padh 1990).

L-ergothioneine has demonstrated its potential as a stable antioxidant. It can remain in the blood circulation for up to four weeks after the ingestion period of 25 mg per day is concluded (Cheah et al. 2017). The elimination rate of L-ergothioneine remains low as only 2% of the overall L-ergothioneine has been removed from the body throughout the study (Cheah et al. 2017). These phenomena could be hypothetically explained via the fact that the liver and the bone marrow are the possible storage sites for L-ergothioneine (Grundemann et al. 2005; Cheah et al. 2017). Studies also showed that the level of L-ergothioneine increases more rapidly in the murine liver than in the whole blood concentration. It was hence suggested that the erythrocytes are responsible for gathering up the L-ergothioneine from the plasma, which may further reduce L-ergothioneine's excretion out of the body (Grundemann et al. 2005; Cheah et al. 2017). Consequently, the L-ergothioneine will be widely distributed throughout the body to minimize the oxidative damage that this COVID-19 infection might have caused through free radical scavenging activities (Tang et al. 2018; Booi et al. 2022).

3.4 Medicinal Mushroom as a Source for Novel Bioactive Compounds

Nature serves as a bioactive reservoir, providing us with abundant resources that are important to promote the discovery of novel bioactive compounds and further advance the currently existing treatment method. For example, the antimicrobial activity of penicillin was discovered in serendipity from a micro fungus, while the lovastatin, a cholesterol-lowering compound, is naturally occurring in a wide array of macrofungi such as the *Pleurotus ostreatus*, *Agaricus bisporus*, and *Cantharellus cibarius* (Dias et al. 2012; Kała et al. 2020).

Therefore, the medicinal mushroom (macrofungi) is recognized as a functional food comprising polysaccharides, proteoglycans, terpenes, phenolic compounds, and other bioactive molecules best known for their abilities to modulate the immunity and protective effects against oxidative stress (Martinez-Medina et al. 2021). These mushroom-derived bioactive components are then being extensively studied at different levels of cellular organization, including those that have successfully progressed into clinical trials, namely the *G. lucidum*, *A. blazei*, *T. versicolor,* and *F. officinalis* that have been mentioned in this chapter. Hence, there is substantial evidence that the medicinal mushroom has the potential to revolutionize the standard therapy for respiratory disorders, especially in the light of minimizing the dependency on corticosteroids due to its comparable anti-inflammatory properties.

3.5 *The Quality, Safety, and Efficacy of Medicinal Mushrooms*

Medicinal mushrooms are either consumed as a delicacy or a supplement. The fruiting bodies of these mushrooms contain bioactive extracts with immense health benefits. The use of medicinal mushrooms in treating respiratory disorders remain largely at the preliminary stages since there are many variations in the clinical parameters and regulations from different batches and sources of mushrooms. These variations could affect the quality, safety and efficacy of the end products processed from the mushrooms. Therefore, it is termed the CAM to distinguish its use from the conventional treatments and promote general well-being.

Food and Drug Administration (FDA) in the US regulates all CAM applications, requiring the CAM to have recorded evidence of use traditionally to prove its safe use. No minimum requirements have been stipulated regarding the clinical evidence to substantiate its safety and efficacy (Food and Drug Administration 2007). In Europe, European Food Safety Authority (EFSA) is responsible for assessing the safety of these CAM using a science-based approach. European Medicine Agency (EMA) regulates the use of CAM as medicines, focusing on the evaluation of safety and quality (European Food Safety Authority 2016). Based on the guidelines in existing regulations, the allowable amounts of bioactive compounds and contaminants vary across similar products. Hence, the standardization of medicinal mushrooms in the CAM industry should be implemented in order for more end-users to receive the benefits of CAM in an optimized and regulated manner.

There has been a significant number of research works ongoing to discover potentially valuable properties of the medicinal mushrooms. Many researchers have begun to focus on studying these mushrooms to disclose more scientific evidence about their medicinal benefits. However, most of the notable research works are mainly in vitro and in vivo studies employing animal models as the test subjects. The use of animal models affects the reliability and reproducibility of the documented outcomes since there are large genetic and physiological differences between animals and human beings. Hence, the probability of successful transitional study from animals to human beings is relatively low (Leenaars et al. 2019). This could be addressed with carefully designed in vivo studies involving human subjects in the future. Before proper guidelines to run human-centric in vivo studies are available, successful animal studies should be encouraged for progression to clinical trials to further improve the credibility of the outcomes of the research works. With adequate laboratory studies and scientific evidence, the safety and effectiveness of the medicinal mushrooms can be substantiated to help make these Asian medications more appealing to the Western as well as the global community (Booi et al. 2022).

4 Conclusion

Medicinal mushrooms in respiratory disorders have been gaining popularity lately due to their histological foundation in complementary and alternative medicine. They invite more rigorous trials and quality control efforts to strengthen further scientific evidence about the benefits of mushrooms to human health. Scientific investigations should be conducted more robustly to focus on understanding the pharmacological effects of the active compounds in the mushroom. These efforts from the scientific community provided a platform for medicinal mushrooms to emerge more prominently as prospective candidates that could help in the management of respiratory conditions.

Acknowledgments Han Ni Booi is a Ph.D. scholarship recipient sponsored by the University of Nottingham Malaysia and the Fundamental Research Grant Scheme (FRGS) Malaysia (Grant code: FRGS/1/2020/SKK0/UNIM/01/2). LiGNO Biotech is the sponsor of the fruiting body powder used in our research. (Project title: Elucidating the relaxant effects of *Ophiocordyceps sinensis* (Berk.) on airways and bladder smooth muscles to substantiate its traditional uses).

References

Ahmed F, Jo DH, Lee SH (2020) Can natural killer cells be a principal player in anti-SARS-CoV-2 immunity? Front Immunol 11:586765. https://doi.org/10.3389/fimmu.2020.586765

Alves C, Robazzi TCV, Mendonça M (2008) Withdrawal from glucocorticosteroid therapy: clinical practice recommendations. J Pediatr 84(3):192–202. https://doi.org/10.2223/JPED.1773

An SS, Bai TR, Bates JHT et al (2007) Airway smooth muscle dynamics: a common pathway of airway obstruction in asthma. Eur Respir J 29(5):834–860. https://doi.org/10.1183/09031936.00112606

Annane D (2021) Corticosteroids for COVID-19. J Intensive Med 1(1):14–25. https://doi.org/10.1016/j.jointm.2021.01.002

Aveyard P, Gao M, Lindson N et al (2021) Association between pre-existing respiratory disease and its treatment, and severe COVID-19: a population cohort study. Lancet Respir Med 9(8): 909–923. https://doi.org/10.1016/S2213-2600(21)00095-3

Barda N, Dagan N, Cohen C et al (2021) Effectiveness of a third dose of the BNT162b2 mRNA COVID-19 vaccine for preventing severe outcomes in Israel: an observational study. Lancet 398(10316):2093–2100. https://doi.org/10.1016/S0140-6736(21)02249-2

Bellan M, Soddu D, Balbo PE et al (2021) Respiratory and psychophysical sequelae among patients with COVID-19 four months after hospital discharge. JAMA Netw Open 4(1):e2036142. https://doi.org/10.1001/jamanetworkopen.2020.36142

Berry MA, Hargadon B, Shelley M et al (2006) Evidence of a role of tumor necrosis factor α in refractory asthma. N Engl J Med 354(7):697–708. https://doi.org/10.1056/NEJMoa050580

Björkström NK, Strunz B, Ljunggren HG (2022) Natural killer cells in antiviral immunity. Nat Rev Immunol 22:112–123. https://doi.org/10.1038/s41577-021-00558-3

Black JL, Panettieri RA, Banerjee A et al (2012) Airway smooth muscle in asthma. Clin Chest Med 33(3):543–558. https://doi.org/10.1016/j.ccm.2012.05.002

Blanchard J, Tozer TN, Rowland M (1997) Pharmacokinetic perspectives on megadoses of ascorbic acid. Am J Clin Nutr 66(5):1165–1171. https://doi.org/10.1093/ajcn/66.5.1165

Booi HN, Lee MK, Fung SY et al (2022) Medicinal mushrooms and their use to strengthen respiratory health during and post-COVID-19 pandemic. Int J Med Mushrooms. (in press)
Borodina I, Kenny LC, McCarthy CM et al (2020) The biology of ergothioneine, an antioxidant nutraceutical. Nutr Res Rev 33(2):190–217. https://doi.org/10.1017/S0954422419000301
Bourdin A, Burgel P-R, Chanez P et al (2009) Recent advances in COPD: pathophysiology, respiratory physiology and clinical aspects, including comorbidities. Eur Respir Rev 18(114): 198–212. https://doi.org/10.1183/09059180.00005509
Bozdogan A, Ulukanlı Z, Bozok F et al (2018) Antioxidant potential of *Lactarius deliciosus* and *Pleurotus ostreatus* from Amanos mountains. Adv Life Sci 5(3):113–120. http://www.als-journal.com/articles/vol5issue3/535.18/567.pdf
Bradding P (2007) Mast cell regulation of airway smooth muscle function in asthma. Eur Respir J 29(5):827–830. https://doi.org/10.1183/09031936.00017707
Brown GD, Gordon S (2003) Fungal β-glucans and mammalian immunity. Immunity 19(3): 311–315. https://doi.org/10.1016/S1074-7613(03)00233-4
Bush A (2019) Pathophysiological mechanisms of asthma. Front Pediatr 7:68. https://doi.org/10.3389/fped.2019.00068
Camilli G, Tabouret G, Quintin J (2018) The complexity of fungal β-glucan in health and disease: effects on the mononuclear phagocyte system. Front Immunol 9:673. https://doi.org/10.3389/fimmu.2018.00673
Canning BJ, Spina D (2009) Sensory nerves and airway irritability. Handb Exp Pharmacol 194: 139–183. https://doi.org/10.1007/978-3-540-79090-7_5
Casas L, Nemery B (2014) Irritants and asthma. Eur Respir J 44(3):562–564. https://doi.org/10.1183/09031936.00090014
Cascella M, Rajnik M, Aleem A et al (2022) Features, evaluation, and treatment of coronavirus (COVID-19). In: StatPearls. StatPearls Publishing, Treasure Island (FL), pp 11–14
Chang Y, Zhang M, Jiang Y et al (2017) Preclinical and clinical studies of *Coriolus versicolor* polysaccharopeptide as an immunotherapeutic in China. Discov Med 23(127):207–219
Cheah IK, Tang RMY, Yew TSZ et al (2017) Administration of pure ergothioneine to healthy human subjects: uptake, metabolism, and effects on biomarkers of oxidative damage and inflammation. Antioxid Redox Signal 26(5):193–206. https://doi.org/10.1089/ars.2016.6778
Chen SY, Ho KJ, Hsieh YJ et al (2012) Contents of lovastatin, γ-aminobutyric acid and ergothioneine in mushroom fruiting bodies and mycelia. LWT-Food Sci Technol 47(2): 274–278. https://doi.org/10.1016/j.lwt.2012.01.019
Chen H, Tian T, Miao H et al (2016) Traditional uses, fermentation, phytochemistry and pharmacology of *Phellinus linteus* : a review. Fitoterapia 113:6–26. https://doi.org/10.1016/j.fitote.2016.06.009
Cheung B, Kwan M, Chan R et al (2016) Potential of Asian natural products for health in aging. In: Malavolta M, Mocchegiani E (eds) Molecular basis of nutrition and aging. Academic Press, Cambridge, pp 659–676
Choe J-H, Yi Y-J, Lee M-S et al (2015) Methyl 9-oxo-(10 e, 12 e)-octadecadienoate isolated from *Fomes fomentarius* attenuates lipopolysaccharide-induced inflammatory response by blocking phosphorylation of STAT3 in murine macrophages. Mycobiology 43(3):319–326. https://doi.org/10.5941/MYCO.2015.43.3.319
Choi YJ, Park JY, Lee HS et al (2021) Effect of asthma and asthma medication on the prognosis of patients with COVID-19. Eur Respir J 57(3):2002226. https://doi.org/10.1183/13993003.02226-2020
Choi YH, Yan GH, Chai OH et al (2006) Inhibition of anaphylaxis-like reaction and mast cell activation by water extract from the fruiting body of *Phellinus linteus*. Biol Pharm Bull 29(7): 1360–1365. https://doi.org/10.1248/bpb.29.1360
Cukic V, Lovre V, Dragisic D et al (2012) Asthma and chronic obstructive pulmonary disease (COPD): differences and similarities. Mater Socio Med 24(2):100–105. https://doi.org/10.5455/msm.2012.24.100-105

Das G, Shin H-S, Leyva-Gómez G et al (2021) Cordyceps spp.: a review on its immune-stimulatory and other biological potentials. Front Pharmacol 11:602364. https://doi.org/10.3389/fphar.2020.602364

Dias DA, Urban S, Roessner U (2012) A historical overview of natural products in drug discovery. Meta 2(2):303–336. https://doi.org/10.3390/metabo2020303

Doeing DC, Solway J (2013) Airway smooth muscle in the pathophysiology and treatment of asthma. J Appl Physiol 114(7):834–843. https://doi.org/10.1152/japplphysiol.00950.2012

Du B, Lin C, Bian Z et al (2015) An insight into anti-inflammatory effects of fungal beta-glucans. Trends Food Sci Technol 41(1):49–59. https://doi.org/10.1016/j.tifs.2014.09.002

Ellertsen LK, Hetland G (2009) An extract of the medicinal mushroom *Agaricus blazei* Murill can protect against allergy. Clin Mol Allergy 7(1):6. https://doi.org/10.1186/1476-7961-7-6

European Food Safety Authority (2016) Botanicals. https://www.efsa.europa.eu/en/topics/topic/botanicals. Accessed 23 May 2021

Fajgenbaum DC, June CH (2020) Cytokine storm. N Engl J Med 383(23):2255–2273. https://doi.org/10.1056/NEJMra2026131

Falandysz J (2008) Selenium in edible mushrooms. J Environ Sci Heal Part C 26(3):256–299. https://doi.org/10.1080/10590500802350086

Feuillet V, Canard B, Trautmann A (2021) Combining antivirals and immunomodulators to fight COVID-19. Trends Immunol 42(1):31–44. https://doi.org/10.1016/j.it.2020.11.003

Firenzuoli F, Gori L, Lombardo G (2008) The medicinal mushroom *Agaricus blazei* Murrill: review of literature and pharmaco-toxicological problems. Evid Based Complement Altern Med 5(1): 3–15. https://doi.org/10.1093/ecam/nem007

Food and Drug Administration (2007) Complementary and alternative medicine products and their regulation by the Food and Drug Administration. https://www.fda.gov/regulatory-information/search-fda-guidance-documents/complementary-and-alternative-medicine-products-and-their-regulation-food-and-drug-administration. Accessed 23 May 2021

Fritz H, Kennedy DA, Ishii M et al (2015) Polysaccharide K and *Coriolus versicolor* extracts for lung cancer. Integr Cancer Ther 14(3):201–211. https://doi.org/10.1177/1534735415572883

Fung KP, Leung PC, Tsui KWS et al (2011) Immunomodulatory activities of the herbal formula Kwan Du Bu Fei dang in healthy subjects: a randomised, double-blind, placebo-controlled study. Hong Kong Med J 17(2):41–43

Gao L, Han J, Si J et al (2017) Cryptoporic acid E from *Cryptoporus volvatus* inhibits influenza virus replication in vitro. Antivir Res 143:106–112. https://doi.org/10.1016/j.antiviral.2017.02.010

Gao L, Sun Y, Si J et al (2014) *Cryptoporus volvatus* extract inhibits influenza virus replication in vitro and in vivo. PLoS One 9(12):e113604. https://doi.org/10.1371/journal.pone.0113604

Girometta C (2019) Antimicrobial properties of *Fomitopsis officinalis* in the light of its bioactive metabolites: a review. Mycology 10(1):32–39. https://doi.org/10.1080/21501203.2018.1536680

Global Initiative for Chronic Obstructive Lung Disease (2019) Global strategy for the diagnosis, management and prevention of chronic pulmonary disease. https://goldcopd.org/wp-content/uploads/2018/11/GOLD-2019-v1.7-FINAL-14Nov2018-WMS.pdf. Accessed 2 Jun 2021

Goldberg Y, Mandel M, Bar-On YM et al (2021) Waning immunity after the BNT162b2 vaccine in Israel. N Engl J Med 385(24):e85. https://doi.org/10.1056/NEJMoa2114228

Goldklang M, Stockley R (2016) Pathophysiology of emphysema and implications. J COPD Found 3(1):454–458. https://doi.org/10.15326/jcopdf.3.1.2015.0175

González A, Atienza V, Montoro A et al (2020) Use of *Ganoderma lucidum* (Ganodermataceae, Basidiomycota) as radioprotector. Nutrients 12(4):1143. https://doi.org/10.3390/nu12041143

Grienke U, Zöll M, Peintner U et al (2014) European medicinal polypores: a modern view on traditional uses. J Ethnopharmacol 154(3):564–583. https://doi.org/10.1016/j.jep.2014.04.030

Grundemann D, Harlfinger S, Golz S et al (2005) Discovery of the ergothioneine transporter. Proc Natl Acad Sci 102(14):5256–5261. https://doi.org/10.1073/pnas.0408624102

Gu Q, Lee L-Y (2022) Neural control of airway smooth muscle. In: Janes SM (ed) Encyclopedia of respiratory medicine, 2nd edn. Academic Press, Cambridge, pp 164–173

Habtemariam S (2020) *Trametes versicolor* (synn. Coriolus versicolor) polysaccharides in cancer therapy: targets and efficacy. Biomedicine 8(5):135. https://doi.org/10.3390/biomedicines8050135

Hall S, Agrawal DK (2014) Key mediators in the immunopathogenesis of allergic asthma. Int Immunopharmacol 23(1):316–329. https://doi.org/10.1016/j.intimp.2014.05.034

Halliwell B, Cheah IK, Tang RMY (2018) Ergothioneine: a diet-derived antioxidant with therapeutic potential. FEBS Lett 592(20):3357–3366. https://doi.org/10.1002/1873-3468.13123

Han B, Baruah K, Cox E et al (2020) Structure-functional activity relationship of β-glucans from the perspective of immunomodulation: a mini-review. Front Immunol 11:658. https://doi.org/10.3389/fimmu.2020.00658

Han J, Li L, Zhong J et al (2016) Officimalonic acids a−h, lanostane triterpenes from the fruiting bodies of *Fomes officinalis*. Phytochemistry 130:193–200. https://doi.org/10.1016/j.phytochem.2016.05.004

Hawksworth DL, Lücking R (2017) Fungal diversity revisited: 2.2 to 3.8 million species. Microbiol Spectr 5(4):79–95. https://doi.org/10.1128/microbiolspec.FUNK-0052-2016

Heffler E, Madeira LNG, Ferrando M et al (2018) Inhaled corticosteroids safety and adverse effects in patients with asthma. J Allergy Clin Immunol Pract 6(3):776–781. https://doi.org/10.1016/j.jaip.2018.01.025

Hemilä H, Chalker E (2019) Vitamin C can shorten the length of stay in the ICU: a meta-analysis. Nutrients 11(4):708. https://doi.org/10.3390/nu11040708

Hetland G, Johnson E, Bernardshaw SV et al (2021) Can medicinal mushrooms have prophylactic or therapeutic effect against COVID-19 and its pneumonic superinfection and complicating inflammation? Scand J Immunol 93(1):e12937. https://doi.org/10.1111/sji.12937

Hetland G, Johnson E, Lyberg T et al (2011) The mushroom *Agaricus blazei* Murill elicits medicinal effects on tumor, infection, allergy, and inflammation through its modulation of innate immunity and amelioration of Th1/Th2 imbalance and inflammation. Adv Pharmacol Sci 2011:157015. https://doi.org/10.1155/2011/157015

Hetland G, Tangen J-M, Mahmood F et al (2020) Antitumor, anti-inflammatory and antiallergic effects of *Agaricus blazei* mushroom extract and the related medicinal basidiomycetes mushrooms, *Hericium erinaceus* and *Grifola frondosa*: a review of preclinical and clinical studies. Nutrients 12(5):1339. https://doi.org/10.3390/nu12051339

Hikichi M, Mizumura K, Maruoka S et al (2019) Pathogenesis of chronic obstructive pulmonary disease (COPD) induced by cigarette smoke. J Thorac Dis 11(Suppl 17):S2129–S2140. https://doi.org/10.21037/jtd.2019.10.43

Hoffmann M, Kleine-Weber H, Schroeder S et al (2020) SARS-CoV-2 cell entry depends on ACE2 and TMPRSS2 and is blocked by a clinically proven protease inhibitor. Cell 181(2):271–280.e8. https://doi.org/10.1016/j.cell.2020.02.052

Hsu E, Bajaj T (2021) Beta 2 Agonists. In: StatPearls. StatPearls Publishing, Treasure Island (FL), pp 4–7

Hu T, Lin Q, Guo T et al (2018) Polysaccharide isolated from *Phellinus linteus* mycelia exerts anti-inflammatory effects via MAPK and PPAR signaling pathways. Carbohydr Polym 200:487–497. https://doi.org/10.1016/j.carbpol.2018.08.021

Hu Y, Wang T, Hu Z et al (2020) Clinical efficacy of glucocorticoid on the treatment of patients with COVID-19 pneumonia: a single-center experience. Biomed Pharmacother 130:110529. https://doi.org/10.1016/j.biopha.2020.110529

Huang G-J, Huang S-S, Deng J-S (2012) Anti-inflammatory activities of inotilone from *Phellinus linteus* through the inhibition of MMP-9, NF-κB, and MAPK activation in vitro and in vivo. PLoS One 7(5):e35922. https://doi.org/10.1371/journal.pone.0035922

Huang Y, Yang C, Xu X et al (2020) Structural and functional properties of SARS-CoV-2 spike protein: potential antivirus drug development for COVID-19. Acta Pharmacol Sin 41(9):1141–1149. https://doi.org/10.1038/s41401-020-0485-4

Hughes-Visentin A, Paul ABM (2020) Asthma and COVID-19: what do we know now. Clin Med Insights Circ Respir Pulm Med 14:1179548420966242. https://doi.org/10.1177/1179548420966242

Hunt C, Chakravorty NK, Annan G et al (1994) The clinical effects of vitamin C supplementation in elderly hospitalised patients with acute respiratory infections. Int J Vitam Nutr Res 64(3): 212–219

Hwang BS, Lee M-S, Lee SW et al (2014) Neuraminidase inhibitors from the fermentation broth of *Phellinus linteus*. Mycobiology 42(2):189–192. https://doi.org/10.5941/MYCO.2014.42.2.189

Inoue A, Kodama N, Nanba H (2002) Effect of maitake (*Grifola frondosa*) D-fraction on the control of the T lymph node Th-1/Th-2 proportion. Biol Pharm Bull 25(4):536–540. https://doi.org/10.1248/bpb.25.536

Jan JT, Cheng TJR, Juang YP et al (2021) Identification of existing pharmaceuticals and herbal medicines as inhibitors of SARS-CoV-2 infection. Proc Natl Acad Sci 118(5):e2021579118. https://doi.org/10.1073/pnas.2021579118

Jesenak M, Majtan J, Rennerova Z et al (2013) Immunomodulatory effect of pleuran (β-glucan from *Pleurotus ostreatus*) in children with recurrent respiratory tract infections. Int Immunopharmacol 15(2):395–399. https://doi.org/10.1016/j.intimp.2012.11.020

Johansson MA, Quandelacy TM, Kada S et al (2021) SARS-CoV-2 transmission from people without COVID-19 symptoms. JAMA Netw Open 4(1):e2035057. https://doi.org/10.1001/jamanetworkopen.2020.35057

Johnathan M, Muhamad SA, Gan SH et al (2021) *Lignosus rhinocerotis* Cooke Ryvarden ameliorates airway inflammation, mucus hypersecretion and airway hyperresponsiveness in a murine model of asthma. PLoS One 16(3):e0249091. https://doi.org/10.1371/journal.pone.0249091

Kała K, Kryczyk-Poprawa A, Rzewińska A et al (2020) Fruiting bodies of selected edible mushrooms as a potential source of lovastatin. Eur Food Res Technol 246(4):713–722. https://doi.org/10.1007/s00217-020-03435-w

Kalač P (2016) Health-stimulating compounds and effects. In: Edible mushrooms. Academic Press, Cambridge, pp 137–153

Kalaras MD, Richie JP, Calcagnotto A et al (2017) Mushrooms: a rich source of the antioxidants ergothioneine and glutathione. Food Chem 233:429–433. https://doi.org/10.1016/j.foodchem.2017.04.109

Kaymakci MA, Guller EM (2020) Promising potential pharmaceuticals from the genus *Cordyceps* for COVID-19 treatment: a review study. Bezmialem Sci 8(3):140–144. https://doi.org/10.14235/bas.galenos.2020.4532

Kieliszek M, Błażejak S (2016) Current knowledge on the importance of selenium in food for living organisms: a review. Molecules 21(5):609. https://doi.org/10.3390/molecules21050609

Kim V, Criner GJ (2013) Chronic bronchitis and chronic obstructive pulmonary disease. Am J Respir Crit Care Med 187(3):228–237. https://doi.org/10.1164/rccm.201210-1843CI

Kim C, Kim W, Jeon JH et al (2021) COVID-19 infection with asymptomatic or mild disease severity in young patients: clinical course and association between prevalence of pneumonia and viral load. PLoS One 16(4):e0250358. https://doi.org/10.1371/journal.pone.0250358

Kim WY, Kweon OJ, Cha MJ et al (2021) Dexamethasone may improve severe COVID-19 via ameliorating endothelial injury and inflammation: a preliminary pilot study. PLoS One 16(7): e0254167. https://doi.org/10.1371/journal.pone.0254167

Kim F, Sakagami H, Tanuma S et al (1990) Stimulation of interferon-gamma-induced human myelogenous leukemic cell differentiation by high molecular weight PSK subfraction. Anticancer Res 10(1):55–58

Kofuji K, Aoki A, Tsubaki K et al (2012) Antioxidant activity of β-glucan ISRN Pharm 2012:1–5. https://doi.org/10.5402/2012/125864

Krammer F, Smith GJD, Fouchier RAM et al (2018) Influenza. Nat Rev Dis Prim 4(1):3. https://doi.org/10.1038/s41572-018-0002-y

Lampejo T (2020) Influenza and antiviral resistance: an overview. Eur J Clin Microbiol Infect Dis 39(7):1201–1208. https://doi.org/10.1007/s10096-020-03840-9

Lam-Sidun D, Peters KM, Borradaile NM (2021) Mushroom-derived medicine? Preclinical studies suggest potential benefits of ergothioneine for cardiometabolic health. Int J Mol Sci 22(6):3246. https://doi.org/10.3390/ijms22063246

Land MH, Wang J (2018) Complementary and alternative medicine use among allergy practices: results of a nationwide survey of allergists. J Allergy Clin Immunol Pract 6(1):95–98.e3. https://doi.org/10.1016/j.jaip.2017.01.017

Lauzon A-M, Martin JG (2016) Airway hyperresponsiveness; smooth muscle as the principal actor. F1000Research 5:306. https://doi.org/10.12688/f1000research.7422.1

Leckie MJ, Brinke A t, Khan J et al (2000) Effects of an interleukin-5 blocking monoclonal antibody on eosinophils, airway hyper-responsiveness, and the late asthmatic response. Lancet 356(9248):2144–2148. https://doi.org/10.1016/S0140-6736(00)03496-6

Lee MK, Li X, Yap ACS et al (2018) Airway relaxation effects of water-soluble sclerotial extract from *Lignosus rhinocerotis*. Front Pharmacol 9:461. https://doi.org/10.3389/fphar.2018.00461

Lee MK, Lim KH, Millns P et al (2018) Bronchodilator effects of *Lignosus rhinocerotis* extract on rat isolated airways is linked to the blockage of calcium entry. Phytomedicine 42:172–179. https://doi.org/10.1016/j.phymed.2018.03.025

Leenaars CHC, Kouwenaar C, Stafleu FR et al (2019) Animal to human translation: a systematic scoping review of reported concordance rates. J Transl Med 17(1):223. https://doi.org/10.1186/s12967-019-1976-2

Leung JM, Niikura M, Yang CWT et al (2020) COVID-19 and COPD. Eur Respir J 56(2):2002108. https://doi.org/10.1183/13993003.02108-2020

Li X-M (2011) Treatment of asthma and food allergy with herbal interventions from traditional chinese medicine. Mt Sinai J Med A J Transl Pers Med 78(5):697–716. https://doi.org/10.1002/msj.20294

Li H, Lee H-S, Kim S-H et al (2014) Antioxidant and anti-inflammatory activities of methanol extracts of *Tremella fuciformis* and its major phenolic acids. J Food Sci 79(4):C460–C468. https://doi.org/10.1111/1750-3841.12393

Li Q, Li W, Jin Y et al (2020) Efficacy evaluation of early, low-dose, short-term corticosteroids in adults hospitalized with non-severe COVID-19 pneumonia: a retrospective cohort study. Infect Dis Ther 9(4):823–836. https://doi.org/10.1007/s40121-020-00332-3

Li RF, Zhou XB, Zhou HX et al (2021) Novel fatty acid in *Cordyceps* suppresses influenza a (H1N1) virus-induced proinflammatory response through regulating innate signaling pathways. ACS Omega 6(2):1505–1515. https://doi.org/10.1021/acsomega.0c05264

Liu Y, Wang J, Wang W et al (2015) The chemical constituents and pharmacological actions of *Cordyceps sinensis*. Evid Based Complement Altern Med 2015:1–12. https://doi.org/10.1155/2015/575063

Liu C, Yang N, Song Y et al (2015) Ganoderic acid C1 isolated from the anti-asthma formula, ASHMI™ suppresses TNF-α production by mouse macrophages and peripheral blood mononuclear cells from asthma patients. Int Immunopharmacol 27(2):224–231. https://doi.org/10.1016/j.intimp.2015.05.018

Lo H-C, Hsieh C, Lin F-Y et al (2013) A systematic review of the mysterious caterpillar fungus *Ophiocordyceps sinensis* in Dong Chong Xia Cao and related bioactive ingredients. J Tradit Complement Med 3(1):16–32. https://doi.org/10.4103/2225-4110.106538

Lu X, Chen T, Wang Y et al (2020) Adjuvant corticosteroid therapy for critically ill patients with COVID-19. Crit Care 24(1):241. https://doi.org/10.1186/s13054-020-02964-w

Lull C, Wichers HJ, Savelkoul HFJ (2005) Anti-inflammatory and immunomodulating properties of fungal metabolites. Mediat Inflamm 2005(2):63–80. https://doi.org/10.1155/MI.2005.63

Ma X, Jiao X, Wu J et al (2018) The function of *Ophiocordyceps sinensis* in airway epithelial cell senescence in a rat COPD model. Can Respir J 2018:6080348. https://doi.org/10.1155/2018/6080348

Ma Y, Zeng H, Zhan Z et al (2020) Corticosteroid use in the treatment of COVID-19: a multicenter retrospective study in Hunan. China Front Pharmacol 11:1198. https://doi.org/10.3389/fphar.2020.01198

Ma Z, Zhang W, Wang L et al (2013) A novel compound from the mushroom *Cryptoporus volvatus* inhibits porcine reproductive and respiratory syndrome virus (PRRSV) in vitro. PLoS One 8(11):e79333. https://doi.org/10.1371/journal.pone.0079333

Mahmood F, Hetland G, Nentwich I et al (2019) *Agaricus blazei* based mushroom extract supplementation to birch allergic blood donors: a randomized clinical trial. Nutrients 11(10): 2339. https://doi.org/10.3390/nu11102339

Mahmud R, Rahman MM, Rassel MA et al (2021) Post-COVID-19 syndrome among symptomatic COVID-19 patients: a prospective cohort study in a tertiary care center of Bangladesh. PLoS One 16(4):e0249644. https://doi.org/10.1371/journal.pone.0249644

Marone PA, Trampota J, Weisman S (2016) A safety evaluation of a nature-identical l-ergothioneine in Sprague Dawley rats. Int J Toxicol 35(5):568–583. https://doi.org/10.1177/1091581816653375

Martinez-Medina GA, Chávez-González ML, Verma DK et al (2021) Bio-funcional components in mushrooms, a health opportunity: ergothionine and huitlacohe as recent trends. J Funct Foods 77:104326. https://doi.org/10.1016/j.jff.2020.104326

Mason RJ (2020) Pathogenesis of COVID-19 from a cell biology perspective. Eur Respir J 55(4): 2000607. https://doi.org/10.1183/13993003.00607-2020

Masterson CH, Murphy E, Major I et al (2019) Purified beta-glucan from the *Lentinus edodes* mushroom attenuates antibiotic resistant *Klebsiella* pneumoniae-induced pulmonary sepsis. In: A28. Bacterial and viral lung infections and pathogenesis. American Thoracic Society, Dallas, p A1222

Mau J-L, Lin H-C, Chen C-C (2002) Antioxidant properties of several medicinal mushrooms. J Agric Food Chem 50(21):6072–6077. https://doi.org/10.1021/jf0201273

Mihailović M, Arambašić Jovanović J, Uskoković A et al (2015) Protective effects of the mushroom *Lactarius deterrimus* extract on systemic oxidative stress and pancreatic islets in Streptozotocin-induced diabetic rats. J Diabetes Res 2015:1–10. https://doi.org/10.1155/2015/576726

Minov J, Bislimovska-Karadzhinska J, Petrova T et al (2017) Effects of pleuran (β–glucan from *Pleurotus ostreatus*) supplementation on incidence and duration of COPD exacerbations. Open Access Maced J Med Sci 5(7):893–898. https://doi.org/10.3889/oamjms.2017.198

Moghadami M (2017) A narrative review of influenza: a seasonal and pandemic disease. Iran J Med Sci 42(1):2–13

Müller C, Klemm S, Fleck C (2021) Bracket fungi, natural lightweight construction materials: hierarchical microstructure and compressive behavior of *Fomes fomentarius* fruit bodies. Appl Phys A Mater Sci Process 127(3):178. https://doi.org/10.1007/s00339-020-04270-2

Murphy EJ, Masterson C, Rezoagli E et al (2020) β-Glucan extracts from the same edible shiitake mushroom *Lentinus edodes* produce differential in-vitro immunomodulatory and pulmonary cytoprotective effects: implications for coronavirus disease (COVID-19) immunotherapies. Sci Total Environ 732:139330. https://doi.org/10.1016/j.scitotenv.2020.139330

Mutanen M (1986) Bioavailability of selenium in mushrooms, *boletus edulis*, to young women. Int J Vitam Nutr Res 56(3):297–301

Naji A, Gatling JW (2021) Muscarinic antagonists. In: StatPearls. StatPearls Publishing, Treasure Island (FL), pp 3–5

Nallathamby N, Phan C-W, Seow SL-S et al (2018) A status review of the bioactive activities of tiger milk mushroom *Lignosus rhinocerotis* (Cooke) Ryvarden. Front Pharmacol 8:998. https://doi.org/10.3389/fphar.2017.00998

Nasserie T, Hittle M, Goodman SN (2021) Assessment of the frequency and variety of persistent symptoms among patients with COVID-19. JAMA Netw Open 4(5):e2111417. https://doi.org/10.1001/jamanetworkopen.2021.11417

National Institute for Health and Care Excellence (2021) Asthma: diagnosis, monitoring and chronic asthma management. https://www.nice.org.uk/guidance/ng80. Accessed 26 Apr 2021

National Institute for Health and Care Excellence (2022) COVID-19 rapid guideline: managing COVID-19. https://app.magicapp.org/#/guideline/L4Qb5n/section/EvybVn. Accessed 10 Jan 2022

National Institute of Health (2021) Mushroom-based product for COVID-19 (MACH19). https://www.clinicaltrials.gov/ct2/show/NCT04667247. Accessed 15 Jul 2021

Navarro-Torné A, Vidal M, Trzaska DK et al (2015) Chronic respiratory diseases and lung cancer research: a perspective from the European Union. Eur Respir J 46(5):1270–1280. https://doi.org/10.1183/13993003.00395-2015

Nguyen TT, Bui TT, Pham BP et al (2021) Identification of *Ganoderma lucidum* (Curtis) p. karst. Species isolated from ironwood (*Erythrophleum fordii oliver*) in Vietnam act through by ITS1 sequence and phylogenetic analysis. Plant Cell Biotechnol Mol Biol 22(17–18):59–70

Nowacka N, Nowak R, Drozd M et al (2015) Antibacterial, antiradical potential and phenolic compounds of thirty-one polish mushrooms. PLoS One 10(10):e0140355. https://doi.org/10.1371/journal.pone.0140355

Ocakli B, Ozmen I, Acarturk Tuncay E et al (2018) A comparative analysis of errors in inhaler technique among COPD versus asthma patients. Int J Chron Obstruct Pulmon Dis 13:2941–2947. https://doi.org/10.2147/COPD.S178951

Ogidi CO, Oyetayo VO, Akinyele BJ (2020) Wild medicinal mushrooms: potential applications in phytomedicine and functional foods. In: Passari AK, Sánchez S (eds) An introduction to mushroom. IntechOpen, London, pp 1–4

Padh H (1990) Cellular functions of ascorbic acid. Biochem Cell Biol 68(10):1166–1173. https://doi.org/10.1139/o90-173

Panda AK, Swain KC (2011) Traditional uses and medicinal potential of *Cordyceps sinensis* of Sikkim. J Ayurveda Integr Med 2(1):9–13. https://doi.org/10.4103/0975-9476.78183

Pandya D, Puttanna A, Balagopal V (2014) Systemic effects of inhaled corticosteroids: an overview. Open Respir Med J 8(1):59–65. https://doi.org/10.2174/1874306401408010059

Pang LY, Tan CS, Ng ST et al (2020) *Ophiocordyceps sinensis*: a prominent source of bioactive components for alleviating overactive bladder. Life Sci Med Biomed 4(9):1519. https://doi.org/10.28916/lsmb.4.9.2020.67

Park MS, Fong JJ, Lee H et al (2014) Determination of coleopteran insects associated with spore dispersal of *Cryptoporus volvatus* (Polyporaceae: Basidiomycota) in Korea. J Asia Pac Entomol 17(4):647–651. https://doi.org/10.1016/j.aspen.2014.06.005

Peintner U, Kuhnert-Finkernagel R, Wille V et al (2019) How to resolve cryptic species of polypores: an example in *Fomes*. IMA Fungus 10(1):17. https://doi.org/10.1186/s43008-019-0016-4

Philip J (2014) The effects of inhaled corticosteroids on growth in children. Open Respir Med J 8(1):66–73. https://doi.org/10.2174/1874306401408010066

Quarma (2003) *Fomitopsis officinalis*. https://commons.wikimedia.org/wiki/File:Fomitopsis_officinalis.JPG. Accessed 3 Jun 2021

Ramirez-Martinez A, Wesolek N, Yadan JC et al (2016) Intake assessment of l-ergothioneine in some European countries and in the United States. Hum Ecol Risk Assess An Int J 22(3):667–677. https://doi.org/10.1080/10807039.2015.1104241

Ren L, Perera C, Hemar Y (2012) Antitumor activity of mushroom polysaccharides: a review. Food Funct 3(11):1118. https://doi.org/10.1039/c2fo10279j

Romanou V, Koukaki E, Chantziara V et al (2021) Dexamethasone in the treatment of COVID-19: primus inter pares? J Pers Med 11(6):556. https://doi.org/10.3390/jpm11060556

Rop O, Mlcek J, Jurikova T (2009) Beta-glucans in higher fungi and their health effects. Nutr Rev 67(11):624–631. https://doi.org/10.1111/j.1753-4887.2009.00230.x

Rózsa S, Măniuțiu D-N, Poșta G et al (2019) Influence of the culture substrate on the *Agaricus blazei* Murrill mushrooms vitamins content. Plan Theory 8(9):316. https://doi.org/10.3390/plants8090316

Ruan Y, Li H, Pu L et al (2018) *Tremella fuciformis* polysaccharides attenuate oxidative stress and inflammation in macrophages through miR-155. Anal Cell Pathol 2018:1–10. https://doi.org/10.1155/2018/5762371

Sánchez C (2010) Cultivation of *Pleurotus ostreatus* and other edible mushrooms. Appl Microbiol Biotechnol 85(5):1321–1337. https://doi.org/10.1007/s00253-009-2343-7

Savjani KT, Gajjar AK, Savjani JK (2012) Drug solubility: importance and enhancement techniques. ISRN Pharm 2012:195727. https://doi.org/10.5402/2012/195727

Sengupta S, Bhattacharyya D, Kasle G et al (2021) Potential immunomodulatory properties of biologically active components of spices against SARS-CoV-2 and pan β-coronaviruses. Front Cell Infect Microbiol 11:729622. https://doi.org/10.3389/fcimb.2021.729622

Shahrajabian MH, Sun W, Shen H et al (2020) Chemical compounds and health benefits of *Tremella*, a valued mushroom as both cuisine and medicine in ancient China and modern era. Amaz J Plant Res 4(3):692–697. https://doi.org/10.26545/ajpr.2020.b00077x

Shreenath AP, Ameer MA, Dooley J (2021) Selenium deficiency. In: StatPearls. StatPearls Publishing, Treasure Island (FL), pp 3–5

Sigdel SR, Rokaya MB, Münzbergová Z et al (2017) Habitat ecology of *Ophiocordyceps sinensis* in western Nepal. Mt Res Dev 37(2):216–223. https://doi.org/10.1659/MRD-JOURNAL-D-16-00075.1

Silva-Filho AGS, Sulzbacher MA, Grebenc T et al (2020) Not every edible orange milkcap is *Lactarius deliciosus*: first record of *Lactarius quieticolor* (sect. Deliciosi) from Brazil. J Appl Bot Food Qual 93:289–299. https://doi.org/10.5073/JABFQ.2020.093.036

Sinyor B, Concepcion Perez L (2021) Pathophysiology of asthma. In: StatPearls. StatPearls Publishing, Treasure Island (FL), pp 2–6

Sletmoen M, Stokke BT (2008) Higher order structure of (1,3)-β-D-glucans and its influence on their biological activities and complexation abilities. Biopolymers 89(4):310–321. https://doi.org/10.1002/bip.20920

Slomski A (2021) Trials test mushrooms and herbs as anti–COVID-19 agents. JAMA 326(20):1997. https://doi.org/10.1001/jama.2021.19388

Soksawatmakhin S, Boonyahotra W (2013) Preliminary study of the applications of *Ganoderma lucidum* in chronic fatigue syndrome. J Asian Assoc Sch Pharm 2:262–268

Spagnuolo V, Guffanti M, Galli L et al (2020) Viral clearance after early corticosteroid treatment in patients with moderate or severe COVID-19. Sci Rep 10(1):21291. https://doi.org/10.1038/s41598-020-78039-1

Spolar MR, Schaffer EM, Beelman RB et al (1999) Selenium-enriched *Agaricus bisporus* mushrooms suppress 7,12-dimethlybenz[a]anthracene bioactivation in mammary tissue. Cancer Lett 138(1–2):145–150. https://doi.org/10.1016/S0304-3835(99)00003-8

Sterlin D, Mathian A, Miyara M et al (2021) IgA dominates the early neutralizing antibody response to SARS-CoV-2. Sci Transl Med 13(577):eabd2223. https://doi.org/10.1126/scitranslmed.abd2223

Sum AYC, Li X, Yeng YYH et al (2020) The immunomodulating properties of tiger milk medicinal mushroom, *Lignosus rhinocerus* TM02® cultivar (Agaricomycetes) and its associated carbohydrate composition. Int J Med Mushrooms 22(8):803–814. https://doi.org/10.1615/IntJMedMushrooms.2020035658

Sun X, Dong Z, Li N et al (2018) Nucleosides isolated from *Ophiocordyceps sinensis* inhibit cigarette smoke extract-induced inflammation via the SIRT1-nuclear factor-κB/p65 pathway in RAW264.7 macrophages and in COPD mice. Int J Chron Obstruct Pulmon Dis 13:2821–2832. https://doi.org/10.2147/COPD.S172579

Sutton I (2010) *Lactarius deliciosus*. https://commons.wikimedia.org/wiki/File:Lactarius_deliciosus_1_(1).jpg. Accessed 3 Jun 2021

Synytsya A, Novák M (2013) Structural diversity of fungal glucans. Carbohydr Polym 92(1):792–809. https://doi.org/10.1016/j.carbpol.2012.09.077

Takimoto H, Kato H, Kaneko M et al (2008) Amelioration of skewed Th1/Th2 balance in tumor-bearing and asthma-induced mice by oral administration of *Agaricus blazei* extracts.

Immunopharmacol Immunotoxicol 30(4):747–760. https://doi.org/10.1080/08923970802279092

Tan ESS, Leo TK, Tan CK (2021) Effect of tiger milk mushroom (*Lignosus rhinocerus*) supplementation on respiratory health, immunity and antioxidant status: an open-label prospective study. Sci Rep 11(1):11781. https://doi.org/10.1038/s41598-021-91256-6

Tang RMY, Cheah IKM, Yew TSK et al (2018) Distribution and accumulation of dietary ergothioneine and its metabolites in mouse tissues. Sci Rep 8(1):1601. https://doi.org/10.1038/s41598-018-20021-z

Tang H, Chen J, Xie Q et al (2003) Effects of polysaccharides of *Cryptoporus volvatus* on bronchial hyperreasponsiveness and inflammatory cells in ovalbumin sensitized rats. J Zhejiang Univ Med Sci 32(4):287–291

Teplow DB (ed) (2019) Progress in molecular biology and translational science, 1st edn. Academic Press, Cambridge

Thau L, Gandhi J, Sharma S (2022) Physiology, cortisol. In: StatPearls. StatPearls Publishing, Treasure Island (FL), pp 2–6

Tsai S-Y, Yang J-H, Tseng Y-H et al (2010) Quality of silver ear steamed bun. J Food Process Preserv 34(4):649–643. https://doi.org/10.1111/j.1745-4549.2009.00383.x

Tuli HS, Sandhu SS, Sharma AK (2014) Pharmacological and therapeutic potential of *Cordyceps* with special reference to Cordycepin. 3 Biotech 4(1):1–12. https://doi.org/10.1007/s13205-013-0121-9

Turner RE, Harrison PF, Swaminathan A et al (2021) Genetic and pharmacological evidence for kinetic competition between alternative poly(a) sites in yeast. elife 10(e65331):35. https://doi.org/10.7554/eLife.65331

Upham JW, Chung LP (2018) Optimising treatment for severe asthma. Med J Aust 209(S2):S22–S27. https://doi.org/10.5694/mja18.00175

Valverde ME, Hernández-Pérez T, Paredes-López O (2015) Edible mushrooms: improving human health and promoting quality life. Int J Microbiol 2015:1–14. https://doi.org/10.1155/2015/376387

van Eeden C, Khan L, Osman MS et al (2020) Natural killer cell dysfunction and its role in COVID-19. Int J Mol Sci 21(17):6351. https://doi.org/10.3390/ijms21176351

Vetter J, Lelley J (2004) Selenium level of the cultivated mushroom *Agaricus bisporus*. Acta Aliment 33(3):297–301. https://doi.org/10.1556/AAlim.33.2004.3.10

Wachtel-Galor S, Yuen J, Buswell JA et al (2011) *Ganoderma lucidum* (Lingzhi or Reishi): a medicinal mushroom. In: Benzie IFF, Wachtel-Galor S (eds) Herbal medicine: biomolecular and clinical aspects, 2nd edn. CRC Press, Boca Raton, pp 1–33

Wadhwa A, Aljabbari A, Lokras A et al (2020) Opportunities and challenges in the delivery of mRNA-based vaccines. Pharmaceutics 12(2):102. https://doi.org/10.3390/pharmaceutics12020102

Wagner JG (1973) A modern view of pharmacokinetics. J Pharmacokinet Biopharm 1(5):363–401. https://doi.org/10.1007/BF01059664

Wang SJ, Bai W, Wang CL et al (2007) Effects of *Cordyceps* sinensis on bleomycin-induced pulmonary fibrosis in mice. Zhongguo Zhong Yao Za Zhi 32(24):2623–2627

Wang N, Li J, Huang X et al (2016) Herbal medicine *Cordyceps sinensis* improves health-related quality of life in moderate-to-severe asthma. Evid Based Complement Alternat Med 2016:6134593. https://doi.org/10.1155/2016/6134593

Wang Z, Lorenzi JCC, Muecksch F et al (2021) Enhanced SARS-CoV-2 neutralization by dimeric IgA. Sci Transl Med 13(577):eabf1555. https://doi.org/10.1126/scitranslmed.abf1555

Wasser SP (2004) Shiitake (*Lentinos edodes*). In: Encyclopedia of dietary supplements, 1st edn. CRC Press, Florida, pp 655–664

Wei W, Zheng L, Yu M et al (2010) Anti-fatigue activity of extract form the submerged fermentation of *Ganoderma lucidum* using *Radix astragali* as substrate. J Anim Plant Sci 6(3):677–684

Wiersinga WJ, Rhodes A, Cheng AC et al (2020) Pathophysiology, transmission, diagnosis, and treatment of coronavirus disease 2019 (COVID-19): a review. JAMA 324(8):782–793. https://doi.org/10.1001/jama.2020.12839

World Health Organization (2017) The global impact of respiratory disease. Forum Int Respir Soc. https://www.who.int/gard/publications/The_Global_Impact_of_Respiratory_Disease.pdf. Accessed 22 Sep 2021

World Health Organization (2019) Global influenza strategy summary 2019–2030 influenza. https://apps.who.int/iris/bitstream/handle/10665/311184/9789241515320-eng.pdf?sequence=18&isAllowed=y. Accessed 6 Jun 2021

World Health Organization (2021) WHO coronavirus (COVID-19) dashboard. World Heal Emerg. https://covid19.who.int/. Accessed 22 Sep 2021

World Health Organization (WHO) (2020) Transmission of SARS-CoV-2: implications for infection prevention precautions. https://www.who.int/news-room/commentaries/detail/transmission-of-sars-cov-2-implications-for-infection-prevention-precautions. Accessed 10 Jul 2021

Wu Y-C, Chen C-S, Chan Y-J (2020) The outbreak of COVID-19: an overview. J Chinese Med Assoc 83(3):217–220. https://doi.org/10.1097/JCMA.0000000000000270

Wu J, Dong F, Wang RA et al (2013) Central role of cellular senescence in TSLP-induced airway remodeling in asthma. PLoS One 8(10):e77795. https://doi.org/10.1371/journal.pone.0077795

Wu S, Wang G, Yang R et al (2016) Anti-inflammatory effects of *boletus edulis* polysaccharide on asthma pathology. Am J Transl Res 8(10):4478–4489

Wu Y, Wei Z, Zhang F et al (2019) Structure, bioactivities and applications of the polysaccharides from *Tremella fuciformis* mushroom: a review. Int J Biol Macromol 121:1005–1010. https://doi.org/10.1016/j.ijbiomac.2018.10.117

Xiao H, Li H, Wen Y et al (2021) *Tremella fuciformis* polysaccharides ameliorated ulcerative colitis via inhibiting inflammation and enhancing intestinal epithelial barrier function. Int J Biol Macromol 180:633–642. https://doi.org/10.1016/j.ijbiomac.2021.03.083

Xie Q-M, Deng J-F, Deng Y-M et al (2006) Effects of cryptoporus polysaccharide on rat allergic rhinitis associated with inhibiting eotaxin mRNA expression. J Ethnopharmacol 107(3):424–430. https://doi.org/10.1016/j.jep.2006.03.040

Yadav M, Schorey JS (2006) The β-glucan receptor dectin-1 functions together with TLR2 to mediate macrophage activation by mycobacteria. Blood 108(9):3168–3175. https://doi.org/10.1182/blood-2006-05-024406

Yan GH, Choi YH (2014a) *Lentinus edodes* suppresses allergen-induced airway hyperresponsiveness and inflammation by downregulating nuclear factor-kappa B activity in a murine model of allergic asthma. Korean J Phys Anthropol 27(2):79. https://doi.org/10.11637/kjpa.2014.27.2.79

Yan GH, Choi YH (2014b) *Phellinus linteus* extract exerts anti-asthmatic effects by suppressing NF-κB and p38 MAPK activity in an ova-induced mouse model of asthma. Immune Netw 14(2):107. https://doi.org/10.4110/in.2014.14.2.107

Yang X, Li Y, He Y et al (2015) Cordycepin alleviates airway hyperreactivity in a murine model of asthma by attenuating the inflammatory process. Int Immunopharmacol 26(2):401–408. https://doi.org/10.1016/j.intimp.2015.04.017

Yang N, Liang B, Srivastava K et al (2013) The *Sophora flavescens* flavonoid compound trifolirhizin inhibits acetylcholine induced airway smooth muscle contraction. Phytochemistry 95:259–267. https://doi.org/10.1016/j.phytochem.2013.07.023

Yao H-Y, Zhang L-H, Shen J et al (2011) Cyptoporus polysaccharide prevents lipopolysaccharide-induced acute lung injury associated with down-regulating toll-like receptor 2 expression. J Ethnopharmacol 137(3):1267–1274. https://doi.org/10.1016/j.jep.2011.07.058

Yasir M, Goyal A, Bansal P et al (2021) Corticosteroid adverse effects. In: StatPearls. StatPearls Publishing, Treasure Island (FL), pp 3–6

Zapora E, Wolkowycki M, Bakier S et al (2016) *Phellinus igniarius*: a pharmacologically active polypore mushroom. Nat Prod Commun 11(7):1934578X1601100. https://doi.org/10.1177/1934578X1601100741

Zhang M, Xie Y, Su X et al (2019) Inonotus sanghuang polyphenols attenuate inflammatory response via modulating the crosstalk between macrophages and adipocytes. Front Immunol 10:286. https://doi.org/10.3389/fimmu.2019.00286

Zhao X, Xie Q, Chen J et al (2004) Inhibitory effects of cryptoporus polysaccharide on airway constriction, eosinophil release, and chemotaxis in Guinea pigs. Acta Pharmacol Sin 25(4): 503–507

Zhong L, Yan P, Lam WC et al (2019) *Coriolus versicolor* and *Ganoderma lucidum* related natural products as an adjunct therapy for cancers: a systematic review and meta-analysis of randomized controlled trials. Front Pharmacol 10:703. https://doi.org/10.3389/fphar.2019.00703

Zhou W, Chen Z, Li W et al (2019) Systems pharmacology uncovers the mechanisms of anti-asthma herbal medicine intervention (ASHMI) for the prevention of asthma. J Funct Foods 52: 611–619. https://doi.org/10.1016/j.jff.2018.11.048

Zhu L-Y, Ni Z-H, Luo X-M et al (2017) Advance of antioxidants in asthma treatment. World J Respirol 7(1):17. https://doi.org/10.5320/wjr.v7.i1.17

Zuyderduyn S, Sukkar MB, Fust A et al (2008) Treating asthma means treating airway smooth muscle cells. Eur Respir J 32(2):265–274. https://doi.org/10.1183/09031936.00051407

Therapeutic and Prophylactic Potential of Medicinal Mushrooms in COVID-19

Rachel Parise, Rishi M. Nadar, Jack DeRuiter, Suhrud Pathak, Sindhu Ramesh, Timothy Moore, Dinesh Chandra Agrawal, and Muralikrishnan Dhanasekaran

Contents

Abstract Medicinal mushrooms have been used for thousands of years as natural medicine. The bioactive constituents found within the mushroom products are responsible for therapeutic effects such as anti-inflammatory, anticancer, antioxidant, and antiviral activities. Since SARS-CoV-2 (COVID-19) became prevalent in December 2019, new research has emerged to explore potential treatment options since there was no FDA-approved product for treatment or prevention. Of the many mushrooms available, those with perhaps the highest medicinal values were discovered to include Reishi, Lion's mane, Cordyceps, Shiitake, and Turkey tail. Each contains one or more bioactive compounds with evidence of therapeutic activity that can aid disease treatment or prevention. Few mushrooms have been studied regarding their activity in treating and preventing COVID-19. The specific medicinal mushrooms impacting COVID-19 infection discussed in this chapter include the following: *Agaricus blazei Murill, Ganoderma lucidum, Hercium erinaceus, Grifola frondose, Inonotus obliquus, Lentinus edodes, Cordyceps militaris, Antrodia cinnamomea, Antrodia salmonea, Pseudoplectania nigrella, Russula paludosa,*

R. Parise · R. M. Nadar · J. DeRuiter · S. Pathak · S. Ramesh · T. Moore · M. Dhanasekaran (✉)
Department of Drug Discovery and Development, Harrison School of Pharmacy, Auburn University, Auburn, AL, USA
e-mail: dhanamu@auburbn.edu

D. C. Agrawal (✉)
Department of Applied Chemistry, Chaoyang University of Technology, Taichung, Taiwan
e-mail: dcagrawal@gm.cyut.edu.tw

D. C. Agrawal, M. Dhanasekaran (eds.), *Mushrooms with Therapeutic Potentials*, https://doi.org/10.1007/978-981-19-9550-7_3

Clitocybe sinopica, Polyozellus multiplex, Psilocybin, Coriolus versicolor, and *Sinomenium acutum.*

Keywords Bioactive constituents · COVID-19 · Medicinal mushrooms · SARS-CoV-2

Abbreviations

7KC	7-ketocholesterol
ACE2	Angiotensin Converting Enzyme 2
ARDS	Acute Respiratory Distress Syndrome
BCE	Before the Common Era
EUA	Emergency Use Authorization
FDA	Food and Drug Administration
HIV-1	Human Immunodeficiency Virus 1
IL	interleukin
NF-kB	Nuclear Factor Kappa B
RAAS	Renin Angiotensin Aldosterone System
RBD	Receptor Binding Domain
RNA	Ribonucleic Acid
SARS-CoV-2	Severe Acute Respiratory Syndrome Coronavirus 2
TNF-α	Tumor Necrosis Factor alpha
TTS	Thrombocytopenia syndrome
WEE	Western Equine Encephalitis

1 Introduction

There are thousands of mushroom species worldwide, but a select few contain active chemical constituents and nutrients responsible for therapeutic activity. Medicinal mushrooms contain vitamins, antioxidants, terpenes, and polysaccharides called beta-glucans, contributing to their potential therapeutic properties (Natural Wellbeing 2021). These compounds can enhance immunity, improve cognitive function, reduce free radical activity, and, in some cases, treat cancer. The most common medicinal mushrooms (Spelman et al. 2017; Natural Wellbeing 2021) are listed in Table 1.

Medicinal mushrooms have been used traditionally in ancient civilizations. Because of their antioxidant components, Reishi mushrooms have been used to boost the immune system and reduce oxidative stress. Lion's mane has a history of use in traditional Chinese medicine due to the many bioactive compounds present in this mushroom. Lion's mane mushrooms have been studied for their ability to inhibit tumor growth in cancer cells. They may also be beneficial for preventing and treating

Table 1 Common medicinal mushrooms

Mushroom	Bioactive constituent	Therapeutic activity
Reishi (*Ganoderma lingzhi*)	• Terpenoids • Antioxidants	• Anti-inflammatory • Boosts immune system • Antioxidant
Lion's mane (*Hercium erinaceus*)	• Beta-glucan polysaccharides • Terpenoids • Sterols • Myconutrients • Isoindolinones	• Antitumor growth • Prevention & treatment of Alzheimer's and Parkinson's disease • Improvement of cognitive performance
Cordyceps	• Antioxidants • Parasitic	• Anticancer • Anti-aging • Reduce fatigue
Shiitake (*Lentinula edodes*)	• Vitamins • Fiber • Amino acids	• Improve immunity • Prevention of cancer • Cholesterol management
Turkey tail (*Trametes versicolor*)	• Polysaccharide K	• Anticancer

neurodegenerative diseases, such as Alzheimer's and Parkinson's disease, by promoting the regeneration of nerve tissue in the brain and possibly exhibiting neuroprotective activity. This mechanism of action may influence cognitive performance, as well. Cordyceps is a parasitic fungus that also displays anticancer activity. Because of its antioxidant components, this mushroom may also reduce fatigue and prevent aging. Shiitake mushrooms were initially discovered in Japan. They contain vitamins, fiber, and amino acids, which may help promote the immune system, prevent cancer development or progression, and manage cholesterol levels, reducing the risk for serious cardiovascular diseases. Also, the turkey tail displays anticancer activity, and the polysaccharide K constituent of this mushroom is currently used as an anticancer drug in Japan.

The use of medicinal mushrooms dates back thousands of years ago. Historically, Indigenous Americans have used various types of mushrooms for wound healing (Calvatia, puffball mushrooms) and tonic use (*Ganoderma lingzhi*, reishi & *Cordyceps sinensis*, caterpillar fungus) by traditional Chinese practitioners (Nomoto 2022). From 100 BCE, ancient Chinese texts include descriptions of mushrooms that are still used today. The texts describe their ability to treat respiratory illnesses and other ailments (Natural Wellbeing 2021). Amadou is a spongy substance prepared from fungi and used as an anti-inflammatory and a wound cauterizer by physician Hippocrates in ancient Greece (Nomoto 2022). Hallucinogenic beverages (*Kykeon*) were consumed by elite Greeks, including Socrates and Plato. The Vikings also consumed hallucinogenic mushrooms before the battle.

Three important medicines were derived from fungi (mushrooms), including penicillin (from *Penicillium notatum*), cyclosporin (from *Trichoderma polysporum* and *Cylindrocarpon lucidum*), and krestin (Fig. 1), extracted from Turkey tail

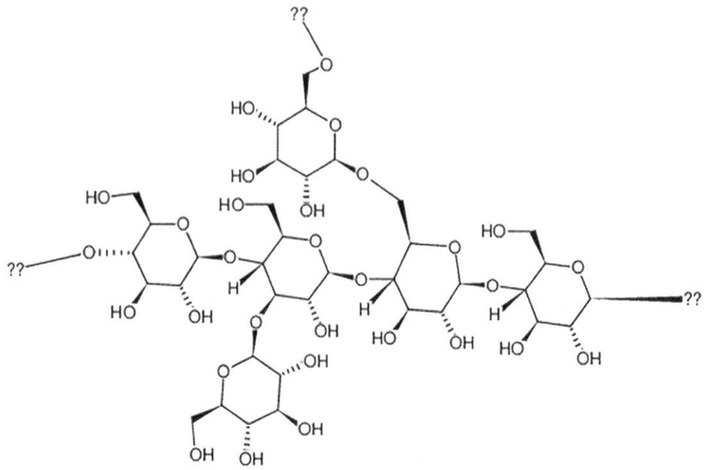

Fig. 1 Krestin

(*Tramates versicolor)* (Nomoto 2022). Lentinan sulfate, shiitake mycelium extract, and Turkey tail have proven efficacy against four human immunodeficiency virus-1 (HIV-1) strains. One strain was human immunodeficiency virus-2 (HIV-2), discovered in 1989 (Nomoto 2022). Reishi, cordyceps, Agaricus, maitake (*Grifola frondosa*), Phellinus, trametes, hericium, and shiitake (*Lentinula edodes*) have shown potential as anticancer agents. In South America, *Agaricus subrufescens* has been used as an immunomodulator and antitumor agent. *Phellinus linteus*, found in Korea, inhibits tumor growth and reduces the frequency of metastases. Japan has used krestin as an anticancer agent and is primarily responsible for producing shiitake, which contains many nutrients and is an extract in Lentinan, a popular anticancer and HIV drug used in Japan and China. A high dose of Oyster mushroom laccase has been shown to inhibit hepatitis C entry into cells and viral reproduction rates. Additionally, evidence shows mushrooms contain constituents responsible for nerve growth factor synthesis, which could directly impact cognitive function. Lion's mane (*Hericium erinaceus*) may also be helpful for the treatment of depression. *Sarcodon scabrosus*, *Ganoderma lucidum*, maitake, and Lion's mane impact nerve and brain health. Tiger milk mushroom (*Lignosus rhinocerotis*) may prevent age-related neurodegenerative diseases.

2 SARS-CoV-2 (COVID-19)

Severe acute respiratory syndrome coronavirus 2 (SARS-CoV-2) is the virus responsible for causing COVID-19. COVID-19 is a mild to severe pulmonary infection that can lead to viral-induced organ complications, hospitalizations, and mortality. The virus was first detected in Wuhan, China, in December 2019 and quickly spread to all parts of the world, causing a global pandemic. The impact of COVID-19 varies significantly among person(s) and has made the pandemic unpredictable and challenging to manage. Other than vaccines, the Food and Drug Administration (FDA) approved only two drugs for COVID-19 treatment to date; Remdesivir (Veklury) for treatment of hospitalized COVID-19 patients, and more recently, issued an emergency use authorization for Paxlovid (nirmatrelvir) tablets for adults and pediatric patients who have tested positive for SARS-CoV-2 and are at high risk for progression to severe COVID-19. The onset of the pandemic stimulated extensive research efforts across the globe to analyze COVID-19 infection, pathology, viral-induced complications, vaccination development, and the creation of drug treatments to battle the worldwide pandemic.

SARS-CoV-2 is transmitted by exposure to infectious respiratory fluids through inhalation, deposition of the virus carried in exhaled droplets or particles exposed on mucous membranes, and touching mucous membranes with hands containing the viral particles (Centers for Disease Prevention and Control 2021). The virus particles bind to angiotensin-converting enzyme 2 (ACE2) receptors located on cellular surfaces, which aids SARS-CoV-2 entry into host cells to cause infection. ACE2 receptors are expressed in the heart, kidneys, and lungs. A disruption in the renin-angiotensin-aldosterone system (RAAS) can occur during this infection process, causing an imbalance of the RAAS with a potential for COVD-19 disease progression, especially among those with comorbid hypertension, diabetes mellitus, and cardiovascular disease (Beyerstedt et al. 2021). As the virus and infection spread in the host, numerous COVID-19-related complications occur by activating inflammatory processes. Acute respiratory distress syndrome (ARDS) is a common severe lung disease in COVID-19 patients. Additionally, cytokine storm is common and is responsible for multi-organ failure or severe life-threatening illnesses.

In the United States, the Pfizer-BioNTech COVID-19 vaccine (Comirnaty) was first used on 11 December 2020 under an emergency use authorization (EUA) and received full FDA approval on 23 August 2021 for ages $\geq$16 years old. The Moderna COVID-19 vaccine (Spikevax) was first used on 18 December 2020 under a EUA and received FDA approval on 31 January 2022 for ages $\geq$18. The third prevalent COVID-19 vaccine is Janssen, uniquely marketed as a single dose versus a double dose, as seen with the other manufactured vaccines. The Janssen COVID-19 vaccine was available on 27 February 2021 under a EUA for >18 years old. As of April 2022, this vaccine is still enacted under the EUA and is not FDA-approved for the prevention of COVID-19. Recently the FDA has limited the authorized use of the Janssen COVID-19 vaccine to individuals 18 years of age and older for whom other authorized or approved COVID-19 vaccines are not accessible or clinically appropriate due to increased reports of the risk of thrombosis with thrombocytopenia syndrome (TTS), associated with the vaccine.

3 Therapeutic and Prophylactic Potential of Medicinal Mushrooms in COVID-19

The mushrooms with current evidence for potential activity against COVID-19 infection are listed in Table 2. Each mushroom's bioactive constituents are responsible for the indicated activities listed in the table and the therapeutic or prophylactic potential against COVID-19.

Hetland et al. (2021) stated that mushrooms have an immunomodulatory mechanism that would stimulate activity against multi-drug-resistant microbes (Hetland et al. 2021). The *Basidiomycetes* family exhibits anti-inflammatory activity that may be used to treat severe lung inflammation caused by SARS-CoV-2. Mycelium extract from *Hericium erinaceus* (HE) and *Grifola frondose* (GF) decreased pro-inflammatory cytokines. It reduced the pro-inflammatory effect in human patients (Johnson et al. 2009; Therkelsen et al. 2016a, b, c; Hetland et al. 2021) The *Agaricus blazei* Murill (AbM) extract reduced pulmonary inflammation in rats (Croccia et al. 2013; Hetland et al. 2021) The AbM extract also reduced TNF-α, IL-6, and IL-1β in mice with cerebral malaria (Val et al. 2015; Hetland et al. 2021) The HE polysaccharide has evidence for downregulating oxidative stress and inflammatory pathways to maintain the intestinal barrier in mice with colitis (Ren et al. 2018; Hetland et al. 2021) The AbM, HE, and GF species have evidence for many antiviral effects against viruses such as the Western equine encephalitis (WEE) virus, poliovirus, hepatitis B and C, herpes simplex virus 1 and 2, influenza virus, and more (Sorimachi et al. 2001; Grinde et al. 2006; Gu et al. 2006, 2007; Faccin et al. 2007; Hsu et al. 2008; Minari et al. 2011; Yamamoto et al. 2013; Cardozo et al. 2013; Eguchi et al. 2017; Hetland et al. 2021) The AbM, HE, and GF species exhibit antimicrobial effects against bacteria and parasites, including *Streptococcus pneumoniae*, *Pseudomonas aeruginosa*, and *Helicobacter pylori* (Bernardshaw et al. 2005; Soković et al. 2014; Liu et al. 2016; Wang et al. 2018; Hetland et al. 2021) The species AbM, HE, and GF could be beneficial as adjunctive therapies in severe COVID-19 pneumonia infections and for excess inflammation, such as cytokine storm.

Inonotus obliquus, also known as Chaga mushroom, has traditionally been used in Asia and Europe to facilitate breathing by reducing nasopharyngeal inflammation (Shahzad et al. 2020). The polysaccharide chemical constituent in this mushroom inhibits the induction of nitric oxide (NO) and other similar cytokines, which could potentially treat COVID-19 by preventing cytokine activation, reducing inflammation, and possibly inflammation preventing cytokine storm. Shahzad et al. (2020) stated that *Inonotus obliquus* is a potential COVID-19 treatment candidate because it can inhibit clathrin-mediated endocytosis and viral infection (Shahzad et al. 2020). The Chaga mushroom is a natural anti-inflammatory and immune booster. There is evidence that components of this mushroom bind to the SARS-CoV-2 carboxy-terminal domain of the receptor-binding domain (RBD), including amino acid residues, previously reported aiding in viral entry through the ACE2 receptor (Eid et al. 2021). They stated that the Chaga mushroom could be a beneficial antiviral

Table 2 Potential use of mushrooms in the treatment and prevention of COVID-19

Family	Species	Bioactive constituents	Therapeutic activity	COVID-19 treatment	COVID-19 prevention	References
Basidiomycetes	*Agaricus blazei* Murill (AbM)	Polysaccharides, Terpenoids, Phenolic compounds, Glycerides, Other low molecular weight molecules	Anti-inflammatory, Antimicrobial, Antiviral	Possible to target severe lung inflammation	No current scientific data	Hetland et al. (2021)
Hericiaceae	*Hericium erinaceus* (HE)	Polysaccharides, Terpenoids, Phenolic compounds, Glycerides, Other low molecular weight molecules,	Anti-inflammatory, Antimicrobial, Antiviral	Possible to target severe lung inflammation	No current scientific data	Hetland et al. (2021)
Meripilaceae	*Grifola frondose* (GF)	Polysaccharides, Terpenoids, Phenolic compounds Glycerides, Other low molecular weight molecules	Anti-inflammatory, Antimicrobial, Antiviral	Possible to target severe lung inflammation	No current scientific data	Hetland et al. (2021)
Hymenochaetaceae	*Inonotus obliquus* (IO)	Polysaccharides	Antiviral, Anti-inflammatory, Anticancer	Possible treatment	No current scientific data	Shahzad et al. (2020); Eid et al. (2021)

(continued)

Table 2 (continued)

Family	Species	Bioactive constituents	Therapeutic activity	COVID-19 treatment	COVID-19 prevention	References
Marasmiaceae	*Lentinus edodes*	Alpha-glucans, Beta-glucans	Reduced pro-inflammatory cytokine, production (TNF-α, IL-8, IL-2, IL-6, IL-22, TGF-β, IL-10), Reduced oxidative stress-induced early apoptosis, Antiviral, Anti-inflammatory	Possible target for cytokine storm & acute respiratory distress syndrome (ARDS)	Protective response	Di Pierro et al. (2020); Murphy et al. (2020); Shahzad et al. (2020)
Cordycipitaceae	*Cordyceps militaris*	No current scientific data	Antiviral, Ergogenic, Antitumor, Antioxidant, Anti-inflammatory, Neuroprotective, Hypolipemic,	Possible inhibition of viral replication	No current scientific data	Jędrejko et al. (2021); Tuli et al. (2014); Zhang et al. (2019)
Fomitopsidaceae	*Antrodia cinnamomea* *Antrodia salmonea*	Antcins	Inhibit ACE2 in epithelial cells, Antioxidant, Anti-inflammatory	No current scientific data	Prevent cell entry	Senthil Kumar et al. (2021)
Sarcosomataceae	*Pseudoplectania nigrella*	Peptides	Antiviral, Antitumor, Antimicrobial, Immune-enhancing agent	No current scientific data	Potential prevention in cellular entrance & viral replication	Oso and Ogidi (2021)
Russulaceae	*Russula paludosa*	Peptides	Antiviral, Antitumor, Antimicrobial, Immune-enhancing agent	No current scientific data	Potential prevention in cellular entrance & viral replication	Oso and Ogidi (2021)

Tricholomataceae	*Clitocybe sinopica*	Peptides	Antiviral, Antitumor, Antimicrobial, Immune-enhancing agent	No current scientific data	Potential prevention in cellular entrance & viral replication	Oso and Ogidi (2021)
Thelephoraceae	*Polyozellus multiplex*	No current scientific data	No current scientific data	Possible SARS-CoV-2 main protease inhibitor	No current scientific data	Sen et al. (2022)
Hymenogastraceae	*Psilocybin*	Psilacetin, Psilocin, Psilocybin	No current scientific data	Potential inhibitor of SARS-CoV-2 main protease Inhibitor of IL-6, Reduction in cytokine storm	No current scientific data	Khan et al. (2021)
Ganodermataceae	*Ganoderma lucidum*	No current scientific data	Antiviral, Anti-inflammatory	Dose-dependent inhibition of SARS-CoV-2 enzyme	No current scientific data	El Sheikha (2022); Fung et al. (2011); Yang et al. (2020); Shahzad et al. (2020)
Polyporaceae	*Coriolus versicolor*	No current scientific data	No current scientific data	Dose-dependent inhibition of SARS-CoV-2 enzyme	No current scientific data	El Sheikha (2022); Fung et al. (2011)
Menispermaceae	*Sinomenium acutum*	No current scientific data	No current scientific data	Dose-dependent inhibition of SARS-CoV-2 enzyme	No current scientific data	El Sheikha (2022); Fung et al. (2011)

used as a supplemental treatment to current drug therapy for COVID-19 (Eid et al. 2021).

Lentinus edodes is commonly known as the shiitake mushroom. The bioactive constituent in this mushroom consists of beta-glucans, known for reducing pro-inflammatory cytokines in in-vitro studies of lung injury (Murphy et al. 2020). The team compared a commercial extract and a novel in-house extract, reducing cytokine-induced nuclear factor kappa B (NF-kB) activation in human alveolar epithelial cells (Murphy et al. 2020). The commercial extract, containing a higher number of alpha glucans vs. beta-glucans, more effectively reduced pro-inflammatory cytokine production, including TNF-α, IL-8, IL-2, IL-6, IL-22, IL-10, and TGF-β. This evidence suggests that this mushroom could target the cytokine storm associated with more severe COVID-19. Beta-glucans are known for their anticancer, antibacterial, and antiviral properties (Mirończuk-Chodakowska et al. 2021). These properties are prevalent based upon the beta-glucan structure, beta-1,3-glucans with short beta-1,6 side chains. Receptors recognize this structure in immune cells, and they form a triple helix conformation that influences the mushrooms' bioactivity. The reported biological activities of beta-glucans in the human body include immunomodulatory properties, antitumor, anti-inflammatory, antioxidant, anti-allergy, antibacterial, antiviral, antifungal, prebiotic, and microbiota modulators, and decrease morbidity and mortality of COVID-19 (Mirończuk-Chodakowska et al. 2021) The alpha-glucan composition has been studied as an immunostimulant and found to promote a protective immune response to viruses, such as West Nile, influenza, avian influenza, hepatitis C, hepatitis B, herpes, and HIV (di Pierro et al. 2020). This evidence may support the use in COVID-19 prevention, but further research is needed.

Cordyceps militaris contains possible antiviral activity against COVID-19. Antiviral activity against influenza, Epstein-Barr virus, herpes simplex virus, and HIV has been confirmed by numerous studies with this mushroom. The mechanism involves inhibiting viral reverse transcriptase and RNA polymerase (Montefiori et al. 1989; Ryu et al. 2014; Jędrejko et al. 2021). An in-silico study showed strong chemical interactions of Cordyceps to COVID-19's RBD at the spike protein and main protease. Such interactions could provide a mechanism to inhibit coronavirus replications (Verma and Aggarwal 2021). Another in-silico study showed Cordycepin might be a potential inhibitor of RNA-dependent RNA polymerase of COVID-19 (Bibi et al. 2022).

The *Fomitopsidaceae* family, consisting of *Antrodia cinnamomea* and *Antrodia salmonea*, contain a steroid compound, antcins (Fig. 2), which exhibit antioxidant

Fig. 2 Antcin

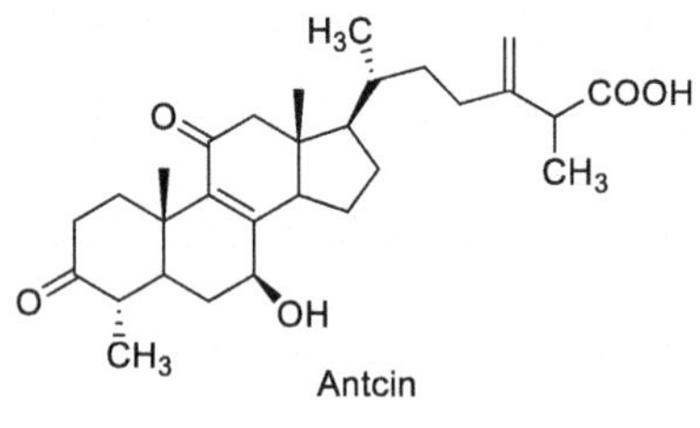

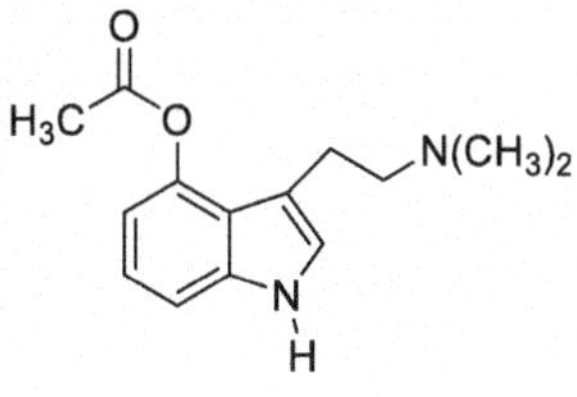

Fig. 3 Psilacetin

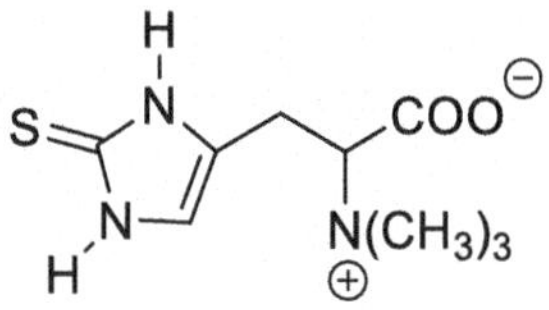

Fig. 4 Ergothioneine

and anti-inflammatory properties. Evidence that the antcin components inhibit ACE2 in epithelial cells may make this mushroom constituent a candidate for prophylactic use against SARS-CoV-2 by preventing viral entry and COVID-19 disease onset (Senthil Kumar et al. 2021).

Pharmaceutical active proteins of mushrooms include lectin, laccases, ribosome-inactivating proteins, nucleases, and fungal immunomodulatory proteins that resemble a natural source of antiviral, antitumor, antimicrobial, and immune-enhancing agents (Zhou et al. 2019; Oso and Ogidi 2021). Oso and Ogidi (2021) claimed that bioactive peptides from medicinal mushrooms, including *Pseudoplectania nigrella*, *Russula paludosa*, and *Clitocybe sinopica,* could be developed as peptide-based drugs against coronavirus infections. These peptides or derivatives could have inhibitory actions on ACE2, COVID-19 HR2 Domain, and the COVID-19 main protease preventing viral cell entry and viral replication.

Sen et al. (2022) reported the potential inhibitory activity of *Polyozellus multiplex* against COVID-19 infection by blocking the SARS-CoV-2 main protease. Additionally, Khan et al. (2021) reported that the psychedelic compounds (psilacetin, psilocin, and psilocybin) of *Psilocybin* mushrooms inhibit SARS-CoV-2 main protease. Psilacetin (Fig. 3) also inhibits IL-6 receptors in humans, reducing cytokine storm in COVID-19 infection. El Sheikha (2022) reported dose-dependent inhibition of the SARS coronavirus enzyme using extracts of *Ganoderma lucidum*, *Coriolus versicolor*, and *Sinomenium acutum* (Fung et al. 2011; El Sheikha 2022). All of these studies cited above indicate the potential of natural active compounds to inhibit coronavirus activity and disease onset or progression (Yang et al. 2020).

Ergothioneine (Fig. 4), an amino acid in edible mushrooms with antioxidant properties, displays a protective effect by reducing inflammation in brain endothelial cells (Koh et al. 2021). This suggests ergothioneine may be beneficial in preventing and treating neurovascular disease. 7-ketocholesterol (7KC) is produced in the central nervous system (CNS) during neuroinflammation, and ergothioneine displays the ability to cross the blood brain barrier to oppose 7KC, which may be useful in COVID-19-related neurologic disease or complications (Koh et al. 2021). An in-silico study on bioactive compounds (colossolactone VIII, colossolactone E, colossolactone G, ergosterol, heliantriol F, and velutin) shows potential activity in the treatment of COVID-19 by inhibiting SARS-CoV-2 main protease (Rangsinth et al. 2021).

4 Conclusion

Many medicinal mushrooms have been used throughout history, and some may potentially inhibit SARS-CoV-2 or COVID-19 infection. This inhibition is related to the bioactive constituents found within the mushroom that can be utilized in various ways, such as inhibition of viral entry by blocking the ACE2 receptor, inhibition of viral replication, and reducing pro-inflammatory cytokines to decrease the severity and prevention of disease onset. The major limitation of using medicinal mushrooms as a treatment or supplemental source for COVID-19 infection is the unavailable evidence of efficacy, safety, and tolerability in humans. Further research is needed to determine the value and benefits of using medicinal mushrooms in COVID-19 treatment, the safety risks to the human patient, and therapeutic dosing while monitoring for intolerability. There is not any current scientific evidence of effectiveness against active COVID-19 infection to prioritize the use of medicinal mushrooms over current COVID-19 drug therapy in those with mild or severe disease.

Acknowledgments We sincerely thank the Department of Drug Discovery and Development at Auburn University Harrison College of Pharmacy.

References

Bernardshaw S, Johnson E, Hetland G (2005) An extract of the mushroom *Agaricus blazei* Murill administered orally protects against systemic *Streptococcus pneumoniae* infection in mice. Scand J Immunol 62:393–398. https://doi.org/10.1111/j.1365-3083.2005.01667.x

Beyerstedt S, Casaro EB, Rangel ÉB (2021) COVID-19: angiotensin-converting enzyme 2 (ACE2) expression and tissue susceptibility to SARS-CoV-2 infection. Eur J Clin Microbiol Infect Dis 40:905–919. https://doi.org/10.1007/s10096-020-04138-6

Bibi S, Hasan MM, Wang YB et al (2022) Cordycepin as a promising inhibitor of SARS-CoV-2 RNA dependent RNA polymerase (RdRp). Curr Med Chem 29:152–162. https://doi.org/10.2174/0929867328666210820114025

Cardozo FTGS, Larsen I, Carballo E et al (2013) *In vivo* anti-herpes simplex virus activity of a sulfated derivative of *Agaricus brasiliensis* mycelial polysaccharide. Antimicrob Agents Chemother 57:2541–2549. https://doi.org/10.1128/AAC.02250-12

Centers for Disease Prevention and Control (2021) Scientific brief: SARS-CoV-2 transmission. In: CDC. https://www.cdc.gov/coronavirus/2019-ncov/science/science-briefs/sars-cov-2-transmission.html Accessed 24 Apr 2022

Croccia C, Lopes AJ, Pinto LFR et al (2013) Royal sun medicinal mushroom *Agaricus brasiliensis* (higher basidiomycetes) and the attenuation of pulmonary inflammation induced by 4-(Methylnitrosamino)-1-(3-Pyridyl)-1-butanone (NNK). Int J Med Mushrooms 15:345–355. https://doi.org/10.1615/IntJMedMushr.v15.i4.20

di Pierro F, Bertuccioli A, Cavecchia I (2020) Possible therapeutic role of a highly standardized mixture of active compounds derived from cultured Lentinula edodes mycelia (AHCC) in patients infected with 2019 novel coronavirus. Minerva Gastroenterol Dietol 66:172–176. https://doi.org/10.23736/S1121-421X.20.02697-5

Eguchi N, Fujino K, Thanasut K et al (2017) *In vitro* anti-influenza virus activity of *Agaricus brasiliensis* KA21. Biocontrol Sci 22(3):171–174. https://doi.org/10.4265/bio.22.171

Eid JI, Das B, Al-Tuwaijri MM, Basal WT (2021) Targeting SARS-CoV-2 with Chaga mushroom: an *in silico* study toward developing a natural antiviral compound. Food Sci Nutr 9:6513–6523. https://doi.org/10.1002/fsn3.2576

El Sheikha AF (2022) Nutritional profile and health benefits of *Ganoderma lucidum* "Lingzhi, Reishi, or Mannentake" as functional foods: current scenario and future perspectives. Foods 11: 1030. https://doi.org/10.3390/foods11071030

Faccin LC, Benati F, Rincão VP et al (2007) Antiviral activity of aqueous and ethanol extracts and of an isolated polysaccharide from *Agaricus brasiliensis* against poliovirus type 1. Letters Appl Microbiol 45:24–28. https://doi.org/10.1111/j.1472-765X.2007.02153.x

Fung KP, Leung PC, Tsui KWS et al (2011) Immunomodulatory activities of the herbal formula Kwan Du Bu Fei dang in healthy subjects: a randomised, double-blind, placebo-controlled study. Hong Kong Med J 17(Suppl 2):41–43

Grinde B, Hetland G, Johnson E (2006) Effects on gene expression and viral load of a medicinal extract from *Agaricus blazei* in patients with chronic hepatitis C infection. Int Immunopharmacol 6:1311–1314. https://doi.org/10.1016/j.intimp.2006.04.005

Gu CQ, Li J, Li JW, Chao FH (2006) Inhibition of hepatitis B virus by D-fraction from *Grifola frondosa*: synergistic effect of combination with interferon-alpha in HepG2 2.2.15. Antivir Res 72:162–165. https://doi.org/10.1016/j.antiviral.2006.05.011

Gu CQ, Li JW, Chao F et al (2007) Isolation, identification and function of a novel anti-HSV-1 protein from Grifola frondosa. Antivir Res 75:250–257. https://doi.org/10.1016/j.antiviral.2007.03.011

Hetland G, Johnson E, Bernardshaw S, Grinde B (2021) Can medicinal mushrooms have prophylactic or therapeutic effect against COVID-19 and its pneumonic superinfection and complicating inflammation? Scand J Immunol 93:e12937. https://doi.org/10.1111/sji.12937

Hsu CH, Hwang KC, Chiang YH, Chou P (2008) The mushroom *Agaricus blazei* Murill extract normalizes liver function in patients with chronic hepatitis B. J Alternat Compl Med 14:299–301. https://doi.org/10.1089/acm.2006.6344

Jędrejko KJ, Lazur J, Muszyńska B (2021) Cordyceps militaris: an overview of its chemical constituents in relation to biological activity. Foods 10(11):2364. https://doi.org/10.3390/foods10112634

Johnson E, Førland DT, Saetre L et al (2009) Effect of an extract based on the medicinal mushroom *Agaricus blazei* Murill on release of cytokines, chemokines and leukocyte growth factors in human blood *ex vivo* and *in vivo*. Scand J Immunol 69:242–250. https://doi.org/10.1111/j.1365-3083.2008.02218.x

Khan FI, Hassan F, Lai D (2021) In silico studies on psilocybin drug derivatives against SARS-CoV-2 and cytokine storm of human interleukin-6 receptor. Front Immunol 12:794780. https://doi.org/10.3389/fimmu.2021.794780

Koh SS, Ooi SC-Y, Lui NMY et al (2021) Effect of ergothioneine on 7-Ketocholesterol-induced endothelial injury. NeuroMolecular Med 23:184–198. https://doi.org/10.1007/s12017-020-08620-4

Liu JH, Li L, Shang XD et al (2016) Anti-helicobacter pylori activity of bioactive components isolated from *Hericium erinaceus*. J Ethnopharmacol 183:54–58. https://doi.org/10.1016/j.jep.2015.09.004

Minari M, Rincão V, Soares A et al (2011) Antiviral properties of polysaccharides from *Agaricus brasiliensis* in the replication of bovine herpesvirus 1. Acta Virol 55:255–259. https://doi.org/10.4149/av_2011_03_255

Mirończuk-Chodakowska I, Kujawowicz K, Witkowska AM (2021) Beta-glucans from fungi: biological and health-promoting potential in the COVID-19 pandemic era. Nutrients 13. https://doi.org/10.3390/nu13113960

Montefiori DC, Sobol RW, Li SW et al (1989) Phosphorothioate and cordycepin analogues of 2′,5′-oligoadenylate: inhibition of human immunodeficiency virus type 1 reverse transcriptase and infection in vitro. Proc Natl Acad Sci U S A 86:7191–7194. https://doi.org/10.1073/pnas.86.18.7191

Murphy EJ, Masterson C, Rezoagli E et al (2020) β-Glucan extracts from the same edible shiitake mushroom *Lentinus edodes* produce differential *in vitro* immunomodulatory and pulmonary cytoprotective effects–implications for coronavirus disease (COVID-19) immunotherapies. Sci Total Environ 732:139330. https://doi.org/10.1016/j.scitotenv.2020.139330

Natural Wellbeing (2021) The fascinating history of medicinal mushrooms. In: Natural Wellbeing. https://blog.naturalwellbeing.com/the-fascinating-history-of-medicinal-mushrooms. Accessed 24 April 2022

Nomoto S (2022) The illustrious history and future potential of medicinal mushrooms. In: TruHavn. https://www.truhavn.com/news/illustrious-history-future-potential-medicinal-mushrooms. Accessed 24 April 2022

Oso BJ, Ogidi CO (2021) Repositioning of fungal-based peptides as modulators of angiotensin-converting enzyme-related carboxypeptidase, SARS-coronavirus HR2 domain, and coronavirus disease 2019 main protease. J Transl Int Med 9:190–199. https://doi.org/10.2478/jtim-2021-0038

Rangsinth P, Sillapachaiyaporn C, Nilkhet S et al (2021) Mushroom-derived bioactive compounds potentially serve as the inhibitors of SARS-CoV-2 main protease: an in silico approach. J Tradit Complement Med 11:158–172. https://doi.org/10.1016/j.jtcme.2020.12.002

Ren Y, Geng Y, Du Y et al (2018) Polysaccharide of *Hericium erinaceus* attenuates colitis in C57BL/6 mice via regulation of oxidative stress, inflammation-related signaling pathways and modulating the composition of the gut microbiota. J Nutr Biochem 57:67–76. https://doi.org/10.1016/j.jnutbio.2018.03.005

Ryu E, Son M, Lee M et al (2014) Cordycepin is a novel chemical suppressor of Epstein-Barr virus replication. Onco Targets Ther 1:866–881. https://doi.org/10.18632/oncoscience.110

Sen D, Debnath B, Debnath P et al (2022) Identification of potential edible mushroom as SARS-CoV-2 main protease inhibitor using rational drug designing approach. Sci Rep 12:1503. https://doi.org/10.1038/s41598-022-05349-x

Senthil Kumar KJ, Gokila Vani M, Hsieh HW et al (2021) Antcins from *Antrodia cinnamomea* and *Antrodia salmonea* inhibit angiotensin-converting enzyme 2 (ACE2) in epithelial cells: can be potential candidates for the development of SARS-CoV-2 prophylactic agents. Plants (Basel) 10. https://doi.org/10.3390/plants10081736

Shahzad F, Anderson D, Najafzadeh M (2020) The antiviral, anti-inflammatory effects of natural medicinal herbs and mushrooms and SARS-CoV-2 infection. Nutrients 12:2573. https://doi.org/10.3390/nu12092573

Soković M, Ćirić A, Glamočlija J et al (2014) *Agaricus blazei* hot water extract shows anti-quorum sensing activity in the nosocomial human pathogen *Pseudomonas aeruginosa*. Molecules 19:4189–4199. https://doi.org/10.3390/molecules19044189

Sorimachi K, Ikehara Y, Maezato G et al (2001) Inhibition by *Agaricus blazei* Murill fractions of cytopathic effect induced by western equine encephalitis (WEE) virus on VERO cells *in vitro*. Biosci Biotechnol Biochem 65:1645–1647. https://doi.org/10.1271/bbb.65.1645

Spelman K, Sutherland E, Bagade A (2017) Neurological activity of lion's mane (*Hericium erinaceus*). J Restor Med 6:19–26. https://doi.org/10.14200/jrm.2017.6.0108

Therkelsen SP, Hetland G, Lyberg T et al (2016a) Cytokine levels after consumption of a medicinal *Agaricus blazei* Murill-based mushroom extract, AndoSan™, in patients with Crohn's disease and ulcerative colitis in a randomized single-blinded placebo-controlled study. Scand J Immunol 84:323–331. https://doi.org/10.1111/sji.12476

Therkelsen SP, Hetland G, Lyberg T et al (2016b) Effect of the medicinal *Agaricus blazei* Murill-based mushroom extract, AndoSan ™, on symptoms, fatigue and quality of life in patients with Crohn's disease in a randomized single-blinded placebo controlled study. PLoS One 11: e0159288. https://doi.org/10.1371/journal.pone.0159288

Therkelsen SP, Hetland G, Lyberg T et al (2016c) Effect of a medicinal *Agaricus blazei* Murill-based mushroom extract, AndoSan™, on symptoms, fatigue and quality of life in patients with ulcerative colitis in a randomized single-blinded placebo controlled study. PLoS One 11: e0150191. https://doi.org/10.1371/journal.pone.0150191

Tuli HS, Sandhu SS, Sharma AK (2014) Pharmacological and therapeutic potential of Cordyceps with special reference to Cordycepin. 3 Biotech 4:1–12. https://doi.org/10.1007/s13205-013-0121-9

Val CH, Brant F, Miranda AS et al (2015) Effect of mushroom *Agaricus blazei* on immune response and development of experimental cerebral malaria. Malar J 14:311. https://doi.org/10.1186/s12936-015-0832-y

Verma AK, Aggarwal R (2021) Repurposing potential of FDA-approved and investigational drugs for COVID-19 targeting SARS-CoV-2 spike and main protease and validation by machine learning algorithm. Chem Biol Drug Des 97:836–853. https://doi.org/10.1111/cbdd.13812

Wang XY, Yin JY, Nie SP, Xie MY (2018) Isolation, purification and physicochemical properties of polysaccharide from fruiting body of *Hericium erinaceus* and its effect on colonic health of mice. Int J Biol Macromol 107:1310–1319. https://doi.org/10.1016/j.ijbiomac.2017.09.112

Yamamoto KA, Galhardi LCF, Rincão VP et al (2013) Antiherpetic activity of an *Agaricus brasiliensis* polysaccharide, its sulfated derivative and fractions. Int J Biol Macromol 52:9–13. https://doi.org/10.1016/j.ijbiomac.2012.09.029

Yang Y, Islam MS, Wang J et al (2020) Traditional Chinese medicine in the treatment of patients infected with 2019-new coronavirus (SARS-CoV-2): a review and perspective. Int J Biol Sci 16: 1708–1717. https://doi.org/10.7150/ijbs.45538

Zhang J, Wen C, Duan Y et al (2019) Advance in *Cordyceps militaris* (Linn) link polysaccharides: isolation, structure, and bioactivities: a review. Int J Biol Macromol 132:906–914. https://doi.org/10.1016/j.ijbiomac.2019.04.020

Zhou R, Liu ZK, Zhang YN et al (2019) Research progress of bioactive proteins from the edible and medicinal mushrooms. Curr Protein Pept Sci 20:196–219. https://doi.org/10.2174/1389203719666180613090710

Influences of Edible Mushrooms on Dynamic and Diversity of Gut Microbiota

Alexandru Stefan Barcan, Laura Dorina Dinu, Vlad Vălu, and Emanuel Vamanu

Contents

Abstract The consumption of mushrooms impacts the immune system and gut microbiota due to the presence of prebiotic and probiotic components found in mushrooms. Mushroom polysaccharides are bioactive macromolecules derived from the fruiting bodies, mycelia, or fermentation broths of edible or medicinal species. Because of their capacity to modulate gut microbiota by lowering pathogen levels and raising friendly microbial strains, mushroom polysaccharides have sparked much interest in the nutraceutical and functional foods businesses. Bacteria can convert several inert chemicals into energy-rich, short-chain fatty acids in the human digestive tract. The immune system is then boosted, which aids in illness prevention. This review highlights the most recent studies on the positive effects of mushroom polysaccharides on the host via targeting gut bacteria. Recent studies on mushroom polysaccharides' impact on the gut microbiota population and the

A. S. Barcan · L. D. Dinu · E. Vamanu (✉)
Faculty of Biotechnology, University of Agronomic Science and Veterinary Medicine, Bucharest, Romania
e-mail: email@emanuelvamanu.ro

V. Vălu
Department of Natural Sciences, Faculty of Science, University of Pitesti, Pitesti, Romania

D. C. Agrawal, M. Dhanasekaran (eds.), *Mushrooms with Therapeutic Potentials*,
https://doi.org/10.1007/978-981-19-9550-7_4

generation of short-chain fatty acids are summarized below. We also discuss the role of the gut microbiome in various disorders and how certain edible mushrooms might affect it. We advocate for the potential use of mushrooms in the near future as an adjuvant treatment for regulating the microbiota in the digestive tract.

Keywords Boletus edulis · Cordyceps militaris · Edible mushrooms · Ganoderma lucidum · Gut microbiota · Polysaccharides · Type 2 diabetes mellitus

List of Abbreviations

ABC	ATP-binding cassette
AMP	antimicrobial peptides
CAZymes	carbohydrate-active enzymes
CEs	carbohydrate esterases
DNA	deoxyribonucleic acid
GI	gastrointestinal
GL	ganoderma lucidum
GLPs	Ganoderma lucidum polysaccharides
GM	gut microbiota
GPC	gel-permeation chromatography
HFD	high-fat diet
HPGPC	high-performance gel-permeation chromatography
HPSEC	high-performance size exclusion chromatography
IL	interleukin
LPs	lipopolysaccharides
MPSs	mushroom polysaccharides
NLRP3	nucleotide-binding oligomerization domain 3
PLs	polysaccharide lyases
PSs	polysaccharides
PULs	polysaccharide utilization loci
SCFAs	short-chain fatty acids
SUS	starch utilization system
T2DM	type 2 diabetes mellitus
TLR	toll-like receptor
TNF	tumor necrosis factors

1 Introduction

The human digestive system harbors complex microbial communities termed gut microbiota (GM). Microbial colonization of the neonatal gut begins postnatally due to the transfer of microbes from the mother to the infant (Milani et al. 2017). It

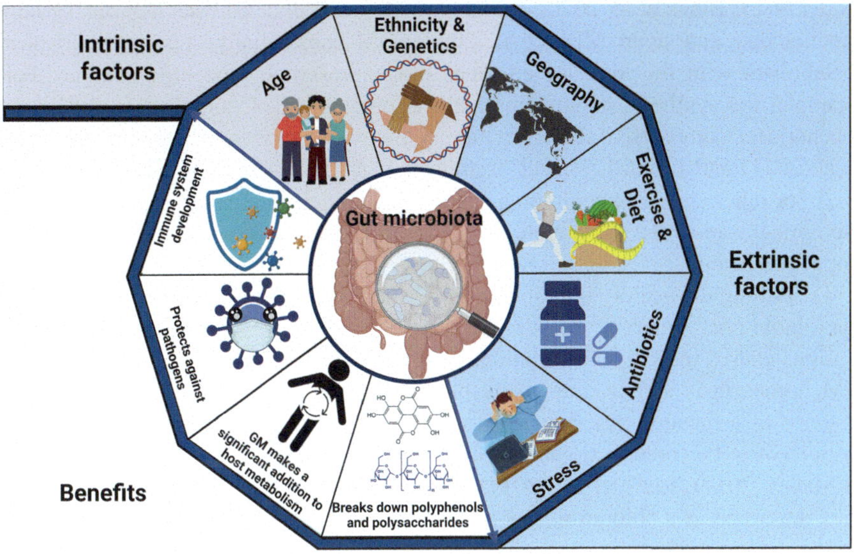

Fig. 1 Factors influencing the composition of GM and its role for human health

proceeds cumulatively from childhood to maturity. After birth, the human gastrointestinal (GI) tract is quickly colonized by microbes from the environment, with events such as sickness, antibiotic therapy, host lifestyle, and dietary choices causing chaotic alteration in the intestinal microbiota (Thursby and Juge 2017).

The dynamic nature of the gut microbiome composition in early life has been widely investigated. Still, more subtle modifications in bacterial variety continue until middle age, when it reaches relative stability (Berg et al. 2020).

GM in adult humans is a dynamic and highly diverse ecosystem of 10–100 trillion microbial cells (primarily bacteria, but also viruses, fungi, and protozoa) (Akobeng et al. 2020). The distal parts of an adult human's gastrointestinal system contain many bacterial communities (Klymiuk et al. 2021). The composition of GM varies among individuals and is influenced by intrinsic factors (Hasan and Yang 2019) (ethnicity, genetics, age) and extrinsic factors such as geographic location (Zafar and Saier Jr. 2021), diet (Jennison and Byrne 2021), level of exercise (Mohr et al. 2020) and medication use including but not limited to antibiotics (Vich Vila et al. 2020) (Fig. 1).

The gut microbial community is generally referred to as our hidden metabolic "organ" due to its massive impact on human wellbeing (Guinane and Cotter 2013). It offers critical advantages in the form of immune system development (Hasan and Yang 2019; D'Amelio and Sassi 2018), supporting resistance to pathogens (Lee et al. 2020), preservation of the structural stability of the intestinal mucosal barrier, maintaining the energy balance and having an effect on intestinal transit, energy intake, and energy expenditure by supplying energy from nondigestible dietary components and modulating intestinal transit (Akobeng et al. 2020; Guinane and

Cotter 2013; Blaak et al. 2020; Zhu et al. 2020), and perhaps even nervous system functionality and brain (Zhu et al. 2020). GM substantially contributes to host metabolism with the input of enzymes not expressed by the host genome. For example, the synthesis of vitamins and the breakdown of polyphenols and polysaccharides. Comparative studies in human microbiota-associated animals (Gruneck et al. 2021) and in vitro research testing human fecal incubation (Bresciani et al. 2020) or other elaborate continuous culture gut models (Harris et al. 2021) confirm the critical role of the microbiota in the assimilation of dietary ingredients and its impact on health.

Describing the healthy microbiota establishes a guideline for comprehending the microbiota-host synergy, including the associations with disease and disorders. Various causes may induce variations in this microbial equilibrium, disrupting the GM homeostasis and leading to dysbiosis (DeGruttola et al. 2016). Due to the absence of a detailed description of a "normal" or healthy microbiota, there is a dispute over the precise definition of dysbiosis (Milani et al. 2017; Hooks and O'Malley 2017). Levy et al. (2017) recognized three forms of dysbiosis, which they described as a "bloom of any potentially pathological organism which, under normal circumstances, lives as a non-harming symbiont, diminishment of commensals, or loss of variety" (Levy et al. 2017), while Vangay et al. (2015) recognized four forms of dysbiosis, namely "reduction of keystone species, diminishment of variety, alterations in metabolic efficiency, and pathogen blooms" (Vangay et al. 2015; Brüssow 2020).

Experimental studies comparing the fecal microbiota of healthy subjects with those of patients strongly suggest that alterations in GM may lead to various GI illnesses and disorders such as irritable bowel syndrome, colon cancer, inflammatory bowel disease, coeliac disease, and antibiotic-associated diarrhea. More recently, evidence has been accumulating that the microbiota may also be involved in obesity and diabetes (Rowland et al. 2018; Guinane and Cotter 2013). More than a thousand different species of bacteria may be found in the gastrointestinal system. These bacteria are classified into five major phyla: Firmicutes, Bacteroidetes, Actinobacteria, Proteobacteria, and Fusobacteria. Bacteria belonging to the phyla Firmicutes and Bacteroidetes comprise most of the gut microbiota, accounting for 90% of its total population (Rinninella et al. 2019).

The abundance and diversity of microbes along the intestine anatomical regions increase from the proximal to the distal intestine, and it is different in terms of regional oxygen level, physiology, nutrient bioavailability, GI transit time (rapid in the mouth to the caecum, slower afterward), pH, mucus, substrate availability, and bile acids. When researchers analyzed 21 healthy people's saliva, mucosal, and fecal samples, they found that each person, and each GI area, has a unique bacterial population, with more richness and diversity in the upper GI tract than in the lower GI tract (Vasapolli et al. 2019).

2 Gut Microbiota's Role in Type 2 Diabetes Mellitus

People's GM changes as they grow older, either because of their nutrition or because of their changing lifestyle. Human lives have altered considerably with a rise in animal products and dietary fats in recent decades. Type 2 diabetes mellitus (T2DM) incidence was linked to these unhealthy lifestyles, implying that dietary variables are important in the beginning and progression of T2DM (Dendup et al. 2018). Evidence shows that a diet rich in fat disrupts the gut's microbial community, which results in an increase in inflammation in type 2 diabetes. In persons with type 2 diabetes, the richness and diversity of the gut microbiota are decreased. In addition, there is an increase in the number of pathogenic microbes, while there is a drop in the number of commensal microorganisms (Aw and Fukuda 2018). The above-mentioned changes in microbiota would cause low-grade inflammation, leading to a decrease in mucous layer and disintegration of the epithelial membrane. Following that, the intestinal permeability increases, allowing the lipopolysaccharide to enter the bloodstream (Honda and Littman 2012).

The equilibrium of the bacteria population that consists of the GM regulates intestinal barrier permeability and maintains local intestinal integrity (Salinas et al. 2021). Any shifts that can alter the gut microbial balance might cause intestinal permeability. The disorder, known as leaky gut or intestinal permeability, damages the integrity of the gut epithelial wall, allowing material from the lumen to translocate into the bloodstream, other organs, or adipose tissue. In addition, as diabetes progresses, there is a reduction in the number of bacteria that produce butyrate, as well as a drop in the production of short-chain fatty acids (SCFAs) by microorganisms (Sharma and Tripathi 2019). These SCFAs are the end products of anaerobic bacterial fermentation of dietary fibers in the large intestine. These molecules are vital for the health of the colon and the rest of the body (den Besten et al. 2013). This decrease in SCFAs reduces the stability of the gut barrier (Hu et al. 2010). The loss of barrier integrity allows lipopolysaccharides (LPs) to translocate, activating inflammatory signaling pathways (Ríos-Covián et al. 2016).

Most research on bacterial translocation and metabolic disorders has focused on LPs measures, resulting in substantial evidence (Salguero et al. 2019; Liaqat et al. 2021; Amar et al. 2011). LPs are substantial, heat-stable components of Gram-negative bacteria's outer membrane that are released into the blood circulation, usually after a bacterium has been destroyed. LPs can get through the gastrointestinal epithelial barrier via chylomicrons or leaky intestinal tight junctions (Ghosh et al. 2020). Excessive HFD consumption alters GM, resulting in higher systemic levels of bacterial products and increased gut permeability for these microbial products (Sircana et al. 2018). Consumption of high-fat, high-cholesterol diets in excess induces an increase in the synthesis of lipopolysaccharide (LPs) from the cell walls of Gram-negative bacteria. This, together with increased chylomicron formation, leads to LPs entering the blood flow (Djuric 2017).

According to the findings of Amar et al. (2011), before the onset of type 2 diabetes, after just 1 week of eating a diet rich in fat, considerable quantities of bacteria DNA are present in adipose tissue and blood, where they can trigger inflammation (Amar et al. 2011). Metabolic endotoxemia is when the blood has an abnormally high number of lipopolysaccharides or endotoxins. One fine example comes from Cani and his collaborators (2007), who observed that high-fat diet-fed mice showed greater blood LPs levels than chow-fed mice, leading to liver and adipose tissue inflammation. Because of this, insulin resistance and nonalcoholic fatty liver disease develop, both of which the investigators categorized as metabolic endotoxemia.

The most significant factors contributing to metabolic endotoxemia development are bacterial endotoxin translocation and increased intestinal permeability (Mohammad and Thiemermann 2021). Both factors play a key part in the pathogenesis of metabolic endotoxemia. Many bacterial communities can stimulate the development of inflammatory T lymphocytes (Belkaid and Hand 2014). In such circumstances, segmented filamentous bacteria invade the gut by direct contact with the epithelium, aided by dendritic cells (Clarke et al. 2010). When LPs get into circulation, it attaches to a protein called plasma LPs-binding protein (LBP), which then triggers the activation of a receptor protein called CD14 in the plasma membrane of macrophages (Lee and Kim 2017). This complex can attach to the Toll-like receptor 4 (TLR4) on macrophage membranes. TLR4 is a type of pattern recognition receptor that helps cells recognize ligands such as endotoxin while also promoting inflammation and immunity.

Several studies have found increased expression and activation of TLRs on cell surfaces in people with diabetes and metabolic syndrome (Redfern and McDermott 2010; Poggi et al. 2007). When TLR4 is activated on innate immune cells, it triggers a proinflammatory cascade that releases cytokines, adhesion molecules, and reactive oxygen species, all of which contribute to chronic inflammation and other obesity-related issues (Liu et al. 2017). During this situation, there is an increase in the expression of proinflammatory molecules such as interleukin-1, interleukin-6, and tumor necrosis factors (TNF-). This, in turn, facilitates the development of Th17 and Th1 cells, both of which have the potential to contribute to autoimmune disease. Several researchers have concluded that certain pro-inflammatory markers, such as IL-1, IL-6, and TNF-, are directly or indirectly connected with insulin resistance (Wang and He 2018; Daniele et al. 2014; Salas-Salvad6 et al. 2006).

As shown in Fig. 2, intestinal immune defense and GM interact to modulate microbial migration-induced tissue inflammation and metabolic regulation. A eubiotic GM comprises a varied bacterial population that may be found inside the gut mucosal layer but preferentially in the lumen. Defensins (such as AMP) and immune cells inhibit luminal bacteria from adhering to and migrating through the mucosa (blue arrows).

When GM dysbiosis occurs, such as when it is induced by consuming a diet high in fat, a dysbiotic mucosal microbiota that is constituted of Proteobacteria and Firmicutes appears. This dysbiotic mucosal microbiota disrupts intestinal epithelial cells' function, resulting in higher intestinal permeability. Throughout the progression of metabolic disorders (gray arrows), mucosal bacteria and fragments such as LPS and peptidoglycan cross through the epithelial layer to the lamina propria,

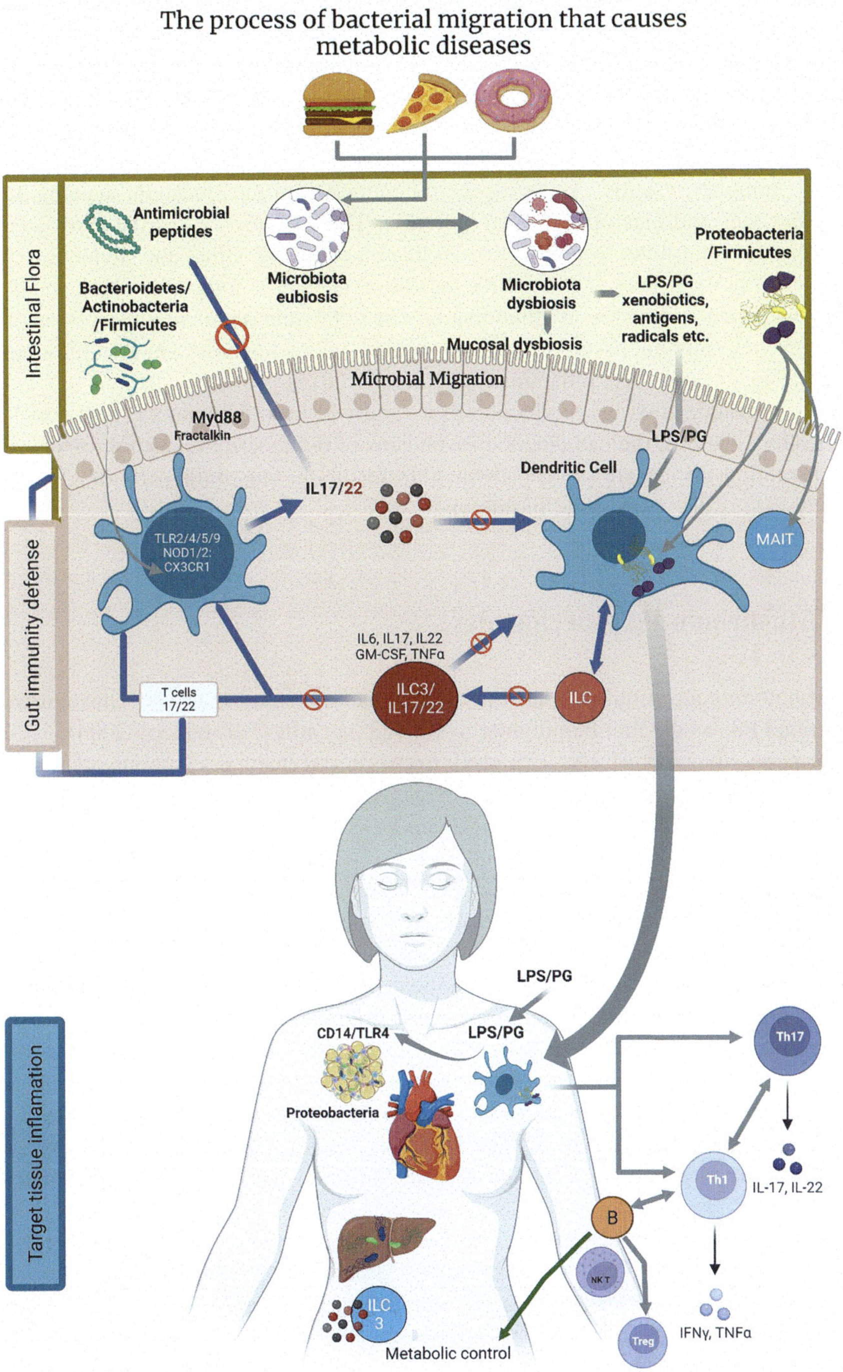

Fig. 2 Microbial migratory pathway that causes metabolic illness

where dendritic cells catch them. The dysfunction of GM interferes with the interaction between phagocytes, ILCs, and T cells. The co-activation that normally occurs between phagocytes and T cells is significantly hampered, which results in decreased production of IL22/17. The liver, the islets of Langerhans, adipose depots, or the heart and vessels are all metabolic tissues that may be targeted by bacteria and bacterial components LPS and peptidoglycan due to the compromised adaptive and innate immune systems. They cause inflammation in the location, leading to preadipocytes and macrophages' proliferation. This, in turn, causes the associated cytokines to contribute to a reduction in insulin signaling. Since the frequency of ILC3 rises in the tissues, a compartmentalization process is triggered, which results in an even greater increase in inflammation due to the release of cytokines. B and T lymphocytes enter the tissues to a greater degree, interacting with newly infiltrating phagocytes to exacerbate inflammation (Burcelin 2016).

This concludes that the production of a variety of proinflammatory mediators is the leading cause of the pathogenic mechanisms of tissue-specific insulin resistance by disrupting insulin signaling pathways. In this case, the equilibrium of glucose metabolism is disrupted, resulting in type 2 diabetes.

3 Mushroom Polysaccharides

Micronutrients are usually found in nature as polysaccharides (PSs). Tolstoguzov described PSs as the first biopolymer to emerge on earth (Tolstoguzov 2004).

A large diversity of PSs extracted from various natural compounds, which include plants, fungi, and polymers biosynthesized by several strains of bacteria, are gaining popularity as health supplements (Maity et al. 2021). Amongst them, mushroom PSs (MPSs) are crucial not only for fungal species' growth and development but also have been discovered to offer enormous promise in biotechnology (Badalyan 2014), immunology (Konusova et al. 2021), molecular biology (Thabthimsuk et al. 2018) and pharmacology (Lawal et al. 2019). Without a doubt, understanding the biological significance of MPSs in public health requires knowledge of their structures. Different mushroom species have polysaccharides of different structures, making it challenging to determine all the structural properties. Due to the diversity of functional mushroom components, such as proteins, peptides, amino acids, phenolic compounds, and other small molecules, purity is essential for accurate structural analysis of MPSs.

The active elements in edible mushrooms fall into two categories: macromolecular compounds and low molecular compounds, regardless of which species they come from (Gong et al. 2020). Polysaccharides, proteins, lipopolysaccharides, glycoproteins, lectins, and polysaccharides protein complexes are examples of high-molecular-weight metabolites; low-molecular-weight metabolites include sterols, phenolic substances, lactones, terpenoids, alkaloids, antibiotics with various active groups, metal chelating agents, and so on. MPSs are structurally distinct from other types in terms of chemical makeup, level of branching, structure length, molecular mass, three-dimensional conformation, glycosidic bond (De Silva et al. 2012), etc.

There is a significant difference between plants and MPSs, mostly due to structural properties. Plant PSs are often acidic, made of pectic PSs with a complex structure of up to 17 monosaccharides with over 20 different linkages. In contrast, the most frequent of MPSs are either α- or β-glucan or a mixed α- and β-glucan (Shi 2016).

One of the most important elements regulating the direction of glycosidic linkage and spatial structure is the α- or β-configuration of monosaccharides.

MPSs are classified as homopolysaccharides or heteropolysaccharides based on their monosaccharide content. Both homopolysaccharides and heteropolysaccharides may have homolinkages and heterolinkages, depending on the configuration and/or location of the connection between them (Wang et al. 2017). Heteropolysaccharides are rich in fucose, galactose, mannose, and xylose. The most common PSs in mushrooms are glycogen-like glucans, which serve as storage components. The cell wall of fungi is made up of two types of PSs: a stiff fibrillar layer made of cellulose or chitin and a matrix-like structure made of α- or β-glucans and glycoproteins (Ghassemi et al. 2022). Fungi contain modest quantities of Alpha-glucans. These include glycogen, dextran, pullulan, and starch. Although beta-D-glucans polymer units are the major polysaccharides and are one of the promising pharmaceutical compounds with excellent nutritional and therapeutic properties. For example, a typical PSs found in many mushrooms is β-D-glucan with a (1 → 3)-linked main chain replaced at O6. According to several published studies, β-D-glucan operates as a biological response modifier, stimulating the formation of phagocytic cells, T cells, cytokines, and NK cells (Liu et al. 2021).

A significant effect on the bioactivity of MPSs is exerted by their molecular weight. MPSs with a larger molecular weight have superior biological activity and anticancer efficacy. This is because only PSs with a molecular weight greater than 90 kDa can form a triple-helix configuration (Zhang et al. 2019). In vivo anticancer experiments conducted by Niu et al. (2021) revealed that Leucopaxillus giganteus PSs (average molecular weight of 1.78×10^6 Da) have outstanding antitumor action and inhibit the growth of H22 solid tumors in a dose-dependent manner. When determining the molecular mass of MPSs, gel-permeation chromatography (GPC) (Fu et al. 2010), high-performance size exclusion chromatography (HPSEC) (Cheong et al. 2015) or high-performance gel-permeation chromatography (HPGPC) (Yang et al. 2008) are often utilized.

4 Modulatory Effects of Mushroom Polysaccharides on Gut Microbiota

MPSs exert various regulatory and protective actions on the GM. For intestinal structure, MPSs promote the growth of beneficial bacteria, which helps to regulate the gastrointestinal tract and make it more balanced (Ma et al. 2021). For intestinal composition, MPSs help to regulate the gastrointestinal tract by treating intestinal mucosal damage (Jayachandran et al. 2017).

Humans and certain animals can only metabolize a limited amount of simple PSs for energy, while complex PSs such as β-glucan, lignin, and cellulose cannot be digested because their DNA does not encode suitable carbohydrate-active enzymes (CAZymes) (Zhang et al. 2022). When consumed, MPSs function as dietary fiber, passing through the digestive tract unaffected by human enzymes and transporters. This allows them to come into contact with the huge microbial population called GM. GM encodes many genes, which is necessary for ensuring complete fiber degradation via the cecal and intestinal fermentation processes (Khursheed et al. 2020).

The only polysaccharide humans can consume without the help of microorganisms is starch, a homopolymer of glucose found in several plants as an energy resource polymer (Kumar et al. 2012). Gut microbes encode a broad array of CAZymes, not found in humans for the breakdown of MPSs and oligosaccharides, making GM a product of the coevolution of the gut environment and the genomes of gut microorganisms (Zhong et al. 2019). This ability to digest MPSs depends on the abundance of glycoside hydrolases (more than 260) present in the vast microbial community, as opposed to just 17 enzymes present in humans, which are capable of breaking down diverse kinds of carbohydrates (Li et al. 2021). The gut microbiota generates several different types of CAZymes required for the degradation of MPSs, including polysaccharide lyases (PLs) which can cleave certain activated glycosidic linkages present in acidic PSs. Glycosyltransferases are specific for the type of linkage (α or β), carbohydrate esterases (CEs) catalyze the de-O or de-N-acylation to eliminate the ester designs from PSs, and glycoside hydrolases can initiate the hydrolysis of the glycosidic linkage in glycosides (Liang et al. 2021).

The phylum Bacteroidetes encodes four times the number of CAZymes than gram-positive bacteria in the Firmicutes phylum (Crost et al. 2013). Furthermore, since the hydrolysis of PSs usually takes place at the surface of the cell membrane of the bacteria, the production of extracellular hydrolytic enzymes is required to facilitate this process (Belkaid and Hand 2014; Wilkes and Aristilde 2017; Ravi et al. 2021). According to bioinformatic research, a signal sequence is found in around 81% of PLs and GHs in Bacteroidetes, but only 19% of those found in Firmicutes contain a signal sequence (Bertucci et al. 2019; Zhu et al. 2016). These data suggest that Bacteroidetes can breakdown a diverse variety of PSs, but Firmicutes have a proclivity to degrade a small number of PSs. To breakdown PSs, the gut microbiota utilizes a variety of transport mechanisms, such as the starch utilization system (SUS) – like transport system, the multi-enzyme complexes process, and the ATP-binding cassette (ABC) transport network (Tingirikari 2018).

For example, all mammalian gut Bacteroides break down PSs through several polysaccharide utilization loci PULs, each encoding a SUS-like mechanism. Glycans, including resistant starch, are captured and transported by these PUL-encoded proteins, which are homologs of SusC and SusD. To transport PSs, SusC-like protein must penetrate the outer membrane (Foley et al. 2016). SusD-like protein is an outer membrane lipoprotein that catches and delivers oligosaccharides to the SusC protein transporter. SusC-like and SusD-like proteins work in synergy with polysaccharide-degrading enzymes and proteins that bind to glycans in the outer membrane of the cell (Schwalm 3rd and Groisman 2017). As the most thoroughly studied PUL-encoded PSs absorption mechanism, the SUS-like transport system of Bacteroides thetaiotaomicron often serves as a basis for comparing or determining the roles of related PUL-encoded proteins. Mammalian gut Firmicutes have fewer CAZymes for extracellular PSs degradation, but they can transfer smaller carbohydrates for intracellular processing through various transporters such as ABC transporters (Hou et al. 2021).

The ABC-transport mechanism for starch degradation in Eubacterium rectale was described by Cockburn and Koropatkin in 2016. It consists of a cell surface substrate-binding protein for specific sugar import in conjunction with ATP hydrolysis (Cockburn and Koropatkin 2016; Yin et al. 2020). Ruminococcus champanellensis, isolated from human fecal samples, has been discovered to have a multi-enzyme complex that breaks down resistant starch and cellulose (Hong et al. 2022).

Bacterial strains belonging to the same genus may have genes with a variety of functions and may differ in their ability to utilize different types of carbohydrates. For instance, the *Lactobacillus bulgaricus* strain prefers lactose. In contrast, the *Lactobacillus helveticus* strain contains genes involved in the biosynthesis pathway and fatty acid metabolism but is not a carrier of dynamic macromolecular carbohydrates (Zeidan et al. 2017). Several researchers have demonstrated that supplementation with MPSs can increase the diversity of gut microbiota species, which might have significant health repercussions (Kanwal et al. 2020; Chen et al. 2020; Amar et al. 2011). Additionally, MPSs intake may inhibit the growth of other gut bacteria species linked with potentially harmful symptoms. One of the causes of the abovementioned phenomena is the usage of MPSs as fermented compounds by GM, with mainly SCFAs as final products, which lowers the pH of the digestive tract and stimulates the prevalence of health-promoting GM in the gastrointestinal tract. In vitro experiments demonstrated that the PSs from *F. velutipes* decreased the pH level from 6.80 to 6.10 after 6 h of fermentation (Li et al. 2021).

Furthermore, GMs with health-promoting properties might absorb the breakdown products of MPSs as carbon and energy sources, allowing them to become the dominant bacterial community in the GI tract. In addition, the bacteria colonizing the gut can stimulate development by engaging in cross-feeding behaviors. In an in vitro study, Turroni et al. (2015) demonstrated that B. bifidum PRL2010 may cross-feed on sugars produced by B. adolescentis 22L, B. breve 12L, and B. thermophilum JCM1207. During fermentation, Bifidobacterium longum can breakdown oligofructose to create free fructose, which Anaerostipes caccae can utilize for its growth (*Rivière et al.* 2016). However, the GM population is highly associated with interindividual differences, such as lifestyle, nutrition, habitat, and age variations. As a result, it is impossible to draw a straight conclusion on the prebiotic effects of MPSs. Modulatory effects of several functional mushrooms on gut microbiota and metabolites synthesis have been listed in Table 1.

Table 1 Modulatory effects of functional mushrooms on gut microbiota and metabolites synthesis

Source	In vitro/in vivo research	Increased metabolites	Gut microbiota modulation	References
Boletus edullis	In vitro gastro-intestinal model	Increased the level of gallic acid	Lactic acid bacteria population growth and diversity ↑	Vamanu and Pelinescu (2017)
Ganoderma lucidum	In vivo model	Increased production of acetate and butyrate	*Allobaculum*, *Bifidobacterium*, and *Christensenellaceae_R-7_group* ↑ *Lachnospiraceae_UCG-001 and Ruminiclostridium* ↓	Sang et al. (2021)
Cordyceps militaris	In vivo model	Induced altered metabolites included lipids, organoheterocyclic compounds, organic acids, benzenoids, phenylpropanoids, organooxygen compounds ↑	Bacteroides, Alloprevotella, Parabacteroides, Butyricimonas, Alistipes, and Akkermansia ↑ while Escherichia-Shigella and Negativebacillus ↓	Huang et al. (2022)
Inonotus obliquus	In vivo model	Butyrate, lysine, 5-aminovaleric acid, and bile acids all showed significant changes ↑	Bacteroidales S24–7 group, *Akkermansia, Holdemanella, Faecalibaculum,* and *Bacteroides* ↑	Yu et al. (2021)
Auricularia auricula-judae	In vivo model	Acetate, propionate and butyrate ↑	*Bacteroides, Paraprevotella, Flavonifractor and Clostridium IV* ↑	Zhang et al. (2020b)
Agaricus bisporus	In vivo model	No significant increase in the concentrations of acetate, propionate, butyrate, isobutyrate, or valerate	*Bacteroides, Parabacteroides, Coprococcus, Sutterella,* and *Anaerostipes* ↑	Hess et al. (2018)
Lentinula edodes	In vitro human salivary, gastric and small intestinal model	Acetic acid and propionic acid followed by butyric acid ↑	*Firmicutes, Bacteroidetes* and *Proteobacteria* ↑	Xue et al. (2020)

5 Short-Chain Fatty Acids' Synthesis by Gut Microbiota

Short-chain fatty acids (SCFAs) and other gases (CO_2, CH_4, H_2S, H_2, etc.) are produced when MPSs are fermented by GM, and these compounds are essential to prebiotic action (Smith et al. 2019).

The primary SCFAs (95%) produced by microbial fermentation of MPSs are acetic acid, propionate, and butyrate. Numerous favorable effects on host health have been attributed to these microbial metabolites, including the suppression of pH-sensitive infections, an enhancement in mineral assimilation, modulation of intestinal transit, and strengthening of the gastrointestinal epithelial barrier (Peredo-Lovillo et al. 2020). However, the possible effects on host physiology of the three principal SCFAs, acetate, propionate, and butyrate, differ substantially. Reductive acetogenesis is another way gut anaerobes make acetate as a net fermentation product (Louis and Flint 2017).

As a result, acetate is the most prevalent SCFA in the large intestinal tract. It is formed by hydrolysis of acetyl-CoA that is converted from pyruvate in most gut bacteria or synthesized through the Wood-Ljungdahl pathway by some acetogenic bacteria. It is transported from the gut to the liver, where it participates in lipogenesis, and it may be processed in muscle, heart, brain, and kidney (Schug et al. 2016). In animal models, acetate stimulates the nucleotide-binding oligomerization domain 3 (NLRP3) inflammasome directly in the gut epithelial cells, enhancing the secretion of IL-18 by engaging the epithelial IL-18 receptor and improving the intestinal barrier integrity (Donovan et al. 2020). Gut bacteria may produce propionate by three separate routes: succinate, acrylate, and propanediol pathways (Reichardt et al. 2014; Louis and Flint 2017).

The succinate pathway is responsible for the metabolism of most hexose and pentose sugars. In contrast, the propanediol pathway is responsible for the metabolism of the deoxy sugars fucose and rhamnose (Alam and Clark 1989). Bacteroidetes and Firmicutes of the Negativicutes class are the most likely to use the succinate route. Lachnospiraceae and Veillonellaceae have been demonstrated to use the acrylate route, whereas Proteobacteria and Lachnospiraceae use the deoxy sugars rhamnose and fucose to synthesize propionate and propanol (Flint et al. 2015). Propionate lowers triglycerides in the liver, enhances insulin sensitivity, and promotes metabolic homeostasis by activating intestinal gluconeogenesis (IGN). It also has anti-cancer, anti-lipogenic, anti-inflammatory, and anti-cholesterol properties (Wang et al. 2019). Butyrate is a short-chain fatty acid that has four carbons and is produced from carbohydrates by joining two molecules of acetyl-CoA to make acetoacetyl-CoA, which is subsequently reduced step by step to butyryl-CoA (Clark and Cronan 2005).

The last stage in the generation of butyrate from butyryl-CoA may occur either via the action of butyryl-CoA: acetate CoA-transferase or through the action of butyrate kinase. Bacteroidetes are the primary producers of butyrate, but clusters IV and XIVa of Clostridia are also important sources (Vital et al. 2014). In addition, Butyrate-producing species are among the most abundant genera of *Faecalibacterium*, *Eubacterium*, and *Roseburia*. *Eubacterium rectale* and the closely related *Roseburia* species comprise a large group of butyrate-producing Firmicutes that share the butyryl-CoA: acetate CoA-transferase pathway for butyrate generation and the same genomic arrangement of their butyrate synthesizing genes from acetyl-CoA to butyryl-CoA (Takahashi et al. 2016). Research has demonstrated that butyrate is beneficial for health because it is a substrate for the colonocytes. It may diffuse into the portal vein and interact with the host body's critical functions, including glucose metabolism, lipogenesis, and gastrointestinal inflammatory processes (Louis and Flint 2009).

Several in vitro and in vivo research have indicated that butyrate exhibits anti-inflammatory, antioxidant, anti-obesity, and metabolic regulatory properties (Jack et al. 2019; Hu et al. 2019; Amiri et al. 2022).

Apart from fiber-derived SCFAs, microbial fermentation forms a wide variety of other compounds, such as polyphenol-derived molecules, which in turn may influence the gut microbial populations and their interactions with the host.

6 Modulatory Effects of *Ganoderma Lucidum* Polysaccharides on Gut Microbiota

The fungus *Ganoderma lucidum* (GL) belongs to the Ganodermataceae family and is known as Lingzhi in China and Reishi in Japan. GL has been utilized as a traditional Chinese medication for thousands of years to promote health, continuous youth, vigor, and lifespan (Sanodiya et al. 2009). Modern medical research has shown that this mushroom has a broad spectrum of pharmacological benefits (Raman et al. 2022). GL contains various components, including glycoproteins, PSs, triterpenoids, meroterpenoids, sesquiterpenoids, steroids, alkaloids, benzopyran derivatives, and benzoic acid derivatives (Pavlik et al. 2020). A few minerals (K, Ca, P, Mg, Se, Fe, Zn, among others) are also found in GL (Papp 2019). In recent years, a growing body of research has shown that the therapeutic characteristics of *Ganoderma* products are influenced by the preparation, culture, and manufacturing processes (Ahmad 2019; Berger et al. 2004; Campi et al. 2021). GL has been grown on various substrates and blended with several additives in the fungi industry.

Additionally, the chemical components of mushrooms often change across species or subspecies that thrive in different environments or under different growth circumstances (Pinya Fernández et al. 2019). The constitution of GL extracts is influenced by whether polar or apolar solvents are utilized in the extraction process. It has recently been identified as a novel form of natural antioxidant due to its enhanced capacity to scavenge free radicals and reduce lipid peroxidation. The most bioactive elements in GL are PSs, ganoderic acid, and adenosine. *Ganoderma lucidum* polysaccharides (GLPs) have long been advocated as multifunctional polysaccharides owing to their several essential biological properties, such as anti-inflammatory, anti-diabetic, anti-ulcer, antioxidant, anti-cancer, anti-bacterial and immunostimulating benefits, amongst other things (Sohretoglu and Huang 2018). Due to GL's rarity in nature, submerged culture offers a viable option for producing bioactive PSs. As a result, PSs derived from farmed GL mycelium might be used to treat antioxidants and regulate GM (Nie et al. 2013).

GLPs are biomacromolecules with a molecular weight distribution of 10 la a 3–10 la 6 Da and contain glycoproteins, (1 → 3), (1 → 6)-α/β-glucans with some strands of monosaccharide residues as their side chains, water-soluble heteropolysaccharides conjugated to galactose, glucose, mannose, arabinose, xylose, and fucose (de Almeida and da Silva 2021).

GLPs have been examined for their ability to enhance cardiovascular function and heart health and maintain healthy blood lipid levels and cellular antioxidant activity (Ferreira et al. 2015). Following the findings of many researchers, it has been hypothesized that the positive effects of GL are strongly associated with the modulation of GM (Chen et al. 2020).

GLPs are often utilized to modulate GM. Examples include studies conducted using fecal transplantation. Chang et al. demonstrated that high molecular weight polysaccharides extracted from *Ganoderma lucidum* might decrease obesity in mice by changing their intestinal flora composition. Moreover, Chang et al. studied the anti-obesity activities of GL in a mouse model of diet-induced obesity. They demonstrated that everyday treatment of a water extract of GL at concentrations ranging from 2% to 8% reduced obesity and weight gain characteristics such as glucose metabolism, fat mass, hepato, and adipose tissue inflammation in high-fat diet mice. Additionally, they discovered that GLPs treatment decreased lipopolysaccharide concentrations and inhibited the activation of the Toll-like receptor 4 pathway in the liver. GLPs significantly enhance the abundance of Lachnospiraceae species, Roseburia, and Lactobacillus, indicating a possible therapeutic method for chronic pancreatitis via microbiota modulation (Li et al. 2016b). GLPs can alter intestinal biological activities by oral administration via boosting microbiota richness, reducing the Firmicutes-to-Bacteroidetes ratio, and generating changes in various gut bacteria (Jin et al. 2017). As part of the therapy of type-2 diabetes, GLPs also significantly raised the anti-inflammatory bacteria Dehalobacterium and Enterococcus in the bloodstream.

The bacteria Dehalobacterium has been shown to inhibit the formation of inflammatory cytokines, while the bacteria Enterococcus has been shown to have anti-inflammatory properties (Ma et al. 2021). Polysaccharide fractions made from G. lucidum significantly reduced obesity caused by a high-fat diet, and the ratio of Firmicutes to Bacteroidetes and the abundance of Proteobacteria. Based on the suppression of apoptosis by reversing microtubule polymerization, Li et al. hypothesized that GLPs might alleviate paclitaxel-induced gut barrier damage (Li et al. 2020). GLPs, one of the most studied and representative PSs, is also recognized as a prebiotic option due to its anti-tumor properties. Compared to galactomannan PSs extracted from guar beans, GLPs supplement consumption reduced CRC symptoms by increasing the amount of Akkermansia (Luo et al. 2020). According to another study, a combination of GLPs and saponins from tetraploid jiaogulan could significantly reduce symptoms associated with CRC, such as inflammatory intestinal barrier, malignant cells growth, and the production of tumorigenic signaling molecules, all of which are associated with an increase of Bacteroidetes bacteria and other microorganisms which can produce SCFAs (Khan et al. 2019). Yao et al. (2021) showed that alcohol extract of GL mycelia promotes sleep through a gut microbiota-dependent and serotonin-associated pathway in mice.

GL extracts have been shown to increase gene expression for fatty acid breakdown and excretion, as well as modify GM for prebiotic and microbiota-regulating transcriptional functions. These systems work together to minimize cholesterol buildup and inflammation (Romero-Córdoba et al. 2021).

GL-related substances have been found in several studies to have the potential to alter the GM and risk factors for cardiovascular disease. Thus, GLPs may help lower the risk of lipid metabolic diseases by changing the GM and the expression of genes involved in hepatic lipid and cholesterol metabolism, respectively (Lv et al. 2019). In research conducted by Meneses et al. (2016), C57BL/6 mice were given a diet rich in cholesterol, and then GL extracts of Mexican origin were administered. Furthermore, the data showed that the composition of GM modulated and increased level of Lactobacillaceae and Lactobacillus genus correlated with a significant decrease in total serum cholesterol, LDL-C, triglyceride concentration, hepatic cholesterol, and hepatic triglycerides, which means that GL genetic resources from Mexico constitute a novel supply of functional compounds with hypocholesterolemic and prebiotic characteristics (Meneses et al. 2016).

On the other hand, Wu et al. (2017) discovered that GL mycelium-fermented liquid was harmful because it increased the population of opportunistic pathogens, such as the Acinetobacter species, while simultaneously decreasing the population of probiotics, such as the Lactococcus genus, in the environment. Furthermore, there was no question that GL mycelium-fermented liquid might lower LDL-C levels in the blood (Wu et al. 2017).

7 Modulatory Effects of *Cordyceps Militaris* Polysaccharides on Gut Microbiota

Cordyceps militaris (CM) is classified as a cultured medicinal and edible fungus. Adaptogenic and tonic properties of Cordyceps spp. belonging to the Ascomycota group have long been used as a natural agent in Asian folkloric medicine because of their potential to reduce tiredness and enhance the human immune system (Zhang and Liang 2013). It has a fruiting body that ranges from orange to red and emerges from the dead pupa of the insects it parasitizes (Mei et al. 2014). CM in their natural state are notoriously difficult to procure and can be quite expensive. It is the primary source of cordycepin, and the mycelium of this fungus is now artificially produced on a culture medium modified to increase the production of this chemical by the fungus (Zhou et al. 2009). *Cordyceps militaris* contains valuable components, including essential fatty acids and amino acids, vitamins B1, B2, B12, E, and K, minerals such as calcium, magnesium, sodium, iron, potassium, and others, enzymes, sterols, and finally, proteins, which give it complexity and a holistic effect on the body. Previous research has demonstrated that CM can boost immunity by stimulating immune cells and certain chemicals already present in the body (Yan et al. 2014). In the case of lung or skin cancers, it may also assist in the fight against tumor cells and reduce the size of tumors (Park et al. 2009; Jordan et al. 2010). Polysaccharides, proteins, pentostatin, cordycepin, adenosine, ergosterol, and myriocin are the main chemical elements responsible for the pharmacological effects of CM. In the past few years, significant research has been conducted on the pharmacological actions and the molecular structure of *Cordyceps militaris* polysaccharides (CMPs). C. militaris possesses a high concentration of polysaccharides, which may vary from 3% to 8% of the entire weight and is often found in the fruiting bodies, the mycelium of solid fermentation submerged cultures, and the broth. Several researchers have shown that CMPs have a wide range of functions, including protecting the intestinal cells from injury and boosting immunostimulatory function (Zhu et al. 2013; Kang et al. 2015). Furthermore, the monosaccharide content and structure of CMPs have been investigated, and these polysaccharides' adhesive capabilities have been studied to preserve starch's structural integrity (Nurmamat et al. 2018). Polysaccharide supplementation from CM is a viable treatment strategy for obesity and metabolic illnesses targeting the gut flora (Yu et al. 2020).

CMPs and cordycepin have recently been used to treat obesity in rats and mice by reducing microbiome dysfunction (Huang et al. 2022). After in vitro digestion and anaerobic fermentation, it has been shown that both *Ophiocordyceps sinensis* and *C. militaris* may change the composition of the microbiota in the intestines by increasing the proportion of Bacteroidetes to Firmicutes. In addition, Ji et al. (2021) demonstrated that *O. sinensis* had more beneficial effects on intestinal health than *C. militaris*. These benefits included, lowering the pH of the environment within the intestinal tract, stimulating the production of beneficial SCFA, and increasing the growth of beneficial genera such as *Phascolarctobacterium* and *Bifidobacterium*. *C. militaris* fruiting body, CMPs, and cordycepin-rich solution displayed varying efficiency in regulating hyperglycemia and GM in HFSD-induced rats (Yu et al. 2015).

Selenium peptides derived from selenium-enriched CM were shown to improve GM dysbiosis by boosting the numbers of *Lactobacillus* and *Alistipes* while simultaneously reducing the numbers of *Akkermansia* and Bacteroides in lipopolysaccharide-injured mice's intestinal mucosa (Wu et al. 2022). Chen et al. (2020) observed that supplementing laying hen diets with 100 and 200 mg CMPs/kg improved the performance of the product and altered the composition of the caecal bacterial population. A study conducted on a pig model showed that CM increased intestinal barrier function, as shown by increased physical, biochemical, and immunological barrier function. Furthermore, CM changed the microbiota composition and raised the levels of SCFAs in pigs.

8 Modulatory Effects of *Boletus Edulis* on Gut Microbiota

Boletus edulis, also known as cep, king bolete, porcini, or penny bun, is one of the most well-known species of wild fungi with a wide range of nutritional and medicinal properties (Zhao et al. 2020). It grows primarily in North America, China, Southern Africa, Eastern Europe, and Italy, where it is consumed in abundance. *B. edulis* has a wide variety of bioactive components, some of which have been identified as polysaccharides, phenolic compounds, and phytosterols (Tan et al. 2022). In addition, it has been demonstrated that bioactive compounds and chemical extracts derived from *B. edulis* have activities that include the prevention of constipation, antioxidant defense, antineoplastic defense, anti-inflammatory defense, hepatoprotective defense, antibacterial defense, and antiviral defense (Ślusarczyk et al. 2021).

In a recently published article, the ameliorating effects of the polysaccharides in the fungus Boletus edulis on dyslipidemia, dysglycemia, and oxidative stress in the liver of rats from the T2DM genetic line were indicated (Mustafa et al. 2022). At the same time, possible pathways or mechanisms for the effect of boletus polysaccharides on diabetic liver disease in rats have been identified. Further studies have shown that *Boletus edulis* polysaccharide treatment could regulate gene expression, decreased liver damage and inflammation while boosting bile acid excretion and cellular oxidative defenses in rats given a high-fat diet and injected with diabetes medication (Xiao et al. 2019). Recent in vitro research demonstrated the potential fermentability of *Boletus edulis* by microbiota from both lean and obese individuals, exhibiting significant increases in fermentation rates and alterations in SCFA synthesis (Gual-Grau et al. 2022). In our previous study, *Boletus edulis* extract administration was linked to changes in the microbiome of patients with cardiovascular disease. Furthermore, the study on the artificial gut system (GIS2 simulator) showed an improvement in the microbiological ecosystem. This was shown due to the presence of certain biomarkers (such as ammonium levels and fingerprints of short-chain fatty acids – SCFAs), and there were more lactic acid bacteria (LAB) and a greater variety of them when porcini mushrooms were consumed (Vamanu and Pelinescu 2017).

9 Mushroom-Derived MicroRNAs: New Multi-Target Bioactive Molecules and Their Perspective for Therapy

Mushrooms have been consumed for thousands of years, have been recognized for their great nutritional value, and are used in traditional medicine (Valverde et al. 2015). Recent advances in food nutrition and health care research have revealed that other "hidden nutrients," such as microRNAs (miRNAs/miRs), are important components that play a significant role in the therapeutic effects. Health regulations based on the food-borne and medicinal miRNAs in humans and other mammals is a new concept that received considerable attention in the last years (Díez-Sainz et al. 2021; Khurana et al. 2021; Zhu et al. 2021; Saiyed et al. 2022). However, many mysteries remain to be explored. Their discovery will facilitate the proper exploitation of edible mushroom resources and open new approaches to preventing or treating human diseases.

MicroRNAs (miRNAs) are a class of small, intracellular, single-stranded, and non-coding RNAs but conserved evolutionary (Spinler et al. 2020; Díez-Sainz et al.

2021). Since 1993, when miRNA was first described as lin-4 in *Caenorhabditis elegans*, thousands of miRNAs have been identified in eukaryotic organisms (Rabuma et al. 2022). In 2010, the first report was published about the presence of fungal miRNA-like RNAs (miRNAs) in Neurospora spp. (Lee et al. 2010). Fungal miRNAs share similarities with miRNAs from other kingdoms but are more closely related to plants than animal miRNAs (Marin et al. 2022). The miRNAs from exogenous species detected in human blood circulation are called xeno-miRNAs, xenomiRs, or simply exogenous miRNAs, compared to endogenous miRNAs from human cells and fluids (Khurana et al. 2021; Díez-Sainz et al. 2021).

Our understanding of the biogenesis of miRNAs and their functional role have improved over the years. MicroRNAs are widely known for their ability to up- or down-regulate multiple target genes at the post-transcriptional and post-translational levels, mainly targeting messenger mRNAs (Ergin and Çetinkaya 2022; Saiyed et al. 2022). Therefore, in all species, miRNAs regulate main biological processes, such as cell growth, differentiation, development, and death or environmental stress response. Moreover, interactive analysis proved that certain miRNAs species, such as miRNA-155, miRNA-168 or miRNA-854 family, may be active in both plant and animal cells, while miRNA-21, miRNA-146a, and miRNA-155 coexist in the gut microbiome and foods (Zhao et al. 2018; Bi et al. 2020). Therefore, an intriguing possibility that exogenous miRNAs are involved in the cross-kingdom movement of miRNAs, regulating gene function in the new host, has emerged as a major research area.

Knowledge of fungal microRNA-like RNAs is relatively low compared to plant and animal miRNA research that received considerable attention. Many reports suggest that fungal miRNAs are crucial mediators of plant-microbe interactions, shaping the rhizosphere microbiota (Marin et al. 2022). Recently, an increased number of fungal miRNAs investigations have been reported to uncover the knowledge gaps on the production, mechanisms, and biological function of fungal miRNAs and their evolutionary origins (Xu et al. 2021; Lau et al. 2020). To date, only a few miRNAs from mushrooms Antrodia cinnamomea (Lin et al. 2015), Ganoderma lingzhi (Li et al. 2016a; Shao et al. 2020; Mu et al. 2015), Coprinus cinereus (Lau et al. 2020), Pleurotus ostreatus (Xu et al. 2021), Agaricus bisporus (Marin et al. 2022) have been proved to have a potential regulatory role in metabolite synthesis, transport, and development in mushrooms. However, more effective researches need to be done based on bioinformatics predictions and high-throughput sequencing technologies to explore their expected functions.

In the case of edible mushrooms, there is an increasing interest in understanding the role of miRNAs in the mechanism of fruiting body development, as they have important nutritional and medicinal value. *Caprinopsis cinerea* is a model mushroom with a well-characterized genome mainly used to study development processes in Basidiomycetes. Therefore, this mushroom was chosen to prove the regulatory role of miRNAs on genes involved in fruiting body development and basidiospores germination and out-growth. Using bioinformatic tools, twenty-two new predicted miRNAs were identified, but only two were validated, and essential genes involved in the miRNAs biogenesis were identified (Lau et al. 2020). Another important edible mushroom is hiratake (*Pleurotus ostreatus*), a delicacy in Asian cuisine, used in medicine and for mycoremediation purposes. Expression profile analysis showed that Po-Mir-1, a novel 22 bp microRNA from *P. ostreatus* expressed differently in distinct development stages: mycelium, primordium, young fruiting body, and mature fruiting body. These miRNAs from oyster mushrooms may perform their function by negatively regulating the hydrophobin POH1 gene, an important gene in fruiting body development (Xu et al. 2021). Worldwide the most consumed mushroom is *Agaricus bisporus* or white button mushroom. It was analyzed using RNA-sequencing, and bioinformatic tools detected 37 de novo miRNAs, including the one already known. Precursors and mature miRNA-like small RNAs were found in edible parts (caps and stipes), and their potential gene targets were involved in the primary and secondary metabolites regulation (Marin et al. 2022).

Medicinal host-specific brown-fungus *Antrodia cinnamomea* is known to produce valuable pharmaceutical compounds with diverse biological activities: antitumor, anti-inflammatory, anti-oxidative, hepatoprotective, and vasorelaxation. Transcriptome analysis and miRNAs prediction tools were helpful in the identification of 4 predicted conserved miRNAs and 63 new miRNAs candidates that regulate important genes involved in the secondary metabolite biosynthesis (Lin et al. 2015). Research about the miRNAs composition of another precious medicinal mushroom, *Ganoderma lucidum* or reishi, added novel miRNAs sequences to the miRBase database. Li et al. (2016a) identified 132 known miRNAs and 34 putative candidate miRNAs in the sporocarp. Another study has noticed that miRNAs were differential expressed at different development stages and indicated the mechanism of interaction between selected miRNAs and target genes (Mu et al. 2015). Recently work has proved miRNAs' crucial regulatory roles in various biological processes of *G. lucidum,* and has identified target genes related to the biosynthesis of triterpenes, polysaccharides and lignin degradation pathway (Shao et al. 2020).

Seven edible commonly consumed mushrooms, including white bottom, Swiss brown, king oyster, shiitake, white beech, brown beech, and oyster mushroom, were used to extract exosome-like nanoparticles (ELNs) (Liu et al. 2020). Mushroom-derived ELNs have sphere-shaped morphology and are 100–140 nm in range. ELNs from three mushrooms (king oyster, shiitake, and white bottom) were analyzed, proving that they are composed of small and large-sized RNAs, proteins, and lipids. The amount and composition of fungal ELNs obtained using the same protocol were species-specific, with the highest yield in the case of white button mushrooms (10^{11}/g). Interestingly, the size and composition of mushroom-derived ELNs were similar to those of dietary plant-derived ELNs (Liu et al. 2020). Moreover, the authors concluded that shiitake mushrooms (*Lentinula edodes*) ELNs at a concentration range of 1–9 × 10^{10}/mL contain undetected active biomolecules and manage to induce a hepatic protective effect in vitro and in vivo experiments. Interestingly, shiitake ELNs inhibited anti-inflammatory cytokines secretion at protein and mRNA levels of the interleukin Il6 and Il1b genes. Also, the expression of NLRP3 and TNF genes was reduced and inhibited the production of macrophage NLRP3 inflammasome (Liu et al. 2020). The authors noted that the well-known polysaccharide lentinan from shiitake mushrooms was not involved in the inhibition process. It has been shown that lentinan has a different target, not on NLRP3 inflammasome. Therefore, there is reason to believe that miRNAs-mediated inhibition of mRNA could play a role in gene regulation (Liu et al. 2020).

Although several mushroom extracts have been reported to modulate endogenous miRNA expression, the mechanism is less investigated (Zhang et al. 2020a; Kang 2019; Gonul et al. 2015; Jin et al. 2016). Moreover, frequently synthetic supplements of important bioactive molecules in the extract do not reveal the same benefits as complex vegetal material. Therefore, it raises the possibility that other hidden multi-target bioactive molecules, such as mushroom-derived miRNAs, could be involved, synergistically acting to enhance the therapeutic effects and/or reduce the toxicity of chemotherapeutic treatments (Vamanu et al. 2021).

Exogenous miRNAs from diet or medicinal preparation should be accessible, stable, absorbed by the host cells, and/or modulate gut microbiota and be active in a cross-kingdom manner. According to the present state of knowledge, there is no information about the ability of mushroom-derived miRNAs to be absorbed and reach their targets or validation studies about their mechanism to regulate host gene expression. It is well known that plant and fungal miRNAs are packed in exosome vesicles that increase their survival in the harsh environment of the digestive tract (Fujita et al. 2018; Qin et al. 2022; Li et al. 2022). However, in the case of mushroom-derived ELNs is not clear if they survive digestion as the anti-inflammatory role of shiitake ELNs were tested using intraperitoneally injected mice (Liu et al. 2020). At the same time, how edible plant and mushroom miRNAs could profile gut microbiota composition by modulating genes involved in microbial growth need to be clarified. Recently, several studies proved that dietary microRNAs might perform as hidden bioactive molecules involved in communication with the human gut microbiome and thus can modulate cellular physiology in the intestinal tract (Liu et al. 2022; Díez-Sainz et al. 2021).

The cross-kingdom regulation based on exogenous miRNAs is a relatively new concept. It gives us hope for exploring the therapeutic perspectives in curing and preventing diseases. The therapeutic potential of shiitake mushroom ELNs in preventing inflammation in acute liver injury mice has been reported (Liu et al. 2020). Investigations on the action mechanism that manages to induce the hepatic protective effect have shown that unknown active biomolecules from ELNs control gene expression at the mRNAs level, similar to the mechanism proposed for miRNAs. In another work, using two different miRNA target prediction tools (psRNATarget v2 and miRanda v3), it was found that miRNA-like small RNAs that regulate basic metabolic pathways in the mushroom *Agaricus bisporus* can be involved in the regulation of 319 human basic metabolic pathways as well as in cancer and infection-related pathways (Marin et al. 2022). This theoretical study provides interesting clues that some pathways in mushrooms and humans could be modulated by miRNAs and are among the top 10 regulated pathways in both organisms. Also, according to the KEGG analysis, abi-miRNAs regulate ~16% of the prostate cancer nodes. Moreover, *Agaricus* sp. inhibited various tumors and was used as an adjuvant in the treatment and prevention of metastases and deaths. Similarly, using target prediction tools and the KEGG Pathway database, Li et al.

concluded in 2016 that *G. lingzhi* miRNAs could target cancer-related genes and human signal transduction pathways (Li et al. 2016a). This analysis paves the way toward identifying mushroom miRNA-like RNAs that might be involved in regulating mammalian genes affecting various physiological and cellular differentiation processes.

Named "treasure in the forest," mushrooms represent a tremendous source of biologically useful and pharmacologically active molecules. There is reason to believe that microRNA-like RNAs are hidden multi-target bioactive molecules with therapeutic perspectives. Recent breakthroughs in nutrition, proteomics, and bioinformatics research have supported us in developing a better understanding of the underlying mechanisms involved in cross-kingdom communication. Therefore, detailed insights into mushroom-derived miRNAs' cross-kingdom gene regulation mechanism will help link our current knowledge of clinically effective therapeutics.

10 Conclusions and Future Prospective

Several types of functional mushrooms have compounds that are very good at fighting inflammation. Since polysaccharides in mushrooms are resistant to hydrolysis by digestive enzymes, they make it to the large intestine, where they are fermented preferentially by gut bacteria and influence absorption of nutrients, metabolic rate, and immune status. The structural characteristics and bioactivities of mushroom polysaccharides can be affected by both the extraction methods and the source of the polysaccharides, which can range from the fruiting body or mycelia to the supernatants of their fermentation or even wastes. Both in vivo and in vitro studies have shown that consuming mushroom polysaccharides influences the composition of GM and the responsiveness of the host immune system. These gut microbiota changes might take several forms, depending on the polysaccharide concentration. However, it seems that the beneficial effects on the host are mediated by metabolites created by the bacteria in the gut. The beneficial interactions between functional mushroom polysaccharides and GM are largely due to SCFAs, which can activate certain cell signaling pathways and target host tissues important in metabolic homeostasis and immunological response. However, more research is needed to determine the whole spectrum of microbial metabolites produced by mushroom polysaccharide ingestion and their impact on host immunity because of the complexity involved in characterizing functional mushroom polysaccharides and variations in GM.

So far, there has been insufficient research on the modification of microbiota by various mushrooms. We have summarized the most recent research on the beneficial effects of prebiotics derived from mushrooms on gut microbiota.

Possible areas for further study of gut microbiota include a genomic exploration of gastrointestinal microbial community, genetic alterations in the microbiome resulting from mushroom consumption, and functional characterization of beneficial gut flora (metagenomics or ecogenomics).

References

Ahmad MF (2019) Ganoderma lucidum: a macro fungus with phytochemicals and their pharmacological properties. In: Plant and human health, vol 2. Springer, Cham, p 491

Akobeng AK, Singh P, Kumar M, Al Khodor S (2020) Role of the gut microbiota in the pathogenesis of coeliac disease and potential therapeutic implications. Eur J Nutr 59(8): 3369–3390. https://doi.org/10.1007/s00394-020-02324-y

Alam KY, Clark DP (1989) Anaerobic fermentation balance of Escherichia coli as observed by in vivo nuclear magnetic resonance spectroscopy. J Bacteriol 171(11):6213–6217. https://doi.org/10.1128/jb.171.11.6213-6217.1989

Amar J, Chabo C, Waget A, Klopp P, Vachoux C, Bermúdez-Humarán LG, Smirnova N, Bergé M, Sulpice T, Lahtinen S, Ouwehand A, Langella P, Rautonen N, Sansonetti PJ, Burcelin R (2011) Intestinal mucosal adherence and translocation of commensal bacteria at the early onset of type 2 diabetes: molecular mechanisms and probiotic treatment. EMBO Mol Med 3(9):559–572. https://doi.org/10.1002/emmm.201100159

Amiri P, Hosseini SA, Ghaffari S, Tutunchi H, Ghaffari S, Mosharkesh E, Asghari S, Roshanravan N (2022) Role of butyrate, a gut microbiota derived metabolite, in cardiovascular diseases: a comprehensive narrative review. Front Pharmacol 12:837509. https://doi.org/10.3389/fphar.2021.837509

Aw W, Fukuda S (2018) Understanding the role of the gut ecosystem in diabetes mellitus. J Diabetes Investig 9(1):5–12. https://doi.org/10.1111/jdi.12673

Badalyan SM (2014) Potential of mushroom bioactive molecules to develop healthcare biotech products. Yugantar Prakashan, New Delhi

Belkaid Y, Hand TW (2014) Role of the microbiota in immunity and inflammation. Cell 157(1): 121–141. https://doi.org/10.1016/j.cell.2014.03.011

Berg G, Rybakova D, Fischer D, Cernava T, Vergès M-CC, Charles T, Chen X, Cocolin L, Eversole K, Corral GH, Kazou M, Kinkel L, Lange L, Lima N, Loy A, Macklin JA, Maguin E, Mauchline T, McClure R, Mitter B, Ryan M, Sarand I, Smidt H, Schelkle B, Roume H, Kiran GS, Selvin J, Souza RSC, van Overbeek L, Singh BK, Wagner M, Walsh A, Sessitsch A, Schloter M (2020) Microbiome definition re-visited: old concepts and new challenges. Microbiome 8(1):103. https://doi.org/10.1186/s40168-020-00875-0

Berger A, Rein D, Kratky E, Monnard I, Hajjaj H, Meirim I, Piguet-Welsch C, Hauser J, Mace K, Niederberger P (2004) Cholesterol-lowering properties of Ganoderma lucidum in vitro, ex vivo, and in hamsters and minipigs. Lipids Health Dis 3:2. https://doi.org/10.1186/1476-511x-3-2

Bertucci M, Calusinska M, Goux X, Rouland-Lefèvre C, Untereiner B, Ferrer P, Gerin PA, Delfosse P (2019) Carbohydrate hydrolytic potential and redundancy of an anaerobic digestion microbiome exposed to acidosis, as uncovered by metagenomics. J Appl Environ Microbiol 85(15):e00895. https://doi.org/10.1128/aem.00895-19

Bi K, Zhang X, Chen W, Diao H (2020) MicroRNAs regulate intestinal immunity and gut microbiota for gastrointestinal health: a comprehensive review. Genes (Basel) 11(9):1075. https://doi.org/10.3390/genes11091075

Blaak EE, Canfora EE, Theis S, Frost G, Groen AK, Mithieux G, Nauta A, Scott K, Stahl B, van Harsselaar J, van Tol R, Vaughan EE, Verbeke K (2020) Short chain fatty acids in human gut and metabolic health. Benef Microbes 11(5):411–455. https://doi.org/10.3920/bm2020.0057

Bresciani L, Favari C, Calani L, Francinelli V, Riva A, Petrangolini G, Allegrini P, Mena P, Del Rio D (2020) The effect of formulation of curcuminoids on their metabolism by human colonic microbiota. Molecules 25(4):940

Brüssow H (2020) Problems with the concept of gut microbiota dysbiosis. Microb Biotechnol 13(2):423–434. https://doi.org/10.1111/1751-7915.13479

Burcelin R (2016) Gut microbiota and immune crosstalk in metabolic disease. Mol Metab 5(9): 771–781. https://doi.org/10.1016/j.molmet.2016.05.016

Campi M, Mancuello C, Ferreira F, Maubet Y, Cristaldo E, Robledo G (2021) Bioactive compounds and antioxidant activity of four native species of the ganodermataceae family (agaricomycetes) from Paraguay. Int J Med Mushrooms 23(8):65–76. https://doi.org/10.1615/IntJMedMushrooms.2021039298

Cani PD, Amar J, Iglesias MA et al (2007) Metabolic endotoxemia initiates obesity and insulin resistance. Diabetes 56:1761–1772. https://doi.org/10.2337/db06-1491

Chen J, Liu J, Yan C, Zhang C, Pan W, Zhang W, Lu Y, Chen L, Chen Y (2020) Sarcodon aspratus polysaccharides ameliorated obesity-induced metabolic disorders and modulated gut microbiota dysbiosis in mice fed a high-fat diet. Food Funct 11(3):2588–2602. https://doi.org/10.1039/C9FO00963A

Cheong KL, Wu DT, Zhao J, Li SP (2015) A rapid and accurate method for the quantitative estimation of natural polysaccharides and their fractions using high performance size exclusion chromatography coupled with multi-angle laser light scattering and refractive index detector. J Chromatogr A 1400:98–106. https://doi.org/10.1016/j.chroma.2015.04.054

Clark DP, Cronan JE (2005) Two-carbon compounds and fatty acids as carbon sources. EcoSal Plus 1(2). https://doi.org/10.1128/ecosalplus.3.4.4

Clarke TB, Davis KM, Lysenko ES, Zhou AY, Yu Y, Weiser JN (2010) Recognition of peptidoglycan from the microbiota by Nod1 enhances systemic innate immunity. Nat Med 16(2): 228–231. https://doi.org/10.1038/nm.2087

Cockburn DW, Koropatkin NM (2016) Polysaccharide degradation by the intestinal microbiota and its influence on human health and disease. J Mol Biol 428(16):3230–3252. https://doi.org/10.1016/j.jmb.2016.06.021

Crost EH, Tailford LE, Le Gall G, Fons M, Henrissat B, Juge N (2013) Utilisation of mucin glycans by the human gut symbiont ruminococcus gnavus is strain-dependent. PLoS One 8(10):e76341. https://doi.org/10.1371/journal.pone.0076341

D'Amelio P, Sassi F (2018) Gut microbiota, immune system, and bone. Calcif Tissue Int 102(4): 415–425. https://doi.org/10.1007/s00223-017-0331-y

Daniele G, Guardado Mendoza R, Winnier D, Fiorentino TV, Pengou Z, Cornell J, Andreozzi F, Jenkinson C, Cersosimo E, Federici M, Tripathy D, Folli F (2014) The inflammatory status score including IL-6, TNF-α, osteopontin, fractalkine, MCP-1 and adiponectin underlies whole-body insulin resistance and hyperglycemia in type 2 diabetes mellitus. Acta Diabetol 51(1): 123–131. https://doi.org/10.1007/s00592-013-0543-1

de Almeida WS, da Silva DA (2021) Does polysaccharide quaternization improve biological activity? Int J Biol Macromol 182:1419–1436. https://doi.org/10.1016/j.ijbiomac.2021.05.012

De Silva DD, Rapior S, Fons F, Bahkali AH, Hyde KD (2012) Medicinal mushrooms in supportive cancer therapies: an approach to anti-cancer effects and putative mechanisms of action. Fungal Divers 55(1):1–35. https://doi.org/10.1007/s13225-012-0151-3

DeGruttola AK, Low D, Mizoguchi A, Mizoguchi E (2016) Current understanding of dysbiosis in disease in human and animal models. Inflamm Bowel Dis 22(5):1137–1150. https://doi.org/10.1097/mib.0000000000000750

den Besten G, van Eunen K, Groen AK, Venema K, Reijngoud DJ, Bakker BM (2013) The role of short-chain fatty acids in the interplay between diet, gut microbiota, and host energy metabolism. J Lipid Res 54(9):2325–2340. https://doi.org/10.1194/jlr.R036012

Dendup T, Feng X, Clingan S, Astell-Burt T (2018) Environmental risk factors for developing type 2 diabetes mellitus: a systematic review. Int J Environ Res Public Health 15(1):78. https://doi.org/10.3390/ijerph15010078

Díez-Sainz E, Lorente-Cebrián S, Aranaz P, Riezu-Boj JI, Martínez JA, Milagro FI (2021) Potential mechanisms linking food-derived microRNAs, gut microbiota and intestinal barrier functions in the context of nutrition and human health. Front Nutr 8:586564. https://doi.org/10.3389/fnut.2021.586564

Djuric Z (2017) Obesity-associated cancer risk: the role of intestinal microbiota in the etiology of the host proinflammatory state. Transl Res 179:155–167. https://doi.org/10.1016/j.trsl.2016.07.017

Donovan C, Liu G, Shen S, Marshall JE, Kim RY, Alemao CA, Budden KF, Choi JP, Kohonen-Corish M, El-Omar EM, Yang IA, Hansbro PM (2020) The role of the microbiome and the NLRP3 inflammasome in the gut and lung. J Leukoc Biol 108(3):925–935. https://doi.org/10.1002/jlb.3mr0720-472rr

Ergin K, Çetinkaya R (2022) Regulation of microRNAs. In: Methods in molecular biology, vol 2257. Humana, New York, NY, pp 1–32. https://doi.org/10.1007/978-1-0716-1170-8_1

Ferreira IC, Heleno SA, Reis FS, Stojkovic D, Queiroz MJ, Vasconcelos MH, Sokovic M (2015) Chemical features of Ganoderma polysaccharides with antioxidant, antitumor and antimicrobial activities. Phytochemistry 114:38–55. https://doi.org/10.1016/j.phytochem.2014.10.011

Flint HJ, Duncan SH, Scott KP, Louis P (2015) Links between diet, gut microbiota composition and gut metabolism. Proc Nutr Soc 74(1):13–22. https://doi.org/10.1017/s0029665114001463

Foley MH, Cockburn DW, Koropatkin NM (2016) The Sus operon: a model system for starch uptake by the human gut Bacteroidetes. Cell Mol Life Sci 73(14):2603–2617. https://doi.org/10.1007/s00018-016-2242-x

Fu L, Chen H, Dong P, Zhang X, Zhang M (2010) Effects of ultrasonic treatment on the physicochemical properties and DPPH radical scavenging activity of polysaccharides from mushroom Inonotus obliquus. J Food Sci 75(4):C322–C327. https://doi.org/10.1111/j.1750-3841.2010.01590.x

Fujita D, Arai T, Komori H, Shirasaki Y, Wakayama T, Nakanishi T, Tamai I (2018) Apple-derived nanoparticles modulate expression of organic-anion-transporting polypeptide (OATP) 2B1 in Caco-2 cells. Mol Pharm 15(12):5772–5780. https://doi.org/10.1021/acs.molpharmaceut.8b00921

Ghassemi N, Poulhazan A, Deligey F, Mentink-Vigier F, Marcotte I, Wang T (2022) Solid-state NMR investigations of extracellular matrixes and cell walls of algae, bacteria, fungi, and plants. Chem Rev 122(10):10036–10086. https://doi.org/10.1021/acs.chemrev.1c00669

Ghosh SS, Wang J, Yannie PJ, Ghosh S (2020) Intestinal barrier dysfunction, LPS translocation, and disease development. J Endocr Soc 4(2):bvz039. https://doi.org/10.1210/jendso/bvz039

Gong P, Wang S, Liu M, Chen F, Yang W, Chang X, Liu N, Zhao Y, Wang J, Chen X (2020) Extraction methods, chemical characterizations and biological activities of mushroom polysaccharides: a mini-review. Carbohydr Res 494:108037. https://doi.org/10.1016/j.carres.2020.108037

Gonul O, Aydin HH, Kalmis E, Kayalar H, Ozkaya AB, Atay S, Ak H (2015) Effects of ganoderma lucidum (higher basidiomycetes) extracts on the miRNA profile and telomerase activity of the MCF-7 breast cancer cell line. Int J Med Mushrooms 17(3):231–239. https://doi.org/10.1615/intjmedmushrooms.v17.i3.30

Gruneck L, Khongphinitbunjong K, Popluechai S (2021) Gut microbiota associated with two species of domesticated honey bees from Thailand. Symbiosis 83:335. https://doi.org/10.1007/s13199-021-00754-8

Gual-Grau A, Guirro M, Crescenti A, Boqué N, Arola L (2022) In vitro fermentability of a broad range of natural ingredients by fecal microbiota from lean and obese individuals: potential health benefits. Int J Food Sci Nutr 73(2):195–209. https://doi.org/10.1080/09637486.2021.1954144

Guinane CM, Cotter PD (2013) Role of the gut microbiota in health and chronic gastrointestinal disease: understanding a hidden metabolic organ. Therap Adv Gastroenterol 6(4):295–308. https://doi.org/10.1177/1756283x13482996

Harris HC, Morrison DJ, Edwards CA (2021) Impact of the source of fermentable carbohydrate on SCFA production by human gut microbiota in vitro - a systematic scoping review and secondary analysis. Crit Rev Food Sci Nutr 61(22):3892–3903. https://doi.org/10.1080/10408398.2020.1809991

Hasan N, Yang H (2019) Factors affecting the composition of the gut microbiota, and its modulation. PeerJ 7:e7502. https://doi.org/10.7717/peerj.7502

Hess J, Wang Q, Gould T, Slavin J (2018) Impact of Agaricus bisporus mushroom consumption on gut health markers in healthy adults. Nutrients 10(10):1402

Honda K, Littman DR (2012) The microbiome in infectious disease and inflammation. Annu Rev Immunol 30:759–795. https://doi.org/10.1146/annurev-immunol-020711-074937

Hong YS, Jung DH, Chung WH, Nam YD, Kim YJ, Seo DH, Park CS (2022) Human gut commensal bacterium Ruminococcus species FMB-CY1 completely degrades the granules of resistant starch. Food Sci Biotechnol 31(2):231–241. https://doi.org/10.1007/s10068-021-01027-2

Hooks KB, O'Malley MA (2017) Dysbiosis and its discontents. mBio 8(5):e01492. https://doi.org/10.1128/mBio.01492-17

Hou L, Yang Y, Sun B, Jing Y, Deng W (2021) Dietary fiber, gut microbiota, short-chain fatty acids, and host metabolism. Am J Life Sci 9(6):162–172

Hu GX, Chen GR, Xu H, Ge RS, Lin J (2010) Activation of the AMP activated protein kinase by short-chain fatty acids is the main mechanism underlying the beneficial effect of a high fiber diet on the metabolic syndrome. Med Hypotheses 74(1):123–126. https://doi.org/10.1016/j.mehy.2009.07.022

Hu Y, Chen D, Zheng P, Yu J, He J, Mao X, Yu B (2019) The bidirectional interactions between resveratrol and gut microbiota: an insight into oxidative stress and inflammatory bowel disease therapy. Biomed Res Int 2019:5403761. https://doi.org/10.1155/2019/5403761

Huang R, Zhu Z, Wu S, Wang J, Chen M, Liu W, Huang A, Zhang J, Wu Q, Ding Y (2022) Polysaccharides from Cordyceps militaris prevent obesity in association with modulating gut microbiota and metabolites in high-fat diet-fed mice. Food Res Int 157:111197. https://doi.org/10.1016/j.foodres.2022.111197

Jack BU, Malherbe CJ, Mamushi M, Muller CJF, Joubert E, Louw J, Pheiffer C (2019) Adipose tissue as a possible therapeutic target for polyphenols: a case for Cyclopia extracts as anti-obesity nutraceuticals. Biomed Pharmacother 120:109439. https://doi.org/10.1016/j.biopha.2019.109439

Jayachandran M, Xiao J, Xu B (2017) A critical review on health promoting benefits of edible mushrooms through gut microbiota. Int J Mol Sci 18(9):1934

Jennison E, Byrne CD (2021) The role of the gut microbiome and diet in the pathogenesis of non-alcoholic fatty liver disease. Clin Mol Hepatol 27(1):22–43. https://doi.org/10.3350/cmh.2020.0129

Ji Y, Tao T, Zhang J et al (2021) Comparison of effects on colitis-associated tumorigenesis and gut microbiota in mice between Ophiocordyceps sinensis and Cordyceps militaris. Phytomedicine 90:153653. https://doi.org/10.1016/j.phymed.2021.153653

Jin M, Zhu Y, Shao D, Zhao K, Xu C, Li Q, Yang H, Huang Q, Shi J (2017) Effects of polysaccharide from mycelia of Ganoderma lucidum on intestinal barrier functions of rats. Int J Biol Macromol 94(Pt A):1–9. https://doi.org/10.1016/j.ijbiomac.2016.09.099

Jin X, Ruiz Beguerie J, Sze DM, Chan GC (2016) Ganoderma lucidum (Reishi mushroom) for cancer treatment. Cochrane Database Syst Rev 4(4):Cd007731. https://doi.org/10.1002/14651858.CD007731.pub3

Jordan JL, Nowak A, Lee TD (2010) Activation of innate immunity to reduce lung metastases in breast cancer. Cancer Immunol Immunother 59(5):789–797. https://doi.org/10.1007/s00262-009-0800-x

Kang H (2019) MicroRNA-mediated health-promoting effects of phytochemicals. Int J Mol Sci 20(10). https://doi.org/10.3390/ijms20102535

Kang HJ, Baik HW, Kim SJ, Lee SG, Ahn HY, Park JS, Park SJ, Jang EJ, Park SW, Choi JY, Sung JH, Lee SM (2015) Cordyceps militaris enhances cell-mediated immunity in healthy Korean men. J Med Food 18(10):1164–1172. https://doi.org/10.1089/jmf.2014.3350

Kanwal S, Aliya S, Xin Y (2020) Anti-obesity effect of dictyophora indusiata mushroom polysaccharide (DIP) in high fat diet-induced obesity via regulating inflammatory cascades and intestinal microbiome. Front Endocrinol (Lausanne) 11. https://doi.org/10.3389/fendo.2020.558874

Khan I, Huang G, Li XA, Liao W, Leong WK, Xia W, Bian X, Wu J, Hsiao WLW (2019) Mushroom polysaccharides and jiaogulan saponins exert cancer preventive effects by shaping the gut microbiota and microenvironment in Apc(Min/+) mice. Pharmacol Res 148:104448. https://doi.org/10.1016/j.phrs.2019.104448

Khurana P, Gupta A, Varshney R (2021) Diet-derived exogenous miRNAs as functional food components: facts and new perspectives

Khursheed R, Singh SK, Wadhwa S, Gulati M, Awasthi A (2020) Therapeutic potential of mushrooms in diabetes mellitus: role of polysaccharides. Int J Biol Macromol 164:1194–1205. https://doi.org/10.1016/j.ijbiomac.2020.07.145

Klymiuk I, Singer G, Castellani C, Trajanoski S, Obermüller B, Till H (2021) Characterization of the luminal and mucosa-associated microbiome along the gastrointestinal tract: results from surgically treated preterm infants and a murine model. Nutrients 13(3):1030

Konusova V, Vorbeychikov E, Shamtsyan M (2021) Potential role of mushroom beta-glucans in immunity and inflammation in viral infections and COVID-19. J Food Bioact 16. https://doi.org/10.31665/JFB.2021.16288

Kumar V, Sinha AK, Makkar HP, de Boeck G, Becker K (2012) Dietary roles of non-starch polysaccharides in human nutrition: a review. Crit Rev Food Sci Nutr 52(10):899–935. https://doi.org/10.1080/10408398.2010.512671

Lau AYT, Xie Y, Cheung MK, Cheung PCK, Kwan HS (2020) Genome-wide mRNA and miRNA analysis in the early stages of germ tube outgrowth in Coprinopsis cinerea. Fungal Genet Biol 142:103416. https://doi.org/10.1016/j.fgb.2020.103416

Lawal TO, Wicks SM, Calderon AI, Mahady GB (2019) Bioactive molecules, pharmacology and future research trends of Ganoderma lucidium as a cancer chemotherapeutic agent. In: New look to phytomedicine. Academic Press, London, pp 159–178

Lee HC, Li L, Gu W, Xue Z, Crosthwaite SK, Pertsemlidis A, Lewis ZA, Freitag M, Selker EU, Mello CC, Liu Y (2010) Diverse pathways generate microRNA-like RNAs and Dicer-independent small interfering RNAs in fungi. Mol Cell 38(6):803–814. https://doi.org/10.1016/j.molcel.2010.04.005

Lee K, Kim D-W, Lee D-H, Kim Y-S, Bu J-H, Cha J-H, Thawng CN, Hwang E-M, Seong HJ, Sul WJ, Wellington EMH, Quince C, Cha C-J (2020) Mobile resistome of human gut and pathogen drives anthropogenic bloom of antibiotic resistance. Microbiome 8(1):2. https://doi.org/10.1186/s40168-019-0774-7

Lee N, Kim W-U (2017) Microbiota in T-cell homeostasis and inflammatory diseases. Exp Mol Med 49(5):e340–e340. https://doi.org/10.1038/emm.2017.36

Levy M, Kolodziejczyk AA, Thaiss CA, Elinav E (2017) Dysbiosis and the immune system. Nat Rev Immunol 17(4):219–232. https://doi.org/10.1038/nri.2017.7

Li B, Cheng X, Zhang T, Liu L, Nie Z, Sheng Q (2016a) The identification of microRNAs in Ganoderma lingzhi sporocarp. Mycoscience 57(4):271–278. https://doi.org/10.1016/j.myc.2016.03.004

Li D, Gao L, Li M, Luo Y, Xie Y, Luo T, Su L, Yong T, Chen S, Jiao C, Su J, Huang S (2020) Polysaccharide from spore of Ganoderma lucidum ameliorates paclitaxel-induced intestinal barrier injury: apoptosis inhibition by reversing microtubule polymerization. Biomed Pharmacother 130:110539. https://doi.org/10.1016/j.biopha.2020.110539

Li K, Zhuo C, Teng C, Yu S, Wang X, Hu Y, Ren G, Yu M, Qu J (2016b) Effects of Ganoderma lucidum polysaccharides on chronic pancreatitis and intestinal microbiota in mice. Int J Biol Macromol 93(Pt A):904–912. https://doi.org/10.1016/j.ijbiomac.2016.09.029

Li Y, Meng Z, Xu Y, Shi Q, Ma Y, Aung M, Cheng Y, Zhu W (2021) Interactions between anaerobic fungi and methanogens in the rumen and their biotechnological potential in biogas production from lignocellulosic materials. Microorganisms 9(1). https://doi.org/10.3390/microorganisms9010190

Li Y, Miao M, Yin F, Shen N, Yu W-Q, Guo S-D (2022) The polysaccharide–peptide complex from mushroom Cordyceps militaris ameliorates atherosclerosis by modulating the lncRNA–miRNA–mRNA axis. Food Funct 13(6):3185–3197. https://doi.org/10.1039/D1FO03285B

Liang J, Zhang M, Wang X, Ren Y, Yue T, Wang Z, Gao Z (2021) Edible fungal polysaccharides, the gut microbiota, and host health. Carbohydr Polym 273:118558. https://doi.org/10.1016/j.carbpol.2021.118558

Liaqat I, Durrani AI, Zafar U, Rubab S, Faheem M, Mubin M, Raza C, Aftab N (2021) Role of modified diet and gut microbiota in metabolic endotoxemia in mice. Int Arch Med Microbiol 203(8):5085–5093. https://doi.org/10.1007/s00203-021-02491-4

Lin Y-L, Ma L-T, Lee Y-R, Lin S-S, Wang S-Y, Chang T-T, Shaw J-F, Li W-H, Chu F-H (2015) MicroRNA-like small RNAs prediction in the development of antrodia cinnamomea. PLoS One 10(4):e0123245. https://doi.org/10.1371/journal.pone.0123245

Liu B, Lu Y, Chen X, Muthuraj PG, Li X, Pattabiraman M, Zempleni J, Kachman SD, Natarajan SK, Yu J (2020) Protective role of shiitake mushroom-derived exosome-like nanoparticles in D-galactosamine and lipopolysaccharide-induced acute liver injury in mice. Nutrients 12(2). https://doi.org/10.3390/nu12020477

Liu T, Zhang L, Joo D, Sun SC (2017) NF-κB signaling in inflammation. Signal Transduct Target Ther 2:17023. https://doi.org/10.1038/sigtrans.2017.23

Liu Y, Tan M-L, Zhu W-J, Cao Y-N, Peng L-X, Yan Z-Y, Zhao G (2022) In vitro effects of Tartary buckwheat-derived nanovesicles on gut microbiota. J Agric Food Chem 70(8):2616–2629. https://doi.org/10.1021/acs.jafc.1c07658

Liu Y, Wu Q, Wu X, Algharib SA, Gong F, Hu J, Luo W, Zhou M, Pan Y, Yan Y, Wang Y (2021) Structure, preparation, modification, and bioactivities of β-glucan and mannan from yeast cell wall: a review. Int J Biol Macromol 173:445–456. https://doi.org/10.1016/j.ijbiomac.2021.01.125

Louis P, Flint HJ (2009) Diversity, metabolism and microbial ecology of butyrate-producing bacteria from the human large intestine. FEMS Microbiol Lett 294(1):1–8. https://doi.org/10.1111/j.1574-6968.2009.01514.x

Louis P, Flint HJ (2017) Formation of propionate and butyrate by the human colonic microbiota. Environ Res Health 19(1):29–41. https://doi.org/10.1111/1462-2920.13589

Luo J, Li T, Xie J, Guo H, Liu L, Zhang G, Peng X (2020) Guar gum different from Ganoderma lucidum polysaccharide in alleviating colorectal cancer based on omics analysis. Food Funct 11. https://doi.org/10.1039/C9FO02786F

Lv X-C, Guo W-L, Li L, Yu X-D, Liu B (2019) Polysaccharide peptides from Ganoderma lucidum ameliorate lipid metabolic disorders and gut microbiota dysbiosis in high-fat diet-fed rats. J Funct Foods 57:48–58. https://doi.org/10.1016/j.jff.2019.03.043

Ma G, Du H, Hu Q, Yang W, Pei F, Xiao H (2021) Health benefits of edible mushroom polysaccharides and associated gut microbiota regulation. Crit Rev Food Sci Nutr:1–18. https://doi.org/10.1080/10408398.2021.1903385

Maity P, Sen IK, Chakraborty I, Mondal S, Bar H, Bhanja SK, Mandal S, Maity GN (2021) Biologically active polysaccharide from edible mushrooms: a review. Int J Biol Macromol 172: 408–417. https://doi.org/10.1016/j.ijbiomac.2021.01.081

Marin FR, Dávalos A, Kiltschewskij D, Crespo MC, Cairns M, Andrés-León E, Soler-Rivas C (2022) RNA-seq, bioinformatic identification of potential microRNA-like small RNAs in the edible mushroom agaricus bisporus and experimental approach for their validation. Int J Mol Sci 23(9). https://doi.org/10.3390/ijms23094923

Mei YX, Yang W, Zhu PX, Peng N, Zhu H, Liang YX (2014) Isolation, characterization, and antitumor activity of a novel heteroglycan from cultured mycelia of Cordyceps sinensis. Planta Med 80(13):1107–1112. https://doi.org/10.1055/s-0034-1382960

Meneses ME, Martínez-Carrera D, Torres N, Sánchez-Tapia M, Aguilar-López M, Morales P, Sobal M, Bernabé T, Escudero H, Granados-Portillo O, Tovar AR (2016) Hypocholesterolemic properties and prebiotic effects of Mexican Ganoderma lucidum in C57BL/6 mice. PLoS One 11(7):e0159631. https://doi.org/10.1371/journal.pone.0159631

Milani C, Duranti S, Bottacini F, Casey E, Turroni F, Mahony J, Belzer C, Delgado Palacio S, Arboleya Montes S, Mancabelli L, Lugli GA, Rodriguez JM, Bode L, de Vos W, Gueimonde M, Margolles A, van Sinderen D, Ventura M (2017) The First microbial colonizers of the human gut: composition, activities, and health implications of the infant gut microbiota. Microbiol Mol Biol Rev 81(4). https://doi.org/10.1128/mmbr.00036-17

Mohammad S, Thiemermann C (2021) Role of metabolic endotoxemia in systemic inflammation and potential interventions. Front Immunol 11. https://doi.org/10.3389/fimmu.2020.594150

Mohr AE, Jäger R, Carpenter KC, Kerksick CM, Purpura M, Townsend JR, West NP, Black K, Gleeson M, Pyne DB, Wells SD, Arent SM, Kreider RB, Campbell BI, Bannock L, Scheiman J, Wissent CJ, Pane M, Kalman DS, Pugh JN, Ortega-Santos CP, ter Haar JA, Arciero PJ, Antonio J (2020) The athletic gut microbiota. J Int Soc Sports Nutr 17(1):24. https://doi.org/10.1186/s12970-020-00353-w

Mu DS, Li C, Shi L, Zhang X, Ren A, Zhao MW (2015) Bioinformatic identification of potential microRNAs and their targets in the Lingzhi or Reishi medicinal mushroom ganoderma lucidum (higher Basidiomycetes). Int J Med Mushrooms 17(8):783–797. https://doi.org/10.1615/intjmedmushrooms.v17.i8.80

Mustafa F, Chopra H, Baig AA, Avula SK, Kumari S, Mohanta TK, Saravanan M, Mishra AK, Sharma N, Mohanta YK (2022) Edible mushrooms as novel myco-therapeutics: effects on lipid level, obesity and BMI. J Fungi (Basel) 8(2). https://doi.org/10.3390/jof8020211

Nie S, Zhang H, Li W, Xie M (2013) Current development of polysaccharides from ganoderma: Isolation, structure and bioactivities. Bioact Carbohydr Diet Fibre 1:10–20. https://doi.org/10.1016/j.bcdf.2013.01.001

Niu L, Wu Y, Liu H et al (2021) The structural characterization of a novel water-soluble polysaccharide from edible mushroom leucopaxillus giganteus and its antitumor activity on H22 tumor-bearing mice. Chem Biodivers. https://doi.org/10.1002/cbdv.202001010

Nurmamat E, Xiao H, Zhang Y, Jiao Z (2018) Effects of different temperatures on the chemical structure and antitumor activities of polysaccharides from Cordyceps militaris. Polymers 10(4): 430

Papp V (2019) Global diversity of the genus Ganoderma: taxonomic uncertainties and challenges. In: Kandikere R, Sridhar SKD (eds) Advances in macrofungi diversity, ecology and biotechnology. Chapman and Hall/CRC, Milton, pp 10–33

Park SE, Yoo HS, Jin CY, Hong SH, Lee YW, Kim BW, Lee SH, Kim WJ, Cho CK, Choi YH (2009) Induction of apoptosis and inhibition of telomerase activity in human lung carcinoma cells by the water extract of Cordyceps militaris. Food Chem Toxicol 47(7):1667–1675. https://doi.org/10.1016/j.fct.2009.04.014

Pavlik M, Zhou S, Zhang J, Tang Q, Feng N, Kurjak D, Pavlík M Jr, Kunca A (2020) Comparative analysis of triterpene composition between *Ganoderma lingzhi* from China and *G. lucidum* from Slovakia under different growing conditions. Int J Med Mushrooms 22(8):793–802. https://doi.org/10.1615/IntJMedMushrooms.2020035662

Peredo-Lovillo A, Romero-Luna HE, Jiménez-Fernández M (2020) Health promoting microbial metabolites produced by gut microbiota after prebiotics metabolism. Food Res Int 136:109473. https://doi.org/10.1016/j.foodres.2020.109473

Pinya Fernández S, Ferriol P, Tejada Gavela S, Sureda A (2019) Mushrooms reishi (Ganoderma lucidum), shiitake (Lentinela edodes), maitake (Grifola frondosa). In: Nonvitamin and nonmineral nutritional supplements. Academic Press, London, pp 517–526. https://doi.org/10.1016/B978-0-12-812491-8.00068-0

Poggi M, Bastelica D, Gual P, Iglesias MA, Gremeaux T, Knauf C, Peiretti F, Verdier M, Juhan-Vague I, Tanti JF, Burcelin R, Alessi MC (2007) C3H/HeJ mice carrying a toll-like receptor 4 mutation are protected against the development of insulin resistance in white adipose tissue in response to a high-fat diet. Diabetologia 50(6):1267–1276. https://doi.org/10.1007/s00125-007-0654-8

Qin X, Wang X, Xu K, Zhang Y, Ren X, Qi B, Liang Q, Yang X, Li L, Li S (2022) Digestion of plant dietary miRNAs starts in the mouth under the protection of coingested food components and plant-derived exosome-like nanoparticles. J Agric Food Chem 70(14):4316–4327. https://doi.org/10.1021/acs.jafc.1c07730

Rabuma T, Gupta OP, Chhokar V (2022) Recent advances and potential applications of cross-kingdom movement of miRNAs in modulating plant's disease response. RNA Biol 19(1): 519–532. https://doi.org/10.1080/15476286.2022.2062172

Raman J, Lakshmanan H, Hyun-Jae S, Jang K-y (2022) The nutritional and pharmacological potential of medicinal mushroom "Ganoderma lucidum (Lingzhi or Reishi)". In: Biology, cultivation and applications of mushrooms. Springer, Singapore, pp 161–183

Ravi A, Troncoso-Rey P, Ahn-Jarvis J, Corbin KR, Harris S, Harris H, Aydin A, Kay GL, Le Viet T, Gilroy R, Pallen MJ, Page AJ, O'Grady J, Warren FJ (2021) Linking carbohydrate structure with function in the human gut microbiome using hybrid metagenome assemblies. bioRxiv:2021.2005.2011.441322. https://doi.org/10.1101/2021.05.11.441322

Redfern RL, McDermott AM (2010) Toll-like receptors in ocular surface disease. Exp Eye Res 90(6):679–687. https://doi.org/10.1016/j.exer.2010.03.012

Reichardt N, Duncan SH, Young P, Belenguer A, McWilliam Leitch C, Scott KP, Flint HJ, Louis P (2014) Phylogenetic distribution of three pathways for propionate production within the human gut microbiota. ISME J 8(6):1323–1335. https://doi.org/10.1038/ismej.2014.14

Rinninella E, Raoul P, Cintoni M, Franceschi F, Miggiano GAD, Gasbarrini A, Mele MC (2019) What is the healthy gut microbiota composition? A changing ecosystem across age, environment, diet, and diseases. Microorganisms 7(1):14. https://doi.org/10.3390/microorganisms7010014

Ríos-Covián D, Ruas-Madiedo P, Margolles A, Gueimonde M, de Los Reyes-Gavilán CG, Salazar N (2016) Intestinal short chain fatty acids and their link with diet and human health. Front Microbiol 7:185. https://doi.org/10.3389/fmicb.2016.00185

Rivière A, Selak M, Lantin D, Leroy F, De Vuyst L (2016) Bifidobacteria and butyrate-producing colon bacteria: importance and strategies for their stimulation in the human gut. Front Microbiol 7:979. https://doi.org/10.3389/fmicb.2016.00979

Romero-Córdoba SL, Salido-Guadarrama I, Meneses ME, Cosentino G, Iorio MV, Tagliabue E, Torres N, Sánchez-Tapia M, Bonilla M, Castillo I, Petlacalco B, Tovar AR, Martínez-Carrera D (2021) Mexican Ganoderma lucidum extracts decrease lipogenesis modulating transcriptional metabolic networks and gut microbiota in C57BL/6 mice fed with a high-cholesterol diet. Nutrients 13(1):38

Rowland I, Gibson G, Heinken A et al (2018) Gut microbiota functions: metabolism of nutrients and other food components. Eur J Nutr 57:1–24. https://doi.org/10.1007/s00394-017-1445-8

Saiyed AN, Vasavada AR, Johar SRK (2022) Recent trends in miRNA therapeutics and the application of plant miRNA for prevention and treatment of human diseases. Future J Pharm Sci 8(1):24. https://doi.org/10.1186/s43094-022-00413-9

Salas-Salvadó J, Bulló M, García-Lorda P, Figueredo R, Del Castillo D, Bonada A, Balanzà R (2006) Subcutaneous adipose tissue cytokine production is not responsible for the restoration of systemic inflammation markers during weight loss. Int J Obes (Lond) 30(12):1714–1720. https://doi.org/10.1038/sj.ijo.0803348

Salguero MV, Al-Obaide MAI, Singh R, Siepmann T, Vasylyeva TL (2019) Dysbiosis of Gram-negative gut microbiota and the associated serum lipopolysaccharide exacerbates inflammation in type 2 diabetic patients with chronic kidney disease. Exp Ther Med 18(5):3461–3469. https://doi.org/10.3892/etm.2019.7943

Salinas E, Reyes-Pavón D, Cortes-Perez NG, Torres-Maravilla E, Bitzer-Quintero OK, Langella P, Bermúdez-Humarán LG (2021) Bioactive compounds in food as a current therapeutic approach to maintain a healthy intestinal epithelium. Microorganisms 9(8):1634. https://doi.org/10.3390/microorganisms9081634

Sang T, Guo C, Guo D, Wu J, Wang Y, Wang Y, Chen J, Chen C, Wu K, Na K, Li K, Fang L, Guo C, Wang X (2021) Suppression of obesity and inflammation by polysaccharide from sporoderm-broken spore of Ganoderma lucidum via gut microbiota regulation. Carbohydr Polym 256:117594. https://doi.org/10.1016/j.carbpol.2020.117594

Sanodiya BS, Thakur GS, Baghel RK, Prasad GB, Bisen PS (2009) Ganoderma lucidum: a potent pharmacological macrofungus. Curr Pharm Biotechnol 10(8):717–742. https://doi.org/10.2174/138920109789978757
Schug ZT, Vande Voorde J, Gottlieb E (2016) The metabolic fate of acetate in cancer. Nat Rev Cancer 16(11):708–717. https://doi.org/10.1038/nrc.2016.87
Schwalm ND 3rd, Groisman EA (2017) Navigating the gut buffet: control of polysaccharide utilization in Bacteroides spp. Trends Microbiol 25(12):1005–1015. https://doi.org/10.1016/j.tim.2017.06.009
Shao J, Wang L, Liu Y, Qi Q, Wang B, Lu S, Liu C (2020) Identification of milRNAs and their target genes in Ganoderma lucidum by high-throughput sequencing and degradome analysis. Fungal Genet Biol 136:103313. https://doi.org/10.1016/j.fgb.2019.103313
Sharma S, Tripathi P (2019) Gut microbiome and type 2 diabetes: where we are and where to go? J Nutr Biochem 63:101–108. https://doi.org/10.1016/j.jnutbio.2018.10.003
Shi L (2016) Bioactivities, isolation and purification methods of polysaccharides from natural products: a review. Int J Biol Macromol 92:37–48. https://doi.org/10.1016/j.ijbiomac.2016.06.100
Sircana A, Framarin L, Leone N, Berrutti M, Castellino F, Parente R, De Michieli F, Paschetta E, Musso G (2018) Altered gut microbiota in type 2 diabetes: just a coincidence? Curr Diab Rep 18(10):98. https://doi.org/10.1007/s11892-018-1057-6
Ślusarczyk J, Adamska E, Czerwik-Marcinkowska J (2021) Fungi and algae as sources of medicinal and other biologically active compounds: a review. Nutrients 13(9):3178
Smith NW, Shorten PR, Altermann EH, Roy NC, McNabb WC (2019) Hydrogen cross-feeders of the human gastrointestinal tract. Gut Microbes 10(3):270–288. https://doi.org/10.1080/19490976.2018.1546522
Sohretoglu D, Huang S (2018) Ganoderma lucidum polysaccharides as an anti-cancer agent. Anticancer Agents Med Chem 18(5):667–674. https://doi.org/10.2174/1871520617666171113121246
Spinler JK, Oezguen N, Runge JK, Luna RA, Karri V, Yang J, Hirschi KD (2020) Dietary impact of a plant-derived microRNA on the gut microbiome. ExRNA 2(1):11. https://doi.org/10.1186/s41544-020-00053-2
Takahashi K, Nishida A, Fujimoto T, Fujii M, Shioya M, Imaeda H, Inatomi O, Bamba S, Sugimoto M, Andoh A (2016) Reduced abundance of butyrate-producing bacteria species in the fecal microbial community in Crohn's disease. Digestion 93(1):59–65. https://doi.org/10.1159/000441768
Tan Y, Zeng N-K, Xu B (2022) Chemical profiles and health-promoting effects of porcini mushroom (Boletus edulis): a narrative review. Food Chem 390:133199. https://doi.org/10.1016/j.foodchem.2022.133199
Thabthimsuk S, Panya N, Owatworakit A, Choksawangkarn W (2018) Acetylcholinesterase inhibition and antioxidant activities of polysaccharide-peptide complexes from edible mushrooms. In: The 6th international conference on biochemistry and molecular biology
Thursby E, Juge N (2017) Introduction to the human gut microbiota. Biochem J 474(11):1823–1836. https://doi.org/10.1042/bcj20160510
Tingirikari JMR (2018) Microbiota-accessible pectic poly- and oligosaccharides in gut health. Food Funct 9(10):5059–5073. https://doi.org/10.1039/c8fo01296b
Tolstoguzov V (2004) Why are polysaccharides necessary? Food Hydrocoll 18:873–877. https://doi.org/10.1016/j.foodhyd.2003.11.011
Turroni F, Özcan E, Milani C et al (2015) Glycan cross-feeding activities between bifidobacteria under in vitro conditions. Front Microbiol. https://doi.org/10.3389/fmicb.2015.01030
Valverde ME, Hernández-Pérez T, Paredes-López O (2015) Edible mushrooms: improving human health and promoting quality life. Int J Microbiol 2015:376387. https://doi.org/10.1155/2015/376387

Vamanu E, Dinu LD, Pelinescu DR, Gatea F (2021) Therapeutic properties of edible mushrooms and herbal teas in gut microbiota modulation. Microorganisms 9(6):1262. https://doi.org/10.3390/microorganisms9061262

Vamanu E, Pelinescu D (2017) Effects of mushroom consumption on the microbiota of different target groups – impact of polyphenolic composition and mitigation on the microbiome fingerprint. LWT - Food Sci Technol 85:262–268

Vangay P, Ward T, Gerber JS, Knights D (2015) Antibiotics, pediatric dysbiosis, and disease. Cell Host Microbe 17(5):553–564. https://doi.org/10.1016/j.chom.2015.04.006

Vasapolli R, Schütte K, Schulz C, Vital M, Schomburg D, Pieper DH, Vilchez-Vargas R, Malfertheiner P (2019) Analysis of transcriptionally active bacteria throughout the gastrointestinal tract of healthy individuals. Gastroenterology 157(4):1081–1092.e1083. https://doi.org/10.1053/j.gastro.2019.05.068

Vich Vila A, Collij V, Sanna S, Sinha T, Imhann F, Bourgonje AR, Mujagic Z, Jonkers DMAE, Masclee AAM, Fu J, Kurilshikov A, Wijmenga C, Zhernakova A, Weersma RK (2020) Impact of commonly used drugs on the composition and metabolic function of the gut microbiota. Nat Commun 11(1):362. https://doi.org/10.1038/s41467-019-14177-z

Vital M, Howe AC, Tiedje JM, Moran MA (2014) Revealing the bacterial butyrate synthesis pathways by analyzing (meta)genomic data. mBio 5(2):e00889-00814. https://doi.org/10.1128/mBio.00889-14

Wang G, Huang S, Wang Y, Cai S, Yu H, Liu H, Zeng X, Zhang G, Qiao S (2019) Bridging intestinal immunity and gut microbiota by metabolites. Cell Mol Life Sci 76(20):3917–3937. https://doi.org/10.1007/s00018-019-03190-6

Wang Q, Wang F, Xu Z, Ding Z (2017) Bioactive mushroom polysaccharides: a review on monosaccharide composition, biosynthesis and regulation. Molecules 22(6). https://doi.org/10.3390/molecules22060955

Wang T, He C (2018) Pro-inflammatory cytokines: the link between obesity and osteoarthritis. Cytokine Growth Factor Rev 44:38–50. https://doi.org/10.1016/j.cytogfr.2018.10.002

Wilkes RA, Aristilde L (2017) Degradation and metabolism of synthetic plastics and associated products by Pseudomonas sp.: capabilities and challenges. J Appl Microbiol 123(3):582–593. https://doi.org/10.1111/jam.13472

Wu Q, Zhang H, Wang P, Chen M (2017) Evaluation of the efficacy and safety of Ganoderma lucidum mycelium-fermented liquid on gut microbiota and its impact on cardiovascular risk factors in human. RSC Adv 7:45093–45100. https://doi.org/10.1039/C7RA08087E

Wu S, Wu Q, Wang J, Li Y, Chen B, Zhu Z, Huang R, Chen M, Huang A, Xie Y, Jiao C, Ding Y (2022) Novel selenium peptides obtained from selenium-enriched cordyceps militaris alleviate neuroinflammation and gut microbiota dysbacteriosis in LPS-injured mice. J Agric Food Chem 70(10):3194–3206. https://doi.org/10.1021/acs.jafc.1c08393

Xiao Y, Chen L, Fan Y, Yan P, Li S, Zhou X (2019) The effect of boletus polysaccharides on diabetic hepatopathy in rats. Chem Biol Interact 308:61–69. https://doi.org/10.1016/j.cbi.2019.05.013

Xu D, Zhou Q, Yan B, Ma A (2021) Identification and physiological function of one microRNA (*Po-MilR-1*) in oyster mushroom *Pleurotus ostreatus*. Mycoscience 62(3):182–188. https://doi.org/10.47371/mycosci.2021.01.004

Xue Z, Ma Q, Chen Y, Lu Y, Wang Y, Jia Y, Zhang M, Chen H (2020) Structure characterization of soluble dietary fiber fractions from mushroom Lentinula edodes (Berk.) Pegler and the effects on fermentation and human gut microbiota in vitro. Food Res Int 129:108870. https://doi.org/10.1016/j.foodres.2019.108870

Yan J-K, Wang W-Q, Wu J-Y (2014) Recent advances in Cordyceps sinensis polysaccharides: mycelial fermentation, isolation, structure, and bioactivities: a review. J Funct Foods 6:33–47. https://doi.org/10.1016/j.jff.2013.11.024

Yang C, He N, Ling X, Ye M, Zhang C, Shao W, Yao C, Wang Z, Li Q (2008) The isolation and characterization of polysaccharides from longan pulp. Sep Purif Technol 63(1):226–230. https://doi.org/10.1016/j.seppur.2008.05.004

Yao C, Wang Z, Jiang H et al (2021) Ganoderma lucidum promotes sleep through a gut microbiota-dependent and serotonin-involved pathway in mice. Sci Rep. https://doi.org/10.1038/s41598-021-92913-6

Yin C, Noratto GD, Fan X, Chen Z, Yao F, Shi D, Gao H (2020) The impact of mushroom polysaccharides on gut microbiota and its beneficial effects to host: a review. Carbohydr Polym 250:116942. https://doi.org/10.1016/j.carbpol.2020.116942

Yu J, Xiang H, Xie Q (2021) The difference of regulatory effect of two Inonotus obliquus extracts on high-fat diet mice in relation to the fatty acid elongation function of gut microbiota. Food Sci Nutr 9(1):449–458. https://doi.org/10.1002/fsn3.2012

Yu M, Hui N, Hayat K, Yang X, Chu S, Zhang D, Wang J, Chen X, Ding J, Chi Y, Zhou P (2020) The regulation effect of crude polysaccharides from cordyceps militaris on obesity control and gut microbiota community via high-fat diet-fed mice model. https://doi.org/10.20944/preprints202012.0052.v1

Yu SH, Chen SY, Li WS, Dubey NK, Chen WH, Chuu JJ, Leu SJ, Deng WP (2015) Hypoglycemic activity through a novel combination of fruiting body and mycelia of Cordyceps militaris in high-fat diet-induced type 2 diabetes mellitus mice. J Diabetes Res 2015:723190. https://doi.org/10.1155/2015/723190

Zafar H, Saier MH Jr (2021) Gut Bacteroides species in health and disease. Gut Microbes 13(1):1–20. https://doi.org/10.1080/19490976.2020.1848158

Zeidan AA, Poulsen VK, Janzen T, Buldo P, Derkx PMF, Øregaard G, Neves AR (2017) Polysaccharide production by lactic acid bacteria: from genes to industrial applications. FEMS Microbiol Rev 41(Supp_1):S168–S200. https://doi.org/10.1093/femsre/fux017

Zhang B, Tian L, Xie J, Chen G, Wang F (2020a) Targeting miRNAs by natural products: a new way for cancer therapy. Biomed Pharmacother 130:110546. https://doi.org/10.1016/j.biopha.2020.110546

Zhang G, Liang Y (2013) Improvement of fruiting body production in Cordyceps militaris by molecular assessment. Int Arch Med Microbiol 195(8):579–585. https://doi.org/10.1007/s00203-013-0904-8

Zhang H, Jiang F, Zhang J, Wang W, Li L, Yan J (2022) Modulatory effects of polysaccharides from plants, marine algae and edible mushrooms on gut microbiota and related health benefits: a review. Int J Biol Macromol 204:169–192. https://doi.org/10.1016/j.ijbiomac.2022.01.166

Zhang J, Wen C, Gu J, Ji C, Duan Y, Zhang H (2019) Effects of subcritical water extraction microenvironment on the structure and biological activities of polysaccharides from Lentinus edodes. Int J Biol Macromol 123:1002–1011. https://doi.org/10.1016/j.ijbiomac.2018.11.194

Zhang T, Zhao W, Xie B, Liu H (2020b) Effects of Auricularia auricula and its polysaccharide on diet-induced hyperlipidemia rats by modulating gut microbiota. J Funct Foods 72:104038. https://doi.org/10.1016/j.jff.2020.104038

Zhao S, Gao Q, Rong C, Wang S, Zhao Z, Liu Y, Xu J (2020) Immunomodulatory effects of edible and medicinal mushrooms and their bioactive immunoregulatory products. J Fungi (Basel) 6(4). https://doi.org/10.3390/jof6040269

Zhao Y, Cong L, Lukiw WJ (2018) Plant and animal microRNAs (miRNAs) and their potential for inter-kingdom communication. Cell Mol Neurobiol 38(1):133–140. https://doi.org/10.1007/s10571-017-0547-4

Zhong L, Ma N, Zheng H, Ma G, Zhao L, Hu Q (2019) Tuber indicum polysaccharide relieves fatigue by regulating gut microbiota in mice. J Funct Foods 63:103580. https://doi.org/10.1016/j.jff.2019.103580

Zhou X, Gong Z, Su Y, Lin J, Tang K (2009) Cordyceps fungi: natural products, pharmacological functions and developmental products. J Pharm Pharmacol 61(3):279–291. https://doi.org/10.1211/jpp/61.03.0002

Zhu N, Yang J, Ji L, Liu J, Yang Y, Yuan H (2016) Metagenomic and metaproteomic analyses of a corn stover-adapted microbial consortium EMSD5 reveal its taxonomic and enzymatic basis for degrading lignocellulose. Biotechnol Biofuels 9(1):243. https://doi.org/10.1186/s13068-016-0658-z

Zhu S, Jiang Y, Xu K, Cui M, Ye W, Zhao G, Jin L, Chen X (2020) The progress of gut microbiome research related to brain disorders. J Neuroinflammation 17(1):25. https://doi.org/10.1186/s12974-020-1705-z

Zhu SJ, Pan J, Zhao B, Liang J, Ze-Yu W, Yang JJ (2013) Comparisons on enhancing the immunity of fresh and dry Cordyceps militaris in vivo and in vitro. J Ethnopharmacol 149(3):713–719. https://doi.org/10.1016/j.jep.2013.07.037

Zhu WJ, Liu Y, Cao YN, Peng LX, Yan ZY, Zhao G (2021) Insights into health-promoting effects of plant microRNAs: a review. J Agric Food Chem 69(48):14372–14386. https://doi.org/10.1021/acs.jafc.1c04737

Mushrooms – From Traditional Remedies to the Modern Therapeutics

Mirjana Stajić, Milica Galić, and Jasmina Ćilerdžić

Contents

Abstract Although the living standard and life expectancy have been increasing significantly, we face numerous arising challenges in modern medicine, such as the presence of increasing exogenous triggers of oxidative stress that lead to the emergence of multiple diseases and disorders, the appearance of an increasing number of resistant microorganisms, an immense number of patients suffering from cardiovascular diseases, cancers, diabetes, neurodegenerative disorders as well as autoimmune and rare diseases. Therefore, we need the help of natural sources of active compounds, among which mushrooms are important. They have been an integral part of traditional medicine for centuries, and modern research has confirmed their bioactivities and given them a scientific basis. Numerous species, primarily from the genera *Ganoderma*, *Lentinus*, *Pleurotus*, *Innonotus*, *Trametes*, *Cordyceps*, *Agaricus*, etc., have shown exceptional immunomodulatory, antioxidative, antihypercholesterolemic, antihypertensive, antitumor,

M. Stajić (✉) · M. Galić · J. Ćilerdžić
Faculty of Biology, University of Belgrade, Belgrade, Serbia
e-mail: stajicm@bio.bg.ac.rs; galic.m@bio.bg.ac.rs; simonicj@bio.bg.ac.rs

D. C. Agrawal, M. Dhanasekaran (eds.), *Mushrooms with Therapeutic Potentials*,
https://doi.org/10.1007/978-981-19-9550-7_5

antineurodegenerative, antidiabetic, antimicrobial as well as numerous other potentials. Studies have shown that the mushrooms' crude extracts, as well as various metabolites, especially polysaccharides, phenolic compounds, terpenoids, and proteins, possess mentioned activities and thereby could be the basis for the development of new, more efficient drugs. However, numerous problems and challenges need to be overcome before mushrooms from the domain of traditional medicine move into the modern one and become part of conventional therapy.

Keywords Bioactivities · Drug development · Functional food · Medicinal mushrooms · Traditional medicine

List of Abbreviations

ACE	angiotensin-converting enzyme
AChE	acetylcholinesterase
AP-1	transcription factor
Bax	a central cell death regulator
Bcl-2	a protein that regulates cell apoptosis
BHA	butylated hydroxyanisole
BHT	butylated hydroxytoluene
CAT	catalase
DPPH•	2-diphenyl-1-picryl-hydrazyl-hydrate free radical
EC_{50}	half maximal effective concentration
GLUT-2	glucose 2 transporter in the liver
GLUT-4	glucose 2 transporter in the muscles
GPx	glutathione peroxidase
HDL	high-density lipoproteins
HMG-CoA reductase	hemoglobin-coenzyme A reductase
IFN-γ	interferon gamma
IL	interleukin
IARC	International Agency for Research on Cancer
LDL	low-density lipoproteins
L-DOPA	amino acid known as l-3,4-dihydroxyphenylalanine
LEP	polysaccharide from *Lentinus edodes*
mRNA	messenger RNA
NF-κB	nuclear transcription factor
NO	nitric oxide
p53	a nuclear transcription factor
PG	propyl gallate
PPAR-γ	peroxisome proliferator-active receptor-γ
PSK	Krestin
PSP	polysaccharide-peptide complex
SOD	superoxide dismutase

TBHQ	tert-butylated hydroxyquinone
Th2	T helper 2 cell
TNF	tumor necrosis factor

1 Introduction

The finding of *Piptoporus betulinus* and *Fomes fomentarius* pieces in the bag of a Neolithic man whose corpse was found in the Alpine Glacier, at an altitude of 3200 m, in 1991 is an excellent indicator that mushrooms have been used as a medicine since prehistoric times. Throughout history, many mushrooms have been used to treat various diseases, but in different regions, preference was given to different species. *Psilocybe* spp. and *Amanita muscaria* were highly prized species by North American Indians who believed that each serious illness was a malfunction of the spirit. *Geastrum* spp. and *Sarcoscypha coccinea* were used by Maya and Cherokee Indians to stop the bleeding and *Fomes officinalis* for the treatment of fever, diarrhea, dysentery, and hepatitis. In sub-Saharan Africa, for about 9000 years, *Calvatia cyathiformis* was used for wound healing, *Phallus aurantiacus* for leprosy, and *Termitomyces microcarpus* for gonorrhea treatment, while ground *Podoxis pistillaris* fruiting bodies have been used for the treatment of patients suffering from cancer as early as the eighteenth century. In India, *Bovista pusilla*, *Geastrum fornicatum,* and *C. cyathiformis* were used to stop bleeding and wound healing, *Cyathus limbatus* and *C. stercoreus* for treating ear diseases, *Phallus rubicundus* for various stomach problems, *T. microcarpus* in paralysis, and *Xylaria polymorpha* for increasing lactation. However, the greatest admirers of mushrooms for thousands of years have been the people of Russia and Far East countries, who used more than 1100 species for the treatment of various diseases. Especially prized species in China, Japan, and Mongolia were *Ganoderma lucidum*, *Lentinus edodes*, *Auricularia auricula*, *Tricholoma matsutake, T. mongolicum*, *Tremella fuciformis*, *Grifola frondosa*, *Cordyceps sinensis,* and even some poisonous species because of the belief that similar treats "similar" and that "there is no such mushroom in nature that could not be used as a medicinal agent." Diet therapy was widely accepted in the Chinese palace and among the population. Mushrooms were used as food by healthy persons, as food supplements for people with mild disorders, and as medicine for sick ones. In Russia, mushrooms have been used not only for treating humans but also for animals. The first clinical studies of the effect of *P. betulinus* against cancers of parotid glands and lips were done as early as the nineteenth century. Figure 1 shows the most commonly used mushrooms in traditional medicine in different parts of the world.

The second half of the twentieth century and the beginning of the 21st one represent the period of the reign of four global evils: (i) intensive industrial and economic development, (ii) poverty, (iii) illnesses, and (iv) wars. The "first evil" attacks nature and humans by increasing air, water, and soil pollution levels,

Fig. 1 The most commonly used mushrooms in traditional medicine in different parts of the world

biodiversity loss, natural resources reduction, the gap between the rich and the poor, and massive world migration. Data from United Nations annual reports can demonstrate the severity of the "second evil" effect. Namely, in 2019 one billion people lived in extreme poverty, and the number of hungry and death rates were extremely high. The greater problem is that the United Nations' idea to eradicate famine in 2015 was unrealizable, and globally the number of hungry people has increased since 2015. The realization of one of the Millennium Development Goals, zero hunger in 2030, presents a huge challenge.

The "third evil" also has a terrifying effect; today, humanity is facing many pandemics. One of the biggest is HIV/AIDS, especially in Africa, where only in Swaziland, 41% of pregnant women are HIV+, and their lifespan is less than 43 years. The latest one is COVID-19, a pandemic of a previously unknown virus, which rapidly raced and became one of the top global killers that took approximately 6,066,000 lives in only 2 years. Obesity is one more pandemic in modern society which becomes a serious problem. Data from 2011 have shown that more than two-thirds of adults and one-third of children in the United States are obese. However, according to a report by the World Health Organization from 2019, ischemic heart disease was the first cause of death; stroke and chronic obstructive pulmonary disease were second and third, while lower respiratory infections and neonatal conditions ranged in the fourth and fifth place, respectively. The sixth place was taken by lung cancer and the seventh by neurodegenerative disorders (Alzheimer's and Parkinson's diseases and various forms of dementia). In the last 20 years, kidney diseases and diabetes have become the top 10 causes of death. However, the leading causes of mortality vary from country to country and depend primarily on the living standard. To this list of diseases, we have to add the disease of crazy cows, foot-and-mouth disease, and other animal diseases, reducing the amount of available food and putting pressure on the world's population to change its diet drastically. Thus, although longevity is a trend in modern society, poverty and the number of diseases and disorders significantly reducing the quality of life are constantly increasing.

Today, on the world market, there are a considerable number of drugs against the mentioned and numerous other diseases, and a large number of new drugs are also tested every year. However, numerous commercial drugs have many side effects and disadvantages besides the potential to repress mentioned diseases and disorders. Thus, butylated hydroxyanisole (BHA) and butylated hydroxytoluene (BHT), the most commonly used synthetic antioxidants, have carcinogenic and hepatotoxic effects. At the same time, conventional therapies against Alzheimer's and Parkinson's diseases are insufficiently efficient, i.e., they cause progression delay for a short time and gastrointestinal problems. Chemo- and radiotherapy give satisfactory results only in the treatment of early cancer development stages, are not effective for some cancer types, and can cause numerous side effects (Chen et al. 2006). Likewise, cytostatics currently available on the world market are not tumor-specific and cause numerous harmful effects in patients. Based on everything mentioned, the twenty-first century should be a century of world education on the importance of disease prevention and supplementing commercial medicines with natural ones. Diseases prevention has particular importance not only because of its positive financial and social impact but also because of maintaining and improving the life quality.

The edible and medicinal mushrooms present a promising and relatively unused source of substances with a high potential for the human diet and disease prevention and treatment. Nowadays, 14,000 mushroom species are described, of which 50% to 70% possess some edibility, and about 700 are medicinal (Wasser 2010; Ayeka 2018). Mushrooms' therapeutic potential is based on the synthesis of numerous biologically active compounds such as proteins, polysaccharides, glycoproteins, lipopolysaccharides, lectins, organic acids, sterols, alkaloids, etc. These metabolites positively affect the immune and cardiovascular systems and possess antioxidative, antitumor, antihypercholesterolemic, antihyperglycemic, antineurodegenerative, antimicrobial, antiparasitic, detoxification, and hepato-protective activities (Table 1). Because mushrooms have high nutritional value and produce numerous biologically active compounds, they are considered functional foods, i.e., food that, besides good nutritional effects, positively affects one or more functions in the body and thereby improves health or reduces the risk of illness. Therefore, dietary mushroom-based supplements help prevent and alleviate diseases. However, mushrooms and their derivates cannot replace commercial drugs but combined with conventional medical treatments, and they can allow patients to feel better. Despite very intensive studies and a considerable amount of results, the overall mushroom medical potential has not yet been fully realized because the biological potentials of many species have not yet been studied, and a large number of species have not yet been discovered and identified. It is estimated that the number of mushroom species on the planet is even ten times higher. Thus, it can be concluded that the mushrooms' medicinal potential is immense.

2 Immunomodulatory Activity

Nowadays, there is a particularly great interest in natural immunomodulators as alternatives to commercial medicines. Mushrooms represent rich natural sources of these compounds that show stimulatory activity on the innate and adaptive immune systems (El Enshasy and Hatti-Kaul 2013). They cause the proliferation and activation of natural killer cells, neutrophils, and macrophages and stimulate the expression and secretion of cytokines that activate adaptive immunity (Table 1). Immunomodulatory lectins, terpenes and terpenoids, proteins, and polysaccharides are the main four groups of mushroom compounds. Some of them are available on the market (lentinan, schizophyllan, grifolan, Krestin, and PSP). They are used simultaneously with radio- and chemotherapy and as auxiliaries to antibiotics and vaccines to minimize harmful effects (Stajić 2015). Besides the purified biologically active compounds, crude extracts of mushroom fruiting bodies and mycelium are also characterized by high immunomodulatory activity. Božić Nedeljković et al. (2022) reported that ethanol extract of *G. lucidum* fruiting bodies cultivated on wheat

Table 1 Bioactive compounds of some mushrooms and their activities

Mushroom	Active compound	Structure	Activity
Agaricus blazei	Agaritin	Hydrazine derivative	Arrest proliferation and induction of apoptosis of leukemia cells
	Blazein	Steroid	Initiation of morphological changes in stomach and lung cancer cells and their apoptosis
Agaricus bisporus	ABL	Lectin	Arrest proliferation and induction of apoptosis of colon cancer cells
Agaricus polytricha	APP	Protein	Immunostimulatory activity
Androdia camphorata	FIP-aca	Immunomodulatory protein	Induction of cytokine and chemokine synthesis
Calvatia utriformis	Calcetine	Ribosome inactivating protein	Reduction of breast cancer cell proliferation
Clitocybe maxima	Laccase	Enzyme	Antimitotic activity against liver and breast cancer cells
Clitocybe nebularis	CNL	Immunomodulatory protein	Antiproliferative activity on leukemia cells
Cordyceps sinensis	Cordycepin	Adenosine derivative	– Inhibition of NF-κB, antiproliferative and proapoptotic activity on leukemia and colon, bladder, and liver cancer cells; – Antidiabetic activity
	Cordlan	Polysaccharide	Induction of dendritic cell maturation
	Cs-HK1	Polysaccharide-protein complex (65%–70%:25%)	Radical neutralization, reduction of Fe^{3+} into Fe^{2+}, and its chelation
Cyathus striatus	Striatins and ciatins	Diterpenoids	Antibacterial, fungicidal, and cytotoxic activity
Flammulina velutipes	Proflamin	Glycoprotein	Cytotoxic activity on melanoma cells
	FIP-fve	Immunomodulatory protein	Stimulation of mitogenesis in human peripheral lymphocytes, stimulation of IL-3 and INF-γ transcription; inhibition of hepatoma growth; anti-HIV activity
	Flammulin	Ribosome inactivating protein	Inhibition of leukemia cell proliferation
Ganoderma spp.	Ganoderic acids F, B, D, H, and Y	Terpens and terpenopids	Antihypertensive activity
	Ganoderic acids T and Me		Inhibition of colon tumor invasion and metastasis

(continued)

Table 1 (continued)

Mushroom	Active compound	Structure	Activity
	Ganoderiol		Anti-HIV activity
	Ganomycin		Antibacterial activity
	Ganoderans A, B, and C		Antidiabetic activity
Ganoderma annulare	Applanoxidic acid A	Terpenoid	Antifungal activity
Ganoderma applanatum	Applanoxidic acid A-H	Terpenoids	Cytotoxic activity on skin tumor cells
Ganoderma australe	Australic acid	Terpenoid	Antibacterial and antifungal activity
Ganoderma atrum	PSG	Polysaccharide-protein complex	Neutralization of DPPH radicals and superoxide anions
Ganoderma lucidum	Ganoderan	B-glucan +4% protein	Stimulation of TNF-α, IL-1, and IFN-γ production
	GLP1	Polysaccharide	Neutralization of free radicals and Fe^{2+} chelation
	GLP2	Polysaccharide	
	LZ-8	Immunomodulatory protein	Stimulation of IL-2, IL-3, IL-4, IFN-γ and TNF-α transcription
	Ganodermin	Protein	Antifungal activity
Ganoderma microsporum	FIP-gmi	Immunomodulatory protein	TNF-α regulation
Ganoderma sinensis	FIP-gsi	Immunomodulatory protein	Stimulation of IL-2, IL-3, IL-4, INF-γ and TNF-α production
Ganoderma tsuge	FIP-gts	Immunomodulatory protein	Induction of cytokine and IFN-γ secretion; proliferation of human peripheral mononuclear cells; antitumor activity against lung adenocarcinoma cells
Grifola frondosa	Grifolan	Polysaccharide	Activation of macrophages and increase of IL-6, IL-1, and TNF-α levels
	GFPPS1b	Polysaccharide-peptide complex	Antiproliferative and pro-apoptotic activity on gastric cancer cells
Hypsizigus marmoreus	Marmorin	Ribosome inactivating protein	Inhibition of hepatoma and breast cancer cell proliferation and HIV-1 reverse transcriptase activity
Lentinus edodes	Lentinan	B-(1 → 3)-glucan with β-(1 → 6) branches	– Induction of non-specific cytotoxicity of macrophages and stimulation of cytokine production; – Increase in SOD and GPS activity;

(continued)

Table 1 (continued)

Mushroom	Active compound	Structure	Activity
			– Reduction of tumor size and inhibition of metastasis; – Vasodilatation
	Lentin	Protein	– Antifungal activity; – Inhibition of HIV-1 reversible transcriptase activity and leukemia cells proliferation
	Eritadenin	Uridine derivative	Antihypercholesterolemic activity
Phellinus linteus	Hispolon	Phenolic compound	Pro-apoptotic effect on breast, bladder, stomach, and lung cancer cells
Pleurotus spp.	Lovastatin	Naphthalene, Polycyclic aromatic hydrocarbon	Antihypercholesterolemic activity
Pleurotus eryngii	Eyingin	Peptide	Antifungal activity
	Ergothioneine	Steroid	Antioxidative activity
	Ribonuclease	Enzyme	Antiviral, immunomodulatory, and antineoplastic activity
Pleurotus nebrodensis	Nebrodeolysin	Triterpenoid	Anti-HIV activity
Pleurotus ostreatus	Pleuran	Polysaccharide	Stimulation of humoral and cellular immunity
	POPS-1	Polysaccharide	Antitumor activity against cervical cancer cells
	Pleurostrin	Peptide	Antifungal activity
Poria cocos	FIP-PCP	Immunomodulatory protein	Increase of IL-1β, IL-6, IL-18, TNF-α and NO production
Russula cyanoxantha	Ergon	Steroid	Antiproliferative and pro-apoptotic activity to liver cancer cells
Shizophyllum commune	Schizophyllan	B-(1 → 3)-glucan with β-(1 → 6) branches	Immunostimulatory and antitumor activity
Suillus placidus	Illudin	Sesquiterpene	Antitumor activity
Trametes versicolor	Krestin	B-(1 → 3)-glucan with β-(1 → 6) branches +25% - 38% protein	Activation of T cells. Induction of IFN-γ, IL-2, TNF-α, IL-1, IL-6, IL-8 production; neutralization of free radicals; arrest of cancer cells cycle and induction of their apoptosis; antiviral activity.
	PSP	B-(1 → 3)-glucan with β-(1 → 6) branches +25% - 38% peptid	
	FIP-tve	Immunomodulatory protein	Proliferation of human peripheral lymphocytes;

(continued)

Table 1 (continued)

Mushroom	Active compound	Structure	Activity
			increasing the production of TNF-α and NO
Tremella fuciformis	Glucuronoxylomannan	1,3 D mannose with xylose and glucuronic acid in the side chains	Improving the immune system and stopping the development of cancer; lowering blood glucose levels
Tremella mesenterica			
Tricholoma giganteum	Trichogin	Peptide	Antifungal activity; inhibition of HIV-1 reverse transcriptase activity
Tricholoma mongolicum	TML-1 and TML-2	Lectins	Immunostimulatory and antitumor activity
Volvariella volvacea	FIP-vvo	Immunomodulatory protein	Stimulation of IL-2, IL-2, IFN-γ and TNF-α production

straw, an alternative substrate, stimulated metabolic and phagocytic activity, adhesion capability, and NO produce ability of peritoneal macrophages initiated the production of certain cytokines and activated monocyte-derived dendritic cells. These authors emphasized that these activities of the extract could be the basis of dendritic cell-based anti-tumor vaccines. However, despite numerous in vitro and in vivo studies of the immunomodulatory potential of mushrooms' extracts, clinical trials have been done only for lentinan, schizophyllan, grifolan, and polysaccharides from *L. edodes*, *Schizophillum commune,* and *G. frondosa*, respectively. Clinical trials showed that immunochemotherapy (cytostatics + lentinan in a daily dose of 0.5–1.0 mg) extended the lifespan of patients with stomach cancer by about 70%. In contrast, in patients with colon cancer, the value was 112% compared to those who received only chemotherapy (Hazama et al. 1995). The addition of schizophyllan to conventional postoperative therapy for patients suffering from cancer also resulted in a longer and more quality of life (Fujimoto et al. 1991). These authors showed that 72.2% of selected patients who began to receive intramuscularly schizophyllan in a dose of 40 mg per week with the cytostatics survived 5 years compared with 61.9% in the control group treated only with cytostatics. Similar results were reported by Mitomi et al. (1992) and Yang (1993), who studied the effect of commercial cytostatic therapy enrichment with Krestin (PSK) or polysaccharide-peptide complex (PSP) on patients who suffered from rectal and esophagus cancer, respectively. *Trametes versicolor*, the producer of these active compounds, was also very efficient in treating breast cancer patients since it caused the production of B cells and reduced the number of IL-2 receptors located on the surface of the malignant cells. Kodama et al. (2002) showed very high efficiency of the mixture of MD-fraction. They milled fruiting bodies of *G. frondosa* in the treatment of patients in later stages (II-IV stage) of liver, lung, and breast cancers (58.3%, 62.5%, and 68.8%, respectively) based on the stimulation of natural killer cell activity.

The antimicrobial efficiency of mushrooms' metabolites and extracts is also based on immune system stimulation. Thus, lentinan is an effective immunostimulator in HIV+ patients. Gordon et al. (1995) observed that combined therapy of these patients with didanosine and lentinan significantly increased the number of auxiliary T cells (CD4+ cells), macrophages, and dendritic cells with specific CD4+ glycoprotein on the surface. Stimulation of humoral and cellular immunity was also reported in patients suffering from respiratory infections after treatment with Immunoglucan P4H, the product based on *Pleurotus ostreatus* polysaccharide pleuran (Jesenak et al. 2013).

A significant decrease in cholesterol level (even 69%) in patients with mild hypercholesterolemia treated with α-glucan originated from *Agaricus bisporus* was based on induction of TNF-α production. At the same time, stimulation of all cytokines' synthesis was the mechanism of action of *A. blazei* fruiting bodies extract (Roupas et al. 2012).

Nowadays, numerous immunomodulators from mushrooms are used in the cosmetics industry as the main ingredients of wound healing and anti-aging creams (Taofiq et al. 2016). Likewise, they are increasingly used as alternatives to antibiotics, which were common additives in feed and whose use has been prohibited in Europe since 2006. Lee et al. (2010) reported that extract of *L. edodes* fruiting bodies improved the immune system of chickens by activation of lymphocytes and macrophages and by increasing the levels of mRNA that encode IL-1β, -6, -12, and -18, while Harikrishnan et al. (2011) and Chang et al. (2013) noted the increase of resistance of cultivated scarps and shrimps fed with feed enriched with *Phellinus linteus* extract, which stimulated the activity of lysozymes and phagocytes, and *Hericium erinaceum* one that increased activity of phenoloxidases, superoxide dismutases, and glutathione peroxidases.

3 Antioxidative Activity

Oxidative stress is one of modern society's most common causal agents of diseases and disorders. It occurs when the level of free radicals overcomes the capacity of the body to neutralize them, i.e., when the capacity of the cellular antioxidative defense system is insufficient (Limón-Pacheco and Gonsebatt 2009). Numerous exogenous and some endogenous factors can be responsible for the formation of increased amounts of free radicals, which in high concentrations can attack nucleic and amino acids, proteins, carbohydrates, lipids, and phospholipids changing their function, and consequently leading to the production of organic radicals, peroxidation of cell membrane lipids, onset and development of various disorders, and cellular death (Burton and Jauniaux 2010; Leopoldini et al. 2011). Different antioxidants can be found on the world market, and mostly used are BHA, BHT, tert-butylated hydroxyquinone (TBHQ), and propyl gallate (PG). Although their main role is organism protection, they can have toxic and mutagenic effects, which unfortunately does not prevent their usage as food stabilizers (Ito et al. 1985; Ćilerdžić et al. 2013). The current trend is a replacement of synthetic antioxidants with natural ones. Mushrooms rich in vitamin C, polyphenols, flavonoids, carotenoids, and low molecular weight peptides present highly effective antioxidants (Table 1). Their regular consumption reduces the risk of cardiovascular diseases, cancers, and stroke (Sarmadi and Ismail 2010; Ćilerdžić et al. 2013). Their antioxidative mechanisms are based on: (i) catalytic removal of free radicals by glutathione peroxidase (GPx), superoxide dismutase (SOD) and catalase (CAT), and thiol-specific antioxidants; (ii) increase of the activity of these antioxidative enzymes by trace elements such as selenium, copper, zinc, and magnesium, presented in the mycelium and fruiting bodies, which act as the enzymes cofactors; (iii) binding proteins to pro-oxidant metal ions (iron and copper); (iv) protection against macromolecular damage by stress or heat shock proteins; (v) reduction of free radicals by electron donors (glutathione, vitamins E and C, β-carotene and bilirubin) (Stajić et al. 2013).

However, despite intensive mycological studies, only approximately 5% of the species are well-studied. Among the genus *Agaricus*, *A. silvaticus* was the strongest antioxidant, with EC_{50} values ranging between 2.08 mg/mL and 5.37 mg/mL (Barros et al. 2008). Methanol extracts of *Macrolepiota procera* var. *procera*, *Amanita rubescens* var. *rubescens*, *Boletus edulis*, *B. pseudosulphureus*, *B. erythropus* var. *erythropus*, and *Suillus luteus* had the significantly lower potential of DPPH• scavenging than commercial antioxidants α-tocopherol, BHA and BHT (77%, 85%, and 97%, respectively). In contrast, the same extracts of *Russula delica*, *Boletus badius*, *Polyporus squamosus*, *P. ostreatus*, *Lepista nuda*, and *Verpa conica* and acetone extract of *B. edulis* neutralized from 97.7% to 99.7% of radicals (Elmastas et al. 2007; Keles et al. 2011). Jayakumar et al. (2007, 2011) and Reis et al. (2012a) showed that *P. ostreatus* ethanol extract was also a highly effective inhibitor of lipid peroxidation level, reducer of Fe^{3+} into Fe^{2+}, as well as a good inducer of vitamins C and E, and SOD, CAT, and GPx activities in aged rats. Similar results were obtained for methanol extracts of *P. ostreatus* and *P. eryngii* mycelia and basidiocarps, as well as for *P. cystidiosus* methanol extract, which chelated 52% of Fe^{2+} (1.0 mg/mL), neutralized 42% of DPPH• (5.0 mg/mL), and inhibited lipid peroxidation by 44% (10.0 mg/mL) (Yang et al. 2002a; Oke and Aslim 2011; Reis et al. 2012a). According to several studies, extracts of *L. edodes*, *Laetiporus sulphureus*, *Hericium erinaceus*, *Agrocybe aegerita*, *G. lucidum*, and *G. applanatum* are effective free radical neutralizers and inhibitors of lipid peroxidation (Yang et al. 2002a; Cheung and Cheung 2005; Mau et al. 2005; Turkoglu et al. 2007; Karaman et al. 2010; Mujić et al. 2010; Carneiro et al. 2013; Ćilerdžić et al. 2014; Milovanović et al. 2015a). Water extract of *G. Tsugae* mycelium is an excellent DPPH• scavenger, even better than basidiocarp one and almost twice as effective as fermentation broth (91.2% vs. 79.3% vs. 58.8%), while *G. lucidum* extracts besides this activity were also good inhibitors of lipid peroxidation (Mau et al. 2002; Ćilerdžić et al. 2014). However, numerous studies showed that the ability of radical neutralization in *L. edodes*, *Flammulina velutipes*, *Lenzites betulinus*, and *G. applanatum* was improved by mycelium enrichment with selenium (Turło et al. 2010; Milovanović et al. 2015a, 2015b, 2015c).

Phenols, including flavonoids, vitamins, polysaccharides, peptides, proteins, organic acids, carotenoids, alkaloids, and nucleotides, are the main carriers of mushroom antioxidative activity (Stajić et al. 2013). Oke and Aslim (2011) and

Vaz et al. (2011) noted significant concentrations of protocatechuic and *p*-hydroxybenzoic acids in *Fistulina hepatica*, while *p*-hydroxybenzoic, gallic and caffeic acids were the major phenolic components in *P. eryngii* and *Auricularia auricula-judae* extracts. Carriers of antioxidative activity in *G. lucidum, G. Applanatum, Meripilus giganteus, L. sulphureus, F. velutipes, Coriolus versicolor, P. Ostreatus,* and *Panus tigrinus* were gallic and protocatechuic acids, which also were dominant in *A. Bisporus* together with *p*-hydroxybenzoic and cinnamic acids (Karaman et al. 2010; Reis et al. 2012a). However, Barros et al. (2008) reported that species of the genus *Agaricus* also synthesize flavonoids, ascorbic acid, β-carotene, and lycopene, but in lower concentrations. Eight ganoderic acids, which belong to phenols, have a key role in the antioxidative activity of *Ganoderma atrum*, while in hot water extracts from *G. tsugae* fruiting body and mycelium, besides phenols, ascorbic acid, α- and δ-tocopherols possessed that function (Mau et al. 2005; Li et al. 2012). Lee et al. (2007) reported that in *P. citrinopileatus*, phenols were the main carriers of antioxidative activity since ascorbic acid, tocopherols, β-carotene, and cysteine were insignificantly present. However, ascorbic acid and β-carotene were important radicals' neutralizers in *P. eryngii* and *P. ostreatus* (Jayakumar et al. 2009; Oke and Aslim 2011; Mishra et al. 2013). In several studies, positive correlations between the high antioxidative capacity of *P. cystidiosus*, *P. eryngii*, *P. ostreatus,* and *P. ferulae* and significant amounts of tocopherols, gallic, protocatechuic, *p*-hydroxybenzoic, *p*-coumaric, and cinnamic acids have been observed (Yang et al. 2002a; Tsai et al. 2009; Oke and Aslim 2011; Reis et al. 2012a). Phenols were the major antioxidant components in *Agrocybe cylindracea* and *A. aegerita* var. *alba*, and *L. sulphureus*, where flavonoids also had an important role (Lo et al. 2005; Huang et al. 2006; Tsai et al. 2006; Turkoglu et al. 2007). However, in the case of the main carriers of *L. edodes* antioxidative activity, opinions differed. According to Yang et al. (2002a) and Cheung et al. (2003); Cheung and Cheung (2005), various phenols were the main DPPH• neutralizers, while according to Carneiro et al. (2013), tocopherols had this function. The high content of tocopherols was also responsible for DPPH• neutralizing activity in *Clitocybe alexandri*, *Laccaria laccata*, *Mycena rosea*, *A. blazei,* and *A. brasiliensis* (Camelini et al. 2005; Tsai et al. 2007; Soares et al. 2009; Heleno et al. 2010). However, in *A. brasiliensis*, β-glucans, phenols, citric, malic, tartaric, oxalic, succinic, lipoic, and phytic acids were also synthesized (Keles et al. 2011; Vaz et al. 2011). Flavonoids were the main antioxidative compounds in numerous species (Stajić et al. 2013), while in *Inonotus* spp. and *Phellinus* spp. highly oxygenated and unsaturated polyphenols-hispidin derivatives (interfungins A, B, and C) had this role (Lee and Yun 2007).

Among polysaccharides, extracellular ones are the leader antioxidants in most mushrooms, and only some, such as *P. eryngii*, *P. cornucopiae,* and *P. nebrodensis* intracellular polysaccharides have this role (Liu et al. 2010a; Stajić et al. 2013). In *G. lucidum,* low molecular weight polysaccharides, as well as free amino acids, peptides, and proteins, were highly effective DPPH• neutralizers, an inhibitor of linoleic acid peroxidation, and stimulators of SOD and CAT activities, even more effective than ascorbic acid (Jia et al. 2009; Saltarelli et al. 2009; Liu et al. 2010b;

Kozarski et al. 2012). On the other hand, polysaccharide-protein complexes from *G. atrum*, *P. ostreatus*, *Phellinus rimosus*, *C. sinensis*, and *Antrodia camphorata*, polysaccharide-peptide complexes from *G. lucidum* and *Grifola umbellata*, as well as krestin from *T. versicolor* were responsible for strong free radical scavenging activities (Song and Yen 2002; Behera et al. 2005; Chen et al. 2008; Tseng et al. 2008; Janardhanan et al. 2009; Leung et al. 2009; Xia et al. 2011). Similar to *G. lucidum*, lentinan and LEP from *L. edodes* had high antioxidative potential based on the increase of SOD and GPx activity, while its high efficiency in lipid peroxidation inhibition was in positive correlation with high content of free amino acids and proteins (Cheung and Cheung 2005; Yu et al. 2009; Feng et al. 2010). However, in this species and *Cantharellus cibarius*, *Calocybe gambosa*, and *Clitocybe odora*, unsaturated fatty acids nucleotides and nucleic acids also significantly contributed to antioxidative activity (Ames et al. 1981; Vaz et al. 2011; Cheng et al. 2012; Carneiro et al. 2013). Likewise, nucleotides and nucleic acids were responsible for the activity of *Agrocybe chaxingu*, *Coprinus comatus*, *A. bisporus*, *Armillariella mellea*, and *F. velutipes* to inhibit lipid peroxidation and convert free radicals to stable forms (Stajić et al. 2013).

4 Effects on Cardiovascular System

Numerous mushroom species have antioxidative, hypocholesterolic, hypotensive, and anti-inflammatory effects. Therefore, they are recommended for the prevention or treatment of cardiovascular diseases, which are the main causal agents of death in most developed countries and countries in transition. The main biomarkers of coronary heart disease have increased low-density lipoproteins (LDL) and triglycerides, reduced high-density lipoproteins (HDL), and high blood pressure.

4.1 Antihypercholesterolemic Activity

Due to the insignificant amount of fats, the dominant presence of polyunsaturated fatty acids, and the absence of trans fatty acids, as well as the high content of soluble fibers, mushrooms are ideal food for reducing the ratio of total and LDL cholesterol in serum (Kalač 2009; Reis et al. 2012b; Wang et al. 2014). Besides that, mushrooms have no side effects contrary to statins, conventional therapy for patients with increased LDL cholesterol levels. Mushrooms' hypocholesterolemic action is based on two mechanisms: (i) increase in the excretion of short-chain bile and fatty acids and inhibition of the cholesterol and triglycerides absorption and (ii) production of inhibitors of hemoglobin-coenzyme A reductase (HMG-CoA reductase), a key enzyme in the hemoglobin synthesis (Schneider et al. 2011; Meneses et al. 2016). Thus, fruiting bodies of *A. auricula* and *Tremella fuciformis*, owing to the presence of soluble dietary fibers, reduce the level of blood cholesterol by the first mechanism, while erythadenin from the *L. edodes* fruiting bodies and mevinolin (lovastatin) from *P. ostreatus*, *P. eryngii* var. *ferulae* and *P. cornucopiae* basidiocarps and *P. sapidus*, *P. saca* and *P. ostreatus* mycelia by the second mechanism (Gunde-Cimerman et al. 1993; Guillamón et al. 2010; Zhang et al. 2020). Chitin and chitosan from the mushrooms' cell walls have functions similar to dietary fibers. Chitosan is now commercialized as a dietary supplement for obese people and people with problems with high cholesterol levels in the blood (Neyrinck et al. 2009).

Hu et al. (2006a) and Alam et al. (2011) showed that the *P. ostreatus* and *P. citrinopileatus* fruiting bodies have a positive effect on rats with hypercholesterolemia induced by the consumption of fatty foods or alcohol, by diabetes, or by a congenital disorder of cholesterol metabolism. The decrease of total lipids, cholesterol, and triglycerides levels in plasma and liver of rabbits with hypercholesterolemia and an increase in HDL/total cholesterol and HDL/LDL cholesterol ratio was caused by a diet enriched with dry *P. florida* fruiting bodies, which increased bile acid excretion (Guillamón et al. 2010). *P. ostreatus* caused a similar effect in humans. Schneider et al. (2011) observed that regular intake of dry fruiting bodies at the daily dose of 30 g for 21 days reduced triglyceride concentrations by about 0.44 mmol/L, oxidized LDL by about 7.2 U/mL, and total cholesterol by about 0.47 mmol/L. Contrary to that, in the control group, which consumed potato soup, the concentration of triglycerides increased significantly. These authors showed that the carriers of the hypocholesterolemic effect of *P. ostreatus* were linoleic acid, mevinolin, ergosterol, and its derivatives, as well as dietary fibers whose content was about sixfold higher than in the potato.

Erythadenin or lentinacin is a carrier of the hypocholesterolemic effect of *L. edodes* and *A. bisporus* basidiocarps (Guillamón et al. 2010). These authors reported that this compound reduced cholesterol levels in rats by 25% after 7 days of consumption at a dose of only 0.005% of the food. The mechanism of its action is based on the modification of phospholipid metabolism in the liver and the change in the fatty acid profile in the liver and plasma. It is well known that body weight reduction leads to a decrease in triglyceride and cholesterol levels in the blood. Enrichment of feed of obesity mice (fat-induced obesity) with 5% chitosan from *A. bisporus* led to a decrease in the level of lipid and adipocytokine absorption in the serum, resulting in a reduction of fat accumulation and triglyceride content in the liver and muscles by 39% and 66%, respectively. Likewise, adding *P. ostreatus* fruiting bodies to the diet of experimental animals, at a dose of about 5% of daily calories, for 6 weeks significantly increased the HDL concentration (Alam et al. 2011). Studies by a group of Baltimore scientists have shown that the replacement of minced beef with *A. bisporus* in only one meal for 4 days significantly reduced fat usage without influencing appetite and satiety (Stajić 2015). If replacement is constantly done once per week, 20,000 kcal can be reduced in a year, and thus obesity can be significantly decreased. The author also reported a reduction in body weight and blood cholesterol level in 90 volunteers who used polysaccharide-protein complexes from *A. blazei* and *L. edodes*.

4.2 Antihypertensive Activity

Hypertension, also known as a "silent killer," presents one more risk of cardiovascular diseases. Because of the high frequency, even at the epidemic level, hypertension is a global threat to modern humans. According to World Health Organization, more than a billion people worldwide suffer from hypertension, and it is assumed that in 2025 this number may increase to 1.6 billion.

The main factors that lead to hypertension are the hyperactivity of the sympathetic nervous system induced by stress, the large production of vasoconstrictors and mineralocorticoids, reduced vasodilators' production, obesity, and diabetes (Mills et al. 2021). Therefore, hypertension should be treated with diuretics, beta-blockers, inhibitors of angiotensin-converting enzyme (ACE), and blockers of angiotensin receptors and calcium channels. However, antihypertensive drugs have various side effects, and therefore great attention is given to finding the natural sources of safe and effective antihypertensive agents. Numerous mushrooms present one of these sources (Table 1). Due to the low sodium and high potassium concentrations in the fruiting bodies, they are used in an antihypertensive diet. Their peptides act as ACE inhibitors and decrease blood pressure without side effects (Wang et al. 2014). Compared with other mushrooms, the water *G. lucidum* extract was the best inhibitor of ACE and the sympathetic nervous system, whose secondary effect was hypotension (Yahaya et al. 2014). Although ganodal A, ganoderols A and B, and ganoderic acids K and S, triterpenoids present in this extract in significant amounts, were slightly weaker ACE inhibitors, regular usage of the extract for 4 weeks led to a significant decrease in blood pressure. Inhibition of this enzyme was also the mechanism of action of *P. ostreatus*, *P. cornucopiae,* and *P. nebrodensis* (Jang et al. 2011; Yahaya et al. 2014). Jang et al. (2011) showed that D-mannitol and two easily absorbed peptides isolated from *P. cornucopiae* were highly effective in the treatment of spontaneous hypertension. Biologically active compounds with a similar structure and mode of action are isolated from the fruiting bodies of *P. eryngii*, *P. flabellatus*, *P. sajor-caju*, *P. cystidiosus,* and *P. florida* (Abdullah et al. 2012). Good ACE inhibitors and, consequently, reducers of systolic blood pressure are also chitin-deacetylated derivatives. Therefore, *L. edodes* fruiting bodies rich in chitin (8.07% of dry weight) present an excellent hypotensive agent (Vetter 2007). However, its activity is also based on high potassium content because it is known that the best way of hypertension prevention is maintaining potassium ions amount at a higher level and the concentration of sodium ions and aldosterone at a lower level (Manzi et al. 1999). Lau et al. (2012) reported that *L. edodes* water extract inhibited the activity of ACE by 90%, while lentinan caused vasodilation and, consequently, blood pressure reduction. *F. velutipes*, *H. erinaceus,* and three peptides isolated from *A. bisporus* basidiocarps were also highly effective ACE inhibitors, i.e., they inhibit the activity by 96%, 90%, and 87%, respectively (Lau et al. 2012). Likewise, *Tricholoma giganteum* and *G. frondosa* showed significant antihypertensive capacity (Lee et al. 2004; Yahaya et al. 2014). Lee et al. (2004) noted that *T. giganteum* extract inhibited the activity of this enzyme by 61%, while low molecular weight peptide isolated from its basidiocarps has shown very high potential in blood pressure reduction, even like some commercial drugs. A significant decrease in systolic blood pressure in rats on a diet enriched with the *G. frondosa* fruiting bodies or extracts (5% of the total food amount) for only 35 days was noted by Yahaya et al. (2014). These authors also observed that peptides isolated from *G. frondosa* water extract significantly inhibited ACE.

5 Antitumor Activity

Nowadays, half of the men and more than a third of women worldwide have cancer, and even a quarter of adult deaths are caused by cancer (Parker 2014). The global situation, as predicted by International Agency for Research on Cancer (IARC), is more pessimistic. Namely, each year until 2030, over 21 million persons worldwide will be new cancer-diagnosed. Besides the high annual rate of new-cancer patients, low curability rate, and high treatment cost represent serious problems, which require good knowledge of cancer causal agents and mechanism of cancer development, creating an effective strategy for cancer prevention, as well as the development of more efficient drugs.

Nowadays, potential carcinogens are numerous, and they can be divided into three groups, chemical, physical and biological. They can cause oxidative stress and/or inflammation that can cause direct and irreversible changes in the genome, cell morphology, polarity, adhesion, communication, mobility, and the synthesis of metalloproteinases and angiogenic factors. All these processes consequently lead to increased cell proliferation, resistance to apoptosis, i.e., the transformation of normal cells into neoplastic ones, and other aggressive tumors neovascularization and metastases (Reuter et al. 2010; Stajić et al. 2019).

The common treatment for cancer patients is surgery combined with chemo- and/or radiotherapy. However, chemo- and radiotherapy have several disadvantages, such as ineffectiveness in the treatment of some cancer types as well as in the treatment of later cancer development stages, absence of tumor-specificity, and causing numerous harmful effects in patients (Chen et al. 2006). Nowadays, preference is given to integrative medicine, presenting a combination of conventional, complementary, and alternative medicines. Preparations based on mushroom extracts or metabolites are important in complementary medicine (Table 1). However, it should be emphasized that mushrooms are not drugs but dietary supplements whose consumption eliminates the accompanying harmful effects of chemo- and radiotherapy (Chang and Wasser 2012).

The anticancer activity of the mushroom extracts and/or metabolites is based on several mechanisms, one of which is the stimulation of the immune system. Numerous studies have shown that species of the genus *Agaricus*, *G. lucidum*, *Cordyceps militaris*, *Phellinus linteus*, *H. erinaceus*, *L. edodes*, *G. frondosa*, *T. versicolor*, *Clitocybe nebulares,* and many others possess the cytotoxic activity and strengthen the immune system (Chihara 1992; Fortes et al. 2009; Patel and Goyal 2012; Ren et al. 2012; Roupas et al. 2012; El Enshasy and Hatti-Kaul 2013). For example, *G. lucidum* extract stimulates TNF-γ synthesis, *C. militaris* one production of IFN-γ and IL-18, *Ph. linteus* extract induces the production of IL-12, IFN-γ, and TNF-α synthesis, as well as macrophage and dendritic cells proliferation, and *H. erinaceus* extract activates natural killer cells and macrophages. In such a way, these species act against stomach cancer, leukemia, hepatoma, and colon cancer, respectively (Patel and Goyal 2012). Lentinan, grifolan, *G. lucidum* polysaccharide, and *C. nebulares* lectin stimulate the production of certain cytokines. At the same time, krestin and *A. bisporus* proteoglycans initiate ILs and INF-γ production and activate natural killer cells, while cordlane from *C. militaris* and *C. nebulares* lectin induce the maturation of dendritic cells (Chihara 1992; Patel and Goyal 2012; Ren et al. 2012; Roupas et al. 2012; El Enshasy and Hatti-Kaul 2013).

Some mushrooms, such as *P. ostreatus* and *Phellinus rimosus*, base their strong cytostatic effect on neutralizing free radicals, i.e., on high antioxidative potential (Stajić et al. 2013). On the other hand, ethanol extracts of *T. versicolor*, *T. hirsuta*, *T. gibbosa*, and Se-enriched *G. lucidum* basidiocarps, methanol extract of *Lactarius vellereus*, as well as water extracts of *A. bisporus*, *G. lucidum* and *Agrocybe cylindracea* possess strong antimutagenic activity, i.e., effectively protect cells against H_2O_2-induced DNA damage and in such a way prevent the transformation of a normal cell to malignant one (Mlinrič et al. 2004; Zhao et al. 2008; Roupas et al. 2012; Ćilerdžić et al. 2016a; Knežević et al. 2018). The same effect was noted for *A. brasiliensis* and *A. blazei* β-glucans, and *A. bisporus* thermolabile protein (Angeli et al. 2006, 2009). Strong anti-inflammatory activity is the basis of the cytotoxic activity of *Ph. rimosus* and *P. ostreatus* (Joseph et al. 2012; El Enshasy and Hatti-Kaul 2013). Namely, these authors observed that *Ph. rimosus* polysaccharide-protein complex increased SOD and GPx activity and decreased the level of reduced glutathione, while *P. ostreatus* β-(1,3/1,6)-D-glucan changed cytokines level in plasma.

Some mushrooms can regulate some cell processes. Roupas et al. (2012) reported the high efficiency of *A. blazei* extracts and its metabolite agaritine in stopping proliferation and inducing apoptosis of some leukemia cell lines, based on induction of cytochrome c release, caspases activation, and Bcl-2 synthesis regulation. High inhibition of breast cancer by *A. bisporus* water extract and colon cancer by its lectin are based on the inhibition of aromatase activity and stimulation of caspase-3 activity, respectively (Grube et al. 2001; Hong et al. 2004). Also, strong antiproliferative activity against breast cancer cells was caused by theanine from *Boletus badius,* which stimulated cytochrome c release and activated caspase (Patel and Goyal 2012). Several studies have demonstrated that *G. lucidum* extract-based cytotoxic activity against stomach cancer cells is caused by caspases activation and inhibition of metalloproteinase expression, while antiproliferative activity against prostate cancer cells is based on inhibition of transcription factor AP-1 (Chen et al. 2010; Patel and Goyal 2012; Roupas et al. 2012). Triterpenoid ganoderic acid Me inhibits colon cancer development by stimulating the expression of p53, Bax, and caspase 3 and the release of cytochrome c (Patel and Goyal 2012). *G. frondosa* β-glucan inhibits bladder cancer cells by activating DNA-dependent protein kinase and arresting the cell cycle. At the same time, its polysaccharide-peptide complex stimulates the synthesis of Bax, inhibits the synthesis of Bcl-2, and activates caspase 3, leading to the apoptosis of stomach cancer cells (Louie et al. 2010; Patel and Goyal 2012). The development of various cancer types can be inhibited by suppression of nuclear transcription factor (NF-κB) activity which could be caused by extracts of *A. brasiliensis, C. sinensis, C. comatus, Sparassis crispa,* and *Phallus impudicus* (Grube et al. 2001; Hong et al. 2004; Petrova et al. 2008).

Numerous studies have demonstrated that arresting cell cycle in a specific phase and induction of rapid cell apoptosis was caused by *L. edodes* extract against leukemia and skin cancer cells, *C. comatus* and *F. velutipes* extracts against prostate and breast cancer, respectively, and *Ph. linteus* extract against liver, bladder, stomach, and lung cancer cells (Gu and Belury 2005; Guo et al. 2007; Zaidman et al. 2008; Patel and Goyal 2012). Arresting cell cycle in G_0/G_1 phase was at the base of melanoma reduction by *T. versicolor* and *Inonotus obliquus* extracts and breast cancer inhibition by *Pleurotus tuber-regium* carboxymethylated polysaccharide (Zhang et al. 2007; Youn et al. 2009; Roupas et al. 2012). On the other hand, arresting cell cycle in the G_2/M phase by *G. tsugae* extract, *G. frondosa* polysaccharide-peptide complex, cordlane from *C. militaris,* and ergon from *Russula cyanoxantha* led to inhibition of colon, stomach, bladder, and liver cancer (Hsu et al. 2008; Patel and Goyal 2012). Vaz et al. (2010) observed that *Clitocybe alexandri* extract had the same effect on colon adenocarcinoma and lung, breast, and stomach cancers but by cell cycle arrest in the S phase and apoptosis induction. Strong cytotoxic activity against breast and pancreas cancer was caused by theanine from *B. badius* and anthraquinones from *A. camphorata*, respectively, which stopped the cell cycle in the G_1 phase (Yu et al. 2012; Patel and Goyal 2012).

B. badius fermentation broth, *G. frondosa* polysaccharide-peptide complex, and ergon from *R. cyanoxantha* inhibited breast, stomach, and liver cancer, respectively, by induction of apoptotic bodies' appearance on cell, cell volume reduction, chromatin condensation, and DNA fragmentation, i.e., by disturbance of DNA synthesis and structure (Cui et al. 2007; Patel and Goyal 2012). On the other hand, some mushrooms cause changes in the morphology and mobility of malignant cells, inhibiting cancer development. Several researchers have demonstrated that the cytotoxic activity of *P. betulinus* fruiting bodies against colon and lung carcinoma and glioma cells and blazein from *A. blazei* against lung and stomach cancers are based on cells morphology and mobility changes, while anthraquinone from *A. camphorata* stimulated degradation of dysfunctional cellular components resulting in the inhibition of proliferation of pancreas cancer cells (Yu et al. 2012; Patel and Goyal 2012; Roupas et al. 2012). Cytotoxic activity of *H. erinaceus* water extract and *G. lucidum* polysaccharide-peptide complex against colon and lung cancer, respectively, were based on angiogenesis inhibition (Kim et al. 2011; Ren et al. 2012).

6 Antineurodegenerative Activity

The seventh leading cause of death is dementia affecting more than 55 million people. World Health Organization estimates that this number will be 78 million in 2030 and even 139 million in 2050. The most common form of dementia is Alzheimer's disease. It is estimated that every ninth man and every fifth woman will suffer from it in 2050. Parkinson's disease is the second most common age-related neurodegenerative disorder, and according to World Health Organization, 7–10 million people worldwide suffer from it, especially men. The probability of Parkinson's disease occurrence and development is 1.5 times higher in men than in women. These two neurodegenerative disorders have physical, psychological, social, and economic impacts. People with these diseases suffer, including their careers, families, and society.

Nearly 10 million new cases of Alzheimer's disease worldwide and about 60,000 with Parkinson's only in the USA are diagnosed yearly. The average cost of Alzheimer's treatment is US$ 20,461 per patient per year, while in the case of Parkinson's disease, the cost is US$2500. However, although numerous commercial antineurodegenerative drugs exist on the world market, their many side effects and disadvantages are known (Phan et al. 2014). Therefore, the current trend in the world is the creation of highly effective natural preparations. Owing to some mushrooms' medicinal properties, they could be efficient antineurodegenerative agents (Table 1). Mushrooms, as excellent antioxidants, can prevent disturbances in the structure of numerous metabolites as well as in the function of organelles and cells; thus, they can prevent the occurrence and development of Alzheimer's and Parkinson's diseases (Asanuma et al. 2003; Zhu et al. 2004; Halliwell 2006; Tessari et al. 2008; Tsang and Chung 2009; Tel et al. 2011; Janjušević et al. 2017). These authors have demonstrated high mushrooms' efficiency in the prevention of (i) abnormal mitochondrial function, (ii) lipid peroxidation and consequently change in cell membrane permeability, (iii) inflammatory responses, (iv) cell apoptosis, (v) neurons' senescence, (vi) inadequate synthesis of acetylcholine, a neurotransmitter that is directly related to increased activity of acetylcholinesterase (AChE) which is a trigger of Alzheimer's disease, (vii) production of highly active tyrosinase that catalyzes the conversion of L-DOPA into a reactive quinone form toxic to dopaminergic neurons, which progressive loss leads to the appearance of Parkinson's disease.

Due to their ability to inhibit neuroinflammation, mushrooms are known as "brain food" (Essa et al. 2012; Phan et al. 2014). However, mushrooms differ in their mechanism of action and efficiency. Numerous studies showed that mushroom can act by one of five the most common mode of action: (i) inhibition of amyloid peptide production or aggregation into amyloid plaques (*G. lucidum* extracts); (ii) inhibition of *p*-tau protein secretion and consequently neurone damage (*A. comphorata*); (iii) free radical neutralization (*Ganoderma* spp. Extracts, hispidin from *Ph. Linteus*, *I. obliquus* protein-bound polysaccharide, etc.); (iv) inhibition of AchE and tyrosinase (*Cortinarius infractus* alkaloids, extracts of *Tricholoma* spp., *Trametes* spp., *G. lucidum*, *P. ostreatus*, *L. sulphureus*, etc.); (v) stimulation of neurotrophins' synthesis and neuronal differentiation (extracts of *Sarcodon* spp., *G. frondosa*, *Pleurotus giganteus*, *C. militaris* and *H. erinaceus*, etc.) (Kawagishi et al. 1997; Wang et al. 2004, 2012; Marcotullio et al. 2006, 2007; Nishina et al. 2006; Jung et al. 2008; Lai et al. 2008; Mori et al. 2008; Dai et al. 2010; Lee et al. 2011; Tel et al. 2011; Phan et al. 2012; Knežević et al. 2018; Ćilerdžić et al. 2019).

The efficiency of mushrooms depends on species/strain, development phase (mycelium/fruiting body), extract type, and concentration of the active metabolites. Thus, for example, *P. ostreatus* extracts showed higher reduction potential than *P. citrinopileatus* ones (Jayakumar et al. 2009; Alam et al. 2010; Lee et al. 2007). Lee et al. (2007) reported that the highest amount of active compounds from *P. citrinopileatus* fruiting bodies was extracted with hot water and that basidiocarp extract was a more efficient reduction agent than mycelium one. On the other hand, Ćilerdžić et al. (2015) obtained the highest extraction yield from *P. ostreatus* with 96% ethanol and similar free radical reduction capacity of basidiocarp and mycelium extracts.

Previous reports showed that extracts of a few fungal species produce compounds that inhibit AChE activity (Patocka 2012; Janjušević et al. 2017). Patocka (2012) and Jamila et al. (2015) emphasized that phenols, terpenoids, and alkaloids were responsible for this activity. *Trametes* species showed higher efficiency in AChE inhibition than *Emericella unguis* (El-Hady et al. 2014a; Knežević et al. 2018), and their potential of tyrosinase activity inhibition was even higher than in commercial inhibitors, i.e., kojic acid (El-Hady et al. 2014b), which can be explained by synergistic interaction of numerous compounds of the crude extracts (Şenol et al. 2010). *P. ostreatus* and *L. sulphureus* extracts were good AChE and tyrosinase inhibitors. However, *P. ostreatus* was a significantly better anti-Alzheimer's agent (Ćilerdžić et al. 2019).

H. erinaceus is another mushroom highly effective in slowing down dementia progression and increasing cognitive function. Ma et al. (2010) found that metabolites hericenones C, D, E, F, G, and H were very effective in patients with dementia and mild cognitive impairment. However, patients returned to the former stage only during treatment, i.e., after the termination of their usage.

7 Antihyperglycemic Activity

According to the World Health Organization report, over 220 million people, or 7.8% of the world's population, suffer from diabetes. The International Diabetes Federation predicts that the number of patients in Europe will likely increase by 20% and in Africa by 98%. This number will be about 366 million in 2030 (Wild et al. 2004). Diabetes is the seventh death causal agent in the United States, sixth in Great Britain, and fifth in Taiwan. Although this is an irreversible disease that cannot be cured, glycemic control is necessary to prevent accompanying complications and reduce mortality. Antihyperglycemic agents delay the absorption of carbohydrates, increase the expression of insulin-sensitive glucose transporters, reduce gluconeogenesis in the liver, and stimulate pancreas beta cells to secrete insulin resulting in increased sensitivity to it (Lo and Wasser 2011).

However, many of the agents can cause side effects in the gastrointestinal tract and should be avoided in patients with heart failure and liver and kidney dysfunctions. Nowadays, numerous national and international programs focus on the prevention or disposal of diabetes occurrence as well as its chronic complications. Diet control, increased physical activity, healthy sleep, and weight reduction are the parts of the most effective strategies. Numerous in vitro and in vivo studies have shown that many mushrooms have a high antihyperglycemic potential, but only a few clinical trials have been done (Table 1). One of them showed that the consumption of Ganopoly (a preparation based on *G. lucidum* polysaccharide) significantly reduced the amount of glycosylated hemoglobin in patients with type 2 diabetes (Wińska et al. 2019). Hsu et al. (2007) observed significantly lower insulin resistance and higher blood adiponectin levels in 536 patients with type 2 diabetes whose conventional therapy was enriched with the *A. brasiliensis* extract at a daily dose of 1500 mg. Usage of *C. comatus* also reduces blood glucose, cholesterol, and triglyceride levels in patients with diabetes without any side effects on the function of the liver and kidneys and changes in body weight (Lo and Wasser 2011).

Mechanisms of mushroom hypoglycemic action can differ depending on how diabetes occurs. Namely, Lo and Wasser (2011) emphasized that diabetes was a result of pancreas beta cells damage, which can occur in several ways: (i) by free radicals which affect the cells inhibiting synthesis and secretion of insulin and inducing their apoptosis and development of accompanying complications; (ii) by infection when NF-κB activates leading to increase the production of inflammatory mediators (cytokines and NO) and at the end to the death of beta cells; (iii) by abnormal fatty acid metabolism. Namely, according to the so-called glucocytotoxic hypothesis, the increased presence of free fatty acids and hyperglycemia act synergistically in causing damage to beta cells.

Numerous studies have shown that compounds originating from mushrooms affect antidiabetic activity on glucose absorption, regeneration of pancreas beta cells, regulation of insulin secretion and metabolisms of carbohydrates and fat, neutralization of free radicals, and anti-inflammatory action (Lo and Wasser 2011). Gray and Flatt (1998) and Yang et al. (2008) observed that water-soluble fibers and polysaccharides of *A. campestris*, *T. versicolor*, *C. sinensis*, and *Fomes fomentarius* increased the viscosity of the gastrointestinal content and reduced the nutrient flow, which led to decrease in glucose absorption and its level in plasma. The high content of dietary fibers in *A. bisporus* and *I. obliquus* can also inhibit the activity of α-amylase and α-glucosidase, enzymes that catalyze the hydrolysis of carbohydrates, thus reducing glucose levels in the blood (Lu et al. 2010). The mushrooms' antihyperglycemic effect can also be based on (i) increase of SOD, CAT and GPx activity (*A. bispotus*, *A. brasiliensis*, *Phellinus baumii*, *T. versicolor* and *Tremella aurantia*) (Wei et al. 1996; Yuan et al. 1996; Hwang et al. 2007; Zhang et al. 2009; Yamac et al. 2010); (ii) protection of beta cells from the cytotoxic effect of hyperglycemic agents (polysaccharides of *P. citrinopileatus* and *Agrocybe chaxingu* and exobiopolymers of *C. sinensis* and *F. fomentarius*) (Hwang et al. 2005); (iii) reparation of beta cells damages to a certain degree (*L. edodes*) (Yang et al. 2002b); (iv) stimulation of the insulin synthesis and secretion (*A. bisporus*, *A. brasiliensis*, *A. campestris*, *C. militaris*, *C. sinensis*, *G. applanatum*, *G. lucidum*, *G. frondosa*, *Ph. linteus* and *Tremella fuciformis*) (Lo and Wasser 2011); (v) inhibition of the production of NO, ILs (1β and 6) and TNF-α in lipopolysaccharide-activated macrophages (*G. frondosa* fraction and cordicepine from *C. militaris*) (Shin et al. 2009); (vi) increase in the content of glucose 2 transporter (GLUT-2) in the liver and GLUT-4 in the muscles (*C. militaris*) (Choi et al. 2004); (vii) increase the activity of glucokinase, hexokinase and glucose-6-phosphate dehydrogenase in the liver (*C. sinensis* and *T. aurantia*) (Kiho et al. 1996, 2000); (viii) stimulation of glucose oxidation and incorporation into glycogen (*A. campestris*) (Gray and Flatt 1998); (ix) increase glycogenesis and reduction of glycogenolysis (*C. militaris*, *C. comatus* and *G. lucidum*) (Choi et al. 2004; Gao et al. 2004; Lv et al. 2009); (x) improving sensitivity to insulin by regulating peroxisome proliferator-active receptor-γ (PPAR-γ) and further lipid metabolism (*Ph. baumii*, *Ph. linteus*, *Poria cocos* and *T. fuciformis*) (Cho et al. 2007; Lee et al. 2008; Li et al. 2011).

Yamac et al. (2010) showed that enrichment of the diet of rats with induced diabetes with *A. bisporus* extract significantly increased the number of beta cells in Langerhans pancreas islands, primarily due to increased activity of antioxidative enzymes, increased insulin level by 78.5%, and reduced level of glucose in serum by 29.7%. Significant reduction in blood glucose level was also caused by extracellular *A. brasiliensis* β-glucans and glycoproteins that increased insulin levels in plasma and antioxidative activity as well as expressed GLUT-4 in the fatty tissue (Oh et al. 2010). Extract of this species in the dose of 400 mg/kg of body weight had a similar efficiency to 500 mg of the commercial drug (metformin) per kg of body weight. *P. eryngii* and *P. citrinopileatus* also cause a significant reduction in blood glucose levels by increasing sensitivity to insulin and reduction of Langerhans islands damage, respectively (Hu et al. 2006b; Kim et al. 2010). Extracellular exopolymer of *L. edodes* (200 mg/kg) also remarkably repaired the damage of pancreas beta cells, increased insulin synthesis by 22.1%, and consequently reduced glucose level in plasma by 21.5% (Yang et al. 2002b). High efficiency in the prevention and treatment of induced diabetes in mice was also shown by neutral water-soluble polysaccharides and the acidic polysaccharides from *A. auricula-judae* fruiting bodies (Yuan et al. 1998).

The exopolysaccharides of *T. fuciformis* fruiting bodies improve the sensitivity to insulin by regulating lipid metabolism and, therefore present good hypoglycemic agents or functional food whose usage is suggested for the treatment of type 2 diabetes (Cho et al. 2007). Acidic heteropolysaccharides from *T. mesenterica* fruiting bodies also can significantly reduce the level of glucose in the blood. At the same time, its fibers and other compounds prevent macrovascular complications in diabetes (Lo et al. 2006). Significant reparation of damaged pancreas beta cells, an increase in insulin secretion, and a reduction in glucose level were noted in rats with induced diabetes after 3 weeks of therapy with *G. applanatum* exopolymer and *G frondosa* fruiting bodies (Kubo et al. 1994; Yang et al. 2007). According to Lo and Wasser (2011), *G. frondosa's* hypoglycemic action was based on antioxidative and immunomodulating activity, i.e., inhibiting macrophage proliferation and decreasing the synthesis of factors destructive to beta cells (NO and IL-1). *I. obliquus,* polysaccharides from *T. versicolor* fruiting bodies and β-glucan-protein complex from its mycelium also base strong antidiabetic activity on high antioxidative capacity, i.e., on the increase of SOD and GPx activities (Wei et al. 1996; Lu et al. 2010). Extremely high efficiency in reduction of glucose level in plasma of rats with induced diabetes, as high as 52.3%, was caused by consumption of *P. baumii* exopolysaccharides, while metabolites of *Ph. linteus* (hispidin and its derivatives) as good antioxidants prevented accompanying complications (Cho et al. 2007; Lee et al. 2008). *C. militaris* metabolites, cordycepin, and acarbose (0.2 mg/kg and 10 mg/kg, respectively), were also highly effective in rats with induced diabetes since they reduced glucose levels in the blood by 48.4% and 37.5%, respectively (Yun et al. 2003).

8 Antimicrobial Activity

Viral, bacterial, and fungal infections are among the most serious threats to human health and quality of life and present a significant challenge to modern medicine. Although the arsenal of antimicrobial drugs constantly expands, it does not meet the increasing requirements for the successful treatment of various infections due to the alarming increase in microbial resistance. Therefore, developing novel natural antimicrobial agents with improved modes of action and higher efficiency are the main requirements of modern society. Mushrooms, as producers of numerous intra- and extracellular antimicrobial metabolites, could be an excellent basis for preparations for successful treatments of human and animal diseases (Table 1). In vitro studies have shown that numerous mushrooms, their extracts, and compounds can potentially inhibit the growth of Gram+ and Gram- bacteria. However, Gram+ bacteria are more susceptible to mushroom extracts than Gram- ones due to the absence of lipoproteins in the cell wall (Kosanić and Ranković 2011). Ćilerdžić et al. (2014) showed that phenols were the main carriers of the mushrooms' antibacterial activity, which was confirmed by the high correlation between phenols content and the inhibitory activity of *G. lucidum* and *G. applanatum* extracts against *Staphylococcus aureus* and *Bacillus* sp. Rare in vivo studies showed the indirect antibacterial effect of mushrooms based on the improvement of the immune system. Thus, *A. brasiliensis* fraction rich in polysaccharides increases host resistance to some infectious agents by stimulating macrophage activity, while *A. blazei* extract by stimulation of TNF-α synthesis (Stajić 2015). This antimicrobial-immunomodulatory activity is also the basis of the antifungal and antiviral effects of mushrooms' extracts and metabolites.

The most potent antifungal compounds are phenols and polysaccharides. However, proteins, peptides, terpenoids, and numerous low molecular mass compounds act as inhibitors of pathogen development and virulence, activators of pathogens' autolytic system, and immunomodulators (Yamaç and Bilgili 2006). Thus,

G. lucidum methanol extract and gallic acid originated from *Clitocybe subconnexa* inhibit virulence of *Aspergillus niger* and *A. fumigatus*, respectively, by demelanization of conidiophores and vesicle (Stajić et al. 2017). Highly effective natural agents against *Aspergillus* spp. were also *I. obliquus* and *G. lucidum* ethanol extracts, which were even better than the commercial fungicide Ketoconazole (Ćilerdžić et al. 2014). *Aspergillus, Penicillium* spp., *Candida albicans, and Trichoderma viride were* sensitive to *Cordyceps militaris* extracts, *G. applanatum,* and *G. carnosum* mycelial extracts and fermentation filtrates (Ćilerdžić et al. 2016b). These are just some examples of mushrooms' antifungal activity that are extensively reviewed (Stajić et al. 2017).

Antiviral activity of mushroom extracts and metabolites is expressed in two modes: (i) direct by inhibition of viral enzymes or virus absorption by the host cell and (ii) indirect by stimulation of the host immune system (Stajić 2015). This author emphasized that proteins, peptides, polysaccharides, and triterpenoids were the carriers of this activity. There are numerous examples of mushroom efficiency in combat against various viruses. Thus, *G. frondosa* is highly effective against hepatitis B virus and herpes simplex virus type 1 owing to its ability to stop their replication, while its activity against influenza A virus is based on virus growth inhibition as well as immune response improvement (Nishihira et al. 2017; Wu et al. 2021). Influenza A and B viruses were also sensitive to triterpenoids from *Ganoderma* spp., phenol compounds from *Inonotus hispidus*, and herpes simplex virus type 1 to krestin and PSP (Stajić 2015). This author reported that inhibition of HIV-1 reverse transcriptase was due to the activity against the HIV-1 virus and can be caused by *G. lucidum* and *G. colossum* triterpenoids, lentinan, *P. ostreatus* protein, *P. nebrodensis* terpenoid, *A. bisporus* lectin, *I. obliquus* water-soluble metabolites, as well as *F. velutipes* ribosome-inactivating protein. On the other hand, krestin and PSP affect anti-HIV-1 activity on immune system stimulation and prevention of virus binding for the cell receptor (Lindequist et al. 2005; Rodríguez-Valentín et al. 2018). Brandler et al. (2020) showed that the activity of some mushrooms against COVID-19 was also based on immunostimulatory and anti-inflammatory effects.

9 Other Activities

Besides the above-mentioned activities of mushroom extracts and compounds, several other less-studied effects have been recorded. For example, the ethanol extract and proteoglycan isolated from *Ph. linteus*, as well as methanol extract of the *Pleurotus florida* fruiting bodies and ganoderic acids A, B, G, and H, isolated from *G. lucidum,* have anti-inflammatory effect in arthritis, which can be stronger than the effects of diclofenac and acetylsalicylic acid (Kim et al. 2003; Jose et al. 2004; Stajić 2015). Numerous studies have shown that this effect is associated with mushrooms' antioxidative and immunomodulatory properties.

Extracts of *C. sinensis* are very effective in the treatment of asthma in children during the remission stage. Their activity is based on the inhibition of proliferation and differentiation of Th2 cells, reduction of the expression of the transcription factor involved in the expression of T cell receptors, and rising IL-10 (Stajić 2015).

Extracts of some mushrooms used to treat allergies can suppress immune responses. For example, using the fruiting bodies of *Tricholoma populinum,* the ethanol extracts of *Hypsizygus marmoreus*, *F. velutipes*, *Pholiota nameko,* and *P. eryngii* cause regression of severe allergic symptoms (Stajić 2015). Chen et al. (2015) showed that ganoderic acids C and D, ganoderiol F, ganodermanontriol, and ganodermonondiol from *G. lucidum* caused inhibition of the complement system and release of histamine from the mast cells. Therefore, their usage is primarily recommended after organ transplantation. In vitro studies showed that extracts of fruiting bodies of *Polyporus badius*, *Lactifluus vellereus*, *Heterobasidion annosum*, *T. versicolor,* and *P. betulinus* inhibit the binding of lipopolysaccharides to CD14 receptors on immune cells and release inflammatory mediators and reactive oxygen species, and thereby prevent the appearance of a complex syndrome known as septic shock (Stajić 2015).

In vitro studies showed that triterpenoids isolated from *G. lucidum* basidiocarps (ganoderic acids R and S and ganosporeric acid A) have hepatoprotective activity, which was later confirmed in vivo (Lin et al. 2002). The addition of two triterpenoid fractions to mice's diet protects them from hepatic necrosis, which is probably associated with the ability to activate antioxidative enzymes. Stajić (2015) reported that by use of Ganopoly, 13% of patients who suffered from chronic hepatitis B did not have serum hepatitis B antigen after 6 months, while in patients with chronic hepatitis C positive effect was observed after 8 weeks of oral administration, twice per day. Also, the hepatoprotective effect was noted in patients with chronic hepatitis B and hepatitis C treated with the *A. blazei* extract.

Obstipation is one of the leading gastrointestinal problems of modern humans. Recently, great attention has been given to the importance of diet and the creation of new dietary supplements with probiotic functions. Mushrooms are potential candidates for prebiotics because they contain chitin, hemicellulose, β- and α-glucans, mannans, xylans, and galactanes (Aida et al. 2009). Nondigestible chitin and β-glucans have the role of dietary fibers and represent sources of prebiotics. The proportion of chitin ranges from 68 mg to 102 mg per gram of dried fruiting body (in *Boletus* spp.), while in the case of β-glucan, the values range between only 0.8 mg/g of dried matter (in *A. bisporus*) and even 548.8 mg/g (in *Boletus* spp.) (Manzi et al. 1999, 2004). Fruiting bodies of *Auricularia* spp. contain about 50% more fibers than other species and significantly improve the state of patients who suffer from functional obstipation without any side effects. Aida et al. (2009) showed that *P. ostreatus* and *P. eryngii* extracts stimulated the growth of 4 strains of *Lactobacillus* sp. (Lac A-D), three strains of *Bifidobacterium* sp. (Bifi A-C), and two strains of *Enterococcus faecium* (Ent A and B). *P. eryngii* was more effective, especially in the Lac B and C and Bifi B strains, while *P. ostreatus* extract primarily stimulated the growth of Bifi A. Results demonstrated that strains Lac B and Lac C were the most effective producers of short-chain fatty acids. These authors divided three fractions from *P. ostreatus* and *P. eryngii* fruiting bodies (water- and bases-soluble fractions and insoluble ones) and showed that they were responsible for the activities.

Several mushroom species' extracts and compounds can be used as analgesics. For example, *P. betulinus*, *G. applanatum*, *Fomitopsis pinicola,* and *Daedaleopsis confragosa* have an inhibitory effect on natural endopeptidases and, thus, activity similar to opiates can be used against pains (Stajić 2015). This author has reported that skutigeral from *Scutiger ovinus* has an affinity for dopamine D1 receptors in the brain and can eliminate pain by acting on vanilloid receptors, non-selective cationic channels that can be activated by various exogenous and endogenous physical and chemical stimulants that cause pain.

Fruiting bodies, extracts, and compounds of some mushrooms can increase bone density and prevent osteoporosis. Shimizu et al. (2006) and Stajić (2015) showed that *G. lucidum* and *P. eryngii* ethanol extracts significantly improved bone density in female rats, which was disrupted by removing the ovaries, i.e., by estrogen deficiency, without significant effect on the uterus. In mice where low bone density was induced by feed that was poor in calcium and vitamin D_2, an increase in femur density and tibia thickness, as well as increased duodenal and renal transport of calcium, was recorded after the addition of *L. edodes* fruiting bodies exposed to UV radiation (Lee et al. 2009). These authors also showed that the bioavailability of vitamin D_2 from *L. edodes* fruiting bodies enriched with this vitamin was high, and the consumption of the species increased the level of the vitamin in humans and improved alkaline phosphatase activity in osteoblasts. The activity of alkaline phosphatase, as well as the level of mineralization, were also significantly increased, in comparison with control cells, by in vitro cultivation of human osteosarcoma cells in the medium enriched with *G. frondosa* water extract, which means that this extract induced bone formation (Chaturvedi et al. 2018). Extracts of *P. eryngii* also increased the activity of alkaline phosphatase and stimulated the expression of osteocalcin mRNA in osteoblasts (Kim et al. 2006).

Some mushrooms can play an important role in wound healing in patients with diabetes which is a major clinical problem. Kwon et al. (2009) showed that wound healing in rats with induced diabetes was accelerated significantly by adding *Sparassis crispa* fruiting bodies to their diet. β-glucan, which is synthesized in significant amounts by this species stimulates the migration of macrophages and fibroblasts and the synthesis of collagen type 1. *H. erinaceus* and fractions of *G. lucidum* and *L. edodes* polysaccharides increase the activity of antioxidative enzymes and the levels of IL-2 and TNF-α. They have shown high efficiency in treating rats with ulcers (Stajić 2015).

Nowadays, cataract is a widespread ophthalmological problem commonly treated by surgery. According to World Health Organization data from 2011, in a sample of 100,000 people, 1100 primarily women, surgically removed the cataract from one or both eyes. Extracts of some mushrooms have been found very effective in preventing cataract emergence both in vitro and in vivo. Isai et al. (2009) reported that incubation of lenses damaged by selenite with *P. ostreatus* extracts caused a decrease in lens blurring and maintenance of antioxidative compounds at an almost normal level, and cataracts did not occur in 75% of treated rats.

It is known that the presence of free radicals is associated with aging and the appearance and progression of various diseases and disorders from which a large part of the world population suffers and dies. DNA is the most susceptible macromolecule to oxidative damage that can be induced by various agents, among which H_2O_2 has significant genotoxic potential. In vitro studies have shown that water extracts of *A. bisporus* (at a temperature of 20 °C), *G. lucidum* (at a temperature of 100 °C), and *Agrocybe cylindracea*, as well as ethanol extracts of *T. versicolor, T. hirsuta,* and *T. gibbosa,* have protective effects against H_2O_2-induced DNA damage (Knežević et al. 2015; Ćilerdžić et al. 2016a). Thermolabile protein isolated from *A. bisporus*, β-glucans from *A. brasiliensis* and *A. blazei,* and extracts of Se-enriched *G. lucidum* basidiocarps have a similar effect. Stajić (2015) showed that *Agaricus* spp. glucan induced the genoprotective effect on the binding of benzopyrene that induced damage or neutralized free radicals, and *G. lucidum* inhibited lipid peroxidation.

10 Scientific Basis, Problems, and Perspectives for Mushroom-Based Drugs Development

Mushrooms are commonly used as prophylactics, i.e., agents that act as preventive or protective against some diseases or infections. Therefore, they are considered dietary supplements or nutraceuticals that can be administered alone or in combination with commercial medicines. On the world market, there are several mushroom-based products: (i) powdered cultivated fruiting bodies, their extracts or extracts mixtures; (ii) dried and milled substrate, mycelium, and primordia after cultivation; (iii) biomass or extracts of mycelium obtained by submerged cultivation in the reactor and (iv) dried fruiting bodies of wild and cultivated species in the form of capsules or tablets. Under strictly controlled conditions, mushrooms' commercial submerged and solid-state cultivation guarantees genetic uniformity, a significant mass of mycelium, fruiting bodies, and certain stable structure and quality metabolites, and allows checking chemical and microbiological correctness.

Until a few decades ago, immunomodulatory and anticancer mushroom-based drugs were primarily based on lentinan, schizophilan, and krestin, i.e., mushrooms' polysaccharides. However, due to the high molecular weight of these metabolites, they are obtained by extraction from fruiting bodies, mycelium, or medium after cultivation. Contemporary pharmaceutical trends in cancer prevention include the development of new drugs based on mushroom low molecular weight metabolites (Wasser 2010). This author listed the mechanisms of the cytotoxic activity of several mushrooms and their metabolites. *Ph. linteus* and *Marasmius oreades* caffeic acid, cordycepin from *C. sinensis*, panepoxidone from *Panus* spp., and *Xylaria* spp. cycloepoxidone can inhibit NF-kB. Some mushroom compounds can inhibit proteinases, matrix metalloproteinases, cyclooxygenase, DNA topoisomerase, or DNA polymerase. Some others can block cell division by binding to specific receptors, interrupt communication between growth-regulation enzymes and the development of tumor cells, or inhibit angiogenesis around them. At the same time, some of them increase vitamin D levels in serum, owing to the synthesis of ergosterol and ergocalciferol, and thus cytostatically affect some malignant cells.

However, the process of mushroom-based drug development still has several disadvantages. International standards and protocols for their production and quality testing are still lacking. According to Wasser (2010), 90–95% of mushrooms' β-glucans on the world market are considered counterfeit, leading to numerous side effects. Various species of the genus *Ganoderma* such as *G. lucidum* and species of the genus *Stereum,* replace species of the genus *Trametes*, and various species of the genus *Cordyceps* are used instead of *C. sinensis*. Further, there are many questions requiring answers. For example, it is still unknown whether mycelium and basidiocarp crude extracts are more effective than isolated compounds. Which is the more efficient? The extract of single species or a mixture of several species? What doses are safe and effective? Can certain mushroom-based preparations be administered to children safely? Can mushroom-based preparations be used during pregnancy and breastfeeding? In the end, a serious disadvantage is that mushroom-based drug development is costly, time-consuming, and can even last several years.

Despite all these limitations, the latest data from the World Health Organization demonstrate that 80% of the world population relies on traditional medicines based on active herbal and mushroom ingredients. In the USA, more than 100 million people use various dietary supplements as a safe and natural way of food enrichment to maintain good health. Today special attention is paid to finding new, highly effective immunomodulators that can be used as both precursors of drugs and prophylactics. The indicator of their importance in modern wellness industries is the budget set aside for their production. In 2012, it was US$ 145.9 billion, which increased to US$ 259.3 billion in 2017.

Presently, a large number of new drugs are developed and tested every year. Pollack (2009) reported that about 860 drugs against various types of cancer had been clinically tested. If several medications for heart disease and stroke, neurodegenerative disorders, AIDS, and other infectious diseases are added, it would add not only to the variety of preparations on the market but also to the budget set aside for each year for these purposes. A good example is Pfizer, the largest world pharmaceutical company with a 2021 research budget of US$ 13.8 billion.

If the discovery of antibiotics marked the twentieth century, the twenty-first century was already marked by the discoveries of the medical potential of mushrooms as well as the beginning of the construction of "a bridge" between eastern, traditional, and western conventional medicine. Solving the above-mentioned problems, standardization and production of mushroom-based preparations, and education of the population present the main tasks and challenges of the scientific community.

Acknowledgments This study was financially supported by the Ministry of Education, Science and Technological Development of the Republic of Serbia (451-03-68/2022-14/200178).

References

Abdullah N, Ismail SM, Aminudin N et al (2012) Evaluation of selected culinary-medicinal mushrooms for antioxidant and ACE inhibitory activities. Evid Based Complement Alternat Med 2012. https://doi.org/10.1155/2012/464238

Aida FMNA, Shuhaimi M, Yazid M, Maaruf AG (2009) Mushroom as a potential source of prebiotics: a review. Trends Food Sci Tech 20:567–575. https://doi.org/10.1016/j.tifs.2009.07.007

Alam N, Yoon KN, Lee KR et al (2010) Antioxidant activities and tyrosinase inhibitory effects of different extracts from *Pleurotus ostreatus* fruiting bodies. Mycobiology 38:295–301. https://doi.org/10.4489/MYCO.2010.38.4.295

Alam N, Yoon KN, Lee TS, Lee UY (2011) Hypolipidemic activities of dietary *Pleurotus ostreatus* in hypercholesterolemic rats. Mycobiology 39:45–51. https://doi.org/10.4489/MYCO.2011.39.1.045

Ames BN, Cathcart R, Schwiers E, Hochstein P (1981) Uric acid provides an antioxidant defense in humans against oxidant and radical-caused aging and cancer: a hypothesis. Proc Natl Acad Sci U S A 78:6858–6862

Angeli JP, Ribeiro LR, Gonzaga ML et al (2006) Protective effects of beta-glucan extracted from *Agaricus brasiliensis* against chemically induced DNA damage in human lymphocytes. Cell Biol Toxicol 22:285–291. https://doi.org/10.1007/s10565-006-0087-z

Angeli JP, Ribeiro LR, Bellini MF, Mantovani MS (2009) Beta-glucan extracted from the medicinal mushroom *Agaricus blazei* prevents the genotoxic effects of benzo[a]pyrene in the human hepatoma cell line HepG2. Arch Toxicol 83:81–86. https://doi.org/10.1007/s00204-008-0319-5

Asanuma M, Miyazaki I, Ogawa N (2003) Dopamine-or L-DOPA-induced neurotoxicity: the role of dopamine quinone formation and tyrosinase in a model of Parkinson's disease. Neurotox Res 5:165–176. https://doi.org/10.1007/BF03033137

Ayeka PA (2018) Potential of mushroom compounds as immunomodulators in cancer immunotherapy: a review. Evid Based Complementary Altern 2018. https://doi.org/10.1155/2018/7271509

Barros L, Falcao S, Baptista P et al (2008) Antioxidant activity of *Agaricus* sp. mushrooms by chemical, biochemical and electrochemical assays. Food Chem 111:61–66. https://doi.org/10.1016/j.foodchem.2008.03.033

Behera BC, Verma N, Sonone A, Makhia U (2005) Evaluation of antioxidant potential of the cultured mycobiont of a lichen *Usnea ghattensis*. Phytother Res 19:58–64. https://doi.org/10.1002/ptr.1607

Božić Nedeljković B, Zmijanjac D, Marković M, Vasilijić S, Ćilerdžić J, Stajić M, Vukojević J, Vučević D (2022) Immunomodulatory effects of extract of Lingzhi or Reishi medicinal mushroom *Ganoderma lucidum* (Agaricomycetes) basidiocarps cultivated on alternative substrate. Int J Med Mushrooms 24:45–59. https://doi.org/10.1615/IntJMedMushrooms.2022044452

Brandler T, Al-Harrasi A, Bauer R et al (2020) Botanical drugs and supplements affecting the immune response in the time of COVID-19: implications for research and clinical practice. Phytother Res 35:3013–3031. https://doi.org/10.1002/ptr.7008

Burton JG, Jauniaux E (2010) Oxidative stress. Best Pract Res Clin Obstet Gynaecol 24:1–13. https://doi.org/10.1016/j.bpobgyn.2010.10.016

Camelini MC, Maraschin M, Mendonça MM et al (2005) Structural characterization of b-glucans of *Agaricus brasiliensis* in different stages of fruiting body maturity and their use in nutraceutical products. Biotechnol Lett 27:1295–1299. https://doi.org/10.1007/s10529-005-0222-6

Carneiro AAJ, Ferreira ICFR, Dueñas M et al (2013) Chemical composition and antioxidant activity of dried powder formulations of *Agaricus blazei* and *Lentinus edodes*. Food Chem 138:2168–2173. https://doi.org/10.1016/j.foodchem.2012.12.036

Chang ST, Wasser SP (2012) The role of culinary-medicinal mushrooms on human welfare with a pyramid model for human health. Int J Med Mushrooms 14:95–134. https://doi.org/10.1615/IntJMedMushr.v14.i2.10

Chang CS, Huang SL, Chen S, Chen S (2013) Innate immune responses and efficacy of using mushroom beta-glucan mixture (MBG) on orange-spotted grouper, *Epinephelus coioides*, aquaculture. Fish Shelfish Immunol 35:115–125. https://doi.org/10.1016/j.fsi.2013.04.004

Chaturvedi VK, Agarwal S, Gupta KK et al (2018) Medicinal mushroom: boon for therapeutic applications. 3 Biotech 8:334. https://doi.org/10.1007/s13205-018-1358-0

Chen X, Hu ZP, Yang XX et al (2006) Monitoring of immune responses to a herbal immunomodulator in patients with advanced colorectal cancer. Int Immunopharmacol 6:499–508. https://doi.org/10.1016/j.intimp.2005.08.026

Chen Y, Xie MY, Nie SP, Li C, Wang YX (2008) Purification, composition analysis and antioxidant activity of a polysaccharide from the fruiting bodies of *Ganoderma atrum*. Food Chem 107:231–241. https://doi.org/10.1016/j.foodchem.2007.08.021

Chen X, Wang W, Li S et al (2010) Optimization of ultrasound-assisted extraction of Lingzhi polysaccharides using response surface methodology an dits inhibitory effect on cervical cancer cells. Carbohyd Polim 80:944–948. https://doi.org/10.1016/j.carbpol.2010.01.010

Chen ML, Hsieh CC, Chiang BL, Lin BF (2015) Triterpenoids and polysaccharide fractions of *Ganoderma tsugae* exert different effects on antiallergic activities. Evid Based Complement Alternat Med. https://doi.org/10.1155/2015/754836

Cheng CL, Wang ZY, Cheng L et al (2012) *In vitro* antioxidant activities of water-soluble nucleotide-extract from edible fungi. Food Sci Technol Res 18:405–412. https://doi.org/10.3136/fstr.18.405

Cheung LM, Cheung PCK, Ooi VEC (2003) Antioxidant activity and total phenolics of edible mushroom extracts. Food Chem 81:249–255. https://doi.org/10.1016/S0308-8146(02)00419-3

Cheung LM, Cheung PCK (2005) Mushroom extracts with antioxidant activity against lipid peroxidation. Food Chem 89:403–409. https://doi.org/10.1016/j.foodchem.2004.02.049

Chihara G (1992) Immunopharmacology of lentinan, a polysaccharide isolated from *Lentinus edodes*: its application as a host defense potentiator. Int J Oriental Med 17:57–77

Cho EJ, Hwang HJ, Kim SW et al (2007) Hypoglycemic effects of exopolysaccharides produced by mycelial cultures of two different mushrooms *Tremella fuciformis* and *Phellinus baumii* in ob/ob mice. Appl Microbiol Biotechnol 75:1257–1265. https://doi.org/10.1007/s00253-007-0972-2

Choi SB, Park CH, Choi MK et al (2004) Improvement of insulin resistance and insulin secretion by water extracts of *Cordyceps militaris*, *Phellinus linteus*, and *Paecilomyces tenuipes* in 90% pancreatectomized rats. Biosci Biotechnol Biochem 68:2257–2264. https://doi.org/10.1271/bbb.68.2257

Cui FJ, Li Y, Xu YY et al (2007) Induction of apoptosis in SGC-7901 cells by polysaccharide-peptide GFPS1b from the cultured mycelia of *Grifola frondosa* GF9801. Toxicol In Vitro 21:417–427. https://doi.org/10.1016/j.tiv.2006.10.004

Ćilerdžić J, Stajić M, Vukojević J, Duletić-Laušević S (2013) Oxidative stress and species of genus *Ganoderma* (Higher Basidiomycetes). Int J Med Mushrooms 15:21–28. https://doi.org/10.1615/intjmedmushr.v15.i1.30

Ćilerdžić J, Vukojević J, Stajić M, Stanojković T, Glamočlija J (2014) Biological activity of *Ganoderma lucidum* basidiocarps cultivated on alternative and commercial substrate. J Ethnopharmacol 155:312–319. https://doi.org/10.1016/j.jep.2014.05.036

Ćilerdžić J, Stajić M, Vukojević J, Milovanović I, Muzgonja N (2015) Antioxidant and antifungal potential of *Pleurotus ostreatus* and *Agrocybe cylindracea* basidiocarps and mycelia. Curr Pharm Biotechnol 16:179–186. https://doi.org/10.2174/1389201015666141202152023

Ćilerdžić J, Stajić M, Živković L, Vukojević J, Bajić V, Spremo-Potparević B (2016a) Genoprotective capacity of alternatively cultivated Lingzhi or Reishi Medicinal Mushroom, *Ganoderma lucidum* (Agaricomycetes), basidiocarps. Int J Med Mushrooms 18:1061–1069. https://doi.org/10.1615/IntJMedMushrooms.v18.i12.10

Ćilerdžić J, Kosanić M, Stajić M, Vukojević J, Ranković B (2016b) Species of genus *Ganoderma* (Agaricomycetes) fermentation broth: a novel antioxidant and antimicrobial agent. Int J Med Mushrooms 18:397–404. https://doi.org/10.1615/intjmedmushrooms.v18.i5.30

Ćilerdžić J, Galić M, Vukojević J, Stajić M (2019) *Pleurotus ostreatus* and *Laetiporus sulphureus* (Agaricomycetes): possible agents against Alzheimer and Parkinson diseases. Int J Med Mushrooms 21:275–289. https://doi.org/10.1615/IntJMedMushrooms.2019030136

Dai YC, Zhou LW, Cui BK, Chen YQ, Decock C (2010) Current advances in *Phellinus sensu lato*: medicinal species, functions, metabolites and mechanisms. Appl Microbiol Biotechnol 87: 1587–1593. https://doi.org/10.1007/s00253-010-2711-3

El Enshasy H, Hatti-Kaul R (2013) Mushroom immunomodulators: unique molecules with unlimited applications. Trends Biotechnol 31:668–677. https://doi.org/10.1016/j.tibtech.2013.09.003

El-Hady FKA, Abdel-Aziz MS, Shaker KH, El-Shahid ZA, Ghani MA (2014a) Coral-derived fungi inhibit acelylcholinesterase, superoxide anion radical and microbial activities. Int J Pharm Sci Rev Res 26:301–308

El-Hady FKA, Abdel-Aziz MS, Shaker KH, El-Shahid ZA (2014b) Tyrosinase, acetylcholinesterase inhibitory potential, antioxidant and antimicrobial activities of sponge derived fungi with correlation to their GC/MS analysis. Int J Pharm Sci Rev Res 26:338–345

Elmastas M, Isildak O, Turkekul I, Temur N (2007) Determination of antioxidant activity and antioxidant compounds in wild edible mushrooms. J Food Compos Anal 20:337–345. https://doi.org/10.4172/2157-7110.1000130

Essa MM, Vijayan RK, Castellano-Gonzales G, Memon MA, Braidy N, Guillemin GJ (2012) Neuroprotective effect of natural products against Alzheimer's disease. Neurochem Res 37: 1829–1842. https://doi.org/10.1007/s11064-012-0799-9

Feng Y, Li W, Wu X, He L, Ma S (2010) Rapid and efficient microwave-assisted sulfate modification of lentinan and its antioxidant and antiproliferative activities in vitro. Carbohydr Polym 82:605–612. https://doi.org/10.1016/j.carbpol.2010.05.025

Fortes RC, Novaes M, Recova VL, Melo AL (2009) Immunological, hematological, and glycemia effects of dietary supplementation with *Agaricus sylvaticus* on patients' colorectal cancer. Exp Biol Med 234:53–62. https://doi.org/10.3181/0806-RM-193

Fujimoto S, Furue H, Kimura T, Kondo T, Orita K, Taguchi T, Yoshida K, Ogawa N (1991) Clinical outcome of postoperative adjuvant immunochemotherapy with sizofiran for patients with resectable gastric cancer—a randomized controlled study. Eur J Cancer 27:1114–1118. https://doi.org/10.1016/0277-5379(91)90306-X

Gao JJ, Nakamura N, Min BS, Hirakawa A, Zuo F, Hattori M (2004) Quantitative determination of bitter principles in specimens of *Ganoderma lucidum* using high-performance liquid chromatography and its application to the evaluation of *Ganoderma* products. Chem Pharm Bull (Tokyo) 52:688–695. https://doi.org/10.1248/cpb.52.688

Gordon M, Guralnik M, Kaneko Y, Mimura T, Goodgame J, DeMarzo C, Pierce D, Baker M, Lang W (1995) A phase II controlled study of a combination of the immune modulator, lentinan, with didanosine (ddI) in HIV patients with CD4 cells of 200-500/mm^3. J Med 26:193–207

Gray AM, Flatt PR (1998) Insulin-releasing and insulin-like activity of *Agaricus campestris* (mushroom). J Endocrinol 157:259–266. https://doi.org/10.1677/joe.0.1570259

Grube BJ, Eng ET, Kao YC, Kwon A, Chen S (2001) White button mushroom phytochemicals inhibit aromatase activity and breast cancer cell proliferation. J Nutr 131:3288–3293. https://doi.org/10.1093/jn/131.12.3288

Gu YH, Belury MA (2005) Selective induction of apoptosis in murine skin carcinoma cells (CH72) by an ethanol extract of *Lentinula edodes*. Cancer Lett 220:21–28. https://doi.org/10.1016/j.canlet.2004.06.037

Guillamón E, García-Lafuente A, Lozano M, D'Arrigo M, Rostagno MA, Villares A, Martínez JA (2010) Edible mushrooms: role in the prevention of cardiovascular diseases. Fitoterapia 81:715–723. https://doi.org/10.1016/j.fitote.2010.06.005

Gunde-Cimerman N, Plemenitas A, Cimerman A (1993) *Pleurotus* fungi produce mevinolin, an inhibitor of HMG CoA reductase. FEMS Microbiol Lett 113:333–337. https://doi.org/10.1111/j.1574-6968.1993.tb06536x

Guo JJ, Zhu TB, Collins L, Xiao ZXJ, Kim SH, Chen CY (2007) Modulation of lung cancer growth arrest and apoptosis by *Phellinus linteus*. Mol Carcinog 46:144–154. https://doi.org/10.1002/mc.20275

Halliwell B (2006) Oxidative stress and neurodegeneration: where are we now? J Neurochem 97: 1634–1658. https://doi.org/10.1111/j.1471-4159.2006.03907.x

Harikrishnan R, Balasundaram C, Heo MS (2011) Diet enriched with mushroom *Phellinus linteus* extract enhances the growth, innate immune response, and disease resistance of kelp grouper, *Epinephelus bruneus* against vibriosis. Fish Shellfish Immunol 30:128–134. https://doi.org/10.1016/j.fsi.2010.09.013

Hazama S, Oka M, Yoshino S, Iizuka N, Wadamori K, Yamamoto K et al (1995) Clinical effects and immunological analysis of intraabdominal and intrapleural injection of lentinan for malignant ascites and pleural effusion of gastric carcinoma. Cancer Chemother 22:1595–1597

Heleno SA, Barros L, Sousa MJ, Martins A, Ferreira ICFR (2010) Tocopherols composition of Portuguese wild mushrooms with antioxidant capacity. Food Chem 119:1443–1450. https://doi.org/10.1016/j.foodchem.2009.09.025

Hong KJ, Dunn DM, Shen CL, Pence BC (2004) Effects of *Ganoderma lucidum* on apoptotic and anti-inflammatory function in HT-29 human colonic carcinoma cells. Phytother Res 18:768–770. https://doi.org/10.1002/ptr.1495

Hsu CH, Liao YL, Lin SC, Hwang KC, Chou P (2007) The mushroom *Agaricus blazei* Murill in combination with metformin and gliclazide improves insulin resistance in type 2 diabetes: a randomized, double-blinded, and placebo-controlled clinical trial. J Altern Complement Med 13:97–102. https://doi.org/10.1089/acm.2006.6054

Hsu SC, Ou CC, Li JW et al (2008) *Ganoderma tsugae* extracts inhibit colorectal cancer cell growth via G2/M cell cycle arrest. J Ethnopharmacol 120:394–401. https://doi.org/10.1016/j.jep.2008.09.025

Hu SH, Liang ZC, Chiu CC, Lien JL (2006a) Antihyperlipidemic and antioxidant effects of extracts from *Pleurotus citrinopileatus*. J Agr Food Chem 54:2103–2110. https://doi.org/10.1021/jf052890d

Hu SH, Wang JC, Lien JL, Liaw ET, Lee MY (2006b) Antihyperglycemic effect of polysaccharide from fermented broth of *Pleurotus citrinopileatus*. Appl Microbiol Biotechnol 70:107–113. https://doi.org/10.1007/s00253-005-0043-5

Huang SJ, Tsai SY, Mau JL (2006) Antioxidant properties of methanolic extracts from *Agrocybe cylindracea*. LWT - Food Sci Technol 39:378–386

Hwang HJ, Kim SW, Lim JM et al (2005) Hypoglycemic effect of crude exopolysaccharides produced by a medicinal mushroom *Phellinus baumii* in streptozotocin-induced diabetic rats. Life Sci 76:3069–3080. https://doi.org/10.1016/j.lfs.2004.12.019

Hwang HJ, Kim SW, Yun JW (2007) Modern biotechnology of *Phellinus baumii* – from fermentation to proteomics. Food Technol Biotechnol 45:306–318

Isai M, Elanchezhian R, Sakthivel M, Chinnakkaruppan A, Rajamohan M, Jesudasan CN, Thomas PA, Geraldine P (2009) Anticataractogenic effect of an extract of the oyster mushroom, *Pleurotus ostreatus*, in an experimental animal model. Curr Eye Res 34:264–273. https://doi.org/10.1080/02713680902774069

Ito N, Fukushima S, Tsuda H (1985) Carcinogenicity and modification of the carcinogenic response by BHA, BHT and other antioxidants. CRC Crit Rev Toxicol 15:109–150

Jamila N, Khairuddean M, Yeong KK, Osman H, Murugaiyah V (2015) Cholinesterase inhibitory triterpenoids from the bark of *Garcinia hombroniana*. J Enzyme Inhib Med Chem 30:133–139. https://doi.org/10.3109/14756366.2014.895720

Janardhanan KK, Meera CR, Nitha B, Vishwakarma RA (2009) Anti-inflammatory and free radical scavening activities of poysaccharide-protein complex isolated from *Phellinus rimosus* (Berk.) Pilal (Aphyllophoromycetideae). Int J Med Mushrooms 11:116

Jang JH, Jeong SC, Jeong-Han K, Lee YH (2011) Characterisation of a new antihypertensive angiotensin I-converting enzyme inhibitory peptide from *Pleurotus cornucopiae*. Food Chem 127:412–418. https://doi.org/10.1016/j.foodchem.2011.01.010

Janjušević L, Karaman M, Šibul F, Tommonaro G, Iodice C, Jakovljević D, Pejin B (2017) The lignicolous fungus *Trametes versicolor* (L.) Lloyd (1920): a promising natural source of antiradical and AChE inhibitory agents. J Enzyme Inhib Med Chem 32:355–362. https://doi.org/10.1080/14756366.2016

Jayakumar T, Thomas PA, Geraldine P (2007) Protective effect of an extract of the oyster mushroom, *Pleurotus ostreatus*, on antioxidants of major organs of aged rats. Exp Gerontol 42:183–191. https://doi.org/10.1016/j.exger.2006.10.006

Jayakumar T, Thomas PA, Geraldine P (2009) *In-vitro* antioxidant activities of an ethanolic extract of the oyster mushroom, *Pleurotus ostreatus*. Innov Food Sci Emerg Technol 10:228–234. https://doi.org/10.1016/j.ifset.2008.07.002

Jayakumar T, Thomas PA, Sheu JR, Geraldine P (2011) *In-vitro* and *in-vivo* antioxidant effects of the oyster mushroom *Pleurotus ostreatus*. Food Res Int 44:851–861. https://doi.org/10.1016/j.foodres.2011.03.015

Jesenak M, Majtan J, Rennerova Z, Kyselovic J, Banovcin P, Hrubisko M (2013) Immunomodulatory effect of pleuran (β-glucan from *Pleurotus ostreatus*) in children with recurrent respiratory tract infections. Int Immunopharmacol 15:395–399. https://doi.org/10.1016/j.intimp.2012.11.020

Jia J, Zhang X, Hua YS, Wua Y, Wang QZ, Li NN, Guo QC, Dong XC (2009) Evaluation of *in vivo* antioxidant activities of *Ganoderma lucidum* polysaccharides in STZ-diabetic rats. Food Chem 115:32–36. https://doi.org/10.1016/J.FOODCHEM2008.11.043

Jose N, Ajith TA, Janardhanan KK (2004) Methanol extract of the oyster mushroom, *Pleurotus florida*, inhibits inflammation and platelet aggregation. Phytother Res 18:43–46. https://doi.org/10.1002/ptr.1355

Joseph J, Panicker SN, Janardhanan KK (2012) Protective effect of polysaccharide-protein complex from a polypore mushroom, *Phellinus rimosus* against radiation-induced oxidative stress. Redox Rep 17:22–27. https://doi.org/10.1179/1351000211Y.0000000018

Jung JY, Lee IK, Seok SJ, Lee HJ, Kim YH, Yun BS (2008) Antioxidant polyphenols from the mycelial culture of the medicinal fungi *Inonotus xeranticus* and *Phellinus linteus*. J Appl Microbiol 104:1824–1832. https://doi.org/10.1111/j.1365-2672.2008.03737.x

Kalač P (2009) Chemical composition and nutritional value of European species of wild growing mushrooms: a review. Food Chem 113:9–16. https://doi.org/10.1016/j.foodchem.2008.07.077

Karaman M, Jovin E, Malbaša R, Matavulj M, Popović M (2010) Medicinal and edible lignicolous fungi as natural sources of antioxidative and antibacterial agents. Phytother Res 24:1473–1481. https://doi.org/10.1002/ptr.2969

Kawagishi H, Ishiyama D, Mori H, Sakamoto H, Ishiquro Y, Furukawa S, Li J (1997) Dictyophorines A and B, two stimulators of NGF-synthesis from the mushroom *Dictyophora indusiata*. Phytochemistry 45:1203–1205. https://doi.org/10.1016/S0031-9422(97)00144-1

Keles A, Koca I, Gencelep H (2011) Antioxidant properties of wild edible mushrooms. J Food Process Technol 2. https://doi.org/10.4172/2157-7110.1000130

Kiho T, Yamane A, Hui J, Usui S, Ukai S (1996) Polysaccharides in fungi. XXXVI. Hypoglycemic activity of a polysaccharide (CS-F30) from the cultural mycelium of *Cordyceps sinensis* and its effect on glucose metabolism in mouse liver. Biol Pharm Bull 19:294–296. https://doi.org/10.1248/bpb.19.294

Kiho T, Morimoto H, Kobayashi T, Usui S, Ukai S, Aizawa K, Inakuma T (2000) Effect of a polysaccharide (TAP) from the fruiting bodies of Tremella aurantia on glucose metabolism in mouse liver. Biosci Biotechnol Biochem 64:417–419. https://doi.org/10.1271/bbb.64.417

Kim JI, Kang MJ, Im J, Seo YJ, Lee YM, Song YH, Lee JH, Kim ME (2010) Effect of king oyster mushroom (*Pleurotus eryngii*) on insulin resistance and dyslipidemia in db/db mice. Food Sci Biotechnol 19:239–242. https://doi.org/10.1007/s10068-010-0033-y

Kim GY, Kim SH, Hwang SY, Kim HY, Park YM, Park SK, Lee MK, Lee SH, Lee TH, Lee JD (2003) Oral administration of proteoglycan isolated from *Phellinus linteus* in the prevention and treatment of collagen-induced arthritis in mice. Biol Pharm Bull 26:823–831. https://doi.org/10.1248/bpb.26.823

Kim SW, Kim HG, Lee BE, Hwang HH, Baek DH, Ko SY (2006) Effects of mushroom, *Pleurotus eryngii*, extracts on bone metabolism. Clin Nutr 25:166–170. https://doi.org/10.1016/j.clnu.2005.10.014

Kim SP, Kang MY, Kim JH, Nam SH, Friedman M (2011) Composition and mechanism of antitumor effects of *Hericium erinaceus* mushroom extracts in tumor-bearing mice. J Agric Food Chem 59:9861–9869. https://doi.org/10.1021/jf201944n

Knežević A, Živković L, Stajić M, Spremo-Potparević B, Vukojević J, Milovanović I (2015) Antigenotoxic effect of *Trametes* spp extracts against DNA damage on human peripheral white blood cells. Sci World J. https://doi.org/10.1155/2015/146378

Knežević A, Stajić M, Sofrenić I, Stanojković T, Milovanović I, Tešević V, Vukojević J (2018) Extracts of *Trametes* species as new antioxidative, antifungal, cytostatic and antineurodegenerative agents. PLoS One 13. https://doi.org/10.1371/journal.pone.0203064

Kodama N, Komuta K, Nanba H (2002) Can maitake MD-fraction aid cancer patients? Altern Med Rev 7:236–239

Kosanić M, Ranković B (2011) Antioxidant and antimicrobial properties of some lichens and their constituents. J Med Food 14:1624–1630. https://doi.org/10.1089/jmf.2010.0316

Kozarski M, Klaus A, Nikšić M, Vrvić MM, Todorović N, Jakovljević D, Van Griensven LJLD (2012) Antioxidative activities and chemical characterization of polysaccharide extracts from the widely used mushrooms *Ganoderma applanatum, Ganoderma lucidum, Lentinus edodes* and *Trametes versicolor*. J Food Comp Anal 26:144–153. https://doi.org/10.1016/j.jfca.2012.02.004

Kubo K, Aoki H, Nanba H (1994) Anti-diabetic activity present in the fruit body of *Grifola frondosa* (Maitake). Biol Pharm Bull 17:1106–1110. https://doi.org/10.1248/bpb.17.1106

Kwon AH, Qiu Z, Hashimoto M, Yamamoto K, Kimura T (2009) Effects of medicinal mushroom (*Sparassis crispa*) on wound healing in streptozotocin-induced diabetic rats. Am J Surg 197:503–509. https://doi.org/10.1016/j.amjsurg.2007.11.021

Lai CSW, Yu MS, Yuen WH, So KF, Zee SY, Chang RC (2008) Antagonizing beta-amyloid peptide neurotoxicity of the anti-aging fungus *Ganoderma lucidum*. Brain Res 1190:215–224. https://doi.org/10.1016/j.brainres.2007.10.103

Lau CC, Abdullah N, Shuib AS, Aminudin N (2012) Proteomic analysis of antihypertensive proteins in edible mushrooms. J Agric Food Chem 60:12341–12348. https://doi.org/10.1021/jf3042159

Lee DH, Kim JH, Park JS, Choi YJ, Lee JS (2004) Isolation and characterization of a novel angiotensin I-converting enzyme inhibitory peptide derived from the edible mushroom *Tricholoma giganteum*. Peptides 25:621–627. https://doi.org/10.1016/j.peptides.2004.01.015

Lee IK, Yun BS (2007) Highly oxygenated and unsaturated metabolites providing a diversity of hispidin class antioxidants in the medicinal mushrooms *Inonotus* and *Phellinus*. Bioorg Med Chem 15:3309–3314. https://doi.org/10.1016/j.bmc.2007.03.039

Lee YL, Huang GW, Liang ZC, Mau JL (2007) Antioxidant properties of three extracts from *Pleurotus citrinopileatus*. LWT - Food Sci Technol 40:823–833. https://doi.org/10.1016/j.lwt.2006.04.002

Lee YS, Kang YH, Jung JY, Lee S, Ohuchi K, Shin KH, Kang IJ, Park JH, Shin HK, Lim SS (2008) Protein glycation inhibitors from the fruiting body of *Phellinus linteus*. Biol Pharm Bull 31:1968–1972. https://doi.org/10.1248/bpb.31.1968

Lee GS, Byun HS, Yoon KH, Lee JS, Choi KC, Jeung EB (2009) Dietary calcium and vitamin D2 supplementation with enhanced *Lentinula edodes* improves osteoporosis-like symptoms and

induces duodenal and renal active calcium transport gene expression in mice. Eur J Nutr 48:75–83. https://doi.org/10.1007/s00394-008-0763-2
Lee SH, Lillehoj HS, Hong YH, Jang SI, Lillehoj EP, Ionescu C, Mazuranok L, Bravo D (2010) *In vitro* effects of plant and mushroom extracts on immunological function of chicken lymphocytes and macrophages. Br Poult Sci 51:213–221. https://doi.org/10.1080/00071661003745844
Lee B, Park J, Park J, Shin HJ, Kwon S, Yeom M, Sur B, Kim S, Kim M, Lee H, Yoon SH, Hahmet DH (2011) *Cordyceps militaris* improves neurite outgrowth in Neuro2a cells and reverses memory impairment in rats. Food Sci Biotechnol 20:1599–1608. https://doi.org/10.1007/s10068-011-0221-4
Leopoldini M, Russo N, Toscano M (2011) The molecular basis of working mechanism of natural polyphenolic antioxidants. Food Chem 125:228–306. https://doi.org/10.1016/j.foodchem.2010.08.012
Leung PH, Zhao S, Ho KP, Wu JY (2009) Chemical properties and antioxidant activity of exopolysaccharides from mycelial culture of *Cordyceps sinensis* fungus Cs-HK1. Food Chem 114:1251–1256. https://doi.org/10.1016/j.foodchem.2008.10.081
Li TH, Hou CC, Chang CL, Yang WC (2011) Anti-hyperglycemic properties of crude extract and triterpenes from *Poria cocos*. Evid Based Complement Alternat Med. https://doi.org/10.1155/2011/128402
Li WJ, Nie SP, Liu XZ, Zhang H, Yang Y, Yu Q, Xie MY (2012) Antimicrobial properties, antioxidant activity and cytotoxicity of ethanol-soluble acidic components from *Ganoderma atrum*. Food Chem Toxicol 50:689–694. https://doi.org/10.1016/j.fct.2011.12.011
Limón-Pacheco J, Gonsebatt ME (2009) The role of antioxidants and antioxidant-related enzymes in protective responses to environmentally induced oxidative stress. Mutat Res 674:137–147. https://doi.org/10.1016/j.mrgentox.2008.09.015
Lin ZB, Wang MY, Liu G, Che QM (2002) Effects of total triterpenoids extract from *Ganoderma lucidum* (Curt.: Fr.) P. Karst. (Reishi Mushroom) on experimental liver injury models induced by carbon tetrachloride or D-Galactosamine in mice. Int J Med Mushrooms 4. https://doi.org/10.1615/IntJMedMushr.v4.i4.70
Lindequist U, Niedermezer THJ, Jülich WD (2005) The pharmacological potential of mushrooms. Evid Based Complement Alternat Med 2:285–299. https://doi.org/10.1093/ecam/neh107
Liu X, Zhou B, Lin R, Jia L, Deng P, Fan K, Wang G, Wang L, Zhang J (2010a) Extraction and antioxidant activities of intracellular polysaccharide from *Pleurotus* sp. mycelium. Int J Biol Macromol 47:116–119. https://doi.org/10.1016/j.ijbiomac.2010.05.012
Liu W, Wang H, Pang X, Yao W, Gao X (2010b) Characterization and antioxidant activity of two low-molecular-weight polysaccharides purified from the fruiting bodies of *Ganoderma lucidum*. Int J Biol Macromol 46:451–457. https://doi.org/10.1016/j.ijbiomac.2010.02.006
Lo KM, Peter CK, Cheung PCK (2005) Antioxidant activity of extracts from the fruiting bodies of *Agrocybe aegerita* var. *alba*. Food Chem 89:533–539. https://doi.org/10.1016/j.foodchem.2004.03.006
Lo HC, Tsai FA, Wasser SP, Yang JG, Huang BM (2006) Effects of ingested fruiting bodies, submerged culture biomass, and acidic polysaccharide glucuronoxylomannan of *Tremella mesenterica* Retz.:Fr. on glycemic responses in normal and diabetic rats. Life Sci 78:1957–1966. https://doi.org/10.1016/j.lfs.2005.08.033
Lo HC, Wasser SP (2011) Medicinal mushrooms for glycemic control in diabetes mellitus: history, current status, future perspectives, and unsolved problems (Review). Int J Med Mushrooms 13:401–426. https://doi.org/10.1615/intjmedmushr.v13.i5.10
Louie B, Rajamahanty S, Won J, Choudhury M, Konno S (2010) Synergistic potentiation of interferon activity with maitake mushroom D-fraction on bladder cancer cells. BJU Int 105:1011–1015. https://doi.org/10.1111/j.1464-410X.2009.08870.x
Lu X, Chen H, Dong P, Fu L, Zhang X (2010) Phytochemical characteristics and hypoglycaemic activity of fraction from mushroom *Inonotus obliquus*. J Sci Food Agric 90:276–280. https://doi.org/10.1002/jsfa.3809

Lv Y, Han L, Yuan C, Guo J (2009) Comparison of hypoglycemic activity of trace elements absorbed in fermented mushroom of *Coprinus comatus*. Biol Trace Elem Res 131:177–185. https://doi.org/10.1007/s12011-009-8352-7

Ma BJ, Shen JW, Yu HY, Ruan Y, Wu TT, Zhao X (2010) Hericenones and erinacines: stimulators of nerve growth factor (NGF) biosynthesis in *Hericium erinaceus*. Mycology 1:92–98. https://doi.org/10.1080/21501201003735556

Manzi P, Gambelli L, Marconi S, Vivanti V, Pizzoferrato L (1999) Nutrients in edible mushrooms: an inter-species comparative study. Food Chem 65:477–482. https://doi.org/10.1016/S0308-8146(98)00212-X

Manzi P, Marconi S, Aguzzi A (2004) Commercial mushrooms: nutritional quality and effect of cooking. Food Chem 84:201–206. https://doi.org/10.1016/S0308-8146(03)00202-4

Marcotullio MC, Pagiott R, Maltese F, Obara Y, Hoshino T, Nakahata N, Curini M (2006) Neurite outgrowth activity of cyathane diterpenes from *Sarcodon cyrneus*, cyrneines A and B. Planta Med 72:819–823. https://doi.org/10.1055/s-2006-946681

Marcotullio MC, Pagiott R, Maltese F, Oball-Mond Mwankie GN, Hoshino T, Obara Y, Nakahata N (2007) Cyathane diterpenes from *Sarcodon cyrneus* and evaluation of their activities of neuritegenesis and nerve growth factor production. Bioorg Med Chem 15:2878–2882. https://doi.org/10.1016/j.bmc.2007.02.019

Mau JL, Lin HC, Chen CC (2002) Antioxidant properties of several medicinal mushrooms. J Agric Food Chem 50:6072–6077. https://doi.org/10.1021/jf0201273

Mau JL, Tsai SY, Tseng YH, Huang SJ (2005) Antioxidant properties of hot water extracts from *Ganoderma tsugae* Murrill. LWT - Food Sci Technol 38:589–597. https://doi.org/10.1016/j.lwt.2004.08.010

Meneses ME, Martínez-Carrera D, Torres N, Sánchez-Tapia M, Aguilar-López M, Morales P, Sobal M, Bernabé T, Escudero H, Granados-Portillo O, Tovar AR (2016) Hypocholesterolemic properties and prebiotic effects of Mexican *Ganoderma lucidum* in C57BL/6 Mice. PLoS One 11. https://doi.org/10.1371/journal.pone.0159631

Mills KT, Stefanescu A, He J (2021) The global epidemiology of hypertension. Nat Rev Nephrol 16:223–237. https://doi.org/10.1038/s41581-019-0244-2

Milovanović I, Stajić M, Stanojković T, Knežević A, Vukojević J (2015a) Effects of selenium presence in mycelia of *Ganoderma* species (Higher Basidiomycetes) on their medicinal properties. Int J Med Mushrooms 17:11–20. https://doi.org/10.1615/intjmedmushrooms.v17.i1.20

Milovanović I, Stanojković T, Stajić M, Vukojević J, Knežević A (2015b) Se effect on biological activity of *Flammulina velutipes*. Ital J Food Sci 27:1–7. https://doi.org/10.14674/1120-1770/IJFS.V27I1.74

Milovanović I, Stanojković T, Stajić M, Brčeski I, Knežević A, Ćilerdžić J, Vukojević J (2015c) Effect of selenium enrichment of *Lenzites betulinus* and *Trametes hirsuta* mycelia on antioxidant, antifungal and cytostatic potential. Curr Pharm Biotechnol 16:920–926. https://doi.org/10.2174/1389201016666150618152531

Mishra KK, Pal RS, Arunkumar R, Chandrashekara C, Jain SK, Bhatt JC (2013) Antioxidant properties of different edible mushroom species and increased bioconversion efficiency of *Pleurotus eryngii* using locally available casing materials. Food Chem 138:1557–1563. https://doi.org/10.1016/j.foodchem.2012.12.001

Mitomi T, Tsuchiya S, Iijima N, Aso K, Suzuki K, Nishiyama K, Amano T, Takahashi T, Murayama N, Oka H, Oya K, Noto T, Ogawa N (1992) Randomized, controlled study on adjuvant immunochemotherapy with PSK in curatively resected colorectal cancer. Dis Colon Rectum 35:123–130

Mlinrič A, Kac J, Fatur T, Filini M (2004) Anti-genotoxic activity of the mushroom *Lactarius vellereus* extract in bacteria and in mammalian cells in vitro. Pharmazie 59:217–221

Mori K, Obara Y, Hirota M, Azumi Y, Kinugasa S, Inatomi S, Nakahata N (2008) Nerve growth factor-inducing activity of *Hericium erinaceus* in 1321N1 human astrocytoma cells. Biol Pharm Bull 31:1727–1732. https://doi.org/10.1248/bpb.31.1727

Mujić I, Zeković Z, Lepojević Ž, Vidović S, Živković J (2010) Antioxidant properties of selected edible mushroom species. J Cent Eur Agric 11:387–392. https://doi.org/10.5513/jcea.v11i4.70

Neyrinck AM, Bindels LB, De Backer F, Pachikian BD, Cani PD, Delzenne NM (2009) Dietary supplementation with chitosan derived from mushrooms changes adipocytokine profile in diet-induced obese mice, a phenomenon linked to its lipid-lowering action. Int Immunopharmacol 9: 767–773. https://doi.org/10.1016/j.intimp.2009.02.015

Nishina A, Kimura H, Sekiguchi A, Fukumoto RH, Nakajima S, Furukawa S (2006) Lysophosphatidylethanolamine in *Grifila frondosa* as a neurotrophic activator via activation of MAPK. J Lipid Res 47:1434–1443. https://doi.org/10.1194/jlr.M600045-JLR200

Nishihira J, Sato M, Tanaka A, Okamatsu M, Azuma T, Tsutsumi N, Yoneyama S (2017) Maitake mushrooms (*Grifola frondosa*) enhances antibody production in response to influenza vaccination in healthy adult volunteers concurrent with alleviation of common cold symptoms. Funct Foods Health Dis 7:462–482. https://doi.org/10.31989/ffhd.v7i7.363

Oh TW, Kim YA, Jang WJ, Byeon JI, Ryu CH, Kim JO, Ha YL (2010) Semipurified fractions from the submerged-culture broth of *Agaricus blazei* Murill reduce blood glucose levels in streptozotocin-induced diabetic rats. J Agric Food Chem 58:4113–4119. https://doi.org/10.1021/jf9036672

Oke F, Aslim B (2011) Protective effect of two edible mushrooms against oxidative cell damage and their phenolic composition. Food Chem 128:613–619. https://doi.org/10.1016/j.foodchem.2011.03.036

Parker L (2014) The impact of the environment on cancer genomics. In: Dellaire G, Berman J, Arceci R (eds) Cancer genomics from bench to personalized medicine. Elsevier, London, pp 449–465

Patel S, Goyal A (2012) Recent developments in mushroom as anti-cancer therapeutics: a review. 3 Biotech 2:1–15. https://doi.org/10.1007/s13205-011-0036-2

Patocka J (2012) Natural cholinesterase inhibitors from mushrooms. Mil Med Sci Lett 81:40–44

Petrova RD, Reznick AZ, Wasser SP, Denchev CM, Nevo E, Mahajna J (2008) Fungal metabolites modulating NF-κB activity: an approach to cancer therapy and chemoprevention (Review). Oncol Rep 19:299–308. https://doi.org/10.3892/or.19.2.299

Pollack A (2009) Drug firms see fortune in treating cancer. International Herald Tribune, pp 15–16

Phan CW, Wong WL, David P, Naidu M, Sabaratnam V (2012) *Pleurotus giganteus* (Berk.) Karunarathna & K.D. Hyde nutritional value and *in vitro* neurite outgrowth activity in rat pheochromocytoma cells. BMC Complement Altern Med 12:102. https://doi.org/10.1186/1472-6882-12-102

Phan CW, David P, Naidu M, Wong KH, Sabaratnam V (2014) Therapeutic potential of culinary-medicinal mushrooms for the management of neurodegenerative diseases: diversity, metabolite, and mechanism. Crit Rev Biotechnol 35:355–368. https://doi.org/10.3109/07388551.2014.887649

Reis FS, Martins A, Barros L, Ferreira ICFR (2012a) Antioxidant properties and phenolic profile of the most widely appreciated cultivated mushrooms: a comparative study between *in vivo* and in vitro samples. Food Chem Toxicol 50:1201–1207. https://doi.org/10.1016/j.fct.2012.02.013

Reis FS, Barros L, Martins A, Ferreira ICFR (2012b) Chemical composition and nutritional value of the most widely appreciated cultivated mushrooms: an inter-species comparative study. Food Chem Toxicol 50:191–197. https://doi.org/10.1016/j.fct.2011.10.056

Ren L, Perera C, Hemar Y (2012) Antitumor activity of mushroom polysaccharides: a review. Food Funct 3:1118–1130. https://doi.org/10.1039/c2fo10279j

Reuter S, Gupta SC, Chaturved MM, Aggarwal BB (2010) Oxidative stress, inflammation, and cancer: how are they linked? Free Radic Biol Med 49:1603–1616. https://doi.org/10.1016/j.freeradbiomed.2010.09.006

Rodríguez-Valentín M, López S, Rivera M, Ríos-Olivares E, Cubano L, Boukli NM (2018) Naturally derived anti-HIV polysaccharide peptide (PSP) triggers a toll-like receptor 4-dependent antiviral immune response. J Immunol Res. https://doi.org/10.1155/2018/8741698

Roupas P, Keogh J, Noakes M, Margetts C, Taylor P (2012) The role of edible mushrooms in health: evaluation of the evidence. J Funct Foods 4:687–709. https://doi.org/10.1016/j.jff.2012.05.003

Saltarelli R, Ceccaroli P, Iotti M, Zambonelli A, Buffalini M, Casadei L, Vallorani L, Stocchi V (2009) Biochemical characterisation and antioxidant activity of mycelium of *Ganoderma lucidum* from Central Italy. Food Chem 116:143–151. https://doi.org/10.1016/j.foodchem.2009.02.023

Sarmadi BH, Ismail A (2010) Antioxidative peptides from food proteins: a review. Peptides 31:1949–1956. https://doi.org/10.1016/j.peptides.2010.06.020

Schneider I, Kressel G, Meyer A, Krings U, Berger RG, Hahn A (2011) Lipid lowering effects of oyster mushroom (*Pleurotus ostreatus*) in humans. J Funct Food 3:17–24. https://doi.org/10.1016/j.jff.2010.11.004

Şenol FS, Orhan I, Celep F, Kahraman A, Doğan M, Yilmaz G et al (2010) Survey of 55 Turkish *Salvia* taxa for their acetylcholinesterase inhibitory and antioxidant activities. Food Chem 120:34–43. https://doi.org/10.1016/j.foodchem.2009.09.066

Shimizu K, Yamanaka M, Gyokusen M, Kaneko S, Tsutsui M, Sato J, Sato I, Sato M, Kondo R (2006) Estrogen-like activity and prevention effect of bone loss in calcium deficient ovariectomized rats by the extract of *Pleurotus eryngii*. Phytother Res 20:659–664. https://doi.org/10.1002/ptr.1927

Shin S, Lee S, Kwon J, Moon S, Lee S, Lee CK, Cho K, Ha NJ, Kim K (2009) Cordycepin suppresses expression of diabetes regulating genes by inhibition of lipopolysaccharide-induced inflammation in macrophages. Immune Netw 9:98–105. https://doi.org/10.4110/in.2009.9.3.98

Soares AA, Marques de Souza CG, Daniel FM, Ferrari GP, Gomes da Costa SM, Peralta RM (2009) Antioxidant activity and total phenolic content of *Agaricus brasiliensis* (*Agaricus blazei* Murril) in two stages of maturity. Food Chem 112:775–781. https://doi.org/10.1016/J.FOODCHEM.2008.05.117

Song TY, Yen GC (2002) Antioxidant properties of *Antrodia camphorata* in submerged culture. J Agric Food Chem 50:3322–3327. https://doi.org/10.1021/jf011671z

Stajić M, Vukojević J, Knežević A, Duletić Lauševič S, Milovanović I (2013) Antioxidant protective effects of mushroom metabolites. Curr Top Med Chem 13:2660–2676. https://doi.org/10.2174/15680266113136660192

Stajić M (2015) Nutritivna svojstva i medicinski potencijal makromiceta. Biološki fakultet Univerziteta u Beogradu (Nutritional properties and medicinal potential of macromycetes. Faculty of Biology, University of Belgrade), p 219

Stajić M, Ćilerdžić J, Vukojević J (2017) Mushrooms as potent sources of new biofungicides. Curr Pharm Biotechnol 18:1055–1066. https://doi.org/10.2174/1389201019666180115145412

Stajić M, Vukojević J, Ćilerdžić J (2019) Mushrooms as potential natural cytostatics. In: Agrawal DC, Dhanasekaran M (eds) Medicinal mushrooms. Recent progress in research and development. Springer, Singapore, pp 143–168

Taofiq O, Heleno SA, Calhelha RC, Alves MJ, Barros L, Barreiro MF, González-Paramás AM, Ferreira ICFR (2016) Development of mushroom-based cosmeceutical formulations with anti-inflammatory, anti-tyrosinase, antioxidant, and antibacterial properties. Molecules 21. https://doi.org/10.3390/molecules21101372

Tel G, Apaydin M, Duru ME, Öztürk M (2011) Antioxidant and cholinesterase inhibition activities of three *Tricholoma* species with total phenolic and flavonoid contents: the edible mushrooms from Anatolia. Food Anal Methods 5:495–504. https://doi.org/10.1007/s12161-011-9275-4

Tessari I, Bisaglia M, Valle F, Samorì B, Bergantino E, Mammi S, Bubacco L (2008) The reaction of α-synuclein with tyrosinase possible implications for Parkinson disease. J Biol Chem 283:16808–16817. https://doi.org/10.1074/jbc.M709014200

Tsai SY, Huang SJ, Mau JL (2006) Antioxidant properties of hot water extracts from *Agrocybe cylindracea*. Food Chem 98:670–677. https://doi.org/10.1016/j.foodchem.2005.07.003

Tsai SY, Tsai HL, Mau JL (2007) Antioxidant properties of *Agaricus blazei*, *Agrocybe cylindracea*, and *Boletus edulis*. LWT - Food Sci Technol 40:1392–1402. https://doi.org/10.1016/j.lwt.2006.10.001

Tsai SY, Huang SJ, Lo SH, Wu TP, Lian PY, Mau JL (2009) Flavour components and antioxidant properties of several cultivated mushrooms. Food Chem 113:578–584. https://doi.org/10.1016/j.foodchem.2008.08.034

Tsang AHK, Chung KKK (2009) Oxidative and nitrosative stress in Parkinson's disease. Biochim Biophys Acta 1792:643–650. https://doi.org/10.1016/j.bbadis.2008.12.006

Tseng YH, Yang JH, Mau JL (2008) Antioxidant properties of polysaccharides from *Ganoderma tsugae*. Food Chem 107:732–738. https://doi.org/10.1016/j.foodchem.2007.08.073

Turkoglu A, Duru ME, Mercan N, Kivrak I, Gezer K (2007) Antioxidant and antimicrobial activities of *Laetiporus sulphureus* (Bull.) Murrill. Food Chem 101:267–273. https://doi.org/10.1016/j.foodchem.2006.01.025

Turło J, Gutkowska B, Herold F (2010) Effect of selenium enrichment on antioxidant activities and chemical composition of *Lentinula edodes* (Berk.) Pegl. mycelial extracts. Food Chem Toxicol 48:1085–1091. https://doi.org/10.1016/j.fct.2010.01.030

Vaz JA, Heleno SA, Martins A, Almeida GM, Vasconcelos MH, Ferreira ICFR (2010) Wild mushrooms *Clitocybe alexandri* and *Lepista inversa*: in vitro antioxidant activity and growth inhibition of human tumour cell lines. Food Chem Toxicol 48:2881–2884. https://doi.org/10.1016/j.fct.2010.07.021

Vaz JA, Barros L, Martins A, Morais JS, Vasconcelos MH, Ferreira ICFR (2011) Phenolic profile of seventeen Portuguese wild mushrooms. LWT - Food Sci Technol 44:343–346. https://doi.org/10.1016/J.LWT.2010.06.029

Vetter J (2007) Chitin content of cultivated mushrooms *Agaricus bisporus*, *Pleurotus ostreatus* and *Lentinula edodes*. Food Chem 102:6–9. https://doi.org/10.1016/j.foodchem.2006.01.037

Wang MF, Chan YC, Wu CL, Wong YC, Hosoda K, Yamamoto S (2004) Effects of *Ganoderma* on aging and learning and memory ability in senescence accelerated mice. Int Congr Ser 1260:399–404

Wang LC, Wang SE, Wang JJ, Tsai TY, Lin CH, Pan TM, Lee CL (2012) *In vitro* and *in vivo* comparisons of the effects of the fruiting body and mycelium of *Androdia comphorata* against amyloid β-protein-induced teurotoxicity and memory impairment. Appl Microbiol Biotechnol 94:1505–1519. https://doi.org/10.1007/s00253-012-3941-3

Wang XM, Zhang J, Wu LH, Zhao YL, Li T, Li JQ, Wang YZ, Liu HG (2014) A mini-review of chemical composition and nutritional value of edible wild-growth mushroom from China. Food Chem 151:279–285. https://doi.org/10.1016/j.foodchem.2013.11.062

Wasser SP (2010) Medicinal mushroom science: history, current status, future trends, and unsolved problems. Int J Med Mushrooms 12:1–16. https://doi.org/10.1615/IntJMedMushr.v12.i1.10

Wei WS, Tan JQ, Guo F, Ghen HS, Zhou ZY, Zhang ZH, Gui L (1996) Effects of *Coriolus versicolor* polysaccharides on superoxide dismutase activities in mice. Acta Pharmacol Sin 17:174–178

Wild S, Roglic G, Green A, Sicree R, King H (2004) Global prevalence of diabetes: estimates for 2000 and projections for 2030. Diabetes Care 27:1047–1053. https://doi.org/10.2337/diacare.27.5.1047

Wińska K, Mączka W, Gabryelska K, Grabarczyk M (2019) Mushrooms of the genus *Ganoderma* used to treat diabetes and insulin resistance. Molecules 24. https://doi.org/10.3390/molecules24224075

Wu JY, Siu KC, Geng P (2021) Bioactive ingredients and medicinal values of *Grifola frondosa* (Maitake). Foods 10:95. https://doi.org/10.3390/foods10010095

Xia F, Fan J, Zhu M, Tong H (2011) Antioxidant effects of a water-soluble proteoglycan isolated from the fruiting bodies of *Pleurotus ostreatus*. J Taiwan Inst Chem Eng 42:402–407. https://doi.org/10.1016/j.jtice.2010.08.012

Yahaya NFM, Rahman MA, Abdullah N (2014) Therapeutic potential of mushrooms in preventing and ameliorating hypertension. Trends Food Sci Technol 39:104–115. https://doi.org/10.1016/j.tifs.2014.06.002

Yamaç M, Bilgili F (2006) Antimicrobial activities of fruit bodies and/or mycelial cultures of some mushroom isolates. Pharm Biol 44:660–667. https://doi.org/10.1080/13880200601006897

Yamac M, Kanbak G, Zeytinoglu M, Senturk H, Bayramoglu G, Dokumacioglu A, van Griensven LJ (2010) Pancreas protective effect of button mushroom *Agaricus bisporus* (J.E. Lange) Imbach (Agaricomycetidae) extract on rats with streptozotocin-induced diabetes. Int J Med Mushrooms 12:379–389. https://doi.org/10.1615/IntJMedMushr.v12.i4.50

Yang QY (1993) A new biological response modifier – PSP. In: Chang ST (ed) Mushroom biology and mushroom products. The Chinese University Press, Hong Kong, pp 247–259

Yang JH, Lin HC, Mau JL (2002a) Antioxidant properties of several commercial mushrooms. Food Chem 77:229–235. https://doi.org/10.1016/S0308-8146(01)00342-9

Yang BK, Kim DH, Jeong SC, Das S, Choi YS, Shin JS, Lee SC, Song CH (2002b) Hypoglycemic effect of a *Lentinus edodes* exo-polymer produced from a submerged mycelial culture. Biosci Biotechnol Biochem 66:937–942. https://doi.org/10.1271/bbb.66.937

Yang BK, Jung YS, Song CH (2007) Hypoglycemic effects of *Ganoderma applanatum* and *Collybia confluens* exo-polymers in streptozotocin-induced diabetic rats. Phytother Res 21: 1066–1069. https://doi.org/10.1002/ptr.2214

Yang BK, Kim GN, Jeong YT, Jeong H, Mehta P, Song CH (2008) Hypoglycemic effects of exo-biopolymers produced by five different medicinal mushrooms in STZ-induced diabetic rats. Mycobiology 36:45–49. https://doi.org/10.4489/MYCO.2008.36.1.045

Youn MJ, Kim JK, Park SY, Kim Y, Park C, Kim ES, Park KI, So HS, Park R (2009) Potential anticancer properties of the water extract of *Inontus obliquus* by induction of apoptosis in melanoma B16–F10 cells. J Ethnopharmacol 121:221–228. https://doi.org/10.1016/j.jep.2008.10.016

Yu ZH, Hua YL, Qian Y, Yan L (2009) Effect of *Lentinus edodes* polysaccharide on oxidative stress, immunity activity and oral ulceration of rats stimulated by phenol. Carbohydr Polym 75: 115–118. https://doi.org/10.1016/j.carpol.2008.07.002

Yu CC, Chiang PC, Lu PH, Kuo MT, Wen WC, Chen P, Guh JH (2012) Antroquinonol, a natural ubiquinone derivative, induces a cross talk between apoptosis, autophagy and senescence in human pancreatic carcinoma cells. J Nutr Biochem 23:900–907. https://doi.org/10.1016/j.jnutbio.2011.04.015

Yuan C, Mei Z, Liu S, Yi L (1996) PSK protects macrophages from lipoperoxide accumulation and foam cell formation caused by oxidatively modified low-density lipoprotein. Atherosclerosis 124:171–181. https://doi.org/10.1016/0021-9150(96)05835-2

Yuan Z, He P, Cui J, Takeuchi H (1998) Hypoglycemic effect of water-soluble polysaccharide from *Auricularia auricula-judae* Quel. on genetically diabetic KK-Ay mice. Biosci Biotechnol Biochem 62:1898–1903. https://doi.org/10.1271/bbb.62.1898

Yun YH, Han SH, Lee SJ, Ko SK, Lee CK, Ha NJ, Kim KJ (2003) Anti-diabetic effects of CCCA, CMESS, and Cordycepin from *Cordyceps militaris* and the immune responses in streptozotocin-induced diabetic mice. Nat Prod Sci 9:291–298

Zaidman BZ, Wasser SP, Nevo E, Mahajan J (2008) *Coprinus comatus* and *Ganoderma lucidum* interfere with androgen receptor function in LNCaP prostate cancer cells. Mol Biol Rep 35:107–117. https://doi.org/10.1007/s11033-007-9059-5

Zhang M, Cui SW, Cheung PCK, Wang Q (2007) Antitumor polysaccharides from mushrooms: a review on their isolation process, structural characteristics and antitumor activity. Trends Food Sci Technol 18:4–19. https://doi.org/10.1016/j.tifs.2006.07.013

Zhang ZC, Lian B, Huang DM, Cui FJ (2009) Compare activities on regulating lipid-metabolism and reducing oxidative stress of diabetic rats of *Tremella aurantialba* broth's extract (TBE) with its mycelia polysaccharides (TMP). J Food Sci 74:H15–H21. https://doi.org/10.1111/j.1750-3841.2008.00989.x

Zhang T, Zhao W, Xie B, Liu H (2020) Effects of *Auricularia auricula* and its polysaccharide on diet-induced hyperlipidemia rats by modulating gut microbiota. J Funct Foods 72. https://doi.org/10.1016/j.jff.2020.104038

Zhao L, Zhao GH, Du M, Zhao ZD, Xiao LX, Hu XS (2008) Effect of selenium on increasing free radical scavenging activities of polysaccharide extracts from a Se-enriched mushroom species of the genus *Ganoderma*. Eur Food Res Technol 226:499–505. https://doi.org/10.1007/s00217-007-0562-7

Zhu X, Raina AK, Lee H, Casadesus G, Smith MA, Perry G (2004) Oxidative stress signaling in Alzheimer's disease. Brain Res 1000:32–39. https://doi.org/10.1016/j.brainres.2004.01.012

Biopharmaceutical Potential of *Ophiocordyceps sinensis* for Human Health

Li-Yin Pang, Sue-Mian Then, Kuan-Hon Lim, Kang-Nee Ting, and Shin-Yee Fung

Contents

Abstract *Ophiocordyceps sinensis* is an entomopathogenic fungus used in traditional Chinese and Tibetan medicine for over 2000 years. It is also commonly consumed in Asia for nutritional purposes. To authenticate the various traditional medicinal claims of *O. sinensis*, researchers have begun to investigate its pharmacological properties and bioactive compounds since the 1990s. Recently, a large proportion of research used the artificially cultivated *O. sinensis* due to the high cost and limited supply of its natural counterpart. Polysaccharides, proteins, amino acids,

L.-Y. Pang · S.-M. Then · K.-H. Lim · K.-N. Ting (✉)
School of Pharmacy, University of Nottingham Malaysia, Jalan Broga, Selangor, Malaysia
e-mail: hsxlp1@nottingham.edu.my; then.sue-mian@nottingham.edu.my; kuanhon.Lim@nottingham.edu.my; kang-nee.ting@nottingham.edu.my

S.-Y. Fung (✉)
Faculty of Medicine, Department of Molecular Medicine, Medicinal Mushroom Research Group (MMRG), Universiti Malaya, Kuala Lumpur, Malaysia
e-mail: syfung@um.edu.my; syfung@ummc.edu.my

D. C. Agrawal, M. Dhanasekaran (eds.), *Mushrooms with Therapeutic Potentials*,
https://doi.org/10.1007/978-981-19-9550-7_6

and nucleosides, such as adenosine and cordycepin, are the main components identified in *O. sinensis*. They are responsible for a wide range of pharmacological effects, including antioxidant, anticancer, immunomodulatory, smooth muscle relaxation, hypoglycemic, cholesterol-lowering, antibacterial, and antiviral properties. The mechanistic pathways that are responsible for eliciting these bioactivities have also been successfully elucidated in many studies. However, more research on the bioactive compounds of *O. sinensis* is needed to translate it into pharmaceutical use, as several components can be responsible for single bioactivity. A deeper understanding of these bioactive compounds that fit the biomolecular target(s) is valuable before developing them into therapeutic agents.

Keywords Bioactive compounds · Medicinal mushrooms · *Ophiocordyceps sinensis* · Pharmacological effects

Abbreviations

ALP	Alkaline Phosphatase
BALF	Bronchoalveolar Lavage Fluids
CAT	Catalase
COPD	Chronic Obstructive Pulmonary Disease
COVID-19	Coronavirus Disease-2019
DNA	Deoxyribonucleic Acid
EBV	Epstein-Barr Virus
FEV	Forced Expiratory Volume
FVC	Forced Vital Capacity
GPx	Glutathione Peroxidase
GSK	Glycogen Synthase Kinase
HIV	Human Immunodeficiency Virus
HMCs	Human Mesangial Cells
HPLC	High-performance Liquid Chromatography
HO-1	Heme oxygenase
IFN	Interferon
Ig	Immunoglobulin
IL	Interleukin
LC	Liquid Chromatography
L-NAME	L-N^G-nitro Arginine Methyl Ester
MAP	Mean Arterial Pressure
MDA	Malondialdehyde
MIC	Minimum Inhibitory Concentration
MMP	Mitochondrial Membrane Potential
M^{pro}	Main Protease
MS/MS	Tandem Mass Spectrometry
MT	Metallothionein

NK	Natural Killer
NO	Nitric Oxide
Nrf2	Nuclear Factor Erythroid-derived 2-like 2
PARP	Poly (ADP-ribose) Polymerase
PEF	Peak Expiratory Flea
PDGF	Platelet-derived Growth Factor
RNA	Ribonucleic Acid
ROS	Reactive Oxygen Species
SARS	Severe Acute Respiratory Syndrome
SOD	Superoxide Dismutase
TRAP	Tartrate-Resistant Acid Phosphatase
TEA	Tetraethylammonium
UTP	Urinary Total Protein
Vif	Viral Infectivity Factor

1 General Introduction to Medicinal Mushrooms

Mushrooms have been used for centuries for their exquisite flavors, nutritional value, and medicinal properties. Out of the 14,000 known species of mushrooms worldwide, only 5% of them have been identified to possess medicinal properties (Hrudayanath and Sameer 2014). Medicinal mushrooms can be defined as macroscopic fungi that are mostly higher Basidiomycetes and some Ascomycetes, and are often used in the form of extracts or powder for the prevention, alleviation of diseases, and/or for nutritional purposes (Wasser 2014). They inevitably represent a largely untapped source of pharmaceutical compounds worthy of scientific investigation.

In the 1960s, Japan began to conduct modern scientific research on medicinal mushrooms to explore compounds with bioactivities and validate their medicinal properties (Jakopovic et al. 2020). The bioactivities of medicinal mushroom-derived polysaccharides, especially β-glucans and polysaccharide-protein complexes, have since received tremendous attention. For instance, Krestin (PSK), a protein-bound glycan isolated from *Trametes versicolor*; lentinan, a purified β-glucan isolated from *Lentinus edodes* in 1985; and sonifilan, which is also a β-glucan, isolated from *Schizophyllum commune*, have been approved for clinical use by the Japanese National Health Registry since 1977, 1985, and 1986, respectively (Taguchi et al. 1985; Fujii 1996; Zhang et al. 2013). Several low-molecular-weight compounds such as triterpenes, lactones, and alkaloids have also been isolated from the fruiting bodies and mycelia of medicinal mushrooms, as well as a number of enzymes that have critical roles in the normal physiology of human health and several disease pathogenesis, such as superoxide dismutase, glucose oxidase, and peroxidase (Wasser 2014).

With growing advances in the biomedical and pharmaceutical industries, more than 130 pharmacological effects of medicinal mushrooms have been identified, including anticancer, antioxidant, immunomodulatory, antibacterial, and antiviral. Some examples of highly valued medicinal mushrooms include *Ganoderma lucidum*, *Agaricus brasiliensis*, *Grifola frondosa*, *Hericium erinaceus*, and *Ophiocordyceps sinensis*. The present chapter provides an overview of *Ophiocordyceps sinensis*, including its traditional uses and the scientific evidence that supports its bioactivities and biopharmaceutical potential for therapeutic application.

2 *Ophiocordyceps sinensis*

Ophiocordyceps sinensis was first described scientifically by a British mycologist Dr. Miles Berkeley in 1843 as *Sphaeria sinensis* Berk. It was renamed *Cordyceps sinensis* by an Italian mycologist Pier Andrea Saccardo in 1878; the name *Cordyceps* originated from two Latin words, "cord" meaning club and "ceps" meaning head, implying a club fungus fruiting out of the head of a caterpillar (Bhandari et al. 2010). Following a molecular phylogenetic classification in 2007, *C. sinensis* was transferred to the genus *Ophiocordyceps* and was renamed *Ophiocordyceps sinensis* (Sung et al. 2007). Its taxonomy is as follows: Fungi (Kingdom), Ascomycota (Phylum), Sordariomycetes (Class), Hypocreales (Order), Ophiocordycipitaceae (Family), and *Ophiocordyceps* (Genus).

O. sinensis is mainly distributed on the Tibetan Plateau alpine between 3000 and 5000 m a.s.l. and its surrounding regions, including eastern Tibet, eastern Qinghai, western Sichuan, northern Yunnan, and southwestern Gansu provinces (Li et al. 2011). It is also found in other Himalayan countries such as Nepal, Bhutan, and India at altitudes above 3800 m. However, due to climate change, the distribution of natural *O. sinensis* has decreased over recent decades and moved towards the central part of the Tibetan Plateau (Yan et al. 2017).

O. sinensis is well-known for its interesting, rather long, and unusual life cycle. In late autumn, the fruiting bodies of *O. sinensis* disperse spores and scatter on the soil surface, developing into infective conidia. These infective conidia then parasitize the underground larvae of ghost moths within the family Hepialidae (Zhang et al. 2012). It is interesting to note that the infection rates are highest in the fourth and fifth instar larvae that are shedding old cuticles and forming new ones, and this stage in the larval life cycle coincides with the release of spores by *O. sinensis*. Once entering the moth's body, the fungus multiplies by yeast-like budding and fills the hemocoel with threadlike hyphae (Guo et al. 2017). Then, each infected larva gradually moves into a position ideal for the growth of the fungus; usually, 2–5 cm below the soil surface, it becomes rigid and dies with the head facing upward. The fungus emerges from the dead host and forms a small stroma bud right before winter. More than 90% of the contents of the larva are eventually replaced by the sclerotium, turning into a nutrient store for the fungus. At the same time, the intact exoskeleton serves as a support for the sclerotium throughout winter (Guo et al. 2017). In the following spring, the fungus ruptures the host larva, forming a sexual sporulating structure known as the fruiting body, and emerges above the soil surface. This interesting life cycle is the origin of its name, "Dong Chong Xia Cao," in Chinese, which means "winter worm, summer grass."

Although it has been used in traditional Chinese medicine for centuries, its demand and the interest in scientific research towards its medicinal properties only surged in the early 1990s after a group of female athletes broke a world record during the 1993 National Games in Beijing, China, and revealed that *O. sinensis* extracts were part of their diet regime. The value of *O. sinensis* was further accelerated following the severe acute respiratory syndrome (SARS) outbreak in China in 2003 due to its traditional use in respiratory diseases. In 2008, the price of natural *O. sinensis* was recorded to be around 30,000 RMB per kg and up to 100,000 RMB per kg, which was 1500 times higher than its price in the 1970s (Dong et al. 2015). Currently, it may cost from 200,000 RMB per kg to 400,000 RMB, depending on its quality, size, and origin.

In the wake of its high cost and being classified as an endangered species, novel approaches such as the artificial cultivation of *O. sinensis* have been taking place since the early 2000s to provide an alternative source for consumers and researchers to substantiate its medicinal value. Artificial cultivation also ensures the quality of the species by producing them in a controlled-environment as wild *O. sinensis* has been found to contain high arsenic content due to soil contamination (Zhou et al. 2017; Xiao et al. 2021). The composition and authenticity of the cultivated *O. sinensis* have been validated in several studies using techniques such as liquid chromatography-mass spectrometry and ultraviolet spectrometry (Zhang et al. 2018; Zhou et al. 2019). Many studies have also revealed the bioactivities and pharmacological actions of the cultivated *O. sinensis*, showing that it is equally potent, if not more, than its natural counterpart in health-promoting functions. For example, Yao et al. (2014) showed that both wild and preparations derived from cultured *O. sinensis* mycelia antagonized the activity of the pro-fibrotic cytokine in renal epithelial cells with similar potency. Then, another study demonstrated that the water extracts of cultured *O. sinensis* have a stronger antioxidant effect as compared to the wild *O. sinensis* (Wang et al. 2015a).

2.1 *Ethnomedicinal Uses of* Ophiocordyceps sinensis

The exotic medicinal properties of *O. sinensis* were discovered over 2000 years ago when local herders in Tibet used *O. sinensis* to improve reproductive capacity and increase the milk production of their livestock (Panda and Swain 2011). Its medicinal values propagated to mainland China during the Qing dynasty and were recorded in the "Essentials of the Materia Medica: *Ben Cao Bei Yao* 1694" by

Wang Ang (Lin and Li 2011). *Ben Cao Bei Yao* was thought to be the first official documentation of *O. sinensis* in the history of traditional Chinese medicine in China. In this piece of record, Wang Ang emphasized its medicinal values in kidney invigoration, hemostasis, lung protection, and its use as an effective antitussive. However, there was no recommendation on dosage and preparation methods. According to the anecdotal record, the uses and methods of preparation of *O. sinensis* were mainly based on the empirical trial-and-error method by the local healers. Thus, they may vary in different areas and countries.

A survey conducted in the Northern Yunnan provinces reported that the local communities used *O. sinensis* to improve eyesight and treat calcium deficiency, indigestion, hypertension, and diabetes (Chen et al. 2010). The methods of preparation include steaming the ground stroma powder with eggs, serving it as a chicken stew, boiling it with *Po Cha* (local butter tea), or having it raw. *O. sinensis* is also consumed as a general tonic to enhance vitality and promote endurance and life expectancy. For this purpose, the Bhutanese community would consume it each morning and evening after brewing it in a cup of *chang* (locally made alcohol) or hot water for an hour. Mixing it with Ginseng root decoction has also been recorded to treat severe diseases such as cancer (Panda and Swain 2011). In North Sikkim, aborigines consumed *O. sinensis* with milk to enhance their libido and sex performance (Panda 2015). Thus, it is commonly regarded as the "Himalayan Viagra" in Tibet and Nepal.

2.2 *Bioactive Substances in* Ophiocordyceps sinensis

Evidence shows that many pharmacological effects of *O. sinensis* can be attributed to the presence of polysaccharides. Depending on the difference in their weighted degree of branching, backbone linkage, side-chain units, and the type of constituent monosaccharides, *O. sinensis*-derived polysaccharides may elicit different pharmacological actions. For example, Yap et al. (2020) showed that the amount of glucose determined the immunomodulating effect of *O. sinensis* extracts. Other monosaccharides in *O. sinensis* include rhamnose, arabinose, ribose, xylose, mannose, galactose, fructose, sorbose, mannitol, and D-mannitol [also known as cordycepic acid (Table 1)]. Several novel polysaccharides have been isolated from *O. sinensis*, such as cordysinocan, which contains glucose, mannose, and galactose in a ratio of 2.4:2:1, cordyglucans with (1→3)-β-D-glucans linkages, AEPS-1 which is composed of glucopyranose and pyrano-glucuronic acid, with an α-D-(1→3)-Glcp backbone structure, and CPS-F which is composed of mannose, glucose, galactose, and ribose in a molar ratio of 37:3:28:20 (Yalin et al. 2005; Cheung et al. 2009; Wang et al. 2011, 2015b).

Besides polysaccharides, *O. sinensis* is also rich in protein, amino acids, and peptides. Most of the proteins in *O. sinensis* are proteases involved in many physiological roles, such as development, physiology, defense, and stress responses. For example, a serine protease with fibrinolytic activity (CSP) was purified from the

Table 1 Some of the pure compounds isolated from *Ophiocordyceps sinensis*

Compound	References
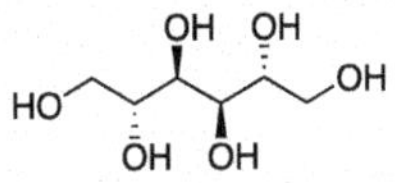 Cordycepic acid (D' mannitol)	Chatterjee et al. (1957)
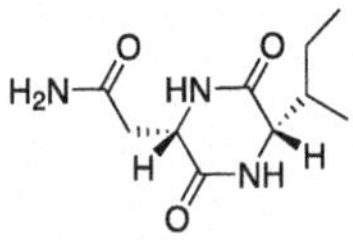 Cordycedipeptide A	Jia et al. (2005)
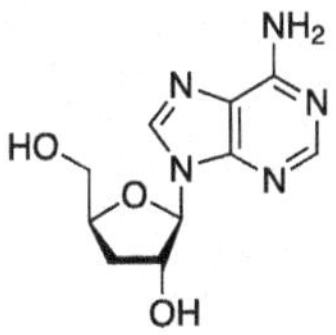 Cordycepin 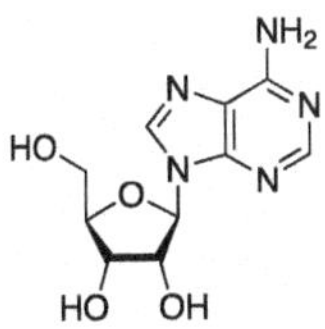 Adenosine	Cheng et al. (2017)
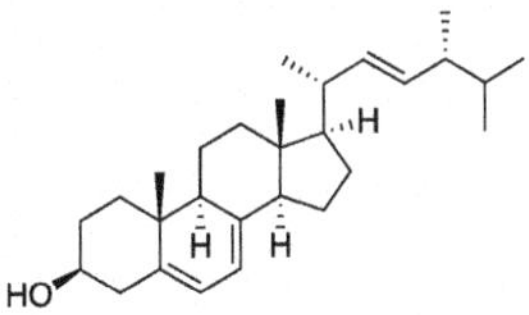 Ergosterol	Yuan et al. (2007)
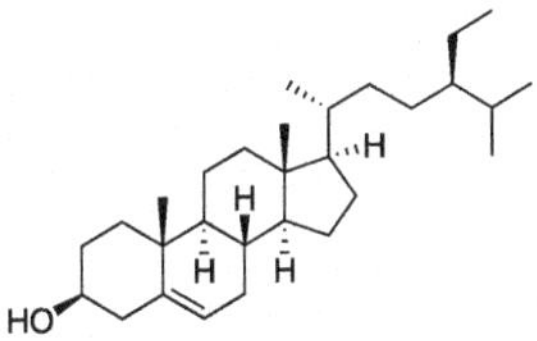 ß-sitosterol	Yang et al. (2009)
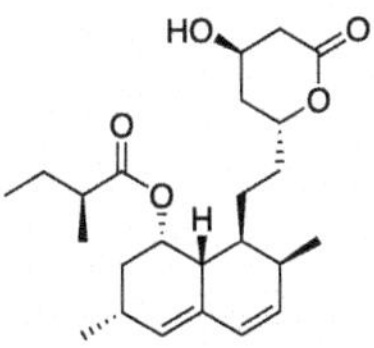 Lovastatin	Chen et al. (2012)

(continued)

Table 1 (continued)

Compound	References
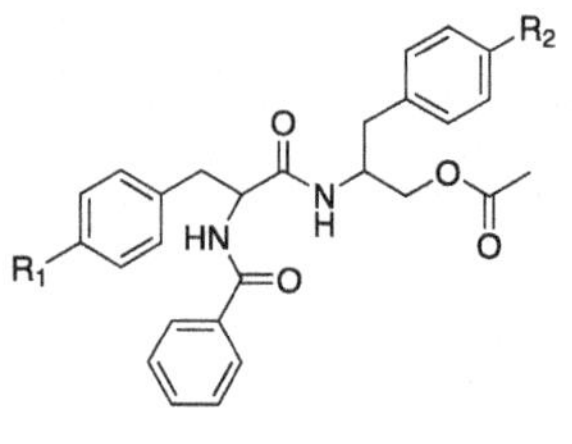 Cordycaemide A: R_1 = OH, R_2 = H Cordycaemide B: R_1 = OH, R_2 = OH	Jia et al. (2009)
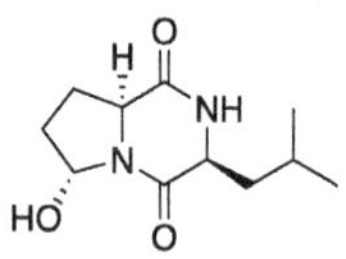 Cordysinin A 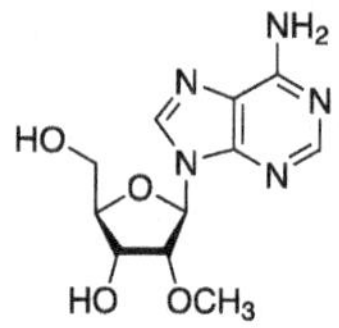 Cordysinin B 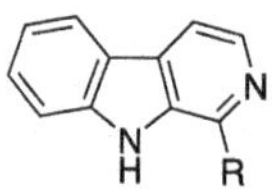 Cordysinin C: R = 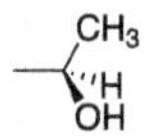 Cordysinin D: R = 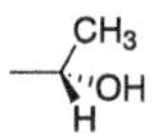 Cordysinin E: R = 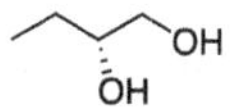	Yang et al. (2011)

culture supernatant of *O. sinensis* (Li et al. 2007). Bi et al. (2011) also purified a novel protease from *O. sinensis* that has a completely new cleavage pattern though the significance of this new cleavage pattern has not been determined. Recently, several potential cytotoxic proteases were identified in a cultivated *O. sinensis* cold water extract (Kong et al. 2021).

To date, 18 amino acids, including L-arginine, glutamic acid, aspartic acid, and tryptophan, have been detected in *O. sinensis*. Hydroxyproline and hydroxylysine were not detected in both wild or cultivated *O. sinensis* (Fung et al. 2018). Novel peptides have also been purified from *O. sinensis*, including cordymin and cordycedipeptide A (Jia et al. 2005; Qian et al. 2012). The chemical structure of cordycedipeptide A is depicted in Table 1. Many antioxidant-rich peptides have also been identified in *O. sinensis* (Mishra et al. 2019; Tong and Guo 2022).

Nucleosides, including adenosine, guanosine, inosine, thymidine, cytidine, and cordycepin, were also detected in both natural and cultured *O. sinensis*, with a higher content in the mycelium than in the fruiting bodies (Cheng et al. 2017). Guanosine was found to be the most abundant nucleoside in *O. sinensis.* Cordycepin is an adenosine-derivative, differing from adenosine by the absence of oxygen in the 3′ position of its ribose entity (Table 1). It was initially isolated from *Cordyceps militaris* and later discovered in *O. sinensis* and other *Cordyceps* (Lin and Li 2011; Kuo et al. 2015). In the past, the amount of cordycepin was much lower in cultured *O. sinensis*. However, several new strains of cultivated *O. sinensis* in recent years have been proven to contain a much higher quantity of cordycepin (Liu et al. 2015; Fung et al. 2018; Zhou et al. 2019). This could be due to the advancements in the cultivation technique.

Three nucleotides, including uridine-5′-monophosphate (UMP), adenosine-5′--monophosphate (AMP), and guanosine-5′-monophosphate (GMP), have been separated by ion-pairing reversed-phase liquid chromatography-mass spectrometry (Yang et al. 2010). AMP, GMP, and UMP can be degraded in the biological system to adenosine, guanosine, and uridine, respectively. Nucleobases, including adenine, guanine, uracil, hypoxanthine, cytosine, and thymine, were also determined in wild and cultivated *O. sinensis* (Cheng et al. 2017).

Like many medicinal mushrooms, *O. sinensis* contains sterols, including ergosterol, the principal sterol in mushrooms. Ergosterol is often present in two forms, either free or esterified. Both forms have been identified in *O. sinensis* via high-performance liquid chromatography (HPLC) (Yuan et al. 2007). Other free sterols, including cholesterol, campesterol, and β-sitosterol, have also been determined in *O. sinensis* using pressurized liquid extraction, trimethylsilyl derivatization, and GC-MS analysis (Yang et al. 2009). The pharmacological effects of these *O. sinensis*-derived sterols have not been actively investigated.

Other compounds such as lovastatin, cordyceamides A–B, and cordysinins A–E have also been identified in *O. sinensis* (Table 1). Lovastatin was discovered in *Aspergillus terreus* and *Monascus ruber* around 50 years ago. It was the first statin approved by the United States Food and Drug Administration as a hypercholesterolemic drug in 1987 (Subhan et al. 2016). Recently, it was also detected in the mycelia of *O. sinensis* at a concentration of 1356 mg/kg (Chen et al. 2012). However, it is not known whether the cholesterol-lowering effects of *O. sinensis* are due to the presence of lovastatin, as the amount of lovastatin was not analyzed in those respective studies. Cordyceamides A–B and cordysinins A–E were reported to possess cytotoxic and anti-inflammatory effects, respectively, but the findings were insufficient to be extrapolated (Yang et al. 2011; Jia et al. 2009). *O. sinensis* also contains a wide range of vitamins and microelements.

3 Biopharmaceutical Potential of *Ophiocordyceps sinensis*

3.1 Antioxidant Activities

Antioxidants are substances capable of preventing or delaying the formation of reactive oxygen species (ROS), such as H_2O_2, hydroxyl radicals, and superoxide radicals, and promoting their decomposition (Santos-Sánchez et al. 2019). Although free radicals are being produced continuously in all cells as part of normal cellular function, excess production of free radicals gives rise to oxidative stress. As a result, cellular structures like proteins, lipids, and nucleic acids may be damaged due to cell injury and apoptosis. Common factors that contribute to the excessive production of free radicals include inflammation, activation of immune cells, exposure to environmental pollutants such as cigarette smoke, consumption of alcohol, and radiation (Pham-Huy et al. 2008). These events create a vicious cycle in the biological system as excessive production of free radicals may initiate or exacerbate several diseases, such as cancer, diabetes, and cardiovascular diseases.

There are several strategies to counteract the effects of free radicals and oxidative stress, including enzymatic and nonenzymatic pathways (Pizzino et al. 2017). Superoxide dismutase (SOD), catalase (CAT), and glutathione peroxidase (GPx) are enzymes that are crucial in the antioxidant defensive system in cells by promoting the breakdown of the reactive molecules. L-arginine, lipoic acid, glutathione, coenzyme Q_{10}, vitamin E, and ascorbic acid are nonenzymatic antioxidant molecules introduced by diet or nutritional supplementation.

The antioxidant activity is one of the most extensively studied properties of *O. sinensis* due to its long-standing use as a tonic and energy booster. Yamaguchi et al. (2000a) pioneered the investigation of the antioxidant properties of *O. sinensis*. They demonstrated that hot water extract of *O. sinensis* displayed a strong inhibitory effect against lipid peroxidation, which results in the inhibition of cholesteryl ester accumulation in macrophages via suppression of low-density lipoprotein oxidation (Yamaguchi et al. 2000a). The group then proceeded with an in vivo study and showed that the administration of the extracts significantly suppressed the serum lipid peroxide level and aortic cholesteryl ester level in mice fed with an atherogenic diet (Yamaguchi et al. 2000b).

Generally, the antioxidant properties of *O. sinensis* are attributable to the presence of polysaccharides, proteins, and peptides. An earlier study showed that compared with the water extracts of *O. sinensis*, the partially purified polysaccharide fractions increased the antioxidant activities by 10–30 times (Li et al. 2001). A pure polysaccharide fraction was then isolated using the antioxidation activity-guided fractionation method, and it exhibited a strong protective effect against H_2O_2-induced oxidative stress in rat pheochromocytoma cells (Li et al. 2003).

Following the administration of Cs-C-Q80 capsules (a formulation of fermented *O. sinensis* that is listed in the Pharmacopoeia of the People's Republic of China, which contains 39.4% of carbohydrate and 14.8% protein) at a dose of 1.5 g/kg per day for 4 weeks, the malondialdehyde (MDA) levels in liver and cardiac tissues of doxorubicin-treated rats were significantly reduced (Wu et al. 2015). Concurrently, the catalase and glutathione peroxidase levels in the tissues were significantly elevated. A recent study highlighted a similar outcome where the exopolysaccharide purified from cultivated *O. sinensis* exhibited a hepatoprotective effect in rats with elevated glutathione levels, and the MDA levels were reduced after exposure to carbon tetrachloride, CCl_4, which promotes oxidation (Nguyen et al. 2021). These results may be considered scientific validation of the traditional use of *O. sinensis* in treating liver diseases, as glutathione plays a critical role in eliminating toxic metabolites. In contrast, the hepatic MDA level represents the degree of membrane lipid peroxidation and oxidative damage in liver tissue.

In a recent breakthrough, over 8000 putative antioxidant peptides have been identified in wild and cultured *O. sinensis* (Tong and Guo 2022). Gene ontology analysis showed that most of these peptides shared similar genes with the typical antioxidant peptides involved in biological processes such as cell redox homeostasis, antioxidative stress, and cellular transport. Some peptides are essential in organizing cellular components (Tong and Guo 2022). Remarkably, these antioxidant peptides in *O. sinensis* are thought to contribute to its ability in high-altitude adaptation. *O. sinensis* has been proved to attenuate hypoxia-induced ROS generation, oxidation of lipids and proteins in human lung epithelial cells via induction of antioxidant genes, including heme oxygenase-1 (HO-1), metallothionein (MT), and nuclear factor erythroid-derived 2-like 2 (Nrf2) (Singh et al. 2013).

O. sinensis also significantly prolonged the life span of fruit flies through an antioxidative stress pathway by upregulating the activity of copper-zinc-containing SOD 1 and catalase and inhibiting the accumulation of lipofuscin (Zou et al. 2015). Due to the complex mechanisms of the oxidative pathway, it is often linked to cancer and immune-mediated diseases. Therefore, the antioxidant properties of *O. sinensis* are commonly investigated along with its anticancer and immunomodulatory properties.

3.2 *Cytotoxic and Anticancer Activities*

Cancer is a leading cause of death worldwide. According to the World Health Organization (WHO), it was responsible for approximately ten million deaths in 2020. Conventional therapies, including surgery, radiotherapy, and chemotherapy, are associated with many serious side effects. It has been well documented that *O. sinensis* possesses significant anticancer activities and has the potential to target multiple cancers. The mechanisms of action by which *O. sinensis* exhibits its anticancer activity include inhibition of proliferation, induction of apoptosis, immunomodulation, and inhibition of metastasis.

The American National Cancer Institute indicated that natural products or compounds must express an IC_{50} of ≤ 30 μg/mL to claim their anticancer activities. IC_{50} measures the potency or effectiveness of a compound in inhibiting a biological function, in this case, cancer cell proliferation (Suffness and Pezzuto 1990). The lower the IC_{50}, the more cytotoxic the compound or drug is. From the cancer cell proliferation assays, the IC_{50} of *O. sinensis* extracts, fractions, and/or isolated bioactive compounds ranged from 14.2 μM to 1 mM (Table 2). The relatively large deviation in the IC_{50} range of *O. sinensis* suggests that it may be more effective against certain cancer types than others, as different cell lines were used in these studies. The extraction method and the origin of *O. sinensis* could also affect its bioactivities. Although the IC_{50} values of *O. sinensis* in some studies did not meet the standard of an anticancer compound defined by the American National Cancer Institute, they have stipulated several plausible anticancer pathways of *O. sinensis* and identified the potential bioactive components. For instance, Kong et al. (2021) identified several proteases that possess cytotoxic effects in the high-molecular-weight fraction of *O. sinensis* using liquid chromatography (LC) and tandem mass spectrometry (MS/MS).

Heteroglycan from *O. sinensis* at concentrations of 50, 150, and 300 μg/mL significantly inhibited the proliferation of S180 murine sarcoma cells by 72.74%, 81.22%, and 90.24%, respectively, after a treatment period of 48 h (Mei et al. 2014). Moreover, the heteroglycan also promoted apoptosis of the cells at a rate higher than that of 5-fluorouracil, indicating that the anticancer effect of *O. sinensis* is exerted via apoptosis (Mei et al. 2014). In another study, *O. sinensis*-derived volatile oil exhibited a greater proliferation inhibitory action against various cancer cell types than 5-fluorouracil (Sang et al. 2020). In particular, the volatile oil exerted an overt inhibitory effect against the paclitaxel-resistant lung cancer cell line A549 and paclitaxel-resistant ovarian cancer cell line A2780, in which the IC_{50} values of the *O. sinensis* volatile oil in these two cell lines were 23.5 and 4.6 times lower than paclitaxel, respectively (Sang et al. 2020). Thus, *O. sinensis* may be great potential as a lead product for discovering therapeutic agents against paclitaxel-resistant tumor cells.

The activation of the adenosine A_3 receptor is thought to be one of the anticancer mechanisms of *O. sinensis*. Yoshikawa et al. (2011) showed that the proliferation inhibitory effect of *O. sinensis* water extract against the B16-BL6 mouse melanoma, mouse Lewis lung carcinoma cells, human fibrosarcoma cells HT1080, and human colon carcinoma cells CW-2 can be inhibited by MRS11191, a selective adenosine A_3 receptor and promoted by an adenosine deaminase inhibitor. Thus, cordycepin which has a high affinity at the adenosine A_3 receptor has been regarded as one of the bioactive components. The authors later demonstrated the anticancer action of cordycepin through the activation of glycogen synthase kinase (GSK)-3β and suppression of cyclin D1 (Nakamura et al. 2015). GSK-3 is a tumor suppressor for several cancers and contributes to the antiapoptotic phenotype of cancer cells, whereas cyclin D1 is an important regulator of cell cycle progression (Alao 2007; Chiara and Rasola 2013). Thus, cordycepin may hold promise in cancer treatment via these mechanisms. Moreover, cordycepin possesses an antimetastatic effect through the inhibition of platelet aggregation, suppression of the invasiveness of cancer cells, and stimulation of the secretion of tissue inhibitors from cancer cells (Nakamura et al. 2015).

Table 2 The antiproliferative activity of *O. sinensis*

Form	Treatment period	Cell line	IC_{50} (μg/mL)	References
Cultivated *O. sinensis* fruiting bodies aqueous extract (high-molecular-weight fraction)	72 h	A549	157.3 ± 10.1	Kong et al. (2021)
Cultivated *O. sinensis* fruiting bodies aqueous extract (medium-molecular-weight fraction)			357.3 ± 54.5	
Cultivated *O. sinensis* fruiting bodies aqueous extract (low-molecular-weight fraction)			>1000	
Cultivated *O. sinensis* fruiting bodies aqueous extract (high-molecular-weight-protein fraction)			107.8 ± 5.9	
Cultivated *O. sinensis* fruiting bodies aqueous extract (high-molecular-weight-non-protein fraction)			213.3 ± 37.5	
Wild *O. sinensis* mycelium ethyl acetate extract	72 h	HL-60	≤25	Zhang et al. (2004)
Cultivated *O. sinensis* mycelium	120 h	A549	235.6 ± 12.6	Lee et al. (2015)
	240 h	HepG2	84.7 ± 2.8	
	120 h	MCF-7	208.23 ± 8.7	
Cultivated *O. sinensis* mycelium Cs-4 volatile oil	24 h	Hela	19.8	Sang et al. (2020)
		HL-60	21.8	
		Daudi	30.7	
		A375	12.3	
		HepG2	44.6	
		Caco-2	14.2	
		MCF-7	43.1	
		A549 (non-resistant cell line)	5.3	
		A549 (paclitaxel-resistant cell line)	1.5	
		A2780 (nonresistant cell line)	2.3	
		A2780 (paclitaxel-resistant cell line)	8.6	

In human pancreatic cancer cells MIAPaCa-2 and Capan-1, cordycepin was found to inhibit proliferation in a dose- and time-dependent manner through a novel mitochondrial-mediated intrinsic apoptotic pathway (Zhang et al. 2022b). Following the exposure to cordycepin, the mitochondrial membrane potential (MMP) was decreased, and the cleaved caspase-3, cleaved caspase-9, and cleaved poly (ADP-ribose) polymerase (PARP) were upregulated. These are all indicators of cell apoptosis via the mitochondrial pathway. Interestingly, Bai et al. (2020) recently showed that *O. sinensis* might provide a therapeutic effect in relieving cerebral ischemia injury via a similar pathway. In the oxygen- and glucose-deprived brain microvascular endothelial cells, *O. sinensis* prevented cellular apoptosis, recovered the reduction of MMP, and reduced the caspase-3, -8, and -9 activities (Bai et al. 2020).

Cordycepin also inhibited the proliferation of colon cancer cells HCT116 and Caco-2 by downregulating MYC mRNA/protein expression and upregulating microRNA-26a (Zhang et al. 2022b). The overexpression of MYC has been linked with aggressive biological behavior and adverse clinical outcome of colon cancer, and its deletion has been found to suppress tumorigenesis (Pan et al. 2020). MicroRNA-26a, on the other hand, is a negative regulator of MYC as it suppresses MYC to inhibit the progression and metastasis of cancer cells (Sander et al. 2008). Therefore, the clinical potential of cordycepin by targeting the MYC/microRNA-26a pathway should be further investigated.

Notwithstanding the great biopharmaceutical potential of *O. sinensis* in multiple cancer types, it is interesting to note that the use of *O. sinensis* may not be suitable for some cancers. The safety of *O. sinensis* consumption by patients with prostate cancer has been questioned in a recent study as administration of *O. sinensis* at a dose of 50 mg/g in mice for 24 days raised the serum testosterone level and caused enlargement of prostate glands (Ma et al. 2018). Besides, the viability of the androgen-responsive prostate cancer cells was enhanced twofold after *O. sinensis* treatment, as compared to the control group. The elevation of testosterone production in mouse Leydig cells following exposure to *O. sinensis* extract of up to 3 mg/mL has also been reported (Huang et al. 2001). These results suggest that *O. sinensis* may promote the growth of prostate cancer cells by increasing the production of testosterone. On the other hand, these studies showed that *O. sinensis* might be useful in some conditions caused by the insufficient secretion of testosterone, such as infertility. The stimulation of testosterone production could also contribute to *O. sinensis*' aphrodisiac effects. These potential health effects, however, have not been investigated.

Taken together, even though *O. sinensis* may not be beneficial in the treatment of prostate cancer and benign prostate hyperplasia, it presents a prominent source of bioactive components with anticancer properties. It deserves attention in the exploration of leads for the development of new anticancer drugs.

3.3 Immunomodulatory Properties

Immunomodulators are substances that may alter the immune response, in a beneficial way, through stimulation, expression, amplification, or inhibition of any portion or stage of the immune response (Abood et al. 2014). Currently, clinical uses of immunomodulators (immunostimulants or immunosuppressants) include the reconstruction of the immune system, such as in AIDS treatment, and the suppression of normal immune response or exaggerated immune response in autoimmune diseases such as Crohn's disease and psoriasis. Many of the currently available immunomodulators are associated with serious side effects. Therefore, there is a growing interest in using natural products to modulate the complex immune system.

Numerous studies have demonstrated the immunomodulatory effects of *O. sinensis*, including promoting macrophages and T-lymphocytes' proliferation, phagocytosis, nitric oxide (NO) production, upregulating or downregulating of inflammatory factors, cytokines, and chemokines. In an earlier study, *O. sinensis* ethanolic fraction reduced the production of IL-1β, IL-6, IL-8, IL-10, and TNF-α in lipopolysaccharide-induced human bronchoalveolar lavage fluids (BALF) cells while increasing the production of IL-12 and INF-γ (Kuo et al. 2001). IL-1β, IL-6, and IL-8 are secreted by macrophages, neutrophils, and fibroblasts and play a vital role in the early inflammatory cells' recruitment to inflamed tissues. These inflammatory cells and cytokines are often elevated in BALF cells of asthmatic patients (Rincon and Irvin 2012). Thus, downregulating these inflammatory factors indicates a potential therapeutic role of *O. sinensis* in the treatment of asthma. In addition, cordysinocan, a polysaccharide of molecular weight around 82 kDa, was shown to enhance the secretion of IL-2, IL-6, and IL-8 and promote the proliferation of T-lymphocytes (Cheung et al. 2009). Its mechanism of action is thought to be mediated by mitogen-activated protein kinases. Cheung et al. (2009) also found that the application of cordysinocan dose-dependently increased the phagocytosis activity of macrophages, suggesting its alleged role in fighting infection. Another purified polysaccharide, AEPS-1, also stimulated the release of IL-1B, IL-6, and IL-10 in macrophages (Wang et al. 2011).

As opposed to Kuo et al. (2001) and Cheung et al. (2009), who demonstrated that *O. sinensis* downregulated IL-6, Yap et al. (2020) recently showed that the cold water extract and fractions of high and medium molecular weight derived from cultured *O. sinensis* promoted the production of IL-6 in murine macrophages. Although mostly regarded as a pro-inflammatory cytokine, IL-6 also plays a role in regenerative and anti-inflammatory activities (Rincon and Irvin 2012). Upregulation of IL-6 is also important in the induction of the differentiation of B cells to secrete antibodies and differentiation of T cells into Th17. Besides, *O. sinensis* extracts upregulated IL-9, which promotes the proliferation and differentiation of mast cells and involves in the anticancer mechanism (Yap et al. 2020).

While the bioactive component(s) of *O. sinensis* that possess an immunomodulating effect is yet to be identified, it is suggested that these immunomodulating actions are mainly due to the presence of polysaccharides and polysaccharide-protein complexes. The cold water extract, high- and medium-molecular-weight fraction of *O. sinensis* reported by Yap et al. (2020), mainly consists of carbohydrates. Pretreatment of a pure polysaccharide fraction of *O. sinensis* (CPS-F) has markedly reduced the expression of TNF-α, TNFR1, and MCP-1 in platelet-derived growth factor-BB (PDGF-BB)-induced inflammation in human mesangial cells (HMCs) (Wang et al. 2015b). This study suggested that *O. sinensis* may be beneficial in the treatment of glomerulonephritis as the production of TNF-α and MCP-1 in the mesangial cells often relates to glomerular injury. Its mechanism of action is likely to involve the ERK1/2 and Akt pathways as *O. sinensis* fraction dose-dependently inhibited the activation of ERK1/2 and Akt, that in turn decreased the expression of pro-inflammatory factors (Wang et al. 2015b). Interestingly, the immunomodulatory properties of *O. sinensis* are reported to only affect cells in pathological conditions as it did not affect the proliferation of normal HMCs (Wang et al. 2015b).

Cordymin, a purified peptide isolated from *O. sinensis*, has exhibited outrageous activities against TNF-α and IL-1β, which diminished inflammatory reactions (Qian et al. 2012). Cordycepin also displayed an immunomodulating effect by downregulating the expression of the TNF-α, IL-6, and IL-17A on cultured murine spleen cells stimulated with lipopolysaccharides (Seo et al. 2013). Other nucleosides found in *O. sinensis*, including adenine, adenosine, 2′-deoxyadenosine, and thymidine, have also been found to inhibit inflammatory mediator expression of inflammatory cytokines, including TNF-α, IL-6, and IL-1ß, on the cigarette smoke-stimulated RAW 264.7 macrophages and in cigarette smoke-induced mice (Sun et al. 2018).

The effect of *O. sinensis* has also been assessed clinically in renal transplant recipients (Li et al. 2009). Adjunct treatment of *O. sinensis* at a dose of 1 g 3 times daily has significantly reduced the serum uric acid and 24-h urinary total protein (24-h UTP) in post-transplantation patients compared to those receiving cyclosporin alone. Incidences of hepatotoxicity and nephrotoxicity within 6 months of transplantation were 11.87 and 14.59% lower in patients who consumed *O. sinensis* concomitantly. The treatment also reduced the incidence of chronic allograft nephropathy by 10.82%. This study showed that *O. sinensis* could potentially be used as an immunosuppressive agent after a renal transplant. In a recent randomized and double-blind clinical trial conducted in South Korea, the administration of cultured *O. sinensis* extract (*Paecilomyces hepiali*) of 1.68 g per day for 8 weeks significantly increased the natural killer (NK) cells activity of the healthy volunteers. While activation of NK cells often promotes the release of cytokines such as IFN-γ and TNF-α, administration of the extract did not affect the cytokine level of the volunteers in this study. This also ensures that the use of *O. sinensis* does not affect cytokine levels and maintains homeostasis in healthy individuals (Jung et al. 2019).

3.4 Smooth Muscle Relaxant Effects

Smooth muscle physiology is responsible for maintaining and preserving every vital sign due to its unique ability to contract and relax involuntarily. As such, smooth muscles have been an important therapeutic target for tackling several clinical conditions. Calcium channel blockers and nitrates have been developed to relax the vascular smooth muscle to lower blood pressure; antimuscarinics such as oxybutynin and tolterodine are used to inhibit bladder contraction to treat an overactive bladder; mebeverine is used to ease colonic muscle spasm; salbutamol and ipratropium are bronchodilators in the treatment of asthma.

Research is still underway to improve the understanding of the physiology of smooth muscle and other components that directly or indirectly affect smooth muscle function. For instance, it was not until the past decade that the role of the urothelium (the innermost layer of the bladder made up of urothelial cells) was established in regulating the contraction and relaxation of urinary bladder (Winder et al. 2014).

The smooth muscle relaxation properties of *O. sinensis* have been demonstrated in several studies. In rat isolated aortic tissues, *O. sinensis* protein extracts exhibited an endothelium-dependent relaxation response that was sensitive to L-N^G-nitro arginine methyl ester (L-NAME, a nitric oxide synthase inhibitor) and tetraethylammonium chloride (TEA, a potassium channel blocker) (Chiou et al. 2000). The results suggested that *O. sinensis* may produce NO and inhibit K_{Ca} channels to promote vasorelaxation. Administration of the extracts also dose-dependently reduced the mean arterial pressure (MAP) of rats (Chiou et al. 2000). At the highest dose of 32 mg/kg, the MAP was reduced from 107 $\pm$ 6 to 49 $\pm$ 3 mmHg.

Following Chiou et al. (2000), a purified *O. sinensis* polysaccharide fraction was found to effectively reduce both systolic and diastolic blood pressure of spontaneously hypertensive rats (SHR) in a dose-dependent manner after a treatment period of 90 days (Xiang et al. 2016). Besides, the administration of *O. sinensis* significantly enhanced NO production and decreased the levels of endothelin-1, epinephrine, norepinephrine, and angiotensin-II in the SHR (Xiang et al. 2016). Endothelin-1 is a potential marker of endothelial dysfunction and is usually elevated in hypertensive subjects, whereas epinephrine and norepinephrine are neurotransmitters that increase blood pressure, heart rate, and blood sugar levels (Floras 1992; Akter et al. 2015). These studies corroborated that *O. sinensis* may provide a beneficial effect on systemic circulation.

Regarding pulmonary circulation, an ex vivo experiment using isolated, ventilated, and buffer-perfused mouse lungs have recently been performed to investigate the possible therapeutic effect of *O. sinensis* in pulmonary hypertension (Luitel et al. 2020). While demonstrating the potent vasodilatory effect of *O. sinensis* in the ex vivo model, the study ruled out cordycepin as the bioactive component of *O. sinensis* in vasodilation. Moreover, a recent study has published the bladder relaxant effect of *O. sinensis* cold water extract (Pang et al. 2022). In line with the results reported by Chiou et al. (2000), the relaxant effect of *O. sinensis* in the bladder was partly urothelium-dependent and sensitive to L-NAME. The study also showed that *O. sinensis* inhibited extracellular calcium influx, thus preventing calcium-induced contractions in the bladder. These findings substantiated the traditional use of *O. sinensis* in frequent urination.

Although the direct relaxant effect of *O. sinensis* in airway smooth muscles has not been established, it is worth mentioning that *O. sinensis* has been proven beneficial in treating respiratory conditions, including asthma and chronic obstructive pulmonary disease (COPD). Airway smooth muscle physiology plays a significant role in the pathophysiology of these conditions, and conventional therapeutic agents such as β-adrenoceptor agonists and muscarinic receptor antagonists promote airway relaxation. In a randomized controlled trial conducted at Xuanwu Hospital, Beijing, from January 2014 to December 2015, 120 asthmatic patients were recruited to study the therapeutic effect of *O. sinensis* (Wang et al. 2016). Half of the recruited patients consumed a Corbin capsule containing 1.2 g of *O. sinensis* 3 times per day, and the other half received a placebo. It was found that *O. sinensis* reduced the asthma onset and significantly increased the mean number of symptom-free days in the treatment group. The lung function, as indicated by the forced expiratory volume in 1 s (FEV1), forced vital capacity (FVC), and mean peak expiratory flow (PEF), was also greatly improved in the treatment group. According to the outcome of Juniper's Asthma Quality of Life Questionnaire (AQLQ), these improvements were accompanied by a better quality of life in the patients. Recently, a systematic review and meta-analysis on the use of *O. sinensis* in patients with stage 2–3 COPD revealed that *O. sinensis* provides beneficial effects in preserving lung function and improving the symptoms (Yu et al. 2019). Treatment with *O. sinensis* also improved the exercise endurance of the life quality of the patients.

3.5 Diabetes and Diabetes-Related Conditions

Diabetes is a highly prevalent, chronic metabolic disorder mainly caused by the lack of insulin or ineffective insulin production by the pancreatic β-cells and insulin resistance. Without proper insulin function and/or the inability of the cells, such as adipose and muscle cells, to respond to circulating insulin due to insulin resistance, glucose builds up in the bloodstream. Eventually, it results in hyperglycemia (Petersen and Shulman 2018). Chronic hyperglycemia can lead to complications such as atherosclerosis, diabetic nephropathy, obesity, hypertension, and hyperlipidemia.

The beneficial effects of *O. sinensis* in lowering blood sugar, cholesterol, and lipids have been established in several studies. It is proposed that *O. sinensis* exerts its antidiabetic effect by promoting glucose metabolism and insulin release via pancreatic β-cells preservation. In an in vivo study, treatment with polysaccharide extract of *O. sinensis* at a dose of 200 mg/kg for 7 days significantly reduced the blood glucose level of streptozotocin-induced diabetic rats and alloxan-induced diabetic mice (Li et al. 2006). At the same time, the serum insulin level in diabetic animals was increased. In another study that administered *O. sinensis* to streptozotocin-induced diabetic rats at a lower dose but for a longer treatment period (100 mg/kg for 21 days), similar effects were observed where the glucose level was significantly reduced, and the reduction was comparable to glibenclamide, a sulfonylurea antidiabetic medication (el Ashry et al. 2012).

Additionally, *O. sinensis* increased serum insulin, HDL-cholesterol, and pancreatic β-cells function. Preservation of pancreatic β-cells is an important aspect in the treatment of diabetes as these cells secrete insulin and are often damaged or become dysfunctional due to high glucose or lipid levels, inflammatory mediators released from the adipose tissue, or oxidative stress. Kan et al. (2012) also showed that *O. sinensis* increased the viability of pancreatic β-cells in addition to decreasing the body weight gain and improving the cholesterol profile in KK/HIJ diabetic mice fed with a high-fat diet (Kan et al. 2012).

In China, *O. sinensis* has been used clinically for diabetic nephropathy since the early 2000s (Luo et al. 2015). Combining the results from 60 randomized controlled trials involving 4288 participants, a meta-analysis showed that *O. sinensis* significantly improved the therapeutic effect of angiotensin-converting enzyme inhibitors and angiotensin receptor blockers in diabetic nephropathy (Luo et al. 2015). The underlying mechanisms of *O. sinensis* in diabetic nephropathy may include inhibition of the purinergic P2X7 receptors, P2X7R, and NLRP3 inflammasome (NLRP3: nucleotide binding and oligomerization domain-like receptor family pyrin domain-containing 3) (Wang et al. 2018). P2X7Rs and NLRP3 have recently received great attention as potential targets in the treatment of diabetic nephropathy due to their various signaling pathways that promote disease progression (Menzies et al. 2017; Nascimento et al. 2020; Yang et al. 2021).

When there is persistently high glucose in the blood, the lining of the arteries gets damaged, thus making them more susceptible to fats and cholesterol build-up, leading to atherosclerosis. Atherosclerosis is the main cause of cardiovascular diseases like myocardial infarction, heart failure, and stroke. A heteropolysaccharide isolated from *O. sinensis* has been found to possess strong inhibitory activity against cholesterol esterase that could prohibit the absorption of dietary cholesterol esters (Kim 2010). *O. sinensis* may also inhibit platelet activation, which is involved in the recruitment of inflammatory cells toward the lesion site in atherosclerosis (Lu et al. 2014; Chang et al. 2015).

3.6 Antibacterial Activities

Antibiotics have always been one of the most notable scientific discoveries in the history of humankind. In addition to treating infectious diseases, antibiotics are essential in many modern medical procedures such as cancer treatment, organ transplants, and surgeries. They are also commonly used for nonmedicinal purposes such as beekeeping, fish farming, horticulture, and food preservation (Meek et al. 2015). Due to the misuse and overuse of antibiotics over time and poor infection prevention and control, antimicrobial resistance has been a global public health threat for the past three decades. Therefore, the research of new alternatives is crucial and has to be escalated to keep pace with the formation of superbugs. Many studies have explored the antibacterial potentials of medicinal mushrooms such as *Ganoderma lucidum*, *Coriolus versicolor*, and *Lignosus rhinocerus* (Mohanarji et al. 2012; Vazirian et al. 2014; Matijaševic et al. 2016). The antibacterial activities of *O. sinensis* are discussed herein.

To assess the antibacterial properties of a compound, methodologies including microdilution, disk diffusion, and agar well diffusion are often used. The antibacterial potency is expressed in MIC (minimum inhibitory concentration) values, which means the lowest concentration of an antibacterial agent at which bacterial growth is completely inhibited. The antibacterial effects of *O. sinensis*-derived extracts and fractions have been examined using these in vitro methods, and the findings are summarized in Table 3. From these studies, *O. sinensis* appears to be effective against both Gram-positive and Gram-negative bacteria. The antibacterial potency may depend on the type of solvent used, extraction method, and assays. For instance, Ren et al. (2014) found that the microdilution method provides higher efficiency and accuracy than the disk diffusion method in measuring the antibacterial activity of high-molecular-weight polysaccharides.

While the antibacterial mechanism of action of *O. sinensis* remains ambiguous, the antibacterial activity of cordycepin has been considerably explored. Recently, a detailed bactericidal dual mechanism of cordycepin against *E. coli* and *B. subtilis* has been reported. The process begins by disrupting the bacterial cell walls and membranes and then, binding to the bacterial genomic DNA to interfere with cellular functions (Jiang et al. 2019). Another study showed that cordycepin is capable of killing mycobacteria (*M. bovis* and *M. tuberculosis*) by competitively inhibiting the activity of adenosine kinase (AdoK), a purine salvage enzyme responsible for the phosphorylation of adenosine to adenosine monophosphate (Huang et al. 2019). Without the formation of AMP, DNA or RNA synthesis would be interrupted, resulting in eventual cell death.

An in vivo study showed that the combination treatment of *O. sinensis* (200 mg/kg) and PA-824 (10 mg/kg), a bicyclic nitroimidazole antituberculosis agent for 8 weeks, caused a more effective reduction of *M. tuberculosis* colony formation in the lungs of infected mice than PA-824 alone. In addition to confirming the antibacterial effect of *O. sinensis*, the study also provided insights into the combinatory use of traditional medicine and modern clinical drugs to possibly improve efficacy in medical treatment (Li and Ren 2017).

3.7 Antiviral Activities

Like a bacterial infection, viral infection remains one of the biggest health threats to humankind. Over the past five decades, the world has been profoundly affected by several viral outbreaks, including Human Immunodeficiency Virus (HIV) in 1981,

Table 3 Antibacterial activity of *O. sinensis* against [a]Gram-positive and [b]Gram-negative bacteria

Form	Test microbes	MIC values	References
Lyophilized wild *O. sinensis* hydroethanolic extract	*Bacillus subtilis*[a]	45–50 μg	Mamta et al. (2015)
	Escherichia coli[b]	93.75–375 μg	
	Pseudomonas aeruginosa[b]	93.75–150 μg	
Cultivated *O. sinensis* polysaccharide extract	*Bacillus subtilis*[a]	938 μg/mL	Ren et al. (2014)
	Staphylococcus epidermidis[a]	469 μg/mL	
Cultivated *O. sinensis* protein extract	*Staphylococcus aureus*[a]	50–75 mg/L	Hu et al. (2006)
	Escherichia coli[b]	75–100 mg/L	
	Salmonella enterica typhi[b]	50 mg/L	
	Proteus vulgaris[b]	75 mg/L	
Cultivated *O. sinensis* aqueous extract	*Bacillus cereus*[a]	0.078 mg/mL	Kaushik et al. (2019)
	Staphylococcus aureus[a]	0.312 mg/mL	
	Listeria monocytogenes[a]	0.625 mg/mL	
	Pseudomonas aeruginosa[b]	0.312 mg/mL	
	Escherichia coli[b]	0.019 mg/mL	
Cultivated *O. sinensis* methanolic extract	*Bacillus cereus*[a]	0.009 mg/mL	
	Staphylococcus aureus[a]	1.25 mg/mL	
	Listeria monocytogenes[a]	0.156 mg/mL	
	Pseudomonas aeruginosa[b]	1.25 mg/mL	
	Escherichia coli[b]	0.078 mg/mL	
Lyophilized wild *O. sinensis* peptide fraction	*Salmonella enterica typhi*[b]	38–45 μg	Mishra et al. (2019)
	Escherichia coli[b]	50–52 μg	

Severe Acute Respiratory Syndrome Coronavirus (SARS-Cov) in 2002, H1N1 influenza virus in 2009, the Middle East Respiratory Syndrome Coronavirus (MERS-Cov) in 2012, Ebola virus in 2013, and the Severe Acute Respiratory Syndrome Coronavirus-2 (SARS-Cov-2) in late 2019 to present (Roychoudhury et al. 2020). Other viruses that cause global health issues include Hepatitis B, Hepatitis C, rabies, Japanese Encephalitis, yellow fever virus, and the occasional epidemic outbreaks of noroviruses, flaviviruses, and influenza viruses. Despite having more than 200 virus species that can infect humans, only 90 antiviral drugs have been approved for human use in the past 50 years, suggesting a considerable gap between demand and supply (Woolhouse et al. 2012; de Clercq and Li 2016). Given the vast therapeutic potential of phytochemicals in viral infections, the antiviral effects of *O. sinensis* have been investigated in several studies.

O. sinensis exhibited potent anti-HIV-1 effects through the inhibition of the reverse-transcriptase activity and the viral infectivity factor (Vif), a protein essential for the replication of HIV-1 (Zhu et al. 2016). This study showed that *O. sinensis* contains bioactive component(s) that may have a potential therapeutic effect on HIV. However, scientific evidence on the anti-HIV effect of *O. sinensis* remains scarce. This occurrence could be due to the challenging and rather time-consuming process of identifying the bioactive component in crude extracts. At the same time, many novel antiretroviral therapies for HIV have successfully been developed in recent years (Morikawa et al. 2020).

Unlike Zhu et al. (2016), who used crude extracts of *O. sinensis*, most studies focused on the antiviral effect of its bioactive component, cordycepin. Ryu et al. (2014) investigated the effect of cordycepin against Epstein-Barr virus (EBV) and found that its mechanism of action is highly related to histone modification. Cordycepin downregulated most of the EBV genes tested, preventing EBV translation, suggesting that cordycepin could act as a strong suppressor of EBV protein synthesis. Moreover, at 125 μM, cordycepin significantly reduced the extracellular and intracellular EBV genome copy numbers by up to 55% and 30%, respectively (Ryu et al. 2014).

Recently, Panya et al. (2021) demonstrated that cordycepin significantly decreased dengue virus protein in the dengue virus-infected Vero cells with an EC_{50} of 26.94 μM. Through in silico molecular docking, it was predicted that cordycepin might bind to the dengue virus nonstructural protein (NS5), which is essential for RNA synthesis (Panya et al. 2021). In order to substantiate this prediction, more detailed molecular elucidation studies are required.

Cordycepin is one of the compounds of interest in combating COVID-19. In an in vitro study, cordycepin exhibited a more potent inhibitory effect against SARS-Cov-2 than remdesivir, a broad-spectrum antiviral medication (Rabie 2022). Rabie (2022) suggested that the inhibitory effect of cordycepin on the replication of SARS-Cov-2 could be due to its chemical similarity with adenosine. As adenosine is one of the four nucleoside building blocks of RNA, the presence of cordycepin may interfere with the action of adenosine and disrupt the formation of the viral RNA. Besides, the anti-inflammatory and immunomodulatory properties of cordycepin which have been reported in previous studies could also be useful in mitigating COVID-19 symptoms. Based on the computational approach, several new studies have also predicted the possible inhibitory affinities of cordycepin against the principal SARS-Cov-2 protein targets such as the main protease (M^{pro}) enzyme and RNA-dependent RNA polymerase enzyme (Suwannarach et al. 2020; Bibi et al. 2021).

3.8 Other Health Benefits

Besides the aforementioned, *O. sinensis* also possesses other therapeutic and health effects. Treatment with high doses of *O. sinensis* (300 and 500 mg/kg/day) was shown to improve body weight, mechanical strength, bone mineral density, and bone mineral content of rats treated with hind limb suspension (Qi et al. 2012). Secondary metabolites may contribute to the osteoprotective effect as administration of isoflavones derived from *O. sinensis* at a dose of 20, 50, and 100 mg/kg/day has significantly increased the serum osteocalcin and estradiol level while decreasing the serum alkaline phosphatase (ALP) and tartrate-resistant acid phosphatase (TRAP) levels in ovariectomized osteopenic rats (Zhang et al. 2014). Given that osteocalcin is essential for bone formation and turnover, ALP promotes bone mineralization, and osteoclasts secrete TRAP during bone resorption, modulation of these markers indicates the potential therapeutic effect of *O. sinensis* on bone health.

Furthermore, *O. sinensis* may be beneficial in reducing the adverse effects caused by radiotherapy and chemotherapy in cancer treatment. Administration of *O. sinensis* water extracts at 200 mg/kg/day significantly prevented the reduction in body weight of mice inoculated with B16 melanoma cells due to methotrexate treatment (Nakamura et al. 2003). In another study, *O. sinensis* showed a radiation protective effect, significantly reducing the production of micronuclei-induced 5-G-γ-whole-body γ-ray irradiation in mice (Lin et al. 2007).

In several studies, *O. sinensis* has also been reported to improve learning memory and possess antidepressant effects. Following administration of *O. sinensis* at a dose of 0.5 g per kg for 60 days, mice with memory impairment showed significant improvement in the water maze test and step-through test compared to the control group (Dong et al. 2014). Recently, seven bioactive compounds in *O. sinensis* have been identified to exert antidepressant effects by regulating the CREB binding protein and antioxidative stress effects (Zhang et al. 2022a).

Lastly, the antifibrotic effect of *O. sinensis* has also been reported, with ergosterol being the bioactive component. In CCl_4-induced hepatic fibrosis mice, the administration of *O. sinensis* significantly alleviated the levels of serum liver functions and attenuated the infiltration of inflammatory cells and collagen deposition (Peng et al. 2014).

4 Conclusion

For a long time, *O. sinensis* has been used in Chinese and Tibetan medicine. Its medicinal properties are believed to be brought by a plethora of bioactive compounds in the fungus that have prompted scientific research over the past few decades. However, cordycepin, a small molecule found in the cocktail of compounds in *O. sinensis*, has been implied in a number of pharmacological actions described above. In addition to the voluminous data obtained from in vitro studies, several in vivo and clinical studies in patients and healthy subjects have manifested promising effects of *O. sinensis* in various medical conditions, especially immune-mediated conditions, cancer, and cardiovascular-related diseases. Although the dosage and administration method used in each study is different and has failed to indicate the recommended dosage for each disease, these studies have provided a great extent of knowledge on the bioactive compounds and the proposed mechanism of action underlying their biological effects. These studies substantiate our current knowledge of *O. sinensis* with its potential as an alternative therapy or complementary treatment for disease modification and maintaining good health.

Acknowledgments This work was supported by LiGNO Biotech Sdn Bhd (LiGNO). Miss Li Yin Pang receives a scholarship for her Ph.D. studies from the University of Nottingham Malaysia.

References

Abood WN, Fahmi I, Abdulla MA, Ismail S (2014) Immunomodulatory effect of an isolated fraction from *Tinospora crispa* on intracellular expression of INF-γ, IL-6 and IL-8. BMC Complement Altern Med 14:205. https://doi.org/10.1186/1472-6882-14-205

Akter S, Jesmin S, Iwashima Y et al (2015) Higher circulatory level of endothelin-1 in hypertensive subjects screened through a cross-sectional study of rural Bangladeshi women. Hypertens Res 38:208–212. https://doi.org/10.1038/hr.2014.160

Alao JP (2007) The regulation of cyclin D1 degradation: roles in cancer development and the potential for therapeutic invention. Mol Cancer 6:24. https://doi.org/10.1186/1476-4598-6-24

Bai X, Tan TY, Li YX et al (2020) The protective effect of *Cordyceps sinensis* extract on cerebral ischemic injury via modulating the mitochondrial respiratory chain and inhibiting the mitochondrial apoptotic pathway. Biomed Pharmacother 124:109834. https://doi.org/10.1016/j.biopha.2020.109834

Bhandari AK, Negi JS, Bisht VK et al (2010) Chemical constituent, inorganic elements and properties of *Cordyceps sinensis*—a review. Nat Sci 8(9):253–256

Bi B, Wang X, Wu H, Wei Q (2011) Purification and characterisation of a novel protease from *Cordyceps sinensis* and determination of the cleavage site motifs using oriented peptide library mixtures. Food Chem 126(1):46–53. https://doi.org/10.1016/j.foodchem.2010.10.057

Bibi S, Hasan MM, Wang YB et al (2021) Cordycepin as a promising inhibitor of SARS-CoV-2 RNA dependent RNA polymerase (RdRp). Curr Med Chem 29(1):152–162. https://doi.org/10.2174/0929867328666210820114025

Chang Y, Hsu WH, Lu WJ et al (2015) Inhibitory mechanisms of CME-1, a novel polysaccharide from the mycelia of *Cordyceps sinensis*, in platelet activation. Curr Pharm Biotechnol 16(5): 451–461. https://doi.org/10.2174/1389201016666150303152237

Chatterjee R, Srinivasan KS, Maiti PC (1957) *Cordyceps sinensis* (Berkeley) Saccardo: structure of cordycepic acid. J Am Pharm Assoc 46(2):114–118. https://doi.org/10.1002/jps.3030460211

Chen J, Lee S, Cao Y et al (2010) Ethnomycological use of medicinal Chinese caterpillar fungus, *Ophiocordyceps sinensis* (Berk.) G. H. Sung et al. (ascomycetes) in Northern Yunnan Province, SW China. Int J Med Mushrooms 12:427–434. https://doi.org/10.1615/IntJMedMushr.v12.i4.100

Chen SY, Ho KJ, Hsieh YJ et al (2012) Contents of lovastatin, γ-aminobutyric acid and ergothioneine in mushroom fruiting bodies and mycelia. LWT 47(2):274–278. https://doi.org/10.1016/j.lwt.2012.01.019

Cheng W, Zhang X, Song Q et al (2017) Determination and comparative analysis of 13 nucleosides and nucleobases in natural fruiting body of *Ophiocordyceps sinensis* and its substitutes. Mycology 8(4):318–326. https://doi.org/10.1080/21501203.2017.1385546

Cheung JKH, Li J, Cheung AWH et al (2009) Cordysinocan, a polysaccharide isolated from cultured *Cordyceps*, activates immune responses in cultured T-lymphocytes and macrophages: signaling cascade and induction of cytokines. J Ethnopharmacol 124(1):61–68. https://doi.org/10.1016/j.jep.2009.04.010

Chiara F, Rasola A (2013) GSK-3 and mitochondria in cancer cells. Front Oncol 3:16. https://doi.org/10.3389/fonc.2013.00016

Chiou WF, Chang PC, Chou CJ, Chen CF (2000) Protein constituent contributes to the hypotensive and vasorelaxant activities of *Cordyceps sinensis*. Life Sci 66:1369–1376. https://doi.org/10.1016/S0024-3205(00)00445-8

de Clercq E, Li G (2016) Approved antiviral drugs over the past 50 years. Clin Microbiol Rev 29(3): 695–747. https://doi.org/10.1128/CMR.00102-15

Dong Y, Yang J, Zhang Y et al (2014) *Cordyceps sinensis* Cs-5 improves memory and learning abilities in a dysmnesia model. FASEB J 28(S1). https://doi.org/10.1096/fasebj.28.1_supplement.684.1

Dong C, Guo S, Wang W, Liu X (2015) *Cordyceps* industry in China. Mycology 6(2):121–129. https://doi.org/10.1080/21501203.2015.1043967

el Ashry FEZZ, Mahmoud MF, el Maraghy NN, Ahmed AF (2012) Effect of *Cordyceps sinensis* and taurine either alone or in combination on streptozotocin induced diabetes. Food Chem Toxicol 50(3–4):1159–1165. https://doi.org/10.1016/j.fct.2011.12.020

Floras JS (1992) Epinephrine and the genesis of hypertension. Hypertension 19(1):1–18. https://doi.org/10.1161/01.hyp.19.1.1

Fujii M (1996) Krestin (PSK). Biotherapy 10(3):315–317

Fung SY, Cheong PCH, Tan NH et al (2018) Nutrient and chemical analysis of fruiting bodies of a cultivar of the Chinese caterpillar mushroom, *Ophiocordyceps sinensis* (Ascomycetes). Int J Med Mushrooms 20(5):459–469. https://doi.org/10.1615/IntJMedMushrooms.2018026252

Guo LX, Hong YH, Zhou QZ et al (2017) Fungus-larva relation in the formation of *Cordyceps sinensis* as revealed by stable carbon isotope analysis. Sci Rep 7:7789. https://doi.org/10.1038/s41598-017-08198-1

Hrudayanath T, Sameer KS (2014) Diversity, nutritional composition and medicinal potential of Indian mushrooms: a review. African J Biotechnol 13(4):523–545. https://doi.org/10.5897/ajb2013.13446

Hu Z, Ye M, Xia L et al (2006) Purification and characterization of an antibacterial protein from the cultured mycelia of *Cordyceps sinensis*. Wuhan Univ J Nat Sci 11:709–714. https://doi.org/10.1007/BF02836695

Huang BM, Hsu CC, Tsai SJ et al (2001) Effects of *Cordyceps sinensis* on testosterone production in normal mouse Leydig cells. Life Sci 69(22):2593–2602. https://doi.org/10.1016/S0024-3205(01)01339-X

Huang F, Li W, Xu H et al (2019) Cordycepin kills *Mycobacterium tuberculosis* through hijacking the bacterial adenosine kinase. PLoS One 14(6):e0218449. https://doi.org/10.1371/journal.pone.0218449

Jakopovic B, Horvatić A, Klobučar M et al (2020) Treatment with medicinal mushroom extract mixture inhibits translation and reprograms metabolism in advanced colorectal cancer animal model as evidenced by tandem mass tags proteomics analysis. Front Pharmacol. 11:1202. https://doi.org/10.3389/fphar.2020.01202

Jia JM, Ma XC, Wu CF et al (2005) Cordycedipeptide A, a new cyclodipeptide from the culture liquid of *Cordyceps sinensis* (BERK.) SACC. Chem Pharm Bull (Tokyo) 53:582–583. https://doi.org/10.1248/cpb.53.582

Jia JM, Tao HH, Feng BM (2009) Cordyceamides A and B from the culture liquid of *Cordyceps sinensis* (Berk.) sacc. Chem Pharm Bull 57(1):99–101. https://doi.org/10.1248/cpb.57.99

Jiang Q, Lou Z, Wang H, Chen C (2019) Antimicrobial effect and proposed action mechanism of cordycepin against *Escherichia coli* and *Bacillus subtilis*. J Microbiol 57:288–297. https://doi.org/10.1007/s12275-019-8113-z

Jung SJ, Jung ES, Choi EK et al (2019) Immunomodulatory effects of a mycelium extract of Cordyceps (*Paecilomyces hepiali*; CBG-CS-2): a randomized and double-blind clinical trial. BMC Complement Alternat Med. 19(1):77. https://doi.org/10.1186/s12906-019-2483-y

Kan WC, Wang HY, Chien CC et al (2012) Effects of extract from solid-state fermented *Cordyceps sinensis* on type 2 diabetes mellitus. Evid Based Complement Alternat Med. 2012:743107. https://doi.org/10.1155/2012/743107

Kaushik V, Arya A, Sindhu A, Singh A (2019) Identification, optimization of culture conditions, and bioactive potential of Chinese caterpillar mushroom *Ophiocordyceps sinensis* (Ascomycetes) mycelium isolated from fruiting body. Int J Med Mushrooms 21(9):931–942. https://doi.org/10.1615/IntJMedMushrooms.2019031841

Kim SD (2010) Isolation, structure and cholesterol esterase inhibitory activity of a polysaccharide, PS-A, from *Cordyceps sinensis*. J Appl Biol Chem 53:784–789. https://doi.org/10.3839/jksabc.2010.118

Kong BH, Yap CSA, Razif MFM et al (2021) Antioxidant and cytotoxic effects and identification of *Ophiocordyceps sinensis* bioactive proteins using shotgun proteomic analysis. Food Technol Biotechnol 59(2):201–208. https://doi.org/10.17113/ftb.59.02.21.7151

Kuo YC, Tsai WJ, Wang JY et al (2001) Regulation of bronchoalveolar lavage fluids cell function by the immunomodulatory agents from *Cordyceps sinensis*. Life Sci 68(9):1067–1082. https://doi.org/10.1016/S0024-3205(00)01011-0

Kuo HC, Huang IC, Chen TY (2015) Cordyceps s.L. (ascomycetes) species used as medicinal mushrooms are closely related with higher ability to produce cordycepin. Int J Med Mushrooms 17(11):1077–1085. https://doi.org/10.1615/IntJMedMushrooms.v17.i11.80

Lee EJ, Jang KH, Im SY et al (2015) Physico-chemical properties and cytotoxic potential of *Cordyceps sinensis* metabolites. Nat Prod Res 29(5):455–459. https://doi.org/10.1080/14786419.2014.948438

Li DG, Ren ZX (2017) *Cordyceps sinensis* promotes immune regulation and enhances bacteriostatic activity of PA-824 via IL-10 in mycobacterium tuberculosis disease. Braz J Med Biol Res 50(9):e6188. https://doi.org/10.1590/1414-431x20176188

Li SP, Li P, Dong TTX, Tsim KWK (2001) Antioxidation activity of different types of natural *Cordyceps sinensis* and cultured *Cordyceps* mycelia. Phytomedicine 8:207–212. https://doi.org/10.1078/0944-7113-00030

Li SP, Zhao KJ, Ji ZN et al (2003) A polysaccharide isolated from *Cordyceps sinensis*, a traditional Chinese medicine, protects PC12 cells against hydrogen peroxide-induced injury. Life Sci 73:2503–2513. https://doi.org/10.1016/S0024-3205(03)00652-0

Li SP, Zhang GH, Zeng Q et al (2006) Hypoglycemic activity of polysaccharide, with antioxidation, isolated from cultured *Cordyceps* mycelia. Phytomedicine 13:428–433. https://doi.org/10.1016/j.phymed.2005.02.002

Li HP, Hu Z, Yuan JL et al (2007) A novel extracellular protease with fibrinolytic activity from the culture supernatant of *Cordyceps sinensis*: purification and characterization. Phytother Res 21(12):1234–1241. https://doi.org/10.1002/ptr.2246

Li Y, Wang XL, Jiao L et al (2011) A survey of the geographic distribution of *Ophiocordyceps sinensis*. J Microbiol 49:913–919. https://doi.org/10.1007/s12275-011-1193-z

Li Y, Xue WJ, Tian PX et al (2009) Clinical application of *Cordyceps sinensis* on Immunosuppressive therapy in renal transplantation. Transplant Proc 41(5):1565–1569. https://doi.org/10.1016/j.transproceed.2009.02.085

Lin BQ, Li SP (2011) *Cordyceps* as an herbal drug. In: Benzie IFF, Wachtel-Galor S (eds) Herbal medicine: biomolecular and clinical aspects, 2nd edn. CRC Press Taylor & Francis, Florida, pp 73–105

Lin CC, Pumsanguan W, Koo MMO et al (2007) Radiation protective effects of *Cordyceps sinensis* in blood cells. Tzu Chi Med J 19(4):226–232. https://doi.org/10.1016/S1016-3190(10)60020-1

Liu Y, Wang J, Wang W et al (2015) The chemical constituents and pharmacological actions of *Cordyceps sinensis*. Evid Based Complement Alternat Med 2015:575063. https://doi.org/10.1155/2015/575063

Lu WJ, Chang NC, Jayakumar T et al (2014) *Ex vivo* and *in vivo* studies of CME-1, a novel polysaccharide purified from the mycelia of *Cordyceps sinensis* that inhibits human platelet activation by activating adenylate cyclase/cyclic AMP. Thromb Res 134(6):1301–1310. https://doi.org/10.1016/j.thromres.2014.09.023

Luitel H, Novoyatleva T, Sydykov A et al (2020) Yarsagumba is a promising therapeutic option for treatment of pulmonary hypertension due to the potent anti-proliferative and vasorelaxant properties. Medicina 56(3):131. https://doi.org/10.3390/medicina56030131

Luo Y, Yang SK, Zhou X et al (2015) Use of *Ophiocordyceps sinensis* (syn. *Cordyceps sinensis*) combined with angiotensin-converting enzyme inhibitors (ACEI)/angiotensin receptor blockers (ARB) versus ACEI/ARB alone in the treatment of diabetic kidney disease: a meta-analysis. Ren Fail 37(4):614–634. https://doi.org/10.3109/0886022X.2015.1009820

Ma MW, Gao XS, Yu HL et al (2018) *Cordyceps sinensis* promotes the growth of prostate cancer cells. Nutr Cancer 70(7):1166–1172. https://doi.org/10.1080/01635581.2018.1504091

Mamta, Mehrotra S, Amitabh et al (2015) Phytochemical and antimicrobial activities of Himalayan *Cordyceps sinensis* (Berk.) Sacc. Indian J Exp Biol 53:36–43

Matijaševic D, Pantic M, Raškovic B et al (2016) The antibacterial activity of *Coriolus versicolor* methanol extract and its effect on ultrastructural changes of *Staphylococcus aureus* and *Salmonella enteritidis*. Front Microbiol 7:1226. https://doi.org/10.3389/fmicb.2016.01226

Meek RW, Vyas H, Piddock LJV (2015) Nonmedical uses of antibiotics: time to restrict their use? PLoS Biol 13:e1002266. https://doi.org/10.1371/journal.pbio.1002266

Mei YX, Yang W, Zhu PX et al (2014) Isolation, characterization, and antitumor activity of a novel heteroglycan from cultured mycelia of *Cordyceps sinensis*. Planta Med 80:1107–1112. https://doi.org/10.1055/s-0034-1382960

Menzies RI, Booth JWR, Mullins JJ et al (2017) Hyperglycemia-induced renal P2X7 receptor activation enhances diabetes-related injury. EBioMedicine 19:73–83. https://doi.org/10.1016/j.ebiom.2017.04.011

Mishra J, Rajput R, Singh K et al (2019) Antioxidant-rich peptide fractions derived from high-altitude Chinese caterpillar medicinal mushroom *Ophiocordyceps sinensis* (Ascomycetes) inhibit bacterial pathogens. Int J Med Mushrooms 21(2):155–168. https://doi.org/10.1615/IntJMedMushrooms.2019030013

Mohanarji S, Dharmalingam S, Kalusalingam A (2012) Screening of *Lignosus rhinocerus* extracts as antimicrobial agents against selected human pathogens. J Pharm Biomed Sci 18(11):1–4

Morikawa T, Tamura S, Wang T (2020) Editorial: discovery and total synthesis of bio-functional natural products from traditional medicinal plants. Front Chem 8:650. https://doi.org/10.3389/fchem.2020.00650

Nakamura K, Konoha K, Yamaguchi Y et al (2003) Combined effects of *Cordyceps sinensis* and methotrexate on hematogenic lung metastasis in mice. Recept Channels. 9(5):329–334. https://doi.org/10.3109/713745176

Nakamura K, Shinozuka K, Yoshikawa N (2015) Anticancer and antimetastatic effects of cordycepin, an active component of *Cordyceps sinensis*. J Pharmacol Sci 127(1):53–56. https://doi.org/10.1016/j.jphs.2014.09.001

Nascimento M, Punaro GR, Serralha RS et al (2020) Inhibition of the P2X7 receptor improves renal function via renin-angiotensin system and nitric oxide on diabetic nephropathy in rats. Life Sci. 251:117640. https://doi.org/10.1016/j.lfs.2020.117640

Nguyen QV, Vu TT, Tran MT et al (2021) Antioxidant activity and hepatoprotective effect of exopolysaccharides from cultivated *Ophiocordyceps Sinensis* against CCl4-induced liver damages. Nat Prof Commun. https://doi.org/10.1177/1934578X21997670

Pan W, Wang W, Huang J et al (2020) The prognostic role of c-MYC amplification in schistosomiasis-associated colorectal cancer. Jpn J Clin Oncol 50(4):446–455. https://doi.org/10.1093/jjco/hyz210

Panda AK (2015) Tracing historical perspective of *Cordyceps sinensis*—an aphrodisiac in Sikkim Himalaya. Proc Indian Natl Sci 45:189–198

Panda AK, Swain KC (2011) Traditional uses and medicinal potential of *Cordyceps sinensis* of Sikkim. J Ayurveda Integr Med 2(1):9–13. https://doi.org/10.4103/0975-9476.78183

Pang LY, Lee MK, Kong BH et al (2022) Chinese caterpillar mushroom, *Ophiocordyceps sinensis* (Ascomycota), cold-water extract relaxes the rat isolated bladder. Int J Med Mushrooms 24(3): 1–14. https://doi.org/10.1615/IntJMedMushrooms.2022043097

Panya A, Songprakhon P, Panwong S et al (2021) Cordycepin inhibits virus replication in dengue virus-infected vero cells. Molecules 26(11):3118. https://doi.org/10.3390/molecules26113118

Peng Y, Tao Y, Wang Q et al (2014) Ergosterol is the active compound of cultured mycelium *Cordyceps sinensis* on antiliver fibrosis. Evid Based Complement Alternat Med. 2014:537234. https://doi.org/10.1155/2014/537234

Petersen MC, Shulman GI (2018) Mechanisms of insulin action and insulin resistance. Physiol Rev 98(4):2133–2223. https://doi.org/10.1152/physrev.00063.2017

Pham-Huy LA, He H, Pham-Huy C (2008) Free radicals, antioxidants in disease and health. Int J Biomed Sci 4(2):89–96

Pizzino G, Irrera N, Cucinotta M et al (2017) Oxidative stress: harms and benefits for human health. Oxidative Med Cell Longev 2017:8416763. https://doi.org/10.1155/2017/8416763

Qi W, Yan YB, Lei W et al (2012) Prevention of disuse osteoporosis in rats by *Cordyceps sinensis* extract. Osteoporos Int 23:2347–2357. https://doi.org/10.1007/s00198-011-1842-4

Qian GM, Pan GF, Guo JY (2012) Anti-inflammatory and antinociceptive effects of cordymin, a peptide purified from the medicinal mushroom *Cordyceps sinensis*. Nat Prod Res. 26(24): 2358–2362. https://doi.org/10.1080/14786419.2012.658800

Rabie AM (2022) Potent inhibitory activities of the adenosine analogue cordycepin on SARS-CoV-2 replication. ACS Omega 7(3):2960–2969. https://doi.org/10.1021/acsomega.1c05998

Ren L, Hemar Y, Perera CO et al (2014) Antibacterial and antioxidant activities of aqueous extracts of eight edible mushrooms. Bioact Carbohydr Diet Fibre 3(2):41–51. https://doi.org/10.1016/j.bcdf.2014.01.003

Rincon M, Irvin CG (2012) Role of IL-6 in asthma and other inflammatory pulmonary diseases. Int J Biol Sci 8(9):1281–1290. https://doi.org/10.7150/ijbs.4874

Roychoudhury S, Das A, Sengupta P et al (2020) Viral pandemics of the last four decades: pathophysiology, health impacts and perspectives. Int J Environ Res Public Health. 17(24): 9411. https://doi.org/10.3390/ijerph17249411

Ryu E, Son M, Lee M et al (2014) Cordycepin is a novel chemical suppressor of Epstein-Barr virus replication. Oncoscience 1(12):866–881. https://doi.org/10.18632/oncoscience.110

Sander S, Bullinger L, Klapproth K et al (2008) MYC stimulates EZH2 expression by repression of its negative regulator miR-26a. Blood 112(10):4202–4212. https://doi.org/10.1182/blood-2008-03-147645

Sang Q, Pan Y, Jiang Z et al (2020) HPLC determination of massoia lactone in fermented *Cordyceps sinensis* mycelium Cs-4 and its anticancer activity in vitro. J Food Biochem 44(9): e13336. https://doi.org/10.1111/jfbc.13336

Santos-Sánchez NF, Salas-Coronado R, Villanueva-Cañongo C, Hernández-Carlos B (2019) Antioxidant compounds and their antioxidant mechanism. In: Shalaby E (ed) Antioxidants. IntechOpen, London. https://doi.org/10.5772/intechopen.85270

Seo MJ, Kim MJ, Lee HH et al (2013) Effect of cordycepin on the expression of the inflammatory cytokines TNF-alpha, IL-6, and IL-17a in C57bL/6 mice. J Microbiol Biotechnol 23:156–160. https://doi.org/10.4014/jmb.1211.11032

Singh M, Tulsawani R, Koganti P et al (2013) *Cordyceps sinensis* increases hypoxia tolerance by inducing heme oxygenase-1 and metallothionein via Nrf2 activation in human lung epithelial cells. Biomed Res Int 2013:1–13. https://doi.org/10.1155/2013/569206

Subhan M, Faryal R, Macreadie I (2016) Exploitation of *Aspergillus terreus* for the production of natural statins. J Fungi 2(2):13. https://doi.org/10.3390/jof2020013

Suffness M, Pezzuto JM (1990) Assays related to cancer drug discovery. In: Hostettamann K (ed) Methods in plant biochemistry: assays for bioactivity. Academic Press, Massachusetts, pp 71–133

Sun X, Dong Z, Li N et al (2018) Nucleosides isolated from *Ophiocordyceps sinensis* inhibit cigarette smoke extract-induced inflammation via the SIRT1–nuclear factor-κB/p65 pathway in RAW264.7 macrophages and in COPD mice. Int J Chron Obstruct Pulmon Dis 13:2821–2832. https://doi.org/10.2147/COPD.S172579

Sung GH, Hywel-Jones NL, Sung JM et al (2007) Phylogenetic classification of *Cordyceps* and the clavicipitaceous fungi. Stud Mycol 57:5–59. https://doi.org/10.3114/sim.2007.57.01

Suwannarach N, Kumla J, Sujarit K et al (2020) Natural bioactive compounds from fungi as potential candidates for protease inhibitors and immunomodulators to apply for coronaviruses. Molecules 25(8):1800. https://doi.org/10.3390/molecules25081800

Taguchi T, Furue H, Kimura T et al (1985) Results of phase III study of lentinan. Jpn J Cancer Chemother 12(2):366–378

Tong X, Guo J (2022) High throughput identification of the potential antioxidant peptides in *Ophiocordyceps sinensis*. Molecules 27(2):438. https://doi.org/10.3390/molecules27020438

Vazirian M, Faramarzi MA, Ebrahimi SES et al (2014) Antimicrobial effect of the Lingzhi or Reishi medicinal mushroom, *Ganoderma lucidum* (higher Basidiomycetes) and its main compounds. Int J Med Mushrooms 16(1):77–84. https://doi.org/10.1615/IntJMedMushr.v16.i1.70

Wang ZM, Peng X, Lee KLD et al (2011) Structural characterisation and immunomodulatory property of an acidic polysaccharide from mycelial culture of *Cordyceps sinensis* fungus Cs-HK1. Food Chem 125:637–643. https://doi.org/10.1016/j.foodchem.2010.09.052

Wang J, Kan L, Nie S et al (2015a) A comparison of chemical composition, bioactive components and antioxidant activity of natural and cultured *Cordyceps sinensis*. Food Sci Tech 63(1):2–7. https://doi.org/10.1016/j.lwt.2015.03.109

Wang Y, Wang Y, Liu D et al (2015b) *Cordyceps sinensis* polysaccharide inhibits PDGF-BB-induced inflammation and ROS production in human mesangial cells. Carbohydr Polym 125: 135–145. https://doi.org/10.1016/j.carbpol.2015.02.012

Wang N, Li J, Huang X et al (2016) Herbal Medicine *Cordyceps sinensis* improves health-related quality of life in moderate-to-severe asthma. Evid Based Complement Alternat Med. 2016: 6134593. https://doi.org/10.1155/2016/6134593

Wang C, Hou XX, Rui HL et al (2018) Artificially cultivated *Ophiocordyceps sinensis* alleviates diabetic nephropathy and its podocyte injury via inhibiting P2X7R expression and NLRP3 inflammasome activation. J Diabetes Res 2018:1390418. https://doi.org/10.1155/2018/1390418

Wasser SP (2014) Medicinal mushroom science: current perspectives, advances, evidences, and challenges. Biomed J 37(6):345–356. https://doi.org/10.4103/2319-4170.138318

Winder M, Tobin G, Zupančič D, Romih R (2014) Signalling molecules in the urothelium. Biomed Res Int 2014:297295. https://doi.org/10.1155/2014/297295

Woolhouse M, Scott F, Hudson Z et al (2012) Human viruses: discovery and emergence. Philos Trans R Soc Lond B Biol Sci. 367(1604):2864–2871. https://doi.org/10.1098/rstb.2011.0354

Wu R, Gao JP, Wang HL et al (2015) Effects of fermented *Cordyceps sinensis* on oxidative stress in doxorubicin treated rats. Pharmacogn Mag 11(44):724–731. https://doi.org/10.4103/0973-1296.165562

Xiang F, Lin L, Hu M, Qi X (2016) Therapeutic efficacy of a polysaccharide isolated from *Cordyceps sinensis* on hypertensive rats. Int J Biol Macromol 82:308–314. https://doi.org/10.1016/j.ijbiomac.2015.09.060

Xiao Y, Li C, Xu W et al (2021) Arsenic content, speciation, and distribution in wild *Cordyceps sinensis*. Evid Based Complement Alternat Med 2021:6651498. https://doi.org/10.1155/2021/6651498

Yalin W, Ishurd O, Cuirong S, Yuanjiang P (2005) Structure analysis and antitumor activity of (1→3)-β-D-glucans (cordyglucans) from the mycelia of *Cordyceps sinensis*. Planta Med 71(4): 381–384. https://doi.org/10.1055/s-2005-864111

Yamaguchi Y, Kagota S, Nakamura K et al (2000a) Antioxidant activity of the extracts from fruiting bodies of cultured *Cordyceps sinensis*. Phytother Res 14(8):647–649. https://doi.org/10.1002/1099-1573(200012)14:8<647::AID-PTR670>3.0.CO;2-W

Yamaguchi Y, Kagota S, Nakamura K et al (2000b) Inhibitory effects of water extracts from fruiting bodies of cultured *Cordyceps sinensis* on raised serum lipid peroxide levels and aortic cholesterol deposition in atherosclerotic mice. Phytother Res 14(8):650–652. https://doi.org/10.1002/1099-1573(200012)14:8<650::AID-PTR675>3.0.CO;2-0

Yan Y, Li Y, Wang WJ et al (2017) Range shifts in response to climate change of *Ophiocordyceps sinensis*, a fungus endemic to the Tibetan Plateau. Biol Conserv 206:143–150. https://doi.org/10.1016/j.biocon.2016.12.023

Yang FQ, Feng K, Zhao J, Li SP (2009) Analysis of sterols and fatty acids in natural and cultured *Cordyceps* by one-step derivatization followed with gas chromatography-mass spectrometry. J Pharm Biomed Anal 49(5):1172–1178. https://doi.org/10.1016/j.jpba.2009.02.025

Yang FQ, Li DQ, Feng K et al (2010) Determination of nucleotides, nucleosides and their transformation products in *Cordyceps* by ion-pairing reversed-phase liquid chromatography-mass spectrometry. J Chromatogr A 1217(34):5501–5510. https://doi.org/10.1016/j.chroma.2010.06.062

Yang ML, Kuo PC, Hwang TL, Wu TS (2011) Anti-inflammatory principles from *Cordyceps sinensis*. J Nat Prod 74:1996–2000. https://doi.org/10.1021/np100902f

Yang M, Wang X, Han Y et al (2021) Targeting the NLRP3 inflammasome in diabetic nephropathy. Curr Med Chem 28(42):8810–8824. https://doi.org/10.2174/0929867328666210705153109

Yao X, Meran S, Fang Y et al (2014) *Cordyceps sinensis*: *in vitro* antifibrotic bioactivity of natural and cultured preparations. Food Hydrocoll 35:444–452. https://doi.org/10.1016/j.foodhyd.2013.06.023

Yap ACS, Li X, Yap YHY et al (2020) Immunomodulatory properties of water-soluble polysaccharides extracted from the fruiting body of Chinese caterpillar mushroom, *Ophiocordyceps sinensis* cultivar OCS02® (Ascomycetes). Int J Med Mushrooms 22(10):967–977. https://doi.org/10.1615/IntJMedMushrooms.2020036351

Yoshikawa N, Nishiuchi A, Kubo E et al (2011) *Cordyceps sinensis* acts as an adenosine A3 receptor agonist on mouse melanoma and lung carcinoma cells, and human fibrosarcoma and colon carcinoma cells. Pharmacol Pharm 2:266–270. https://doi.org/10.4236/pp.2011.24034

Yu X, Mao Y, Shergis JL et al (2019) Effectiveness and safety of oral *Cordyceps sinensis* on stable COPD of GOLD Stages 2–3: systematic review and meta-analysis. Evid Based Complement Alternat Med 2019:4903671. https://doi.org/10.1155/2019/4903671

Yuan JP, Wang JH, Liu X et al (2007) Simultaneous determination of free ergosterol and ergosteryl esters in *Cordyceps sinensis* by HPLC. Food Chem 105(4):1755–1759. https://doi.org/10.1016/j.foodchem.2007.04.070

Zhang Q, Wu J, Hu Z, Li D (2004) Induction of HL-60 apoptosis by ethyl acetate extract of *Cordyceps sinensis* fungal mycelium. Life Sci 75(24):2911–2919. https://doi.org/10.1016/j.lfs.2004.05.029

Zhang Y, Li E, Wang C et al (2012) *Ophiocordyceps sinensis*, the flagship fungus of China: terminology, life strategy and ecology. Mycology 3:2–10. https://doi.org/10.1080/21501203.2011.654354

Zhang Y, Kong H, Fang Y et al (2013) Schizophyllan: a review on its structure, properties, bioactivities and recent developments. Bioact Carbohydr Diet Fibre 1(1):53–71. https://doi.org/10.1016/j.bcdf.2013.01.002

Zhang DW, Wang ZL, Qi W, Zhao GY (2014) The effects of *Cordyceps sinensis* phytoestrogen on estrogen deficiency-induced osteoporosis in Ovariectomized rats. BMC Complement Alternat Med. 14:484. https://doi.org/10.1186/1472-6882-14-484

Zhang P, Li S, Li J et al (2018) Identification of *Ophiocordyceps sinensis* and its artificially cultured *Ophiocordyceps* mycelia by ultra-performance liquid chromatography/orbitrap fusion mass spectrometry and chemometrics. Molecules 23(5):1013. https://doi.org/10.3390/molecules23051013

Zhang X, Wang M, Qiao Y et al (2022a) Exploring the mechanisms of action of *Cordyceps sinensis* for the treatment of depression using network pharmacology and molecular docking. Ann Transl Med 10(6):282. https://doi.org/10.21037/atm-22-762

Zhang Z, Li K, Zheng Z, Liu Y (2022b) Cordycepin inhibits colon cancer proliferation by suppressing MYC expression. BMC Pharmacol Toxicol 23(1):12. https://doi.org/10.1186/s40360-022-00551-z

Zhou L, Hao QX, Wang S et al (2017) Study on distribution of five heavy metal elements in different parts of *Cordyceps sinensis* by microwave digestion ICP-MS. Zhongguo Zhongyao Zazhi 42(5):2934–2938. https://doi.org/10.19540/j.cnki.cjcmm.20170714.016

Zhou Y, Wang M, Zhang H et al (2019) Comparative study of the composition of cultivated, naturally grown *Cordyceps sinensis*, and stiff worms across different sampling years. PLoS One. 14(12):e0225750. https://doi.org/10.1371/journal.pone.0225750

Zhu Y, Ma L, Hu Q et al (2016) In vitro anti-HIV-1 activity of *Cordyceps sinensis* extracts. Bing Du Xue Bao 32(4):417–422

Zou Y, Liu Y, Ruan M et al (2015) *Cordyceps sinensis* oral liquid prolongs the lifespan of the fruit fly, *Drosophila melanogaster*, by inhibiting oxidative stress. Int J Mol Med 36(4):939–946. https://doi.org/10.3892/ijmm.2015.2296

Anticancer Potential of *Ganoderma lucidum* and Its Underlying Mechanisms

Seren Gündoğdu and Nadire Özenver

Contents

Abstract *Ganoderma lucidum* (Curtis) P. Karst. is a medicinal mushroom with extensive pharmacological and therapeutical effects and outstanding industrial value. It has been used for diverse health benefits in Asian countries for millennia. Emerging scientific research has pointed out the potential of *G. lucidum* in the management of various cancers. In this chapter, we focused on the anticancer capacity of *G. lucidum*-based products and their relevant modes of action. We performed a literature search using PubMed, Google Scholar, Web of Science, and Scopus, mainly focusing on the studies conducted in the last years. During the search, we used keywords such as *Ganoderma lucidum*, anticancer, mechanism, pharmacology, bioactive compounds, etc. We present an overview of *G. lucidum* and its probable anticancer mechanisms and draw attention to the use of natural products alone or chemotherapy in the management of cancer. Possible anticancer mechanisms against various cancer cells are indicated based on preclinical and clinical findings. Furthermore, some essential points are highlighted to ensure *G. lucidum*-based products come to the clinic.

S. Gündoğdu · N. Özenver (✉)
Faculty of Pharmacy, Department of Pharmacognosy, Hacettepe University, Ankara, Turkey
e-mail: nadire@hacettepe.edu.tr

D. C. Agrawal, M. Dhanasekaran (eds.), *Mushrooms with Therapeutic Potentials*,
https://doi.org/10.1007/978-981-19-9550-7_7

Keywords *Ganoderma lucidum* · Anticancer · Mechanism · *G. lucidum* composition

Abbreviations

CAM	Chorioallantoic Membranes
DNA	Deoxyribonucleic Acid
EMSA	Electrophoretic
ESAC	Ethanol-Soluble and Acidic Component
FDA	the United States Food and Drug Administration
GLE	*Ganoderma lucidum* Extract
HIV	Human Immunodeficiency Virus
HPLC	High-Performance Liquid Chromatography
IARC	International Agency for Research on Center
MTT	(3-(4,5-dimethylthiazol-2-yl)-2,5-diphenyl)/tetrazolium bromide
NCI	National Cancer Institute
NTF	Neutral Triterpene Fraction
SRB	Sulforhodamine B
TCM	Traditional Chinese Medicine
TB	Trypan Blue
TNBC	Triple Negative Breast Cancer
TNF-α	Tumor Necrosis Factor
TUNEL	Terminal deoxynucleotidyl transferase-mediated dUTP Nick end Labeling
US	United States
WHO	World Health Organization

1 Introduction

Cancer is unique among life-threatening diseases because it has become a thriving health issue worldwide. Cancer is the second foremost reason for death (WHO 2022). Based on the reports published by the International Agency for Research on Center (IARC), the estimated number of new cases from 2020 to 2040 will reach 30.2 million from 19.3 million, drawing attention to the situation (IARC 2022). Cancer prevention is mediated mainly by sustaining a healthy lifestyle, prohibiting cancer-causing agents, and getting medicines or vaccines that can avoid cancer occurrence. When it comes to treatment, modern medical approaches are available such as surgery, chemotherapy, immunotherapy, targeted therapy, radiotherapy, etc., and/or their combinations (NCI 2022). At this point, discovering new agents is crucial to combat cancer. Nature provides an undeniable resource of potential drug candidates with improved efficacy and side effect profiles. For example, only 13% of the anticancer drugs approved by the United States Food and Drug Administration (FDA) between 1981 and 2014 were reported to be fully synthetic drugs, and the rest originated from natural sources (Newman and Cragg 2016). Besides, the uses of natural products in traditional medicine provide the basis for the fact that they are relatively safe (Atanasov et al. 2021).

Ganoderma lucidum (Curtis) P. Karst. is an oriental fungus from the Ganodermataceae family with a history of long-term traditional use for longevity and health improvement in Asian countries, including China and Japan. In China, it is called "lingzhi," while the corresponding name is "reishi" or "mannentake" in Japan (Wachtel-Galor et al. 2011; Ahmad 2020). *G. lucidum* has been broadly applied in traditional Chinese medicine (TCM) for decades. Asian populations have used it for medical purposes, not only for typical well-being conditions but also during cancer intervention (Gordan et al. 2011). *G. lucidum* has also been widely used in managing various conditions such as obesity, hypertension, diabetes, asthma, etc. (Kim et al. 2004; Wang et al. 2020a; Wachtel-Galor et al. 2011).

G. lucidum is precious due to its pharmaceutical aspects. Traditional uses, anecdotal grounds, and cultural sides indicate health-associated advantages of *G. lucidum*, most of which were scientifically proven by preclinical and clinical experiments. Emerging evidence has demonstrated anticancer (Sohretoglu and Huang 2018; Wang et al. 2020b; Zhang et al. 2019; Yuen and Gohel 2005), antimetastatic (Xian et al. 2021; Kimura et al. 2002; Cheng and Sliva 2015), antioxidative (Lin and Deng 2019; Kozarski et al. 2012), neuroprotective (Zhou et al. 2012; Lu et al. 2019; Ren et al. 2019; Wu et al. 2022), immunomodulatory (Rubel et al. 2018; Li et al. 2020a), antimicrobial (Savin et al. 2020; Kaur et al. 2015), hypoglycemic (Huang et al. 2021; Adeyi et al. 2021; Shao et al. 2022), hypotensive (Adeyi et al. 2021; Shevelev et al. 2018), anti-HIV activity (el-Mekkawy et al. 1998; Min et al. 1998; Cör et al. 2018), and other biological activities (Adeyi et al. 2021; Shevelev et al. 2018; Cör et al. 2018) of *G. lucidum* and *G. lucidum*-obtained compounds/extracts. *G. lucidum* includes a variety of molecules from diverse chemical groups in the composition of their fruiting bodies and mycelia, including proteins, polysaccharides, fatty acids, enzymes, triterpenoids, amino acids, nucleosides, alkaloids, steroids, lactones, and lectins. Among those, triterpenoids and polysaccharides are substantial bioactive compounds with noteworthy pharmacological properties. Various nutraceuticals, nutriceuticals, and pharmaceuticals have been formulated based on *G. lucidum* or its bioactive substances (Bulam et al. 2019; Wachtel-Galor et al. 2011; Wang et al. 2020a).

G. lucidum has broad distribution worldwide and draws attention due to its medicinal, ornamental, and economic value. *G. lucidum* products have been increasingly marketed worldwide, and growing demand exists, leading to its cultivation (Roy et al. 2015; Wang et al. 2020a). Various commercial *G. lucidum* products exist in different forms, including tea, dietary supplements, or powders that include assorted mushroom parts (Wachtel-Galor et al. 2011). *G. lucidum*-associated products are sold in contemporary markets for decorative or health care purposes. *G. lucidum* has been manufactured in different forms: substantially as tablet or capsule form in which fruiting bodies are crushed into powder. Other "non-extracted" products are obtained from entire fungal spores, dried and crushed mycelia, dried and crushed combinations of substrate, mycelia, and mushroom primordia (Wachtel-Galor et al. 2011). Extracted products are prepared with substances in the composition of fruiting bodies or mycelia of *G. lucidum* that can be encapsulated individually or together. Moreover, the combined products of *G. lucidum* with other mushrooms or medicinal plants are also available in the modern market (Bishop et al. 2015; Wachtel-Galor et al. 2011).

Throughout history, *G. lucidum* has held great interest in possessing medicinal records in nations' cultures or exhibiting critical pharmacological activities. In the present chapter, we plan to focus on the anticancer potential of *G. lucidum* and its relevant modes of action based on the current scientific data. In this context, we intend to render a certain theoretical basis for the rational use of *G. lucidum* products and medicinal historical records associated with *G. lucidum*. The potential value of *G. lucidum* products in the pharmaceutical field was also highlighted.

2 *Ganoderma lucidum* as an Anticancer Agent

Cancer is an important public health problem among the leading causes of death worldwide. In addition to modern treatment approaches, the use of semisynthetic/synthetic drugs and advanced personalized medicine have substantially contributed to our fight against cancer. It has been reported that there has been little change in the prevalence of cancer during the past 10 years (Siegel et al. 2019). Therefore, it is imperative to develop new active compounds to fight against cancer by humanity. Phytochemicals obtained from natural sources and known for their benefits to human health for centuries are significant sources for anticancer drug discovery (Graziose et al. 2010; Gullett et al. 2010).

Mushrooms, which are consumed as a delicious and nutritious food by many societies due to their rich protein content, vitamins, and minerals, have recently become interesting natural sources for developing functional foods and drugs beneficial for human health. Studies have shown that bioactive chemicals extracted from medicinal mushrooms have the potential to serve as anticancer agents, in particular (Sullivan et al. 2006; Ferreira et al. 2010; Singh 2017).

Ganoderma is one of the few mushroom species used as nutraceuticals worldwide and is probably the most common mushroom of medicinal origin. *G. lucidum* (Ganodermataceae), commonly known as Lingzhi (in China) and Reishi (in Japan) mushrooms in the Far East, is one of the wood-degrading basidiomycetous fungi and has been used for longevity and health promotion since antiquity (Sliva 2003; Bishop et al. 2015). *G. lucidum* has been mainly distributed in Asian countries and listed in the pharmacopeia of China since 2005 (Zhou et al. 2007). It has been reported to have therapeutic efficacy in many human diseases such as allergies, chronic bronchitis, hypertension, inflammation, thrombosis, cardiovascular disorders, viral infection (e.g., HIV), tumors, and cancer. Moreover, it has been recognized as an alternative adjuvant for treating carcinoma, leukemia, and hepatitis. In

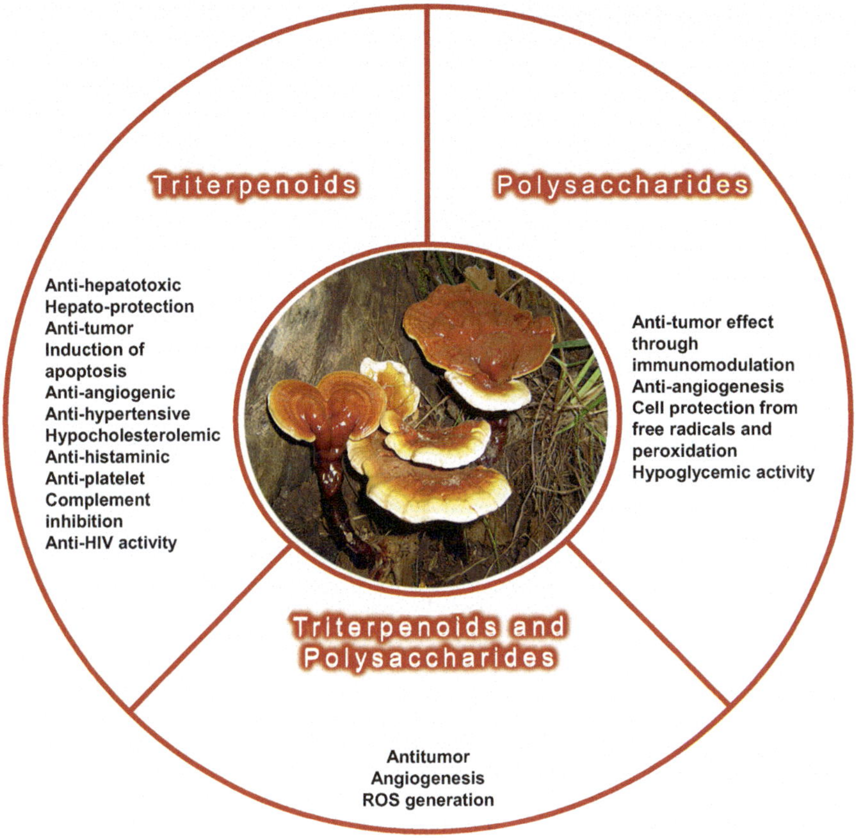

Fig. 1 Biological activities and pharmacological functions reported for *G. lucidum*. The picture of *G. lucidum* was taken from Wikimedia Commons (https://commons.wikimedia.org/wiki/File:GANODERMA_LUCIDUM_(Curt._Fr.)_Karsten_(6006072051).jpg)

addition, it has numerous pharmacological effects, including antioxidant, radical scavenging activity, modulating the immune system, hepatoprotective, and prolonging the life span (Fig. 1) (Mau et al. 2001; Sliva 2003; Sanodiya et al. 2009; Cao et al. 2012). *G. lucidum* has been a valuable ancient "herbal" nutraceutical for over 2000 years, giving the industry a market share of $2.5 billion (US). There are commercial *G. lucidum* preparations in the industry in many forms, such as tea and food supplements, based on their benefits to human health. These products mostly contain water/ethanol extracts prepared from different parts of the plant, such as fruiting bodies, mycelia, and spores (Bishop et al. 2015).

2.1 Bioactive Constituents

Almost all fungi on earth consist of 90% water by weight. A study on nonvolatile parts of *G. lucidum* reported that the remaining 10% of this weight consisted of 7–8% protein (rich in lysine and leucine), 26–28% carbohydrates, 3–5% fat, and fiber. It is also rich in potassium, calcium, phosphorus, magnesium, selenium, iron, zinc, and copper (Borchers et al. 1999; Mau et al. 2001). In addition to this rich nutritional content, *G. lucidum* is known to contain active metabolites such as terpenes, alkaloids (choline and betaine), sterols, steroids, phenols, nucleotides (adenosine and guanosine) and derivatives, glycoproteins or peptides, and polysaccharides (Borchers et al. 1999; Paterson 2006; Boh et al. 2007; Zhou et al. 2007; Sanodiya et al. 2009). Over 400 different bioactive compounds have been isolated/purified or detected from several parts of *G. lucidum* as mycelia, spores, and the fruit body (Sanodiya et al. 2009). Many researchers attribute the major pharmacological activities of *G. lucidum* to its active components known as triterpenoids, polysaccharides, and peptidoglycans (Boh et al. 2007; Zhou et al. 2007). Approximately 150 triterpenoids, such as ganoderic and lucidenic acids, were identified in extracts of *G. lucidum* spores prepared with solvents of different polarities, such as methanol, ethanol, acetone, and chloroform. Also, they can be purified by various normal or reverse phase chromatographic methods, including high-performance liquid chromatography (HPLC). These identified triterpenes have shown significant therapeutic and pharmacological efficacy on many human diseases, including cancer (Su et al. 2001; Yue et al. 2010). Various polysaccharides and polyglycans, macromolecules with high molecule weights naturally synthesized by fungi, have been the subject of many past studies (Zhou et al. 2007). Over 200 polysaccharides have been identified from *G. lucidum* (Bao et al. 2001, 2002; Huie and Di 2004). The chemical structures of some biologically active triterpenoids related to the anticancer activity of *G. lucidum* are shown in Table 1.

2.2 In Vitro Studies

Scientific data has unraveled that *G. lucidum* exerts anticancer potential in vitro via diverse mechanisms in a variety of cell lines (e.g., breast at G0/G1 phase; lung at G1 phase; liver at G1/G2 phase; and bladder, prostate, and leukemia at G2 phase), and it possesses anticancer potential by regulating the expression of various signals (Wachtel-Galor et al. 2011). Some of the biologically active components/extracts related to in vitro anticancer activity of *G. lucidum* are summarized in Table 2.

In addition to the biologically active compounds summarized in Table 2, some other triterpenes and polysaccharides in the composition of *G. lucidum* are also known to have cytotoxic activity. These are triterpene derivatives ganoderic acid (F, U, V, W, X, Y), ganoderic acid (A, D), and polysaccharide derivatives (1→3)-β-D-glucans (Sliva 2003; Martínez-Montemayor et al. 2019).

Table 1 The chemical structures of some biologically active triterpenoids related to the anticancer activity of *G. lucidum*

Compound name	Molecular formula	Chemical structures
Ganoderic acid A	$C_{30}H_{44}O_7$	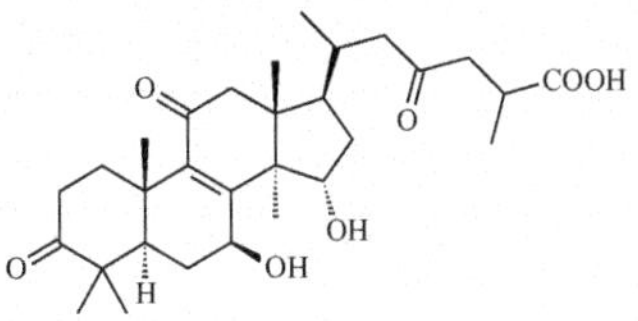
Ganoderic acid D	$C_{30}H_{42}O_7$	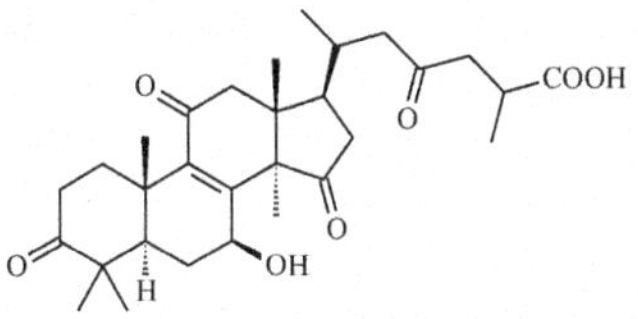
Ganoderic acid DM	$C_{30}H_{39}O_5$	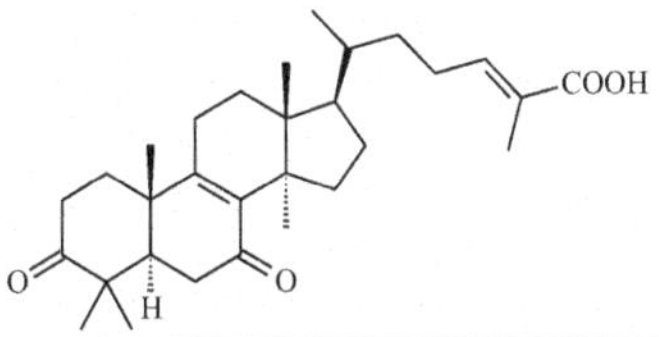
Ganoderic acid Me	$C_{34}H_{50}O_6$	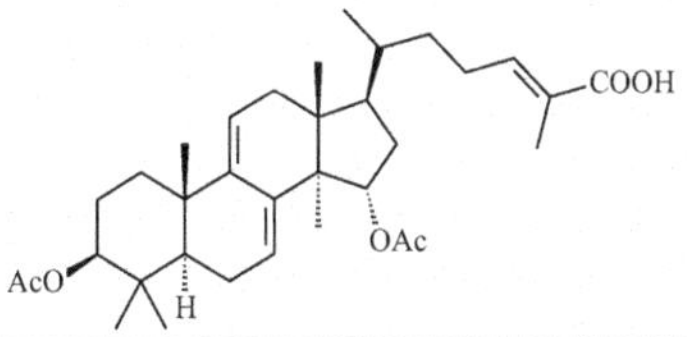
Ganoderic acid S	$C_{32}H_{48}O_5$	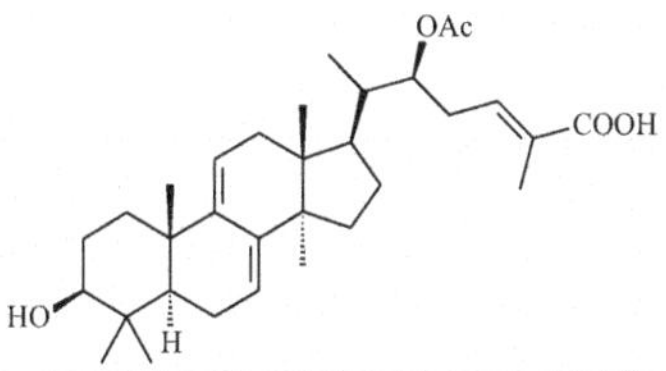
Ganoderic acid T	$C_{36}H_{53}O_7$	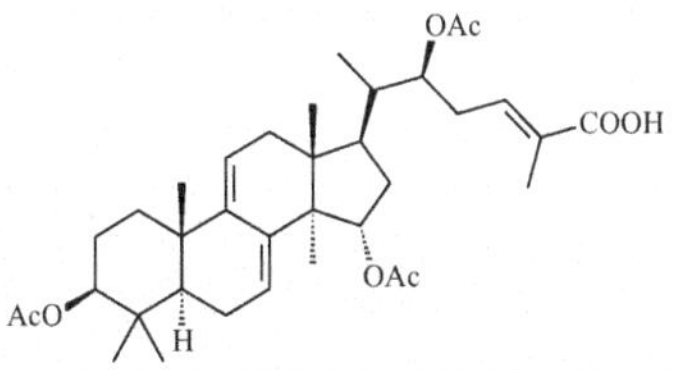
Ganoderiol F	$C_{30}H_{46}O_3$	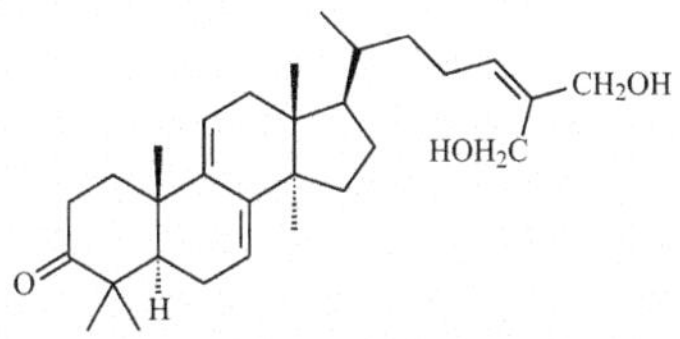

(continued)

Table 1 (continued)

Compound name	Molecular formula	Chemical structures
Ganodermanotriol	$C_{30}H_{49}O_4$	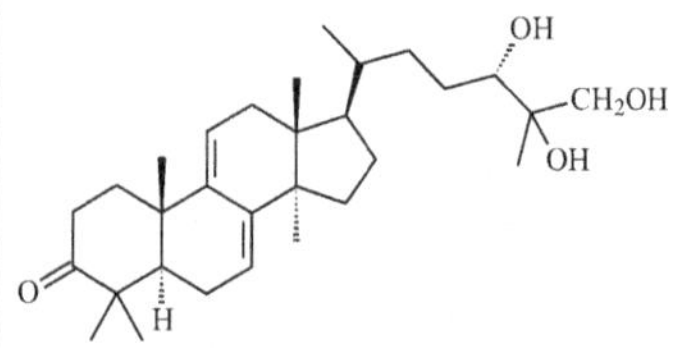
Lucialdehyde A	$C_{30}H_{46}O_2$	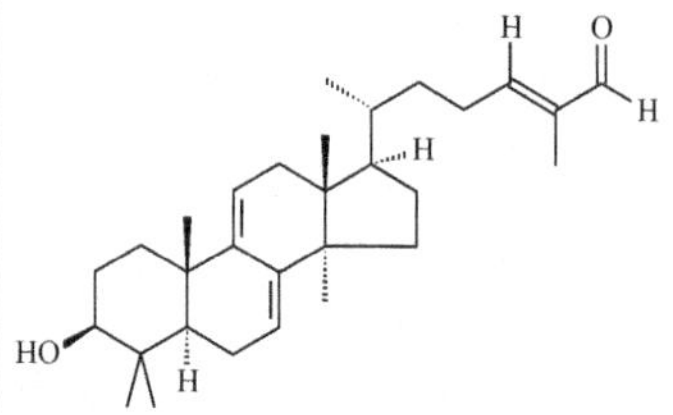
Lucialdehyde B	$C_{30}H_{44}O_3$	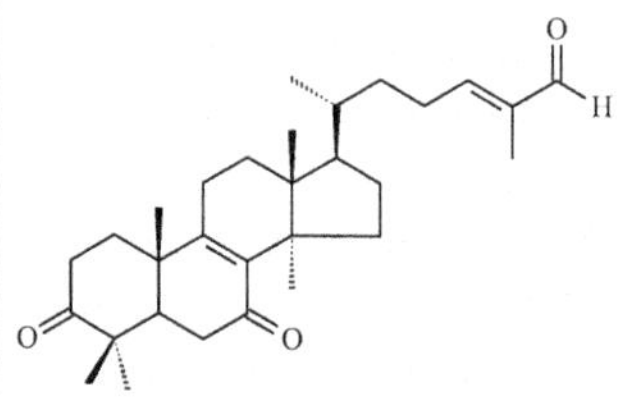

2.3 *In Vivo Studies*

In vivo studies of the possible antitumor effects of *G. lucidum* date back to the early 1990s. Some of those are indicated as below.

A new medium-term model was tested using benzo(a)pyrene (BP)-induced pulmonary adenocarcinoma in newborn mice. It has been suggested that the dose showing 50% tumor incidence was 0.5 mg BP, the incidence of lung tumors decreased with this model, which lasted for approximately 9 weeks, and that *G. lucidum* could be a chemopreventive agent (Yun et al. 1995; Yun 1999). When assessed with chick chorioallantoic membranes (CAM) and a mouse dorsal air-sac model, a polysaccharide mixture combined with genistein (an isoflavone aglycon) from cultured *G. lucidum* mycelia inhibited angiogenesis in vivo (Miura et al. 2002). In another study, oral administration of lipids extracted from germinating spores of *G. lucidum* and bio-enhanced sporoderm-fractured germinating spores inhibited the growth of mice hepatoma, sarcoma S-180, and reticulocyte sarcoma L-II cells by 80–90%. It has been noted that these spores were dose-dependent antitumor activity. It has also been suggested that sporoderm-fractured germinating spores can significantly reduce the toxic and side effects of radiotherapy and chemotherapy in some cancer patients (Liu et al. 2002). Water-soluble mycelial extract from *G. lucidum* culture given to 80 male F344/Du Crj rats divided into different groups can be a

Table 2 In vitro anticancer activity of *G. lucidum* extracts or isolated compounds

G. lucidum extract/ pure compound	Cancer species	Cell lines	Analysis methods	IC_{50} (μg/ ml)	LD_{50} (μg/ml)	Function/mechanism/outcome	Refs.
Extracts							
Ethanol extract	Breast	MCF-7 MDA-MB-231	Flow cytometry, MTT and Comet, Western blot	219.5 71.6	–	Inducing G1 arrest, apoptosis and DNA damage	Liu et al. (2009) and Wu et al. (2012)
	Myeloid leukemia	HL-60	MTT	136.3	–	Antiproliferative activity	Liu et al. (2009)
	Hepatoma	Huh-7 Hep-G2 Hep-3B	ACP, Western blot	100–450 81.2–162.6 18.1	450 (Huh-7)	Inducing G2/M arrest and apoptosis, downregulating of JNK, PKC and p38 MAPK	Lin et al. (2003), Shiao (2003), and Liu et al. (2009)
	Colon	HT-29	Caspase-3 activity assay, flow cytometry and Western blot	–	–	Inducing G0/G1 arrest, apoptosis, autophagy, suppressing	Hong et al. (2004) and Thyagarajan et al. (2010)
Fruiting body ethanol extract	Gastric carcinoma	AGS	Caspase-3 activity assay, Flow cytometry and MTT	–	–	Inducing apoptosis, inactivating AKT signal pathway	Jang et al. (2010)
	Bladder	HUC-PC MTC-11	MTT	325 129.3	–	Inducing G2/M arrest	Lu et al. (2004)
Spore ethanol extract	Bladder	HUC-PC MTC-11	MTT	521 274.7	–	Inducing G2/M arrest	Lu et al. (2004)
Fruiting body aqueous and methanol extract	Leukemia	NB4	Flow cytometry and Western blot	–	–	Inducing apoptosis	Calvino et al. (2010)
Sporoderm- broken or nonbroken spores ethanol extract	Cervical	HeLa	TB, MTT and Flow cytometry	750	4700	Inducing G1 arrest, decreasing the level of intracellular calcium	Zhu et al. (2000)
		RAW264.7		14.8	–		

(continued)

Table 2 (continued)

G. lucidum extract/ pure compound	Cancer species	Cell lines	Analysis methods	IC_{50} (μg/ml)	LD_{50} (μg/ml)	Function/mechanism/outcome	Refs.
Ethanolic triterpene extract	Murine macrophage cell		MTT, EMSA, Western blot			Inducing G0/G1-G2/M arrest, activation NF-kB and AP-1 signalings	Dudhgaonkar et al. (2009)
Crude methanol extract	Murine lymphocytic leukemia	L1210 3LL	MTT	15 10	–	Cytotoxic activity	Tomasi et al. (2004)
Pure compounds							
Ergosta-7,22-diene-2β,3α,9α-triol	Hepatoma	PLC/PRF/5 KB	MTT	1.2 0.9	–	–	Lin and Tome (1991)
Ganoderic acid A	Breast and Hepatoma	MDA-MB-231 Hep-G2	MTT, Flow cytometry, Western-blot	–	–	Inhibition of NF-kB AP-1/uPA and JAK-STAT3 signalings, Anti-invasion, anti-proliferation	Jiang et al. (2008) and Yao et al. (2012)
Ganoderic acid D	Cervical	HeLa	Flow cytometry	17.3 μM	–	G_2/M cell cycle arrest and apoptosis	Yue et al. (2008)
Ganoderic acid (G, γ, ε, θ)	Mouse lung sarcoma	Meth-A LLC	SRB	–	5.7–15.6 > 20	Cytotoxic activity	Min et al. (2000)
Lucidumol A~B	Mouse lung sarcoma	Meth-A LLC	SRB	–	4.2–8.5 2.3–16.6	Cytotoxic activity	Min et al. (2000)
Ganodermanontriol	Colon	Meth-A	MTT, Western blot	–	5.4	Inhibition of ß-catenin signaling and protein expression	Min et al. (2000) and Jedinak et al. (2011)
		LLC			9.6		
		HCT-116			–		
		HT-29			–		

potent chemopreventive agent for azoxymethane-induced colon carcinogenesis (Lu et al. 2003). The study by Nonaka et al. (2006) stated the antitumor and immunostimulant activity of *G. lucidum*. Oral dry powder of *G. lucidum* has been shown to inhibit tumor growth and has prolonged survival in both ddY mice with allogeneic sarcoma-180 and C3H/He mice with syngeneic MM-46 mammary tumors (Nonaka et al. 2006). It has also been reported that *G. lucidum* has a chemopreventive effect on prostate cancer by suppressing ventral prostate growth induced by testosterone through its triterpenes (Liu et al. 2007).

2.4 *Clinical Perspective*

Many preclinical studies have revealed that *G. lucidum* polysaccharide fractions show potent antitumor activity through their immunostimulatory effects. For this reason, many commercial preparations containing *G. lucidum* extracts rich in polysaccharides have emerged. A randomized, double-controlled, multicenter clinical trial with 34 patients of different cancer origins used *G. lucidum* capsule supplement (a patented over-the-counter product, Ganopoly) at a dose of 1800 mg given for 12 weeks. As a result of the study, IL-2, IL-6, and IFN-y plasma concentrations significantly increased ($p < 0.05$) in the majority of these patients (80%), whereas IL-1 and tumor necrosis factor (TNF-α) plasma concentrations significantly ($p < 0.05$) decreased. In another study evaluating the efficacy and safety of the same preparation on 68 patients with lung cancer, the evaluation of the quality of life was measured by Karnofsky scores, and it was reported that approximately 65% improvement was observed. There was also an increase in enhanced host immune function (e.g., enhanced natural killer cell activity) in the short preparation group compared to the placebo group (Gao et al. 2003a, b). These clinical data, which support preclinical studies, show that the antitumor effect of *G. lucidum* occurs through its various effects on the immune system (Gao and Zhou 2003).

3 Possible Mechanisms of *Ganoderma lucidum* Against Cancer

G. lucidum is a favored supplementation received by cancer patients along with standard cures. The anticancer potential of *G. lucidum* on different cancer types was evidenced by preclinical and clinical findings as indicated in above sections (Loganathan et al. 2014; Cao and Lin 2006; Zolj et al. 2018; Jiao et al. 2020; Ahmad et al. 2021; Ahmad 2020; Sohretoglu and Huang 2018). Emerging scientific data has unraveled relevant modes of action of *G. lucidum*-based products summarized below.

It has been investigated that the antitumor effect of an alcohol extract of *G. lucidum on* MCF-7 cells, dose-dependently repressed cell proliferation through upregulation of p21/Waf1 and downregulation of cyclin D1 in addition to its apoptotic activity through upregulating pro-apoptotic Bax protein (Hu et al. 2002).

In another study, *G. lucidum* extracts markedly alleviated cell viability of MDA-MB 231 and B16F10 cell lines in a time- and concentration-dependent manner by lessening the releases of matrix metalloproteases (Barbieri et al. 2017).

Ethanol-soluble and acidic component (ESAC) prepared from *G. lucidum* concentration-dependently eased the cell viability of MCF-7 and MDA-MB-231 cells by inducing G1 cell cycle arrest, DNA damage, and apoptosis (Wu et al. 2012).

G. lucidum extract (GLE) adversely affected breast cancer stem cells in vitro and triple-negative breast cancer (TNBC) animal tumor models in vivo. It significantly decreased TNBC cell viability by downregulating the STAT3 pathway (Rios-Fuller et al. 2018).

GLE inhibited breast cancer cell viability, migration, and invasion capacity by impairing Rac activity and inhibiting expressions of Lamellipodin, ENA/VASP, p-FAK, Cdc42, and c-Myc. Thus, GLE lessened lamellipodia formation, suggesting GLE as a potential therapeutic in breast cancer metastasis (Acevedo-Díaz et al. 2019).

G. lucidum spore oil, a lipid substance isolated from the sporoderm-broken spore of *G. lucidum*, inhibited the proliferation of MDA-MB-231 cells and tumors in vivo, leading to apoptosis mediated by the mitochondrial apoptotic pathway (Jiao et al. 2020).

G. lucidum caused oxidative DNA damage in colorectal cancer cells while it preserved nonmalignant cells from the accumulation of reactive oxygen species. Furthermore, DNA damage sensitized cancer cells to 5-Fluorouracil, providing a ground for the potential application of a conventional chemotherapeutic in combination with natural compounds (Opattova et al. 2019).

G. lucidum polysaccharide reduced the aggressiveness of cervical cancer cells, inhibited the cell cycle, and stimulated apoptosis via suppression of epithelial-mesenchymal and JAK/STAT5 pathways (Jin et al. 2020).

Neutral Triterpene Fraction (NTF) obtained from the dry fruiting body of *G. lucidum* acted against colon cancer cells stimulating mitochondrial-dependent pathways, and ganoderic alcohols were considered as bioactive constituents (Li et al. 2020b).

Ergosterol peroxide gained from *G. lucidum* exhibited antiproliferative effects on triple negative and inflammatory breast cancer cells through G1 phase cell cycle arrest, inducing apoptosis via caspase 3/7 activation and PARP cleavage, suggesting ergosterol peroxide as a promising anticancer agent (Martínez-Montemayor et al. 2019).

A glucose-rich polysaccharide (WSG) extracted from *G. lucidum* suppressed the proliferation of lung cancer cells. WSG alleviated phosphorylation of ERK1/2 in EGF or TGFβ stimulated cells. WSG also inhibited the phosphorylation of diverse molecules such as FAK, AKT, and Smad2, presenting a potential therapeutic intervention in lung cancer (Hsu et al. 2020).

Emerging scientific data have pointed out the cytotoxic effects and relevant mechanisms associated with *G. lucidum*-based products. We exemplified some of those indicated above to provide an overview of possible mechanisms for their anticancer profiles. You can find further information about *G. lucidum*-associated anticancer mechanisms in Table 2.

4 Future Perspectives

Due to gaining in popularity, many investigations are being accomplished concerning the composition, cultivation, and pharmacological activity of *G. lucidum*. Emerging evidence has proved the ranging activities of *G. lucidum*, among which cancer is of particular interest within the context of our present chapter. Despite the presence of *G. lucidum*-associated anticancer potential based on the outcomes of cell culture and animal models in scientific literature, clinical data are insufficient and need further research. At present, reliable experimental data with clinical significance from well-designed human trials are required, holding principal eminence. Another point of critical importance is the standardization of *G. lucidum* formulations and their quality for determining *G. lucidum*-associated mechanisms and bioactive substances. Thereby, standardized *G. lucidum* formulations whose effects were proven with sufficient preclinical and clinical data will enable the emergence of approved *G. lucidum* products in cancer management.

5 Conclusion

G. lucidum is a well-established Asian herbal medication with a variety of applications, particularly in the field of cancer. It has unraveled noteworthy impacts on cancer management due to in vitro and in vivo experimental data. However, it cannot be applied as first-line therapy in cancer interventions due to insufficient characterization of the extract and being short of purified substances that hold vital importance to ensure uniform formulations. The consumption of *G. lucidum* is increasing globally; therefore, growing demand for patented and commercially accessible products containing *G. lucidum* as a principal ingredient exists as food supplements. These include isolated compounds and extracts in the form of numerous formulations.

The composition of *G. lucidum* may vary based on the cultivation environment, *G. lucidum* strain, and extraction method (Peng et al. 2015; Peksen et al. 2011; Kim et al. 2006; Liu et al. 2017). Future clinical studies require particular amounts of characterized extracts or bioactive compounds. Thus, using standardized extracts for in vitro and in vivo assessment is a rational approach to unraveling the *G. lucidum*-specific activity (Cheng and Sliva 2015).

G. lucidum has been used as an adjuvant therapy along with drugs used in cancer therapy. In the joint administration of *G. lucidum* products with chemotherapeutic agents, the extent of possible synergistic interactions should be investigated to foresee their synergistic potential in cancer intervention. The relation between *G. lucidum* consumption and cancer risks should be revealed via preclinical and clinical research. Determination of toxicity, a suitable dose, efficacy, and safety are required either alone or with chemotherapeutics (Ahmad et al. 2021; Ahmad 2020; Sharma et al. 2019; Lawal et al. 2019).

Natural products exert modest side effects holding the capacity as anticancer agents that cover the principal human need of current chemotherapeutics to surmount cancer. This chapter attempts to accumulate evidence of bioactive substances of *G. lucidum* and/or *G. lucidum* itself, demonstrating anticancer potential either alone or along with chemotherapy. Their probable modes of action regarding their anticancer effects on numerous cancer cells, clinical trials, and chemotherapy-associated toxicity challenges. They hold natural alternatives to fight cancer, particularly in conventional combination regimens.

Prospective clinical trials are urgently desired to ensure that the outcomes monitored in animal studies are also available in humans. On the other hand, one of the principal stumbling issues to efficient clinical trials is the quality control process of these products. Modern perspectives, including chemometrics and biochemometrics, may enable scientists to surmount these drawbacks as these approaches allow the correlation of chemistry with biological activities. To ensure reproducible effects, the development of standardized extracts holds primary significance for clinical studies. When such parameters are determined as a result of preclinical and clinical research, *Ganoderma* will come to the forefront as an adjunct therapy for cancer in clinical trials, and then the perspective for *Ganoderma* products will be tremendous globally.

References

Acevedo-Díaz A, Ortiz-Soto G, Suárez-Arroyo IJ et al (2019) *Ganoderma lucidum* extract reduces the motility of breast cancer cells mediated by the RAC–lamellipodin axis. Nutrients 11(5): 1116. https://doi.org/10.3390/nu11051116

Adeyi AO, Awosanya SA, Adeyi OE et al (2021) *Ganoderma lucidum* ethanol extract abrogates metabolic syndrome in rats: *in vivo* evaluation of hypoglycemic, hypolipidemic, hypotensive and antioxidant properties. Obes Med 22:100320. https://doi.org/10.1016/j.obmed.2021.100320

Ahmad MF (2020) *Ganoderma lucidum*: a rational pharmacological approach to surmount cancer. J Ethnopharmacol 260:113047. https://doi.org/10.1016/j.jep.2020.113047

Ahmad R, Riaz M, Khan A et al (2021) *Ganoderma lucidum* (Reishi) an edible mushroom; a comprehensive and critical review of its nutritional, cosmeceutical, mycochemical, pharmacological, clinical, and toxicological properties. Phytother Res 35(11):6030–6062. https://doi.org/10.1002/ptr.7215

Atanasov AG, Zotchev SB, Dirsch VM et al (2021) Natural products in drug discovery: advances and opportunities. Nat Rev Drug Discov 20(3):200–216. https://doi.org/10.1038/s41573-020-00114-z

Bao X, Liu C, Fang J et al (2001) Structural and immunological studies of a major polysaccharide from spores of *Ganoderma lucidum* (Fr.) Karst. Carbohydr Res 332:67–74. https://doi.org/10.1016/S0008-6215(01)00075-1

Bao XF, Wang XS, Dong Q et al (2002) Structural features of immunologically active polysaccharides from *Ganoderma lucidum*. Phytochemistry 59:175–181. https://doi.org/10.1016/S0031-9422(01)00450-2

Barbieri A, Quagliariello V, Del Vecchio V et al (2017) Anticancer and anti-inflammatory properties of *Ganoderma lucidum* extract effects on melanoma and triple-negative breast cancer treatment. Nutrients 9(3):210. https://doi.org/10.3390/nu9030210

Bishop KS, Kao CH, Xu Y et al (2015) From 2000 years of *Ganoderma lucidum* to recent developments in nutraceuticals. Phytochemistry 114:56–65. https://doi.org/10.1016/j.phytochem.2015.02.015

Boh B, Berovic M, Zhang J, Zhi-Bin L (2007) *Ganoderma lucidum* and its pharmaceutically active compounds. Biotechnol Annu Rev 13:265–301. https://doi.org/10.1016/S1387-2656(07)13010-6

Borchers AT, Stern JS, Hackman RM et al (1999) Minireview: mushrooms, tumors and immunity. Proc Soc Exp Biol Med 221:281–293. https://doi.org/10.1046/j.1525-1373.1999.d01-86.x

Bulam S, Üstün NŞ, Pekşen A (2019) Health benefits of *Ganoderma lucidum* as a medicinal mushroom. Turk Tarim Gida Bilim Teknol Derg 7(1):84–93. https://doi.org/10.24925/turjaf.v7isp1.84-93.2728

Calvino E, Manjon JL, Sancho P et al (2010) *Ganoderma lucidum* induced apoptosis in NB4 human leukemia cells: involvement of Akt and Erk. J Ethnopharmacol 128:71–78. https://doi.org/10.1016/j.jep.2009.12.027

Cao QZ, Lin ZB (2006) *Ganoderma lucidum* polysaccharides peptide inhibits the growth of vascular endothelial cell and the induction of VEGF in human lung cancer cell. Life Sci 78(13):1457–1463. https://doi.org/10.1016/j.lfs.2005.07.017

Cao Y, Wu SH, Dai YC (2012) Species clarification of the prize medicinal *Ganoderma* mushroom "Lingzhi". Fungal Divers 56:49–62. https://doi.org/10.1007/s13225-012-0178-5

Cheng S, Sliva D (2015) *Ganoderma lucidum* for cancer treatment: we are close but still not there. Integr Cancer Ther 14(3):249–257. https://doi.org/10.1177/1534735414568721

Cör D, Knez Ž, Knez Hrnčič M (2018) Antitumour, antimicrobial, antioxidant and antiacetylcholinesterase effect of *Ganoderma lucidum* terpenoids and polysaccharides: a review. Molecules 23(3):649. https://doi.org/10.3390/molecules23030649

Dudhgaonkar S, Thyagarajan A, Sliva D (2009) Suppression of the inflammatory response by triterpenes isolated from the mushroom *Ganoderma lucidum*. Int Immunopharmacol 9:1272–1280. https://doi.org/10.1016/j.intimp.2009.07.011

el-Mekkawy S, Meselhy MR, Nakamura N et al (1998) Anti-HIV-1 and anti-HIV-1-protease substances from *Ganoderma lucidum*. Phytochemistry 49(6):1651–1657. https://doi.org/10.1016/s0031-9422(98)00254-4

Ferreira I, Vaz JA, Vasconcelos MH et al (2010) Compounds from wild mushrooms with antitumor potential. Anti Cancer Agents Med Chem 10(5):424–436

Gao Y, Zhou S (2003) Cancer prevention and treatment by *Ganoderma*, a mushroom with medicinal properties. Food Rev Int 19:275–325. https://doi.org/10.1081/FRI-120023480

Gao YH, Dai XH, Chen GL et al (2003a) A randomized, placebo-controlled, multicenter study of *Ganoderma lucidum* (W.Curt.: Fr.) Lloyd (Aphyllophoromycetideae) polysaccharides (Ganopoly®) in patients with advanced lung cancer. Int J Med Mushrooms 5:369. https://doi.org/10.1615/InterJMedicMush.v5.i4.40

Gao YH, Zhou SF, Jiang WQ et al (2003b) Effects of Ganopoly® (a *Ganoderma lucidum* polysaccharide extract) on the immune functions in advanced-stage cancer patients. Immunol Investig 32:201–215. https://doi.org/10.1081/IMM-120022979

Gordan JD, Chay WY, Kelley RK et al (2011) And what other medications are you taking? J Clin Oncol 29(11):288–291. https://doi.org/10.1200/JCO.2010.32.8054

Graziose R, Ann Lila M, Raskin I (2010) Merging traditional Chinese medicine with modern drug discovery technologies to find novel drugs and functional foods. Curr Drug Discov Technol 7(1):2–12

Gullett NP, Amin AR, Bayraktar S et al (2010) Cancer prevention with natural compounds. Semin Oncol 37(3):258–281. https://doi.org/10.1053/j.seminoncol.2010.06.014

Hong KJ, Dunn DM, Shen CL et al (2004) Effects of *Ganoderma lucidum* on apoptotic and anti-inflammatory function in HT-29 human colonic carcinoma cells. Phytother Res 18:768–770. https://doi.org/10.1002/ptr.1495

Hsu WH, Qiu WL, Tsao SM et al (2020) Effects of WSG, a polysaccharide from *Ganoderma lucidum*, on suppressing cell growth and mobility of lung cancer. Int J Biol Macromol 165: 1604–1613. https://doi.org/10.1016/j.ijbiomac.2020.09.227

Hu H, Ahn NS, Yang X et al (2002) *Ganoderma lucidum* extract induces cell cycle arrest and apoptosis in MCF-7 human breast cancer cell. Int J Cancer 102(3):250–253. https://doi.org/10.1002/ijc.10707

Huang CH, Lin WK, Chang SH et al (2021) Evaluation of the hypoglycaemic and antioxidant effects of submerged *Ganoderma lucidum* cultures in type 2 diabetic rats. Mycology 12(2): 82–93. https://doi.org/10.1080/21501203.2020.1733119

Huie CW, Di X (2004) Chromatographic and electrophoretic methods for Lingzhi pharmacologically active components. J Chromatogr B 812:241–257. https://doi.org/10.1016/j.jchromb.2004.08.038

IARC (2022) Estimated number of new cases from 2020 to 2040, Both sexes, age [0-85+]. https://gco.iarc.fr/tomorrow/en/dataviz/isotype?types=0&single_unit=500000. Accessed 24 Jul 2022

Jang KJ, Han MH, Lee BH et al (2010) Induction of apoptosis by ethanol extracts of *Ganoderma lucidum* in human gastric carcinoma cells. J Acupunct Meridian Stud 3:24–31. https://doi.org/10.1016/S2005-2901(10)60004-0

Jedinak A, Thyagarajan-Sahu A, Jiang J et al (2011) Ganodermanontriol, a lanostanoid triterpene from *Ganoderma lucidum*, suppresses growth of colon cancer cells through ss-catenin signaling. Int J Oncol 38:761–767. https://doi.org/10.3892/ijo.2011.898

Jiang J, Grieb B, Thyagarajan A et al (2008) Ganoderic acids suppress growth and invasive behavior of breast cancer cells by modulating AP-1 and NF-kappaB signaling. Int J Mol Med 21:577–584. https://doi.org/10.3892/ijmm.21.5.577

Jiao C, Chen W, Tan X et al (2020) *Ganoderma lucidum* spore oil induces apoptosis of breast cancer cells *in vitro* and *in vivo* by activating caspase-3 and caspase-9. J Ethnopharmacol 247: 112256. https://doi.org/10.1016/j.jep.2019.112256

Jin H, Song C, Zhao Z, Zhou G (2020) *Ganoderma lucidum* polysaccharide, an extract from *Ganoderma lucidum*, exerts suppressive effect on cervical cancer cell malignancy through mitigating epithelial-mesenchymal and jak/stat5 signaling pathway. Pharmacology 105(7-8): 461–470. https://doi.org/10.1159/000505461

Kaur H, Sharma S, Khanna P et al (2015) Evaluation of *Ganoderma lucidum* strains for the production of bioactive components and their potential use as antimicrobial agents. J Appl Nat Sci 7(1):298–303. https://doi.org/10.31018/jans.v7i1.605

Kim JH, Lee DH, Lee SH et al (2004) Effect of *Ganoderma lucidum* on the quality and functionality of Korean traditional rice wine, yakju. J Biosci Bioeng 97(1):24–28. https://doi.org/10.1016/S1389-1723(04)70160-7

Kim HM, Park MK, Yun JW (2006) Culture pH affects exopolysaccharide production in submerged mycelial culture of *Ganoderma lucidum*. Appl Biochem Biotechnol 134(3):249–262. https://doi.org/10.1385/ABAB:134:3:249

Kimura Y, Taniguchi M, Baba K (2002) Antitumor and antimetastatic effects on liver of triterpenoid fractions of *Ganoderma lucidum*: mechanism of action and isolation of an active substance. Anticancer Res 22:3309–3318

Kozarski M, Klaus A, Nikšić M et al (2012) Antioxidative activities and chemical characterization of polysaccharide extracts from the widely used mushrooms *Ganoderma applanatum, Ganoderma lucidum, Lentinus edodes* and *Trametes versicolor*. J Food Compos Anal 26(1): 144–153. https://doi.org/10.1016/j.jfca.2012.02.004

Lawal TO, Wicks SM, Calderon AI et al (2019) Bioactive molecules, pharmacology and future research trends of *Ganoderma lucidium* as a cancer chemotherapeutic agent. In: Ahmad Khan MS, Ahmad I, Chattopadhyay D (eds) New look to phytomedicine. Academic Press, London, pp 159–178. https://doi.org/10.1016/B978-0-12-814619-4.00007-0

Li J, Gu F, Cai C et al (2020a) Purification, structural characterization, and immunomodulatory activity of the polysaccharides from *Ganoderma lucidum*. Int J Biol Macromol 143:806–813. https://doi.org/10.1016/j.ijbiomac.2019.09.141

Li P, Liu L, Huang S et al (2020b) Anticancer effects of a neutral triterpene fraction from *Ganoderma lucidum* and its active constituents on sw620 human colorectal cancer cells. Anti Cancer Agents Med Chem 20(2):237–244. https://doi.org/10.2174/1871520619666191015102442

Lin Z, Deng A (2019) Antioxidative and free radical scavenging activity of *Ganoderma* (Lingzhi). Adv Exp Med Biol 1182:271–297. https://doi.org/10.1007/978-981-32-9421-9_12

Lin CN, Tome WP (1991) Novel cytotoxic principles of Formosan *Ganoderma lucidum*. J Nat Prod 54:998–1002. https://doi.org/10.1021/np50076a012

Lin SB, Li CH, Lee SS et al (2003) Triterpene-enriched extracts from *Ganoderma lucidum* inhibit growth of hepatoma cells via suppressing protein kinase C, activating mitogen-activated protein kinases and G2-phase cell cycle arrest. Life Sci 72:2381–2390. https://doi.org/10.1016/S0024-3205(03)00124-3

Liu X, Yuan JP, Chung CK et al (2002) Antitumor activity of the sporoderm-broken germinating spores of *Ganoderma lucidum*. Cancer Lett 182:155–161. https://doi.org/10.1016/S0304-3835(02)00080-0

Liu J, Shimizu K, Konishi F et al (2007) Anti-androgenic activities of the triterpenoids fraction of *Ganoderma lucidum*. Food Chem 100:1691–1696. https://doi.org/10.1016/j.foodchem.2006.01.003

Liu YW, Gao JL, Guan J et al (2009) Evaluation of antiproliferative activities and action mechanisms of extracts from two species of *Ganoderma* on tumor cell lines. J Agric Food Chem 57: 3087–3093. https://doi.org/10.1021/jf900011f

Liu SR, Ke BR, Zhang WR et al (2017) Breeding of new *Ganoderma lucidum* strains simultaneously rich in polysaccharides and triterpenes by mating basidiospore-derived monokaryons of two commercial cultivars. Sci Hortic 216:58–65. https://doi.org/10.1016/j.scienta.2016.12.016

Loganathan J, Jiang J, Smith A et al (2014) The mushroom *Ganoderma lucidum* suppresses breast-to-lung cancer metastasis through the inhibition of pro-invasive genes. Int J Oncol 44(6): 2009–2015. https://doi.org/10.3892/ijo.2014.2375

Lu H, Kyo E, Uesaka T et al (2003) A water-soluble extract from cultured medium of *Ganoderma lucidum* (Rei-shi) mycelia suppresses azoxymethane-induction of colon cancers in male F344 rats. Oncol Rep 10:375–379. https://doi.org/10.3892/or.10.2.375

Lu QY, Jin YS, Zhang QF et al (2004) *Ganoderma lucidum* extracts inhibit growth and induce actin polymerization in bladder cancer cells *in vitro*. Cancer Lett 216(1):9–20. https://doi.org/10.1016/j.canlet.2004.06.022

Lu SY, Peng XR, Dong JR et al (2019) Aromatic constituents from *Ganoderma lucidum* and their neuroprotective and anti-inflammatory activities. Fitoterapia 134:58–64. https://doi.org/10.1016/j.fitote.2019.01.013

Martínez-Montemayor MM, Ling T, Suárez-Arroyo IJ (2019) Identification of biologically active *Ganoderma lucidum* compounds and synthesis of improved derivatives that confer anticancer activities *in vitro*. Front Pharmacol 10:115. https://doi.org/10.3389/fphar.2019.00115

Mau JL, Lin HC, Chen CC (2001) Non-volatile components of several medicinal mushrooms. Food Res Int 34(6):521–526. https://doi.org/10.1016/S0963-9969(01)00067-9

Min BS, Nakamura N, Miyashiro H et al (1998) Triterpenes from the spores of *Ganoderma lucidum* and their inhibitory activity against HIV-1 protease. Chem Pharm Bull (Tokyo) 46(10): 1607–1612. https://doi.org/10.1248/cpb.46.1607

Min BS, Gao JJ, Nakamura N (2000) Triterpenes from the spores of *Ganoderma lucidum* and their cytotoxicity against Meth-A and LLC tumor cells. Chem Pharm Bull 48:1026–1033. https://doi.org/10.1248/cpb.48.1026

Miura T, Yuan L, Sun B et al (2002) Isoflavone aglycon produced by culture of soybean extracts with basidiomycetes and its anti-angiogenic activity. Biosci Biotechnol Biochem 66:2626–2631. https://doi.org/10.1271/bbb.66.2626

NCI (2022) Types of cancer treatment. Accessed 24 Jul 2022

Newman DJ, Cragg GM (2016) Natural products as sources of new drugs from 1981 to 2014. J Nat Prod 79(3):629–661. https://doi.org/10.1021/acs.jnatprod.5b01055

Nonaka Y, Shibata H, Nakai M et al (2006) Antitumor activities of the antlered form of *Ganoderma lucidum* in allogeneic and syngeneic tumor-bearing mice. Biosci Biotechnol Biochem 70:2028–2034. https://doi.org/10.1271/bbb.50509

Opattova A, Horak J, Vodenkova S (2019) *Ganoderma lucidum* induces oxidative DNA damage and enhances the effect of 5-Fluorouracil in colorectal cancer *in vitro* and *in vivo*. Mutat Res Genet Toxicol Environ Mutagen 845:403065. https://doi.org/10.1016/j.mrgentox.2019.06.001

Paterson RRM (2006) *Ganoderma*–a therapeutic fungal biofactory. Phytochemistry 67(18): 1985–2001. https://doi.org/10.1016/j.phytochem.2006.07.004

Peksen A, Yakupoglu G, Yakupoglu T et al (2011) Changes in chemical compositions of substrates before and after *Ganoderma lucidum* cultivation. World J Microbiol Biotechnol 27(3):637–642. https://doi.org/10.1007/s11274-010-0500-x

Peng L, Qiao S, Xu Z et al (2015) Effects of culture conditions on monosaccharide composition of *Ganoderma lucidum* exopolysaccharide and on activities of related enzymes. Carbohydr Polym 133:104–109. https://doi.org/10.1016/j.carbpol.2015.07.014

Ren ZL, Wang CD, Wang T et al (2019) *Ganoderma lucidum* extract ameliorates MPTP-induced parkinsonism and protects dopaminergic neurons from oxidative stress via regulating mitochondrial function, autophagy, and apoptosis. Acta Pharmacol Sin 40(4):441–450. https://doi.org/10.1038/s41401-018-0077-8

Rios-Fuller TJ, Ortiz-Soto G, Lacourt-Ventura M et al (2018) *Ganoderma lucidum* extract (GLE) impairs breast cancer stem cells by targeting the STAT3 pathway. Oncotarget 9(89): 35907–35921. https://doi.org/10.18632/oncotarget.26294

Roy S, Jahan M, Das K et al (2015) Artificial cultivation of *Ganoderma lucidum* (Reishi medicinal mushroom) using different sawdusts as substrates. Am J BioSci 3:178–182. https://doi.org/10.11648/j.ajbio.20150305.13

Rubel R, Santa HSD, Dos Santos LF et al (2018) Immunomodulatory and antitumoral properties of *Ganoderma lucidum* and agaricus brasiliensis (agaricomycetes) medicinal mushrooms. Int J Med Mushrooms 20(4):393–403. https://doi.org/10.1615/IntJMedMushrooms.2018025979

Sanodiya BS, Thakur GS, Baghel RK et al (2009) *Ganoderma lucidum*: a potent pharmacological macrofungus. Curr Pharm Biotechnol 10:717–742. https://doi.org/10.2174/138920109789978757

Savin S, Craciunescu O, Oancea A et al (2020) Antioxidant, cytotoxic and antimicrobial activity of chitosan preparations extracted from *Ganoderma lucidum* mushroom. Chem Biodivers 17(7): 2000175. https://doi.org/10.1002/cbdv.202000175

Shao W, Xiao C, Yong T et al (2022) A polysaccharide isolated from *Ganoderma lucidum* ameliorates hyperglycemia through modulating gut microbiota in type 2 diabetic mice. Int J Biol Macromol 197:23–38. https://doi.org/10.1016/j.ijbiomac.2021.12.034

Sharma P, Tulsawani R, Agrawal U (2019) Pharmacological effects of *Ganoderma lucidum* extract against high-altitude stressors and its subchronic toxicity assessment. J Food Biochem 43(12): 13081. https://doi.org/10.1111/jfbc.13081

Shevelev OB, Seryapina AA, Zavjalov EL et al (2018) Hypotensive and neurometabolic effects of intragastric Reishi (*Ganoderma lucidum*) administration in hypertensive ISIAH rat strain. Phytomedicine 41:1–6. https://doi.org/10.1016/j.phymed.2018.01.013

Shiao MS (2003) Natural products of the medicinal fungus *Ganoderma lucidum*: occurrence, biological activities, and pharmacological functions. Chem Rec 3:172–180. https://doi.org/10.1002/tcr.10058

Siegel RL, Miller KD, Jemal A (2019) Cancer statistics, 2019. CA Cancer J Clin 69(1):7–34. https://doi.org/10.3322/caac.21551

Singh R (2017) A review on different benefits of mushroom. IOSR J Pharm Biol Sci 12(1): 107–111. https://doi.org/10.9790/3008-120102107111

Sliva D (2003) *Ganoderma lucidum* (Reishi) in cancer treatment. Integr Cancer 2:358–364. https://doi.org/10.1177/1534735403259066

Sohretoglu D, Huang S (2018) *Ganoderma lucidum* polysaccharides as an anticancer agent. Anti Cancer Agents Med Chem 18(5):667–674. https://doi.org/10.2174/1871520617666171113121246

Su CH, Yang YZ, Ho C et al (2001) High-performance liquid chromatographic analysis for the characterization of triterpenoids from *Ganoderma*. J Chromatogr Sci 39(3):93–100. https://doi.org/10.1093/chromsci/39.3.93

Sullivan R, Smith JE, Rowan NJ (2006) Medicinal mushrooms and cancer therapy: translating a traditional practice into Western medicine. Perspect Biol Med 49(2):159–170. https://doi.org/10.1353/pbm.2006.0034

Thyagarajan A, Jedinak A, Nguyen H et al (2010) Triterpenes from *Ganoderma lucidum* induce autophagy in colon cancer through the inhibition of p38 mitogen-activated kinase (p38 MAPK). Nutr Cancer 62:630–640. https://doi.org/10.1080/01635580903532390

Tomasi S, Lohezic-Le Devehat F, Sauleau P et al (2004) Cytotoxic activity of methanol extracts from Basidiomycete mushrooms on murine cancer cell lines. Pharmazie 59:290–293

Wachtel-Galor S, Yuen J, Buswell JA et al (2011) *Ganoderma lucidum* (Lingzhi or Reishi): a medicinal mushroom. In: Benzie IFF, Wachtel-Galor S (eds) Herbal medicine: biomolecular and clinical aspects, 2nd edn. CRC Press, Boca Raton, FL

Wang L, Li JQ, Zhang J (2020a) Traditional uses, chemical components and pharmacological activities of the genus *Ganoderma* P. Karst.: a review. RSC Adv 10(69):42084–42097. https://doi.org/10.1039/D0RA07219B

Wang X, Wang B, Zhou L et al (2020b) *Ganoderma lucidum* put forth antitumor activity against PC-3 prostate cancer cells via inhibition of Jak-1/STAT-3 activity. Saudi J Biol Sci 27(10): 2632–2637. https://doi.org/10.1016/j.sjbs.2020.05.044

WHO (2022) Cancer. https://www.who.int/health-topics/cancer#tab=tab_1. Accessed 24 Jul 2022

Wu G, Qian Z, Guo J et al (2012) *Ganoderma lucidum* extract induces g1 cell cycle arrest, and apoptosis in human breast cancer cells. Am J Chin Med 40(3):631–642. https://doi.org/10.1142/s0192415x12500474

Wu KJ, Wang SQ, Shi RP et al (2022) Neuroprotective effect of *Ganoderma lucidum* polysaccharides in an epileptic rat model. Curr Top Nutraceutical Res 20(2):416–423. https://doi.org/10.37290/ctnr2641-452X.20:416-423

Xian H, Li J, Zhang Y et al (2021) Antimetastatic effects of *Ganoderma lucidum* polysaccharide peptide on b16-f10-luc-g5 melanoma mice with sleep fragmentation. Front Pharmacol 12: 650216. https://doi.org/10.3389/fphar.2021.650216

Yao X, Li G, Xu H et al (2012) Inhibition of the JAK-STAT3 signaling pathway by ganoderic acid a enhances chemosensitivity of hepg2 cells to cisplatin. Planta Med 78:1740–1748. https://doi.org/10.1055/s-0032-1315303

Yue QX, Cao ZW, Guan SH et al (2008) Proteomics characterization of the cytotoxicity mechanism of ganoderic acid D and computer-automated estimation of the possible drug target network. Mol Cell Proteomics 7:949–961. https://doi.org/10.1074/mcp.M700259-MCP200

Yue QX, Song XY, Ma C et al (2010) Effects of triterpenes from *Ganoderma lucidum* on protein expression profile of HeLa cells. Phytomedicine 17(8-9):606–613. https://doi.org/10.1016/j.phymed.2009.12.013

Yuen JWM, Gohel MDI (2005) Anticancer effects of *Ganoderma lucidum*: a review of scientific evidence. Nutr Cancer 53(1):11–17. https://doi.org/10.1207/s15327914nc5301_2

Yun TK (1999) Update from Asia. Asian studies on cancer chemoprevention. Ann NY Acad Sci 889:157–192. https://doi.org/10.1111/j.1749-6632.1999.tb08734.x

Yun TK, Kim SH, Lee YS (1995) Trial of a new medium-term model using benzo(a)pyrene induced lung tumor in newborn mice. Anticancer 15:839–845

Zhang QH, Hu QX, Xie D et al (2019) *Ganoderma lucidum* exerts an anticancer effect on human osteosarcoma cells via suppressing the wnt/β-catenin signaling pathway. Integr Cancer Ther 18: 1534735419890917. https://doi.org/10.1177/1534735419890917

Zhou X, Lin J, Yin Y et al (2007) Ganodermataceae: natural products and their related pharmacological functions. Am J Chin Med 35(4):559–574. https://doi.org/10.1142/S0192415X07005065

Zhou Y, Qu ZQ, Zeng YS et al (2012) Neuroprotective effect of preadministration with *Ganoderma lucidum* spore on rat hippocampus. Exp Toxicol Pathol 64(7):673–680. https://doi.org/10.1016/j.etp.2010.12.011

Zhu HS, Yang XL, Wang LB et al (2000) Effects of extracts from sporoderm-broken spores of *Ganoderma lucidum* on HeLa cells. Cell Biol Toxicol 16:201–206. https://doi.org/10.1023/A:1007663006548

Zolj S, Smith MP, Goines JC et al (2018) Antiproliferative effects of a triterpene-enriched extract from lingzhi or reishi medicinal mushroom, *Ganoderma lucidum* (Agaricomycetes), on human lung cancer cells. Int J Med Mushrooms 20(12). https://doi.org/10.1615/IntJMedMushrooms.2018028823

Cognitive Enhancing Effects of Medicinal Mushrooms: A Potential Neuroprotective Implication in Dementias

Phillip McCain, Jack Deruiter, Suhrud Pathak, Keyi Liu, Rishi M. Nadar, Sindhu Ramesh, Timothy Moore, Dinesh Chandra Agrawal, and Muralikrishnan Dhanasekaran

Contents

Abstract Dementia, a form of neurodegenerative disease, is a chronic brain disorder that affects memory and learning. Oxidative stress and inflammation are believed to contribute to the pathogenesis of dementia, as an imbalance of oxidant-antioxidant/pro-inflammatory-anti-inflammatory cytokines homeostasis causes neuroinflammation and an increased rate of reactive oxygen species generation. Natural, medicinal mushrooms, enriched with antioxidants and anti-inflammatory properties, are growing popular as agents to provide beneficial neuronal health effects, such as inhibiting oxidative stress, protecting neuronal excitotoxicity, and boosting the human immune response. Mushrooms such as *Hericium erinaceus* and *Coriolus versicolor* have been demonstrated to improve cognitive function and prevent and reduce Alzheimer's and Parkinson's disease. The bioactive compounds in these mushrooms include erinacines, polyphenols, polysaccharopeptides, and terpenoids. Erinacine A, one of the main bioactive components, effectively reduces neurodegeneration and dementia and has no known adverse side effects. The

P. McCain · J. Deruiter · S. Pathak · K. Liu · R. M. Nadar · S. Ramesh · T. Moore · M. Dhanasekaran (✉)
Department of Drug Discovery and Development, Harrison School of Pharmacy, Auburn University, Auburn, AL, USA
e-mail: dhanamu@auburbn.edu

D. C. Agrawal (✉)
Department of Applied Chemistry, Chaoyang University of Technology, Taichung, Taiwan
e-mail: dcagrawal@gm.cyut.edu.tw

D. C. Agrawal, M. Dhanasekaran (eds.), *Mushrooms with Therapeutic Potentials*,
https://doi.org/10.1007/978-981-19-9550-7_8

mycelia are safe for oral consumption over an extended period and can be supplemented in a diet. This chapter summarizes current scientific information on the health properties of *H. erinaceus* and *C. versicolor* to contribute to prophylactic and therapeutic findings regarding neuroprotection against dementia.

Keywords Alzheimer's Disease · *Coriolus versicolor* · Dementia · *Hericium erinaceus* · Mushrooms · Neurodegeneration

Abbreviations

Ach	Acetylcholine
AD	Alzheimer's Disease
APP	Amyloid Precursor Protein
CASI	Cognitive Abilities Screening Instrument
ChAT	Choline Acetyltransferase
COX-2	Cyclooxygenase-2
EAHE	Erinacine A-Enriched *H. erinaceus*
HE	*Hericium erinaceus*
HO-1	Heme oxygenase 1
Hsp	Heat Shock Protein
IADL	Instrumental Activities of Daily Living
IkBa	I-Kappa-B-Alpha
JNK	c-jun N-terminal Kinase
LXA4	Lipoxin A4
MMSE	Mini-Mental State Examination
NFkb	Nuclear Factor Kappa B
NGF	Nerve Growth Factors
NPI	Neuropsychiatric Inventory
PD	Parkinson's Disease
PSP	Polysaccharopeptides
ROS	Reactive Oxygen Species
SOD	Superoxide Dismutase
Trx	Thioredoxin

1 Introduction

Neurodegenerative diseases are fatal and enfeebling conditions caused by increased neuronal cell death and deterioration of the nervous system. Common neurodegenerative diseases include Alzheimer's (AD), Parkinson's disease (PD), and Huntington's disease. AD mainly affects the population of those 65 years or older and is projected to affect 12.7 million individuals by 2050 (Alzheimer Association

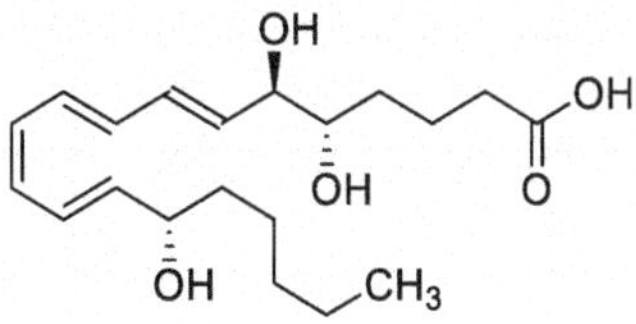

Fig. 1 Chemical structure of lipoxin A4 (LXA4)

2022). AD is responsible for most cases of disability and morbidity in the elderly and is the most prevalent cause of dementia worldwide. Parkinson's disease (PD) is the second most typical neurodegenerative disorder characterized by progressive loss of both motor and non-motor functions (Ziegler-Graham et al. 2008). Motor symptoms include tremors, postural instability, and slowing of movement. At the same time, non-motor symptoms can present as mood disorders, anxiety, depression, and cognitive impairment, which occur before motor symptoms (Mhyre et al. 2012). While some cases of the neurologic disease are reversible and are due to depression, drug intoxication, and hormone imbalances, 80% of cases of the neurodegenerative disorder are irreversible (Bello and Schultz 2011).

A number of hypotheses have been advanced to describe the pathogenesis of neurodegenerative diseases, with one of the main theories being excessive oxidative stress. It is suggested that reactive oxygen species (ROS) have a role in the cause of major neurodegenerative diseases (Sperling et al. 2011). Oxyradical damage, memory impairment, and aging result from an imbalance between ROS generation and antioxidant enzyme activities (Valko et al. 2007). The depletion of antioxidants in the brain or a reduction in complexes I and IV of the mitochondrial electron transport chain causes an increased rate of formation of ROS, which consequently leads to more lipid and protein peroxidation, inflammation, mitochondrial and nuclear damage, and apoptosis, eventually affecting the normal functions of the brain (Talalay and Zimmerman 2015). AD progression can be linked to an accumulation of abnormal proteins, namely, amyloid-β plaques, phosphorylated tau, and neurofibrillary tangles, which lead to neuronal death and synapse reduction (Calabrese et al. 2010; Trovato et al. 2016).

To prevent such pathology, brain cells utilize heat shock proteins, sirtuins, thioredoxin (Trx), and lipoxin A4 (LXA4) to control and respond to various stresses (Kirstein et al. 2015). LXA4 (Fig. 1), an anti-inflammatory eicosanoid, prevents the further production of pro-inflammatory radical oxygen species and represents a signal stop in the inflammatory response. Its neuroprotective features are associated with reduced AD in the brain and other neurodegenerative disorders in the nervous system. However, with dementia, there is a deficit in stress tolerance and protective proteins. During progressive neurodegeneration, redox homeostasis is flawed, and the subsequent chronic neuroinflammation can worsen progression.

Despite this, there is currently no cure for these serious health illnesses that affect the elderly globally. Medications only slow down the progression of neurodegenerative disease. The neurodegenerative process of AD is theorized to begin significantly before the onset of symptoms and manifest into a progressive cascade of reactions over time (Wang et al. 2016). Studies have shown that the density of aggregated tau tangles remains unchanged after the removal of amyloid-β plaques. While treatment of AD in later stages may not hinder progression, therapies used before plaque formation have proven effective (Demattos et al. 2012).

Mushrooms, found in everyday diets and medicines, are an excellent candidate for enhancing cognition in those diagnosed with dementia and other age-based neurodegenerative diseases (Heneka 2017). The compounds within these natural mushrooms have few side effects and have potent anti-inflammatory properties, leading to improved cognitive functions and survival of neurons. Mushrooms are abundant in polysaccharides, such as β-glucans, which act as strong immune stimulators and incite cytokine responses (Yang et al. 2014). Consistent consumption of antioxidant-rich mushrooms decreases the risk of NDs and enhances host immune responses. The bioactive compounds and secondary metabolites of edible mushrooms inhibit the antioxidant and oxidant balance disruption, consequently preventing ROS generation, neuroinflammation, and progression of AD and PD (Phan et al. 2017).

2 *Hericium erinaceus*

Among most edible mushrooms, *Hericium erinaceus* (HE), also known as lion's mane mushroom, is one of the most studied and the best examples of medicinal mushrooms with the potential for immunotherapy and improvement of cognitive impairment (Friedman 2015). Compounds isolated from this mushroom include erinacines, hericenones, sterols, polysaccharides, and glycoproteins (Table 1).

Another compound in HE, dilinoleoyl-phosphatidylethanolamine (Fig. 2), reduces endoplasmic reticulum stress, hindering neuronal cell loss (De Silva et al. 2013). Erinacine A (Fig. 3), derived from the mushroom's mycelia, is a key component in promoting nerve growth factors (NGF) and treating AD and PD. Studies have found that NGF is synthesized in astrocytes and is transported to nerve cells, where NGF assists in the necessary maintenance, regeneration, and repair of neuronal stability (Lai et al. 2010). According to one study, erinacine A successfully enhanced the NGF expression in the hippocampus and locus coeruleus of rats after administration, consequently improving the behaviors of the animal models (Shimbo et al. 2005). HE bioactivity results in the potent promotion of NGF mRNA expression in the hippocampus with the c-jun N-terminal kinase (JNK) pathway in human astrocytoma cells. Other studies found that dried HE's fruiting body powder enhanced visual recognition, and spatial short-term memory improved after 16 weeks (Mori et al. 2009). In a randomized, double-blind, placebo-controlled study, 31 participants, averaging 61.3 years, consumed either 0.8 g of HE's fruiting body powder or placebo cookies for 12 weeks. The HE group scored higher in the Mini-Mental State Examination (MMSE) than the placebo group, confirming that HE improved cognitive function and prevented the deterioration of short-term memory (Saitsu et al. 2019).

Table 1 Bioactives and mechanisms of *Hericium erinaceus* and *Coriolus versicolor*

Mushrooms	Bioactives	Mechanisms
Hericium erinaceus  https://unsplash.com/photos/icmAHvzMuPY (free to use under the Unsplash License)	• Erinacines (A) • Dilinoleoyl-phosphatidyl-ethanolamine • Hericenones • Sterols • Polysaccharides glycoproteins	• Promotes nerve growth factors (NGF) • Affects the c-Jun N-terminal kinase (JNK) pathway • Helps in the stability, maintenance, regeneration, and repair of neurons. • Antioxidant (reduce ROS formation) • Blocks Ca^{2+} overloading • Revives cholinergic signaling • Increases the LXA4 gene expression • Initiates the Hsp pathway and HO-1 • Slows the degeneration of dopaminergic neurons • Anti-inflammatory activity (decreased expression of Fas and Bax and phosphorylation of JNK, p38, and activated B cell pathways) • Reduces endoplasmic reticulum stress
Coriolus versicolor  https://unsplash.com/photos/o3QdK-LTfRY (free to use under the Unsplash license)	• Polysaccharopeptides • Polyphenolic compounds (gallic acid, syringic acid, protocatechuic acid, caffeic acid, chlorogenic acid)	• Inhibits oxidative stress • Bolsters the immune system. • Stimulates the assembly of IFNs, macrophages, IgG, and T lymphocytes. • Promotes neuronal resilience • Anti-inflammatory • Exhibits resistance to proteotoxic proteins • Inhibits cholinergic neurodegenerative degradation • Increases the LXA4 gene expression (increases the number of cytoprotective proteins responding to cellular stress and inhibiting the degradation of neuronal cells) • Enhances the quality of mitochondria to maintain energy homeostasis

One of the first events of AD is the loss of function of cholinergic neurons. With HE, increased NGF affects the survival and regeneration of cholinergic neurons, reducing ROS formation and blocking Ca^{2+} overloading. Cholinergic signaling is revived in the cortex and hippocampus, and acetylcholine (Ach) and choline

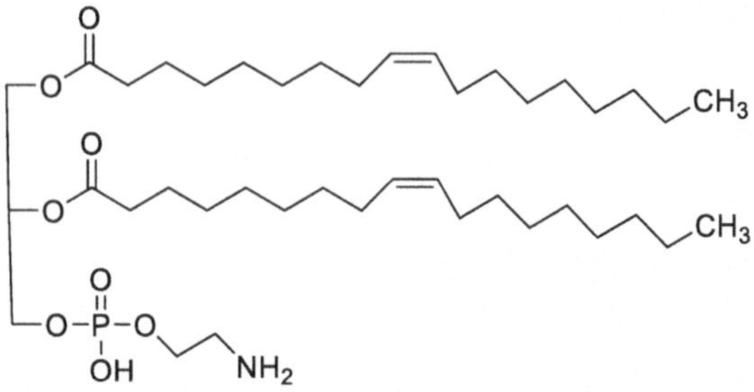

Fig. 2 Chemical structure of dilinoleoyl-phosphatidylethanolamine

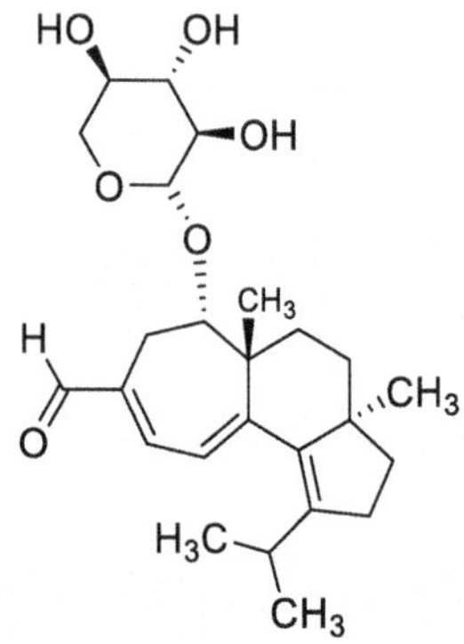

Fig. 3 Chemical structure of erinacine A

acetyltransferase (ChAT) concentrations are raised in the hypothalamus (Iulita and Cuello 2014). In an AD animal model, treatment with HE overcomes behavior abnormalities in the water maze and autonomic activity tests, therefore exhibiting neuronal protection in AD.

Additionally, the treatment of HE significantly improves the expression of LXA4 in cytoprotective proteins involved in stress response, namely, Trx, Hsp 70, and heme oxygenase 1 (HO-1) (Wu et al. 2011). With the initiation of the Hsp pathway and HO-1, there is an increase in vital cellular stress response and a defensive effect against oxidative stress, respectively. In a study of oral administration of HE mushrooms in rats, LXA4 was upregulated to a maximum in specific brain regions such as the hippocampus, cortex, striatum, and cerebellum. Treatment of AD with HE promotes the upregulation of protective proteins, which would balance the redox state and combat several neurodegenerative diseases.

In a trial observing the lifestyle of 633 elderly Chinese citizens in Singapore between 2011 and 2017, it was found that multiple mushrooms, including HE, exhibited therapeutic in AD with their antioxidant properties (Feng et al. 2019). Furthermore, a clinical study of mildly cognitively impaired Japanese men and women aged 50 to 80 found an improvement in cognitive performance after taking HE tablets for 16 weeks (Mori et al. 2009). However, the primary chemical concentrations and active constituents were not effectively assessed, raising concerns about whether hericenones in HE exhibits neuronal protective properties. As for erinacine A, a study of amyloid precursor protein mice was evaluated after oral administration of EAHE for 30 days (Chen et al. 2016; Tsai et al. 2016). It was found that the mycelia could decrease the amyloid- β plaque accumulation, promote insulin-degrading enzyme expression, and enhance the generation of neuron progenitors. After 81 days of the administration, the treated mice were shown to have additional improvements in other impaired brain regions, raising the possibility that EAHE could be an effective therapy for managing AD.

In a human clinical study, Li et al. investigated the effects of HE on brain function and the nervous system. Participants were randomly placed either in a placebo group or an erinacine A-enriched *H. erinaceus* (EAHE) group. The EAHE group received three mycelia capsules per day for 49 weeks (Li et al. 2020). The clinical effects were evaluated by using the Cognitive Abilities Screening Instrument (CASI), MMSE, Neuropsychiatric Inventory (NPI), and Instrumental Activities of Daily Living (IADL). At 25 weeks, the placebo scored significantly lower in CASI than the baseline. Also, the EAHE group showed significantly higher MMSE scores at 49 weeks than the placebo group, achieving better cognitive sensitivity. There was a significant difference in IADL scores at 49 weeks, representing a lower level of dependence for the EAHE group (Chiu et al. 2016).

Mushrooms rich in supplements, such as vitamins, phytochemicals, and minerals, have efficiently prevented and alleviated clinical symptoms of various neurodegenerative disorders. Polysaccharides, polyphenols, carotenoids, and vitamins are abundant in mushrooms, providing potent antioxidant and anti-inflammatory activity. The bioactivity in mushrooms slows the degeneration of dopaminergic neurons in those affected by PD. In one study, the drug 1-methyl-4-phenyl-1,2,3,6-tetrahydropyridine (MPTP) was introduced to a brain to cause nigral dopaminergic neuronal death, mimicking the pathology of PD (Meredith and Rademacher 2011). Mechanisms associated with the EAHE included lowered expression of Fas and Bax and phosphorylation of JNK, p38, and activated B cell pathways. With the combination of these factors, EAHE is shown to have the potential to act as a therapeutic agent for treating PD and other neurodegenerative disorders. Additionally, it is expected that the injection of NGF in brain cells is expected to improve symptoms of PD, further supporting the safety of HE oral intake, as NGF synthesis in astrocytes is confirmed to be an effective therapeutic agent for dementia (Aloe et al. 2012; Apfel et al. 2000; Apfel 2002).

3 *Coriolus versicolor*

Coriolus versicolor (Cv), also known as *Trametes versicolor,* is an edible mushroom that possesses polysaccharopeptides (PSP) (Fig. 4, Table 1) and polyphenolic compounds (Fig. 5) that can treat dementia and bolster the immune system. These components of the mycelia stimulate the assembly of interferons, macrophages, immunoglobulin G, and T lymphocytes. Additionally, Cv induces the production of superoxide dismutase (SOD), an enzyme to recover leftover oxygen free radicals and lower active oxygen toxicity and other free radicals (Cui and Chisti 2003).

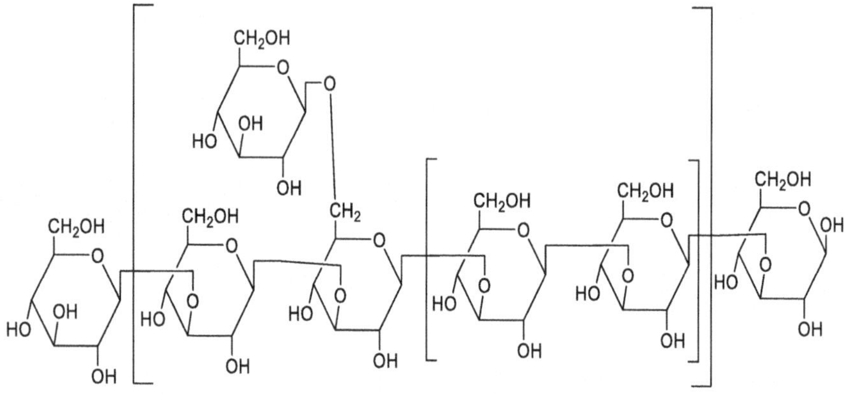

Fig. 4 Polysaccharopeptides in *Coriolus versicolor*

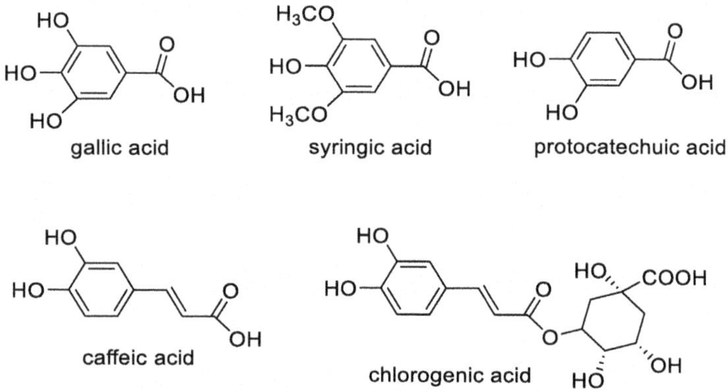

Fig. 5 Polyphenols in *Coriolus versicolor*

Similar to HE, Cv can be supplemented in a diet to treat AD through the equalizing of oxidants and pro-inflammatory cytokines, as well as increasing the expression of LXA4 in rat models. Cv enhances the quality of mitochondria to maintain energy homeostasis, as dysfunction is attributed to the lacking energy equivalents. In these conditions, Cv biomass promotes neuronal resilience and resistance to proteotoxic proteins and inhibits cholinergic neurodegenerative degradation.

Homeostasis of the neurological system enforces a biphasic dose response, in which high doses are toxic to humans and cause ROS formation. In contrast, a low dose inhibits oxidative stress and subsequent dementia (Calabrese and Mattson 2017). The polyphenols of Cv exhibit properties that can activate the Hsp pathway. Furthermore, after administration of Cv, LXA4 levels were consistently high in the cortex and hippocampus, increasing the number of cytoprotective proteins responding to cellular stress and inhibiting the degradation of neuronal cells (Le et al. 2002).

4 Toxicology Studies

Currently, either HE or Cv consumption is safe from toxic effects with no reports of mortality or significant adverse effects. In one study, the acute oral LD_{50} of EAHE mycelia was higher than 5 g/kg in rats, indicating that the mushroom is reasonably safe (Li et al. 2018). Furthermore, reoccurring daily HE doses have been used in rats without adverse effects. HE was found not mutagenic in a bacterial reverse mutation test, and it was not teratogenic in Sprague-Dawley rats with doses up to 2625 mg/kg (Li et al. 2018).

One study showed that EAHE demonstrates no mortality or toxic effects with prolonged use and has no significant difference in adverse effects between the treated and placebo groups. There were no noticeable differences between the placebo and treated groups in rat body weights, histopathological parameters, consumption patterns, and biochemical parameters. Therefore, it can be suggested that EAHE can be supplemented in diets for an extended period. Another study of EAHE products launched in the Taiwanese markets found no adverse effects (Li et al. 2018).

Four patients dropped out in the double-blind 49-week study of EAHE with 50 patients due to adverse effects, such as abdominal discomfort, nausea, and skin rashes. Overall, the incidence was 8.2% over the entire study period. However, it cannot be determined if these effects were caused by EAHE consumption, as there is an increasing trend in prescribing elderly patients with multiple medications that could also cause similar adverse effects (Poleksic and Xie 2019).

5 Conclusions

Neurodegenerative diseases result from progressive deterioration of the neurons in a specific brain region due to oxidative stress and inflammation. The imbalance of ROS and antioxidant enzymes and the cascade of amyloid-β plaques, NFTs, and tau proteins is responsible for neuronal cell deaths and dementia. Mushrooms, such as *Hericium erinaceus* and *Coriolus versicolor*, exhibit abundant antioxidant and anti-inflammatory properties, promote neuron regeneration, and repair and inhibit neuroinflammation (Fig. 6). The mushrooms are reportedly safe for extended supplementation in diets and do not cause adverse side effects or mortality. Erinacine A, prominent in the mycelia of HE, is evidenced to delay neuronal death and provide healthful benefits when consumed orally. In random human trials, treated groups scored significantly higher in cognitive function tests than the control group, confirming the protective effects of mushrooms on neural networks and the promotion of the human immune response.

Further studies are needed on humans and the differing bioactive components of mushrooms to determine the efficiency of an application. In conclusion, it can be affirmed that mushrooms can be used as supplements to prevent the initiation and

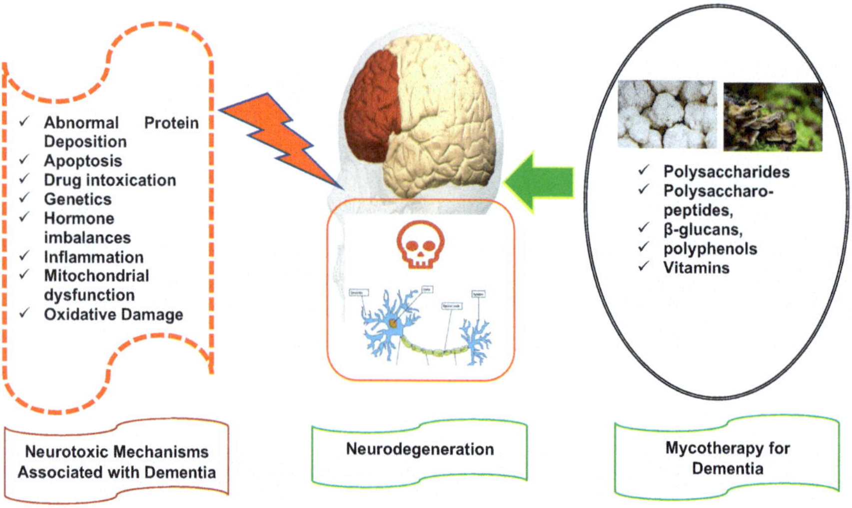

Fig. 6 Neuroprotective effects of *Hericium erinaceus* and *Coriolus versicolor* mushrooms. Polygon data were generated by Database Center for Life Science (DBCLS) [2], CC BY-SA 2.1 JP <https://creativecommons.org/licenses/by-sa/2.1/jp/deed.en>, via Wikimedia Commons. BrunelloN, CC BY-SA 4.0 <https://creativecommons.org/licenses/by-sa/4.0>, via Wikimedia Commons. Injurymap, CC BY 4.0 <https://creativecommons.org/licenses/by/4.0>, via Wikimedia Commons. User:PatilRohan0, CC BY-SA 4.0 <https://creativecommons.org/licenses/by-sa/4.0>, via Wikimedia Commons

progression of neurodegenerative diseases (Fig. 6). The fruiting body and mycelia components, polysaccharides, erinacines, polysaccharopeptides, β-glucans, and polyphenolic compounds, balance homeostasis, promote "neuro health," and have an implication of being therapeutic agents for AD and PD.

References

Aloe L, Rocco ML, Bianchi P, Manni L (2012) Nerve growth factor: from the early discoveries to the potential clinical use. J Transl Med 10:239. https://doi.org/10.1186/1479-5876-10-239

Alzheimer Association (2022) 2022 Alzheimer's disease facts and figures. Alzheimer Dement 18(4):700–789. https://doi.org/10.1002/alz.12638

Apfel SC (2002) Nerve growth factor for the treatment of diabetic neuropathy: what went wrong, what went right, and what does the future hold? Int Rev Neurobiol 50:393–413. https://doi.org/10.1016/s0074-7742(02)50083-0

Apfel SC, Schwartz S, Adornato BT et al (2000) Efficacy and safety of recombinant human nerve growth factor in patients with diabetic polyneuropathy: a randomized controlled trial. rhNGF Clinical Investigator Group. JAMA 284(17):2215–2221. https://doi.org/10.1001/jama.284.17.2215

Bello VME, Schultz RR (2011) Prevalence of treatable and reversible dementias: a study in a dementia outpatient clinic. Dement Neuropsychol 5(1):44–47. https://doi.org/10.1590/s1980-57642011dn05010008

Calabrese EJ, Mattson MP (2017) How does hormesis impact biology, toxicology, and medicine? NPJ Aging Mech Dis 3(1):13. https://doi.org/10.1038/s41514-017-0013-z

Calabrese V, Cornelius C, Dinkova-Kostova AT et al (2010) Cellular stress responses, the hormesis paradigm, and vitagenes: novel targets for therapeutic intervention in neurodegenerative disorders. Antioxid Redox Signal 13(11):1763–1811. https://doi.org/10.1089/ars.2009.3074

Chen CC, Tzeng TT, Chen CC et al (2016) Erinacine S, a rare sesterterpene from the mycelia of *Hericium erinaceus*. J Nat Prod 79(2):438–441. https://doi.org/10.1021/acs.jnatprod.5b00474

Chiu PY, Tsai CT, Chen PK et al (2016) Neuropsychiatric symptoms in Parkinson's disease dementia are more similar to Alzheimer's disease than dementia with lewy bodies: a case-control study. PLoS One 11(4):e0153989. https://doi.org/10.1371/journal.pone.0153989

Cui J, Chisti Y (2003) Polysaccharopeptides of *Coriolus versicolor*: physiological activity, uses, and production. Biotechnol Adv 21(2):109–122. https://doi.org/10.1016/S0734-9750(03)00002-8

De Silva DD, Rapior S, Sudarman E et al (2013) Bioactive metabolites from macrofungi: ethnopharmacology, biological activities and chemistry. Fungal Divers 62(1):1–40. https://doi.org/10.1007/s13225-013-0265-2

Demattos RB, Lu J, Tang Y et al (2012) A plaque-specific antibody clears existing β-amyloid plaques in Alzheimer's disease mice. Neuron 76(5):908–920. https://doi.org/10.1016/j.neuron.2012.10.029

Feng L, Cheah IK, Ng MM, Li J et al (2019) The association between mushroom consumption and mild cognitive impairment: a community-based cross-sectional study in Singapore. J Alzheimers Dis 68(1):197–203. https://doi.org/10.3233/jad-180959

Friedman M (2015) Chemistry, nutrition, and health-promoting properties of *Hericium erinaceus* (Lion's Mane) mushroom fruiting bodies and mycelia and their bioactive compounds. J Agricul Food Chem 63(32):7108–7123. https://doi.org/10.1021/acs.jafc.5b02914

Heneka MT (2017) Inflammasome activation and innate immunity in Alzheimer's disease. Brain Pathol 27(2):220–222. https://doi.org/10.1111/bpa.12483

Iulita MF, Cuello AC (2014) Nerve growth factor metabolic dysfunction in Alzheimer's disease and Down syndrome. Trends Pharmacol Sci 35(7):338–348. https://doi.org/10.1016/j.tips.2014.04.010

Kirstein J, Morito D, Kakihana T et al (2015) Proteotoxic stress and ageing trigger the loss of redox homeostasis across cellular compartments. EMBO J 34(18):2334–2349. https://doi.org/10.15252/embj.201591711

Lai CL, Lin RT, Liou LM, Liu CK (2010) The role of event-related potentials in cognitive decline in Alzheimer's disease. Clin Neurophysiol 121(2):194–199. https://doi.org/10.1016/j.clinph.2009.11.001

Le Y, Murphy PM, Wang JM (2002) Formyl-peptide receptors revisited. Trends Immunol 23(11):541–548. https://doi.org/10.1016/S1471-4906(02)02316-5

Li IC, Chen WP, Chen YP et al (2018) Acute and developmental toxicity assessment of erincine A-enriched Hericium erinaceus mycelia in Sprague–Dawley rats. Drug Chem Toxicol 41(4):459–464. https://doi.org/10.1080/01480545.2017.1381110

Li IC, Chang HH, Lin CH et al (2020) Prevention of early Alzheimer's disease by erinacine A-enriched *Hericium erinaceus* mycelia pilot double-blind placebo-controlled study. Front Aging Neurosci 12. https://doi.org/10.3389/fnagi.2020.00155

Meredith GE, Rademacher DJ (2011) MPTP mouse models of Parkinson's disease: an update. J Parkinsons Dis 1(1):19–33. https://doi.org/10.3233/jpd-2011-11023

Mhyre TR, Boyd JT, Hamill RW, Maguire-Zeiss KA (2012) Parkinson's disease. Subcell Biochem 65:389–455. https://doi.org/10.1007/978-94-007-5416-4_16

Mori K, Inatomi S, Ouchi K, Azumi Y, Tuchida T (2009) Improving effects of the mushroom Yamabushitake (*Hericium erinaceus*) on mild cognitive impairment: a double-blind placebo-controlled clinical trial. Phytother Res 23(3):367–372. https://doi.org/10.1002/ptr.2634

Phan C-W, David P, Sabaratnam V (2017) Edible and medicinal mushrooms: emerging brain food for the mitigation of neurodegenerative diseases. J Med Food 20(1):1–10. https://doi.org/10.1089/jmf.2016.3740

Poleksic A, Xie L (2019) Database of adverse events associated with drugs and drug combinations. Sci Rep 9(1):20025. https://doi.org/10.1038/s41598-019-56525-5

Saitsu Y, Nishide A, Kikushima K et al (2019) Improvement of cognitive functions by oral intake of *Hericium erinaceus*. Biomed Res 40(4):125–131. https://doi.org/10.2220/biomedres.40.125

Shimbo M, Kawagishi H, Yokogoshi H (2005) Erinacine A increases catecholamine and nerve growth factor content in the central nervous system of rats. Nutr Res 25(6):617–623. https://doi.org/10.1016/j.nutres.2005.06.001

Sperling RA, Jack CR, Aisen PS (2011) Testing the right target and right drug at the right stage. Sci Transl Med 3(111):111cm33. https://doi.org/10.1126/scitranslmed.3002609

Talalay P, Zimmerman AW (2015) Reply to Scahill: behavioral outcome measures in autism. Proc Nat Acad Sci 112(4):E350–E350. https://doi.org/10.1073/pnas.1422473112

Trovato A, Siracusa R, Di Paola R et al (2016) Redox modulation of cellular stress response and lipoxin A4 expression by *Hericium erinaceus* in rat brain: relevance to Alzheimer's disease pathogenesis. Immun Ageing 13(1):23. https://doi.org/10.1186/s12979-016-0078-8

Tsai TT, Chin CC, Li YL et al (2016) Erinacine A-enriched *Hericium erinaceus* mycelium ameliorates Alzheimer's disease-related pathologies in APPswe/PS1dE9 transgenic mice. J Biomed Sci 23(1):49. https://doi.org/10.1186/s12929-016-0266-z

Valko M, Leibfritz D, Moncol J et al (2007) Free radicals and antioxidants in normal physiological functions and human disease. Int J Biochem Cell Biol 39(1):44–84. https://doi.org/10.1016/j.biocel.2006.07.001

Wang L, Benzinger TL, Su Y et al (2016) Evaluation of Tau imaging in staging Alzheimer disease and revealing interactions between β-amyloid and tauopathy. JAMA Neurol 73(9):1070–1077. https://doi.org/10.1001/jamaneurol.2016.2078

Wu J, Wang A, Min Z et al (2011) Lipoxin A4 inhibits the production of proinflammatory cytokines induced by β-amyloid in vitro and in vivo. Biochem Biophys Res Commun 408(3):382–387. https://doi.org/10.1016/j.bbrc.2011.04.013

Yang F, Xie J, Wang W et al (2014) Regional arterial infusion with lipoxin A4 attenuates experimental severe acute pancreatitis. PLoS One 9(9):e108525. https://doi.org/10.1371/journal.pone.0108525

Ziegler-Graham K, Brookmeyer R, Johnson E, Arrighi HM (2008) Worldwide variation in the doubling time of Alzheimer's disease incidence rates. Alzheimer Dement 4(5):316–323. https://doi.org/10.1016/j.jalz.2008.05.2479

Potential of Medicinal Mushrooms in Human Health and Welfare: An Overview

Itika Sharma, Malvi Choudhary, Sanjana Kaul, Dinesh Chandra Agrawal, and Manoj K. Dhar

Contents

Abstract Escalating progression in human diseases and problems has led to an intensive search for new drug sources and effective metabolites. In this scenario, the

I. Sharma · M. Choudhary · S. Kaul (✉)
Fungal Biotechnology Laboratory, School of Biotechnology, University of Jammu, Jammu, Jammu and Kashmir, India

D. C. Agrawal
Department of Applied Chemistry, Chaoyang University of Technology, Taichung, Taiwan
e-mail: dcagrawal@gm.cyut.edu.tw

M. K. Dhar
Genome Research Laboratory, School of Biotechnology, University of Jammu, Jammu, Jammu and Kashmir, India

D. C. Agrawal, M. Dhanasekaran (eds.), *Mushrooms with Therapeutic Potentials*,
https://doi.org/10.1007/978-981-19-9550-7_9

therapeutic potential of macrofungi is worth mentioning at the frontier of developing new nutraceuticals. Mushrooms have been treasured and appreciated as traditional medicine by human civilization for ages. They represent a potent source of bioactive metabolites, including polysaccharides, phenolics, terpenoids, peptides, minerals, and vitamins. The health augmenting properties of these biocomponents are extended to their structural variability, imparting biological and functional attributes. Medicinal mushrooms and their competent secondary metabolites effectively treat human disorders such as anticancer, anti-inflammatory, antiviral, antidiabetic, antioxidant, immune-stimulatory, hepatoprotective, antimicrobial, nephroprotective, hypocholesterolemic, hypoglycemic, etc. Therefore, this chapter documents and summarizes the knowledge on biomolecules of mushrooms in human health and highlights their healing traits against major lifestyle-related diseases. It also discloses the capacity of these bioactive metabolites in developing innovative functional food products with pharmaceutical potential.

Keywords Bioactive metabolites · Human health · Macrofungi · Medicinal mushrooms · Nutraceuticals · Pharmaceutical potential

Abbreviations

ACE	Angiotensin-converting enzyme
ACHN	Renal carcinoma cell line
A549	Human lung carcinoma cell line
Caco 2	Colon cancer cell line
Cy	Cyclophosphamide
DNA	Deoxyribonucleic acid
DPPH	1,1-Diphenyl-2-Picrylhydrazyl
GHz	Gigahertz
HCl	Hydrochloric acid
HIV	Human immunodeficiency virus
HPLC-PDAD	High-Performance Liquid Chromatography-Photodiode Array Detector
HT-22	Neuronal cell line
HT-29	Human intestinal cell line
iNOS	Inducible nitric oxide synthase
KDa	Kilodalton
KOH	Potassium hydroxide
LOVO/DX, LOVO	Human colon cancer cell lines
MAPK	Mitogen-activated protein kinase
MBPs	Mushroom bioactive peptides
MCF-7/DX, MCF-7, MDA-MB-231	Human breast cancer cell line
MHz	Megahertz

MRC5	Human lung cell line
MW	Molecular weight
NaOH	Sodium hydroxide
NF-κB	Nuclear factor kappa B
NK	Natural killer cells
PLA2	Phospholipase A2
PSP	Polysaccharide protein
PSK	Polysaccharopeptides
RNA	Ribonucleic acid
ROS	Reactive oxygen species
SHE	Ethyl alcohol fraction
SHP	Polysaccharide
SiHa, HeLa, and CaSki	Human cervical cell lines
Spp.	Species
STZ	Streptozotocin
SW620	Human colorectal cancer cell line
25 (OH)D	25-Hydroxyvitamin D
UV	Ultraviolet

1 Introduction

Fungi, the third food group, is the most diverse and evolutionary advanced member constituting the ethnomycological group. Filamentous macrofungi growing above or beneath the top soil form large sporocarps, known as mushrooms, distinctly conspicuous to the unaided eye (Martinez-Medina et al. 2021). These are exceptionally beneficial basidiomycetes which are considered "superfoods" and have been a part of culinary delicacies in the ancient civilization in Asia (Bamigboye et al. 2022).

Out of 2000 mushrooms described as safe species, around 700 have been documented to be bestowed with pharmacological potential (Motta et al. 2021). Medicinal and palatable mushrooms embody a large number of mycochemicals with specific health-promoting functionalities. Many species of medicinal mushrooms, for instance, *Lentinula edodes, Cordyceps, Russula virescens, Flammulina velutipes, Termitomyces eurhizus, Grifola frondosa, Pleurotus tuber-regium, Agaricus bisporus, Boletus edulis, Cantharellus cibarius, Hericium erinaceum, Agaricus blazei, Ganoderma lucidum, Coriolus versicolor, Phellinus linteus,* etc. possess unique biological properties (Jeitler et al. 2020). The demand and utilization of mushroom is growing in different parts of the world owing to their unique biochemical composition, aroma, low-calorie content, ease of availability, and high protein and fiber content. Lectins, proteins, phenolic compounds, polysaccharides, steroids, vitamins, terpenoids, unsaturated fatty acids, and indole compounds are major biochemicals imparting medicinal, nutraceutical, psychotropic, and dietary potentialities (Niego et al. 2021; Rizzo et al. 2021). However, many factors like the developmental stage, mushroom type, growing environmental conditions, and geographical location significantly manipulate the activity and content of these biomolecules (Martinez-Medina et al. 2021).

The active ingredients in mushroom extract work synergistically to furnish promising defensive and therapeutic properties. Traditional knowledge and modern studies have demonstrated the ameliorative functions of these mycochemicals such as cytotoxic, antioxidant, antidiabetic, antimicrobial, antihypertensive, antihyperglycemic, antiviral, immune-stimulation, neuroprotective, hepatoprotective, and renoprotective functions. All these prospective applications are attributed to the regulation of apoptosis, signaling pathways in tumor progression, activation of immune cells, cytokines, interleukins, and angiogenesis (Blagodatski et al. 2018). Hence, in all certainty, medicinal mushrooms have increasingly attracted consumers' attention for being nutritious, safe, and rich in bioactive and therapeutic properties (Das et al. 2021).

Therefore, this chapter systematically details the curative potential of medicinal mushroom biocomponents in ongoing pathologies and invigorates further investigation in the search for novel therapeutic candidates. Basic techniques in processing the extraction of bioactive components in different species of mushroom have been given in Table 1.

2 Biofunctional Components in Medicinal Mushrooms

Mushrooms have been considered over-the-counter health supplements across the globe, especially for their organoleptic merit and the synergistic effects of medicinal biocomponents (Yang et al. 2019). High nutritional value, unique flavor, and medicinal properties have substantiated their usage in gourmet cuisines since their earliest history (Valverde et al. 2015). The varied physiological and health-promoting effects have been ascribed to several bioactive compounds, namely, polysaccharides, proteins, minerals, vitamins, terpenes, phenolic and lactones in medicinal mushrooms (Elkhateeb et al. 2019; Venturella et al. 2021). These biofunctional compounds with tremendous potential have aroused scientific interest in mycotherapy for producing safe and genetically pure health supplements.

2.1 *Polysaccharides*

The mushroom polysaccharides are the most structurally diverse macromolecules among all the bioactive compounds. They are naturally occurring high molecular weight polymers of glucans forming the structural component of the mushroom cell wall (Yin et al. 2020). α-, β-, and mixed glucans are the most common polysaccharides found in edible mushrooms. Their structural buildup characteristics, like molecular mass, polymerization, branching degree, chain length, linkage,

Table 1 Basic techniques in mushroom processing for the extraction of bioactive components

S. no.	Technique	Methodology	Examples	
	Conventional extraction techniques			
1.	Hot water extraction	The most commonly used method for extraction of water-soluble polysaccharides. In this method, hot water is used at high temperatures (50–100 °C for a certain period (1.5–5 h)	*Termitomyces eurhizus*, *Grifola frondosa*, *Pleurotus tuber-regium*, *Agaricus bisporus*	Patel et al. (2021)
2.	Acid/alkali extraction	This step is usually accompanied by hot water extraction to enhance mushroom polysaccharides' release. In this method, acid (HCl) or alkaline (NaOH, KOH) treatments are given, which further facilitates the destruction of cell walls and linkages, causing the release of bound polysaccharides out of the cell	*Pleurotus ostreatus*	Baeva et al. (2019)
3.	Organic solvent extraction	Extraction is performed using low-boiling organic solvents such as hexane, chloroform, etc. The mushroom mycelia or fermentation broth is treated with appropriate organic solvent for a week, filtered, and concentrated under a vacuum in a rotary evaporator for further analysis	*Grifola frondosa*, *Hericium erinaceum*, *Agaricus blazei*, *Ganoderma lucidum*, *Coriolus versicolor*, *Phellinus linteus*	Patel et al. (2021)
4.	Soxhlet extraction	This is the most extensively used method for extracting polyphenolic components using organic solvents. In this method, the boiling solvent is reflux through the Soxhlet apparatus, which causes the vapors to aspire through the vapor duct and siphon tube and results in the condensation of the vapors in the extraction flask attached	*Cantharellus cibarius*	Heleno et al. (2016), Sevindik (2019)
	Latest extraction techniques			
1.	Ultrasound-mediated extraction	This is the most effective method based on generating a bubble within the solvent in an ultrasound environment. The bubble formation	*Agaricus bisporus*	Alves Filho et al. (2021)

(continued)

Table 1 (continued)

S. no.	Technique	Methodology	Examples	
		facilitates the interruption and penetration of solvents in the cell cytoplasm, resulting in the movement of bioactive components out of the cell.		
2.	Supercritical fluid extraction	It is the most promising, effective, and eco-friendly extraction method for isolating natural oils, oleoresins, fats, and other nonpolar compounds. This method uses inert, noncorrosive, and nontoxic carbon dioxide as the extraction solvent under supercritical conditions to preserve the quantity and quality of heat-labile compounds.	*Hericium erinaceus* and *Lentinula edodes*	Joradon et al. (2022), Morales et al. (2017)
3.	Accelerated solvent extraction	It is a relatively new automatic extraction technique operative under high pressure and temperature. This technique has been categorized as a greener approach for the extraction of various bioactive metabolites by offering advantages like simplicity in procedures, high product quality, low extraction cost, high efficiency, and less time-consuming	*Agrocybe aegerita*	Du et al. (2022)
4.	Enzyme-assisted extraction	This method involves the use of enzymes for the hydrolysis and degradation of the mushroom cell wall for the release of bioactive compounds. As this method operates under room temperature and does not involve any harsh conditions, therefore, the three-dimensional structure and bioactivity of the extracted compounds remain intact	*Auricularia auricula*	Li et al. (2019b)
5.	Microwave-assisted extraction	The use of microwave (frequency from 300 MHz to 300 GHz) energy causes the	*Lentinula edodes*	Xiaokang et al. (2020)

(continued)

Table 1 (continued)

S. no.	Technique	Methodology	Examples	
		direct heating of the material by dipole polarization and iconic conduction, thereby causing internal heating. This liquid vaporization generates pressure within the cells, as a result of which cell wall ruptures and bioactive metabolites are recovered		

three-dimensional conformation, etc., distinguish them from their plant counterparts (Gong et al. 2020). Fruiting bodies of mushrooms, mycelia, and fermentation broth are exceptionally rich in β-D glucans, linked by β-type glycosidic bonds involving C1–C3 and C1–C6 (Pandya et al. 2019). These β-linked carbohydrate-based polymers are the most efficacious substances derived from mushrooms with miraculous biological properties (Cerletti et al. 2021). Nontoxicity, biocompatibility, and biodegradability are few therapeutic advantages mushroom polysaccharides offer. Due to their comprehensive structural features and variability, these biopolymers possess antitumor, antidiabetic, antioxidative, immunomodulatory, anti-inflammatory, antiviral effects, etc. (Chakraborty et al. 2021).

Many studies have endorsed the vast potential of β-D-glucans in modern medicine, immunology, biotechnology, and the pharmaceutical industry. For instance, Lentinan, an approved biological response modifier, is a highly purified polysaccharide extracted from the fruit body of *Lentinus edodes* (Berk). It is the second most relished culinary delicacy globally, valued for its unique flavor, nutritional benefits, and therapeutic capabilities. It is reported to have remarkable antioxidant, antitumor, antiaging, immunoprotective, anti-inflammatory, antiviral, nephroprotective, and hepatoprotective effects by modulating the immune system through cytokine production (Sheng et al. 2021). Another similar β-glucan claimed to be efficacious against emerging health concerns is Schizophyllan, isolated from the *Schizophyllan commune*. It has various bioactivities, especially anticancer and immunomodulatory effects (Saetang et al. 2022). Likewise, Ganoderan (*Ganoderma lucidum*) and Maitake d-fraction from *Grifola frondosa* possess similar immunostimulatory and antitumoral mechanisms (Garcia et al. 2022).

It has been ascertained that in some mushroom species (*Ganoderma lucidum, Tricholoma lobayense, Coriolus versicolor*), the polysaccharides are associated with protein or peptide moiety forming polysaccharide-protein complex (PSP) or the polysaccharopeptides (PSK/Krestin) complex. PSP and PSK are two clinically authorized adjuvants in China, Europe, and Japan for cancer treatment (Neergheen et al. 2020). The processing and extraction of mycelium or fermentation broth of medicinal mushrooms yield these two important bioactive components (Dou et al. 2019). Consumption of these heteroglycans has been proven to improve the innate and adaptive immune system via maintenance of redox potential, reduced lipid peroxidation, activation of immune cells like macrophages, natural killer cells, etc. (Maity et al. 2021).

Bioactive polysaccharides have also been advocated as a great source of prebiotics. Moumita and Das (2022) assessed the prebiotic potential of β-glucan and inulin isolated from *Pleurotus florida*. The results of their study highlighted the role of these polysaccharides in lowering blood pressure and cholesterol and can come up as a functional food with hypercholesteremic and antihypertensive activity. Further, mushroom polysaccharides extend their biotechnological application in the field of the cosmetic industry to design formulations for oral and topical administration. The bioactive polysaccharides with antityrosinase, antihyaluronidase, antielastase, anti-pigmentation, anti-collagenase, antiaging, etc. encourage the development of organic nutraceuticals, cosmeceuticals, and nutracosmetics (Badalyan et al. 2022). So, these biofunctional polysaccharides may lead to the development of innovative functional food products with detailed health-related effects.

2.2 *Vitamins and Minerals*

Vitamin D deficiency is a subject of debate globally concerned with skeletal deformities, chronic diseases, and acute conditions. As per the guidelines of the Endocrine Society Task Force, a cut-off level of 50 nmol/L of serum/plasma 25 (OH)D (25-hydroxyvitamin D) concentration is considered Vitamin D-deficient (Amrein et al. 2020). There is growing evidence that UV-treated mushrooms are the only natural vegetarian sources of Vitamin D (Leung and Cheung 2021). Synthesis of Vitamin D occurs by converting ergosterol to ergocalciferol after UV irradiation of mushrooms. Ergosterol is distributed unevenly throughout the body of mushrooms, acting as a precursor of Vitamin D_2 (ergocalciferol). Photoirradiation of ergosterol generates various photolysis products, namely, previtamin D2, lumisterol, and tachysterol. The previtamin D_2 undergoes spontaneous rearrangement to vitamin D_2 (Nowak et al. 2022). Besides, they are exceptionally rich in B-complex vitamins like cobalamin (B_{12}), niacin (B_3), pantothenic acid (B_5), riboflavin (B_2), thiamine (B_1), pyridoxine (B_6), and biotin (B_7) (Alzand et al. 2019). The relevance of mushroom consumption is accompanied by the association of these molecules in energy homeostasis, regulation of membrane permeability, fluidity, immune system, endocrine system, and cytoskeletal organization.

Moreover, mushrooms are also reported to be enriched with various minerals such as calcium, potassium, selenium, copper, phosphorus, and magnesium. The edible mushroom contains less sodium and high potassium concentration, making it safe for consumption by hypertensive and diabetic individuals (Yadav and Negi 2021). So, mushrooms' combined vitamin and mineral composition may help develop mushroom-based medicament products.

2.3 Proteins

Mushrooms are an excellent source of proteins, essential and nonessential amino acids. Mushroom bioactive peptides (MBPs) are desirable peptides extracted from either mushrooms or mycelia. These are defined as small and bioactive fragments of proteins with encouraging beneficial effects on human health and disease prevention (Zhou et al. 2020). MBPs serve as vital macronutrients covering dietetics requirements with a complete essential amino acid profile and low cost. Their multifaceted applications vary significantly with their physicochemical properties, bioavailability, amino acid composition, processing, digestibility, and purity (González et al. 2020). Antioxidant, anticancer, antihypertensive, and antimicrobial activities are few of mushroom proteins' reputed physiological properties (Sosalagere et al. 2022). The scientific literature has revealed the radical scavenging potential of mushroom peptides. One such report was presented by Wongaem et al. (2021) in assessing the free radical scavenging potential of protein hydrolysates isolated from split gill mushrooms (*Schizophyllum commune*). The protein hydrolysate was subjected to fractionation, and the fraction with molecular weight < 0.65 KDa was refractionated via Reverse Phase High-Performance Liquid Chromatography to obtain five antioxidant peptides. Moreover, the fraction with molecular weight < 0.65 KDa also represented outstanding antioxidant potential in HT-29 cancer cell lines.

Recently, Kaprasob et al. (2022) reported the isolation and characterization of novel bioactive peptides from King Boletus mushrooms, followed by an evaluation of their antioxidant and antihypertensive activities. A total of four fractions were obtained from protein hydrolysate on treatment with Bromelain. The fraction KBMPHF4 (MW < 1 KDa) possessed the highest antioxidative and ACE (angiotensin-converting enzyme) inhibitory capacity. Furthermore, six novel peptides (DLDLLEKGIRKT, NGGNAPI, VSWNVLQEP, DTGRGLGASH, IDDNLDNLIIKL, and LIYAQGFSK) were isolated from the fraction mentioned above, exhibiting the highest antihypertensive activity and moderate free radical scavenging capability. These studies widen the horizon of mushroom peptides as a natural source of antihypertensives and antioxidants by necessitating their utilization as functional food and nutraceuticals.

2.4 Phenolics

Several phenolic compounds, such as flavonoids, oxidized polyphenols, lignans, tannins, phenolic acids, hydroxybenzoic acids, hydroxycinnamic acids, and stilbenes, are present in mushrooms. The peel and gills portions of the mushroom body significantly contribute to phenolic and flavonoid compounds. These minor metabolites offer many biological activities, including antioxidant, anti-tyrosinase, anti-inflammatory, antitumor, osteoprotective, etc. Recently, large observational data has been compiled on the role of phenolic compounds in the prophylaxis of many human diseases such as cancer, diabetes, atherosclerosis, Alzheimer's disease, Parkinson's disease, chronic kidney disease, chronic obstructive pulmonary disease, and cardiovascular disorders (Podkowa et al. 2021). Çayan et al. (2021) assessed the phenolic and antioxidant activity of five mushroom species (*Chondrostereum purpureum, Hymenochaete rubiginosa, Macrolepiota procera, Phaeolus schweinitzii,* and *Phellinus igniarius*) via HPLC-DAD (High-Performance Liquid Chromatography-Photodiode Array Detector) analysis. The study revealed the presence of phenolic compounds (fumaric acid, protocatechuic acid, 6,7-dihydroxy coumarin, and β-carotene-linoleic acid) in the methanolic and hexane extracts of the mushroom species and suggested their potential use as a source of antioxidant and versatile biological agents. Several such exploratory reports based on the quantification of phenols and phenolic compounds in mushrooms have already been documented in the literature (Çayan et al. 2020; Rašeta et al. 2020; Darmasiwi et al. 2022; Vazquez-Armenta et al. 2022). Depending on their disease prevention and bioactive profile, they are an outstanding source of functional food, nutraceuticals, and cosmeceuticals.

2.5 Terpenes

Terpenes, the most structurally diverse group of naturally occurring compounds. These belong to volatile unsaturated hydrocarbons contributing primarily to the organoleptic properties of mushrooms (Cox-Georgian et al. 2019). At the beginning of the twenty-first century, mushroom-derived therapeutic terpenes have gained momentum for being highly versatile from a pharmaceutical perspective. It is imperative to disclose the capacity of bioactive terpenoids in several microbial, neurodegenerative, and viral diseases and tumor progression (Yadav and Negi 2021). Dasgupta and Acharya (2019) have reviewed the detailed data on mushroom terpenoids' for their healing and curing properties. Several bioactive terpenes have been isolated from the culture filtrate and solid culture substrate of *Flammulina velutipes* with a detailed explanation of their health functionalities (Fukushima-Sakuno 2020). These reports initiate mounting interest in the therapeutic potential and significance of terpenoids in unraveling the gray areas of medical research.

3 Therapeutic Activities of Mushrooms

Mushrooms have been used as foods, nutraceuticals, and medicines for ages due to the bioactive agents such as phenols, polysaccharides, flavonoids, alkaloids, terpenes, proteins, vitamins, and low fat content present in them. These compounds attribute to various biological applications of mushrooms like antimicrobial activity, antioxidant activity, anti-inflammatory activity, immunomodulatory activity, anticancer activity, and many others (Fig. 1).

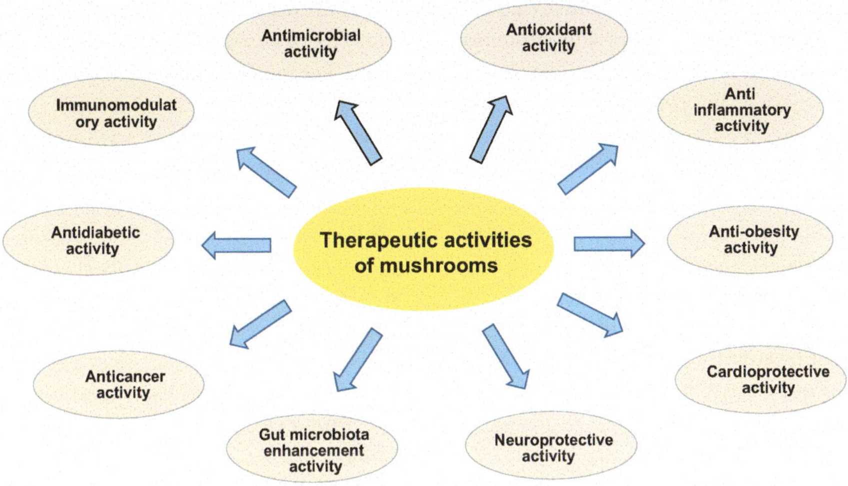

Fig. 1 Pharmacological properties of mushrooms

3.1 Antimicrobial Activity

Food security is a major problem due to continuous health threats by foodborne pathogens (Shen et al. 2017). Also, synthetic antimicrobial drugs have developed multidrug resistance among microorganisms. Hence, there is a requirement for more sustainable antimicrobials against pathogenic microorganisms. Mushrooms have emerged as the most reliable antimicrobials due to their natural defense strategies, as they are exposed to several microorganisms in their natural environment (Ogidi et al. 2020; Sevindik 2021). Of 1,40,000 known mushroom species, 2000 are safe to be consumed as food, and only 158 possess antimicrobial properties (Shen et al. 2017; Assemie and Abaya 2022). The antimicrobial potential of edible mushrooms is attributed to their significant amounts of phenols and alkaloids. Different solvents like water, methanol, chloroform, acetone, etc. are used to extract these active compounds from mushrooms. However, these extracts show variations in their antimicrobial properties (Shen et al. 2017). Secondary compounds like triterpene in mushrooms act as antiviral agents, mainly in HIV infections (Assemie and Abaya 2022). Widely cultivated edible mushrooms that possess antimicrobial activity are *Sarcodon imbricatus, Lactarius deliciosus, Tricholoma portentosum, Agaricus bisporus, Lentinus edodes, Pleurotus* spp., and *Flammulina velutipes* (Roman et al. 2020; Assemie and Abaya 2022). Shen et al. (2017) reported 88 genera of mushrooms possessing antimicrobial properties, majorly dispersed in various regions of the world (specifically Bangladesh, India, China, Turkey, Portugal, Korea, Northern Serbia, and Spain). Three kinds of mushrooms, namely, naturally occurring wild mushrooms (*Panus fulvus, Agrocybe* spp., and *Auricularia auricula-judae*), commercial varieties (*Pleurotus* spp. and *Agaricus bisporus*), and medicinal mushrooms (*Lentinus edodes, Ganoderma spp.,* and *Cordyceps militaris*) were listed. Bach et al. (2019) assessed the antimicrobial properties of five mushroom phenolic extracts comprising *Flammulina velutipes, Agaricus bisporus* (Champignon and Portobello), *Agaricus brasiliensis,* and *Lentinula edodes* against pathogenic bacteria, *Bacillus cereus, Staphylococcus aureus, Escherichia coli,* and *Salmonella enteritidis*. The phenolic extract of *Agaricus brasiliensis* was observed to possess maximum antibacterial activity. Similarly, three *Macrolepiota* species (*Macrolepiota rhacodes, Macrolepiota mastoidea,* and *Macrolepiota procera*) were examined for their antimicrobial potential. Methanolic extract of *Macrolepiota mastoidea* showed the best antibacterial activity, while the lowest activity has been shown by methanolic extract of *Macrolepiota procera*. Regarding antifungal activity, *Macrolepiota mastoidea* and *Macrolepiota procera* extracts have shown the most pronounced effects (Ciric et al. 2019).

Further, Alkin et al. (2021) evaluated the antimicrobial properties of nine mushroom genera (*Agaricus bisporus, Hericium erinaceus, Lactarius deliciosus, Morchella spp., Lentinus edodes, Cantharellus cibarius, Ganoderma lucidum, Pleurotus ostreatus,* and *Boletus edulis*) from Turkey. *Boletus edulis* and *Cantharellus cibarius* showed the most potent antimicrobial activity. Another study was conducted by Suliaman et al. (2021) in which the inhibitory effects of methanol extracts of *Pleurotus nebrodensis, Trametes trogii, Boletus luridus,* and *P. ostreatus* have been reported against bacterial and yeast pathogens. Of all the wild mushroom species, *Trametes trogii* possesses the strongest antimicrobial activity against most microorganisms. In contrast, *Boletus luridus* produced inhibitory effects only against *Salmonella typhi, Candida parapsilosis,* and *Saccharomyces cerevisiae.* Another mushroom, *Poletus ostreatus,* showed antimicrobial activity against *Staphylococcus aureus, Saccharomyces cerevisia,* and *Shigella* spp. However, no activity has been shown by *P. nebrodensis*. In addition, the high antifungal activity of a wild mushroom, *Laeticutis cristata,* was reported by Sevindik and Bal (2021). Also, Soliman and El-Sayed (2021) conducted a study to illustrate the antimicrobial potential of some wild mushrooms (*Cyclocybe cylindracea, Chlorophyllum molybdites, Bjerkandera adusta, Lentinus squarrosulu,* and *Agrocybe aegerita*). Amongst these, *Cyclocybe cylindracea* and *Bjerkandera adusta* have been stated to possess strong antimicrobial effects against *Pseudomonas aeruginosa, Micrococcus luteus, Escherichia coli, Staphylococcus aureus, Candida albicans,* and *Streptococcus pneumonia.* Moreover, gram-positive bacteria were more sensitive to their ethanolic extracts than gram-negative bacteria. However, *Agrocybe aegerita* exhibited only a weak antimicrobial effect against *Escherichia coli* and *Pseudomonas aeruginosa*. Similarly, Sevindik (2021) reported the effect of *Melanoleuca melaleuca* ethanolic extract on tested microorganisms (*Staphylococcus aureus, Enterococcus faecalis, Pseudomonas aeruginosa, Escherichia coli, Candida albicans, Acinetobacter baumannii, Candida krusei,* and *Candida glabrata*). *Melanoleuca melaleuca* has shown maximum antimicrobial activity against gram-negative bacteria and fungal strains. Antimicrobial properties have been explored for edible Chilean mushrooms. Dichloromethane extracts of mushrooms are more active against bacteria and fungi than methanol extracts (Jacinto-Azevedo et al. 2021). Further, a study by Krishnamoorthi et al. (2022) has depicted the strong antimicrobial activity of an edible mushroom, *Agaricus bisporus,* against some clinically important pathogens (*Staphylococcus aureus* and *Candida albicans*). Since mushrooms have been eaten safely for years and have antimicrobial potential, they can be used as an alternative to currently used antimicrobials. More strategies need to be developed for their mass production.

3.2 Antioxidant Activity

The oxidative stress environment is the major cause of the production of reactive oxygen species in cellular entities. Excessive ROS generation damages macromolecules such as DNA, RNA, lipids, and proteins, further leading to tissue death. These reactive oxygen species include hydrogen peroxide, superoxide and hydroxyl radicals, etc., and are involved in causing various diseases such as cancer, aging, cataracts, and many others (Abdelshafy et al. 2022). Hence, finding natural substances that can scavenge free radicals is important. One such reliable natural source is medicinal mushrooms, as they contain a significant amount of flavonoids, cytokines, polysaccharides, terpenoids, etc., that can reduce oxidative stress (Shaffique et al. 2021).

Yan et al. (2019) explored the antioxidant potential of four water-soluble polysaccharides, isolated from *Pleurotus ostreatus, Flammulina velutipes*, *Pleurotus eryngii*, and white *Hypsizygus marmoreus,* and purified their acidic and neutral fractions. On screening, it was found that the acidic polysaccharides from *Pleurotus ostreatus* exhibited maximum activity, which might have contributed to the presence of β-(1-6)-glucan isolated from the strongest acidic fraction. Garrab et al. (2019) further examined the antioxidant properties of three mushrooms: *Agaricus silvaticus Schaeff., Hydnum rufescens* Pers., and *Meripilus giganteus* (Pers.) Karst by using the DPPH (1,1-Diphenyl-2-Picrylhydrazyl) method, and *Hydnum rufescens* has shown the highest antioxidant activity. Similarly, the antioxidant activity of two wild mushrooms *Melanoleuca cognata* and *Melanoleuca stridula* was investigated by Bahadori et al. (2019). Water extracts of both species have shown maximum antiradical activity, followed by methanolic extracts. Ethyl acetate extracts have shown the least activity. Furthermore, an experiment was done to study the antioxidant effects of five mushrooms: *Flammulina velutipes, Agaricus bisporus* (Champignon and Portobello), *Agaricus brasiliensis,* and *Lentinula edodes. Agaricus brasiliensis* was discovered to have the maximum antioxidant potential (Bach et al. 2019). In addition, Ciric et al. (2019) illustrated the antioxidant properties of three different species of the genus *Macrolepiota,* namely, *Macrolepiota mastoidea, Macrolepiota rhacodes,* and *Macrolepiota procera.* The maximum antioxidant effect has been shown by *Macrolepiota procera*. Besides, a comparison has been made by Contato et al. (2020) between the mycelium and basidioma extracts of an eatable mushroom *Pleurotus pulmonarius* to examine their antioxidant potential, and even small concentrations of basidioma extracts have shown better antioxidant activity as compared to mycelium extracts.

One more approach has been followed by Kruzselyi et al. (2020) to compare the radical scavenging activity of different tissues (inner cap, stipe, peel, and gills) of eight edible mushrooms (*Cyclocybe cylindracea, Pleurotus eryngii,* white and brown *Agaricus bisporus, Flammulina velutipes, Pleurotus ostreatus, Leccinum duriusculum,* and *Lentinula edodes*). The results indicated that the peel and gills displayed maximum radical scavenging activity compared to the inner cap and stipe. Further, *n*-hexane, ethyl acetate, chloroform, acetone, ethanol, and pure water extracts of five edible mushrooms *Tricholoma scalpturatum*, *Neolentinus cyathiformis*, *Chlorophyllum agaricoides*, *Tricholoma populinum* and *Lycoperdon utriforme* were evaluated to determine their antioxidant efficiency. *L. utriforme*, *C. agaricoides*, and *T. populinum mushrooms* showed maximum radical scavenging properties (Sezgin et al. 2019).

Sevindik (2021) demonstrated the antioxidant potential of an edible mushroom *Melanoleuca melaleuca,* which has shown positive results. Besides, Jacinto-Azevedo et al. (2021) have analyzed the biological properties of 24 Chilean edible mushrooms, and *Ramaria flava* has been shown to possess maximum antioxidant properties. In addition, Radhika et al. (2021) examined the antioxidant properties of aqueous and ethanolic extracts of Ganoderma lucidum and reported significant antioxidant activities.

Recently, de Menezes Filho et al. (2022) evaluated the DPPH free radical reduction activity of 70% ethanol extract of *Scleroderma verrucosum*. The strong activity of the extract was attributed to the presence of flavonoids and phenolic compounds that exhibit commendable antioxidant properties. Similarly, strong antioxidant effects of *Ganoderma lucidum*, *Ganoderma applanatum*, and *Ranunculus cajanderi* have been elucidated by Hossen et al. (2022), and methanol extracts of tested mushrooms have been observed to possess noteworthy antioxidant properties. Also, the radical scavenging activity of an edible mushroom *Agaricus bisporus* has been explored by Krishnamoorthi et al. (2022), and it showed remarkable results. The outcomes of various studies reported suggest utilizing mushrooms as substantial sources of natural antioxidants.

3.3 Anti-Inflammatory Activity

Inflammation is the response of body immunity against numerous factors such as pathogens, tissue injury, and toxic metabolites (Chen et al. 2017; Abdelshafy et al. 2022). The usual characteristic features of inflammation are redness, loss of function, pain, and swelling. Harmful stimuli can lead to chronic inflammatory responses in various organs of the body, such as the liver, heart, kidney, brain, lungs, etc., further leading to tissue damage or diseases like obesity, cancer, diabetes, and cardiovascular disorders (Rowaiye et al. 2022). Immune cells secrete a large number of inflammatory mediators such as TNF-α (tumor necrosis factor-α), interleukins, ICAM-1 (inducible type cyclooxygenase-2), etc. that regulate various mechanisms like chemotaxis, phagocytosis, vasodilation, and fever (Shapouri-Moghaddam et al.

2018; Li et al. 2019c). The excessive production of these mediators leads to tissue damage and causes many diseases like rheumatoid arthritis, cancer, multiple sclerosis, bronchitis, diabetes, and many others (Rowaiye et al. 2022; Ruksiriwanich et al. 2022). Nonsteroidal inflammatory drugs are commonly used against inflammation, but many studies have depicted that these drugs have long-term side effects (Wongrakpanich et al. 2018). Hence, the focus has been shifted towards natural anti-inflammatory compounds. One such reliable source is mushrooms, which have emerged as a potential alternative to these chemical compounds. Extracts from many mushroom species and their bioactive molecules act as anti-inflammatory agents and downregulate inflammatory mediators' gene expression (Muszynska et al. 2018; Rowaiye et al. 2022). Various literature reports are available that describe the anti-inflammatory potential of mushrooms. Souilem et al. (2017) analyzed two wild mushrooms *Suillus bellinii* and *Pleurotus eryngii* for their anti-inflammatory properties, and *P. eryngii* has shown more anti-inflammatory activity as compared to *S. bellinii.* Further, Saad et al. (2018) evaluated the anti-inflammatory activity of five culinary-medicinal mushrooms: *Ganoderma lucidum, Agaricus bisporus* (brown and white), *Pleurotus pulmonarius, Pleurotus floridanus,* and *Hypsizygus marmoreus* to elucidate their use in cosmeceuticals. Amongst all the mushrooms tested, *A. bisporus* has shown the best results, which indicates its use as a skincare product for soothing the inflammatory reactions on the skin. Similarly, Yamada et al. (2019) screened the anti-inflammatory properties of 44 mushrooms (wild varieties) from Japan, and ten mushroom samples from five species (*Pholiota mixta, Naematoloma fasciculare, C. triumphans, Cortinarius balteatocumatilis,* and *Russula rosacea*) have shown noteworthy results.

Stastny et al. (2022) explored the anti-inflammatory potential of five medicinal mushrooms from the genus *Pleurotus*: *P. pulmonarius*, *P. flabellatus*, *P. opuntiae*, *P. ostreatus* Sylvan Ivory, and *P. ostreatus* Florida, and found that chloroform extract of *P. flabellatus* had significant anti-inflammatory activity. Also, Shashikant et al. (2022) studied the anti-inflammatory properties of the ethanolic extract of *Calocybe indica* and obtained favorable results. Furthermore, the anti-inflammatory activity of bioactive compounds present in ethanolic extracts of the edible mushroom *Dictyophora indusiata* was evaluated by Ruksiriwanich et al. (2022), and the extract has shown strong anti-inflammatory activity.

Various studies have also been done to analyze the compounds present in medicinal mushrooms that contribute to their anti-inflammatory potential. Jiang et al. (2018) studied 19 chemical constituents from *Phellinus igniarius* and suggested that ingredients from *P. igniarius* could be used as anti-inflammatory agents for future studies.

Zuo et al. (2021) examined the anti-inflammatory response of an extracellular polysaccharopeptide isolated from *Sanghuangporus lonicericola* and found it to be a potent anti-inflammatory compound. This further shows that *Sanghuangporus lonicericola* is a suitable option for the extraction of bioactive nutraceuticals. Further, Khalilov et al. (2022) isolated four triterpenoids (Sulphurenoids A-D) from the fruiting bodies of a medicinal mushroom *Laetiporus sulphureus* and investigated their anti-inflammatory properties. The results showed that the compounds exhibited high anti-inflammatory activity, suggesting the medicinal use of *Laetiporus sulphureus* fruiting bodies. The presence of these anti-inflammatory compounds recommends the use of mushrooms as a better alternative to nonsteroidal inflammatory drugs.

3.4 Immunomodulatory Activity

Immunomodulators are usually categorized into three classes: immunosuppressants, immunoadjuvants, and immunostimulants. Their use has swiftly increased over the years due to the medical condition of patients that require immunomodulators. Although most immunomodulators are synthetic compounds, the demand for natural products is increasing daily. Mushrooms are potent natural immunomodulators. Terpenes, terpenoids, polysaccharides, and lectins are some compounds that contribute to their immunomodulatory properties. The immune modulation effects of mushrooms depend upon the distribution of these compounds in the fractions used for the study (Zhao et al. 2020). Mushroom extracts are usually prepared in water and ethanol. Water extracts trigger the release of cytolytic proteins, perforin, and granulysin by stimulating intracellular pathways in NK cells. On the other hand, ethanol extracts produce contrary effects on NK cells by inhibiting intracellular pathway activation and lowering cytolytic protein secretion. These effects can be attributed to the potency and solubility of compounds present in these extracts. Polysaccharides, dissolved in water extracts, activate immune responses.

In contrast, hydrophobic compounds, such as flavonoids and terpenoids in ethanol extracts, act as immune system inhibitors, representing the potent immunomodulatory compounds in these extracts (Martel et al. 2017). To analyze the immunomodulatory properties of polysaccharides, Hu et al. (2017) extracted four polysaccharide fractions from *Lignosus rhinocerotis* sclerotia and evaluated their immunomodulatory activities by cyclophosphamide (Cy)-induced immunosuppression model. All the fractions were found to improve immune organs by showing protective effects against immunosuppression in mice and stimulating cytokine release. They further suggested their utilization as effective immunostimulants in food industries. Also, Smith et al. (2017) evaluated the immunostimulatory effects of five mushrooms (*Trichaptum abietinum, Leucocybe connate, Hericium coralloides, Hydnellum* spp., and *Gyromitra esculenta*), and they all exhibited prominent immunostimulatory activity. In addition, Wang et al. (2018) examined the polysaccharide (14,942 Da) from *Collybia radicata*. They found that it stimulated macrophages and induced the cells to secrete nitric oxide, tumor necrosis factor, and interleukins. Therefore, it could be used as a novel immunopotentiator.

Similarly, Li et al. (2018) isolated a bioactive polysaccharide GFP-22 from the fruiting bodies of *Grifola frondosa*. It has been seen that its administration could reverse cyclophosphamide-induced immunosuppression and considerably increase the proliferation of spleen lymphocytes and the production of cytokines in splenocytes. These outcomes suggest that GFP-22 could be discovered as a natural immunomodulatory agent. Lectins also play a significant role in immunoregulation by activating macrophages. Wang et al. (2019) purified the lectin LSL from *Latiporus sulphureus* and elucidated that it shows appreciable immunomodulating effects; hence, it could be used in pharmacology and food sectors.

Pan et al. (2019) illustrated that the polysaccharides from *Amauroderma rude* stimulate the production of cytokines and trigger PLA2-AA, MAPK, and iNOS pathways involved in the immunomodulatory process. Li et al. (2020) isolated a polysaccharide ECIP-1A from *Eurotium cristatum.* They demonstrated its immunomodulatory activity, and it has been observed to considerably stimulate the proliferation of RAW264.7 cells and incite the production of interleukin-6, tumor necrosis factor-α, and nitric oxide by RAW264.7 cells. Hence, their findings recommended it to be an efficient immunomodulator. In addition to this, Chen et al. (2020) compared the immunomodulatory effect of three polysaccharide fractions (F1, F2, F3) from *Lentinula edodes,* and the first fraction (F1) has been witnessed to enhance cellular immunity, while the other fractions have been shown to improve the innate and adaptive immunity. They also inferred that the F3 fraction, due to its highest molecular weight, possessed the maximum activity, followed by F2 and F1, further highlighting the role of molecular weight in determining their bioactivities.

Also, spores of *Ganoderma lucidium* serve as a prominent immunomodulatory food in Asia. As the polysaccharides present in it are responsible for the immunomodulatory activity, Sheng et al. (2022) isolated and purified the main water-soluble polysaccharide (GLSP-I) from *G. lucidum.* Further analysis indicated that it was a glucan and could activate macrophages, showing notable immunomodulatory activity. Elhusseiny et al. (2022) demonstrated the immunostimulatory effect of aqueous extracts of five edible mushrooms (*Agaricus bisporus, Pleurotus columbinus, Lentinula edodes, Pleurotus ostreatus,* and *Pleurotus sajor-caju)* in vivo using Wistar albino rats. Their results have shown that all the mushroom extracts substantially boosted the white blood cells and lymphocyte counts along with the increase in NO concentration, lysozyme activity, and cytokine production at a 400 mg/kg dose, which was considered the effectual dose. Hence, their study recommended the use of these mushrooms as effective immunostimulators. Zhang et al. (2022) further characterized the immunomodulatory potential of a polysaccharide, mannan (HLP-1), from *Helvella leucopus.* It significantly induced the activation of macrophages through *the* NF-κB pathway, suggesting its role as an active immune enhancement agent. Considering mushrooms' immunomodulatory properties, exploring their compounds for more productive use is important.

3.5 Anticancer Activity

Although the health care sector has grown immensely in recent years, still the world is struggling with the menace of cancer. Cancer is one of the foremost causes of death and, hence, the most significant public health problem worldwide (Kiddane and Kim 2021). The compromised immune system is a major barrier to successfully executing recently developed anticancer drugs. New therapies like nano-formulation and gene therapy have also been launched, but these techniques are too costly for ordinary people. Hence, at this point, the interest of scientists is heading toward the natural source of medicines that not only regulate cancer cells but also boost the immune system. Mushrooms are considered the best dietary food for immunosuppressed patients as they have many immunostimulatory properties, as discussed in our previous section (Pathak et al. 2022). *Pleurotus* spp., also known as oyster mushrooms, are the most cultivated edible mushrooms with different therapeutic properties like antitumor, immunomodulatory, antidiabetic, etc. Extracts from various *Pleurotus* spp. have been observed to incite apoptosis and cell cycle and restrict the growth of cancer cells. Even the nanoparticles prepared from *Pleurotus* extracts have shown potent anticancer properties with negligible side effects. However, more studies are needed in this sector (Mishra et al. 2021).

Secme et al. (2018) examined the anticancer potential of *Macrolepiota procera* on the A549 lung cancer cell line and witnessed the reduction in invasion in cancer cells, highlighting its antiproliferative properties. Also, Datta et al. (2019) isolated a water-soluble polysaccharide from *Marasmiellus palmivorus,* and it was found to possess favorable immunomodulatory properties that were correlated with their anticancer effects. Sairi et al. (2020) further explored the anticancer activity of two aqueous extracts (hot and cold extracts) of *Donkioporiella mellea* on the A549 cancer cell line and MRC5 normal cell line. Both hot and cold extracts are cytotoxic to cancer cell lines. However, they showed no effects on MRC5 cells, illustrating it as a safe option for human consumption. Similarly, Ghosh et al. (2020) tested the anticancer activity of different solvent (ethanolic, water, and ethyl acetate) extracts of *Hexogonia glabra* against SiHa, HeLa, and CaSki (cervical cancer) cell lines. All extracts were active in arresting the growth of cell lines, but ethanolic extract had the maximum effect.

Kolniak-Ostek et al. (2022) evaluated the anticancer potential of bioactive compounds of the fruiting body of *Ganoderma lucidum.* Antiproliferative studies were done on colorectal cancer cell line (SW620), breast cancer cell lines (MCF-7/DX, MCF-7, MDA-MB-231), and colon cancer cell lines (LOVO/DX, LOVO) and significant results were obtained. Dixon et al. (2022) further investigated the anticancer effect of *Poria* mushroom extract, and it exhibited significant results on ACHN cells, causing a decrease in cell viability. In addition, Yamac et al. (2022) characterized the antiproliferative effects of ethyl acetate extracts of fruiting body, culture liquid, and mycelial biomass of *Omphalotus olearius* in cell culture, human cancer, and healthy cell lines by WST-1 assay. A549 (lung cancer cell line) treated with mycelial biomass extract and Caco 2 (colon cancer cell line) treated with culture liquid extract have shown maximum activities. Although mushrooms represent a promising alternative to anticancer therapies, most studies have been performed on only five mushroom species (*Ganoderma lucidium, Lentinula edodes, Agaricus bisporus, Coriolus versicolor,* and *Grifola frondosa*). Only three cancer lines (breast cancer followed by colorectal and lung cancer cell lines) are explored for such studies. Hence, more scientific efforts need to be made to illustrate the value of mushrooms in cancer treatments (Panda et al. 2022).

3.6 Antidiabetic Activity

Diabetes mellitus is a biological syndrome caused by an increase in glucose levels in the blood due to a lack of insulin, affecting millions of people. It can further lead to organ failure and can be fatal if not treated properly. Although many commercial drugs are available in the market for their control, they have long-term side effects. This has highlighted the importance of natural alternatives such as mushrooms. They are rich in fibers, polysaccharides, and phenolics, involved in antidiabetic and antihyperlipidemic activities. Also, they do not contain any fats and calories. Hence, they are healthy foods (Khursheed et al. 2020). Polysaccharides in mushrooms alter insulin secretion and reduce blood glucose levels, which therefore help in regulating insulin metabolism (Anusiya et al. 2021). The antidiabetic potential of aqueous extracts of three medicinal mushrooms (*Volvariella volvacea, Pleurotus ostreatus,* and *Calocybe indica*) has been evaluated by Singh et al. (2017). *C. indica* extract has been seen as the best α-amylase inhibitor in vitro and paved the way for its in vivo studies. It also showed significant results in mice and reduced blood glucose levels. These results suggested that it could be developed as an antidiabetic drug for future use. Prabu and Kumuthakalavalli (2017) further investigated the antidiabetic properties of *Pleurotus florida* (oyster mushroom), and it showed prominent α-amylase and α-glucosidase inhibitory activities (both in vitro and in vivo). Similarly, Ekowati et al. (2018) determined the antidiabetic effects of *Agaricus bisporus* on the blood glucose levels of alloxan-induced diabetic rats, and its extract has shown notable antidiabetic effects activity. In addition, Stojkovic et al. (2019) assessed the antidiabetic properties of methanolic extracts of six mushrooms (*Coprinus comatus, Inonotus obliquus, Agaricus blazei, Cordyceps militaris, Phellinus linteus,* and *Morchella conica*). They found that *Coprinus comatus* had the most potent inhibitory potential on α-amylase while *Inonotus obliquus* extract has shown the best inhibitory effect on α-glucosidase. Also, Kaewnarin et al. (2020) characterized the antidiabetic properties of bioactive polysaccharides from three edible mushrooms (*Phlebopus portentosus, Rugibolutus extremiorientalis,* and *Russula emetica*). The results indicated that the polysaccharides isolated from *R. emetica* had the highest antidiabetic potential. Furthermore, Liu et al. (2020) examined the antidiabetic activity of a polysaccharide, ARPs-p, from *Anoectochilus roxburghii,* and it showed appreciable antihyperglycemic activity.

Hou et al. (2021) explored two fractions, SHP (polysaccharide) and SHE (ethyl alcohol fraction), obtained from the fermentation broth of *Sanghuangporus fenghuang* for their antidiabetic properties. Amongst both extracts, SHP possessed strong antidiabetic properties, suggesting its application in functional foods. Similarly, Deveci et al. (2021) tested the antidiabetic activities of 24 mushrooms' hexane and methanol extracts, and significant results were obtained. The highest inhibitory effect on α-amylase and α-glucosidase was shown by *Cryptoblepharus rutilus* and *Pleurotus ostreatus* hexane extract, respectively. Recently, Ratnaningtyas et al. (2022) studied the antidiabetic effect of *Coprinus comatus* ethanol extract on streptozotocin-induced diabetic rats and observed that its administration considerably reduced glucose levels and increased insulin concentration in rats. Considering the potent antidiabetic effects of mushrooms, it is necessary to explore more strategies to convert their extracts into commercial pharmaceutical products so that these nutraceuticals could reach the bedside of patients who have diabetes.

3.7 Cardioprotective Activity

Cardiovascular disorders, such as heart attacks and stroke, are the major reasons for most mortality cases. Raised blood pressure and increased cholesterol and blood glucose levels are risk factors for cardiovascular diseases (Chugh et al. 2022). Hence, scientists are showing much interest in cardiac research. It is always advisable to consume naturally available food for a healthy heart, and mushroom is one such functional food comprising many valuable bioactive metabolites, contributing to its physiological properties. The well-known bioactive compound that contributes to the biological properties of mushrooms, including cardioprotective properties, is β-glucan. It decreases cholesterol absorption and increases the excretion of bile acids, and thus, helps in the prevention and treatment of cardiovascular disorders (Khan et al. 2018). Hypertension (due to increased blood pressure) is a common clinical condition affecting the heart. As mushrooms comprise low amounts of sodium and high potassium concentration, they can be viewed as the right dietary supplement to combat hypertension. One of the mechanisms shown by mushrooms is attributed to their antihypertensive properties is ACE (angiotensin I-converting enzyme) inhibition as it converts angiotensin-I to angiotensin-II, which has potent hypertensive properties (Shibu et al. 2017).

Bhushan and Kulshreshtha (2019) assessed the cardioprotective activity of hydroalcoholic extract of *Agaricus bisporus* against isoproterenol-induced myocardial infarction in Wistar albino rats and found positive results. Sathishkumar et al. (2020) further explored the cholesterol esterase inhibitory potential of the purified phenolic content of *Agaricus bisporus* and observed remarkable results, further suggesting the cardioprotective feature of *Agaricus bisporus*. Another mushroom species that possess significant cardioprotective properties is *Ganoderma lucidum*. It reduces blood cholesterol, blood pressure, and triglyceride levels (Ahmad 2018). Shaher et al. (2020) examined the hyperglycemia-mediated cardiomyopathy protection of *Ganoderma lucidum* spores in streptozotocin (STZ)-induced diabetic rats. The treatment of *G. lucidum* spores has reduced cardiomyopathy by decreasing hyperglycemia, apoptosis, inflammation, and oxidative stress.

Similarly, Liu et al. (2021) analyzed the cardioprotective activity of extracts from *G. lucidum* spores in trimethylamine-N-oxide-induced cardiac dysfunction in rats. They illustrated the importance of polysaccharides and lipophilic components present in extracts in the functioning of the heart and in reducing the risk of cardiovascular diseases by regulating the expression of proteins involved in causing disorders and affecting the gut microbiota. All these findings reinforce the utilization of mushrooms in managing cardiovascular diseases, but much more clinical confirmation is still required for further proceedings.

3.8 Anti-Obesity Activity

Obesity is a metabolic syndrome caused by various reasons such as diet, heredity, lifestyle, and environment. It reflects an inequity between energy uptake and expenditure, ultimately leading to body fat storage. Excessive storage of fats in the adipose tissue is the major reason for metabolic disorders such as diabetes, cardiovascular diseases, pulmonary diseases, and many others, which can be fatal to life. Diet, exercise, and sometimes surgical interventions can help manage obesity. However, the drugs used for their treatment cause many complications such as insomnia, headache, constipation, etc. (Ganesan and Xu 2018). Numerous in vitro and in vivo studies have been performed to explore the anti-obesity potential of polysaccharides from mushrooms. Kanwal et al. (2020) investigated the effect of polysaccharides from *Dictyophora indusiata* in a high-fat diet-induced obesity mice model. They observed that it significantly reversed the obesity parameters, indicating its high therapeutic value. Nagaraj et al. (2021) further studied the anti-obesity effect of *Calocybe indica* using the diet-induced obese Zebrafish model, and it has been seen to reduce the levels of cholesterol and triglycerides and also lowered the fat accumulation in the liver, hence showing a beneficial effect on obesity. In addition, Babac et al. (2021) assessed the effects of ethanolic extract of mycelia of *Lentinus strigosus* on the intake of food and locomotion of *Caenorhabditis elegans*. The outcomes have depicted that the extract considerably reduced the food intake and increased the roaming activities of *C. elegans*. Therefore, it is concluded that the mycelium of *L. strigosus* has obesity management properties. Although studies have suggested that the intake of mushrooms reduces obesity, this practice must be combined with physical exercise and lifestyle modifications for better results.

3.9 Neuroprotective Activity

Oxidative stress is responsible for various neurodegenerative disorders like dementia, Alzheimer's disease, etc. The antioxidant potential of mushrooms helps reduce stress in neurons, hence contributing to neuroprotection. Bioactive compounds in mushrooms like erinacines, dictyophorines, hericenones, and scabronines hinder the formation of beta-amyloid and acetylcholinesterase, which, therefore, help in protection from neurodegeneration. Some mushrooms commonly used to treat neural disorders include *Ganoderma, Antrodia, Hericium, Lignosus, Pleurotus,* and *Grifola* (Anusiya et al. 2021).

Sun et al. (2017) evaluated the role of polysaccharides of *Ganoderma lucidum* in inhibiting neuronal apoptosis, further suggesting their importance as neuroprotective components. Similarly, Lew et al. (2020) studied the neuroprotective effects of aqueous extract of *Hericium erinaceus* against a depression mimicking cellular model and observed appreciable effects, highlighting its role as a potent antidepressant. Sillapachaiyaporn et al. (2021) further analyzed the neuroprotective property of *Auricularia polytricha* mushroom extracts against glutamate-induced HT-22 neural damage by assessing the cytotoxicity and accumulation of reactive oxygen species. They found that the extracts reduced the cytotoxicity and ROS accumulation, indicating their role as neuroprotective agents. In addition, Kittimongkolsuk et al. (2021) examined the neuroprotective effects of ethanol, cold and hot water extracts of *Lignosus rhinocerus* against glutamate-induced oxidative stress in mouse hippocampal cells in vitro and in the case of *Caenorhabditis elegans* in vivo. Amongst these extracts, only ethanol extracts have shown neuroprotective effects. Hence, compounds in ethanol extracts of *L. rhinocerus* may be used as neuroprotectants in clinical studies. Also, Sevindik et al. (2021) investigated the neuroprotective activity of *Octaviania asterosperma* and concluded that it could be considered a functional food for the treatment of neurodegenerative disorders owing to its phenolic content and antioxidant properties. More studies need to be conducted in the future to illustrate the clear mechanism of action behind the neuroprotective effects of mushrooms. Also, further clinical trials are required for authenticated results (Rai et al. 2021).

3.10 Gut Microbiota Enhancement Activity

The human gut contains a diverse microflora that contributes to a healthy digestive system, modulates the immune system, and thus influences a person's overall health (Jayachandran et al. 2017). Apart from possessing various medicinal properties, edible mushrooms can also act as prebiotics and enhance the gut microbiota. Commonly described mushrooms that enhance the gut microflora are *H. erinaceus, G. frondosa, G. lucidum,* and *L. edodes*. Gut microbiota considerably changes in older ages. Hence, mushrooms can restore the gut microbial balance by increasing beneficial bacteria and decreasing harmful microbes (Li et al. 2021). The polysaccharides present in mushrooms contribute a lot to their gut-enhancing properties. Kanwal et al. (2018) examined the polysaccharides from *Dictyophora indusiata* for their gut microflora enhancement properties. They found that it efficiently restored the gut microbiota and enhanced the diversity of beneficial microbes deteriorated by antibiotic administration.

Similarly, Li et al. (2019a) explored the potential effects of polysaccharides from an edible mushroom *Grifola frondosa* on gut microbiota dysbiosis and lipid metabolic syndromes. Results showed that polysaccharides favorably molded the gut microflora and antihyperlipidemic activities. Sun et al. (2019) reported that the water-insoluble polysaccharide isolated from the sclerotium of *Poria cocos* improved hyperglycemia and hyperlipidemia by modulating gut microbes. In addition, Su et al. (2019) characterized the polysaccharides from *Flammulina velutipes* and observed the modulation of gut microflora. They significantly increased the levels of *Bacteroidaceae* and *Bifidobacteriaceae* and reduced *Enterococcaceae* and *Lachnospiraceae* species.

Mitsou et al. (2020) studied the effects of mushrooms rich in β-glucans on aging gut microbiota. They have been found to exert beneficial results, further suggesting β-glucans as potent prebiotic agents. Additionally, Diling et al. (2020) observed that the administration of alcohol extract of *Ganoderma lucidum* in high fat and sugar-induced obese mice reversed the damage to gut microflora by reducing harmful microbes and also stimulated the growth of beneficial microbiota. Furthermore, Hu et al. (2022) investigated the effect of *Pleurotus ostreatus* in obese mice to analyze its role in gut modulation. It positively molded the gut microflora and helped reduce obesity. Future studies need to be done to gain more understanding of mushrooms' gut microflora-enhancing properties and to find out the mechanisms of action between gut microbiota and other organs. Also, mushroom product commercialization must be carried out for their utilization as prebiotics.

4 Conclusion

Several edible and medicinal mushroom species are a source of biologically active compounds and dietary fiber. This chapter portrays the bioactive profile of mushrooms, various extraction procedures, and the pharmaceutical potential of bioactive metabolites. Mushrooms are rich sources of β-glucans, phenolics, unsaturated fatty acids, lectins, flavonoids, Vitamin D, B-complex vitamins, terpenes, peptides, inulin, and dietary fibers. These bioconstituents exhibit many pharmacological actions such as antidiabetic, cytotoxic, immunomodulatory, antioxidant, anticholesterolemic, antiviral, antimicrobial, neuroprotective, antihyperlipidemic, and hypotensive. Hence, the inclusion of mushrooms may provide an everlasting anecdotal advantageous outcome in preventing ongoing life-threatening pathologies. In this regard, mushroom-based formulations/supplementation may further strengthen the mushroom industry in increasing the overall efficacy of delivering health benefits as a functional food. Further analyses of the underlying molecular mechanism via cellular and experimental models would hopefully add scientific-based knowledge to the future development of mushrooms as functional food with nutritional and pharmaceutical aspects.

References

Abdelshafy AM, Belwal T, Liang Z et al (2022) A comprehensive review on phenolic compounds from edible mushrooms: occurrence, biological activity, application and future prospective. Crit Rev Food Sci Nutr 62:6204–6224. https://doi.org/10.1080/10408398.2021.1898335

Ahmad MF (2018) *Ganoderma lucidum*: persuasive biologically active constituents and their health endorsement. Biomed Pharmacother 107:507–519. https://doi.org/10.1016/j.biopha.2018.08.036

Alkin M, Söğüt E, Seydim AC (2021) Determination of bioactive properties of different edible mushrooms from Turkey. Food Measure 15:3608–3617. https://doi.org/10.1007/s11694-021-00941-7

Alves Filho EG, Lima M, Silva L et al (2021) Green ultrasound-assisted extraction of bioactive compounds from button mushrooms, potatoes, and onion peels. ACS Food Sci Technol 1:1274–1284. https://doi.org/10.1021/acsfoodscitech.1c00153

Alzand KI, Bofaris MS, Ugis A (2019) Chemical composition and nutritional value of edible wild growing mushrooms: a review. World J Pharm Res 8:31–46. https://doi.org/10.20959/wjpr20193-14261

Amrein K, Scherkl M, Hoffmann M et al (2020) Vitamin D deficiency 2.0: an update on the current status worldwide. Eur J Clin Nutr 74:1498–1513. https://doi.org/10.1038/s41430-020-0558-y

Anusiya G, Gowthama Prabu U, Yamini NV et al (2021) A review of the therapeutic and biological effects of edible and wild mushrooms. Bioengineered 12:11239–11268. https://doi.org/10.1080/21655979.2021.2001183

Assemie A, Abaya G (2022) The effect of edible mushrooms on health and their biochemistry. Int J Microbiol 2022:1–7. https://doi.org/10.1155/2022/8744788

Babac BD, Dulay RM, Calpito RA et al (2021) In-vitro activity of ethanolic extract of *Lentinus strigosus* mycelia in N2 wild strain *Caenorhabditis elegans* – an animal model for obesity and its chemical composition. J Appl Biol Biotechnol 9:41–46. https://doi.org/10.7324/JABB.2021.9106

Bach F, Zielinski AAF, Helm CV et al (2019) Bio compounds of edible mushrooms: in vitro antioxidant and antimicrobial activities. LWT 107:214–220. https://doi.org/10.1016/j.lwt.2019.03.017

Badalyan SM, Barkhudaryan A, Rapior S (2022) Medicinal macrofungi as cosmeceuticals: a review. Int J Med Mushrooms 24:1–13. https://doi.org/10.1615/IntJMedMushrooms.2022043124

Baeva E, Bleha R, Lavrova E et al (2019) Polysaccharides from basidiocarps of cultivating mushroom *Pleurotus ostreatus*: isolation and structural characterization. Molecules 24:2740. https://doi.org/10.3390/molecules24152740

Bahadori MB, Sarikurkcu C, Yalcin OU, Cengiz M, Gungor H (2019) Metal concentration, phenolics profiling, and antioxidant activity of two wild edible *Melanoleuca* mushrooms (*M. cognata* and *M. stridula*). Microchem J 150:1–6. https://doi.org/10.1016/j.microc.2019.104172

Bamigboye CO, Oloke JK, Adebayo EA, Omomowo IO (2022) Mushrooms in Nigeria: neglected but beneficial nutraceutical and functional food. Pan Afr J Life Sci 6:395–340. https://doi.org/10.36108/pajols/2202/60.0140

Bhushan A, Kulshreshtha M (2019) Cardioprotective activity of *Agaricus bisporus* against isoproterenol-induced myocardial infarction in laboratory animals. Curr Nutr Food Sci 15:401–417. https://doi.org/10.2174/1573401314666180427161119

Blagodatski A, Yatsunskaya M, Mikhailova V et al (2018) Medicinal mushrooms as an attractive new source of natural compounds for future cancer therapy. Oncotarget 9:29259–29274. https://doi.org/10.18632/oncotarget.25660

Çayan F, Deveci E, Tel-Çayan G, Duru ME (2020) Identification and quantification of phenolic acid compounds of twenty-six mushrooms by HPLC–DAD. Food Measure 14:1690–1698. https://doi.org/10.1007/s11694-020-00417-0

Çayan F, Tel-Çayan G, Deveci E, Duru ME (2021) A comprehensive study on phenolic compounds and bioactive properties of five mushroom species via chemometric approach. J Food Process Preserv 45:1–31. https://doi.org/10.1111/jfpp.15695

Cerletti C, Esposito S, Iacoviello L (2021) Edible mushrooms and beta-glucans: impact on human health. Nutrients 13:2195. https://doi.org/10.3390/nu13072195

Chakraborty N, Banerjee A, Sarkar A, Ghosh S, Acharya K (2021) Mushroom polysaccharides: a potent immune-modulator. Biointerface Res Appl Chem 11:8915–8930. https://doi.org/10.33263/BRIAC112.89158930

Chen L, Deng H, Cui H et al (2017) Inflammatory responses and inflammation-associated diseases in organs. Oncotarget 9:7204–7218. https://doi.org/10.18632/oncotarget.23208

Chen S, Liu C, Huang X et al (2020) Comparison of immunomodulatory effects of three polysaccharide fractions from *Lentinula edodes* water extracts. J Funct Foods 66:1–6. https://doi.org/10.1016/j.jff.2020.103791

Chugh RM, Mittal P, Namratha MP et al (2022) Fungal mushrooms: a natural compound with therapeutic applications. Front Pharmacol 13:1–16. https://doi.org/10.3389/fphar.2022.925387

Ciric A, Kruljevic I, Stojkovic D et al (2019) Comparative investigation on edible mushrooms *Macrolepiota mastoidea*, *M. rhacodes* and *M. procera*: functional foods with diverse biological activities. Food Funct 10:7678–7686. https://doi.org/10.1039/C9FO01900F

Contato AG, Inácio FD, de Araujo CA et al (2020) Comparison between the aqueous extracts of mycelium and basidioma of the edible mushroom *Pleurotus pulmonarius*: chemical composition and antioxidant analysis. Food Measure 14:830–837. https://doi.org/10.1039/C9FO01900F

Cox-Georgian D, Ramadoss N, Dona C, Basu C (2019) Therapeutic and medicinal uses of terpenes. In: Joshee N, Dhekney S, Parajuli P (eds) Medicinal plants. Springer, Cham, pp 333–359

Darmasiwi S, Aramsirirujiwet Y, Kimkong I (2022) Antibiofilm activity and bioactive phenolic compounds of ethanol extract from the *Hericium erinaceus* basidiome. J Adv Pharm Technol Res 13:111–116. https://doi.org/10.4103/japtr.japtr_1_22

Das AK, Nanda PK, Dandapat P et al (2021) Edible mushrooms as functional ingredients for development of healthier and more sustainable muscle foods: a flexitarian approach. Molecules 26:2463–2488. https://doi.org/10.3390/molecules26092463

Dasgupta A, Acharya K (2019) Mushrooms: an emerging resource for therapeutic terpenoids. 3 Biotech 9:369. https://doi.org/10.1007/s13205-019-1906-2

Datta HK, Das D, Koschella A et al (2019) Structural elucidation of a heteropolysaccharide from the wild mushroom *Marasmiellus palmivorus* and its immune-assisted anticancer activity. Carbohydr Polym 211:272–280. https://doi.org/10.1016/j.carbpol.2019.02.011

de Menezes Filho AC, Ventura MV, Alves I et al (2022) Phytochemical prospection, total flavonoids and total phenolic and antioxidant activity of the mushroom extract *Scleroderma verrucosum* (Bull.) Pers. Braz J Sci 1:1–4. https://doi.org/10.1016/j.carbpol.2019.02.011

Deveci E, Çayan F, Tel-Çayan G, Duru ME (2021) Inhibitory activities of medicinal mushrooms on α-amylase and α-glucosidase-enzymes related to type 2 diabetes. S Afr J Bot 137:19–23. https://doi.org/10.1016/j.sajb.2020.09.039

Diling C, Yinrui G, Longkai Q et al (2020) Metabolic regulation of *Ganoderma lucidum* extracts in high sugar and fat diet-induced obese mice by regulating the gut-brain axis. J Funct Foods 65:1–22. https://doi.org/10.1016/j.jff.2019.103639

Dixon A, Elyaguov J, Choudhury M, Konno S (2022) Anticancer effect of medicinal mushroom extract on renal cell carcinoma: alternative therapeutic implication. World J Nephrol Urol 11:1–9. https://doi.org/10.14740/wjnu433

Dou H, Chang Y, Zhang L (2019) *Coriolus versicolor* polysaccharopeptide as an immunotherapeutic in China. Prog Mol Biol Transl Sci 163:361–381. https://doi.org/10.1016/bs.pmbts.2019.03.001

Du B, Peng F, Xie Y, Wang H, Wu J, Liu C, Yang Y (2022) Optimization extraction and antioxidant activity of crude polysaccharide from chestnut mushroom (*Agrocybe aegerita*) by accelerated solvent extraction combined with response surface methodology (ASE-RSM). Molecules 27:2380. https://doi.org/10.3390/molecules27082380

Ekowati N, Yuniati NI, Hernayanti H, Ratnaningtyas NI (2018) Antidiabetic potentials of button mushroom (*Agaricus bisporus*) on alloxan-induced diabetic rats. Biosaintifika 10:655–662. https://doi.org/10.15294/biosaintifika.v10i3.17126

Elhusseiny SM, El-Mahdy TS, Elleboudy NS et al (2022) Immunomodulatory activity of extracts from five edible basidiomycetes mushrooms in Wistar albino rats. Sci Rep 12:1–11. https://doi.org/10.1038/s41598-022-16349-2

Elkhateeb WA, Daba GM, Thomas PW, Wen TC (2019) Medicinal mushrooms as a new source of natural therapeutic bioactive compounds. Egypt Pharm J 18:88–101

Fukushima-Sakuno E (2020) Bioactive small secondary metabolites from the mushrooms *Lentinula edodes* and *Flammulina velutipes*. J Antibiot 73:687–696. https://doi.org/10.1038/s41429-020-0354-x

Ganesan K, Xu B (2018) Anti-obesity effects of medicinal and edible mushrooms. Molecules 23: 2880–2906. https://doi.org/10.3390/molecules23112880

Garcia J, Rodrigues F, Saavedra MJ et al (2022) Bioactive polysaccharides from medicinal mushrooms: a review on their isolation, structural characteristics and antitumor activity. Food Biosci 14:101955. https://doi.org/10.1016/j.fbio.2022.101955

Garrab M, Edziri H, El Mokni R et al (2019) Phenolic composition, antioxidant and anticholinesterase properties of the three mushrooms *Agaricus silvaticus* Schaeff., *Hydnum rufescens* Pers. and *Meripilus giganteus* (Pers.) Karst. in Tunisia. S Afr J Bot 124:359–363. https://doi.org/10.1016/j.sajb.2019.05.033

Ghosh SK, Sanyal T, Bera T (2020) Anticancer activity of solvent extracts of *Hexogonia glabra* against cervical cancer cell lines. Asian Pac J Cancer Prev 21:1977–1986. https://doi.org/10.31557/APJCP.2020.21.7.1977

Gong P, Wang S, Liu M et al (2020) Extraction methods, chemical characterizations and biological activities of mushroom polysaccharides: a mini-review. Carbohydr Res 494:108037. https://doi.org/10.1016/j.carres.2020.108037

González A, Cruz M, Losoya C et al (2020) Edible mushrooms as a novel protein source for functional foods. Food Funct 11:7400–7414

Heleno SA, Diz P, Prieto MA et al (2016) Optimization of ultrasound-assisted extraction to obtain mycosterols from *Agaricus bisporus* L. by response surface methodology and comparison with conventional Soxhlet extraction. Food Chem 197:1054–1063. https://doi.org/10.1016/j.foodchem.2015.11.108

Hossen SM, Hossain MS, Yusuf AT et al (2022) Profiling of phytochemical and antioxidant activity of wild mushrooms: evidence from the in vitro study and phytoconstituent's binding affinity to the human erythrocyte catalase and human glutathione reductase. Food Sci Nutr 10:88–102. https://doi.org/10.1002/fsn3.2650

Hou R, Zhou L, Fu Y et al (2021) Chemical characterization of two fractions from *Sanghuangporus sanghuang* and evaluation of antidiabetic activity. J Funct Foods 87:1–10. https://doi.org/10.1016/j.jff.2021.104825

Hu T, Huang Q, Wong K, Yang H (2017) Structure, molecular conformation, and immunomodulatory activity of four polysaccharide fractions from *Lignosus rhinocerotis* sclerotia. Int J Biol Macromol 94:423–430. https://doi.org/10.1016/j.ijbiomac.2016.10.051

Hu Y, Xu J, Sheng Y et al (2022) *Pleurotus Ostreatus* ameliorates obesity by modulating the gut microbiota in obese mice induced by high-fat diet. Nutrients 14:1868–1886. https://doi.org/10.3390/nu14091868

Jacinto-Azevedo B, Valderrama N, Henríquez K, Aranda M, Aqueveque P (2021) Nutritional value and biological properties of Chilean wild and commercial edible mushrooms. Food Chem 356: 129651–129659. https://doi.org/10.1016/j.foodchem.2021.129651

Jayachandran M, Xiao J, Xu B (2017) A critical review on health promoting benefits of edible mushrooms through gut microbiota. Int J Mol Sci 18:1934–1946. https://doi.org/10.3390/ijms18091934

Jeitler M, Michalsen A, Frings D et al (2020) Significance of medicinal mushrooms in integrative oncology: a narrative review. Front Pharmacol 11:580656. https://doi.org/10.3389/fphar.2020.580656

Jiang Z, Jin M, Zhou W et al (2018) Anti-inflammatory activity of chemical constituents isolated from the willow bracket medicinal mushroom *Phellinus Igniarius* (Agaricomycetes). Int J Med Mushrooms 20:119–128. https://doi.org/10.1615/IntJMedMushrooms.201802553

Joradon P et al (2022) A comparative study of conventional and supercritical carbon dioxide extraction methods for the recovery of bioactive compound from Lion's Mane mushroom (*Hericium erinaceus*). In: E3S web of conferences, vol 355. EDP Sciences, Les Ulis, p 02015

Kaewnarin K, Suwannarach N, Kumla J et al (2020) Characterization of polysaccharides from wild edible mushrooms from Thailand and their antioxidant, antidiabetic, and antihypertensive activities. Int J Med Mushrooms 22:221–233. https://doi.org/10.1615/IntJMedMushrooms.2020034092

Kanwal S, Joseph TP, Owusu L et al (2018) A polysaccharide isolated from *Dictyophora indusiata* promotes recovery from antibiotic-driven intestinal dysbiosis and improves gut epithelial barrier function in a mouse model. Nutrients 10(8):1003–1027. https://doi.org/10.3390/nu10081003

Kanwal S, Aliya S, Xin Y (2020) Anti-obesity effect of *Dictyophora indusiata* mushroom polysaccharide (DIP) in high fat diet-induced obesity via regulating inflammatory cascades and intestinal microbiome. Front Endocrinol 11:1–18. https://doi.org/10.3389/fendo.2020.558874

Kaprasob R, Khongdetch J, Laohakunjit N et al (2022) Isolation and characterization, antioxidant, and antihypertensive activity of novel bioactive peptides derived from hydrolysis of King Boletus mushroom. LWT 160:1–11. https://doi.org/10.1016/j.lwt.2022.113287

Khalilov Q, Numonov S, Sukhrobov P et al (2022) New triterpenoids from the fruiting bodies of *Laetiporus sulphureus* and their anti-inflammatory activity. ACS Omega 7:27272–27277. https://doi.org/10.1021/acsomega.2c02165

Khan AA, Gani A, Khanday FA, Masoodi FA (2018) Biological and pharmaceutical activities of mushroom β-glucan discussed as a potential functional food ingredient. Bioact Carbohydr Dietary Fibre 16:1–3. https://doi.org/10.1016/j.bcdf.2017.12.002

Khursheed R, Singh SK, Wadhwa S, Gulati M, Awasthi A (2020) Therapeutic potential of mushrooms in diabetes mellitus: role of polysaccharides. Int J Biol Macromol 164:1194–1205. https://doi.org/10.1016/j.ijbiomac.2020.07.145

Kiddane AT, Kim GD (2021) Anticancer and immunomodulatory effects of polysaccharides. Nutr Cancer 73:2219–2231. https://doi.org/10.1080/01635581.2020.1861310

Kittimongkolsuk P, Pattarachotanant N, Chuchawankul S, Wink M, Tencomnao T (2021) Neuroprotective effects of extracts from tiger milk mushroom *Lignosus rhinocerus* against glutamate-induced toxicity in ht22 hippocampal neuronal cells and neurodegenerative diseases in *Caenorhabditis elegans*. Biology 10:30–46. https://doi.org/10.3390/biology10010030

Kolniak-Ostek J, Oszmiański J, Szyjka A, Moreira H, Barg E (2022) Anticancer and antioxidant activities in *Ganoderma lucidum* wild mushrooms in Poland, as well as their phenolic and triterpenoid compounds. Int J Mol Sci 23:9359–9374. https://doi.org/10.3390/ijms23169359

Krishnamoorthi R, Srinivash M, Mahalingam PU, Malaikozhundan B (2022) Dietary nutrients in edible mushroom, *Agaricus bisporus* and their radical scavenging, antibacterial, and antifungal effects. Process Biochem 121:10–17. https://doi.org/10.1016/j.procbio.2022.06.021

Kruzselyi D, Moricz AM, Vetter J (2020) Comparison of different morphological mushroom parts based on the antioxidant activity. LWT 127:1–5. https://doi.org/10.1016/j.lwt.2020.109436

Leung MF, Cheung PC (2021) Vitamins D and D2 in cultivated mushrooms under ultraviolet irradiation and their bioavailability in humans: a mini-review. Int J Med Mushrooms 23:1–15. https://doi.org/10.1615/IntJMedMushrooms.2021040390

Lew SY, Lim SH, Lim LW, Wong KH (2020) Neuroprotective effects of *Hericium erinaceus* (Bull.: Fr.) Pers. against high-dose corticosterone-induced oxidative stress in PC-12 cells. BMC Complement Altern Med 20:1–6. https://doi.org/10.1186/s12906-020-03132-x

Li Q, Zhang F, Chen G, Chen Y et al (2018) Purification, characterization and immunomodulatory activity of a novel polysaccharide from *Grifola frondosa*. Int J Biol Macromol 111:1293–1303. https://doi.org/10.1016/j.ijbiomac.2018.01.090

Li L, Guo WL, Zhang W et al (2019a) *Grifola frondosa* polysaccharides ameliorate lipid metabolic disorders and gut microbiota dysbiosis in high-fat diet fed rats. Food Funct 10:2560–2572. https://doi.org/10.1039/C9FO00075E

Li L, Yang X, Pan L, Su Y, Wang Y (2019b) Comparing three methods of extraction of *Auricularia auricula* polysaccharides. Curr Top Nutraceutical Res 17:7–10. https://doi.org/10.1016/j.foodchem.2015.11.108

Li Q, Ye T, Long T, Peng X (2019c) Ginkgetin exerts anti-inflammatory effects on cerebral ischemia/reperfusion-induced injury in a rat model via the TLR4/NF-κB signaling pathway. Biosci Biotechnol Biochem 83:675–683. https://doi.org/10.1080/09168451.2018.1553608

Li H, Mi Y, Duan Z, Ma P, Fan D (2020) Structural characterization and immunomodulatory activity of a polysaccharide from *Eurotium cristatum*. Int J Biol Macromol 162:609–617. https://doi.org/10.1016/j.ijbiomac.2020.06.099

Li M, Yu L, Zhao J et al (2021) Role of dietary edible mushrooms in the modulation of gut microbiota. J Funct Foods 83:1–12. https://doi.org/10.1016/j.jff.2021.104538

Liu Y, Tang T, Duan S et al (2020) The purification, structural characterization and antidiabetic activity of a polysaccharide from *Anoectochilus roxburghii*. Food Funct 11:3730–3740. https://doi.org/10.1039/C9FO00860H

Liu Y, Lai G, Guo Y et al (2021) Protective effect of *Ganoderma lucidum* spore extract in trimethylamine-N-oxide-induced cardiac dysfunction in rats. J Food Sci 86:546–562. https://doi.org/10.1111/1750-3841.15575

Maity P, Sen IK, Chakraborty I et al (2021) Biologically active polysaccharide from edible mushrooms: a review. Int J Biol Macromol 172:408–417. https://doi.org/10.1016/j.ijbiomac.2021.01.081

Martel J, Ko YF, Ojcius DM et al (2017) Immunomodulatory properties of plants and mushrooms. Trends Pharmacol Sci 38:967–981. https://doi.org/10.1016/j.tips.2017.07.006

Martinez-Medina GA, Chávez-González ML, Verma DK et al (2021) Bio-functional components in mushrooms, a health opportunity: ergothionine and huitlacohe as recent trends. J Funct Foods 77:104326. https://doi.org/10.1016/j.jff.2020.104326

Mishra V, Tomar S, Yadav P, Singh MP (2021) Promising anticancer activity of polysaccharides and other macromolecules derived from oyster mushroom (*Pleurotus* sp.): an updated review. Int J Biol Macromol 182:1628–1637. https://doi.org/10.1016/j.ijbiomac.2021.05.102

Mitsou EK, Saxami G, Stamoulou E et al (2020) Effects of rich in β-glucans edible mushrooms on aging gut microbiota characteristics: an in vitro study. Molecules 25:2806–2831. https://doi.org/10.3390/molecules25122806

Morales D, Gil-Ramirez A, Smiderle FR et al (2017) Vitamin D-enriched extracts obtained from shiitake mushrooms (*Lentinula edodes*) by supercritical fluid extraction and UV-irradiation. Innov Food Sci Emerg Technol 41:330–336. https://doi.org/10.1016/j.ifset.2017.04.008

Motta F, Gershwin ME, Selmi C (2021) Mushrooms and immunity. J Autoimmun 117:102576. https://doi.org/10.1016/j.jaut.2020.102576

Moumita S, Das B (2022) Assessment of the prebiotic potential and bioactive components of common edible mushrooms in India and formulation of synbiotic microcapsules. LWT 156: 113050. https://doi.org/10.1016/j.lwt.2021.113050

Muszynska B, Grzywacz-Kisielewska A, Kała K, Gdula-Argasinska J (2018) Anti-inflammatory properties of edible mushrooms: a review. Food Chem 243:373–381. https://doi.org/10.1016/j.foodchem.2017.09.149

Nagaraj A, Wilson SA, Vaidyanathan L (2021) Anti-obesity properties of *Calocybe Indica* in zebra fishes with short-term high-fat diet induction. Biomed Pharmacol J 14:411–423. https://doi.org/10.13005/bpj/2141

Neergheen VS, Kam AH, Pem Y, Ramsaha S, Bahorun T (2020) Regulation of cancer cell signaling pathways as key events for therapeutic relevance of edible and medicinal mushrooms. Semin Cancer Biol 1–46:145. https://doi.org/10.1016/j.semcancer.2020.03.004

Niego AG, Rapior S, Thongklang N et al (2021) Macrofungi as a nutraceutical source: promising bioactive compounds and market value. J Fungi 7:397. https://doi.org/10.3390/jof7050397

Nowak R, Nowacka-Jechalke N, Pietrzak W, Gawlik-Dziki U (2022) A new look at edible and medicinal mushrooms as a source of ergosterol and ergosterol peroxide-UHPLC-MS/MS analysis. Food Chem 369:1–9. https://doi.org/10.1016/j.foodchem.2021.130927

Ogidi CO, Oyetayo VO, Akinyele BJ (2020) Wild medicinal mushrooms: potential applications in phytomedicine and functional foods. In: Passari AK, Sánchez S (eds) An introduction to mushroom. IntechOpen, London, pp 118–126

Pan H, Zhao X, Lei S et al (2019) The immunomodulatory activity of polysaccharides from the medicinal mushroom *Amauroderma rude* (Agaricomycetes) is mediated via the iNOS and PLA2-AA pathways. Int J Med Mushrooms 21:793–804. https://doi.org/10.1615/IntJMedMushrooms.2019031543

Panda SK, Sahoo G, Swain SS, Luyten W (2022) Anticancer activities of mushrooms: a neglected source for drug discovery. Pharmaceuticals 15:176–201. https://doi.org/10.3390/ph15020176

Pandya U, Dhuldhaj U, Sahay NS (2019) Bioactive mushroom polysaccharides as antitumor: an overview. Nat Prod Res 33:2668–2680. https://doi.org/10.1080/14786419.2018.1466129

Patel DK, Dutta SD, Ganguly K, Cho SJ, Lim KT (2021) Mushroom-derived bioactive molecules as immunotherapeutic agents: a review. Molecules 26:1359. https://doi.org/10.3390/molecules26051359

Pathak MP, Pathak K, Saikia R et al (2022) Immunomodulatory effect of mushrooms and their bioactive compounds in cancer: a comprehensive review. Biomed Pharmacother 149:112901. https://doi.org/10.1016/j.biopha.2022.112901

Podkowa A, Kryczyk-Poprawa A, Opoka W, Muszyńska B (2021) Culinary–medicinal mushrooms: a review of organic compounds and bioelements with antioxidant activity. Eur Food Res Technol 247:513–533. https://doi.org/10.1007/s00217-020-03646-1

Prabu M, Kumuthakalavalli R (2017) Antidiabetic potential of the oyster mushroom *Pleurotus florida* (Mont.) Singer. Int J Curr Pharm Res 9:51–54. https://doi.org/10.22159/ijcpr.2017v9i4.20765

Radhika R, Sundar SK, Anuradha R (2021) Antioxidant property of fruiting bodies of *Ganoderma Lucidum*. Int J Mod Agric 10:3508–3514

Rai SN, Mishra D, Singh P, Vamanu E, Singh MP (2021) Therapeutic applications of mushrooms and their biomolecules along with a glimpse of in silico approach in neurodegenerative diseases. Biomed Pharmacother 137:1–14. https://doi.org/10.1016/j.biopha.2021.111377

Rašeta M, Popović M, Knežević P et al (2020) Bioactive phenolic compounds of two medicinal mushroom species *Trametes versicolor* and *Stereum subtomentosum* as antioxidant and antiproliferative agents. Chem Biodivers 17:1–11. https://doi.org/10.1002/cbdv.202000683

Ratnaningtyas NI, Hernayanti H, Ekowati N, Husen F (2022) Ethanol extract of the mushroom *Coprinus comatus* exhibits antidiabetic and antioxidant activities in streptozotocin-induced diabetic rats. Pharm Biol 60:1126–1136. https://doi.org/10.1080/13880209.2022.2074054

Rizzo G, Goggi S, Giampieri F, Baroni L (2021) A review of mushrooms in human nutrition and health. Trends Food Sci Technol 117:60–73. https://doi.org/10.1016/j.tifs.2020.12.025

Roman MP, Mantilla NB, Flórez SA et al (2020) Antimicrobial and antioxidant potential of wild edible mushrooms. In: Passari AK, Sánchez S (eds) An introduction to mushroom. IntechOpen, London, pp 1–8

Rowaiye A, Wilfred OI, Onuh OA et al (2022) Modulatory effects of mushrooms on the inflammatory signaling pathways and pro-inflammatory mediators. Clin Complement Med Pharmacol 2:1–11. https://doi.org/10.1016/j.ccmp.2022.100037

Ruksiriwanich W, Khantham C, Linsaenkart P et al (2022) Anti-inflammation of bioactive compounds from ethanolic extracts of edible bamboo mushroom (*Dictyophora indusiata*) as functional health promoting food ingredients. Int J Food Sci Technol 57:110–122. https://doi.org/10.1111/ijfs.15338

Saad HM, Sim KS, Tan YS (2018) Antimelanogenesis and anti-inflammatory activity of selected culinary-medicinal mushrooms. Int J Med Mushrooms 20:141–153. https://doi.org/10.1615/IntJMedMushrooms.2018025463

Saetang N, Rattanapot T, Manmai N et al (2022) Effect of hot water extraction process on schizophyllan from split gill mushroom. Biomass Conv Bioref 2022:1–10. https://doi.org/10.1007/s13399-021-02286-z

Sairi AM, Ismail SI, Sukor A et al (2020) Cytotoxicity and anticancer activity of *Donkioporiella mellea* on MRC5 (normal human lung) and A549 (human lung carcinoma) cells lines. Evid Based Complementary Altern Med 2020:1–10. https://doi.org/10.1155/2020/7415672

Sathishkumar T, Suriyakala V, Lakshmanabharathy M et al (2020) In vitro cholesterol esterase inhibitory activity of some purified phenolic acids from *Agaricus bisporus*: an investigation of cardioprotective properties. Jundishapur J Nat Pharm Prod 15:1–8. https://doi.org/10.5812/jjnpp.84839

Secme M, Kaygusuz O, Eroglu C et al (2018) Potential anticancer activity of the parasol mushroom, *Macrolepiota procera* (Agaricomycetes), against the A549 human lung cancer cell line. Int J Med Mushrooms 20:1075–1086. https://doi.org/10.1615/IntJMedMushrooms.2018028589

Sevindik M (2019) Wild edible mushroom *Cantharellus cibarius* as a natural antioxidant food. Turk J Agric For Sci Tech 7:1377–1381. https://doi.org/10.24925/turjaf.v7i9.1377-1381.2644

Sevindik M (2021) Phenolic content, antioxidant and antimicrobial potential of *Melanoleuca melaleuca* edible mushroom. J Anim Plant Sci 31:824–830

Sevindik M, Bal C (2021) Antioxidant, antimicrobial, and antiproliferative activities of wild mushroom, *Laeticutis cristata* (Agaricomycetes), from Turkey. Int J Med Mushrooms 23:85–90. https://doi.org/10.1615/IntJMedMushrooms.2021040415

Sevindik M, Akgul H, Selamoglu Z, Braidy N (2021) Antioxidant, antimicrobial and neuroprotective effects of *Octaviania asterosperma* in vitro. Mycology 12:128–138. https://doi.org/10.1080/21501203.2020.1816584

Sezgin S, Dalar A, Uzun Y (2019) Determination of antioxidant activities and chemical composition of sequential fractions of five edible mushrooms from Turkey. J Food Sci Technol 57:1866–1876. https://doi.org/10.1007/s13197-019-04221-7

Shaffique S, Kang SM, Kim AY et al (2021) Current knowledge of medicinal mushrooms related to anti-oxidant properties. Sustainability 13:7948–7964. https://doi.org/10.3390/su13147948

Shaher F, Wang S, Qiu H et al (2020) Effect and mechanism of *Ganoderma lucidum* spores on alleviation of diabetic cardiomyopathy in a pilot in vivo study. Diabetes Metab Syndr Obes 13:4809–4822. https://doi.org/10.2147/DMSO.S281527

Shapouri-Moghaddam A, Mohammadian S, Vazini H et al (2018) Macrophage plasticity, polarization, and function in health and disease. J Cell Physiol 233:6425–6440. https://doi.org/10.1002/jcp.26429

Shashikant M, Bains A, Chawla P et al (2022) In-vitro antimicrobial and anti-inflammatory activity of modified solvent evaporated ethanolic extract of *Calocybe indica*: GCMS and HPLC characterization. Int J Food Microbiol 376:109741. https://doi.org/10.1016/j.ijfoodmicro.2022.109741

Shen HS, Shao S, Chen JC, Zhou T (2017) Antimicrobials from mushrooms for assuring food safety. Compr Rev Food Sci Food Saf 16:316–329. https://doi.org/10.1111/1541-4337.12255

Sheng K, Wang C, Chen B et al (2021) Recent advances in polysaccharides from *Lentinus edodes* (Berk.): isolation, structures and bioactivities. Food Chem 358:1–14. https://doi.org/10.1016/j.foodchem.2021.129883

Sheng Z, Wen L, Yang B (2022) Structure identification of a polysaccharide in mushroom Lingzhi spore and its immunomodulatory activity. Carbohydr Polym 278:118939. https://doi.org/10.1016/j.carbpol.2021.118939

Shibu MA, Agrawal DC, Huang CY (2017) Mushrooms: a pandora box of cardioprotective phytochemicals. In: Agrawal DC et al (eds) Medicinal plants and fungi: recent advances in research and development. Springer, Singapore, pp 337–362

Sillapachaiyaporn C, Rangsinth P, Nilkhet S et al (2021) Neuroprotective effects against glutamate-induced HT-22 hippocampal cell damage and *Caenorhabditis elegans* lifespan/healthspan enhancing activity of *Auricularia polytricha* mushroom extracts. Pharmaceuticals 14:1001–1020. https://doi.org/10.3390/ph14101001

Singh V, Bedi GK, Shri R (2017) In vitro and in vivo antidiabetic evaluation of selected culinary-medicinal mushrooms (Agaricomycetes). Int J Med Mushrooms 19:17–25. https://doi.org/10.1615/IntJMedMushrooms.v19.i1.20

Smith A, Javed S, Barad A et al (2017) Growth-inhibitory and immunomodulatory activities of wild mushrooms from North-Central British Columbia (Canada). Int J Med Mushrooms 19:485–497. https://doi.org/10.1615/IntJMedMushrooms.v19.i6.10

Soliman ER, El-Sayed H (2021) Molecular identification and antimicrobial activities of some wild Egyptian mushrooms: *Bjerkandera adusta* as a promising source of bioactive antimicrobial phenolic compounds. J Genet Eng Biotechnol 19:1–11. https://doi.org/10.1186/s43141-021-00200-8

Sosalagere C, Kehinde BA, Sharma P (2022) Isolation and functionalities of bioactive peptides from fruits and vegetables: a reviews. Food Chem 366:1–24. https://doi.org/10.1016/j.foodchem.2021.130494

Souilem F, Fernandes A, Calhelha RC et al (2017) Wild mushrooms and their mycelia as sources of bioactive compounds: antioxidant, anti-inflammatory and cytotoxic properties. Food Chem 230:40–48. https://doi.org/10.1016/j.foodchem.2017.03.026

Stastny J, Marsik P, Tauchen J et al (2022) Antioxidant and anti-inflammatory activity of five medicinal mushrooms of the genus *Pleurotus*. Antioxidants 11:1569–1585. https://doi.org/10.3390/antiox11081569

Stojkovic D, Smiljkovic M, Ciric A et al (2019) An insight into antidiabetic properties of six medicinal and edible mushrooms: inhibition of α-amylase and α-glucosidase linked to type-2 diabetes. S Afr J Bot 120:100–103. https://doi.org/10.1016/j.sajb.2018.01.007

Su A, Ma G, Xie M et al (2019) Characteristic of polysaccharides from *Flammulina velutipes* in vitro digestion under salivary, simulated gastric and small intestinal conditions and fermentation by human gut microbiota. Int J Food Sci Technol 54:2277–2287. https://doi.org/10.1111/ijfs.14142

Suliaman SQ, Al-Abbasi SH, Mahmood YH, Al-Azzawi HA (2021) Antimicrobial activity of four selected wild mushrooms in Iraq. Biochem Cell Arch 21:4533–4537

Sun SS, Wang K, Ma K, Bao L, Liu HW (2019) An insoluble polysaccharide from the sclerotium of *Poria cocos* improves hyperglycemia, hyperlipidemia and hepatic steatosis in ob/ob mice via modulation of gut microbiota. Chin J Nat Med 17:3–14. https://doi.org/10.1016/S1875-5364(19)30003-2

Sun XZ, Liao Y, Li W, Guo LM (2017) Neuroprotective effects of *Ganoderma lucidum* polysaccharides against oxidative stress-induced neuronal apoptosis. Neural Regen Res 12:953–958. https://doi.org/10.4103/1673-5374.208590

Valverde ME, Hernández-Pérez T, Paredes-López O (2015) Edible mushrooms: improving human health and promoting quality life. Int J Microbiol 2015. https://doi.org/10.1155/2015/376387

Vazquez-Armenta FJ, Leyva JM, Mata-Haro V et al (2022) Phenolic compounds of *Phellinus* spp. with antibacterial and antiviral activities. Braz J Microbiol 53:1187–1197. https://doi.org/10.1007/s42770-022-00745-x

Venturella G, Ferraro V, Cirlincione F, Gargano ML (2021) Medicinal mushrooms: bioactive compounds, use, and clinical trials. Int J Mol Sci 22:634. https://doi.org/10.3390/ijms22020634

Wang Y, Tian Y, Shao J et al (2018) Macrophage immunomodulatory activity of the polysaccharide isolated from *Collybia radicata* mushroom. Int J Biol Macromol 108:300–306. https://doi.org/10.1016/j.ijbiomac.2017.12.025

Wang Y, Zhang Y, Shao J et al (2019) Potential immunomodulatory activities of a lectin from the mushroom *Latiporus sulphureus*. Int J Biol Macromol 130:399–406. https://doi.org/10.1016/j.ijbiomac.2019.02.150

Wongaem A, Reamtong O, Srimongkol P et al (2021) Antioxidant properties of peptides obtained from the split gill mushroom (*Schizophyllum commune*). J Food Sci Technol 58:680–691. https://doi.org/10.1007/s13197-020-04582-4

Wongrakpanich S, Wongrakpanich A, Melhado K, Rangaswami J (2018) A comprehensive review of non-steroidal anti-inflammatory drug use in the elderly. Aging Dis 9:143–150. https://doi.org/10.14336/AD.2017.0306

Xiaokang W, Lyng JG, Brunton NP et al (2020) Monitoring the effect of different microwave extraction parameters on the recovery of polyphenols from shiitake mushrooms: comparison with hot-water and organic-solvent extractions. Biotechnol Rep 27:1–9. https://doi.org/10.1016/j.btre.2020.e00504

Yadav D, Negi PS (2021) Bioactive components of mushrooms: processing effects and health benefits. Food Res Int 148:1–23. https://doi.org/10.1016/j.foodres.2021.110599

Yamac M, Ozturk SE, Canturk Z, Dikmen M (2022) Comparative analysis of antioxidant and anticancer activities of fruiting bodies, mycelial biomass and culture liquid of *Omphalotus olearius* OBCC 2503 (Agaricomycetes) from Turkey. Int J Med Mushrooms 24:61–72. https://doi.org/10.1615/IntJMedMushrooms.2022045033

Yamada S, Tanaka M, Miura R et al (2019) Anti-inflammatory and antimicrobial activities of aqueous extracts of wild mushrooms from Japan. Int J Med Mushrooms 21:469–486. https://doi.org/10.1615/IntJMedMushrooms.2019030637

Yan J, Zhu L, Qu Y et al (2019) Analyses of active antioxidant polysaccharides from four edible mushrooms. Int J Biol Macromol 123:945–956. https://doi.org/10.1016/j.ijbiomac.2018.11.079

Yang D, Zhou Z, Zhang L (2019) An overview of fungal glycan-based therapeutics. In: Zhang L (ed) Progress in molecular biology and translational science. Academic Press, New York, pp 135–163

Yin C, Noratto GD, Fan X et al (2020) The impact of mushroom polysaccharides on gut microbiota and its beneficial effects to host: a review. Carbohydr Polym 250:1–17. https://doi.org/10.1016/j.carbpol.2020.116942

Zhang WN, Gong LL, Zhou ZB et al (2022) Structural characterization and immunomodulatory activity of a mannan from *Helvella leucopus*. Int J Biol Macromol 212:495–507. https://doi.org/10.1016/j.ijbiomac.2022.05.132

Zhao S, Gao Q, Rong C et al (2020) Immunomodulatory effects of edible and medicinal mushrooms and their bioactive immunoregulatory products. J Fungus 6:269–307. https://doi.org/10.3390/jof6040269

Zhou J, Chen M, Wu S et al (2020) A review on mushroom-derived bioactive peptides: preparation and biological activities. Food Res Int 134:1–61. https://doi.org/10.1016/j.foodres.2020.109230

Zuo K, Tang K, Liang Y et al (2021) Purification and antioxidant and anti-inflammatory activity of extracellular polysaccharopeptide from sanghuang mushroom, *Sanghuangporus lonicericola*. J Sci Food Agric 101:1009–1020. https://doi.org/10.1002/jsfa.10709

Mycotherapeutics Reduce Nephrotoxicity and Renal Diseases

Rishi M. Nadar, Keyi Liu, Jack DeRuiter, Suhrud Pathak, Sindhu Ramesh, Timothy Moore, Dinesh Chandra Agrawal, and Muralikrishnan Dhanasekaran

Contents

R. M. Nadar · K. Liu · J. DeRuiter · S. Pathak · S. Ramesh · T. Moore · M. Dhanasekaran (✉)
Department of Drug Discovery and Development, Harrison College of Pharmacy, Auburn University, Auburn, AL, USA
e-mail: dhanamu@auburbn.edu

D. C. Agrawal (✉)
Department of Applied Chemistry, Chaoyang University of Technology, Taichung, Taiwan
e-mail: dcagrawal@gm.cyut.edu.tw

D. C. Agrawal, M. Dhanasekaran (eds.), *Mushrooms with Therapeutic Potentials*,
https://doi.org/10.1007/978-981-19-9550-7_10

Abstract Acute/chronic exposure to endogenous and exogenous toxins, aging, and comorbidities simultaneously increase the prevalence of nephronal dysfunction substantially. Mycotherapeutics are becoming progressively more popular as a viable alternative therapy for treating nephrotoxicity and aging. As a result, mycotherapeutics is being seriously validated by numerous worldwide natural bioactive authorities. Mycotherapeutics can be pharmaceutical and nutraceutical inventions to prevent and treat kidney illnesses since they have various innovative pharmacodynamic mechanisms and advantageous pharmacokinetic mechanisms to lessen nephrotoxicity without significant adverse effects and hypersensitivity reactions. This chapter focuses on the nephroprotective properties of mushrooms against nephrotoxins and aging.

Keywords Alternative natural bioactives · Kidney disease · Medicinal mushrooms · Nephrotoxicity · Nephroprotection · Renal damage

Abbreviations

apoA	Apolipoprotein A
apoB	Apolipoprotein B
Bax	BCL2-associated X protein
BCL-2	B-cell lymphoma-2
BMI	Body mass index
CCl_4	Carbon tetrachloride
CDC	Centers for disease control
CKD	Chronic kidney disease
CYP2E1	Cytochrome P450 family 2 subfamily E member 1
DKD	Diabetic kidney disease
EPO	Erythropoietin
GFR	Glomerular filtration rate
HDL	High-density lipoprotein
IgA	Immunoglobulin A
LDL	Low-density lipoprotein
MAPK	Mitogen-activated protein kinase
NF-κB	Nuclear factor kappa-light-chain-enhancer of activated B cells
NPDS	National poison data system
PAA	Poricoic acid
PMPS	Phosphorylated mycelia polysaccharides
TGF-β1	Transforming growth factor beta 1
Vitamin B-12	Cobalamin
VLDL	Very-low-density lipoprotein

1 Introduction

The prevalence of nephronal dysfunction drastically increases with aging, acute/chronic exposure to endogenous and exogenous toxins, and comorbidities. Nephrotoxicity can occur due to excessive alcohol (ethanol consumption), renal anemia, chronic diseases, fibrosis, hypercholesterolemia/hyperlipidemia, diet (high-fat diet), obesity, hyperglycemia, increased blood pressure, hyperuricemia, iatrogenesis/drug-induced, inflammation, oxidative stress, heavy metals, oxidative stress, senescence, and urolithiasis (kidney stones). Mycotherapies have been used as a conventional treatment for centuries. There is growing interest in using mushroom bioactives for treating the symptoms of central nervous system disorders such as anxiety, addiction, and major depressive disorder. The pharmacodynamic effect, combined with the safety profile of certain mushrooms, has increased the health beneficial interest (Lee et al. 2017a; Ugbogu et al. 2019; Tsai et al. 2016; Kornsakulkarn et al. 2020). Mycotherapeutics target the renal and immune systems implying a potential for the treatment of renal pathology. As a result, mycotherapeutics are being seriously validated by numerous worldwide natural bioactive authorities. The "Quadruplex Helix" concept (Carayannis and Campbell 2009) is associated with new and novel drug discovery and development, which bonds and integrates academia, nutraceutical/pharmaceutical companies (healthcare industry), strategies (policy), and customers/society (Fig. 1). This concept facilitates and validates authentication, analysis, bio-digging/bio-exploring, and the development of mycotherapeutics to slow down aging and reduce age-related disorders. This chapter reviews scientific studies relating to mycotherapeutic bioactives in treating renal diseases and the nephroprotective effects.

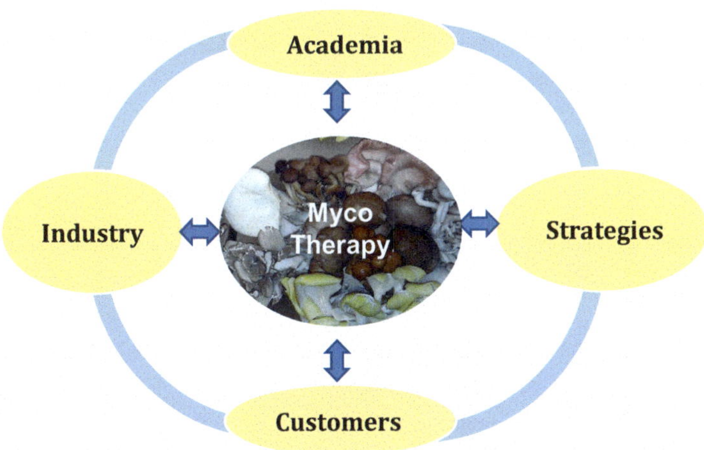

Fig. 1 "Quadruplex Helix" concept of new drug discovery and development

2 Factors that Induce Nephronal Dysfunction and Renal Damage

Acute and chronic kidney diseases can influence the ability of the human body capability in controlling acid-base, water balance, and electrolytes balance, remove toxins and waste products, regulate blood pressure, produce erythropoietin, and activate vitamin D. In renal pathological conditions, waste products and fluid can build up that can lead to systemic effects including shortness of breath, nausea, weakness, edema, sleep disorders and can be life-threatening. (Table 1).

The most common causes of renal damage are auto-immune renal disorders, dehydration, iatrogenic (drug-induced), urinary tract obstruction, and uncontrolled systemic cardiac and hepatic diseases. The other causes of renal diseases are due to other renal etiologies, including comorbidities (diabetes mellitus or lupus nephritis) and genetic disorders.

The most common approaches to treating renal diseases are dialysis, surgical procedures, kidney transplantation, and non-pharmacological and pharmacological approaches. Dialysis (synonym = hemodialysis, artificial kidney) is a process that uses an instrument to filter and remove waste products (extra salt) and excess fluid (water) from the blood when the kidneys do not function appropriately. However, this procedure has several adverse effects, and numerous patients believe a kidney transplant can provide a better quality of life than dialysis. The non-pharmacological

Table 1 Factors that induce nephronal dysfunction

Factors that induce nephronal dysfunction
Alcohol (excessive)
Anemia (renal anemia)
Chronic kidney disease
Endogenous nephrotoxin
Exogenous nephrotoxin
Fibrosis
Hypercholesterolemia/hyperlipidemia/high-fat diet
Hyperglycemia
Hypertension
Hyperuricemia
Iatrogenesis/drug-induced
Immune-based: IgA glomerulonephritis
Infection
Inflammation
Kidney cancer
Metal-induced (arsenic and chromium)
Nephritis
Oxidative stress
Pesticide (carbofuran)
Senescence
Urolithiasis (kidney stone)

approaches include watching the diet (appropriate BMI), exercise (routine physical activity in everyday life), proper sleep, quitting smoking and alcohol, and maintaining good mental health (coping with stress and depression in everyday life). The current pharmacological interventions are effective but can induce several adverse drug effects, contraindications, hypersensitivity reactions, and drugs to treat hypertension, hyperglycemia, and hypercholesteremia. Therefore, mycotherapeutics can be a novel alternative prophylactic and therapeutic avenue to prevent nephrotoxicity and protect the kidney from the physiological process of aging and various forms of renal damage and disease states.

3 Nephroprotective Effects of Mushrooms

Mycotherapeutics have demonstrated significant nephroprotective effects by exhibiting novel pharmacodynamic action, as detailed in the following sections.

3.1 Nephroprotective Effects of Mycotherapeutics Against Excessive Alcohol (Ethanol Consumption)-Induced Nephrotoxicity and Renal Damage

The Centers for Disease Control (CDC) has documented that a substantial number of American adults (nearly 70%) consume ethanol (alcohol) (CfDCa 2022). Binge drinking is a highly prevalent, expensive, and fatal form of excessive ethanol consumption in the United States. Binge drinking can result in detrimental pathological scenarios or conditions (Table 2). "Binge" drinking has detrimental impacts on renal structure and function and can result in kidney failure. Binge drinking is a critical but preventable public health issue.

"Binge" drinking can cause a toxic effect on the nephrons, leading to reversible or irreversible renal damage. Treatment for binge drinking is specific to an individual patient and can be a non-pharmacological and pharmacological approach. In an animal model of alcoholism (ethanol abuse), *Phellinus rimosus*, a polyspore from the family Hymenochaetaceae, protects the liver and kidney due to its antioxidant activity (Ajith and Janardhanan 2015) (Fig. 2). The beneficial effect of *Phellinus rimosus* can be attributed to its reactive oxygen species scavenging action and the CYP2E1 inhibitory activity. By these mechanisms, *Phellinus rimosus* affected the activity of the antioxidant enzymes (catalase and superoxide dismutase) and decreased lipid peroxidation.

Table 2 Factors that induce nephronal dysfunction associated with binge drinking

Binge drinking	Factors that induce nephronal dysfunction
Consuming • Four or more drinks at an event for women • Five or more alcoholic beverages/drinks at an event for men  Ukko.de, CC BY-SA 3.0 https://creativecommons.org/licenses/by-sa/3.0, via Wikimedia commons	(i) Accidents: Motor vehicle/crashes (ii) Acute and chronic kidney damage (iii) Alcohol poisoning (iv) Burns (v) Cancer (breast, among females), colon, esophagus, kidney, larynx, liver, mouth, pharynx, and rectum (vi) Chronic pathological conditions: Hypertension, cardiovascular problems, stroke, and hepatic dysfunction (vii) Cognitive defects (viii) Falls (ix) Serious pregnancy consequences (miscarriage, stillbirth, fetal alcohol spectrum disorders) (x) Sexual problems: Assaults, sexually transmitted diseases, unintended pregnancy (xi) Sudden infant death syndrome (xii) Violence (homicide, assault)

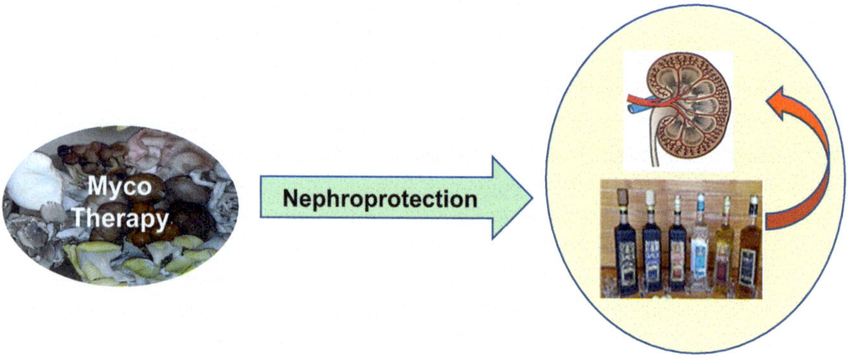

Fig. 2 Mycotherapy protects against alcohol-induced renal damage

3.2 *Nephroprotective Effects of Mycotherapeutics Against Anemia (Renal Anemia)-Induced Nephrotoxicity and Renal Damage*

Anemia is a pathological disorder resulting from decreased production and function of red blood cells. In anemia, the body (cells/tissues/organs) does not receive sufficient oxygenated red blood cells (oxyhemoglobin). Due to the lack of oxygen, an anemic patient's symptoms are fatigue (tired or weak), arrhythmia (irregular heartbeat), shortness of breath, dizziness, or headache. According to the Centers for Disease Control and Prevention, nearly three million people in the United States are diagnosed with anemia. There are specific types of anemia, including iron-deficiency anemia, hemolytic anemia, and Vitamin B12 deficiency anemia. Most individuals with renal damage or kidney disease also develop anemia. Decreased iron due to intestinal hemorrhage and lowered iron absorption can lead to anemia. Erythropoietin (EPO) is a glycoprotein hormone synthesized in the kidney by the peritubular cells. EPO is involved in the process of hematopoiesis, especially erythropoiesis, because EPO stimulates the production of erythrocytes (red blood cells). The synthesis of EPO is reduced/decreased due to renal diseases associated with diabetes mellitus, chronic kidney damage – stage 3 or 4, and kidney failure – stage 5. Anemia occurs early during renal pathology and escalates as kidneys fail and can no longer synthesize EPO. There is a flourishing interest in more natural iron-rich products, such as algae and fungus/mushrooms, that blend antioxidant and anti-inflammatory effects in addition to the presence of iron. A number of fungi, including *Pleurotus ostreatus*, *Schizophyllum commune*, *Agaricus subrufescens*, and *Ganoderma lucidum,* contain a considerable amount of iron and consequently may be used to prevent and treat renal anemia (Almeida et al. 2015).

3.3 Nephroprotective Effects of Mycotherapeutics in the Prevention of Chronic Kidney Damage

Chronic kidney disease (CKD) is a pathological disorder where the kidneys are damaged and cannot function appropriately. The primary risk factors that increase kidney disease are hyperglycemia, hypertension, cardiovascular diseases, and a family history of renal failure. Patients with kidney disease can prevent damage or insult to the nephron by an appropriate healthy lifestyle, monitoring for kidney disease, being more active, and targeting a healthy weight. The bioactive polysaccharides from *Auricularia polytricha* exhibit antioxidant (increase the activities of antioxidant enzymes), antiapoptotic (downregulate Bax and Caspase-3 and upregulate Bcl-2 expressions), and anti-inflammatory (reduce pro-inflammatory cytokines) activities and prevent against CKD (Song et al. 2021).

3.4 Nephroprotective Effects of Mycotherapeutics Against Fibrosis

Renal fibrosis refers to the scarring or thickening of the renal tissue, where the normally thin, lacy walls of the kidney become hard/scarred and thick, resulting in decreased renal function. Renal fibrosis is manifested by tubulointerstitial fibrosis and glomerulosclerosis, resulting in the final stage of chronic kidney disease. Renal fibrotic signaling includes renal apoptosis, initiation, and activation of transforming growth factor beta, fibroblast, and angiotensin II, significant accumulation of extracellular matrix components due to epithelial-to-mesenchymal changes, and

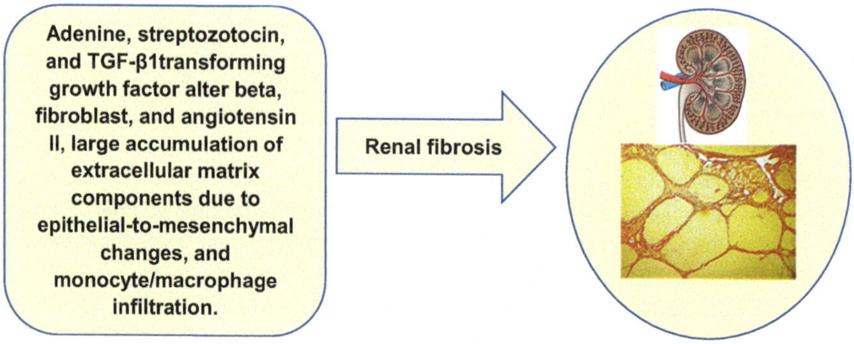

Fig. 3 Renal fibrosis. Durgesh1104, CC BY-SA 4.0 https://creativecommons.org/licenses/by-sa/4.0, via Wikimedia Commons

monocyte/macrophage infiltration. Adenine, streptozotocin, and TGF-β1 have been shown to induce renal fibrosis. Adenine is a potent nephrotoxin that induces nephrotoxicity by inducing oxidative stress and inflammation, resulting in renal fibrosis. The phosphorylated mycelia polysaccharides (PMPS) from *Pleurotus djamor* exhibited several nephroprotective activities, such as antioxidant, anti-inflammatory, and anti-fibrosis effects against adenine-induced chronic renal failure in animals. The PMPS contains α-pyranose structured bioactive compounds such as galacturonic acid and glucose (Li et al. 2021a). Transforming growth factor-β1 (TGF-β1) activates the Smad3 and mitogen-activated protein kinase (MAPK) pathways resulting in extracellular matrix accumulation and renal fibrosis. The poricoic acid A (PAA), an active bioactive in *Poria cocos,* inhibits the suppression of TGF-β1-induced renal fibroblast. PAA blocked the ECM accumulation and fibrosis formation by modulating the Smad3 and MAPK pathways.

Furthermore, PAA also exhibited an antioxidant effect and protected against acute renal injury (Li et al. 2021b). Diabetes mellitus is a complex endocrine metabolic disorder characterized by hyperglycemia leading to the toxicity of several cells/tissues resulting in various organ failures. Streptozotocin can cause hyperglycemia and lead to nephrotoxicity. Streptozotocin has been shown to affect several markers of oxidative stress, like the activities of superoxide dismutase, glutathione peroxidase, and catalase leading to lipid peroxidation. Thus, the oxidative stress induced by hyperglycemia caused by streptozotocin can result in severe nephrotoxicity. Edible mushrooms belonging to the *Pleurotus* genus have nutritional value and antioxidant activity. Natural bioactives with antioxidant actions have been shown to protect against hyperglycemia-induced nephrotoxicity. *Pleurotus djamor* possesses zinc polysaccharides that can suppress the oxidative stress induced by hyperglycemia (Zhang et al. 2015). Thus, *Pleurotus djamor and Poria cocos* have been shown to protect against endogenous and exogenous nephrotoxin-induced renal fibrosis and nephrotoxicity (Li et al. 2019, 2021b; Zhang et al. 2015) (Fig. 3).

3.5 *Nephroprotective Effects of Mycotherapeutics in Hypercholesterolemia (Hyperlipidemia/High-Fat Diet) and Obesity*

Hypercholesterolemia refers to excessive amounts of cholesterol in the blood (Fig. 4). Abnormal amounts of total cholesterol and high low-density lipoprotein (LDL) induce significant renal injury. Increased triglycerides primarily characterize dyslipidemia with high VLDL, apoB, and pre-β HDL and low levels of HDL and apoA in the serum. Interestingly, the amount of cholesterol is exceedingly high in proteinuric patients. Hypercholesterolemia induces macrophage infiltration and foam cell formation resulting in glomerulosclerosis.

The various key markers used to assess the effect of natural bioactives and synthetic drugs on body weight/abdominal fat weight and renal and hepatic functions are appetite (food intake), body mass (weight gain), hepatic function, and serum lipid profiles (aminotransferases [alanine aminotransferase and aspartate aminotransferase]), creatinine, and urea levels. A hypercholesterolemic and/or hyperglycemic diet induces a significant increase in total and non-HDL cholesterol, affects the atherogenic index, and increases body weight. Mycotherapeutics can decrease the cholesterol or lipid content which in turn decreases the body weight. Mycotherapeutics have significantly improved the lipid profile in hypercholesterolemic diet animal models. Various polysaccharides present in the *Pleurotus ostreatus and Pholiota nameko* exerted hypocholesterolemia effect as seen by the significant reduction of triglyceride and cholesterol content in plasma, the inhibiting activity of glutamate–oxaloacetate transaminase and glutamate–pyruvate transaminase in blood, and the content of thiobarbituric acid reactive substance (de Miranda et al. 2014, Oh et al., 2010). *Agaricus blazei, Pleurotus ostreatus, Pleurotus abalonus, Pholiota nameko, Inonotus obliquus,* and *Pleurotus citrinopileatus* have been shown to protect against hypercholesterolemia/hyperlipidemia/high-fat diet and obesity-induced nephrotoxicity (de Miranda et al. 2014; Oh et al. 2010; Alam et al. 2009; Li et al. 2007, 2010; Chou et al. 2016; Kim et al. 2019).

Fig. 4 Cholesterol

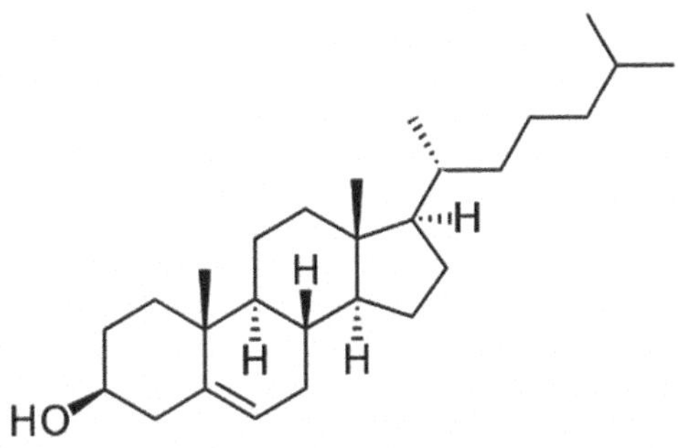

3.6 Nephroprotective Effects of Mycotherapeutics in Hyperglycemia

Chronic exposure to high blood sugar due to diabetes mellitus can cause direct nephronal insult and damage to the renal blood vessels. Diabetes mellitus is the leading cause of renal disorders. Also, many people with diabetes mellitus have hypertension, which can affect the structure and function of the kidneys (see the section below). Hyperglycemia can also severely damage the nerves. Diabetic nephropathy affects the kidneys' ability to remove waste products and extra fluid from the body. Renal damage can progress to kidney failure (end-stage kidney disease), a life-threatening pathological condition. Treatment for this pathological condition is dialysis or a kidney transplant. Drugs such as streptozotocin, alloxan, streptozotocin-nicotinamide, and diet-induced obesity have also been shown to induce hyperglycemia in rodents and rabbits. Natural bioactives in mushrooms, polysaccharides, steroids, and phenolic compounds have reduced hyperglycemia and nephrotoxicity (Fig. 5). Mycotherapeutics can reduce oxidative stress and inflammation (inflammasome, NF-κB signaling) and prevent renal morphopathological alterations, nucleotide-binding oligomerization domain-like receptor protein 3 modification, renal fibrosis, intestinal microbiota dysbiosis, colonic inflammation, and barrier dysfunction, tyrosine phosphatase 1B, and insulin resistance.

Polysaccharides present in *Armillariella tabescens* mycelia exhibited substantial anti-inflammatory effects. These polysaccharides significantly decreased several serum pro-inflammatory mediators, interleukin (IL)-1β and IL-18, and hence can

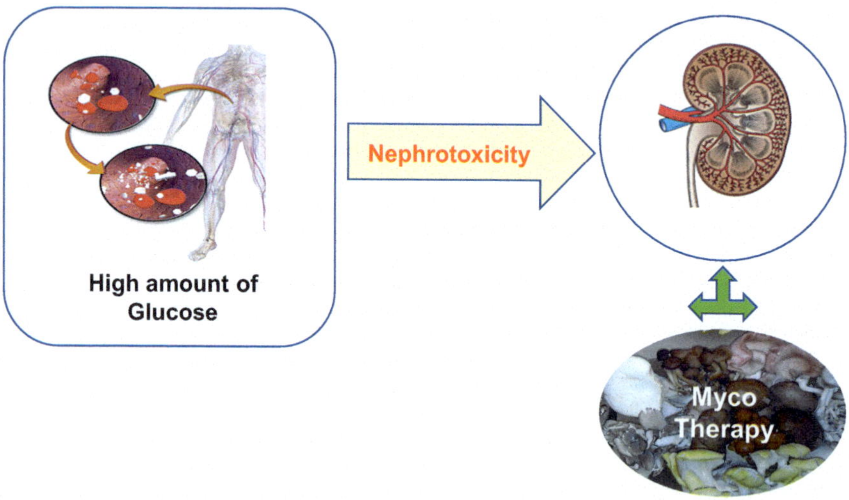

Fig. 5 Mycotherapy protects against hyperglycemia-induced nephrotoxicity. Blausen.com staff (2014). "Medical gallery of Blausen Medical 2014". WikiJournal of Medicine 1 (2). DOI:10.15347/wjm/2014.010. ISSN 2002-4436, CC BY 3.0 https://creativecommons.org/licenses/by/3.0, via Wikimedia Commons

be used to prevent and treat diabetic kidney disease (DKD). Furthermore, the bioactives of various mushrooms have shown to decrease blood glucose and hence protect against hyperglycemia-mediated nephrotoxicity (Yang et al. 2020; Liu et al. 2013a, b, 2017, 2018; Gao et al. 2021; Zahid et al. 2020; Lee et al. 2008; Zhang et al. 2021). Additionally, the bioactives present in the mushrooms can also decrease the total cholesterol, triglycerides, low-density lipoprotein-cholesterol, and maleic dialdehyde resulting in the nephroprotective effects against hyperlipidemic-induced nephrotoxicity (Yadav et al. 2020; Chi et al. 2017; Zhang et al. 2018; Gao et al. 2018a; Balaji et al. 2020; Yang et al. 2019a; Hwang et al. 2005; Lin et al. 2016; Jiang et al. 2020, 2022; Kou et al. 2019). Interestingly, the polysaccharides and the sterols present in the mushrooms can scavenge free radicals and increase the content and activities of the antioxidant molecules and enzymes. Consequently, the bioactives of mushrooms can decrease the inflammatory pathways and modulate the PTEN/PI3K/Akt and Wnt-1/β-catenin pathway and reverse kidney-related injuries mediated by various nephronal toxins.

A wide variety of mushrooms, including *Agaricus blazei, Armillaria tabescens, Catathelasma ventricosum, Coprinus comatus, Fomitopsis pinicola, Hypsizygus marmoreus, Inonotus obliquus, Morchella conica Pers., Ophiocordyceps sinensis, Pleurotus ostreatus, Pleurotus citrinopileatus, Pleurotus eryngii, Pleurotus pulmonarius, Pleurotus tuber-regium, Suillus luteus, Termitornyces albuminosus, Phellinus baumii, Flammulina velutipes, and Grifola frondose*, has shown to protect against hyperglycemia-induced nephrotoxicity (Fig. 5) (Yang et al. 2020; Liu et al. 2013a, b, 2017, 2018; Gao et al. 2018a, 2021; Zahid et al. 2020; Lee et al. 2008; Zhang et al. 2018, 2021; Yadav et al. 2020; Chi et al. 2017; Balaji et al. 2020; Yang et al. 2019a; Hwang et al. 2005; Lin et al. 2016; Jiang et al. 2020, 2022; Kou et al. 2019).

3.7 Nephroprotective Effects of Mycotherapeutics in Hypertension

High-sodium diets can increase blood pressure and are considered a good and valid hypertension model. This animal model is an accepted and valid model for studying the effect of synthetic drugs and natural bioactives for their antihypertensive effects. Hypertension induces the blood vessels (arteries) around the kidneys to constrict, weaken, or harden and induce nephrotoxicity. β-glucans in the mushrooms induce Corin expression, atrial natriuretic peptide production, and sodium excretion, leading to hypotensive effects. *Phellinus baumii* and *Pleurotus sajor-caju* have shown to protect against hypertension-mediated nephrotoxicity (Fig. 6) (Tam et al. 1986).

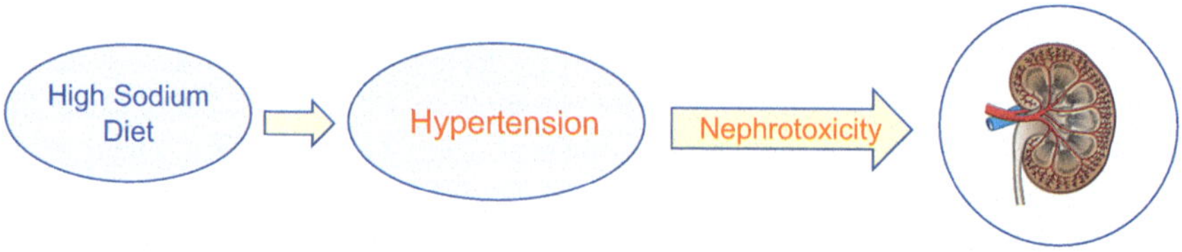

Fig. 6 Mycotherapy protects against hypertension-induced nephrotoxicity. https://www.myupchar.com/en, CC BY-SA 4.0. https://creativecommons.org/licenses/by-sa/4.0, via Wikimedia Commons

3.8 Nephroprotective Effects of Mycotherapeutics in Hyperuricemia

Hyperuricemia can be induced by administering potassium oxonate and hypoxanthine and by nephrectomy. Ethanol extracts and triterpenoids from mushrooms such as *Inonotus obliquus* and *Taiwanofungus camphorates* have been shown to suppress xanthine oxidase (XOD) activity in serum and liver and also downregulate renal uric acid transporter 1 (URAT1). These actions reduce serum uric acid levels and hyperuricemia (Luo et al. 2022; Yong et al. 2018; Wang et al. 2016).

3.9 Nephroprotective Effects of Mycotherapeutics in Iatrogenesis/Drug-Induced Nephrotoxicity

The common drugs that induce nephrotoxicity are cisplatin, cyclophosphamide, gentamicin, adenine, and paracetamol. *Cordyceps cicadae, Grifola frondose, Hypoxylon truncatum, Ophiocordyceps lanpingensis, Pleurotus cornucopiae, Auricularia polytricha, Agaricus brasiliensis,* and *Pleurotus porrigens* have shown to protect against the above therapeutic drugs-induced nephrotoxicity (Yang et al. 2019b; Li et al. 2018a; Masuda et al. 2009; Degen et al. 2013; Hwang et al. 2018; Zhou et al. 2021; Lee et al. 2017b). Furthermore, the methanolic extract of *G. lucidum* rendered a significant preventive effect against cisplatin-induced nephrotoxicity. The extract did not interfere with the antitumor activity of cisplatin. The findings suggest the potential therapeutic use of South Indian *G. lucidum* in cancer chemotherapy.

3.10 Nephroprotective Effects of Mycotherapeutics Against Immune-Based Toxicity

The immune system and the kidneys are intimately connected to maintaining normal health by providing immune homeostasis. However, in certain disease states, the elements of the immune system facilitate several acute forms of kidney disorders and play a key part in the advancement of chronic kidney disease. An altered immune system drastically impacts the kidney directly or indirectly. Direct immune-facilitated renal damage occurs due to autoimmune (autoantibodies) mechanisms, where the specific antibodies target a specific renal antigen. Indirect immune-facilitated renal damage occurs due to systemic autoimmunity with the immune complex formation and uncontrolled activation of the complement pathways. Failure of immune homeostasis in kidney disorders results in unending immune cell recruitment, leading to nephrotoxicity.

Inadequate efforts to repair the kidney cells and tissue after immune-facilitated disorders or nonimmune facilitated injury result in renal fibrosis of structures essential for kidney physiology, resulting in nephrotoxicity and renal malfunction. Lentinan ((1-6,1-3)-beta-glucan) is a potent bioactive of *Lentinus edodes* (Shiitake) which has shown to exert immune-modulatory effects on the immunological system. Lentinan stimulates the nonspecific (innate immunity) system and upregulated Interleukin-1beta, immunoglobulins, and other transcription factors' expression in the nephrons. Additionally, bioactives of *Pleurotus pulmonarius* and *Lentinula edodes* amplify the effect on B cells and facilitate humoral immunity (Ching et al. 2021; Gaullier et al. 2011; Li et al. 2018b). Thus, *Pleurotus pulmonarius* and *Lentinula edodes* have been shown to protect against immune-mediated nephrotoxicity (Ching et al. 2021; Gaullier et al. 2011; Li et al. 2018b).

3.11 Nephroprotective Effects of Mycotherapeutics in Inflammation/Nephritis

Nephritis is a pathological ailment in which the nephrons are inflamed due to the increase in the pro-inflammatory cytokines and/or decrease in the anti-inflammatory cytokines in the kidney. Nephritis can affect the anatomy and physiology of the nephrons, leading to decreased filtration and clearance of the waste materials from the blood. The etiopathology of nephritis can be due to comorbidities (lupus), genetic conditions, infection, inflammation, auto-immune disorders, exposure to drugs/medicines, or certain exogenous toxins/chemicals. The major signs and symptoms of nephritis are the accumulation of blood and protein in the urine, hypertension, accumulation of fluid in certain parts of the body (swelling of the face, hands, feet, and legs); lethargy/fatigue; and anemia. The small molecular fraction of various mushrooms (*Lentinula edodes, Oudemansiella radicata, Pleurotus ostreatus, Pleurotus djamor, Lentinula edodes, Oudemansiella radicata, Pleurotus ostreatus,* and *Irpex lacteus)* has protected against acute and chronic nephritis induced by various endogenous and exogenous nephrotoxins, The small molecular fraction of various mushrooms significantly decreased the proteins, blood urea nitrogen, creatinine, tumor necrosis factor-α, and maleic dialdehyde. Consequently, *Lentinula edodes, Oudemansiella radicata, Pleurotus ostreatus, Pleurotus djamor, Lentinula edodes, Oudemansiella radicata, Pleurotus ostreatus,* and *Irpex lacteus* protected against inflammation-mediated nephrotoxicity (Song et al. 2020; Han et al. 2020).

3.12 Nephroprotective Effects of Mycotherapeutics in Kidney Cancer

Kidney cancer or renal cell carcinoma is the most common type of kidney cancer. According to the American Cancer Society, kidney cancer is a pathological condition where the cells in the kidney begin to grow excessively without control leading to decreased kidney function and mortality. The diverse types of cancer include renal cell carcinoma, clear cell renal cell carcinoma, non-clear cell renal cell carcinoma, transitional/urothelial carcinoma, Wilms' tumor (nephroblastoma), and renal sarcoma. Usually, renal cell carcinomas are cancer and are usually fatal. Numerous traditional remedial choices are available to treat cancers, such as chemotherapeutics, immunotherapy, hormone therapy, radiotherapy, and surgery. Even though these traditional therapeutic approaches exhibit optimal response initially, severe adverse effects, drug interactions, hypersensitivity reactions, and mortality occur due to recurrence or ensuing relapse. However, the water-soluble polysaccharide of mushrooms (*Phellinus linteus* and *Pleurotus eryngii*) activated the immune response and possessed potent anticancer activities. Moreover, mycotherapeutics can induce apoptosis and reduce pathologies associated with renal cancer. *Phellinus linteus* and *Pleurotus eryngii* have been shown to protect against kidney cancer (Fig. 7) (Yang et al. 2013).

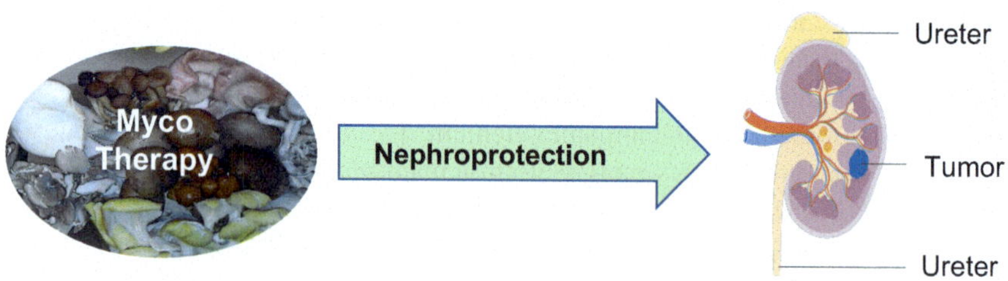

Fig. 7 Mycotherapy protects against kidney cancer. Cancer Research UK, CC BY-SA 4.0 <https://creativecommons.org/licenses/by-sa/4.0>, via Wikimedia Commons

3.13 Nephroprotective Effects of Mycotherapeutics in Metal-Induced Nephrotoxicity

Persistent and excessive exposure to heavy metals results in systemic disorders, damaging many organs. A kidney is the major target organ that is affected due to heavy metal exposure. The ability of the kidney to filter, reabsorb, and concentrate urine is lost due to the exposure to heavy metals, and it also induces nephrotoxicity. The magnitude of and the manifestation of nephrotoxicity are based on the type of metals, age, sex, comorbidities, and the dose and duration of exposure. Essentially, acute renal damage differs from chronic kidney damage due to the toxicokinetic and toxicodynamic effects and the extent of the nephrotoxic consequences. The alkaloids and flavonoids of *Pleurotus tuber-regium* have been shown to protect against heavy metals, arsenic, and cadmium-induced nephrotoxicity (Ogbomida et al. 2018).

3.14 Nephroprotective Effects of Mycotherapeutics in Oxidative Stress

Agaricus bisporus, Agrocybe cylindracea, Lactarius deliciosus, Pleurotus ostreatus, Pleurotus tuber-regium, Lentinula edodes, Oudemansiella radicata, Pleurotus ostreatus, Phellinus rimosus, Pleurotus djamor, Phellinus baumii, Pleurotus abalonus, Wolfiporia cocos, and *Morchella conica Pers.* have shown to exhibit antioxidant activity and protect against oxidative stress (Gao et al. 2018b; Jayakumar et al. 2008; Sainkhuu et al. 2016; Schulman et al. 2016; You et al. 2011; Li et al. 2018c).

3.15 Nephroprotective Effects of Mushrooms Against Anatomical and/or Physiological Nephronal Insults Related to Senescence

Senescence is an organic phenomenon that is a natural, progressive, and predictable biological process revealed by continuing distinct deterioration of cellular/tissue/organ structure (anatomy) and function (physiology) in many central and peripheral human organs. Like the brain and other peripheral organs, the nephrons and other renal tissues also proceed to the common senescence pathway, as seen by the decline in the structure (anatomy) and function. This age-related anatomical and physiological nephronal deterioration is an incredibly predictable change that illustrates the progression of aging. Nevertheless, the progress and changes in aging development differ significantly from those induced by a pathological condition. However, the irreversible age-related structural damage and the functional rate are complicated to comprehend. Occasionally, it is complicated to characterize the anatomical and physiological changes in renal senescence and kidney pathologies in an aged human (Denic et al. 2016). The kidney contains 700,000 to 1.8 million functional nephrons/kidney. The anatomical alteration during senescence in the aging kidney can be classified as micro and macrostructural-based changes.

The micro-renal structural changes are the following:

(i) Nephrosclerosis
  (a) arteriosclerosis
  (b) focal and global glomerulosclerosis
  (c) interstitial fibrosis
  (d) tubular atrophy
(ii) Nephron hypertrophy
(iii) Tubular diverticuli
(iv) Simple cysts

The macro-renal structural changes are the following:

(i) Renal volume
(ii) Kidney cysts
(iii) Tumors
(iv) Parenchymal alterations:
  (a) Cortical focal scars
  (b) Fibromuscular dysplasia
  (c) Renal artery atherosclerosis

Similarly, the physiological alterations during senescence in the aging kidney can be categorized based on the following parameters:

(i) Glomerular filtration rate (GFR)
(ii) Age-related acute kidney damage
(iii) Age-related chronic kidney damage

Thus, the senescence-associated renal damage with loss of nephron physiology irrefutably triggers an aged adult to be highly vulnerable to acute renal damage, leading to chronic kidney disease. Therefore, senescence-associated renal damage can increase the risk of morbidities and mortality (Fig. 8). Mycotherapeutics (*Agaricus bisporus*, *Pleurotus djamor*, and *Lentinula edodes*) exhibited significant effects to protect against senescence-induced nephrotoxicity. D-galactose administration induces oxidative stress and inflammation leading to renal damage and physiological decline and is also implicated in senescence/aging. Polysaccharides of *Agaricus bisporus exhibited antioxidant activity and* protected against renal damage in D-galactose-induced aging in rodents. Similarly, *Pleurotus djamor* increased the survival and life expectancy and decreased the mortality of *Anastrepha ludens* (Mexican fruit fly) and D-galactose-induced aging in rodents (Li et al. 2019; Sánchez et al. 2015). *Drosophila melanogaster* is an appropriate model system for

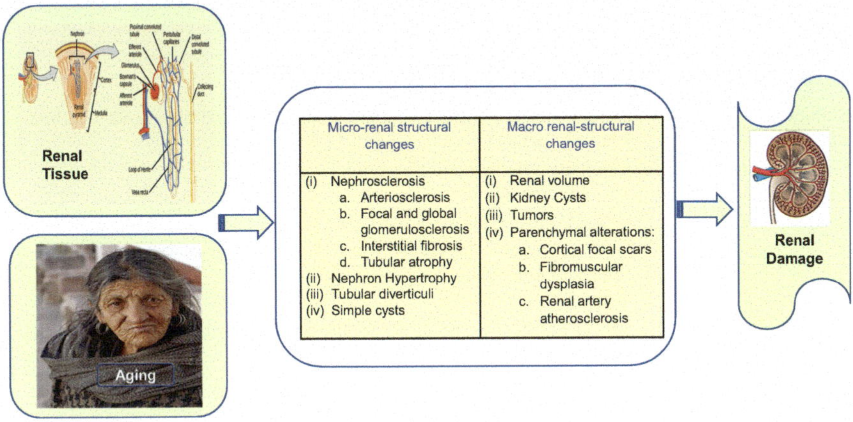

Fig. 8 Aging-related renal damage. CNX OpenStax, CC BY 4.0 https://creativecommons.org/licenses/by/4.0, via Wikimedia Commons. Tomas Castelazo, CC BY 3.0 https://creativecommons.org/licenses/by/3.0, via Wikimedia Commons. Artwork by Holly Fischer, CC BY 3.0 https://creativecommons.org/licenses/by/3.0, via Wikimedia Commons

investigating the molecular signaling of the antiaging properties. Shiitake mushroom (*Lentinula edodes*) has exhibited nephroprotective and antiaging effects (Matjuskova et al. 2014).

3.16 *Nephroprotective Effects of Mycotherapeutics Against Toxins (Exogenous and Endogenous Nephrotoxins)*

The American Association of Poison Control Centers' National Poison Data System (NPDS) reports that nearly 17 million toxin exposures are frequently occurring. Approximately 17,000 single chemical exposures can lead to nephronal damage renal effects, of which about 50–55% of the cases had severe renal damage and kidney complications. These toxins (plant, animal, or from the environment, Table 3) can cause induce renal complications as observed by increased creatinine, anuria, oliguria, and kidney failure.

The short- or long-term exposure or intoxication of various nephrotoxins can induce acute or chronic toxicity based on the toxicokinetic and toxicodynamic actions of the specific nephrotoxin and the severity of renal impairment. The nephrotoxins can affect the kidney due to oral ingestion, inhalation, and entry through the dermis (skin). Thus, exposure to renal toxins can cause nephronal damage and kidney failure and affect other organs. The nonrenal presentation is manifested as the following clinical symptoms of neurological signs (crumps, unconsciousness, dizziness, seizures, coma), cardiovascular issues (decreased blood pressure – hypotension), gastrointestinal problems (diarrhea, dehydration, nausea, salivation, vomiting), liver toxicity with jaundice and pulmonary failure.

Table 3 List of nephrotoxins

Endogenous nephrotoxins	Endogenous nephrotoxins
• Glucose • Lipids • Angiotensin • Hydrogen peroxide •Salt	• Medicines (antibiotics, chemotherapeutics, contrast agents – Iodine) • Heavy metals • Botanical/herbs • Animal venom • Fungi/mushroom • Environmental chemicals • Agricultural chemicals-pesticides • Alcohols • Illicit substances of abuse (amphetamine, cocaine, heroin, synthetic cannabinoids)

Patients exposed to endogenous and exogenous nephrotoxins with nephrotoxicity and kidney failure had higher rates of morbidities and mortality.

Endogenous and exogenous nephrotoxins have been shown to increase prooxidants and decrease antioxidants leading to increased lipid peroxidation that can lead to severe nephronal damage. Mycotherapeutics have been shown to protect against hydrogen peroxide (Xu et al. 2018), carbon tetrachloride – CCl_4, and pesticide (Carbofuran)-induced nephrotoxicity (Hossen et al. 2018; Hasar et al. 2020; Dogan et al. 2022; Jayakumar et al. 2008; Nworu et al. 2014). The polysaccharides from *Morchella conica Pers. and Morchella conica* and the phenolic bioactives from *Ganoderma lucidum* and *Auricularia polytricha* exhibit antioxidant effects by decreasing the reactive oxygen species generation and increasing the mitochondria membrane potential levels. In addition, the phenolic bioactives of mushrooms significantly improved the lipid profile, blocked oxidative stress biomarkers, and protected the nephrons.

3.17 Nephroprotective Effects of Mycotherapeutics in Urolithiasis

Kidney stones (synonyms: renal calculi, nephrolithiasis, urolithiasis) are solid/hard deposits formed within the kidneys and are composed of minerals and salts. Kidney stones are formed due to the urine concentration that causes the minerals to crystallize and stick together. Improper lifestyle (unhealthy diet, higher BMI, comorbidities, and drugs/medicines) can increase the risk of kidney stones. Pathologically, kidney stones influence the renal system. Excreting (passing) kidney stones is usually extremely painful, and these stones generally induce no irreversible damage if they are diagnosed or identified early. The standard remedies are analgesics to suppress pain and intake of water (excessive) to excrete the kidney stone. However, kidney stones in the urinary tract can increase the risk of urinary infection and cause renal and nonrenal complications. Regarding the in vitro and animal

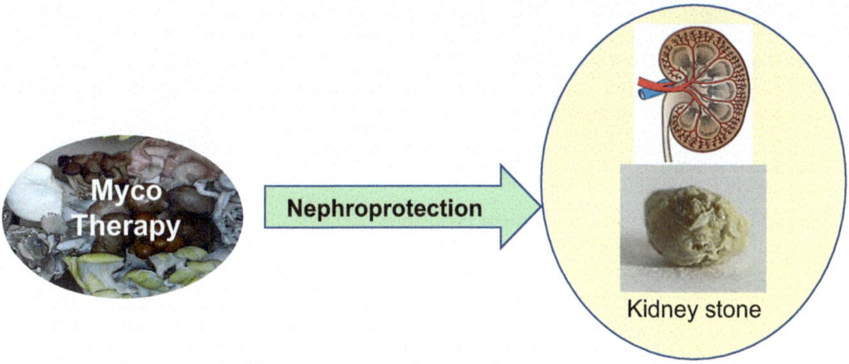

Fig. 9 Mycotherapy protects against kidney stones. Fvasconcellos (talk contribs), Public domain, via Wikimedia Commons. Jacek Proszyk, CC0, via Wikimedia Commons

models, ethylene glycol administration results in metabolism to oxalic acid, which sequesters calcium. It precipitates in the renal tubular system, forming kidney stones obstructing the nephron (urolithiasis). *Agaricus bisporus* and *Pleurotus ostreatus* (Ahmed et al. 2020; Li et al. 2018c; Ismaya et al. 2017; Walton et al. 1997) have nephroprotective effects by altering the NF-κB activation, intrinsic and extrinsic apoptotic signaling, and antioxidant action (Fig. 9).

4 Conclusion

Mycotherapeutics are gradually growing as a feasible alternative therapy to counteract aging and nephrotoxicity. Therefore, various international natural bioactive agencies are seriously validating the use of mycotherapeutics. Mushrooms possess several novel pharmacodynamic mechanisms and favorable pharmacokinetic mechanisms to reduce nephrotoxicity and thus can be pharmaceutical and nutraceutical formulations to prevent and treat kidney diseases.

References

Ahmed OM, Ebaid H, El-Nahass ES et al (2020) Nephroprotective effect of Pleurotus ostreatus and Agaricus bisporus extracts and carvedilol on ethylene glycol-induced urolithiasis: roles of NF-κB, p53, Bcl-2, Bax and Bak. Biomol Ther 10(9). https://doi.org/10.3390/biom10091317

Ajith TA, Janardhanan KK (2015) Medicinal mushroom cracked-cap polypore, Phellinus rimosus (higher basidiomycetes) attenuates acute ethanol-induced lipid peroxidation in mice. Int J Med Mushrooms 17(11):1061–1067. https://doi.org/10.1615/intjmedmushrooms.v17.i11.60

Alam N, Amin R, Khan A et al (2009) Comparative effects of oyster mushrooms on lipid profile, liver and kidney function in hypercholesterolemic rats. Mycobiology 37(1):37–42. https://doi.org/10.4489/myco.2009.37.1.037

Almeida SM, Umeo SH, Marcante RC et al (2015) Iron bioaccumulation in mycelium of Pleurotus ostreatus. Braz J Microbiol 46(1):195–200. https://doi.org/10.1590/S1517-838246120130695

Balaji P, Madhanraj R, Rameshkumar K et al (2020) Evaluation of antidiabetic activity of Pleurotus pulmonarius against streptozotocin-nicotinamide induced diabetic wistar albino rats. Saudi J Biol Sci 27(3):913–924. https://doi.org/10.1016/j.sjbs.2020.01.027

Carayannis EG, Campbell DFJ (2009) Mode 3′ and 'Quadruple Helix': toward a 21st century fractal innovation ecosystem. Int J Technol Manag 46:201–234. https://doi.org/10.1504/IJTM.2009.023374

CfDCa (2022) Binge drinking. Division of population health, National center for chronic disease prevention and health promotion, Centers for disease control and Prevention. https://www.cdc.gov/alcohol/fact-sheets/binge-drinking.htm

Chi Q, Wang G, Sheng Y et al (2017) Ethanolic extract of the Golden oyster mushroom, Pleurotus citrinopileatus (Agaricomycetes), alleviates metabolic syndrome in diet-induced obese mice. Int J Med Mushrooms 19(11):1001–1008. https://doi.org/10.1615/IntJMedMushrooms.2017024486

Ching JJ, Shuib AS, Abdullah N et al (2021) Hot water extract of Pleurotus pulmonarius stalk waste enhances innate immune response and immune-related gene expression in red hybrid tilapia Oreochromis sp. following challenge with pathogen-associated molecular patterns. Fish Shellfish Immunol 116:61–73. https://doi.org/10.1016/j.fsi.2021.06.005

Chou YJ, Kan WC, Chang CM et al (2016) Renal protective effects of low molecular weight of Inonotus obliquus polysaccharide (LIOP) on HFD/STZ-induced nephropathy in mice. Int J Mol Sci 17(9). https://doi.org/10.3390/ijms17091535

de Miranda AM, Ribeiro GM, Cunha AC et al (2014) Hypolipidemic effect of the edible mushroom Agaricus blazei in rats subjected to a hypercholesterolemic diet. J Physiol Biochem 70(1): 215–224. https://doi.org/10.1007/s13105-013-0295-y

Degen M, Alexander B, Choudhury M et al (2013) Alternative therapeutic approach to renal-cell carcinoma: induction of apoptosis with combination of vitamin K3 and D-fraction. J Endourol 27(12):1499–1503. https://doi.org/10.1089/end.2013.0207

Denic A, Glassock RJ, Rule AD (2016) Structural and functional changes with the aging kidney. Adv Chronic Kidney Dis 23(1):19–28. https://doi.org/10.1053/j.ackd.2015.08.004

Dogan A, Uyar A, Hasar S et al (2022) The protective effects of the Lactarius deliciosus and Agrocybe cylindracea mushrooms on histopathology of carbon tetrachloride induced oxidative stress in rats. Biotech Histochem 97(2):143–151. https://doi.org/10.1080/10520295.2021.1918349

Gao Z, Lai Q, Yang Q et al (2018a) The characteristic, antioxidative and multiple organ protective of acidic-extractable mycelium polysaccharides by Pleurotus eryngii var. tuoliensis on high-fat emulsion induced-hypertriglyceridemic mice. Sci Rep 8(1):17500. https://doi.org/10.1038/s41598-018-35706-8

Gao Z, Zhang C, Tian C et al (2018b) Characterization, antioxidation, anti-inflammation and renoprotection effects of selenized mycelia polysaccharides from Oudemansiella radicata. Carbohydr Polym 181:1224–1234. https://doi.org/10.1016/j.carbpol.2017.12.007

Gao Z, Kong D, Cai W et al (2021) Characterization and anti-diabetic nephropathic ability of mycelium polysaccharides from Coprinus comatus. Carbohydr Polym 251:117081. https://doi.org/10.1016/j.carbpol.2020.117081

Gaullier JM, Sleboda J, Øfjord ES et al (2011) Supplementation with a soluble β-glucan exported from Shiitake medicinal mushroom, Lentinus edodes (Berk.) singer mycelium: a crossover, placebo-controlled study in healthy elderly. Int J Med Mushrooms 13(4):319–326. https://doi.org/10.1615/intjmedmushr.v13.i4.10

Han Y, Ma L, Bao H et al (2020) Preventive effect of small molecular fraction of Irpex lacteus (Agaricomycetes) fruiting body against chronic nephritis in mice and identification of active compounds. Int J Med Mushrooms 22(4):313–323. https://doi.org/10.1615/IntJMedMushrooms.2020034220

Hasar S, Dogan A, Demirel K (2020) Determination of tissue protective and antioxidant activities of two medicinal mushrooms from Turkey against CCl4-induced experimental oxidative stress in rats. Int J Med Mushrooms 22(7):671–681. https://doi.org/10.1615/IntJMedMushrooms.2020035054

Hossen MS, Billah Prince MM, Tanvir EM et al (2018) Ganoderma lucidum and Auricularia polytricha mushrooms protect against Carbofuran-induced toxicity in rats. Evid Based Complement Alternat Med 2018:6254929. https://doi.org/10.1155/2018/6254929

Hwang HJ, Kim SW, Lim JM et al (2005) Hypoglycemic effect of crude exopolysaccharides produced by a medicinal mushroom Phellinus baumii in streptozotocin-induced diabetic rats. Life Sci 76(26):3069–3080. https://doi.org/10.1016/j.lfs.2004.12.019

Hwang BS, Lee D, Choi P et al (2018) Renoprotective effects of Hypoxylonol C and F isolated from Hypoxylon truncatum against cisplatin-induced cytotoxicity in LLC-PK1 cells. Int J Mol Sci 19(4). https://doi.org/10.3390/ijms19040948

Ismaya WT, Efthyani A, Tjandrawinata RR et al (2017) Biological responses in Balb/c mice after long-term parenteral administration of the light subunit of mushroom tyrosinase. J Biochem Mol Toxicol 31(11). https://doi.org/10.1002/jbt.21958

Jayakumar T, Sakthivel M, Thomas PA et al (2008) Pleurotus ostreatus, an oyster mushroom, decreases the oxidative stress induced by carbon tetrachloride in rat kidneys, heart and brain. Chem Biol Interact 176(2–3):108–120. https://doi.org/10.1016/j.cbi.2008.08.006

Jiang T, Wang L, Ma A et al (2020) The hypoglycemic and renal protective effects of Grifola frondosa polysaccharides in early diabetic nephropathy. J Food Biochem 44(12):e13515. https://doi.org/10.1111/jfbc.13515

Jiang T, Shen S, Wang L et al (2022) Grifola frondosa polysaccharide ameliorates early diabetic nephropathy by suppressing the TLR4/NF-κB pathway. Appl Biochem Biotechnol 194:4093. https://doi.org/10.1007/s12010-022-03976-8

Kim SH, Thomas MJ, Wu D et al (2019) Edible mushrooms reduce atherosclerosis in Ldlr$^{-/-}$ mice fed a high-fat diet. J Nutr 149(8):1377–1384. https://doi.org/10.1093/jn/nxz075

Kornsakulkarn J, Palasarn S, Choowong W et al (2020) Antimalarial 9-methoxystrobilurins, Oudemansins, and related polyketides from cultures of basidiomycete Favolaschia species. J Nat Prod 83(4):905–917. https://doi.org/10.1021/acs.jnatprod.9b00647

Kou L, Du M, Liu P et al (2019) Anti-diabetic and anti-nephritic activities of Grifola frondosa mycelium polysaccharides in diet-streptozotocin-induced diabetic rats via modulation on oxidative stress. Appl Biochem Biotechnol 187(1):310–322. https://doi.org/10.1007/s12010-018-2803-6

Lee SI, Kim JS, Oh SH et al (2008) Antihyperglycemic effect of Fomitopsis pinicola extracts in streptozotocin-induced diabetic rats. J Med Food 11(3):518–524. https://doi.org/10.1089/jmf.2007.0155

Lee SS, Tan NH, Pailoor J et al (2017a) Safety evaluation of sclerotium from a medicinal mushroom, Lignosus cameronensis (cultivar): preclinical toxicology studies. Front Pharmacol 8:594. https://doi.org/10.3389/fphar.2017.00594

Lee SR, Lee D, Lee HJ et al (2017b) Renoprotective chemical constituents from an edible mushroom, Pleurotus cornucopiae in cisplatin-induced nephrotoxicity. Bioorg Chem 71:67–73. https://doi.org/10.1016/j.bioorg.2017.01.012

Li L, Ng TB, Song M et al (2007) A polysaccharide-peptide complex from abalone mushroom (Pleurotus abalonus) fruiting bodies increases activities and gene expression of antioxidant enzymes and reduces lipid peroxidation in senescence-accelerated mice. Appl Microbiol Biotechnol 75(4):863–869. https://doi.org/10.1007/s00253-007-0865-4

Li H, Zhang M, Ma G (2010) Hypolipidemic effect of the polysaccharide from Pholiota nameko. Nutrition 26(5):556–562. https://doi.org/10.1016/j.nut.2009.06.009

Li Q, Chen G, Chen H et al (2018a) Se-enriched G. frondosa polysaccharide protects against immunosuppression in cyclophosphamide-induced mice via MAPKs signal transduction pathway. Carbohydr Polym 196:445–456. https://doi.org/10.1016/j.carbpol.2018.05.046

Li Q, Chen G, Wang W et al (2018b) A novel Se-polysaccharide from Se-enriched G. frondosa protects against immunosuppression and low Se status in Se-deficient mice. Int J Biol Macromol 117:878–889. https://doi.org/10.1016/j.ijbiomac.2018.05.180

Li S, Liu M, Zhang C et al (2018c) Purification, in vitro antioxidant and in vivo anti-aging activities of soluble polysaccharides by enzyme-assisted extraction from Agaricus bisporus. Int J Biol Macromol 109:457–466. https://doi.org/10.1016/j.ijbiomac.2017.12.108

Li H, Zhao H, Gao Z et al (2019) The antioxidant and anti-aging effects of acetylated mycelia polysaccharides from Pleurotus djamor. Molecules 24(15). https://doi.org/10.3390/molecules24152698

Li H, Feng Y, Sun W et al (2021a) Antioxidation, anti-inflammation and anti-fibrosis effect of phosphorylated polysaccharides from Pleurotus djamor mycelia on adenine-induced chronic renal failure mice. Int J Biol Macromol 170:652–663. https://doi.org/10.1016/j.ijbiomac.2020.12.159

Li Q, Ming Y, Jia H et al (2021b) Poricoic acid A suppresses TGF-β1-induced renal fibrosis and proliferation via the PDGF-C, Smad3 and MAPK pathways. Exp Ther Med 21(4):289. https://doi.org/10.3892/etm.2021.9720

Lin L, Cui F, Zhang J et al (2016) Antioxidative and renoprotective effects of residue polysaccharides from Flammulina velutipes. Carbohydr Polym 146:388–395. https://doi.org/10.1016/j.carbpol.2016.03.071

Liu Y, Sun J, Rao S et al (2013a) Antidiabetic activity of mycelia selenium-polysaccharide from Catathelasma ventricosum in STZ-induced diabetic mice. Food Chem Toxicol 62:285–291. https://doi.org/10.1016/j.fct.2013.08.082

Liu Y, Sun J, Rao S et al (2013b) Antihyperglycemic, antihyperlipidemic and antioxidant activities of polysaccharides from Catathelasma ventricosum in streptozotocin-induced diabetic mice. Food Chem Toxicol 57:39–45. https://doi.org/10.1016/j.fct.2013.03.001

Liu M, Song X, Zhang J et al (2017) Protective effects on liver, kidney and pancreas of enzymatic- and acidic-hydrolysis of polysaccharides by spent mushroom compost (Hypsizigus marmoreus). Sci Rep 7:43212. https://doi.org/10.1038/srep43212

Liu M, Lan Y, Tian C et al (2018) The characterization, renoprotection and antioxidation of enzymatic and acidic exopolysaccharides from Hypsizigus marmoreus. Sci Rep 8(1):2048. https://doi.org/10.1038/s41598-018-20440-y

Luo LS, Wang Y, Dai LJ et al (2022) Triterpenoid acids from medicinal mushroom Inonotus obliquus (Chaga) alleviate hyperuricemia and inflammation in hyperuricemic mice: possible inhibitory effects on xanthine oxidase activity. J Food Biochem 46(3):e13932. https://doi.org/10.1111/jfbc.13932

Masuda Y, Inoue M, Miyata A et al (2009) Maitake beta-glucan enhances therapeutic effect and reduces myelosupression and nephrotoxicity of cisplatin in mice. Int Immunopharmacol 9(5):620–626. https://doi.org/10.1016/j.intimp.2009.02.005

Matjuskova N, Azena E, Serstnova K et al (2014) The influence of the hot water extract from shiitake medicinal mushroom, Lentinus edodes (higher basidiomycetes) on the food intake, life span, and age-related locomotor activity of Drosophila melanogaster. Int J Med Mushrooms 16(6):605–6015. https://doi.org/10.1615/intjmedmushrooms.v16.i6.100

Nworu CS, Ihim SA, Ugwu LE et al (2014) Hepato- and nephroprotective activities of a Nigerian local king tuber oyster mushroom, Pleurotus tuberregium (higher basidiomycetes), in chemically induced organ toxicities in rats. Int J Med Mushrooms 16(4):305–318. https://doi.org/10.1615/intjmedmushrooms.v16.i4.10

Ogbomida ET, Omofonmwan K, Aganmwonyi I et al (2018) Bioactive profiling and therapeutic potential of mushroom (Pleurotus tuberregium) extract on Wistar albino rats (Ratus norvegicus) exposed to arsenic and chromium toxicity. Toxicol Rep 5:401–410. https://doi.org/10.1016/j.toxrep.2018.03.004

Oh TW, Kim YA, Jang WJ et al (2010) Semipurified fractions from the submerged-culture broth of Agaricus blazei Murill reduce blood glucose levels in streptozotocin-induced diabetic rats. J Agric Food Chem 58(7):4113–4119. https://doi.org/10.1021/jf9036672

Sainkhuu B, Park BS, Kim HW (2016) Induction of Flavin-containing monooxygenase in mice by oral administration of Phellinus baumii (Agaricomycetes) extract. Int J Med Mushrooms 18(9): 793–806. https://doi.org/10.1615/IntJMedMushrooms.v18.i9.40

Sánchez JE, Jiménez-Pérez G, Liedo P (2015) Can consumption of antioxidant rich mushrooms extend longevity?: antioxidant activity of Pleurotus spp. and its effects on Mexican fruit flies' (Anastrepha ludens) longevity. Age (Dordr) 37(6):107. https://doi.org/10.1007/s11357-015-9847-0

Schulman A, Chaimowitz M, Choudhury M et al (2016) Antioxidant and renoprotective effects of mushroom extract: implication in prevention of nephrolithiasis. J Clin Med Res 8(12):908–915. https://doi.org/10.14740/jocmr2781w

Song X, Ren Z, Wang X et al (2020) Antioxidant, anti-inflammatory and renoprotective effects of acidic-hydrolytic polysaccharides by spent mushroom compost (Lentinula edodes) on LPS-induced kidney injury. Int J Biol Macromol 151:1267–1276. https://doi.org/10.1016/j.ijbiomac.2019.10.173

Song X, Pang H, Cui W et al (2021) Renoprotective effects of enzyme-hydrolyzed polysaccharides from Auricularia polytricha on adenine-induced chronic kidney diseases in mice. Biomed Pharmacother 135:111004. https://doi.org/10.1016/j.biopha.2020.111004

Tam SC, Yip KP, Fung KP et al (1986) Hypotensive and renal effects of an extract of the edible mushroom Pleurotus sajor-caju. Life Sci 38(13):1155–1161. https://doi.org/10.1016/0024-3205(86)90169-4

Tsai MY, Hung YC, Chen YH et al (2016) A preliminary randomised controlled study of short-term Antrodia cinnamomea treatment combined with chemotherapy for patients with advanced cancer. BMC Complement Altern Med 16(1):322. https://doi.org/10.1186/s12906-016-1312-9

Ugbogu EA, Akubugwo EI, Ude VC et al (2019) Safety evaluation of an aqueous extract of Termitomyces robustus (Agaricomycetes) in Wistar rats. Int J Med Mushrooms 21(2): 193–203. https://doi.org/10.1615/IntJMedMushrooms.2018029737

Walton K, Coombs MM, Catterall FS et al (1997) Bioactivation of the mushroom hydrazine, agaritine, to intermediates that bind covalently to proteins and induce mutations in the Ames test. Carcinogenesis 18(8):1603–1608. https://doi.org/10.1093/carcin/18.8.1603

Wang SC, Yang CH, Grumezescu AM et al (2016) Renoprotective effects of shout camphor medicinal mushroom (Taiwanofungus camphorates, basidiomycetes) mycelia on several media in mice with chronic kidney disease. Int J Med Mushrooms 18(12):1105–1114. https://doi.org/10.1615/IntJMedMushrooms.v18.i12.50

Xu N, Lu Y, Hou J et al (2018) A polysaccharide purified from Morchella Conica Pers. prevents oxidative stress induced by H_2O_2 in human embryonic kidney (HEK) 293T cells. Int J Mol Sci 19(12). https://doi.org/10.3390/ijms19124027

Yadav SK, Ir R, Jeewon R et al (2020) A mechanistic review on medicinal mushrooms-derived bioactive compounds: potential mycotherapy candidates for alleviating neurological disorders. Planta Med 86(16):1161–1175

Yang Z, Xu J, Fu Q et al (2013) Antitumor activity of a polysaccharide from Pleurotus eryngii on mice bearing renal cancer. Carbohydr Polym 95(2):615–620. https://doi.org/10.1016/j.carbpol.2013.03.024

Yang C, Feng Q, Liao H et al (2019a) Anti-diabetic nephropathy activities of polysaccharides obtained from termitornyces albuminosus via regulation of NF-κB signaling in db/db mice. Int J Mol Sci 20(20). https://doi.org/10.3390/ijms20205205

Yang CH, Su CH, Liu SC et al (2019b) Isolation, anti-inflammatory activity and physico-chemical properties of bioactive polysaccharides from fruiting bodies of cultivated cordyceps cicadae (ascomycetes). Int J Med Mushrooms 21(10):995–1006. https://doi.org/10.1615/IntJMedMushrooms.2019031922

Yang R, Li Y, Mehmood S et al (2020) Polysaccharides from Armillariella tabescens mycelia ameliorate renal damage in type 2 diabetic mice. Int J Biol Macromol 162:1682–1691. https://doi.org/10.1016/j.ijbiomac.2020.08.006

Yong T, Chen S, Liang D et al (2018) Actions of inonotus obliquus against hyperuricemia through XOD and bioactives screened by molecular modeling. Int J Mol Sci 19(10). https://doi.org/10.3390/ijms19103222

You R, Wang K, Liu J et al (2011) A comparison study between different molecular weight polysaccharides derived from Lentinus edodes and their antioxidant activities in vivo. Pharm Biol 49(12):1298–1305. https://doi.org/10.3109/13880209.2011.621960

Zahid MT, Idrees M, Abdullah I et al (2020) Antidiabetic properties of the Red Belt conk medicinal mushroom Fomitopsis pinicola (Agaricomycetes) extracts on streptozotocin-induced diabetic rats. Int J Med Mushrooms 22(8):731–741. https://doi.org/10.1615/IntJMedMushrooms.2020035472

Zhang J, Meng G, Zhang C et al (2015) The antioxidative effects of acidic-, alkalic-, and enzymatic-extractable mycelium zinc polysaccharides by Pleurotus djamor on liver and kidney of streptozocin-induced diabetic mice. BMC Complement Altern Med 15:440. https://doi.org/10.1186/s12906-015-0964-1

Zhang C, Zhang L, Liu H et al (2018) Antioxidation, anti-hyperglycaemia and renoprotective effects of extracellular polysaccharides from Pleurotus eryngii SI-04. Int J Biol Macromol 111:219–228. https://doi.org/10.1016/j.ijbiomac.2018.01.009

Zhang J, Lu Y, Zhao L et al (2021) Hypoglycemic activity of total flavonoids from slippery Jack mushroom, Suillus luteus (Agaricomycetes), in vitro and in vivo. Int J Med Mushrooms 23(11):17–26. https://doi.org/10.1615/IntJMedMushrooms.2021040384

Zhou S, Zhou Y, Yu J et al (2021) A neutral polysaccharide from Ophiocordyceps lanpingensis restrains cisplatin-induced nephrotoxicity. Food Sci Nutr 9(7):3602–3616. https://doi.org/10.1002/fsn3.2317

Proteomic Research on the Therapeutic Properties of Medicinal Mushrooms

Boris Jakopovic, Nada Oršolić, Sandra Kraljević Pavelić, and Ivan Jakopovich

Contents

Abstract According to estimates, more than 800 mushroom species have been confirmed to possess various pharmacological properties. While the primary

B. Jakopovic (✉) · I. Jakopovich
Dr Myko San–Health from Mushrooms Co., Zagreb, Croatia
e-mail: borisjakopovic@yahoo.com; ivan.jakopovic@mykosan.com

N. Oršolić
Division of Animal Physiology, Faculty of Science, University of Zagreb, Zagreb, Croatia
e-mail: nada.orsolic@biol.pmf.hr

S. Kraljević Pavelić
Faculty of Health Studies, University of Rijeka, Rijeka, Croatia
e-mail: sandrakp@uniri.hr

D. C. Agrawal, M. Dhanasekaran (eds.), *Mushrooms with Therapeutic Potentials*,
https://doi.org/10.1007/978-981-19-9550-7_11

research focus has been on immunomodulatory and, more recently, on direct antitumor effects, many other therapeutic properties have also become important research subjects. Systems biology approaches, which include various omics methods coupled with bioinformatics, have begun to enter the field of medicinal mushroom research. Proteomics is a well-established hypothesis-generating discipline focused on the systemic study of proteins, used in discovering new pathways and processes resulting from physiological and pathological states or pharmacological interventions. It is particularly powerful in new potential therapeutic target discovery of "druggable proteins" or disease biomarkers of prognostic or therapeutic significance. Despite challenges in proteomics' extensive data analysis and interpretation, proteomics remains indispensable for the detailed understanding and characterization of pharmacological effects.

In comparison with genomics, proteomics generates more accurate data since several protein products may result from only one gene. This results from differential splicing and more than 200 posttranslational modifications, which affect protein function, stability, protein-protein, and other interactions. This chapter provides a current overview of proteomic research on medicinal mushrooms concerning anticancer, immunomodulatory, antidiabetic, hypolipidemic, antiatherosclerotic, hepatoprotective, and neuroprotective properties.

Keywords Anticancer · Antidiabetic · Hepatoprotective · Hypolipidemic · Medicinal mushrooms · Neuroprotective · Proteomics

Abbreviations

2-DE	Two-dimensional gel electrophoresis
ALT	Alanine transaminase or alanine aminotransferase
AST	Aspartate transaminase or aspartate aminotransferase
Cy	Cyclophosphamide
DEP	Differentially expressed protein
DIGE	Difference gel electrophoresis
ECM	Extracellular matrix
ESI	Electrospray ionization
GL-SP	*Ganoderma lucidum* spore polysaccharides
GSH	Glutathione
HPLC	High-performance liquid chromatography
IARC	International agency for research on cancer
IL-2	Interleukin 2
iTRAQ	Isobaric tag for relative and absolute quantitation
KEGG	Kyoto encyclopedia of genes and genomes
LC-MS	Liquid chromatography-mass spectrometry
LPS	Lipopolysaccharide
MALDI	Matrix-assisted laser desorption/ionization

MAPK	Mitogen-activated protein kinase
NMR	Nuclear magnetic resonance
NSCLS	Non-small-cell lung carcinoma
PPI	Protein-protein interactions
PRDX	Peroxiredoxins
ROS	Reactive oxygen species
SOD	Superoxide dismutase
SWATH	Sequential windowed acquisition of all theoretical mass spectra
TCA cycle	Tricarboxylic acid cycle or citric acid cycle
TMT	Tandem mass tag
UPR	Unfolded protein response
WHO	World Health Organization

1 Introduction

Pharmacologically, medicinal mushrooms are a rich and complex system of chemical constituents interacting with numerous physiological and pathological pathways and processes in various eukaryotic organisms. Moreover, their active compounds also interact with viruses and prokaryotes. This results in more than 130 noted therapeutic properties which have been researched extensively, including anticancer, immunomodulatory, immunosuppressive, antibacterial, antifungal, anti-inflammatory, antioxidant, antiviral, hepatoprotective, hypocholesterolemic, hypoglycemic, and others (Zaidman et al. 2005). A multitude of high- and low molecular weight compounds, including polysaccharides, polysaccharide-peptides (proteins), peptidoglycans, alkaloids, lectins, lipids, phenolics, polyketides, proteins, steroids, terpenoids, ribosomal, and non-ribosomal peptides are being extensively researched and characterized by various analytical methods, ranging from general constituent quantification to NMR spectroscopy characterization. More recently, the bioactive proteins from medicinal mushrooms have been characterized by proteomics methods of ever-increasing resolution as one of the omics methods for mining novel bioactive compounds from natural products.

On the other hand, systems biology approaches, including omics platforms and various bioinformatic tools, have been proven invaluable for researching multiple complex diseases, most notably cancer. Proteomics is a large-scale study of proteins on a cellular, tissue, or organismal level via two-dimensional gel electrophoresis (2-DE), mass spectrometry, multidimensional protein identification techniques (advanced HPLC systems coupled with mass spectrometry), isobaric tags method for relative and absolute quantitation (iTRAQ) and protein microarrays. Proteomics approaches characterize and quantify proteins and proteomes in physiological or pathological states and identify new biomarkers and/or processes affected by a disease or various pharmacological interventions. This approach is particularly relevant in studying medicinal mushrooms, which are well-studied and acknowledged as nonselective, systemic therapeutics since they possess diverse bioactive components with established pharmacological properties. Furthermore, the results obtained from such research could be crucial in further using medicinal mushrooms in treating various pathologies for which current pharmacology has not found satisfactory solutions.

2 Proteomic Research on Anticancer Properties

2.1 Colorectal Cancer

According to WHO, in 2020, colorectal cancer (CRC) was third by incidence (1.93 million cases) and second by mortality (916,000 deaths) worldwide (Ferlay et al. 2020). While it is one of the best-studied cancers in terms of pathophysiology, late diagnosis is still one of the main challenges, with 35% of patients being diagnosed at stage IV, that is, in the metastatic stage of the disease (Zacharakis et al. 2010). Recent large-scale proteomic research on colorectal cancer in patient tumor samples clarified qualitative, quantitative, and temporal (dynamic) changes in the tumor proteome during colorectal cancer progression (Peng et al. 2016; Vasaikar et al. 2019; Zhang et al. 2019). This further enables the potential therapeutics to be studied regarding their effect on the various known processes (which might be modulated by only several proteins or large protein clusters with known PPI interactions) altered in different tumor stages. Many new biomarkers of the disease have thus been established, with further benefits in disease and therapy monitoring.

Erinacine A is a diterpenoid derivative from mycelium of *Hericium erinaceus* (lion's mane mushroom), belonging to a class of compounds with a range of proven anti-inflammatory and neuroprotective effects (Bailly and Gao 2020). It is also one of the most significant anticancer cyathanes, its main mechanisms being the induction of apoptosis through ROS production and disruption of cytoskeleton organization. Erinacine A showed specific effects on the proteome of colorectal cancer cell lines HCT-116 and DLD-1, which include upregulation of proteins profilin-1 and cofilin-1, thereby inducing ROS generation and thus directly controlling the actin cytoskeleton pathways and dynamics (Lee et al. 2017). These perturbations induce apoptosis and inhibit colon cancer cell motility through ROCK1/LIMK2/cofilin cascades (Maekawa et al. 1999). Erinacine A treatment also led to the downregulation of nucleophosmin (NPM), which negatively correlates with cancer development, including CRC (Yu et al. 2021). This protein is implicated in numerous cellular functions with potential tumor-promoting effects, such as ribosome biogenesis and DNA repair. Another protein relevant for CRC development and cancer cell migration, the hepatoma-derived growth factor (HDGF), was downregulated.

The apoptotic effect of fruiting body extracts of *Pleurotus sajor-caju* obtained by various solvents on wild type as well as Bax, p21, and p53 knockout HCT-116 cells was investigated by Finimundy et al. (2018). While all the extracts demonstrated antiproliferative activity mediated by apoptosis and G2/M cell cycle arrest, this was most prominent with the n-hexane extract of *P. sajor-caju* (PSC-hex). While the antiproliferative effect was evident in wild-type cells, it was not observed in knock-out cells, confirming the known requirement of these proteins in the apoptosis of HCT-116 cells (Wang and Youle 2012). A closer insight into the proteome dynamics performed by use of the Proteome Profiler Array pointed to changes underlying apoptosis and corroborated by the involvement of apoptosis-related proteins, that is, tumor-suppressor p53, Bim, Bid, Bad, and Bax, cytochrome-c, caspase-3, Smac, Survivin, HTRA, Xiap, heat-shock proteins HSP70, HSP60, and Fas.

HCT-116 colorectal cancer model was also used to investigate the antitumor properties of *Sporisorium reilianum*, a known plant pathogen studied as a sorghum head smut inducer (Kan et al. 2020). *S. reilianum* isolated polysaccharide WM-NP-60 reduced cell viability and increased the G1 cell population, which indicated a G1 cell arrest. TMT-labeling detected 369 differentially expressed proteins after treating cells with 4 mg/mL of WM-NP-60 for 48 h. The protein thrombospondin 1 (THBS1) was downregulated. Its perturbations directly influence the PI3K-Akt, p53, Rap1 (Ras-related protein 1), and TGF-β (transforming growth factor-beta) signaling pathways. It affects the focal adhesions and interactions of the ECM receptors as well. Bioinformatic analysis using the STRING database revealed protein-protein interactions (PPI) between upregulated TGFβR1, P107 (cell proliferation), and DP1 (transcription), and downregulated THBS1. The upregulation of TGFβR1 by WM-NP-60 results in increased TGF-β binding. In summary, the HCT-116 cells' proliferation was halted in the cell cycle phase G1, followed by apoptosis induction.

Xiong and coworkers investigated the effects of a specific selenium-chelating tripeptide RLA, isolated from the protein hydrolysate of *Grifola frondosa* mushroom (Xiong et al. 2021). While Se deficiency has been found to correlate with infertility states, cardiovascular disease, or even gestational diabetes, inorganic Se compounds are toxic at a relatively small lethal dose (Ying and Zhang 2019). Therefore, the *Grifola* polypeptide RLA-Se chelate could be used for selenium deficiency since it can be well absorbed, unlike inorganic selenium. This Se-chelating peptide, characterized by UV spectroscopy, scanning electron microscopy (SEM), X-ray diffraction (XDR), and NMR spectroscopy, directly halted the proliferation of human colorectal adenocarcinoma Caco-2 cells in a dose-dependent manner. iTRAQ-labeled tandem mass spectrometry revealed 40 differentially regulated proteins in Caco-2 cells compared to the control group. The upregulated proteins in the treated group, RHO guanine nucleotide exchange factor 2 (ARHGEF2), cyclin-dependent kinase 9 (CDK9), tumor protein p53 binding protein 2 (TP53BP2), and Aurora kinase A (AURKA), are regulators of the cell cycle and promote cell growth. This is in line with the research substantiating Se′s antiproliferative role and/or proapoptotic effect (Zeng and Botnen 2007). RLA-treated Caco-2 cells also had elevated expression of E3 proteins and ubiquitin-conjugating enzymes, which indicates an increase in cysteine levels. These are correlated with a heightened level of glutathione peroxidase (GSH-Px), with crucial roles in antioxidation and tumor prevention (Jain et al. 2014).

While individual isolates of either low molecular weight or high molecular weight bioactive substances from single mushroom species are predominantly studied, various research indicates that combining multiple medicinal mushroom species may have synergistic effects regarding their pharmacological action (Kurashige et al. 1997; Shamtsyan et al. 2004). This applies to both direct antitumor (by modulating various oncogenic pathways) and indirect, that is, immunomodulating effects, since concerted activation of multiple immune cell receptors may result in a more prominent immune response. Jakopovic et al. (2020a) performed large-scale proteomics analysis of the CT26.WT colorectal cancer in mice treated with registered preparation Agarikon.1 (AG.1), a medicinal mushroom extract mixture consisting of *Lentinus edodes*, *Ganoderma lucidum*, *Agaricus brasiliensis* (=*blazei* ss. Heinem), *Grifola frondosa*, *Pleurotus ostreatus*, and *Trametes versicolor*, which also contains the usual excipients (Jakopovich 2011). This preparation was already found to bear antiproliferative and proapoptotic properties in HCT-116 and SW620 colorectal cancer cell lines, along with strong immunostimulatory effects through induction of macrophage M1 polarization and tumor angiogenesis inhibition in vivo (Jakopovic et al. 2020a, b). The effects of this preparation (AG.1) were also studied in combination with 5-fluorouracil (5-FU). Since the treatment started only after the tumor size was ≥700 mm^3, that is, 14 days after tumor transplantation, this represents a very advanced, grade IV metastatic tumor model, unlike most current preclinical research. Tumor tissue analysis by tandem mass tag (TMT) uncovered a set of 95 dysregulated proteins. Treatment with AG.1 and a combination of AG.1 and 5-FU resulted in the downregulation of proteins involved in ribosome biogenesis, assembly, and translation. These processes are known to be progressively upregulated during the CRC progression, as well as mRNA splicing and influenza life cycle pathway, which were likewise downregulated as a result of treatment with AG.1 and AG.1 + 5-FU. Concomitantly, some key processes involved in CRC development, that is, unfolded protein response (UPR), lipid metabolism, and TCA cycle pathways were upregulated due to treatment. These processes are usually increasingly downregulated during colorectal cancer progression (Peng et al. 2016). This data provides further mechanistic clarification of the antitumor effects conferred by Agarikon.1 preparation.

2.2 Lung Cancer

With 1.80 million deaths in 2020 alone, lung cancer is the first in terms of cancer mortality worldwide (Ferlay et al. 2020). *Nectria haematococca*, a common soil Ascomycota fungus, a known plant and human pathogen, has been previously studied as an antitumor agent in lung cancer (Xie et al. 2018). Fungal immunomodulatory proteins (FIPs) are a class of small molecular weight bioactive substances, many of which have been isolated and characterized (LZ-8 or FIP-glu from *Ganoderma lucidum*, FIP-gts from *Ganoderma tsugae*, FIP-fve from *Flammulina velutipes* and FIP-nha from *Nectria haematococca*) (Ko et al. 1995; Wu et al. 2011;

Li et al. 2014). FIP-nha possesses superior growth-suppressing efficacy on A549 lung adenocarcinoma cells compared with FIP-fve or LZ-8. Particularly, it led to tumor volume decrease in a nude Balb/c mouse model, which was equal to those observed with the chemotherapy drug doxorubicin. Furthermore, the iTRAQ analysis of the same cells followed by functional bioinformatic enrichment analysis revealed that identified upregulated proteins play a significant role in processes relevant to tumor cell development that include cell migration, cell apoptosis, and autophagy, or cytoskeleton-related pathways, PI3K/Akt signaling, and extracellular matrix organization signaling. Downregulated proteins were mainly enriched in functions such as ubiquitination, telomere maintenance, cell proliferation, and G1/S or G2/M cell cycle arrest.

Although cisplatin, as the first-line chemotherapy in advanced NSCLC, shows some benefits in terms of life expectancy, its nonselective effect on healthy tissues causes acute damage to normal cells (Hesketh 2008). Additionally, cisplatin resistance is a major clinical problem, resulting in treatment failure. Cordycepin is a purine nucleoside antimetabolite and antibiotic from the entomopathogenic Ascomycete *Cordyceps militaris*, which has proven to have dose-dependent antitumor effects (apoptosis induction, inhibition of cell proliferation, and migration) (Tian et al. 2015). Jeong et al. (2019) studied the effects of *Cordyceps militaris* on A549/CR lung cancer cells resistant to cisplatin. CME (*Cordyceps militaris* extract) induced apoptosis in A549/CR cells (up to only 17.63% of viable cells). A protein chip-based antibody array identified H-Ras as the only cell-cycle protein out of 42 analyzed to be significantly downregulated. H-Ras regulates protein whose downregulation may overcome NSCLC cisplatin resistance and is thus an important molecule for cell cycle progression.

Lin et al. (2021) further studied the antitumor effects of Ling-Zhi-8 (LZ-8), a fungal immunomodulatory protein from *Ganoderma lucidum*. C57BL/6 mice were inoculated with LLC1 lung cancer cells. This study was performed on tumor tissue profiled by using ESI-MS/MS. The shotgun proteomics approach showed 21 differentially expressed proteins compared to the control after treatment with LZ-8. The consequent bioinformatic analysis done by Ingenuity Pathway Analysis pointed to a significant overlap of 15 canonical pathways, including protein ubiquitination pathways, 14-3-3-mediated, and aldosterone signaling. Moreover, 4 of these 21 proteins, namely HSP70, HSP90, binding immunoglobulin protein GRP78 (Bip), and protein disulfide isomerase (PDI)-related proteins, are part of the protein processing/endoplasmic reticulum stress pathway. Importantly, the LZ-8 treatment successfully downregulated the levels of heat shock proteins with a known role in stabilizing oncoprotein levels in cancer cells and inhibiting apoptosis (Rong and Yang 2018). These oncoproteins are ALK, KRAS, EGFR, and HER2.

2.3 Breast Cancer

The epidemiological data indicate that breast cancer was the first in terms of incidence in women (2.26 million cases) but fifth in terms of cancer deaths (685,000 cases) in 2020 (Ferlay et al. 2020). This indicates improved screening strategies and better clinical management of patients, although the mortality rate is still too high (Ahmad 2019). *Cordyceps sinensis* polysaccharide (WECS) was investigated on a 4T1 breast cancer model. WECS reduced the number of metastatic lung nodules and significantly increased survival in mice (Cai et al. 2018). The protein array of the lung tissue homogenates showed that out of 111 analyzed cytokines, osteopontin (OPN), IL-33, CCL6, CCL17, CCL12 (chemokine (C-C motif) ligand 12), and matrix metallopeptidase 9 (MMP-9) were upregulated more than twofold. The significantly reduced cytokines due to WECS treatment included OPN, IL-33, CCL17, and MMP-9, which are known to have various roles in lung metastasis and immunosuppression (IL-33) (Colombo and Chiodoni 2013; Liu et al. 2014). Accordingly, *C. sinensis* might reduce DNA damage and, consequently, the DNA damage response (DDR). This mechanism of action, which involves immune response, leads to reduced inflammatory reactions, particularly in advanced tumors (Kang et al. 2015).

One of the most widely consumed mushrooms, *Agaricus bisporus* or "champignon," also has antitumor properties, as evidenced in a breast cancer model. MCF-7 breast cancer cells were accordingly treated with a lectin from *A. bisporus* (ABL-*A. bisporus* lectin) conjugated with $CaCO_3$ nanoparticles (ABL-$CaCO_3$NPs) (Mahmood et al. 2021). It is thought that the known antioxidant, antitumor, and immunostimulating properties of fungal lectins might be enhanced through nanotechnology, enabling better drug efficiency through increased drug delivery (Hassan et al. 2015; Yan et al. 2020). Proteomic analysis revealed that the treatment of MCF-7 cells for 24 h dysregulated 13 proteins. Identified downregulated proteins were cytoskeleton and associated proteins β-catenin, tropomyosin alpha, and cytoplasmic actin 1, metabolic enzyme triosephosphate isomerase (a glycolysis pathway enzyme), and membrane-associated proteins (annexin A2, human serum albumin, and V-set and immunoglobulin domain). Interestingly, the heat shock protein disulfide isomerase (PDI) was upregulated, which may correlate with UPR and ER stress induction in tumor cells.

2.4 Liver Cancer

Primary liver cancer, of which hepatocellular carcinoma (HCC) constitutes about 80% of cases, is the third by mortality worldwide (Ferlay et al. 2020). There is a significant geographic variation in HCC incidence, with 5.1 per 100,000 person-years in Europe to 17.7 per 100,000 person-years in eastern Asia (Dasgupta et al. 2020). This depends on the regional variability of risk factors, such as the prevalence of viral hepatitis infections, which account for 60% of cases worldwide. Other risk factors include high alcohol consumption, smoking, and dietary aflatoxin exposure. This cancer type is third in cancer mortality, with 830,000 deaths in 2020, with the median survival of advanced HCC cases being 1–1.5 years for cases treated with systemic therapies (Ferlay et al. 2020; Llovet et al. 2021).

The effects of single fraction polysaccharides from *Phellinus linteus* (PL), *Ganoderma lucidum* (GL), and *Auricularia auricula* (AA) on hepatocellular carcinoma cells (HepG2) tumor markers were studied after it was confirmed that these polysaccharides induce apoptosis and G1- or S-phase cell cycle arrest (Ouyang et al. 2013). The proteomics method used for the study, MALDI-TOF-MS mass spectrometry, showed 59 differentially expressed proteins after treatment of hepatocellular carcinoma HepG2 cells with 1 mg/mL of PL, GF, and AA. The identified proteins fall into 78 enriched metabolic pathways (Chai et al. 2016). For example, the 14-3-3 protein, which has a role in many signaling pathways activated in cancer cells, such as the PI3K-AKT pathway, was upregulated, suggesting this as a potential resistance mechanism of polysaccharide-treated HepG2 cells. The identified DJ-1 protein (protein deglycase DJ-1 or PARK7) was downregulated due to treatment. This protein has proliferative and antiapoptotic properties and is often upregulated in HCC, indicating its prognostic significance and pharmacological value.

The effects of *Phellinus linteus* alone on HepG2 cells have also been studied (Li et al. 2013). Novel proteoglycan P1, purified from *P. linteus* fresh fruiting bodies, inhibited HepG2 cell proliferation without apoptosis and led to a slower increase in tumor volume and weight in nude Balb/c mice. MALDI-TOF/TOF proteomic analysis of HepG2 cells after treatment with P1 revealed that a calreticulin (CRT) precursor was the only significantly downregulated protein. Calreticulin, upregulated in gastric and breast cancers, has a role in many important cellular processes such as cell motility, cell cycle progression, apoptosis, protein folding, and metabolism (Eric et al. 2009; Chen et al. 2009; Michalak et al. 2009).

The effects of black fungus or *Auricularia auricula* alone on HCC have also been studied (Kang et al. 2020). Of three *A. auricula* isolates, bioconverted *A. auricula* extract (BS) exhibited the most significant antiproliferative and apoptotic effects in Huh-7 cells and hepatocellular carcinoma cells with p53 point mutations. Out of seven dysregulated proteins in Huh-7 cells after treatment with BS, further research was focused on peroxiredoxin-1 (PRDX1), which was downregulated. This is an antioxidant protein whose downregulation reduces the cancer cells' capacity to eliminate elevated ROS levels, leading to apoptosis. By silencing PRDX1 in Huh-7 cells or BS-treated Huh-7 cells, the authors observed an apparent decrease in GSH and SOD levels, respectively, which corroborates the inactivation of antioxidant enzymes in treated cells.

Lentinan is a β-(1→3)-D-glucan, which has been used as a potent anticancer drug in Japan since 1985 (Chihara et al. 1987; Chen et al. 2013). The effects of this polysaccharide on hepatocellular carcinoma were presented in several studies. In an H22 liver cancer model, lentinan demonstrated a dose-dependent antiproliferative effect, which was considered specific since cytotoxicity was not observed in the normal human liver line HL7720 (Wang et al. 2017). Variable lentinan doses were also studied in a KM mouse model, where it was noted that the group treated with 0.4 mg/kg of lentinan i.p. once a week for three weeks showed a 20% increase in life span (ILS) and an improvement in immunological parameters (thymus and spleen indices). The groups treated with 0.02 mg/kg and 1 mg/kg of lentinan i.p. once a week for three weeks had no change in life span compared to untreated animals (control). They had lower macrophage phagocytic indices in comparison with the control group. A shotgun proteomics study using LC-MS/MS on H22 cells incubated with 1.28 mg/mL lentinan yielded six potential protein targets affected by lentinan treatment. These include the annexin A5, PDZ, and LIM domain protein 1, 60S acidic ribosomal protein P2, peroxiredoxin-2, cortactin, and moesin. These are all tumor-promoting proteins with proliferative effects, antioxidative defense, membrane dynamics, tumor invasion, and metastasis. Increased levels of these proteins point to a certain lentinan potential in liver cancer immunoprophylaxis but closely monitoring the applied dosage.

In another study, the effects of lentinan were monitored in ascites and solid H22 liver cancer models (Yang et al. 2020—preprint). The immunogen, that is, liver cancer vaccine LHA was obtained by coculturing H22 cells with lentinan (170 μL/mg). In both SPF KM mouse groups (ascites and solid tumor models), body weight was lower than control, indicating an antitumor effect. The analysis of the antigen protein sample revealed six proteins with an antitumor effect. Their main functions are apoptosis mediated by TNF-α-related apoptosis-inducing ligand, inhibition of angiogenesis (brain-specific angiogenesis inhibitor 1), and direct inhibition of invasion and metastasis (septin-7, cortactin).

Lentinan was also studied as an immunogenic cell death (ICD) inducer in a H22 hepatocellular carcinoma model (Wang et al. 2022). One of the more recent findings was that chemotherapy and radiotherapy, besides their previously known mechanisms of action, are characterized by the enhancement of the transformation of tumor cells in the apoptosis stage from non-immunogenic to immunogenic cells, that is, by causing ICD (Krysko et al. 2012). Tumor cells undergoing ICD are characterized by the expression of various proteins called DAMPs (danger-associated molecular patterns), which activate the adaptive immune response (Casares et al. 2005). H22 mouse hepatoma cells were obtained from the ascites a week after tumor inoculation. Treatment of such cells with several lentinan concentrations showed an apparent dose- and time-dependent cell viability decrease accompanied by apoptosis induction. By use of the HPLC-MS approach, ICD marker proteins (DAMPs) were identified upon lentinan treatment; calreticulin (CRT), high mobility group protein B1 (HMGB1), various heat shock proteins (HSP 70, HSP 90-alpha, HSP 90-beta), ER membrane protein complex subunit 1, ATP (ADP/ATP translocase 2), and annexin 1, confirming lentinan to be an ICD inducer in this model.

Liu et al. (2018) investigated the antitumor properties of a newly isolated natural triterpene-farnesyl hydroquinone hybrid GL22 obtained from *Ganoderma leucocontextum* fruiting bodies. This compound showed to induce a strong antiproliferative effect on liver cancer cell line Huh7.5 in vitro or in vivo in Huh7.7-derived tumor xenografts. Proteomic analysis revealed 128 and

141 differentially expressed proteins after treatment of Huh7.5 with GL22 for 12 and 24 h, respectively. Multiple proteins involved in fatty acid metabolism were downregulated. Some downregulated proteins in GL22-treated xenograft tumors included fatty acid-binding proteins FABP1, FABP4, and FABP5, as well as PPARα and PPARγ (peroxisome proliferator-activated receptors with a crucial role in lipid metabolism). This points to the PPAR-FABPs signaling pathway as a potential target in liver cancer treatment. In addition, GL22 treatment decreased cardiolipin levels, a mitochondrial function regulator, adding to another GL22 antitumor mechanism of action.

Proteomic methods have also been used to study *Cordyceps*' antitumor effects. In particular, lyophilized hot water *Cordyceps cicadae* extract had a dose-dependent inhibition of hepatocellular carcinoma MGCC97H cells. This effect was accompanied by G2/M cell accumulation (Wang et al. 2014). In addition, the 2-DE analysis revealed 28 proteins with dysregulated levels in tumor cells treated with the extract at 500 μg/mL for 48 h. Some of these proteins have important roles in cell growth and cell cycle regulation. 14-3-3 was downregulated, which could account for the observed G2/M phase arrest. Proteins with known functions in spindle checkpoint and mitosis regulation, such as dynactin subunit 2 (DCTN2), microtubule-associated protein RP/EB family member 1 (MAPRE1), and mitotic checkpoint protein BUB3 isoform A (BUB3), were found to be deregulated. Some of the downregulated proteins have important roles in cancer progression. WD-40 repeat protein (STRAP) regulates pre-mRNA splicing, while peroxiredoxin (PRDX1) plays a significant role in tumor cells' antioxidative systems. Both these proteins were downregulated in treated cells.

Cordyceps sinensis powder extract was analyzed in a hepatocellular carcinoma rat model where diethylnitrosamine (DEN) was used for shifting hepatocytes into a state of oxidative stress through chaperone protein and enzyme carbonylation induction, required for liver tumor progression (Paula Santos et al. 2014; Wang et al. 2016). The hypothesis that *C. sinensis* has hepatoprotective and antioxidative effects was confirmed by the almost complete prevention of the increase in the level of hepatic enzymes alanine transaminase (ALT) and aspartate transaminase (AST), as well as by the reversal of histopathological changes at 2 and 8 weeks such as fibrosis. The approach based on high-resolution 2-DE and MALDI-MS analysis revealed that the *C. sinensis* treatment group attenuated DEN-induced ubiquitin-proteasomal proteolysis cascades responsible for cellular stress responses inhibited c-Myc, AKT, p53, and NF-κB levels while activating PPARγ. PPARγ upregulation has a known antitumor effect by regulating lipid metabolism and cell inflammation (Kim et al. 2007). The upregulation of nuclear factor erythroid-derived 2-like 2 (Nrf-2), an important transcription factor that regulates the expression of antioxidant proteins, is also an important factor in *C. sinensis* HCC-preventive efficacy.

The effects of *Antrodia cinnamomea* were investigated in a HepG2 and C3A liver cancer model (Chen et al. 2020). A 2-DIGE approach using fluorescent labeling, which enables a broader dynamic range of protein detection, was used, followed by MALDI-TOF/MS. This approach showed that *A. cinnamomea* ethanol extract (EEAC) induced TRAIL-mediated apoptosis through the downregulation of mRNA processing proteins and heterogenous nuclear ribonucleoproteins A2/B1 and K (Chen et al. 2010). EEAC-downregulated glycolytic enzymes (PKM1, PGAM1, LDHA, GAPDH, G6PD, ENO1, and ALDOA) might result from elevated oxidative stress that shunts metabolism to NADPH production or from upregulation of Bad, a proapoptotic protein (Shenton and Grant 2003).

2.5 *Gastric Cancer*

Gastric (stomach) cancer is sixth in terms of incidence, with 1.09 million cases worldwide in 2020, while being fifth in terms of cancer mortality (685,000 cases) (Ferlay et al. 2020). The incidence of gastric cancer varies by up to 15- to 20-fold with geographic location, with high-risk regions being East Asia, Eastern Europe, and Central and South America (Wong et al. 2021). Although various recognized causes of stomach cancer include obesity, smoking, gastric ulcer disease, and gastroesophageal reflux disease, the most common cause is infection by the bacterium *Helicobacter pylori* (Ahmed 2005).

Grifola frondosa (maitake or hen-of-the-woods mushroom) is a well-known medicinal mushroom with maitake D-fraction, a mixed β-D-glucan fraction of the fruiting bodies being the most notable. However, the effects of *G. frondosa* glycoprotein GFG-3a from fermented mycelia have been analyzed by proteomics on human gastric tumor cells (Cui et al. 2016). While fungal polysaccharides exert antitumor effects primarily through immunostimulation, glycoproteins have a direct antitumor effect (Shoji-Kawata et al. 2013). Treatment of SGC-7901 cells showed a dose-dependent inhibition in cell proliferation and apoptosis. 2-DE analysis revealed downregulation of cell-cycle proteins, that is, RuvB-like 1, histone-binding protein RBBP4, important in cell replication, which could induce S-phase arrest and p53 activation (Gorynia et al. 2011). Another group of downregulated proteins included heat shock protein 90-beta (HSP90B) and 78 kDa glucose-regulated protein (GRP78), molecular chaperones, and stress proteins that might contribute to p53-regulated cell apoptosis. Nucleic-acid-related proteins nucleophosmin and heterogeneous nuclear ribonucleoprotein F were also downregulated. For example, nucleophosmin was already acknowledged as a potential gastric, colon, ovarian, and prostate cancer marker as it inhibits AKT and activates tumor-suppressor p53 protein (Yu et al. 2021).

The mechanism of action of erinacine A, a diterpenoid derivative from *Hericium erinaceus* mycelium, on the inhibition of gastric carcinoma cell line TSGH9201 was investigated by Kuo et al. (2017). Immunoblotting confirmed the induction of apoptosis, while the inhibition of invasiveness and motility was revealed by matrigel assay. Microtubule-associated tumor suppressor candidate 2 (MTUS2) was upregulated, and 14-3-3 protein sigma was involved in G2/M checkpoint control (Mhawech 2005). MTUS2 protein controls microtubules and the cytoskeleton depolarization pathway. Further results confirmed increased ROS generation after the treatment, and it was concluded that erinacine exerts its effects through ROS→p-

FAK→p-AKT→p-p70mTOR→p-PAK→1433S/MTUS2 pathway. Downregulated proteins included 14-3-3 zeta/delta, with known roles in signal transduction, apoptosis, and cell migration and nucleophosmin (B23), important in ribosome biogenesis, genome stability and repair, and cell cycle (Lindstrom 2011).

2.6 *Cervical Cancer*

Cervical cancer is both the fourth most common cancer by incidence and the fourth most common cause of cancer deaths in women worldwide. In 2020 alone, more than 600,000 new cases were registered, with 340,000 deaths (Sung et al. 2021). Infection with human papillomavirus (HPV) is the single most significant risk factor for cervical cancer, of which HPV types 16 and 18 cause 75% of cases (Clifford et al. 2006). Other major causes include tobacco use, long-term use of oral contraceptives, and multiple pregnancies (Gadducci et al. 2011). Although cervical cancer screening and preventive HPV vaccines have significantly reduced the incidence of cervical cancer, it still represents a major health issue, especially in areas with low access to these preventative measures (Jemal et al. 2013).

Ganoderic acids are a group of highly oxygenated lanostane-type triterpenoids from *Ganoderma lucidum* with a wide range of pharmacological properties (Xu et al. 2010). Ganoderic acid has been recognized as a modulator of various cancer-promoting pathways, such as IR, IGFR-1, IGFR-2, VEGFR-1, VEFGR-2, and EGFR. It primarily targets PI3K/Akt/mTOR, RAS-MAPK, NF-κB, and cell cycle, which results in apoptosis (Gill et al. 2018). Recent research shows that it also alleviates chemotherapy-induced fatigue and improves 5-fluorouracil-induced cognitive dysfunction in mice (Abulizi et al. 2021a, b). The effects of 99% pure ganoderic acid D (GAD) on the proteome of human cervical carcinoma HeLa cells were studied by Yue et al. (2008a). MALDI-TOF MS/MS analysis revealed that proteins with various cellular functions were altered in treated cells. Proteins eIF5A (eukaryotic translation initiation factor 5A-1) and spermidine synthase are important in cell survival and proliferation, so their downregulation indicates the mechanism for the observed cytotoxicity as G2/M cell cycle arrest and apoptosis. GAD also induced protein expression changes of metabolic enzymes and their regulators, that is, protein-disulfide isomerase, thioredoxin-dependent peroxide reductase mitochondrial precursor, ubiquinol cytochrome c reductase core protein 1, and an activator of heat shock 90-kDa protein ATPase homolog 1. One of the indicated antitumor mechanisms on HeLa cells was protein degradation, mediated by the upregulated 26 S proteasome subunit p40.5. Furthermore, GAD-induced apoptosis could also be induced by the downregulation of several cytoskeletal proteins, such as cytokeratin 1 and 19, calumenin, and microtubule-associated protein RP/EB family member 1.

The effects and interactions of *Ganoderma lucidum* triterpenes (GTS) with doxorubicin (DOX) were investigated in cervical carcinoma cells (HeLa) (Yue et al. 2008b). The combination index (CI) revealed that GTS and DOX synergistically affect cytotoxicity, apoptosis, and G2/M cell cycle arrest. The potential targets of GTS were studied by proteomics, and 14 proteins were found to be dysregulated compared to the untreated cells. It was revealed that some of the same proteins were downregulated after treatment with GTS, confirming previous research (Yue et al. 2008a). These included eIF5A, PRDX2 with roles in antioxidative defense, protein phosphatase 2 (PP2A), an enzyme with a phosphatase function in oncogenic signaling cascades such as AKT, Raf, MEK, and cytokeratin 19. Proteins involved in energy metabolism, including ATP synthase F0 subunit d, enoyl CoA hydratase chain 1, and LDH B were also downregulated due to GTS treatment, which can inhibit tumor cell proliferation. Chain B of the Ku heterodimer (Ku80) protein, involved in DNA repair, was downregulated. The downregulation of Ku80 has been shown to increase the response of cancer cells to DNA-damaging agents and sensitizes HeLa cells to chemotherapy (Ayene et al. 2005).

In another research, HeLa cells were treated with various purified ganoderic acids: ganoderic acid AM1 (GAAM1), ganoderic acid B (GAB), ganoderic acid D (GAD), ganoderic acid F (GAF), and ganoderic acid K (GAK) (Yue et al. 2010). After treating the cells with ganoderic acids IC50 value (15 μM), proteomic analyses were done by MALDI-TOF MS/MS. Twelve proteins with a similar change tendency in their expression across all ganoderic acid-treated groups were identified and subsequently classified into four biological function categories. eIF5A, an elongation factor (downregulated), 14-3-3 beta/alpha (upregulated), and protein phosphatase 2 (PP2A) subunit (downregulated) belong to the group related to cell proliferation and/or cell death. In the second group, carcinogenesis-related proteins included heterogeneous nuclear ribonucleoprotein K (HNRPK) (mRNA splicing and processing) and interleukin-17E (T-cell-mediated angiogenesis), which were both downregulated. Downregulated proteins from the third group, DJ-1 protein chain A and peroxiredoxin (PRDX2), have an antioxidative function, so their dysfunction contributes to the inability of cancer cells to resolve ROS (Trachootham et al. 2009). The last group of dysregulated proteins included nucleobindin-1 and reticulocalbin-1, with ER stress and calcium signaling functions, respectively (Ozawa and Muramatsu 1993; Tsukumo et al. 2007).

2.7 *Ovarian Cancer*

Ovarian cancer accounts for 3.4% of all cancers in females and is the cause of 4.7% of overall cancer deaths in women (Sung et al. 2021). The prognosis is poor because there is no established screening test, which, in addition to often vague symptoms, leads to late diagnosis in advanced tumor stages (Jayson et al. 2014). Malignant ascites, which entails a poorer prognosis, is most common in ovarian cancer (Barni et al. 2011). The effects of lentinan (LNT)-functionalized Selene nanoparticles (Selene) were studied in OVCAR-3 human ovarian cancer cells, along with the Ehrlich ascites (EAT) tumor model (Liu et al. 2020). Selene was obtained using Na_2SeO_3 and LNT under reduction conditions. Selene inhibited ascites volume in the EAC model and decreased Balb/c mice body weight and EAC cell numbers.

Also, Selene reduced ascites by decreasing vascular leakage and induced apoptosis of OVCAR-3 cells in an orthotopic nude mice model. Selene reduced the expression of various inflammatory cytokines, such as IL-1β, IL-6, and TNF-α. This is of great importance since it proved that Selene could induce the immunologically silent process of apoptosis, unlike pyroptosis, an inflammatory type of cell death that sometimes results from chemotherapy drugs. Pyroptosis is characterized by the induction of various cytokines that may stimulate ascites production (Kampan et al. 2017; Zhang and Zhang 2018). Further research revealed that Selene internalizes into mitochondria, causing apoptosis through induction of the TLR4/TRAF3/MFN1 pathway, unlike Selene nanoparticles (SeNPs), which are internalized into lysosomes, leading to pyroptosis. Proteomic analysis of Selene-treated OVCAR-3 cells showed that Selene acts on mitochondrion and mitochondrion translation, oxidative phosphorylation, apoptosis, adhesion, and endocytosis.

2.8 Prostate Cancer

With over 1.4 million new cases and over 375,000 deaths in 2020, prostate cancer is the second cancer by incidence in men, surpassed only by lung cancer (Sung et al. 2021). The anticancer activities of MPSSS, a novel polysaccharide from *Lentinus edodes*, were studied on cancer-associated fibroblasts (CAFs), known for their immunosuppressive function, and on PC-3, androgen insensitive human prostate cancer cell line (Zhang et al. 2021). PC-3 growth was promoted in the regular CAF medium but was reduced in a dose-dependent manner in the MPSSS-conditioned CAF medium. MPSSS untreated or treated prostate CAFs fractions were separated into high (>100 kDa) (hmwCAFS/MT-hmwCAFS) or low molecular weight fractions (3–100 kDa) (lmwCAFS/MT-lmwCAFS). The lmwCAFs promoted and MT-lmwCAFs inhibited the growth of PC-3 cells, and a comparative secretome/proteome analysis was performed to identify the proteins underlying the observed effect. TMT LC-MS/MS, followed by heatmap analysis, revealed that 73 differentially expressed proteins were enriched for chaperone binding and transforming growth factor-beta (TGF-β). TGF-β3 was upregulated in MT-lmwCAFS compared to lmwCAFS and was correlated to the inhibition of PC-3 proliferation (Laverty et al. 2009). Exactly 188 differentially expressed proteins were observed between lmwCAFS-treated PC-3 cells and MT-lmwCAFS-treated PC-3 cells. Heatmap analysis revealed that the cell cycle was the most prominently altered biological process. Moreover, other enriched dysregulated processes included response to stress, response to growth factors, and regulating lipid metabolism. KEGG analysis showed that Forkhead box O (FoxO), regulated by AMPK, insulin, and TGF-β, altered the most prominent biological process. Further STRING analysis confirmed the interaction between TGF-β3, which arrested the cell cycle in PC-3 cells, and FoxO pathway proteins IL-6, SMAD6, and TGFBR2 (TGF beta receptor 2). This confirmed other research indicating therapeutic interventions upregulate TGF-β expression and cause cancer cell cycle arrest (Chung et al. 2013).

2.9 Melanoma

Skin melanoma accounted for over 324,000 new cases and over 57,000 deaths in 2020 (Sung et al. 2021). Sleep disorders are listed as IARC Group 2A carcinogens, with several epidemiological and clinical studies confirming that various sleep disorders, such as sleep fragmentation, significantly increase the risk of cancer (Blask 2009). Xian et al. (2021) performed a study on the effects of chemically characterized *Ganoderma lucidum* water-soluble polysaccharopeptide (GL-pp) in Balb/c melanoma model (B16-F10-luc-G5 luciferase expressing cells) under conditions of sleep fragmentation (SF). Although the survival rate after 15 days was 100% in all groups, in vivo imaging using luciferase demonstrated that sleep fragmentation (T + SF) in the tumor-bearing group exerted an elevated tumor burden compared with either the T or GL-pp group. GL-pp treated group, which was likewise subjected to sleep fragmentation, exhibited a lower tumor burden and fewer lung metastases, which confirmed GL-pp anticancer and antimetastatic effects in conditions of sleep fragmentation. Lung tissue label-free quantitative proteomics followed by global gene network KEGG analysis detected differentially expressed 43 key regulatory genes between T + SF and GL-pp groups. Focal adhesion, glycerophospholipid and purine metabolism, and ECM-receptor interactions were also identified by KEGG pathway analysis as upregulated mechanisms after GL-pp treatment, while mRNA processing was among the downregulated pathways. Cytoscape analysis revealed that Lama2 (laminin subunit alpha-2) appeared in the "focal adhesion" and "response to hormone" clusters and that the degrees of the Lama2 were the highest. Lama2 downregulation has been observed in liver, ovarian, lung, breast, colorectal cancer, and laryngeal squamous cell carcinoma (Jhunjhunwala et al. 2014). PTK2 protein tyrosine kinase 2 (PTK2) or focal adhesion kinase (FAK) and growth factor receptor-bound protein 2 (Grb2) also had tight correlations in the focal adhesion pathway. Accordingly, it was hypothesized that GL-pp exerts antimetastatic effects by disturbing the FAK function and the pathway involving Ptk2 and Grb2. In this research, Xian et al. have also shown that GL-pp treatment had a significant impact on gut microbiota by decreasing the microbial taxa ratio Firmicutes: Bacteroidetes (F: B), which, if elevated, may result in inflammatory processes and poor prognosis in diverse pathologies (Ismail et al. 2011; Spychala et al. 2018).

2.10 Sarcoma

Sarcomas are a large group of malignant tumors originating from mesenchymal transformation, that is, connective tissue such as bone, cartilage, vascular, fat, or hematopoietic tissues. These tumors are relatively rare, accounting for approximately 1% of all malignant tumors in adults and up to 15% in children aged 0–4 (Toro et al. 2006). The effects of *Gl*PS, a *Ganoderma lucidum* polysaccharide peptide, were investigated in a murine sarcoma 180 (S180) model (Li et al. 2008). *Gl*PS, obtained by hot extraction of the *G. lucidum* fruiting body, was used for ten days after tumor inoculation in Balb/c mice, and a dose-dependent tumor growth inhibition was observed (up to 45.24%). Proteomic analysis (SDS-PAGE) of the animal serum revealed that three proteins were dysregulated due to the treatment. Serum amyloid A (SAA) and haptoglobin were upregulated, while only apolipoprotein A-II was downregulated. SAA is a major acute-phase serum protein important in host immunity, stimulating leukocyte recruitment (Badolato et al. 1994). Moreover, SAA has a role in tumor cell adhesion and migration by inhibiting malignant cell attachment to ECM and inducing the expression of enzymes degrading the ECM (Gutfeld et al. 2006). The antitumor effects of *Gl*PS-treated serum on tumor cell adhesion were tested and corroborated in PC-3M prostate cancer cells, where it significantly inhibited adhesion ability to HUVEC endothelial cells.

2.11 Lymphoma

Lymphomas are tumors that develop from lymphocytes, a white blood cell type. According to WHO 2017 classification, more than 80 mature lymphoma types are recognized, which can be grouped into three major categories: B-cell neoplasms, T-cell and NK-cell neoplasms, and Hodgkin lymphomas (HLs). This classification includes both lymphomas and lymphoid leukemias (De Leval and Jaffe 2020). The effects of fractionated proteins from *Agrocybe aegerita* (AA-f2-MNC-CM) were studied by proteomics on the U937 human myeloid leukemia cell line (Wang et al. 2004). *Agrocybe* (or *Cyclocybe*) *aegerita* (poplar or chestnut mushroom) is an edible, high-quality mushroom. Proteins from the cold-water extract of this mushroom have been shown to reduce the proliferation of U937 cells incubated with human blood mononuclear cells (MNC) by about 80% (Ou et al. 2005). The proteomic analysis between the mononuclear cell-conditioned medium (MNC-CM) *vs.* MNC-CM treated with 5, and 25 μg of AA-f2-MNC-CM revealed that all identified proteins were downregulated. The downregulated proteins have a role in DNA and RNA synthesis, various catabolism pathways (citric acid cycle, glycolysis pathway, uric acid synthesis, and pentose phosphate pathway), and protein conformation. According to these results, the inhibition of proliferation and differentiation of U937 cells by AA-f2-MNC-CM may be associated with the inhibition of disulfide bond formation and proteolysis of cellular proteins. Moreover, processes involved in cell proliferation (inhibition of RNA and DNA replication and catabolism pathways inhibition) along with the accumulation of toxic intermediates may be associated with the observed effects.

A summary of the proteomic research on the antitumor properties of medicinal mushrooms is presented in Fig. 1.

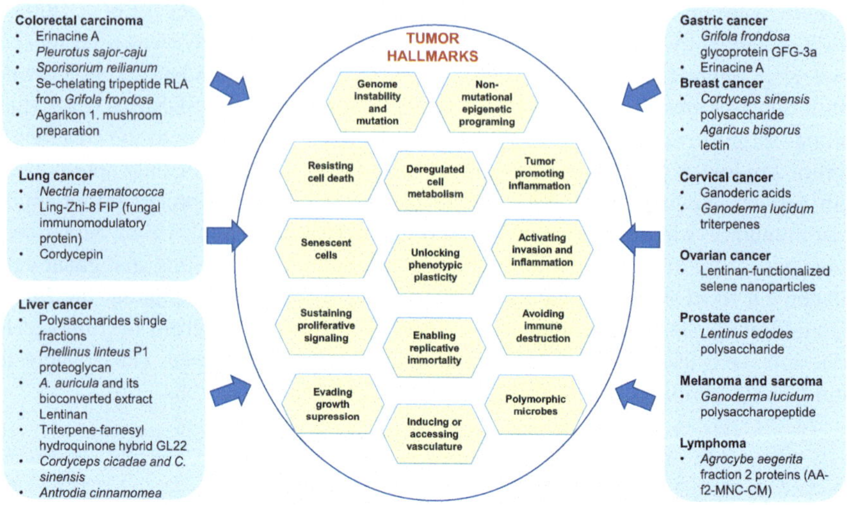

Fig. 1 Mushroom constituents tested on various tumor models using proteomics methods show different molecular mechanisms directed towards major tumor hallmarks. Tumor hallmarks encompass a wide range of biological processes showing a need for an integrative and combined treatment. Cancer hallmarks are described in detail in Hanahan (2022)

3 Proteomic Research on Immunomodulatory Properties

In the last decades, significant discoveries have led to the conclusion that tumor biology can not only be understood by characterizing the tumor cells themselves but must also encompass the various effects of the tumor microenvironment in tumorigenesis. This allowed for a conceptual shift in understanding cancer systemically and not only as a purely genetic disease. The tumor immune microenvironment is a very complex and dynamic system of a tumor, immune, and stromal, that is, support cells that include tumor-associated fibroblasts and endothelial cells (Fridman et al. 2012). There are multiple active compounds in medicinal mushrooms, from high molecular weight ones, such as polysaccharides (especially beta-glucans), polysacharopeptides, and fungal immunomodulatory proteins (FIPs) to low molecular weight compounds, such as lectins, which have important immunostimulatory effects. These properties enable anticancer action in all three tumor immunoediting stages: elimination, equilibrium, and escape. This is a consequence of the known inherent capability of the immune system to affect tumors quantitatively and qualitatively by modulating tumor immunogenicity.

Ganoderma lucidum spores are one of the sources of polysaccharides with immunomodulatory functions since they induce MAPK pathway and spleen tyrosine kinase Syk-dependent TNF-α and interleukin-6 secretion (Guo et al. 2009). The treatment of murine splenic mononuclear cells (MNCs) with *Ganoderma lucidum* spores (GL-SP) resulted in a dose-dependent increase in the proliferation of these cells, as well as a rise in IL-2 and TNF-α production (Ma et al. 2008). Proteomics revealed ten proteins that were dysregulated in MNCs after treatment, which can be classified into three categories based on their biological function. 14–3-3 tau protein is in a group of dysregulated proteins involved in cell viability, and its downregulation increases the total number of viable mononuclear cells (Hermeking and Benzinger 2006). Among the five proteins involved in cell activation and motility, the upregulation of phosphatidylinositol transfer protein α (PITP alpha), which modulates cellular responses of lymphocytes to LPS and other mCD14 ligands, and of Rho, GDP dissociation inhibitor beta, which has important roles in the maintenance of marginal zone T cells and retention of mature T cells in thymic medulla, clearly indicates the mechanisms of GL-SP immunomodulating effects (Wang et al. 1998; Ishizaki et al. 2006). The third group of proteins, downregulated (tubulin alpha, beta-actin, and gamma actin), have roles in cytoskeleton structure and maintenance of cell shape and motility, so their dysregulation could indicate cytoskeletal remodeling during lymphocyte activation (Miletic et al. 2003).

The effects of *Ganoderma lucidum* spore polysaccharides (GL-SP) on the potential reversal of cyclophosphamide (Cy) induced immunosuppression in a KM mice model was studied by Ma et al. (2009). Immunosuppression is one of the important side effects of this alkylating agent and is driven by increased free radical production and apoptosis of thymic immune cells (Fraiser et al. 1991). The study showed Cy decreased body and thymus weight due to its toxic and immunosuppressive effect. Contrarily, the application of GL-SP partially prevented a thymus Cy-induced injury. This combined use of GL-SP with Cy showed those proteins that may be considered GL-SP's possible target proteins. One of these proteins is glutathione peroxidase (GSH-Px), one of the primary antioxidant enzymes that scavenge organic hydroperoxides and hydrogen peroxide. Another such protein is the platelet-activating factor (PAF) acetylhydrolase, with a prominent role in immunity and previously identified as dysregulated in pathological inflammation states (sepsis, shock, traumatic injury) (Prescott et al. 2000). Further on, NADH-ubiquinone oxidoreductase, involved in electron transport chain in mitochondria and the main source of reactive oxygen species (ROS) in mitochondria, is also a possible GL-SP's target protein (Zickermann et al. 2008). These results show that GL-SP impacts the Cy-induced toxicity and immunosuppressive processes mediated by oxidative stress and apoptosis (Bischoff et al. 2000).

A proteomics study was further performed in murine macrophage cell line RAW 264.7 treated with PEP 1b, a novel *Pleurotus eryngii* immunoregulatory protein, by Ma et al. (2020). This protein enhances cellular immune response, particularly by induction of the cytokine and NO (nitric oxide) secretion mediated by Toll-like receptor 4 (TLR4) (Hu et al. 2018). The iTRAQ-based quantitative proteomic analysis revealed a dose-response increase in the expression level of some proteins with the rising concentration of PEP 1b (50–200 μg/mL). These proteins include cyclooxygenase 2 (Cox2), Ras-related protein 1b (Rap1b), and sequestosome (Sqstm 1), macrophage migration inhibitory factor (Mif), and interferon-induced transmembrane protein 3 (Ifitm3). KEGG analysis pointed to PEP 1b's effect on the macrophages and their immunoregulation function. Further on, the KEGG analysis showed that NF-κB, VEGF, and TNF pathways were among the multiple pathways upregulated due to PEP 1b treatment. PEP 1b modulated the MAPK pathway through Mif, superoxide dismutase (Cu-Zn), and peroxiredoxin 2 (Prdx2) upregulation. Moreover, PEP 1b affected nitric oxide biosynthesis by upregulating Cox2, heat shock protein (Hsp90aa1), protein tyrosine kinase 2 beta (Pyk2), important in signaling, as well as integrin-beta 2 (Itgb2).

The role of mushroom compounds as immunosuppressors, particularly important in transplantation and autoimmune diseases, has also been researched. The proteome of human T lymphocytes after treatment with cyclosporine A (CsA) and polysaccharopeptide (PSP) from *Trametes versicolor* has been analyzed by Lee et al. (2007). Cyclosporine A from the fungus *Tolypocaldium inflatum* is used in the clinic for the prevention of graft-versus-host disease as well as in certain autoimmune disorders (Morris 1991). Further on, PSP is a potent immunomodulatory agent used for neoplasms treatment and infection control. It also has immunosuppressive effects that are mediated by the suppression of IL-2 (Cui and Chisti 2003; Lee et al. 2007). CsA and PSP have both shown a reduction of the stimulation index in T cells of ex vivo healthy human male donors. MALDI-TOF MS analysis revealed a change in regulating multiple proteins important in immune cell metabolism after CsA and PSP treatment. These include the upregulated Rho GDP dissociation inhibitor β, which affects the GDP/GTP conversion, and alpha-enolase, which acts on the metabolism of glucose and migration of immune cells (Olofsson 1999). Proteasome-mediated degradation pathway, which is required for IL-2 signaling, was downregulated after PSP or CsA treatment (Yu and Burakoff 1997). Galectin-1, an anti-inflammatory agent and an inhibitor of lymphocyte effector functions, was upregulated after CsA treatment (Rabinovich et al. 2000). Ferritin was one of the proteins affected by PSP only, its upregulation reducing the availability of free iron necessary for immune function. These results indicate a possibility of including PSP in immunosuppressive therapies.

4 Proteomic Research on Antidiabetic Properties

With 537 million adults living with diabetes, and 6.7 million deaths in 2021 alone, this is the world's most common endocrine disease, which involves metabolic disorders of carbohydrates, fats, and protein (Karvonen et al. 2000; IDF Diabetes Atlas 2021). While the main characteristic of the disease is heightened blood sugar level (hyperglycemia), long-term metabolic disturbances lead to complications such as cardiovascular disease, stroke, chronic kidney disease, cognitive deficit, and nerve and eye damage (WHO 2022a). Besides the pancreas, diabetes is closely linked to liver dysfunction, which involves changes in glycogen, lipid metabolism, and antioxidant status (McLennan et al. 1991; O'Meara et al. 1991; Chatila and West 1996). The need for novel medicines from natural sources remains, given the various side effects and toxicity observed with current synthetic antidiabetic drugs (Zhang and Lin 2004; Feng et al. 2022).

The effects of orally administered extracellular mushroom polysaccharides (EPS) from *Phellinus baumii* on the liver proteome in streptozotocin (STZ)-induced diabetic rats were investigated by Hwang et al. (2007). *P. baumii* ETS given 48 h after STZ treatment exhibited an excellent hypoglycemic effect, lowering the average plasma glucose level in EPS-fed rats to 55.1% of STZ-treated rats. All 69 dysregulated proteins in the diabetic rats were noted to be partially or fully restored to healthy (i.e., nondiabetic control) levels after EPS treatment. These dysregulated proteins were relevant in five functional categories: carbohydrate metabolism, lipid metabolism, amino acid metabolism, energy metabolism, and oxidative stress response. In this study, proteomics revealed that two carbohydrate metabolism-related proteins were restored to healthy levels after EPS administration: L-type pyruvate kinase (glycolytic enzyme) and fructose-1,6-bisphosphatase (FBTASE) (gluconeogenesis). Six proteins involved in lipid and fatty acid metabolism, known to be altered in hyperglycemia, were found to be altered in diabetic rats. Cholesterol esterase, known to be increased in diabetes, which results in higher total cholesterol and triglyceride concentrations, was downregulated by EPS. On the other hand, apolipoprotein A-I (apo A-I), which transports cholesterol from tissues to the liver, was decreased in diabetic rats and upregulated to higher than control levels after EPS treatment. 14–3-3 proteins were among the ones involved in the amino acid metabolism normalized by EPS. 14-3-3 gamma was important in glucose transporter type 4 (GLUT4) trafficking and was found to be upregulated after EPS treatment (Ramm et al. 2006). Dysregulated energy metabolism-associated proteins included 10-formyltetrahydrofolate (FDH), reduced in *lep/lep* mice (obese mice) and upregulated by EPS. Proteins relevant to maintaining intracellular ATP levels decreased in alloxan-treated insulin-secreting cells, ATP synthase β chain, and mitochondrial H^+-ATP synthase α subunit, normalized by EPS treatment (Sakurai et al. 2001). Oxidative stress response proteins dysregulated by diabetes induction and normalized by EPS treatment included various heat shock proteins (HSPs) and enzymes (catalase, glutathione S-transferase, peroxiredoxin I and II).

The follow-up research by Kim et al. (2008) analyzed the transcriptome and proteome of rat pancreas after diabetes induction and after mycelial culture of *Phellinus baumii* EPS treatment. Immunohistochemistry revealed that EPS treatment increased the insulin antigenicity of diabetic islet β-cells, suggesting a proliferative or regenerative effect of EPS treatment. Proteomic results confirmed that EPS prevents dysregulation of oxidative stress proteins (Cu/Zn-SOD and PRX IV), which results in antioxidant and β-cell proliferation. Diabetes increases the activity of pancreatic lipase (PL) while decreasing amylase activity. EPS treatment reversed these activities, restoring them to normal levels completely. Further, proteins normalized by EPS treatment included those with functions in fatty acid oxidation, namely, acetyl-coenzyme A dehydrogenase (ACAD), 3-hydroxy-3-methylglutaryl-CoA synthase 2 (HMGCS), triacylglycerol lipase (PL), with the potential of resolving diabetic ketoacidosis. Processes altered after EPS treatment also involved pancreatic cell differentiation (NDK; nucleoside diphosphate kinase, PDI; protein disulfide isomerase, Reg; islet of Langerhans regenerating protein; and PRX; peroxiredoxin) and insulin biosynthesis and release (MAWDBP, NDK, and PDI).

Dyslipidemia, characterized by high triglyceride, low high-density lipoprotein (HDL), and small and dense low-density lipoprotein (LDL) in the blood, is one of the main features of insulin resistance (Gordon and Rifkind 1989). Proteins of plasma from *ob/ob* mice were analyzed by 2-DE followed by MALDI-TOF after treatment with extracellular polysaccharides (EPS) from *Tremella fuciformis* (Kim et al. 2009). The hypoglycemic activity of these EPS from the mycelial culture has been established. Various proteins involved in lipid metabolism were found to be altered after treatment in comparison with the control. Apolipoprotein A-I (Apo A-I) is a protein involved in lipid metabolism, which was upregulated after EPS treatment, which indicated an increase in muscle glucose uptake and a decrease in gluconeogenesis in the liver, mediated by AMP-activated protein kinase (AMPK). Apo A IV, which correlates with triglyceride levels and is increased in diabetic patients, was downregulated. EPS-upregulated adiponectin stimulates glucose metabolism through AMPK and promotes insulin sensitivity. Retinol-binding protein 4 (RBP4), associated with many metabolic syndromes, such as obesity and diabetes mellitus type 2 (T2DM), is also a predictor of insulin resistance and cardiovascular risk. Hence, its downregulation indicates its antidiabetic effect (Wolf 2007). Resistin, also known as an adipose tissue-specific secretory factor (ADSF), has a role in the development of insulin resistance in T2DM patients and is positively correlated with triglycerides, waist circumference, systolic blood pressure, and apo A-I/apo B ratio was downregulated (Norata et al. 2007).

5 Proteomic Research on Hypolipidemic and Antiatherosclerotic Properties

Hyperlipidemia is a form of dyslipidemia characterized by elevated levels of total cholesterol (TC), low-density lipoprotein cholesterol (LDL-C), and triacylglycerols (TG), accompanied by a reduced concentration of high-density lipoprotein cholesterol (HDL-C) in the blood. This metabolic disorder is a primary risk factor for the development of arteriosclerosis and cardiovascular disease, which are the leading causes of death globally (WHO 2022b). Furthermore, hyperlipidemia is a common comorbidity in diabetes (Zhang et al. 2009). Statins are currently the most used hypolipidemic drugs, which reduce serum TC levels in patients by inhibiting hepatic 3-hydroxy-3-methyl-glutaryl-coenzyme A reductase (HMGCR) (Farmer 1998). Various medicinal mushrooms, such as *Pleurotus ostreatus*, *Craterellus cornucopioides*, *Amanita ponderosa*, *Auricularia auricula*, and *Lentinus edodes*, were shown to inhibit HMGCR in vitro or in vivo at high percentages (52–76%) (Ding et al. 2016).

Ding et al. (2016) investigated a rat liver proteome after treatment with *Grifola frondosa* fruiting bodies. The hyperlipidemic group was fed a hypercholesterolemic diet. Administration of *Grifola frondosa* to the hyperlipidemic group did not significantly reduce body weight gain compared to the untreated hyperlipidemic rats.

However, it significantly lowered the serum TG, TC, and LDL-C. The lowering of high-serum LDL is generally accepted as a factor in preventing and reversing atherosclerosis and cardiovascular disease, thus defining *G. frondosa* as food with a potential lipid-lowering effect (Li et al. 2010). Proteomic analysis revealed that 8 out of 20 differentially expressed proteins were relevant in hepatic lipid metabolism. HMGCR was found to be downregulated, which inhibits cholesterol synthesis and decreases serum TC levels. The treatment also resulted in the downregulation of long-chain-fatty-acid—CoA ligase 1 (ACSL1), required for fatty acid biosynthesis, and in the upregulation of alpha-methylacyl-CoA racemase (AMACR), which is necessary for fatty acid oxidation, that is, catalysis (Mobley et al. 2003). Antioxidative enzymes SOD1, PRDX-1, and hemopexin (HPX), a plasma protein with antioxidant and immunoregulatory functions, were upregulated, which can indicate an increased resistance to oxidized LDL (oxLDL) and foam cells formation, which are recognized as early atherosclerosis events (Augusti et al. 2012). S-adenosylmethionine synthase isoform type-1 (MAT1A) and glutamate-cysteine ligase regulatory subunit (GCLM) were the dysregulated proteins relevant to amino acid metabolism. These proteins are important in glutathione (GSH) synthesis and, therefore, indirectly in lipid peroxidation.

In another study, the lipolysis effects of the *Ganoderma lucidum* protein hydrolysates on 3T3-L1 differentiated adipocytes were studied by proteomics (Krobthong et al. 2021). The protein hydrolysates were loaded on nanoscale liposome-based carriers since liposome encapsulation is a delivery system that optimizes the active compound's stability and bioactivity (Allen and Cullis 2013). The viability of control fibroblasts and 3 T3-L1 cells was not significantly affected, up to a concentration of 52.34 μg/mL, which indicates the no-observed-adverse-effect level. Triglyceride breakdown was increased in loaded liposome-treated adipocytes, which was evident by increased glycerol, a byproduct of lipolysis release, and reduced lipid accumulation. Oil Red O (ORO) staining showed a reduced lipid accumulation in the liposome-treated adipocytes. The proteomic analysis detected a fivefold suppression of fatty acid synthase (FAS), which has significant functions in lipogenesis by synthesizing long-chain fatty acids from acetyl-CoA, malonyl-CoA, and NADPH. A protein that elongates long-chain fatty acids, PUFA elongase (Elov5), was downregulated.

6 Proteomic Research on Hepatoprotective Properties

Liver fibrosis, called cirrhosis or end-stage liver disease, is caused by most chronic liver diseases, including viral hepatitis (B and C), alcoholic liver disease, and biliary diseases (Gines et al. 2022). It is a consequence of long-term hepatic inflammation, which can, in its advanced irreversible stage, lead to hepatic encephalopathy and liver cancer (Yang et al. 2022).

Phellinus linteus, known as meshimakobu in Japanese, song gen in Chinese, or sanghwang in Korean, is a medicinal mushroom known for its antioxidative, antiangiogenic, and anticancer properties, including a spontaneous regression of hepatocellular carcinoma in patients with multiple metastases (Nam et al. 2005; Zhu et al. 2008). Wang et al. (2012) conducted proteomic research in a thioacetamide (TAA)-induced rat model to study the hepatoprotective mechanisms of lyophilized hot water extract of wild-type *P. linteus* (PLP). The induction of liver fibrosis in TAA-treated rats was confirmed by histological assessment. In contrast, a marked reduction of fibrosis and inflammation in TAA-treated rats was observed after oral PLP administration (50 mg/kg twice daily for two weeks). Proteomics revealed 13 differentially expressed hepatic proteins in TAA-induced fibrosis rats treated with PLP. Dysregulated proteins with antioxidative effects included hemopexin, preprohaptoglobin, and glutathione S-transferase alpha-4 (GSTA4). PLP treatment upregulated preprohaptoglobin while downregulating hemopexin, while opposite tendencies have been recognized as potential markers of fibrosis because of their involvement in liver iron homeostasis (Cheung et al. 2009). The increase in GSTA4 after PLP treatment is in line with the increased GSH-related antioxidative detoxification, in which this protein plays a central role. The reduced expression of betaine-homocysteine S-methyltransferase 1 (BHMT1), and mitochondrial 2-oxoisovalerate dehydrogenase subunit alpha (BCKDHA), involved in branched-chain amino acid (BCAA) catabolism indicate the promotion of accumulation of substrates for GSH synthesis, cysteine, and glutamate, respectively (Wu et al. 2004). The upregulation of BCKDHA, as well as of quinoid dihydropteridine reductase (QDPR), dihydrofolate reductase (DHFR), and ribonuclease UK114 indicate an elevated amino acid and nucleic acid metabolism, with the potential of relieving protein malnutrition in liver disease and thus aiding liver regeneration (Holecek 2010). A highly upregulated (6.5-fold) galectin-5 expression after PLP administration indicates the promotion of erythropoiesis, inflammation regulation, and liver regeneration.

In another study, the effects of *Ganoderma lucidum* ethanol extracts (GLE) on alcohol-induced liver injury in mice were studied using an iTRAQ-based proteomics approach (Zhao et al. 2019). *Ganoderma lucidum* triterpenoids and polysaccharides have shown hepatoprotective efficacy in cadmium-, α-amanitin, and tetrachloride (CCl_4)-induced hepatocyte damage (Jin et al. 2013; Liu et al. 2015; Wu et al. 2016). GLE restored hepatic histology, reduced serum AST and ALT levels, and enhanced antioxidant activity by increasing GSH and SOD levels. Exactly 457 differentially expressed proteins were found between the EtOH group and GLE/EtOH group, with their biological functions being mitochondrial respiration, calcium metabolism, energy metabolism, enzyme metabolism, and lipid metabolism. Protein-protein interaction (PPI) analysis revealed that cytochrome p4590 2E1 (CYP2E1) and alcohol dehydrogenase (ADH1) interacted with most other proteins. Prolonged ethanol consumption increases oxidative stress in the liver and induces the CYP2E1 enzyme, which metabolizes and activates ethanol to more reactive and toxic products such as acetaldehyde and 1-hydroxyethyl radical (Wu et al. 2010). GLE treatment reduced the CYP2E1 level significantly compared with the EtOH untreated group, while ADH1 increased. ADH1 converts toxic alcohols into aldehydes and ketones, which are used in various metabolic processes.

Arsenic (As) is an environmental toxin and human carcinogen; exposure is known to damage various organ systems, especially the cardiovascular system and liver (Goudarzi et al. 2018). The hot water extract of the fruiting body of *Dictyophora indusiata* (*Dictyophora* polysaccharide, DIP) has been shown to antagonize the arsenic cytotoxicity to L-02 human hepatic cells and inhibit sodium arsenite ($NaAsO_2$)-induced apoptosis (Hu et al. 2021). Western blot analysis revealed that this apoptosis inhibition was accompanied by a downregulation of proapoptotic Bax and an upregulation of antiapoptotic Bcl-2. iTRAQ proteomic analysis showed that among the identified proteins, 60, 71, and 13 proteins were identified as DEPs in As/Ctrl group, DIP + As/As a group, and DIP + As/Ctrl groups, respectively. Moreover, cluster analysis revealed that the expression changes in As-treated cells were largely reversed in As-exposed cells that were pretreated with DIP, with the main changes observed in ribosomal proteins, apoptosis-related, mitochondria-related, and metabolism-related proteins. Sixteen proteins that are involved in apoptotic mitochondrial changes, mitochondrial morphogenesis, mitochondrial cytochrome c release, and transcription, such as clusterin (CLU), mitochondrial fission factor (MFF), peptidyl-prolyl cis-trans isomerase F, mitochondrial (PPIF) and eukaryotic translation initiation factor 5A-1 (EIF5A) which were dysregulated by As, were reversed by DIP. The effect of arsenic on reducing the levels of various metabolic-related proteins, thus decreasing cellular metabolism (glycolysis), has also been noted. Their expression was found to be largely normalized by DIP treatment. Cytoplasmic ribosomal proteins (RPs) were largely downregulated after the As treatment. The reversal in their expression in the DIP + As group indicated a restoration of their functions, including various adaptive responses and apoptosis activation (Chen and Ioannou 1999). On the other hand, mitochondrial ribosomal proteins (MRPs) were largely upregulated after As treatment, while DIP reversed these changes.

7 Proteomic Research on Neuroprotective Properties

Neurotoxicity results from exposure to a biological, chemical, or physical agent, leading to reversible or permanent damage to nervous tissue (Cunha-Oliveira et al. 2008). Arsenic is a common environmental and occupational contaminant that affects 200 million people globally (Duan et al. 2020). It is known to cross the blood-brain barrier, and its prooxidative and apoptotic effects in the striatum and hippocampus can lead to decreased learning and memory capacity (Pandey et al. 2017). Zhang et al. (2022) investigated the neuroprotective effects of *Dictyophora indusiata* polysaccharide (DIP) on arsenic-exposed rats. Sodium arsenite ($NaAsO_2$)-exposed rats treated with DIP showed an improvement in spatial learning and memory, as measured by Morris Water Maze (MWM) test. SWATH-MS analysis revealed 172, 75, and 82 DEPs in the As/Ctrl, DIP + As/As, and DIP + As/Ctrl groups, respectively.

Compared to the As/Ctrl group, most DEPs in the As/Ctrl group returned to previous expression levels or reversed in the As + DIP/As group. Hub gene analysis by Cytohubba revealed that caveolin 1 (Cav1) and Hsp90aa1 were the common hub genes among the As/Ctrl and DIP + As/As groups which belonged to a group with opposite expression trends (reversed proteins). Proteomics revealed that exposure to arsenic led to a differential expression of proteins involved in the expression of synapses and neuron-related proteins. Certain downregulated proteins, such as protein unc-13 homolog A (Unc13a) and metabotropic glutamate receptor 7 (GRM7), are involved in excitatory glutaminergic signaling, which is thought to result in learning and memory deficits observed in As-treated rats. The expression of these proteins was reversed or restored after DIP treatment. Also, various proteins contributing to As-induced neuronal apoptosis were dysregulated. These included keratin, type I cytoskeletal 18 (Krt18), and caveolin-1 (Cav1), identified as a common hub protein in the As/Ctrl and DIP + As/As group. It is known that As damages the mitochondrial respiratory chain, leading to ROS-induced apoptosis (Rezaei et al. 2019). In this study, 29 mitochondrial-associated proteins (including Hsp90aa1) in the hippocampus were altered after As treatment, but their expression reversed in the DIP + As/As group. Most energy metabolism proteins, specifically pyruvate metabolism, ascorbate, aldarate metabolism, ribonucleotide catabolic process, and glycosaminoglycan degradation, were downregulated in arsenic-treated groups. However, their expression was reversed after DIP treatment. Energy deficiency is a known mechanism of As-associated nerve injury, so restoring energy metabolism, reducing apoptosis, and relieving oxidative stress are the main mechanisms of DIP neuroprotective activities (Panchal and Tiwari 2019).

8 Conclusions

The use of the system biology approach that is now increasingly beginning to reexamine the historical reductionist approach has begun in the medicinal mushroom field, where large-scale proteomic research has been conducted into their anticancer, immunomodulatory, antidiabetic, hypolipidemic and antiatherosclerotic as well as hepatoprotective and neuroprotective properties. The previous research into therapeutic properties could be regarded as partial or preliminary given this new approach. Still, various challenges remain. The analytical proteomic and bioinformatic methods have rapidly developed over a short period and vary in their sensitivity, influencing result interpretation. The variability of the results obtained can also be a consequence of a model which is being researched in terms of scale (cells, tissues, organisms), disease stage (early vs. late models of disease), and other disease-specific parameters (e.g., immunocompetent vs. nude xenografts in tumor models). Also, the importance of many proteins for which the function is not yet known must not be underestimated. Recently, an Understudied Protein Initiative has been proposed to reduce the annotation gap by systematically associating uncharacterized proteins with proteins of known biological function (Kustatscher et al. 2022). This should result in casting a wider net through mechanistic studies in the fight against various diseases since out of ~3000 druggable proteins, only 5–10% are currently targeted by current pharmaceuticals (Oprea et al. 2018).

References

Abulizi A, Hu L, Ma A et al (2021a) Ganoderic acid alleviates chemotherapy-induced fatigue in mice bearing colon tumor. Acta Pharmacol Sin 42:1703–1713. https://doi.org/10.1038/s41401-021-00669-6

Abulizi A, Ran J, Ye Y et al (2021b) Ganoderic acid improves 5-fluorouracil-induced cognitive dysfunction in mice. Food Funct 12:12325–12337. https://doi.org/10.1039/d1fo03055h

Ahmad A (2019) Breast cancer statistics: recent trends in advances in experimental medicine and biology. In: Ahmad A (ed) Breast cancer metastasis and drug resistance: challenges and progress, 2nd edn. Springer, New York. https://doi.org/10.1007/978-3-030-20301-6_1

Ahmed N (2005) 23 years of the discovery of Helicobacter pylori: is the debate over? Ann Clin Microbiol Antimicrob 4:17. https://doi.org/10.1186/1476-0711-4-17

Allen TM, Cullis PR (2013) Liposomal drug delivery systems: from concept to clinical applications. Adv Drug Deliv Rev 65:36–48. https://doi.org/10.1016/j.addr.2012.09.037

Augusti PR, Ruviaro AR, Quatrin A et al (2012) Imbalance in superoxide dismutase/thioredoxin reductase activities in hypercholesterolemic subjects: relationship with low density lipoprotein oxidation. Lipids Health Dis 11:79. https://doi.org/10.1186/1476-511X-11-79

Ayene IS, Ford LP, Koch CJ (2005) Ku protein targeting by Ku70 small interfering RNA enhances human cancer cell response to topoisomerase II inhibitor and gamma radiation. Mol Cancer Ther 4:529–536. https://doi.org/10.1158/1535-7163.MCT-04-0130

Badolato R, Wang JM, Murphy WJ et al (1994) Serum amyloid A is a chemoattractant: induction of migration, adhesion, and tissue infiltration of monocytes and polymorphonuclear leukocytes. J Exp Med 180:203–209. https://doi.org/10.1084/jem.180.1.203

Bailly C, Gao JM (2020) Erinacine A and related cyathane diterpenoids: molecular diversity and mechanisms underlying their neuroprotection and anticancer activities. Pharmacol Res 159: 104953. https://doi.org/10.1016/j.phrs.2020.104953

Barni S, Cabiddu M, Ghilardi M et al (2011) A novel perspective for an orphan problem: old and new drugs for the medical management of malignant ascites. Crit Rev Oncol Hematol 79:144–153. https://doi.org/10.1016/j.critrevonc.2010.07.016

Bischoff PL, Holl V, Coelho D et al (2000) Apoptosis at the interface of immunosuppressive and anticancer activities: the examples of two classes of chemical inducers, oxysterols and alkylating agents. Curr Med Chem 7:693–713. https://doi.org/10.2174/0929867003374769

Blask DE (2009) Melatonin, sleep disturbance and cancer risk. Sleep Med Rev 13:257–264. https://doi.org/10.1016/j.smrv.2008.07.007

Cai H, Li J, Gu B et al (2018) Extracts of Cordyceps sinensis inhibit breast cancer cell metastasis via down-regulation of metastasis-related cytokines expression. J Ethnopharmacol 214:106–112. https://doi.org/10.1016/j.jep.2017.12.012

Casares N, Pequignot MO, Tesniere A et al (2005) Caspase-dependent immunogenicity of doxorubicin-induced tumor cell death. J Exp Med 202:1691–1701. https://doi.org/10.1084/jem.20050915

Chai Y, Wang G, Fan L et al (2016) A proteomic analysis of mushroom polysaccharide-treated HepG2 cells. Sci Rep 6:23565. https://doi.org/10.1038/srep23565

Chatila R, West AB (1996) Hepatomegaly and abnormal liver tests due to glycogenosis in adults with diabetes. Medicine (Baltimore) 75:327–333. https://doi.org/10.1097/00005792-199611000-00003

Chen FW, Ioannou YA (1999) Ribosomal proteins in cell proliferation and apoptosis. Int Rev Immunol 18:429–448. https://doi.org/10.3109/08830189909088492
Chen CN, Chang CC, Su TE et al (2009) Identification of calreticulin as a prognosis marker and angiogenic regulator in human gastric cancer. Ann Surg Oncol 16:524–533. https://doi.org/10.1245/s10434-008-0243-1
Chen LC, Chung IC, Hsueh C et al (2010) The antiapoptotic protein, FLIP, is regulated by heterogeneous nuclear ribonucleoprotein K and correlates with poor overall survival of nasopharyngeal carcinoma patients. Cell Death Differ 17:1463–1473. https://doi.org/10.1038/cdd.2010.24
Chen YW, Hu DJ, Cheong KL et al (2013) Quality evaluation of lentinan injection produced in China. J Pharm Biomed Anal 78–79:176–182. https://doi.org/10.1016/j.jpba.2013.02.012
Chen JF, Tsai YT, Lai YH et al (2020) Proteomic analysis of Antrodia Cinnamomea-induced ER stress in liver cancer cells. J Pharm Biomed Anal 187:113142. https://doi.org/10.1016/j.jpba.2020.113142
Cheung KJ, Tilleman K, Deforce D et al (2009) The HCV serum proteome: a search for fibrosis protein markers. J Viral Hepat 16:418–429. https://doi.org/10.1111/j.1365-2893.2009.01083.x
Chihara G, Hamuro J, Maeda YY et al (1987) Antitumor and metastasis-inhibitory activities of lentinan as an immunomodulator: an overview. Cancer Detect Prev Suppl 1:423–443
Chung KS, Cho SH, Shin JS et al (2013) Ginsenoside Rh2 induces cell cycle arrest and differentiation in human leukemia cells by upregulating TGF-beta expression. Carcinogenesis 34:331–340. https://doi.org/10.1093/carcin/bgs341
Clifford G, Franceschi S, Diaz M et al (2006) Chapter 3: HPV type-distribution in women with and without cervical neoplastic diseases. Vaccine 24(Suppl 3):26–34. https://doi.org/10.1016/j.vaccine.2006.05.026
Colombo MP, Chiodoni C (2013) Osteopontin produced by myeloid cells determines the outcome of breast cancer metastases. Mol Cancer Res 11. https://doi.org/10.1158/1557-3125.ADVBC-A102
Cui T, Chisti Y (2003) Polysaccharopeptides of Coriolus versicolor: physiological activity, uses, and production. Biotechnol Adv 21:109–122. https://doi.org/10.1016/S0734-9750(03)00002-8
Cui FJ, Zan XY, Li YH et al (2016) Grifola frondosa glycoprotein GFG-3a arrests S phase, alters proteome, and induces apoptosis in human gastric cancer cells. Nutr Cancer 68:267–279. https://doi.org/10.1080/01635581.2016.1134599
Cunha-Oliveira T, Rego AC, Oliueira CR (2008) Cellular and molecular mechanisms involved in the neurotoxicity of opioid and psychostimulant drugs. Brain Res Rev 58:192–208. https://doi.org/10.1016/j.brainresrev.2008.03.002
Dasgupta P, Henshaw C, Youlden DR et al (2020) Global trends in incidence rates of primary adult liver cancers: a systematic review and meta-analysis. Front Oncol 10:171. https://doi.org/10.3389/Fonc.2020.00171
De Leval L, Jaffe ES (2020) Lymphoma classification. Cancer J 26:176–185. https://doi.org/10.1097/Ppo.0000000000000451
Ding YR, Xiao C, Wu QP et al (2016) The mechanisms underlying the hypolipidaemic effects of grifola frondosa in the liver of rats. Front Microbiol 7:1186. https://doi.org/10.3389/Fmicb.2016.01186
Duan TX, Hu T, Wu CY et al (2020) PINK1/Parkin-mediated mitophagy is involved in NaAsO2-induced apoptosis of human hepatic cells through activation of ERK signaling. Toxicol In Vitro 66:104857. https://doi.org/10.1016/J.Tiv.2020.104857
Eric A, Juranic Z, Milovanovic Z et al (2009) Effects of humoral immunity and calreticulin overexpression on postoperative course in breast cancer. Pathol Oncol Res 15:89–90. https://doi.org/10.1007/s12253-008-9112-2
Farmer JA (1998) Aggressive lipid therapy in the statin era. Prog Cardiovasc Dis 41:71–94. https://doi.org/10.1016/S0033-0620(98)80006-6

Feng J, Wang XH, Ye XC et al (2022) Mitochondria as an important target of metformin: the mechanism of action, toxic and side effects, and new therapeutic applications. Pharmacol Res 177:106114. https://doi.org/10.1016/J.Phrs.2022.106114

Ferlay J, Ervik M, Lam F, Colombet M, Mery L, Pineros M (2020) Global cancer observatory: cancer today. https://gco.iarc.fr/today

Finimundy TC, Abreu RMV, Bonetto N et al (2018) Apoptosis induction by Pleurotus sajor-caju (Fr.) Singer extracts on colorectal cancer cell lines. Food Chem Toxicol 112:383–392. https://doi.org/10.1016/j.fct.2018.01.015

Fraiser LH, Kanekal S, Kehrer JP (1991) Cyclophosphamide toxicity. Characterising and avoiding the problem. Drugs 42:781–795. https://doi.org/10.2165/00003495-199142050-00005

Fridman WH, Pages F, Sautes-Fridman C et al (2012) The immune contexture in human tumours: impact on clinical outcome. Nat Rev Cancer 12:298–306. https://doi.org/10.1038/nrc3245

Gadducci A, Barsotti C, Cosio S et al (2011) Smoking habit, immune suppression, oral contraceptive use, and hormone replacement therapy use and cervical carcinogenesis: a review of the literature. Gynecol Endocrinol 27:597–604. https://doi.org/10.3109/09513590.2011.558953

Gill BS, Navgeet MR et al (2018) Ganoderic acid, lanostanoid triterpene: a key player in apoptosis. Investig New Drugs 36:136–143. https://doi.org/10.1007/s10637-017-0526-0

Gines P, Castera L, Lammert F et al (2022) Population screening for liver fibrosis: toward early diagnosis and intervention for chronic liver diseases. Hepatology 75:219–228. https://doi.org/10.1002/hep.32163

Gordon DJ, Rifkind BM (1989) High-density lipoprotein--the clinical implications of recent studies. N Engl J Med 321:1311–1316. https://doi.org/10.1056/NEJM198911093211907

Gorynia S, Bandeiras TM, Pinho FG et al (2011) Structural and functional insights into a dodecameric molecular machine - the RuvBL1/RuvBL2 complex. J Struct Biol 176:279–291. https://doi.org/10.1016/j.jsb.2011.09.001

Goudarzi M, Fatemi I, Siahpoosh A et al (2018) Protective effect of ellagic acid against sodium arsenite-induced cardio- and hematotoxicity in rats. Cardiovasc Toxicol 18:337–345. https://doi.org/10.1007/s12012-018-9446-2

Guo L, Xie J, Ruan Y, Zhou L, Zhu H, Yun X, Jiang Y, Lü L, Chen K, Min Z et al (2009) Characterization and immunostimulatory activity of a polysaccharide from the spores of Ganoderma lucidum. Int Immunopharmacol 9:1175–1182

Gutfeld O, Prus D, Ackerman Z et al (2006) Expression of serum amyloid A, in normal, dysplastic, and neoplastic human colonic mucosa: implication for a role in colonic tumorigenesis. J Histochem Cytochem 54:63–73. https://doi.org/10.1369/jhc.5A6645.2005

Hanahan D (2022) Hallmarks of cancer: new dimensions. Cancer Discov 12:31–46. https://doi.org/10.1158/2159-8290.CD-21-1059

Hassan MA, Rouf R, Tiralongo E et al (2015) Mushroom lectins: specificity, structure and bioactivity relevant to human disease. Int J Mol Sci 16:7802–7838. https://doi.org/10.3390/ijms16047802

Hermeking H, Benzinger A (2006) 14-3-3 proteins in cell cycle regulation. Semin Cancer Biol 16: 183–192. https://doi.org/10.1016/j.semcancer.2006.03.002

Hesketh PJ (2008) Chemotherapy-induced nausea and vomiting. N Engl J Med 358:2482–2494. https://doi.org/10.1056/NEJMra0706547

Holecek M (2010) Three targets of branched-chain amino acid supplementation in the treatment of liver disease. Nutrition 26:482–490. https://doi.org/10.1016/j.nut.2009.06.027

Hu Q, Du H, Ma G et al (2018) Purification, identification and functional characterization of an immunomodulatory protein from Pleurotus eryngii. Food Funct 9:3764–3775. https://doi.org/10.1039/c8fo00604k

Hu T, Shen L, Huang Q et al (2021) Protective effect of dictyophora polysaccharides on sodium arsenite-induced hepatotoxicity: a proteomics study. Front Pharmacol 12:749035. https://doi.org/10.3389/fphar.2021.749035

Hwang HJ, Baek YM, Kim SW et al (2007) Differential expression of kidney proteins in streptozotocin-induced diabetic rats in response to hypoglycemic fungal polysaccharides. J Microbiol Biotechnol 17:2005–2017

International Diabetes Federation (2021) IDF Diabetes Atlas, 10th edn. International Diabetes Federation, Brussels, Belgium

Ishizaki H, Togawa A, Tanaka-Okamoto M et al (2006) Defective chemokine-directed lymphocyte migration and development in the absence of Rho guanosine diphosphate-dissociation inhibitors alpha and beta. J Immunol 177:8512–8521. https://doi.org/10.4049/jimmunol.177.12.8512

Ismail NA, Ragab SH, Abd Elbaky A et al (2011) Frequency of Firmicutes and Bacteroidetes in gut microbiota in obese and normal weight Egyptian children and adults. Arch Med Sci 7:501–507. https://doi.org/10.5114/aoms.2011.23418

Jain SK, Micinski D, Huning L et al (2014) Vitamin D and L-cysteine levels correlate positively with GSH and negatively with insulin resistance levels in the blood of type 2 diabetic patients. Eur J Clin Nutr 68:1148–1153. https://doi.org/10.1038/ejcn.2014.114

Jakopovic B, Horvatic A, Klobucar M et al (2020a) Treatment with medicinal mushroom extract mixture inhibits translation and reprograms metabolism in advanced colorectal cancer animal model as evidenced by tandem mass tags proteomics analysis. Front Pharmacol 11:1202. https://doi.org/10.3389/fphar.2020.01202

Jakopovic B, Orsolic N, Kraljevic Pavelic S (2020b) Antitumor, immunomodulatory and antiangiogenic efficacy of medicinal mushroom extract mixtures in advanced colorectal cancer animal model. Molecules 25. https://doi.org/10.3390/molecules25215005

Jakopovich I (2011) New dietary supplements from medicinal mushrooms: Dr Myko San: a registration report. Int J Med Mushrooms 13:307–313. https://doi.org/10.1615/intjmedmushr.v13.i3.110

Jayson GC, Kohn EC, Kitchener HC et al (2014) Ovarian cancer. Lancet 384:1376–1388. https://doi.org/10.1016/S0140-6736(13)62146-7

Jemal A, Simard EP, Dorell C et al (2013) Annual report to the Nation on the Status of Cancer, 1975-2009, featuring the burden and trends in human papillomavirus(HPV)-associated cancers and HPV vaccination coverage levels. J Natl Cancer Inst 105:175–201. https://doi.org/10.1093/jnci/djs491

Jeong MK, Yoo HS, Kang IC (2019) The extract of cordyceps militaris inhibited the proliferation of cisplatin-resistant A549 lung cancer cells by downregulation of H-Ras. J Med Food 22:823–832. https://doi.org/10.1089/jmf.2018.4232

Jhunjhunwala S, Jiang Z, Stawiski EW et al (2014) Diverse modes of genomic alteration in hepatocellular carcinoma. Genome Biol 15:436. https://doi.org/10.1186/s13059-014-0436-9

Jin H, Jin F, Jin JX et al (2013) Protective effects of Ganoderma lucidum spore on cadmium hepatotoxicity in mice. Food Chem Toxicol 52:171–175. https://doi.org/10.1016/j.fct.2012.05.040

Kampan NC, Madondo MT, Mcnally OM et al (2017) Interleukin 6 present in inflammatory ascites from advanced epithelial ovarian cancer patients promotes tumor necrosis factor receptor 2-expressing regulatory T cells. Front Immunol 8:1482. https://doi.org/10.3389/fimmu.2017.01482

Kan L, Cui D, Chai Y et al (2020) TMT-based quantitative proteomic analysis of antitumor mechanism of Sporisorium reilianum polysaccharide WM-NP-60 against HCT116 cells. Int J Biol Macromol 165:1755–1764. https://doi.org/10.1016/j.ijbiomac.2020.10.056

Kang C, Xu Q, Martin TD et al (2015) The DNA damage response induces inflammation and senescence by inhibiting autophagy of GATA4. Science 349:aaa5612. https://doi.org/10.1126/science.aaa5612

Kang MA, Jeon YK, Nam MJ (2020) Auricularia auricula increases an apoptosis in human hepatocellular carcinoma cells via a regulation of the peroxiredoxin1. J Food Biochem. https://doi.org/10.1111/jfbc.13373

Karvonen M, Viik-Kajander M, Moltchanova E et al (2000) Incidence of childhood type 1 diabetes worldwide. Diabetes Mondiale (DiaMond) Project Group. Diabetes Care 23:1516–1526. https://doi.org/10.2337/diacare.23.10.1516

Kim KR, Choi HN, Lee HJ et al (2007) A peroxisome proliferator-activated receptor gamma antagonist induces vimentin cleavage and inhibits invasion in high-grade hepatocellular carcinoma. Oncol Rep 18:825–832

Kim SW, Hwang HJ, Baek YM et al (2008) Proteomic and transcriptomic analysis for streptozotocin-induced diabetic rat pancreas in response to fungal polysaccharide treatments. Proteomics 8:2344–2361. https://doi.org/10.1002/pmic.200700779

Kim SW, Hwang HJ, Baek YM et al (2009) Proteomic analysis in ob/ob mice before and after hypoglycemic polysaccharide treatments. J Microbiol Biotechnol 19:1109–1121

Ko JL, Hsu CI, Lin RH et al (1995) A new fungal immunomodulatory protein, FIP-fve isolated from the edible mushroom, Flammulina velutipes and its complete amino acid sequence. Eur J Biochem 228:244–249

Krobthong S, Yingchutrakul Y, Visessanguan W et al (2021) Study of the lipolysis effect of nanoliposome-encapsulated ganoderma lucidum protein hydrolysates on adipocyte cells using proteomics approach. Foods 10. https://doi.org/10.3390/foods10092157

Krysko DV, Garg AD, Kaczmarek A et al (2012) Immunogenic cell death and DAMPs in cancer therapy. Nat Rev Cancer 12:860–875. https://doi.org/10.1038/nrc3380

Kuo HC, Kuo YR, Lee KF et al (2017) A comparative proteomic analysis of erinacine A's inhibition of gastric cancer cell viability and invasiveness. Cell Physiol Biochem 43:195–208. https://doi.org/10.1159/000480338

Kurashige S, Akuzawa Y, Endo F (1997) Effects of Lentinus edodes, Grifola frondosa and Pleurotus ostreatus administration on cancer outbreak, and activities of macrophages and lymphocytes in mice treated with a carcinogen, N-butyl-N-butanolnitrosoamine. Immunopharmacol Immunotoxicol 19:175–183. https://doi.org/10.3109/08923979709007657

Kustatscher G, Collins T, Gingras AC et al (2022) Understudied proteins: opportunities and challenges for functional proteomics. Nat Methods 19:774–779. https://doi.org/10.1038/s41592-022-01454-x

Laverty HG, Wakefield LM, Occleston NL et al (2009) TGF-beta3 and cancer: a review. Cytokine Growth Factor Rev 20:305–317. https://doi.org/10.1016/j.cytogfr.2009.07.002

Lee CL, Jiang PP, Sit WH et al (2007) Proteome of human T lymphocytes with treatment of cyclosporine and polysaccharopeptide: analysis of significant proteins that manipulate T cells proliferation and immunosuppression. Int Immunopharmacol 7:1311–1324. https://doi.org/10.1016/j.intimp.2007.05.013

Lee KC, Kuo HC, Shen CH et al (2017) A proteomics approach to identifying novel protein targets involved in erinacine A-mediated inhibition of colorectal cancer cells' aggressiveness. J Cell Mol Med 21:588–599. https://doi.org/10.1111/jcmm.13004

Li YB, Wang R, Wu HL et al (2008) Serum amyloid A mediates the inhibitory effect of Ganoderma lucidum polysaccharides on tumor cell adhesion to endothelial cells. Oncol Rep 20:549–556

Li H, Zhang M, Ma G (2010) Hypolipidemic effect of the polysaccharide from Pholiota nameko. Nutrition 26:556–562. https://doi.org/10.1016/j.nut.2009.06.009

Li YG, Ji DF, Zhong S et al (2013) Polysaccharide from Phellinus linteus induces S-phase arrest in HepG2 cells by decreasing calreticulin expression and activating the P27kip1-cyclin A/D1/E-CDK2 pathway. J Ethnopharmacol 150:187–195. https://doi.org/10.1016/j.jep.2013.08.028

Li S, Nie Y, Ding Y et al (2014) Recombinant expression of a novel fungal immunomodulatory protein with human tumor cell antiproliferative activity from Nectria haematococca. Int J Mol Sci 15:17751–17764. https://doi.org/10.3390/ijms151017751

Lin TY, Hua WJ, Yeh H et al (2021) Functional proteomic analysis reveals that fungal immunomodulatory protein reduced expressions of heat shock proteins correlates to apoptosis in lung cancer cells. Phytomedicine 80:153384. https://doi.org/10.1016/j.phymed.2020.153384

Lindstrom MS (2011) NPM1/B23: a multifunctional chaperone in ribosome biogenesis and chromatin remodeling. Biochem Res Int 2011:195209. https://doi.org/10.1155/2011/195209

Liu J, Shen JX, Hu JL et al (2014) Significance of interleukin-33 and its related cytokines in patients with breast cancers. Front Immunol 5:141. https://doi.org/10.3389/fimmu.2014.00141

Liu YJ, Du JL, Cao LP et al (2015) Anti-inflammatory and hepatoprotective effects of Ganoderma lucidum polysaccharides on carbon tetrachloride-induced hepatocyte damage in common carp (Cyprinus carpio L.). Int Immunopharmacol 25:112–120. https://doi.org/10.1016/j.intimp.2015.01.023

Liu G, Wang K, Kuang S et al (2018) The natural compound GL22, isolated from Ganoderma mushrooms, suppresses tumor growth by altering lipid metabolism and triggering cell death. Cell Death Dis 9:689. https://doi.org/10.1038/s41419-018-0731-6

Liu HJ, Qin Y, Zhao ZH et al (2020) Lentinan-functionalized selenium nanoparticles target tumor cell mitochondria via TLR4/TRAF3/MFN1 pathway. Theranostics 10:9083–9099. https://doi.org/10.7150/thno.46467

Llovet JM, Kelley RK, Villanueva A et al (2021) Hepatocellular carcinoma. Nat Rev Dis Primers 7: 6. https://doi.org/10.1038/s41572-020-00240-3

Ma C, Guan SH, Yang M et al (2008) Differential protein expression in mouse splenic mononuclear cells treated with polysaccharides from spores of Ganoderma lucidum. Phytomedicine 15:268–276. https://doi.org/10.1016/j.phymed.2007.11.015

Ma C, Yue QX, Guan SH et al (2009) Proteomic analysis of possible target-related proteins of cyclophosphamide in mice thymus. Food Chem Toxicol 47:1841–1847. https://doi.org/10.1016/j.fct.2009.04.041

Ma N, Du H, Ma G et al (2020) Characterization of the immunomodulatory mechanism of a Pleurotus eryngii protein by isobaric tags for relative and absolute quantitation proteomics. J Agric Food Chem 68:13189–13199. https://doi.org/10.1021/acs.jafc.0c00219

Maekawa M, Ishizaki T, Boku S et al (1999) Signaling from Rho to the actin cytoskeleton through protein kinases ROCK and LIM-kinase. Science 285:895–898. https://doi.org/10.1126/science.285.5429.895

Mahmood RI, Abbass AK, Razali N et al (2021) Protein profile of MCF-7 breast cancer cell line treated with lectin delivered by CaCO3NPs revealed changes in molecular chaperones, cytoskeleton, and membrane-associated proteins. Int J Biol Macromol 184:636–647. https://doi.org/10.1016/j.ijbiomac.2021.06.144

Mclennan SV, Heffernan S, Wright L et al (1991) Changes in hepatic glutathione metabolism in diabetes. Diabetes 40:344–348. https://doi.org/10.2337/diab.40.3.344

Mhawech P (2005) 14-3-3 proteins: an update. Cell Res 15:228–236. https://doi.org/10.1038/sj.cr.7290291

Michalak M, Groenendyk J, Szabo E et al (2009) Calreticulin, a multi-process calcium-buffering chaperone of the endoplasmic reticulum. Biochem J 417:651–666. https://doi.org/10.1042/BJ20081847

Miletic AV, Swat M, Fujikawa K et al (2003) Cytoskeletal remodeling in lymphocyte activation. Curr Opin Immunol 15:261–268. https://doi.org/10.1016/s0952-7915(03)00054-2

Mobley JA, Leav I, Zielie P et al (2003) Branched fatty acids in dairy and beef products markedly enhance alpha-methylacyl-CoA racemase expression in prostate cancer cells in vitro. Cancer Epidemiol Biomark Prev 12:775–783

Morris PJ (1991) Cyclosporine, FK-506 and other drugs in organ transplantation. Curr Opin Immunol 3:748–751. https://doi.org/10.1016/0952-7915(91)90107-c

Nam SW, Han JY, Kim JI et al (2005) Spontaneous regression of a large hepatocellular carcinoma with skull metastasis. J Gastroenterol Hepatol 20:488–492. https://doi.org/10.1111/j.1440-1746.2005.03243.x

Norata GD, Ongari M, Garlaschelli K et al (2007) Plasma resistin levels correlate with determinants of the metabolic syndrome. Eur J Endocrinol 156:279–284. https://doi.org/10.1530/eje.1.02338

Olofsson B (1999) Rho guanine dissociation inhibitors: pivotal molecules in cellular signalling. Cell Signal 11:545–554. https://doi.org/10.1016/s0898-6568(98)00063-1

O'Meara NM, Devery RA, Owens D et al (1991) Serum lipoproteins and cholesterol metabolism in two hypercholesterolaemic rabbit models. Diabetologia 34:139–143. https://doi.org/10.1007/BF00418266

Oprea TI, Bologa CG, Brunak S et al (2018) Unexplored therapeutic opportunities in the human genome. Nat Rev Drug Discov 17:317–332. https://doi.org/10.1038/nrd.2018.14

Ou HT, Shieh CJ, Chen JY, Chang HM (2005) The antiproliferative and differentiating effects of human leukemic U937 cells are mediated by cytokines from activated mononuclear cells by dietary mushrooms. J Agric Food Chem 53:300–305

Ouyang F, Wang G, Guo W et al (2013) AKT signalling and mitochondrial pathways are involved in mushroom polysaccharide-induced apoptosis and G1 or S phase arrest in human hepatoma cells. Food Chem 138:2130–2139. https://doi.org/10.1016/j.foodchem.2012.10.047

Ozawa M, Muramatsu T (1993) Reticulocalbin, a novel endoplasmic reticulum resident Ca(2+)-binding protein with multiple EF-hand motifs and a carboxyl-terminal HDEL sequence. J Biol Chem 268:699–705

Panchal K, Tiwari AK (2019) Mitochondrial dynamics, a key executioner in neurodegenerative diseases. Mitochondrion 47:151–173. https://doi.org/10.1016/j.mito.2018.11.002

Pandey R, Rai V, Mishra J et al (2017) From the cover: arsenic induces hippocampal neuronal apoptosis and cognitive impairments via an up-regulated BMP2/Smad-dependent reduced BDNF/TrkB signaling in rats. Toxicol Sci 159:137–158. https://doi.org/10.1093/toxsci/kfx124

Paula Santos N, Colaco A, Gil Da Costa RM et al (2014) N-diethylnitrosamine mouse hepatotoxicity: time-related effects on histology and oxidative stress. Exp Toxicol Pathol 66:429–436. https://doi.org/10.1016/j.etp.2014.07.002

Peng F, Huang Y, Li MY et al (2016) Dissecting characteristics and dynamics of differentially expressed proteins during multistage carcinogenesis of human colorectal cancer. World J Gastroenterol 22:4515–4528. https://doi.org/10.3748/wjg.v22.i18.4515

Prescott SM, Zimmerman GA, Stafforini DM et al (2000) Platelet-activating factor and related lipid mediators. Annu Rev Biochem 69:419–445. https://doi.org/10.1146/annurev.biochem.69.1.419

Rabinovich GA, Alonso CR, Sotomayor CE et al (2000) Molecular mechanisms implicated in galectin-1-induced apoptosis: activation of the AP-1 transcription factor and downregulation of Bcl-2. Cell Death Differ 7:747–753. https://doi.org/10.1038/sj.cdd.4400708

Ramm G, Larance M, Guilhaus M et al (2006) A role for 14-3-3 in insulin-stimulated GLUT4 translocation through its interaction with the RabGAP AS160. J Biol Chem 281:29174–29180. https://doi.org/10.1074/jbc.M603274200

Rezaei M, Keshtzar E, Khodayar MJ et al (2019) SirT3 regulates diabetogenic effects caused by arsenic: an implication for mitochondrial complex II modification. Toxicol Lett 301:24–33. https://doi.org/10.1016/j.toxlet.2018.10.025

Rong B, Yang S (2018) Molecular mechanism and targeted therapy of Hsp90 involved in lung cancer: new discoveries and developments (review). Int J Oncol 52:321–336. https://doi.org/10.3892/ijo.2017.4214

Sakurai K, Katoh M, Someno K et al (2001) Apoptosis and mitochondrial damage in INS-1 cells treated with alloxan. Biol Pharm Bull 24:876–882. https://doi.org/10.1248/bpb.24.876

Shamtsyan M, Konusova V, Maksimova Y et al (2004) Immunomodulating and anti-tumor action of extracts of several mushrooms. J Biotechnol 113:77–83. https://doi.org/10.1016/j.jbiotec.2004.04.034

Shenton D, Grant CM (2003) Protein S-thiolation targets glycolysis and protein synthesis in response to oxidative stress in the yeast Saccharomyces cerevisiae. Biochem J 374:513–519. https://doi.org/10.1042/BJ20030414

Shoji-Kawata S, Sumpter R, Leveno M et al (2013) Identification of a candidate therapeutic autophagy-inducing peptide. Nature 494:201–206. https://doi.org/10.1038/nature11866

Spychala MS, Venna VR, Jandzinski M et al (2018) Age-related changes in the gut microbiota influence systemic inflammation and stroke outcome. Ann Neurol 84:23–36. https://doi.org/10.1002/ana.25250

Sung H, Ferlay J, Siegel RL et al (2021) Global cancer statistics 2020: GLOBOCAN estimates of incidence and mortality worldwide for 36 cancers in 185 countries. CA Cancer J Clin 71:209–249. https://doi.org/10.3322/caac.21660

Tian X, Li Y, Shen Y et al (2015) Apoptosis and inhibition of proliferation of cancer cells induced by cordycepin. Oncol Lett 10:595–599. https://doi.org/10.3892/ol.2015.3273

Toro JR, Travis LB, Wu HJ et al (2006) Incidence patterns of soft tissue sarcomas, regardless of primary site, in the surveillance, epidemiology and end results program, 1978-2001: an analysis of 26,758 cases. Int J Cancer 119:2922–2930. https://doi.org/10.1002/ijc.22239

Trachootham D, Alexandre J, Huang P (2009) Targeting cancer cells by ROS-mediated mechanisms: a radical therapeutic approach? Nat Rev Drug Discov 8:579–591. https://doi.org/10.1038/nrd2803

Tsukumo Y, Tomida A, Kitahara O et al (2007) Nucleobindin 1 controls the unfolded protein response by inhibiting ATF6 activation. J Biol Chem 282:29264–29272. https://doi.org/10.1074/jbc.M705038200

Vasaikar S, Huang C, Wang X et al (2019) Proteogenomic analysis of human colon cancer reveals new therapeutic opportunities. Cell 177(1035–1049):e1019. https://doi.org/10.1016/j.cell.2019.03.030

Wang C, Youle RJ (2012) Predominant requirement of Bax for apoptosis in HCT116 cells is determined by Mcl-1's inhibitory effect on Bak. Oncogene 31:3177–3189. https://doi.org/10.1038/onc.2011.497

Wang PY, Kitchens RL, Munford RS (1998) Phosphatidylinositides bind to plasma membrane CD14 and can prevent monocyte activation by bacterial lipopolysaccharide. J Biol Chem 273:24309–24313. https://doi.org/10.1074/jbc.273.38.24309

Wang YT, Huang ZJ, Chang HM (2004) Proteomic analysis of human leukemic U937 cells incubated with conditioned medium of mononuclear cells stimulated by proteins from dietary mushroom of Agrocybe aegerita. J Proteome Res 3:890–896. https://doi.org/10.1021/pr049922h

Wang H, Wu G, Park HJ et al (2012) Protective effect of Phellinus linteus polysaccharide extracts against thioacetamide-induced liver fibrosis in rats: a proteomics analysis. Chin Med 7:23. https://doi.org/10.1186/1749-8546-7-23

Wang H, Zhang J, Sit WH et al (2014) Cordyceps cicadae induces G2/M cell cycle arrest in MHCC97H human hepatocellular carcinoma cells: a proteomic study. Chin Med 9:15. https://doi.org/10.1186/1749-8546-9-15

Wang PW, Hung YC, Li WT et al (2016) Systematic revelation of the protective effect and mechanism of Cordycep sinensis on diethylnitrosamine-induced rat hepatocellular carcinoma with proteomics. Oncotarget 7:60270–60289. https://doi.org/10.18632/oncotarget.11201

Wang Y, Han X, Li YD et al (2017) Lentinan dose dependence between immunoprophylaxis and promotion of the murine liver cancer. Oncotarget 8:95152–95162. https://doi.org/10.18632/oncotarget.19808

Wang W, Yang X, Li C et al (2022) Immunogenic cell death (ICD) of murine H22 cells induced by lentinan. Nutr Cancer 74:640–649. https://doi.org/10.1080/01635581.2021.1897632

WHO (2022a) Diabetes. https://www.who.int/news-room/fact-sheets/detail/diabetes. Accessed 5 June 2022

WHO (2022b) The top 10 causes of death. https://www.who.int/news-room/fact-sheets/detail/the-top-10-causes-of-death. Accessed 6 June 2022

Wolf G (2007) Serum retinol-binding protein: a link between obesity, insulin resistance, and type 2 diabetes. Nutr Rev 65:251–256. https://doi.org/10.1111/j.1753-4887.2007.tb00302.x

Wong MCS, Huang J, Chan PSF et al (2021) Global incidence and mortality of gastric cancer, 1980-2018. JAMA Netw Open 4:e2118457. https://doi.org/10.1001/jamanetworkopen.2021.18457

Wu G, Fang YZ, Yang S et al (2004) Glutathione metabolism and its implications for health. J Nutr 134:489–492. https://doi.org/10.1093/jn/134.3.489

Wu YW, Fang HL, Lin WC (2010) Post-treatment of Ganoderma lucidum reduced liver fibrosis induced by thioacetamide in mice. Phytother Res 24:494–499. https://doi.org/10.1002/ptr.2949

Wu CT, Lin TY, Hsu HY et al (2011) Ling Zhi-8 mediates p53-dependent growth arrest of lung cancer cells proliferation via the ribosomal protein S7-MDM2-p53 pathway. Carcinogenesis 32: 1890–1896. https://doi.org/10.1093/carcin/bgr221

Wu H, Tang S, Huang Z et al (2016) Hepatoprotective effects and mechanisms of action of triterpenoids from Lingzhi or Reishi medicinal mushroom ganoderma lucidum (agaricomycetes) on alpha-amanitin-induced liver injury in mice. Int J Med Mushrooms 18:841–850. https://doi.org/10.1615/IntJMedMushrooms.v18.i9.80

Xian H, Li J, Zhang Y et al (2021) Antimetastatic effects of Ganoderma lucidum polysaccharide peptide on B16-F10-luc-G5 melanoma mice with sleep fragmentation. Front Pharmacol 12: 650216. https://doi.org/10.3389/fphar.2021.650216

Xie Y, Li S, Sun L et al (2018) Fungal immunomodulatory protein from nectria haematococca suppresses growth of human lung adenocarcinoma by inhibiting the PI3K/Akt pathway. Int J Mol Sci 19. https://doi.org/10.3390/ijms19113429

Xiong Y, Chen ZH, Zhang FL et al (2021) A specific selenium-chelating peptide isolated from the protein hydrolysate of Grifola frondosa. RSC Adv 11:10272–10284. https://doi.org/10.1039/d0ra10886c

Xu JW, Zhao W, Zhong JJ (2010) Biotechnological production and application of ganoderic acids. Appl Microbiol Biotechnol 87:457–466. https://doi.org/10.1007/s00253-010-2576-5

Yan L, Shen J, Wang J et al (2020) Nanoparticle-based drug delivery system: a patient-friendly chemotherapy for oncology. Dose Response 18:1559325820936161. https://doi.org/10.1177/1559325820936161

Yang X, Li Y, Wang W, Li C, Zhao D, Hu G, Rao H, Hao J, Han X (2020) Antigens from H22 cell mutation induced by lentinan are correlated with HCC immunoprophylaxis. https://www.researchsquare.com/article/rs-43542/v1. Accessed 12 May 2022

Yang J, Pan G, Guan L et al (2022) The burden of primary liver cancer caused by specific etiologies from 1990 to 2019 at the global, regional, and national levels. Cancer Med 11:1357–1370. https://doi.org/10.1002/cam4.4530

Ying H, Zhang Y (2019) Systems biology of selenium and complex disease. Biol Trace Elem Res 192:38–50. https://doi.org/10.1007/s12011-019-01781-9

Yu CL, Burakoff SJ (1997) Involvement of proteasomes in regulating Jak-STAT pathways upon interleukin-2 stimulation. J Biol Chem 272:14017–14020. https://doi.org/10.1074/jbc.272.22.14017

Yu ACY, Chern YJ, Zhang P et al (2021) Inhibition of nucleophosmin 1 suppresses colorectal cancer tumor growth of patient -derived xenografts via activation of p53 and inhibition of AKT. Cancer Biol Ther 22:112–123. https://doi.org/10.1080/15384047.2020.1839278

Yue QX, Cao ZW, Guan SH et al (2008a) Proteomics characterization of the cytotoxicity mechanism of ganoderic acid D and computer-automated estimation of the possible drug target network. Mol Cell Proteomics 7:949–961. https://doi.org/10.1074/mcp.M700259-MCP200

Yue QX, Xie FB, Guan SH et al (2008b) Interaction of Ganoderma triterpenes with doxorubicin and proteomic characterization of the possible molecular targets of Ganoderma triterpenes. Cancer Sci 99:1461–1470. https://doi.org/10.1111/j.1349-7006.2008.00824.x

Yue QX, Song XY, Ma C et al (2010) Effects of triterpenes from Ganoderma lucidum on protein expression profile of HeLa cells. Phytomedicine 17:606–613. https://doi.org/10.1016/j.phymed.2009.12.013

Zacharakis M, Xynos ID, Lazaris A et al (2010) Predictors of survival in stage IV metastatic colorectal cancer. Anticancer Res 30:653–660

Zaidman BZ, Yassin M, Mahajna J et al (2005) Medicinal mushroom modulators of molecular targets as cancer therapeutics. Appl Microbiol Biotechnol 67:453–468. https://doi.org/10.1007/s00253-004-1787-z

Zeng H, Botnen JH (2007) Selenium is critical for cancer-signaling gene expression but not cell proliferation in human colon Caco-2 cells. Biofactors 31:155–164. https://doi.org/10.1002/biof.5520310302

Zhang HN, Lin ZB (2004) Hypoglycemic effect of Ganoderma lucidum polysaccharides. Acta Pharmacol Sin 25:191–195

Zhang X, Zhang H (2018) Chemotherapy drugs induce pyroptosis through caspase-3-dependent cleavage of GSDME. Sci China Life Sci 61:739–740. https://doi.org/10.1007/s11427-017-9158-x

Zhang L, Qiao Q, Tuomilehto J et al (2009) Blood lipid levels in relation to glucose status in seven populations of Asian origin without a prior history of diabetes: the DECODA study. Diabetes Metab Res Rev 25:549–557. https://doi.org/10.1002/dmrr.994

Zhang Y, Wu W, Qu H (2019) Integrated analysis of the gene expression changes during colorectal cancer progression by bioinformatic methods. J Comput Biol 26:1168–1176. https://doi.org/10.1089/cmb.2019.0056

Zhang T, Chen X, Sun L et al (2021) Proteomics reveals the function reverse of MPSSS-treated prostate cancer-associated fibroblasts to suppress PC-3 cell viability via the FoxO pathway. Cancer Med 10:2509–2522. https://doi.org/10.1002/cam4.3825

Zhang J, Hu T, Wang Y et al (2022) Investigating the neurotoxic impacts of arsenic and the neuroprotective effects of dictyophora polysaccharide using SWATH-MS-based proteomics. Molecules 27. https://doi.org/10.3390/molecules27051495

Zhao C, Fan J, Liu Y et al (2019) Hepatoprotective activity of Ganoderma lucidum triterpenoids in alcohol-induced liver injury in mice, an iTRAQ-based proteomic analysis. Food Chem 271: 148–156. https://doi.org/10.1016/j.foodchem.2018.07.115

Zhu T, Kim SH, Chen CY (2008) A medicinal mushroom: Phellinus linteus. Curr Med Chem 15: 1330–1335. https://doi.org/10.2174/092986708784534929

Zickermann V, Drose S, Tocilescu MA et al (2008) Challenges in elucidating structure and mechanism of proton pumping NADH:ubiquinone oxidoreductase (complex I). J Bioenerg Biomembr 40:475–483. https://doi.org/10.1007/s10863-008-9171-9

Cordyceps militaris (L.): Medicinal Aspects in Terms of Ethnobotanical and Pharmacological Perspectives

Nuriye Hilal Taştekin and Nadire Özenver

Contents

Abstract *Cordyceps militaris* (L.) (*Clavicipitaceae*) is a favorite member of the genus *Cordyceps* and has been used for various purposes in traditional medicine applications as a functional food, an energetic, an aphrodisiac, and a remedy for a variety of ailments, such as chronic bronchitis, asthma, inflammation, hemoptysis, allergy, epilepsy, anemia, arrhythmia, and cancer. As evidenced by scientific literature, *C. militaris* or its containing molecules have demonstrated various pharmacological properties ranging from renal to neurological diseases. Within the context of the present chapter, we focused on *C. militaris* in terms of ethnobotanical and pharmacological perspectives. We initially pointed out its traditional uses and then

N. H. Taştekin
Department of Pharmacognosy, Faculty of Pharmacy, Ankara Medipol University, Ankara, Turkey
e-mail: hilal.tastekin@ankaramedipol.edu.tr

N. Özenver (✉)
Department of Pharmacognosy, Faculty of Pharmacy, Hacettepe University, Ankara, Turkey
e-mail: nadire@hacettepe.edu.tr

D. C. Agrawal, M. Dhanasekaran (eds.), *Mushrooms with Therapeutic Potentials*,
https://doi.org/10.1007/978-981-19-9550-7_12

mentioned relevant medicinal aspects proved by scientific research worldwide. In this regard, we aimed to emphasize the therapeutic potential of mushrooms with a special interest in *C. militaris*.

Keywords *Cordyceps militaris* · Ethnobotany · Pharmacology · Fungus · Traditional medicine

Abbreviations

AECM	Aqueous extract of *C. militaris*
ALT	Alanine aminotransferase
ALP	Alkaline phosphatase
AST	Aspartate aminotransferase
cAMP	cyclic Adenosine monophosphate
CE	*Cordyceps militaris* extract
CMα	Cordycepin-rich *C. militaris*
CME	70% Ethanolic extracts of cultured mycelia
CMP	*C. militaris* polysaccharides
CNS	Central nervous system
CXCR2	CXC chemokine receptor 2
DC	Dendritic cell
FBE	70% Ethanolic extracts of fruiting bodies
FDA	Food and drug administration
GGT	γ-glutamyl transferase
GSH	Glutathione
HCV	Hepatitis C virus
HFD	High-fat diet
Ig	Immunoglobulin
MAPK	Mitogen-activated protein kinase
mTORC1	Autophagy-related mammalian target of rapamycin complex 1
NAFLD	Non-alcoholic fatty liver disease
NF-κB	Nuclear factor kappa B
Nrf2	Nuclear factor erythroid 2-related factor 2
RSV	Respiratory syncytial virus
SIRT1	Histone deacetylating sirtuin 1
SREBP-2	Sterol regulatory element-binding proteins
TGF-β1	Transforming growth factor-beta 1
TCM	Traditional chinese medicine
T1DM	Type 1 diabetes mellitus
Th2	T helper type 2
TLR4	Toll-like receptor 4
TNF-α	Tumor necrosis factor-alpha

1 Introduction

Medicinal mushrooms have held great importance for millennia because they generate biometabolites used or investigated as probable conventions for various diseases. The genus *Cordyceps* has been used for medicinal purposes in different parts of Asia throughout history (Ying et al. 1987). Cordyceps is a general term identifying a class of ascomycetous fungi that have constituted a hollow as endoparasites, particularly of arthropods, and in the meantime as symbionts of the ascomycete genus *Elaphomyces* (Dong et al. 2015). More than 500 species have been described and documented as Cordyceps (Fungorum 2022).

The genus *Cordyceps*, a favorite fungal traditional Chinese medicine (TCM) species, has been gaining interest for the last several years. Numerous *Cordyceps*-based commercial products exist in Western countries as an over-the-counter medicine for various purposes (Li et al. 2006; Paterson 2008). They are believed to alleviate humans' anxiety about inhabiting technologically developed countries (Lakhanpal and Rana 2005). Moreover, they are known for their ameliorative capacity toward various disorders and their enhancement impacts on cellular immunity, energy level, and aerobic capacity (Yadav 2020; Hirsch et al. 2017; Shashidhar et al. 2013). Many *Cordyceps* products are considered by the Food and Drug Administration (FDA) as dietary supplements (FDA 2022). Therefore, an increase in market demand has been observed recently.

Cordyceps militaris (L.) (Fig. 1) (known as Bei Chong Cao or Yong Chong Cao, *Clavicipitaceae* family), being a favorite member of *Cordyceps*, has been applied for various conditions for years in China (Paterson 2008). It has been formally classified

Fig. 1 *Cordyceps militaris* in nature. The picture was taken from Wikimedia Commons. https://commons.wikimedia.org/wiki/File:Puppenkernkeule.jpg

as a drug in Chinese medicine since 1964 in *Chinese Pharmacopoeia* (China 2015). *C. militaris* has been applied thoroughly in traditional medicine or against pests (Gu et al. 2022; Zhang et al. 2019).

The natural *C. militaris* is costly in the local market, and it is quite a rare species in nature because it requires particular hosts and strict growth conditions (Das et al. 2010). Currently, numerous modern technology approaches focus on culturing *C. militaris* to explore its potential for health-related conditions (Mizuno 1999).

A variety of compounds have been isolated, and their structures were identified. Cordycepin; adenosine; macrolides; cephalosporolides C, E, and F; 2-carboxymethyl-4-(3′-hydroxybutyl)furan; and pyridine-2, 6-dicarboxylic acid are among the compounds obtained from *C. militaris* (Rukachaisirikul et al. 2004; Singpoonga et al. 2020), most of which were reported to attribute *C. militaris*-associated biological activity (Yu et al. 2004; Ma et al. 2015; Parunyakul et al. 2021; Xiong et al. 2013).

In the present chapter, we focused on *C. militaris*, one of the valuable mushrooms, in terms of ethnobotanical and pharmacological context. Primarily, we mentioned its traditional uses in history, followed by its relevant medicinal aspects, proven by scientific research globally. In this regard, we aimed to draw attention to the therapeutic potential of mushrooms with a special interest in *C. militaris*. Future perspectives of discovering its health potential and incorporating it into the pharmaceutical industry have been discussed.

2 Ethnobotanical Investigations on *Cordyceps militaris*

Although the first inscribed medicinal usage of *Cordyceps* spp. dates to the time of the Chinese Qing dynasty in 1757 (Devkota 2006), these species have been utilized for centuries due to the high mountain conditions of the Himalayas by the Chinese, Tibetans, Nepalis, and Indians, typically as tonics and stimulants to boost energy and for their adaptogenic characteristics (Kim et al. 2014; Jędrejko et al. 2021). Among others, *C. sinensis* having therapeutic properties similar to *C. militaris* has been traditionally used to treat various diseases, including spermatorrhea, hemorrhoids, tuberculosis, gastralgia, and impotence in China (Ji et al. 2004). *C. sinensis* is known to be consumed in Nepal as a cardiac tonic, sexual stimulant, immune system booster, liver and heart fortifier, as well as against influenza, cough, cold, headache, diarrhea, rheumatism, and respiratory disorders (Das et al. 2012; Tiwari et al. 2009; Tuli et al. 2014).

In Asian ethnomedicine, primarily applied in China, Japan, India, Korea, Nepal, Taiwan, and Tibet, *C. militaris* has also long been used as a functional food, an energetic, an aphrodisiac, and a remedy for a variety of ailments, such as chronic bronchitis, asthma, inflammation, hemoptysis, allergy, epilepsy, anemia, arrhythmia, and cancer (Ramgir et al. 2022; Fung et al. 2011; Oh et al. 2011; Lin et al. 2021b; Chiu et al. 2016; Hsu et al. 2008; Olatunji et al. 2018; Rao et al. 2006). TCM suggests that root deficiency, namely, the inefficacy of both lung and kidney, causes several diseases (Gao et al. 2009). Thus, due to its beneficial effects on the lung and the kidney, hyperlipidemia, renal failure, and various liver diseases have frequently been cured using *C. militaris* (Zhu et al. 2013). As stated in the Compendium of Materia Medica, *C. militaris* has also historically been employed to lessen convulsions, anxiety, and terror (Lin et al. 2021b).

Scientific knowledge indicates that the influences of *C. militaris* depend mainly on its chemical constituents (e.g., cordycepin and polysaccharides) (Mehra et al. 2017; Yang et al. 2021; Yu et al. 2021; Ma et al. 2015). Table 1 provides an overview of *C. militaris* ethnobotanical usage related to the pharmacological properties of the mushroom and its constituents.

3 Pharmacological Aspects of *Cordyceps militaris*

Various pharmacological effects of *C. militaris* have been documented, specifically linked to various polysaccharides, proteins, adenosine, cordycepin, myriocin, and ergosterol in *C. militaris* (Zhang et al. 2019).

Despite reports of numerous *C. militaris*-associated effects in preclinical models, its possible modes of action remain unclear and need to be explored by experimental, epidemiological, and clinical studies. For instance, the anticancer properties of *C. militaris* were demonstrated in different murine and human cancer models despite the lack of sufficient data about their related mechanisms.

We have classified and provided details about the pharmacological effects of *C. militaris* on different diseases (Fig. 2) in the following sections below.

3.1 Respiratory Diseases

Recent pharmacological studies have demonstrated substantial therapeutic effects of *C. militaris* on the respiratory system, supporting its historical application.

In vitro experiments demonstrated that the secretion by human airway epithelial cells was stimulated when given hot water extract of *C. militaris* (Yue et al. 2008). Elucidating mechanistically, the extract promoted Cl^- secretion via two major intracellular signaling pathways, i.e., cAMP and Ca^{2+}, in a derived human bronchial surface epithelium cell line (Fung et al. 2011). Furthermore, cordycepin from *C. militaris* acted in vitro on sodium, potassium, and chloride ion transportation in the respiratory epithelium (Jędrejko et al. 2021).

The proliferation of infected HEK293 cells was inhibited by different extracts of *C. militaris* on the respiratory syncytial virus (RSV), a prominent cause of pediatric pneumonia in children under 2 years of age. Results indicated that hydrophilic constituents of *C. militaris* were associated with the effects on RSV (Wei et al. 2015).

Table 1 Ethnobotanical uses of *C. militaris* with supporting pharmacological activity

System	Ethnobotanical use	Pharmacological activity	Experimental category	Reference
Cardiovascular	*Hyperlipidemia*	Decreased serum lipid levels	In vivo	(Lee et al. 2021)
	Arrhythmia	Antiarrhythmic specifics[a]	In vivo	(Mei et al. 1989)
Nervous	*Aphrodisiac*	Increased testosterone levels	In vivo	(Nguyen et al. 2021)
	Epilepsy	Reduced release of excitatory neurotransmitters	In vivo	(Cui 2014)
	Anxiety	Reduced release of excitatory neurotransmitters	In vivo	(Cui 2014)
Respiratory	*Asthma*	Decreased respiratory tract hyperresponsiveness	In vivo	(Zheng et al. 2020)
	Bronchitis	Improved lung functions, relief of the symptoms	Clinical trial	(Gai et al. 2004)
	Expectorant	Promoted secretion of airway epithelial cells	In vitro	(Yue et al. 2008)
Immune	*Allergy*	Immunomodulatory activity	In vivo/clinical trial	(Liu et al. 2016a; Jędrejko et al. 2021)
	Anti-inflammatory	Decreased inflammatory response	In vivo	(Jeong et al. 2010)
Endocrine	*Diabetes*	Decreased blood glucose levels, inhibition of the occurrence of diabetes-associated kidney and spleen damage	In vivo	(Ma et al. 2015)
	Liver diseases	Modulatory effects on liver enzymes	In vivo	(Wang et al. 2018)
Neoplasm	*Cancer*	Cytotoxic activity Antiproliferative effects	In vitro	(Dong et al. 2014); (Wong et al. 2011)
Urinary	*Renal dysfunction*	Ameliorated renal parameters	In vivo	(Yu et al. 2016)

[a] *C. sinensis*

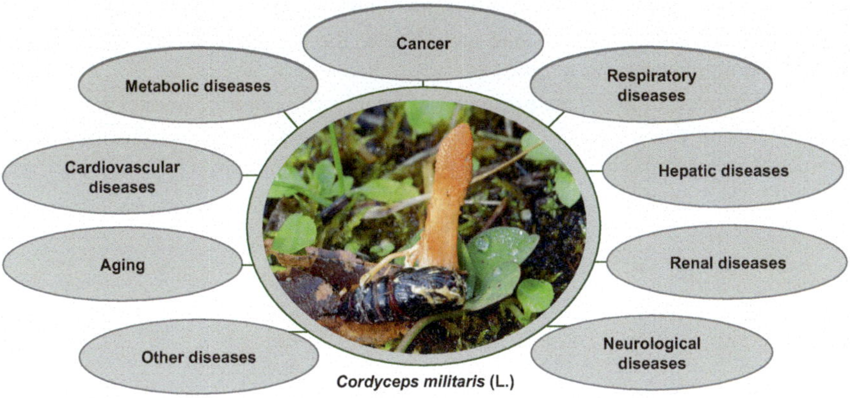

Fig. 2 Pharmacological effects of *C. militaris* on numerous diseases. The picture of *C. militaris* was taken from Wikimedia Commons. https://commons.wikimedia.org/wiki/File:Puppenkernkeule.jpg

An in vivo study conducted with an ovalbumin-induced mouse asthma model demonstrated that the water extract of dry fruiting bodies of *C. militaris* was insufficient to control pulmonary inflammation compared to montelukast and prednisolone. Still, histologic findings revealed that the extract relatively reduced the inflammatory cells and the thickness of the smooth muscle layer in the airway tract (Hsu et al. 2008). However, a related investigation evaluating airway inflammation in asthmatic mice revealed that the fraction of *C. militaris* rich in polysaccharides alleviated the inflammation and hyperresponse of the respiratory tract suppressing the activation of the TGF-β1/Smad pathway (Zheng et al. 2020). In the same pattern, traditionally utilized medicine in China, consisting of Astragali Radix and *C. militaris* (10:3), showed beneficial effects on asthma airway remodeling via inhibitory effects on TGF-β1/Smad pathway activation resulting in collagen deposition decline (Gao et al. 2009). Antiasthmatic specifics of cordycepin were demonstrated in a mouse asthma model. The underlying mechanisms were identified as suppressor effects on releasing IgE- and Th2-driven cytokines and chemokines in the lung (Yang et al. 2015).

DBA/2 mice infected with the pandemic H1N1 virus, which led to lung lobe inflammation via hemorrhagic pleural effusion and cellular invasion, were treated with *C. militaris* extract. Following the administration, TNF-α levels responsible for lung damage significantly decreased, in addition to constant body weight and lower mortality (Lee et al. 2014).

Administration of *C. militaris* on mice having induced acute lung injury relieved the severity of the condition and lung CXCR2 levels inhibiting the expression of CXCR2 through the two miRNAs of *C. militaris* (Liu et al. 2015).

Researchers operating a randomized, double-blind, placebo-controlled clinical study attending 100 healthy subjects evaluated the symptoms or frequency of upper airway infections. They measured the levels of cytokines, IgA, and natural killer cells following *C. militaris* supplementation for 12 weeks. Ultimately, protective effects on upper respiratory tract infections could not be seen. However, NK cell activity and IgA secretion were stimulated due to the immunostimulatory activity of the mushroom (Jędrejko et al. 2021).

C. militaris capsules were evaluated clinically in a multi-center, randomized, single-blind study involving 425 patients with chronic bronchitis. Results indicated a significant improvement in the symptoms and lung functions in *C. militaris* and *C. sinensis* groups against the control (Gai et al. 2004).

3.2 Renal Diseases

Traditional uses of *Cordyceps* prompted investigators to survey the renal effects of these fungi. In this context, research on *C. sinensis* and *C. militaris* came to the fore. Preclinical studies evidenced the renoprotective effects of *C. sinensis*, which was effective on splenic lymphocytes isolated from rats with chronic renal failure. In vivo experiments also inferred the protective effects of *C. sinensis* against acute kidney injury (Lin and Li 2011). The therapeutic effects of *C. sinensis* were further substantiated clinically by improving renal functions (Ng and Wang 2005). The use of *C. sinensis* in renal disease has been thoroughly investigated. On the other hand, as discussed in the following paragraphs, comprehensive research is needed for *C. militaris*.

Human mesangial cells isolated from glomeruli were administrated with *C. sinensis* and *C. militaris* and low-density lipoproteins (LDL), a factor in the hypertrophy or proliferation of mesangial cells that contributes to the progress of glomerular sclerosis. The proliferation of the cells was suppressed by both fungi (Zhao-Long et al. 2000).

In a mouse model of diabetic nephropathy, fruiting bodies and mycelia of *C. militaris* evidenced protective effects on the renal system; particularly, *C. militaris* positively impacted renal parameters related to renal damage, such as serum creatinine, kidney to body weight ratio, pathological changes of renal tissue, collagenous deposition, and TGF-β1 release (Yu et al. 2016).

Fruiting bodies of *C. militaris* were beneficial in rats in preventing membranous glomerulonephritis, a major contributor to nephrotic syndrome, by decreasing NF-κB activity. In addition to restraining oxidative damage, *C. militaris* regulated the parameters associated with chronic kidney disease, including creatinine, serum albumin, lipid levels, and also inflammatory cytokines (Song et al. 2016). More preclinical and clinical research is needed to understand how *C. militaris* affects the urinary system.

3.3 Hepatic Diseases

The widespread use of *C. militaris* in TCM to treat a variety of liver conditions, including hepatic cirrhosis and its alleged regulator effects on the liver, led to several scientific studies associated with hepatic effects of the fungus, which are detailed below (Oh et al. 2011; Wang et al. 2018).

In vitro tests on *C. militaris* capsule and liquid formulations on the hepatitis C virus (HCV) revealed that only the capsule form had a modest effect on HuH-7- and Li23-derived HCV RNA-replicating cells. Additionally, cordycepin was identified as the molecule in charge of the activity. Researchers also noted the additive activity when the *C. militaris* capsule's use was combined with interferon-α or ribavirin (Ueda et al. 2014).

The extract and intra–/extracellular biopolymers of *C. militaris* were investigated in a rat model of biliary fibrosis/cirrhosis caused by bile duct ligation and scission. All the samples given an oral daily dose of 30 mg/kg decreased remarkably serum levels of alanine aminotransferase (ALT), aspartate aminotransferase (AST), and alkaline phosphatase (ALP) compared to the control group. However, neither the extract nor the intracellular biopolymers demonstrated any antifibrotic properties. The extracellular biopolymers hindered the formation of collagen deposition and lipids' oxidation, indicating antifibrotic properties through suppression of hepatic stellate cell activation (Nan et al. 2001).

Beneficial effects of culture broth of *C. militaris* against liver toxicity have been demonstrated on HaM/ICR rats following the intraperitoneal injection of benzo (a)pyrene. The damage of benzo(a)pyrene-induced hepatotoxicity resulted in elevated levels of serum enzymes (ALT, AST) and lipid peroxidation, which were later ameliorated by *C. militaris* (Jo et al. 2009).

When *C. militaris* and cordycepin-enriched *C. militaris* were administered in an in vivo model of CCl_4-induced hepatotoxicity, high blood GSH that was a consequence of CCl_4 application was later brought back to normal. Results further indicated that the cordycepin enrichment had a strong antioxidant impact with greater GSH and lower rates of lipid peroxidation, as well as hepatoprotective properties (Ahn et al. 2013). Both *C. militaris* supplements were evaluated for preventing alcoholic hepatotoxicity compared to silymarin. All samples affected serum markers and decreased the high concentrations of blood alcohol and acetaldehyde through the mediation of enzyme activities such as alcohol dehydrogenase and acetaldehyde dehydrogenase. Comparable to silymarin, cordycepin-enriched *C. militaris* was found to have a greater protective function than *C. militaris* toward alcoholic toxicity (Cha et al. 2013).

Selenium-enriched *C. militaris* demonstrated beneficial outcomes on Wistar rats intoxicated with triptolide. To explain protective effects on acute hepatotoxicity, serum markers of liver injury, including ALT, AST, ALP, γ-glutamyl transferase (GGT), and total cholesterol levels were measured, and histopathological analyses were conducted. Findings indicated that Se-enriched and regular *C. militaris* possess antioxidative capabilities via promoting the expression of downstream genes for

Nrf2 signaling pathway enzymes such as NQO1, GCLC, and HO-1 in addition to Nrf2 translocation (Wang et al. 2018).

The prophylactic effects of *C. militaris* were exhibited in a randomized, double-blind clinical study carried out in Korea on 57 patients with mild hepatic impairment. Over 8 weeks, *C. militaris* (1.5 g/day) reduced the lipid formation in hepatocytes that caused chronic fibrosis or cirrhosis. This study evidenced the safety of *C. militaris* as a functional food and its efficacy (Heo et al. 2015).

The study investigating the possible protective function of *C. militaris* against nonalcoholic fatty liver disease (NAFLD) in an obese mouse model uncovered that the consumption of *C. militaris* markedly reduced serum glucose levels and enhanced insulin sensitivity. Besides, *C. militaris* improved hepatic oxidative stress and declined serum proinflammatory cytokine levels. Thus, *C. militaris* exhibited protective influences against the development of NAFLD (Choi et al. 2014).

3.4 Neurological Diseases

C. militaris is used in Asia to heal neurological conditions like seizures and anxiety (Lin et al. 2021b). To date, various neurological effects of *Cordyceps* spp. have been linked to some common compounds of these fungi, as addressed in the following paragraphs.

As implicated in several pharmacological activities, adenosine, a fundamental nucleoside component of *C. sinensis* and *C. militaris*, has neurological effects by reducing the release of excitatory neurotransmitters spontaneously and in response to evoked stimuli in the central nervous system (CNS). This controls several factors, such as sleep quality, arousal degree, seizure vulnerability, and pain sensation (Ribeiro 1995; Gu et al. 2007; Nxumalo et al. 2020; Cui 2014). In an experimental study conducted on neuronal cells, adenosine obtained from *C. cicadae* protected the cells against apoptotic effects of glutamate by decreasing reactive oxygen species, Ca^{2+} influx, Bax/Bcl-2 ratio, and phosphorylation of MAPKs like ERK, p38, and JNK (Olatunji et al. 2016).

It was reported that cordycepin protects the CNS against the toxicity of lipopolysaccharide-activated microglial BV2 cells, which release NO, PGE2, IL-1β, and TNF-α. Particularly, cordycepin hindered the activation of NF-κB, Akt, and MAPK pathways, reducing inflammatory mediators (Jeong et al. 2010).

In another study, *C. militaris* and its primary component cordycepin exhibited potential neuroprotective benefits in a cerebral ischemia model of gerbil by inhibiting the release of mediators linked to glial cells and protecting hippocampal pyramidal neurons from ischemic injury, which is a key contributor to permanent brain damage (Hwang et al. 2008). Rats were administered ischemic neuronal injury to assess the cerebroprotective potential of the peptide molecule cordymin derived from *C. sinensis*. Cordymin, also a natural constituent of *C. militaris,* dramatically reduced the emergence of behavioral abnormalities demonstrating neuroprotective potential through anti-inflammatory and antioxidant capabilities (Wang et al. 2012).

C. militaris also demonstrated neuroprotective potential in rats with cerebral ischemia via preventing neuronal death in the hippocampus and enhancing memory deficits (Kim et al. 2019).

C. militaris reversed the scopolamine-induced memory impairment in rats and enhanced the immunoreactivity of the hippocampus via compensating the neuronal loss, along with advanced cognitive skills. The ability of the extract to promote neurite outgrowth was tested using the murine neuroblastoma cell line, and the results were comparable to those of the positive control retinoic acid (Lee et al. 2011). Likewise, *C. militaris* remarkably enhanced learning and memory on a scopolamine-induced amnesic murine model. The polypeptide fraction was, however, given credit for the activity. Research on mechanistic pathways of neuroprotective effects revealed the involvement of endogen antioxidant mechanisms and altered expressions of the genes, such as Pik3r5, IL-1β, and Slc18a2, relevant to the nervous system (Yuan et al. 2018).

The value of *C. militaris* as a supportive therapy of duloxetine for insomnia, the primary symptom of major depressive disorder, was evaluated clinically. Even though the treatment was deemed safe, it was ineffective regarding clinical outcomes at the prescribed dose and application period. Extensive trials are required due to the study's other restrictions, including limited sample size and the subjectivity of sleep evaluation (Zhou et al. 2021).

3.5 Metabolic Disorders (Diabetes, Obesity, Etc.)

Many studies confirm the prominent role of *C. militaris* or its active substances against metabolic disorders such as diabetes, obesity, etc. Despite the diversity of experimental studies on *C. militaris* and its active substances, most of the investigations emphasized the therapeutic potential of cordycepin, one of the substantial compounds in the composition of *C. militaris*, in the management of metabolic diseases exemplified below.

Examining the role of *Cordyceps militaris* extract (CE) on high-fat diet-induced (HFD) metabolic disorders, CE was demonstrated to attenuate body weight gain in an in vivo mice model receiving HFD. The quantity of epididymal fat and the size of adipocytes was further lowered in the case of CE administration. Besides, fat accumulation in the liver and liver weight were markedly lessened in the CE-treated group, in addition to the advantageous effects of CE on plasma parameters associated with lipid profiles. Thus, the possible beneficial function of CE in HFD-induced metabolic disorders was disclosed (Kim et al. 2014).

Evaluating the influences of *C. militaris* and its fractions on modifying metabolic syndrome in vivo, high-fat/high-sucrose diet-given mice under the treatment of various *C. militaris* samples were significantly affected. Particularly, the polysaccharide fraction of *C. militaris* lessened blood sugar and serum lipid levels. Besides, *C. militaris* polysaccharide administration ameliorated intestinal dysbiosis through the stimulation of the population of next-generation probiotic *Akkermansia muciniphila* in the gut. Thus, the polysaccharide fraction holds a vital role as a probable dietary supplement through modulating gut microbiota to amend the metabolic syndrome (Lee et al. 2021).

C. militaris has been used in TCM for the convention of metabolic syndrome. Cordycepin, the substantial active substance of *C. militaris*, significantly decreased blood glucose levels and promoted oral glucose tolerance when administered intraperitoneally in diabetic mice. Furthermore, it lessened characteristic symptoms of diabetes and prohibited the occurrence of diabetes-associated kidney and spleen damage, all indicating cordycepin as a therapeutic candidate in managing diabetes (Ma et al. 2015). Likewise, cordycepin downregulated type 2 diabetes mellitus regulating genes, including 11β-HSD1 and PPARγ in lipopolysaccharide-activated macrophages (Shin et al. 2009). Another study investigating the impact of cordycepin on streptozotocin-induced type 1 diabetes mellitus (T1DM) mice in terms of the expression of liver proteins demonstrated that cordycepin lowered final body weight and food intake. It also enhanced the tricarboxylic acid cycle, one of the major energy generation pathways in the liver. It sustained the energy metabolism by adjusting AMPK activity, presenting cordycepin as a drug lead for T1DM therapy (Parunyakul et al. 2021).

Exploring the impacts of cordycepin on obesity-induced NAFLD indicated that it alleviated body weight in addition to the weight of adipose and the liver by decreasing assorted markers associated with lipid metabolism and inflammatory responses. Thus, in vivo mice models treated with a high-fat diet and cordycepin markedly attenuated the expression levels of relevant proteins involved in lipid anabolism and enhanced the levels of relevant proteins associated with β-oxidation in those treated with a high-fat diet. Thus, the potential of cordycepin in managing NAFLD was highlighted (Gong et al. 2021). Similarly, cordycepin protected palmitic acid-induced hepatic lipid collection by stimulating autophagy via the PKA/mTOR pathway, indicating its therapeutic function for the treatment of NAFLD (Li et al. 2019). Investigating the effect of cordycepin on in vivo mice models receiving a high-fat diet, a remarkable reduction in serum total cholesterol levels, triglycerides, and low-density lipoprotein cholesterol was documented. The high-fat diet-linked increase was markedly repressed in relative retroperitoneal fat by activating AMP-activated protein kinase, suggesting cordycepin to prohibit hyperlipidemia and promote insulin sensitivity (Guo et al. 2010).

3.6 Cardiovascular Diseases

Emerging evidence points out that mushrooms possess ranging beneficial activities in the body. Heart-associated diseases comprise one of those affected by mushrooms, particularly by their constituents, with a special emphasis on polysaccharides. Recently, many investigations have unraveled the favorable function of those against cardiovascular diseases (Amirullah et al. 2018; Guillamón et al. 2010; Rahman et al. 2018). In this context, various findings exist about *C. militaris* and its included constituents attributing their behavior to heart diseases exemplified below.

A water-soluble polysaccharide from *C. militaris*, CM1, markedly lessened total plasma cholesterol, triglyceride, and epididymal fat index in vivo. It also suppressed preadipocyte differentiation in vitro by decreasing key gene expression levels in lipid droplet generation (Yu et al. 2021). Likewise, another study demonstrated that CM1 remarkably lowered the total cholesterol and triglyceride in the plasma of mice and reduced lipid accumulation and organization of atherosclerotic plaque dose dependently, possibly affecting diverse signaling pathways, all suggesting the likelihood of polysaccharides in *C. militaris* as a potential therapeutic agent for repressing hyperlipidemia and the occurrence of atherosclerotic cardiovascular diseases (Lin et al. 2021a).

Another polysaccharide CM3II revealed strong anti-atherosclerotic effects via lessening plasma lipids in apolipoprotein *E*-deficient mice by promoting liver X receptor α and inhibiting sterol regulatory element-binding proteins (SREBP-2). Thus, the potential application of CM3II against atherosclerosis was highlighted (Yang et al. 2021).

The ethanol extract of *C. militaris* obstructed platelet accumulation and displayed antithrombotic activity in a rat model of thrombosis. Thus, it exhibited a favorable effect on promoting blood flow and retrieving vessel damage (Choi et al. 2020).

Comparing antioxidant and antihyperlipidemic influences of *Paecilomyces japonica*, *C. militaris*, and cordycepin-rich *C. militaris* (CMα), cordycepin-rich CMα displayed the most efficient antioxidant and antihyperlipidemic effect in rats. This induced alcoholic hyperlipidemia and oxidative stress, indicating that CMα might be efficient in suppressing the oxidation and hyperlipidemia probably due to possible anti-oxidative and antihyperlipidemic properties of cordycepin (Ahn et al. 2020).

3.7 **Cordyceps militaris** *against Tumors*

Epidemiological studies, confirmed by preclinical evidence from in vitro and in vivo experiments and by clinical outcomes, have illuminated the fact that there is a link between diet and avoiding cancer (Greenwald et al. 2001; Zhou et al. 2016). As a popular part of natural dietary products, mushrooms may prohibit or alleviate cancer-associated deaths, and recent findings have emphasized edible fungi's significance in cancer management. Various case control studies have documented that mushroom consumption may be linked to the decreased risk of certain cancers despite the absence of sufficient epidemiologic studies (Li et al. 2014; Khan et al. 2019; Shin et al. 2010; Lee et al. 2019).

Methanol extracts from fruiting bodies and fermented mycelia of *C. militaris* displayed potent cytotoxic activities against the A549 tumor cell line (Dong et al. 2014).

In another investigation, aqueous extract of *C. militaris* (AECM) inhibited the proliferation of human leukemia U937 cells dose dependently, linked to the morphological alterations and apoptosis through caspase protease activity and modulation of various growth regulatory gene products. Thereby, the therapeutic capacity of AECM in managing human leukemia was disclosed (Park et al. 2005).

C. militaris hindered the proliferation and metastasis of Lewis pneumonic cancer, enhanced cortisol and testosterone levels in the plasm, and exhibited a male hormonelike impact in vivo. Thereby, it repressed tumor proliferation and extended the survival of mice (Liu et al. 1997).

The polysaccharide gained from *C. militaris* (cordlan) caused the maturation of dendritic cells (DC) through the toll-like receptor 4 (TLR4) signaling pathway (Kim et al. 2010). The defects occurring during DC maturation in the tumor microenvironment led to enhanced expressions of co-stimulatory molecules and defective cytokine generation impeding cancer immunotherapy (Rabinovich et al. 2007). Thus, cordlan provided an alternative to inducing the phenotypic and functional maturation of DCs and contributing to the promotion of DC-based cancer immunotherapy (Kim et al. 2010).

The peptide cordymin obtained from *C. militaris* displayed distinct antiproliferative effects. In particular, cordymin inhibited the growth of MCF-7 breast cancers. However, it was ineffective against HT-29 colon cancer cells (Wong et al. 2011).

An investigation in human renal cell carcinoma indicated that when cordycepin was compared to the *C. militaris* extract, the latter was more efficient at inhibiting proliferation, inducing apoptosis, disrupting the cell cycle, and activating the extracellular signal-regulated kinase (Yamamoto et al. 2015).

3.8 **Cordyceps militaris** *in Aging*

Another noteworthy field *C. militaris* is used is the promotion of health and longevity (Lee et al. 2020; Liu et al. 2016a). Despite the folkloric knowledge related to its use as an invigorating agent for longevity, *C. militaris* was short of information about its antiaging activity. Thereby, emerging evidence has proved the possible related antiaging activity. For instance, Li et al. (2010) unraveled that the polysaccharides from cultivated fruiting bodies of *C. militaris* (CMP) obstructed Fe^{2+}-L-cysteine-induced mitochondrial damage and swelling. It also scavenged superoxide anion remarkably and enhanced the effects of antioxidants. CMP was a potential pharmaceutic with antiaging and mitochondrial protective activity (Li et al. 2010). Similarly, Liu et al. (2016a, b) found that homogeneous polysaccharide CP2-c2-s2 from *C. militaris* markedly encouraged the growth of T and B lymphocytes, prolonged the typical life span without side effects on fertility, and postponed the age-related decline in *Caenorhabditis elegans*, indicating remarkable immunomodulatory and antiaging activities of CMP (Liu et al. 2016b).

3.9 Other Conditions

Numerous preclinical findings exist disclosing numerous impacts of *C. militaris* in ranging conditions exemplified in the following. The anti-inflammatory and immune-enhancing aspects identified may give a chance to manage other related diseases.

CMP, when assessed in terms of their immune roles and reactive oxygen species-scavenging activity in vivo, enhanced the immune activity in mice and markedly improved the spleen and thymus markers, the levels of white blood cells, the spleen lymphocyte function, and IgG activity in mice serum. Furthermore, CMP exerted strong antioxidant capacity, which confirmed improved immune function through blockage of oxidative stress, all presenting CMP as a probable leader in developing new drugs and health supplements (Liu et al. 2016a).

Methanol extract from the fruiting bodies of *C. militaris* displayed broad antimicrobial activity against all the bacteria and fungi microorganisms tested. In contrast, methanol extract from the fermented mycelia demonstrated selective activity and significant antioxidant properties (Dong et al. 2014).

The polysaccharides obtained from the fruiting bodies of cultured *C. militaris* (WCBP50, WCBP50 I, and WCBP50 II) presented potent antioxidant activity (Chen et al. 2013). In another research among the three polysaccharide fractions called P50–1, P70–1, and P70–2, P70–1 possessed hydroxyl radical-scavenging activity, confirming the folkloric uses of *C. militaris* (Yu et al. 2007).

The antioxidant activity of *C. sinensis* and *C. militaris* dose dependently scavenged the $ABTS^{\bullet+}$ free radical as strong hydrogen donators, possibly due to the presence of polysaccharides. Therefore, the importance of *C. sinensis* and *C. militaris* consumption has been emphasized to improve therapeutic effects in human diseases linked to oxidative stress (Yu et al. 2006).

The immunomodulatory function of *C. militaris*, a prominent or critical point in human health, has been discussed in several studies. Various extracts, polysaccharides, and bioactive compounds of *C. militaris* have been shown to possess immunomodulatory impacts affecting different types of immunity (Lee et al. 2020). The balance between type 1 and type 2 immunity is significant in maintaining healthy conditions (Gieseck et al. 2018). Extracts obtained using water or 50% ethanol and polysaccharides from *C. militaris* enhance type 1 immunity, while total extracts obtained using 70–80% ethanol and cordycepin from *C. militaris* mostly improved type 2 immunity. The constituents of *C. militaris* exhibited different influences on immunity based on their chemical profile, providing knowledge about its utilization in folk medicine and a guide for its use in individuals (Lee et al. 2020). On the other hand, one of its active constituents, cordycepin, hindered T-cell activation both in vitro and in vivo and downregulated the immune system representing a potential as a beneficial immunosuppressive agent in curing unwanted immune responses (Xiong et al. 2013).

The 70% ethanolic extracts of cultured mycelia (CME) and fruiting bodies (FBE) of *C. militaris* displayed anti-inflammatory, antinociceptive activities associated with antioxidant, antiangiogenic, and nitric oxide generation-suppressive action. CME repressed the NO generation and downregulated iNOS expression in lipopolysaccharide-induced RAW 264.7 in a dose-dependent manner, in addition to the potent inhibitory effects of CME and FBE on the chick embryo chorioallantoic membrane angiogenesis. Cordycepin was shown to account for their associated anti-inflammatory and antiangiogenic responses (Won and Park 2005).

Oral administration of *C. militaris* mycelium powder dramatically elevated testosterone levels and sexual activity, suppressed by the induction of diabetes in rats (Nguyen et al. 2021). Kopalli et al. (2019) assessed the influences of *C. militaris*-derived cordycepin in a rat model with senile testicular dysfunction. Cordycepin markedly amended sperm motility, progressiveness, and the modifications in spermatogenesis-associated protein and mRNA expression (e.g., histone deacetylating sirtuin 1 (SIRT1)). Also, the autophagy-related mammalian target of rapamycin complex 1 (mTORC1) proposes cordycepin as a compound with a therapeutic value in ameliorating age-linked male sexual dysfunctions (Kopalli et al. 2019). In another study, Sohn et al. (2012) examined cordycepin's impact on age-induced testicular function decrease. It notably improved sperm motility and the progressiveness of sperm migration in the case of long-term administration, indicating cordycepin as an agent antagonizing testicular function loss (Sohn et al. 2012).

4 Conclusion

Nature is an undeniable source providing a significant number of products/molecules with therapeutic potential. People worldwide use natural products because they believe they are safe and have been used in traditional medicine throughout history. Medicinal mushrooms and their constituents are well-known natural resources holding prominence in maintaining health with reasonable safety. Despite the well-known folkloric use of *C. militaris*, inadequate scientific data limits its recognition and usage worldwide. Therefore, there is a need for further studies regarding the preclinical and clinical effects of *C. militaris*. On the other hand, its similar chemical composition and biological characteristics to *C. sinensis*, a species difficult to cultivate with significant therapeutic potential, suggests that *C. militaris* may be a suitable substitution for *C. sinensis* (Chen et al. 2018; Shrestha et al. 2012). Hence, it is imperative to intensify research on large-scale cultivation and therapeutic effects of *C. militaris*.

References

Ahn HY, Park KR, Kim YR, Cha JY, Cho YS (2013) Effect of Cordycepin-increased *Cordyceps militaris* powder on tissues lipid peroxidation and antioxidative activity in carbon tetrachloride-induced hepatic damage in rats. J Life Sci 23(7):904–912. https://doi.org/10.5352/jls.2013.23.7.904

Ahn HY, Cho HD, Cho YS (2020) Anti-oxidant and anti-hyperlipidemic effects of cordycepin-rich *Cordyceps militaris* in a Sprague–Dawley rat model of alcohol-induced hyperlipidemia and oxidative stress. Bioresour Bioprocess 7(1):33. https://doi.org/10.1186/s40643-020-00323-9

Amirullah NA, Zainal Abidin N, Abdullah N (2018) The potential applications of mushrooms against some facets of atherosclerosis: a review. Food Res Int 105:517–536. https://doi.org/10.1016/j.foodres.2017.11.023

Cha JY, Ahn HY, Cho YS, Je JY (2013) Protective effect of cordycepin-enriched *Cordyceps militaris* on alcoholic hepatotoxicity in Sprague-Dawley rats. Food Chem Toxicol 60:52–57. https://doi.org/10.1016/j.fct.2013.07.033

Chen X, Wu G, Huang Z (2013) Structural analysis and antioxidant activities of polysaccharides from cultured *Cordyceps militaris*. Int J Biol Macromol 58:18–22. https://doi.org/10.1016/j.ijbiomac.2013.03.041

Chen L, Liu Y, Guo Q, Zheng Q, Zhang W (2018) Metabolomic comparison between wild *Ophiocordyceps sinensis* and artificial cultured *Cordyceps militaris*. Biomed Chromatogr 32(9):e4279. https://doi.org/10.1002/bmc.4279

China (2015) NPCPotPsRo, part one. China Medical Science and Technology Press, China

Chiu CP, Liu SC, Tang CH, Chan Y, El-Shazly M, Lee CL, Du YC, Wu TY, Chang FR, Wu YC (2016) Anti-inflammatory cerebrosides from cultivated *Cordyceps militaris*. J Agric Food Chem 64(7):1540–1548. https://doi.org/10.1021/acs.jafc.5b05931

Choi HN, Jang YH, Kim MJ, Seo MJ, Kang BW, Jeong YK, Kim JI (2014) *Cordyceps militaris* alleviates non-alcoholic fatty liver disease in Ob/Ob mice. Nutr Res Pract 8(2):172–176. https://doi.org/10.4162/nrp.2014.8.2.172

Choi E, Oh J, Sung G-H (2020) Antithrombotic and antiplatelet effects of *Cordyceps militaris*. Mycobiology 48(3):228–232. https://doi.org/10.1080/12298093.2020.1763115

Cui JD (2014) Biotechnological production and applications of *Cordyceps militaris*, a valued traditional Chinese medicine. Crit Rev Biotechnol 35(4):475–484. https://doi.org/10.3109/07388551.2014.900604

Das SK, Masuda M, Hatashita M, Sakurai A, Sakakibara M (2010) Optimization of culture medium for cordycepin production using *Cordyceps milita*ris mutant obtained by ion beam irradiation. Process Biochem 45(1):129–132. https://doi.org/10.1016/j.procbio.2009.08.008

Das T, Mishra SB, Saha D, Agarwal S (2012) Ethnobotanical survey of medicinal plants used by ethnic and rural people in Eastern Sikkim Himalayan Region. Afr J Basic Appl Sci 4(1):16–20. https://doi.org/10.5829/idosi.ajbas.2012.4.1.61133

Devkota S (2006) Yarsagumba [*Cordyceps sinensis* (Berk.) Sacc.]; traditional utilization in Dolpa District Western Nepal. Our Nature 4:48–52. https://doi.org/10.3126/on.v4i1.502

Dong C, Yang T, Lian T (2014) A comparative study of the antimicrobial, antioxidant, and cytotoxic activities of methanol extracts from fruit bodies and fermented mycelia of caterpillar medicinal mushroom *Cordyceps militaris* (Ascomycetes). Int J Med Mushrooms 16(5): 485–495. https://doi.org/10.1615/IntJMedMushrooms.v16.i5.70

Dong C, Guo S, Wang W, Liu X (2015) Cordyceps industry in China. Mycology 6(2):121–129. https://doi.org/10.1080/21501203.2015.1043967

Food and Drug Administration (FDA) (2022). https://search.usa.gov/search?query=cordyceps&affiliate=fda1. Accessed 02 Jun 2022

Fung JC, Yue GG, Fung KP, Ma X, Yao XQ, Ko WH (2011) *Cordyceps militaris* extract stimulates Cl (−) secretion across human bronchial epithelia by both Ca2+− and cAMP-dependent pathways. J Ethnopharmacol 138(1):201–211. https://doi.org/10.1016/j.jep.2011.08.081

Fungorum (2022). http://www.indexfungorum.org/names/Names.asp?pg=3. Accessed 9 Jun 2022

Gai GZ, Jin SJ, Wang B (2004) The efficacy of *Cordyceps militaris* capsules in treatment of chronic bronchitis in comparison with Jinshuibao capsules. Chin J New Drugs 13:169–170

Gao GX, Li QM, Shen HH (2009) Effect of Astragali-*Cordyceps mixtura* on TGF-beta/Smad signal pathway in the lung of asthma airway remodeling. J Ethnopharmacol 125(1):68–74. https://doi.org/10.1016/j.jep.2009.06.012

Gieseck RL, Wilson MS, Wynn TA (2018) Type 2 immunity in tissue repair and fibrosis. Nat Rev Immunol 18(1):62–76. https://doi.org/10.1038/nri.2017.90

Gong X, Li T, Wan R, Sha L (2021) Cordycepin attenuates high-fat diet-induced non-alcoholic fatty liver disease via down-regulation of lipid metabolism and inflammatory responses. Int Immunopharmacol 91:107173. https://doi.org/10.1016/j.intimp.2020.107173

Greenwald P, Clifford CK, Milner JA (2001) Diet and cancer prevention. Eur J Cancer 37(8): 948–965. https://doi.org/10.1016/S0959-8049(01)00070-3

Gu YX, Wang ZS, Li SX, Yuan QS (2007) Effect of multiple factors on accumulation of nucleosides and bases in *Cordyceps militaris*. Food Chem 102(4):1304–1309. https://doi.org/10.1016/j.foodchem.2006.07.018

Gu C, Zhang D, Zhai W, Zhang H, Wang S, Lv S, Bao Y, Zhu D, Feng S, Guo S, Wang Z (2022) Research progress on *Cordyceps militaris* polysaccharides. Food Biosci 45:101503. https://doi.org/10.1016/j.fbio.2021.101503

Guillamón E, García-Lafuente A, Lozano M, D'Arrigo M, Rostagno MA, Villares A, Martínez JA (2010) Edible mushrooms: role in the prevention of cardiovascular diseases. Fitoterapia 81(7): 715–723. https://doi.org/10.1016/j.fitote.2010.06.005

Guo P, Kai Q, Gao J, Lian Z-q, Wu C-m, Wu C-a, Zhu H-b (2010) Cordycepin prevents hyperlipidemia in hamsters fed a high-fat diet via activation of AMP-activated protein kinase. J Pharmacol Sci 113(4):395–403. https://doi.org/10.1254/jphs.10041FP

Heo JY, Baik HW, Kim HJ, Lee JM, Kim HW, Choi YS, Won JH, Kim HM, Park WI, Kim CY (2015) The efficacy and safety of *Cordyceps militaris* in Korean adults who have mild liver dysfunction. J Clin Nutr 7(3):81–86. https://doi.org/10.15747/jcn.2015.7.3.81

Hirsch KR, Smith-Ryan AE, Roelofs EJ, Trexler ET, Mock MG (2017) *Cordyceps militaris* improves tolerance to high-intensity exercise after acute and chronic supplementation. J Diet Suppl 14(1):42–53. https://doi.org/10.1080/19390211.2016.1203386

Hsu CH, Sun HL, Sheu JN, Ku MS, Hu CM, Chan Y, Lue KH (2008) Effects of the immunomodulatory agent *Cordyceps militaris* on airway inflammation in a mouse asthma model. Pediatr Neonatol 49(5):171–178. https://doi.org/10.1016/S1875-9572(09)60004-8

Hwang IK, Lim SS, Yoo KY, Lee YS, Kim HG, Kang IJ, Kwon HJ, Park J, Choi SY, Won MH (2008) A Phytochemically characterized extract of *Cordyceps militaris* and cordycepin protect hippocampal neurons from ischemic injury in gerbils. Planta Med 74(2):114–119. https://doi.org/10.1055/s-2008-1034277

Jędrejko KJ, Lazur J, Muszyńska B (2021) *Cordyceps militaris*: an overview of its chemical constituents in relation to biological activity. Foods 10(11):2634. https://doi.org/10.3390/foods10112634

Jeong JW, Jin CY, Kim GY, Lee JD, Park C, Kim GD, Kim WJ, Jung WK, Seo SK, Choi IW, Choi YH (2010) Anti-inflammatory effects of cordycepin via suppression of inflammatory mediators in BV2 microglial cells. Int Immunopharmacol 10(12):1580–1586. https://doi.org/10.1016/j.intimp.2010.09.011

Ji H, Shengji P, Chunlin L (2004) An ethnobotanical study of medicinal plants used by the Lisu people in Nujiang, Northwest Yunnan, China. Econ Bot 58:253–264. https://doi.org/10.1663/0013-0001(2004)58[S253:AESOMP]2.0.CO;2

Jo SJ, Lee TH, Kim JM, Han YH (2009) Effect of culture broth of *Cordyceps militaris* on recovery of mice hepatic damage caused by benzo(a)pyrene treatment. Korean J Microbiol 45(4): 416–418

Khan I, Huang G, Li XA, Liao W, Leong WK, Xia W, Bian X, Wu J, Hsiao WLW (2019) Mushroom polysaccharides and jiaogulan saponins exert cancer preventive effects by shaping the gut microbiota and microenvironment in ApcMin/+ mice. Pharmacol Res 148:104448. https://doi.org/10.1016/j.phrs.2019.104448

Kim HS, Kim JY, Kang JS, Kim HM, Kim YO, Hong IP, Lee MK, Hong JT, Kim Y, Han S-B (2010) Cordlan polysaccharide isolated from mushroom *Cordyceps militaris* induces dendritic cell maturation through toll-like receptor 4 signalings. Food Chem Toxicol 48(7):1926–1933. https://doi.org/10.1016/j.fct.2010.04.036

Kim SB, Ahn B, Kim M, Ji H-J, Shin S-K, Hong IP, Kim CY, Hwang BY, Lee MK (2014) Effect of *Cordyceps militaris* extract and active constituents on metabolic parameters of obesity induced by high-fat diet in C58BL/6J mice. J Ethnopharmacol 151(1):478–484. https://doi.org/10.1016/j.jep.2013.10.064

Kim YO, Kim HJ, Abu-Taweel GM, Oh J, Sung GH (2019) Neuroprotective and therapeutic effect of *Cordyceps militaris* on ischemia-induced neuronal death and cognitive impairments. Saudi J Biol Sci 26(7):1352–1357. https://doi.org/10.1016/j.sjbs.2018.08.011

Kopalli SR, Cha K-M, Lee S-H, Hwang S-Y, Lee Y-J, Koppula S, Kim S-K (2019) Cordycepin, an active constituent of nutrient powerhouse and potential medicinal mushroom *Cordyceps militaris* Linn., ameliorates age-related testicular dysfunction in rats. Nutrients 11(4):906. https://doi.org/10.3390/nu11040906

Lakhanpal TN, Rana M (2005) Medicinal and nutraceutical genetic resources of mushrooms. Plant Genetic Resources 3(2):288–303. https://doi.org/10.1079/PGR200581

Lee B, Park J, Park J, Shin H-J, Kwon S, Yeom M, Sur B, Kim S, Kim M, Lee H, Yoon SH, Hahm D-H (2011) *Cordyceps militaris* improves neurite outgrowth in Neuro2A cells and reverses memory impairment in rats. Food Sci Biotechnol 20(6):1599–1608. https://doi.org/10.1007/s10068-011-0221-4

Lee HH, Park H, Sung GH, Lee K, Lee T, Lee I, Park MS, Jung YW, Shin YS, Kang H, Cho H (2014) Anti-influenza effect of *Cordyceps militaris* through immunomodulation in a DBA/2 mouse model. J Microbiol 52(8):696–701. https://doi.org/10.1007/s12275-014-4300-0

Lee DH, Yang M, Keum N, Giovannucci EL, Sun Q, Chavarro JE (2019) Mushroom consumption and risk of total and site-specific cancer in two large U.S. prospective cohorts. Cancer Prev Res (Phila) 12(8):517–526. https://doi.org/10.1158/1940-6207.CAPR-19-0101

Lee CT, Huang KS, Shaw JF, Chen JR, Kuo WS, Shen G, Grumezescu AM, Holban AM, Wang Y-T, Wang J-S, Hsiang Y-P, Lin Y-M, Hsu H-H, Yang C-H (2020) Trends in the immunomodulatory effects of *Cordyceps militaris*: total extracts, polysaccharides and cordycepin. Front Pharmacol 11:575704. https://doi.org/10.3389/fphar.2020.575704

Lee BH, Chen CH, Hsu YY, Chuang PT, Shih MK, Hsu WH (2021) Polysaccharides obtained from *Cordyceps militaris* alleviate hyperglycemia by regulating gut microbiota in mice fed a high-fat/sucrose diet. Foods 10(8):1870. https://doi.org/10.3390/foods10081870

Li SP, Yang FQ, Tsim KW (2006) Quality control of *Cordyceps sinensis*, a valued traditional Chinese medicine. J Pharm Biomed Anal 41(5):1571–1584. https://doi.org/10.1016/j.jpba.2006.01.046

Li XT, Li HC, Li CB, Dou DQ, Gao MB (2010) Protective effects on mitochondria and anti-aging activity of polysaccharides from cultivated fruiting bodies of *Cordyceps militaris*. Am J Chinese Med 38(06):1093–1106. https://doi.org/10.1142/S0192415X10008494

Li J, Zou L, Chen W, Zhu B, Shen N, Ke J, Lou J, Song R, Zhong R, Miao X (2014) Dietary mushroom intake may reduce the risk of breast cancer: evidence from a meta-analysis of observational studies. PLoS One 9(4):e93437. https://doi.org/10.1371/journal.pone.0093437

Li T, Wen L, Cheng B (2019) Cordycepin alleviates hepatic lipid accumulation by inducing protective autophagy via PKA/mTOR pathway. Biochem Biophys Res Commun 516(3):632–638. https://doi.org/10.1016/j.bbrc.2019.06.108

Lin BQ, Li SP (2011) Cordyceps as an herbal drug. In: Benzie IFF, Wachtel-Galor S (eds) Herbal medicine: biomolecular and clinical aspects. CRC Press, Hoboken

Lin P, Yin F, Shen N, Liu N, Zhang B, Li Y, Guo S (2021a) Integrated bioinformatics analysis of the anti-atherosclerotic mechanisms of the polysaccharide CM1 from *Cordyceps militaris*. Int J Biol Macromol 193:1274–1285. https://doi.org/10.1016/j.ijbiomac.2021.10.175

Lin YE, Chen YC, Lu KH, Huang YJ, Panyod S, Liu WT, Yang SH, Lu YS, Chen MH, Sheen LY (2021b) Antidepressant-like effects of water extract of *Cordyceps militaris* (Linn.) link by modulation of ROCK2/PTEN/Akt signaling in an unpredictable chronic mild stress-induced animal model. J Ethnopharmacol 276:114194. https://doi.org/10.1016/j.jep.2021.114194

Liu J, Yang S, Yang X, Chen Z, Li J (1997) Anticarcinogenic effect and hormonal effect of *Cordyceps militaris* link. Zhongguo Zhong Yao Za Zhi 22(2):111–113

Liu S, Tang J, Huang L, Xu Q, Ling X, Liu J (2015) *Cordyceps militaris* alleviates severity of murine acute lung injury through miRNAs-mediated CXCR2 inhibition. Cell Physiol Biochem 36(5):2003–2011. https://doi.org/10.1159/000430168

Liu JY, Feng CP, Li X, Chang MC, Meng J-l, Xu LJ (2016a) Immunomodulatory and antioxidative activity of *Cordyceps militaris* polysaccharides in mice. Int J Biol Macromol 86:594–598. https://doi.org/10.1016/j.ijbiomac.2016.02.009

Liu X, Huang Y, Chen Y, Cao Y (2016b) Partial structural characterization, as well as immunomodulatory and anti-aging activities of CP2-c2-s2 polysaccharide from *Cordyceps militaris*. RSC Adv 6(106):104094–104103. https://doi.org/10.1039/C6RA23612J

Ma L, Zhang S, Du M (2015) Cordycepin from *Cordyceps militaris* prevents hyperglycemia in alloxan-induced diabetic mice. Nutr Res 35(5):431–439. https://doi.org/10.1016/j.nutres.2015.04.011

Mehra A, Zaidi KU, Mani A, Thawani V (2017) The health benefits of *Cordyceps militaris*—a review. Kavaka 48(1):27–32

Mei Q, Tao J, Gao S, Xu G, Chen L, Su J (1989) Antiarrhythmic effects of *Cordyceps sinensis* (Berk.) Sacc. Zhongguo Zhong Yao Za Zhi 14(10):616–618

Mizuno T (1999) Medicinal effects and utilization of Cordyceps (Fr.) link (ascomycetes) and Isaria Fr. (Mitosporic fungi) Chinese caterpillar fungi, "Tochukaso" (review). Int J Med Mushrooms 1:251–261. https://doi.org/10.1615/IntJMedMushrooms.v1.i3.80

Nan JX, Park EJ, Yang BK, Song CH, Ko G, Sohn DH (2001) Antifibrotic effect of extracellular biopolymer from submerged mycelial cultures of *Cordyceps militaris* on liver fibrosis induced by bile duct ligation and scission in rats. Arch Pharm Res 24(4):327–332. https://doi.org/10.1007/BF02975101

Ng TB, Wang HX (2005) Pharmacological actions of Cordyceps, a prized folk medicine. J Pharm Pharmacol 57(12):1509–1519. https://doi.org/10.1211/jpp.57.12.0001

Nguyen TV, Chumnanpuen P, Parunyakul K, Srisuksai K, Fungfuang W (2021) A study of the aphrodisiac properties of *Cordyceps militaris* in streptozotocin-induced diabetic male rats. Vet World 14(2):537–544. https://doi.org/10.14202/vetworld.2021.537-544

Nxumalo W, Elateeq AA, Sun Y (2020) Can Cordyceps cicadae be used as an alternative to *Cordyceps militaris* and *Cordyceps sinensis*?—a review. J Ethnopharmacol 257:112879. https://doi.org/10.1016/j.jep.2020.112879

Oh JY, Choi WS, Lee CH, Park HJ (2011) The ethyl acetate extract of *Cordyceps militaris* inhibits IgE-mediated allergic responses in mast cells and passive cutaneous anaphylaxis reaction in mice. J Ethnopharmacol 135(2):422–429. https://doi.org/10.1016/j.jep.2011.03.030

Olatunji OJ, Feng Y, Olatunji OO, Tang J, Ouyang Z, Su Z, Wang D, Yu X (2016) Neuroprotective effects of adenosine isolated from *Cordyceps cicadae* against oxidative and ER stress damages induced by glutamate in PC12 cells. Environ Toxicol Pharmacol 44:53–61. https://doi.org/10.1016/j.etap.2016.02.009

Olatunji OJ, Tang J, Tola A, Auberon F, Oluwaniyi O, Ouyang Z (2018) The genus Cordyceps: an extensive review of its traditional uses, phytochemistry and pharmacology. Fitoterapia 129:293–316. https://doi.org/10.1016/j.fitote.2018.05.010

Park C, Hong SH, Lee JY, Kim GY, Choi BT, Lee YT, Park DI, Park YM, Jeong YK, Choi YH (2005) Growth inhibition of U937 leukemia cells by aqueous extract of *Cordyceps militaris* through induction of apoptosis. Oncol Rep 13(6):1211–1216. https://doi.org/10.3892/or.13.6.1211

Parunyakul K, Srisuksai K, Charoenlappanit S, Phaonakrop N, Roytrakul S, Fungfuang W (2021) Metabolic impacts of cordycepin on hepatic proteomic expression in streptozotocin-induced type 1 diabetic mice. PLoS One 16(8):e0256140. https://doi.org/10.1371/journal.pone.0256140

Paterson RRM (2008) Cordyceps—a traditional Chinese medicine and another fungal therapeutic biofactory? Phytochemistry 69(7):1469–1495. https://doi.org/10.1016/j.phytochem.2008.01.027

Rabinovich GA, Gabrilovich D, Sotomayor EM (2007) Immunosuppressive strategies that are mediated by tumor cells. Annu Rev Immunol 25:267–296. https://doi.org/10.1146/annurev.immunol.25.022106.141609

Rahman MA, Abdullah N, Aminudin N (2018) Evaluation of the antioxidative and hypocholesterolemic effects of lingzhi or reishi medicinal mushroom, *Ganoderma lucidum* (Agaricomycetes), in ameliorating cardiovascular disease. Int J Med Mushrooms 20(10): 961–969. https://doi.org/10.1615/IntJMedMushrooms.2018028370

Ramgir SS, Renu K, Vellingiri B, George A, Tirupapuliyur D, Thiagarajan P, Valsala Gopalakrishnan A (2022) Phytomedicinal therapeutics for male infertility: critical insights and scientific updates. J Nat Med 76(3):546–573. https://doi.org/10.1007/s11418-022-01619-0

Rao YK, Chou CH, Tzeng YM (2006) A simple and rapid method for identification and determination of cordycepin in *Cordyceps militaris* by capillary electrophoresis. Anal Chim Acta 566(2):253–258. https://doi.org/10.1016/j.aca.2006.02.071

Ribeiro JA (1995) Purinergic inhibition of neurotransmitter release in the central nervous system. Pharmacol Toxicol 77(5):299–305. https://doi.org/10.1111/j.1600-0773.1995.tb01031.x

Rukachaisirikul V, Pramjit S, Pakawatchai C, Isaka M, Supothina S (2004) 10-membered macrolides from the insect pathogenic fungus *Cordyceps militaris* BCC 2816. J Nat Prod 67(11):1953–1955. https://doi.org/10.1021/np0401415

Shashidhar MG, Giridhar P, Udaya Sankar K, Manohar B (2013) Bioactive principles from *Cordyceps sinensis*: a potent food supplement—a review. J Funct Foods 5(3):1013–1030. https://doi.org/10.1016/j.jff.2013.04.018

Shin S, Lee S, Kwon J, Moon S, Lee S, Lee C-K, Cho K, Ha N-J, Kim K (2009) Cordycepin suppresses expression of diabetes regulating genes by inhibition of lipopolysaccharide-induced inflammation in macrophages. Immune Netw 9(3):98–105. https://doi.org/10.4110/in.2009.9.3.98

Shin A, Kim J, Lim S-Y, Kim G, Sung M-K, Lee E-S, Ro J (2010) Dietary mushroom intake and the risk of breast cancer based on hormone receptor status. Nutr Cancer 62(4):476–483. https://doi.org/10.1080/01635580903441212

Shrestha B, Zhang W, Zhang Y, Liu X (2012) The medicinal fungus *Cordyceps militaris*: research and development. Mycol Prog 11(3):599–614. https://doi.org/10.1007/s11557-012-0825-y

Singpoonga N, Rittiron R, Seang-on B, Chaiprasart P, Bantadjan Y (2020) Determination of adenosine and cordycepin concentrations in *Cordyceps militaris* fruiting bodies using near-infrared spectroscopy. ACS Omega 5(42):27235–27244. https://doi.org/10.1021/acsomega.0c03403

Sohn S-H, Lee S-C, Hwang S-Y, Kim S-W, Kim I-W, Ye M, Kim S-K (2012) Effect of long-term administration of Cordycepin from *Cordyceps militaris* on testicular function in middle-aged rats. Planta Med 78:1620–1625. https://doi.org/10.1055/s-0032-1315212

Song J, Wang Y, Liu C, Huang Y, He L, Cai X, Lu J, Liu Y, Wang D (2016) *Cordyceps militaris* fruit body extract ameliorates membranous glomerulonephritis by attenuating oxidative stress and renal inflammation via the NF-kappaB pathway. Food Funct 7(4):2006–2015. https://doi.org/10.1039/c5fo01017a

Tiwari UL, Kotia A, Rawat GS (2009) Medico-ethnobotany of the Monpas in Tawang and West Kameng districts of Arunachal Pradesh, India. Pleione 3(1):1–8

Tuli HS, Sandhu SS, Sharma AK (2014) Pharmacological and therapeutic potential of *Cordyceps* with special reference to Cordycepin. 3 Biotech 4(1):1–12. https://doi.org/10.1007/s13205-013-0121-9

Ueda Y, Mori K, Satoh S, Dansako H, Ikeda M, Kato N (2014) Anti-HCV activity of the Chinese medicinal fungus *Cordyceps militaris*. Biochem Biophys Res Commun 447(2):341–345. https://doi.org/10.1016/j.bbrc.2014.03.150

Wang J, Liu YM, Cao W, Yao KW, Liu ZQ, Guo JY (2012) Anti-inflammation and antioxidant effect of Cordymin, a peptide purified from the medicinal mushroom *Cordyceps sinensis*, in middle cerebral artery occlusion-induced focal cerebral ischemia in rats. Metab Brain Dis 27(2): 159–165. https://doi.org/10.1007/s11011-012-9282-1

Wang L, Huang QH, Huang YF, Xie JH, Qu C, Chen JP, Zheng L, Yi TG, Zeng HF, Li H-l (2018) Comparison of protective effect of ordinary *Cordyceps militaris* and selenium-enriched *Cordyceps militaris* on triptolide-induced acute hepatotoxicity and the potential mechanisms. J Funct Foods 46:365–377. https://doi.org/10.1016/j.jff.2018.05.016

Wei M, Zheng S, Jiang C, Xiao M, Lu Y (2015) Inhibitory effects of *Cordyceps militaris* (Bombyx mori) extract against respiratory syncytial virus. Agric Biotechnol 4(1):62–64

Won SY, Park EH (2005) Anti-inflammatory and related pharmacological activities of cultured mycelia and fruiting bodies of *Cordyceps militaris*. J Ethnopharmacol 96(3):555–561. https://doi.org/10.1016/j.jep.2004.10.009

Wong JH, Ng TB, Wang H, Sze SC, Zhang KY, Li Q, Lu X (2011) Cordymin, an antifungal peptide from the medicinal fungus *Cordyceps militaris*. Phytomedicine 18(5):387–392. https://doi.org/10.1016/j.phymed.2010.07.010

Xiong Y, Zhang S, Xu L, Song B, Huang G, Lu J, Guan S (2013) Suppression of T-cell activation *in vitro* and *in vivo* by cordycepin from *Cordyceps militaris*. J Surg Res 185(2):912–922. https://doi.org/10.1016/j.jss.2013.06.057

Yadav R (2020) Entomopathogenic mushroom (*Cordyceps* sp.) as immunity booster. Biotica Research Today 2(7):690–692

Yamamoto K, Shichiri H, Uda A, Yamashita K, Nishioka T, Kume M, Makimoto H, Nakagawa T, Hirano T, Hirai M (2015) Apoptotic effects of the extracts of *Cordyceps militaris* via erk phosphorylation in a renal cell carcinoma cell line. Phytother Res 29(5):707–713. https://doi.org/10.1002/ptr.5305

Yang X, Li Y, He Y, Li T, Wang W, Zhang J, Wei J, Deng Y, Lin R (2015) Cordycepin alleviates airway hyperreactivity in a murine model of asthma by attenuating the inflammatory process. Int Immunopharmacol 26(2):401–408. https://doi.org/10.1016/j.intimp.2015.04.017

Yang X, Lin P, Wang J, Liu N, Yin F, Shen N, Guo S (2021) Purification, characterization and anti-atherosclerotic effects of the polysaccharides from the fruiting body of *Cordyceps militaris*. Int J Biol Macromol 181:890–904. https://doi.org/10.1016/j.ijbiomac.2021.04.083

Ying J, Ying C, Wang Y (1987) Icons of medicinal fungi from China. Science Press, China

Yu R, Song L, Zhao Y, Bin W, Wang L, Zhang H, Wu Y, Ye W, Yao X (2004) Isolation and biological properties of polysaccharide CPS-1 from cultured *Cordyceps militaris*. Fitoterapia 75(5):465–472. https://doi.org/10.1016/j.fitote.2004.04.003

Yu HM, Wang BS, Huang SC, Duh PD (2006) Comparison of protective effects between cultured *Cordyceps militaris* and natural *Cordyceps sinensis* against oxidative damage. J Agric Food Chem 54(8):3132–3138. https://doi.org/10.1021/jf053111w

Yu R, Yang W, Song L, Yan C, Zhang Z, Zhao Y (2007) Structural characterization and antioxidant activity of a polysaccharide from the fruiting bodies of cultured *Cordyceps militaris*. Carbohydr Polym 70(4):430–436. https://doi.org/10.1016/j.carbpol.2007.05.005

Yu SH, Dubey NK, Li WS, Liu MC, Chiang HS, Leu SJ, Shieh YH, Tsai FC, Deng WP (2016) *Cordyceps militaris* treatment preserves renal function in type 2 diabetic nephropathy mice. PLoS One 11(11):e0166342. https://doi.org/10.1371/journal.pone.0166342

Yu WQ, Yin F, Shen N, Lin P, Xia B, Li YJ, Guo SD (2021) Polysaccharide CM1 from *Cordyceps militaris* hinders adipocyte differentiation and alleviates hyperlipidemia in LDLR(+/−) hamsters. Lipids Health Dis 20(1):178. https://doi.org/10.1186/s12944-021-01606-6

Yuan G, An L, Sun Y, Xu G, Du P (2018) Improvement of learning and memory induced by cordyceps polypeptide treatment and the underlying mechanism. Evid Based Complement Alternat Med 2018:9419264. https://doi.org/10.1155/2018/9419264

Yue GG, Lau CB, Fung KP, Leung PC, Ko WH (2008) Effects of *Cordyceps sinensis*, *Cordyceps militaris* and their isolated compounds on ion transport in Calu-3 human airway epithelial cells. J Ethnopharmacol 117(1):92–101. https://doi.org/10.1016/j.jep.2008.01.030

Zhang J, Wen C, Duan Y, Zhang H, Ma H (2019) Advance in *Cordyceps militaris* (Linn) link polysaccharides: isolation, structure, and bioactivities: a review. Int J Biol Macromol 132:906–914. https://doi.org/10.1016/j.ijbiomac.2019.04.020

Zhao-Long W, Xiao-Xia W, Wei-Ying C (2000) Inhibitory effect of *Cordyceps sinensis* and *Cordyceps militaris* on human glomerular mesangial cell proliferation induced by native LDL. Cell Biochem Funct 18(2):93–97. https://doi.org/10.1002/(SICI)1099-0844(200006)18:2<93::AID-CBF854>3.0.CO;2-#

Zheng Y, Li L, Cai T (2020) Cordyceps polysaccharide ameliorates airway inflammation in an ovalbumin-induced mouse model of asthma via TGF-beta1/Smad signaling pathway. Respir Physiol Neurobiol 276:103412. https://doi.org/10.1016/j.resp.2020.103412

Zhou Y, Li Y, Zhou T, Zheng J, Li S, Li H-B (2016) Dietary natural products for prevention and treatment of liver cancer. Nutrients 8(3):156. https://doi.org/10.3390/nu8030156

Zhou J, Chen X, Xiao L, Zhou J, Feng L, Wang G (2021) Efficacy and safety of *Cordyceps militaris* as an adjuvant to duloxetine in the treatment of insomnia in patients with depression: a 6-week double-blind, randomized, placebo-controlled trial. Front Psych 12:754921. https://doi.org/10.3389/fpsyt.2021.754921

Zhu SJ, Pan J, Zhao B, Liang J, Wu ZY, Yang JJ (2013) Comparisons on enhancing the immunity of fresh and dry *Cordyceps militaris in vivo* and *in vitro*. J Ethnopharmacol 149(3):713–719. https://doi.org/10.1016/j.jep.2013.07.037

Mycotherapeutics Affecting Dopaminergic Neurotransmission to Exert Neuroprotection

Shannon Kim, Rishi M. Nadar, Jack DeRuiter, Suhrud Pathak, Sindhu Ramesh, Timothy Moore, Dinesh Chandra Agrawal, and Muralikrishnan Dhanasekaran

Contents

Abstract Alternative or integrative medicine is a holistic approach to healthcare that has been used for centuries in many different cultures. Today, there is a significant trend of a growing body of research related to mushrooms to suggest that integrative therapies have much to offer in the prevention and treatment of disease, as well as in rehabilitation and promoting overall health and wellness. Mycotherapeutics are primarily preferred by health-conscious individuals who lead a "wellness-oriented" healthy lifestyle and are dedicated to nutrition, fitness, stress relief, and their environment. Although there is an expanding acceptance of the value of holistic approaches to mycotherapeutics, it has been long overlooked due to the dogma of Western medicine. Naturally, the origins of mycotherapeutics are deeply

---

Shannon Kim and Rishi M. Nadar contributed equally with all other contributors.

S. Kim · R. M. Nadar · J. DeRuiter · S. Pathak · S. Ramesh · T. Moore · M. Dhanasekaran (✉)
Department of Drug Discovery and Development, Harrison School of Pharmacy, Auburn University, Auburn, AL, USA
e-mail: dhanamu@auburbn.edu

D. C. Agrawal (✉)
Department of Applied Chemistry, Chaoyang University of Technology, Taichung, Taiwan
e-mail: dcagrawal@gm.cyut.edu.tw

D. C. Agrawal, M. Dhanasekaran (eds.), *Mushrooms with Therapeutic Potentials*,
https://doi.org/10.1007/978-981-19-9550-7_13

rooted in a culture's religious and philosophical beliefs. In the past few years, the natural bioactives of mushrooms have been isolated, and their chemical properties and pharmacodynamic actions have been well characterized. Bioactives of certain mushrooms can significantly affect the dopaminergic neurotransmission in the central and peripheral nervous systems. Thus, mycotherapeutics can yield a potentiating/synergistic pharmacodynamic impact which can be used to treat various diseases associated with impaired dopaminergic neurotransmission.

Keywords Alternative or integrative medicine · Dopamine · Holistic approach · Mushrooms · Mycotherapeutics · Neuroprotection

Abbreviations

5-HT receptors	5-hydroxytryptamine receptors
Akt	Protein kinase B (PKB)
AMPA	α-amino-3-hydroxy-5-methyl-4-isoxazole propionic acid
BDNF	Brain-derived neurotrophic factor
CES-D	Center for epidemiologic studies depression scale
CNS	Central nervous system
COX-2	Cyclooxygenase-2
D2	Dopamine receptor 2
GABA	γ-Aminobutyric acid
GADD45	Growth arrest and DNA damage-inducible 45
GMP	Good manufacturing practice
GSK-3β	Glycogen synthase kinase-3 beta
HIV	Human immunodeficiency virus
ICARS	International cooperative ataxia rating scale
ICI	Indefinite complaints index
MPTP	1-Methyl-4-phenyl-1,2,3,6-tetrahydropyridine
NCCIH	National center for complementary and integrative health
NF-κB	Nuclear factor kappa B
NGF	Nerve growth factor
NMDA	N-methyl-D-aspartate
PAK1	Serine/threonine-protein kinase
PI3Ks	Phosphoinositide 3-kinases
PTSD	Post-traumatic stress disorder
SARA	Scale for the assessment and rating of ataxia
SK channels	Small-conductance calcium-activated potassium channels
TBI	Traumatic brain injury

1 Introduction

Despite the new and novel synthetic and biological drug discoveries, there is a prominent need for a comprehensive integrative holistic healing approach to current and future global healthcare. An integrative holistic healing approach is a caring and well-planned treatment technique that is used in healthcare in some parts of the world which focuses on alternative and complementary medicine, ethical facets, familial and societal values, psychological phenomena, and religious as well as biological aspects for vigor and treatment of ailments (Gannotta et al. 2018). The enlightenment philosophy in the sixteenth century indicated the "human mind" as distinct from the physical body (Thibaut 2018). In the past few years, healthcare professionals have endeavored to theoretically and systematically "tie up" the human mind and body to enhance and recognize this as a complete system (Gannotta et al. 2018). The fundamental concepts associated with the holistic approach of integrative medicine encompass a relationship between an individual and the healthcare team of professionals by utilizing the holistic approach to improvise and enhance the mind and body's innate healing response (Gannotta et al. 2018).

As per the National Center for Complementary and Integrative Health (NCCIH), "most integrative therapies fall into one of two subgroups, the natural products, and mind and body practices" (Rosenthal et al. 2018). These practices embrace the involvement of nutritional strategies (dietary supplements, natural bioactives, antioxidants, vitamins, individualized/specialized diets, medicinal/dietary mycotherapeutics, natural bioactive supplements (herbs/botanicals), homeopathy, mind-body remedy, lifestyle modifications, physiotherapy, and cognitive/behavioral counseling). The novel holistic mind-body (such as meditation, mindfulness, and guided imagery approaches) involves music therapy, creative arts therapy, hypnosis, yoga, tai chi, and qigong, along with physical exercises and spiritual practices (Rosenthal et al. 2018). Physical therapies include manipulative, massage, chiropractic, and energy-based practices (reiki) (Rosenthal et al. 2018).

An integrative holistic approach highlights the exclusivity of an individual with a specific group of healthcare professionals, the responsiveness of an individual patient-healthcare professional relationship, the individual's responsibility for their healthcare, and society's emphasis on the prophylactic medicinal approaches and enhancement of healthcare. Alternative and complementary medicines can provide direct beneficial and synergistic pharmacological effects to the existing prescription-based and nonprescription natural/synthetic drugs. Due to their scientific validity, widespread acceptance, and use, integrative holistic approaches have become an alternative therapeutic tactic and practice for healing (Gannotta et al. 2018). The major principle of integrative holistic approaches to healthcare is its emphasis that each individual with a pathological condition should be appropriately diagnosed and cared for as an inimitable individual. They also consider that a pathological condition comprises the body (anatomy and physiology), psyche (mind), and spirit, for which the treatment also should include the living condition (Gannotta et al. 2018). Accordingly, the integrative holistic approach to patient care is a blend of the bionomical compassion of traditional treatment methods and the meticulousness of current knowledge of anatomy, physiology, pharmacology, and pathology. Traditional treatment methods have been preserved for centuries, and their usefulness is widely accepted (Bell et al. 2002).

Moreover, nutraceutical, pharmacognosy, and alternative medicine-based techniques endure and have gained advocates and believers globally (Tabish 2008). Numerous physicians worldwide gradually accept and recognize the beneficial value of the integrative holistic tactic for the patient's wellness (Patel et al. 2017). The major advantage of this scientific approach is its emphasis on prophylactic and therapeutic actions. Currently, an effective cumulative extent of systematic evidence is developing regarding the beneficial activities of traditional natural bioactives, specifically mushrooms (Valverde et al. 2015).

The fungi (mushrooms) are effective synthesizers of several natural bioactives with potent beneficial effects for human and animal welfare. Mushrooms can contain bioactives with pharmacodynamic effects which can affect the central nervous system (CNS) and peripheral nervous system, ophthalmic system, cardiovascular system, respiratory system, gastrointestinal system, renal system, reproductive system, endocrine and exocrine system, integumentary system, and the immune system. Thus, "mycotherapy" is considered an important integrative healing approach that uses medicinal mushrooms in healthcare to prevent and treat pathological conditions alone or in combination with other natural or synthetic agents (Venturella et al. 2021).

Conventionally, medicinal mushrooms were separated from the toxic mushroom species and have been valued for their healing properties by numerous cultures and religions for centuries (Valverde et al. 2015). For example, *Ganoderma* species have been used for over 2000 years as a tonic to prevent and treat acute/chronic liver diseases, cardiovascular conditions (hypertension), neurological diseases (anxiety, insomnia, neurasthenia), respiratory diseases (bronchitis), gastrointestinal pathologies (gastric ulcer), endocrinological diseases (hyperglycemia), and cancer in Asian countries (China, Korea, and Japan) and some other countries (Wachtel-Galor et al. 2011). Interestingly, Ayurveda (Ayu = life, Veda = science) reveres mycotherapeutics for their holistic nutraceutical and therapeutic efficacy (Panda and Swain 2011). This book chapter reviews the effect of mycotherapeutics on dopaminergic neurotransmission and treating diseases associated with dopaminergic neurotransmission.

2 Mycotherapy for the Current and Future Human and Animal Healthcare: Treats Acute and Chronic Disease States

Due to the presence of pharmacodynamic bioactives (beta-glucans, minerals, prebiotics, proteins, terpenoids, and vitamins) with diverse pharmacological effects, mycotherapeutics (chaga, Coriolus, *Ganoderma*, and maitake) have been found to treat various neurological diseases states, lethargy, and gastrointestinal pathological conditions (Zou et al. 2020; Ferreiro et al. 2018; Guo et al. 2021; Bai et al. 2019). Moreover, *Cordyceps* and reishi mushrooms have been used in athletics to protect sportspersons from nonfunctional overreaching and overtraining by modulating the cortisol and testosterone content (Rossi et al. 2014). The bioactives of mushrooms can cross the blood-brain barrier and induce a pharmacological impact with neuroprotective effects (neurological and physiological effects) (Venturella et al. 2021). *H. erinaceus* extracts were incorporated into cookies and were found to lower the Indefinite Complaints Index (ICI) and the Center for Epidemiologic Studies Depression Scale (CES-D) (Venturella et al. 2021). These results suggest that *H. erinaceus* contains lipophilic bioactives that can readily cross the blood-brain barrier and interact with targets resulting in a constructive influence on the emotional health aspect of humans. Thus, physicians, nurses, or pharmacists suggesting integrative holistic therapies will focus on the whole body instead of seemingly separate, organ-specific complaints.

By considering all the variables that could impact a patient's well-being, such as their environment, activity, emotional well-being, and lifestyle, the practitioner can gain a comprehensive view of their health. Thus, this integrative holistic approach represents a supplemental approach to managing patients' existing pathologies and overall mental and physical well-being. Compared to the existing prescription or nonprescription-based synthetic drugs (allopathic medicines), the integrative holistic approach focuses on and emphasizes prophylactic efficacy, which makes it optimal for a healthy individual's well-being by preventing the cellular damage associated with the etiology of a pathological condition (Snyderman and Weil 2002). Interestingly, an integrative holistic approach is beneficial in reducing the risks and treating chronic disease states (arthritis, asthma, diabetes mellitus, multiple sclerosis) and infectious diseases (hepatitis C and HIV) (Roy 2010; Marzio and Fenkel 2014; Littlewood and Vanable 2008). Therefore, an integrative holistic approach, including mushrooms' bioactives, can potentially reinforce the innate physiological capability to reconcile, restore, and utilize lifestyle alterations, mind-body remedies, and patient education to achieve therapeutic efficacy with minimal adverse effects and hypersensitivity reactions. Particularly interesting case studies about integrative medicine come from oncology. Thus, an integrative holistic approach using mushrooms is commonly practiced complementing the present prescription-based and nonprescription natural/synthetic drugs (Patel and Goyal 2012).

3 Designing Mycotherapeutics-Based Strategies to Target Dopaminergic Neurotransmission to Prevent and Treat Neurological Diseases

Dopamine is a monoaminergic neurotransmitter in the central nervous system (CNS) and peripheral nervous system. The precursor for the synthesis of dopamine is the amino acid tyrosine (Fig. 1). Synthesizing enzymes tyrosine hydroxylase and amino

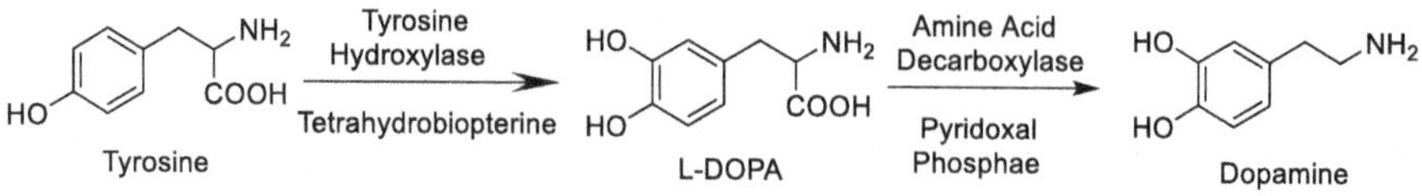

Fig. 1 Dopamine biosynthesis

acid decarboxylase convert tyrosine to the precursor, levodopa, which is then converted to dopamine in cofactors tetrahydrobiopterin, iron, and pyridoxal phosphate.

In the CNS, dopamine is present in the four major neuronal pathways (Rasheed and Alghasham 2012). These dopaminergic neuronal tracts are the nigrostriatal, mesolimbic, mesocortical, and tuberoinfundibular pathways. There are also other minor pathways in the brain that contain dopamine. Neurophysiologically, the major actions of dopamine in the brain are to control movement and mental functions and modify the reward system and prolactin secretion (Juárez et al. 2016). Alteration of the dopaminergic neurotransmission (increased or decreased) can lead to severe neuropathological conditions (Juárez et al. 2016). Parkinson's disease and other movement disorders occur due to the nigral dopaminergic neurodegeneration resulting in movement disorders (Rasheed and Alghasham 2012).

In contrast, the positive symptoms associated with schizophrenia (psychosis) occur due to increased dopaminergic neurotransmission in the mesolimbic/ mesocortical tract (Hany et al. 2022). Therefore, synthetic drugs or natural bioactives that augment or diminish the dopaminergic neurotransmission are used to treat various dopaminergic-associated neuronal disorders (Juárez et al. 2016). Thus, in this book chapter, the role of mycotherapeutics in dopaminergic neuronal diseases, such as ataxia, and autism spectrum disorders is discussed (Table 1).

3.1 Ataxia

Ataxia is a common manifestation of cerebellar dysfunction characterized by impaired coordination of voluntary movements and issues with stance, gait, speech, and precise motions (Mariotti et al. 2005). On rare occasions, it can be hereditary (Ashizawa and Xia 2016). The term "ataxia" has been used to describe or explain diseases as far back as the nineteenth century (Klockgether and Paulson 2011). The prevalence rate of ataxia in children is approximately 26 out of 100,000 individuals. The prevalence rate of dominant hereditary cerebellar ataxia is 2.7 out of 100,000 individuals, and the prevalence rate of recessive hereditary cerebellar ataxia is 3.3 out of 100,000 individuals (Salman 2018).

Regarding the etiology, lesions in the cerebellum commonly cause ataxia (Ashizawa and Xia 2016). This is likely because these lesions interfere with sensory neurotransmissions to the cerebellum (Hafiz and De Jesus 2022). Patients display different symptomatic presentations of the condition based on the location of the

Table 1 The role of mycotherapeutics in dopaminergic neuronal diseases

Neurological disease	Dopaminergic drugs	Mycotherapy	Dopaminergic neuroprotective signaling
• Ataxia • Hereditary • Cerebellar ataxia • Depression	• Bupropion	• *Pleurotus giganteus* • *Ganoderma lucidium* • *Hericium erinaceus*	• Activation of the ubiquitin-proteasome system • Suppresses oxidative stress • Maintenance of intracellular calcium homeostasis • Regulation of chaperones
• Autism spectrum disorder	• Typical / Atypical anti-psychotics • Partial dopamine agonist	• *Antioxidant mushrooms*	• Mushrooms scavenge and suppress pro-oxidants
• Neuropsychiatric disorders: • Bipolar • Psychosis/ • Schizophrenia	• Lithium • Dopamine antagonist (Typical and Atypical anti-psychotics)	• *Agaricus bisporus* • *Agrocybe cylindracea* • *Ganoderma lucidum* • *Hericium erinaceus* • *Lentinus crinitus* • *Pleurotus eryngii* • *Pleurotus ostreatus*	• Lithium biofortification
• Depression	• Norepinephrine-Dopamine reuptake inhibitor (NDRI) • Monoamine oxidase inhibitors (MAOI) • COMT inhibitors	• *Armillaria mellea* • *Cordyceps militaris* • *Ganoderma lucidum* • *Hericium erinaceus* • *Inonotus* sp. • *Marasmius androsaceus* • *Poria cocos* • *Psilocybe* genus	• Psilocybin (increased both the extracellular dopamine and 5-HT) • Modulate neurotrophic and neurogenic mechanism (Stimulate NGF), Modulating BDNF/PI3K/Akt/GSK-3β) • Block the NF-κB signaling • Upregulated BDNF expression • Reduce stress • Shortened the duration of immobility in both tail suspension and forced swimming tests • Increase dopamine and γ-aminobutyric acid (GABA) and decreased glutamate
• Parkinson's disease	• Dopamine Agonist • MAOI • COMT inhibitors • Dopamine Precursor	• *Coriolus versicolor* • *Hericium erinaceus* • *Agaricus muscarius*	• Affect tyrosine hydroxylase, tyrosinase, and dopamine transporter • Prevent neuroinflammation • Affect redox stress response (reduce oxidative

(continued)

Table 1 (continued)

Neurological disease	Dopaminergic drugs	Mycotherapy	Dopaminergic neuroprotective signaling
		• *Pleurotus ostreatus* • *Laetiporus sulphureus* • *Agaricus blazei* • *Amauroderma rugosum*	stress) • Reduce apoptosis • Reduce excitotoxicity • Regulating the p21/GADD45 cell death pathways and PAKalpha, p21activated kinase 1 (PAK1) survival pathways
• Post-Traumatic Stress Disorders (PTSD)	• Dopamine Antagonist (Typical and Atypical antipsychotics)	• *Magic mushrooms* • Over 100 mushrooms species contain psilocybin	• Stimulating nerve cell regrowth in parts of the brain responsible for emotion and memory • Psilocybin stimulates hippocampal neurogenesis
• Restless Leg Syndrome	• Dopamine agonist	• *Reishi & Agaricusus*	• Soothe RLS
• Smoking cessation	• Nicotine agonist • Partial nicotine agonist • NDRI	• *Magic Mushrooms* • Psilocybin	• Increased both the extracellular dopamine and 5-HT • Increase dopamine and GABA) • Decreases glutamate
• Tourette syndrome/Tics	• Dopamine Antagonist	• *Magic mushroom*	
• Traumatic brain injury (TBI)	• Dopamine Antagonist	• *Coriolus versicolor* • *Hericium erinaceus*	• Prevent neuroinflammation • Affect redox stress response

lesions. For example, lesions in the midline cerebellar typically cause gait and truncal ataxia, lesions in the unilateral cerebellar hemisphere typically cause ipsilateral cerebellar ataxia, and lesions in the posterior lobe typically cause postural instability and gait ataxia. Damage to other brain areas or parts of the body, such as the brain stem and spinal cord, can also result in motor-sensory deficits associated with ataxia (Ashizawa and Xia 2016). Genetic ataxia may be caused by X-linked, mitochondrial, or autosomal dominant or recessive inheritance (de Silva et al. 2019). Impairment in the dopaminergic neurotransmission has also been significantly involved in the etiology and pathology of ataxia.

There are various approaches to the treatment of ataxia. One method focuses on alleviating the condition's symptoms using medication such as riluzole (Rilutek™, Tiglutik™), which improves speech and gait by partially returning neuronal firing patterns to proper function through modulation of calcium-activated potassium channels-SK channels. These ion channels are small-conductance calcium-activated potassium channels commonly expressed in the CNS and the neurons. Physiologically, the SK channels are involved in facilitating one process of the after-hyperpolarization event that comes after an action potential. Techniques involving neuromodulation, such as transcranial direct current stimulation, have also been shown to improve patients' Scale for the Assessment and Rating of Ataxia (SARA) and International Cooperative Ataxia Rating Scale (ICARS) scores (Kwei and Kuo 2020).

Bioactives of the *Ganoderma lucidum*, *Hericium erinaceus*, and *Pleurotus giganteus* have been shown to affect dopaminergic neurotransmission and reduce ataxia (Yu et al. 2017; Chong et al. 2020). The neuromolecular signaling mechanisms involved in the neuroprotective mechanisms of this mycotherapeutics are activating the ubiquitin-proteasome system, preventing oxidative stress, maintaining intracellular calcium homeostasis, and regulating chaperones (Liang et al. 2012; Lew et al. 2020; Ma et al. 2007). Ataxia patients had diminished dopaminergic neurotransmission due to inhibition of dopamine synthesis and dopamine D_2 receptor dysfunction. The pathology of ataxia may not be confined to the cerebellum but is distributed across various brain regions, including presynaptic and postsynaptic dopaminergic neurons. Thus, mycotherapeutics have been shown to alter the dopaminergic neurotransmission in cerebellum and other brain regions and thus can be used in treating ataxia. Conclusively, bupropion (a dopamine reuptake inhibitor) in ataxia validated the dopaminergic therapeutic effects of mycotherapeutics (Li et al. 2011).

3.2 *Autism Spectrum Disorders*

Autism, also known as an autism spectrum disorder, is a classification of neurodevelopmental conditions (Sharma et al. 2018). In 1977, the first genetics study was conducted by Susan Folstein and Michael Rutter (Thapar and Rutter 2021). Autism's prevalence has increased over the last two decades, estimated at 1 out of 36 children (Sharma et al. 2018). Genetics appears to have a major influence on autism spectrum disorder. Studies have shown that de novo copy mutations and rare variant mutations lead to atypical alleles, which ultimately have a neuropathological effect on neurological, anatomical, and behavioral characteristics in autism patients. Major genetic abnormalities of autism include the dysregulation of genes associated with synapse function and genes related to the dopaminergic signal transduction mechanism of synapse formation. In addition, the transmembrane and scaffolding proteins involved in the maintenance of synaptogenesis display abnormal assembly and structure (Samsam et al. 2014).

The pharmacological therapies for autism spectrum disorder include psychostimulants such as methylphenidate, which enhance the release of monoamines—dopamine, serotonin, and norepinephrine. These therapies have decreased hyperactivity and impulsivity, mainly in autistic children. Furthermore, atypical or second-generation antipsychotic drugs such as risperidone, which function as dopamine and serotonin receptor antagonists, have decreased aggression, anxiety, and repeating behaviors in adult autistic patients. Thus, these therapeutic approaches have indicated the role of dopamine in the etiology and pathophysiology of autism spectrum disorders. The various non-pharmacological therapies shown to help children with autism include music therapy. Music therapy activates bilateral temporal brain networks similar to the control group of neurotypical children and maintains functional front-temporal connectivity (Sharma et al. 2018).

Furthermore, oxidative stress has been shown to affect dopaminergic neurotransmission in autism (Pangrazzi et al. 2020). Mushrooms with antioxidant properties can significantly decrease the pathologies associated with autism. The mushrooms with potent antioxidant effects include (Vishvakarma and Mishra 2019; Zhao et al. 2017; Erbiai et al. 2021; Zhang et al. 2021; Seweryn et al. 2021; Li et al. 2017; Contato et al. 2020; Umeo et al. 2015; Song et al. 2016; Erdem Guzel et al. 2021; Elhusseiny et al. 2021; Wu et al. 2020; Nkadimeng et al. 2020):

- *Agaricus bisporus*
- *Agrocybe cylindracea*
- *Armillaria mellea*
- *Cordyceps militaris*
- *Ganoderma lucidum*
- *Hericium erinaceus*
- *Inonotus* sp.
- *Lentinus crinitus*
- *Marasmius androsaceus*
- *Pleurotus eryngii*
- *Pleurotus ostreatus*
- *Poria cocos*
- *Psilocybe* genus.

Thus, mycotherapeutics can decrease the oxidative stress-related dopaminergic neuronal insult and exert neuroprotection in autism. By protecting the dopaminergic neurons, the mycotherapeutics can increase the dopaminergic neurotransmission and decrease the symptoms of autism.

3.3 Neuropsychiatric Disorders

3.3.1 Schizophrenia

Schizophrenia is a mental disorder that causes hallucinations, delusional beliefs, and disturbances in thought and behavior. This word was first termed in 1908 by Eugen Bleuler (Hany et al. 2022). Although relatively low in prevalence at about 0.28% of the global population, the mental condition is associated with significant personal and healthcare burdens (Charlson et al. 2018). Psychosis denotes the presence of hallucinations, delusions, and impairment in the patient's perception of reality (Arciniegas 2015). Schizophrenia's symptoms are classified into two categories—positive symptoms, which include hallucinations, delusions, and formal thought disorders, and negative symptoms, which include lack of motivation, anhedonia, and scarcity of speech. One hypothesis regarding the development of the disorder identifies extreme activation of D_2 pathways by the mesolimbic pathway as the cause for its positive symptoms, low mesocortical dopamine levels due to the mesocortical pathway as the cause for its negative symptoms, and a scarcity of dopamine in the nigrostriatal pathway as the cause for its motor symptoms. In addition, issues in the tuberoinfundibular pathway may lead to decreased supply of tuberoinfundibular dopamine, causing the elevation of prolactin levels and resulting in other symptoms of schizophrenia, including low libido and amenorrhea (Hany et al. 2022).

Pharmacological treatments for schizophrenia most often involve antipsychotic drugs aimed at mitigating symptoms that compose the condition of psychosis. First-generation antipsychotics primarily act as antagonists for dopamine D_2 receptors in the mesolimbic pathway, lessening positive schizophrenic symptoms. However, they also block D_2 receptors in the nigrostriatal pathway resulting in undesirable motor effects such as extrapyramidal effects and in the tuberoinfundibular pathway causing increased prolactin secretion and reduced pleasure. Furthermore, they may even worsen schizophrenia's negative symptoms (Stępnicki et al. 2018). Second-generation antipsychotics function as antagonists at dopamine D_2 receptors and serotonin 2A receptors. The net effect of their pharmacologic actions results in a reduced disturbance of motor function or prolactin secretion.

Psilocybin mushrooms may have potential use as a treatment for schizophrenia. They have been associated with a reduction in the expression of the serotonin 5-HT_{2A} receptor (Mahmood et al. 2022). An imbalance in the ratio of 5-HT_{1A} and 5-HT_{2A} receptors is a critical factor in developing schizophrenia-related abnormalities in the brain (Mahmood et al. 2022). Furthermore, there is evidence that, through regulation of striatum and cortex dopamine release, 5-HT_{2A} leads to the motor and cognitive defects associated with schizophrenia patients (Raote et al. 2007). Thus, psilocybin's ability to reduce the expression of the serotonin 5-HT_{2A} receptor may help alleviate some of the disorder's symptoms.

3.3.2 Bipolar Disorder

Bipolar disorder is a long-term episodic condition characterized by periodic incidences of depressive or manic symptoms (Culpepper 2014). It was formerly known as manic depression (Miklowitz and Johnson 2006). The disorder has been estimated to have a prevalence range of anywhere between 0.5 and 5% of the population (Clemente et al. 2015). Dysfunction in certain regions of the brain is strongly linked with bipolar disorder. Hyperactivity and above-average volume of the amygdala, the section of the brain that processes both positive and negative emotional stimuli, have been shown in studies to be tied to the condition, as has diminished activity and below-average volume of the hippocampus and prefrontal cortex. Below-average volume in the basal ganglia and the anterior cingulate have also been observed in patients with bipolar disorder. During manic episodes, the experimental group's activation of the amygdala and subgenual anterior cingulate cortex was diminished in patients with bipolar disorder compared to the control group (Miklowitz and Johnson 2006).

Pharmacological treatments for bipolar disorder typically treat manic, depressive, and mixed episodes. Lithium is the most well-documented medication for the treatment of mania and appears to work by inhibiting protein kinase C as well as calcium, G-proteins, and glutamate, which play a part in the intracellular signaling cascade. However, it has been noted to have numerous adverse effects in many patients involving the GI tract (nausea, vomiting, diarrhea, and constipation), the renal system (polydipsia, polyuria, kidney failure), and the CNS (confusion, vertigo, memory loss), such as stomach irritation and kidney clearance problems. For the treatment of depression, antidepressants, which reduce depressive symptoms through serotonergic, dopaminergic, and noradrenergic mechanisms, are used in conjunction with an atypical antipsychotic agent (Miklowitz and Johnson 2006).

Concerning the mycotherapeutics, *Agaricus bisporus*, *Agrocybe cylindracea*, *Ganoderma lucidum*, *Hericium erinaceus*, *Lentinus crinitus*, *Pleurotus eryngii*, and *Pleurotus ostreatus* have been used to treat neuropsychiatric disorders (Naeem et al. 2021). These mycotherapeutics have been shown to affect lithium biofortification and dopaminergic neurotransmission and have fewer adverse drug reactions than synthetic drugs used in the treatment (Naeem et al. 2021).

3.3.3 Depression

Depression is a mood disorder that causes a persistent feeling of sadness and loss of interest. Also called a major depressive disorder, clinical depression affects emotions, thoughts, and behavior and can lead to various emotional and physical problems. Those with depression may have trouble doing normal day-to-day activities and sometimes feel like life isn't worth living (Chand and Arif 2022). Impacting about 17% of the population across different stages of life, it is associated with immense economic ramifications and great personal distress (Duman 2014). Depression is related to alterations in the activity of certain brain regions. Alterations in the functional connections between these regions have also been shown to correlate with behavior characteristics of depression. Decreased activity in the prefrontal cortex and its connections to the limbic and subcortical structures associated with depression has been observed in patients with depression. It is also possible that a disruption in the link between the prefrontal cortex and the amygdala plays a role in causing some depressive symptoms, such as decreased motivation and reward (Duman 2014). On a broader scale, it has also been evidenced through clinical and preclinical trials that a disruption of the activity of CNS serotonin contributes to the disorder (Chand and Arif 2022). Antidepressants such as tricyclics, selective serotonin reuptake inhibitors, and serotonin-norepinephrine reuptake inhibitors, which alter serotonergic and noradrenergic neurotransmission in the CNS, are one of the most effective treatments for chronic and severe depression (Swedish Council on Health Technology 2004). Also, the drug ketamine in small doses can rapidly improve treatment-resistant depression, including symptoms of suicidal ideation, and has long-lasting effects as an antidepressant. In animals, ketamine has been observed to increase the amount and function of spine synapses in the medial prefrontal cortex and reverse the negative effect of chronic unpredictable stress on the number of synapses (Duman 2014).

In addition to these synthetic drugs, the following mushrooms have been used to relieve depression, *Armillaria mellea*, *Cordyceps militaris*, *Ganoderma lucidum*, *Hericium erinaceus*, *Inonotus* sp., *Marasmius androsaceus*, *Poria cocos*, and *Psilocybe* genus (Sakashita et al. 2015). These mycotherapeutics reportedly increased the extracellular monoamine dopamine and 5-HT, modulated neurotrophic and neurogenic mechanisms (stimulated NGF), modulated BDNF/PI3K/Akt/GSK-3β, blocked the NF-κB signaling, upregulated BDNF expression, reduced stress, increased γ-aminobutyric acid (GABA), and decreased glutamate. *Ganoderma lucidum* exhibits antidepressant effects by restricting the expression of IL-1β and TNF-α, both proinflammatory cytokines; increasing the expression of IL-10, an anti-inflammatory cytokine; and hindering microglia activation and astrocyte proliferation in the hippocampus (Li et al. 2021b). *Armillaria mellea* has alleviated stress-related rises in IL-1β and TNF-α in the cerebrum and serum. It also decreases the protein expression of the ionized calcium-binding adaptor molecule 1, thereby reducing stress-related brain inflammation (Lin et al. 2021).

3.3.4 Parkinson's Disease

Parkinson's disease is an age-related neurodegenerative neurological condition that results in uncontrolled movements (tremor), slowness of movement (bradykinesia), and posture and gait abnormality (Thrash et al. 2009). The first clear medical description of the disorder was written in 1817 by James Parkinson (Goetz 2011). Among those over 65 years old, it is one of the most common causes of disability, affecting approximately 800 out of 1000 individuals (Thrash et al. 2009). Parkinson's disease causes selective neuronal degeneration of the striatal dopaminergic neurons in the nigrostriatal pathway, leading to dopamine depletion in the striatum. Dopamine is necessary for striatal neurons to properly regulate the actions of GABA and glutamate, which ultimately control movement. Parkinson's disease ultimately leads to delay in the execution of voluntary movements, stooped posture and righting reflex, increased muscle tone, and increased resistance to movement, all of which can be evidenced by muscular rigidity, bradykinesia, resting tremors, and other such motor deficits (Thrash et al. 2009). Treatment for Parkinson's disease involves increasing dopaminergic neurotransmission and protecting the nigrostriatal dopaminergic neurons from neurodegeneration. Dopaminergic neurotransmission is enhanced by treatment with the dopamine precursor L-DOPA, dopamine agonists, and monoamine oxidase inhibitors. Another approach uses NMDA receptor antagonists, such as memantine, which blocks excitotoxicity. The coenzyme Q10 and melatonin display neuroprotective properties against MPTP and rotenone. A third method of protecting neurons from neurodegeneration involves anti-inflammatory therapeutic agents. For example, aspirin can modify COX-2 to produce anti-inflammatory lipoxins instead of the natural inflammatory prostanoids that the COX-2 enzyme has (Thrash et al. 2009).

Coriolus versicolor, *Hericium erinaceus*, *Agaricus muscarius*, *Pleurotus ostreatus*, *Laetiporus sulphureus*, *Agaricus blazei*, and *Amauroderma rugosum* have been shown to exhibit neuroprotection against nigrostriatal neurodegeneration and thus has been used to prevent or treat Parkinson's disease. The above mycotherapeutics have demonstrated dopaminergic neuroprotection by the following mechanisms (D'Amico et al. 2021; Agunloye et al. 2021; Venkatesh et al. 2019; Ziaja et al. 2018; Sam et al. 2022).

Mycotherapeutics can reduce dopaminergic neurodegeneration by the following key mechanisms:

- Prevent neuroinflammation in dopaminergic neuron.
- Affect tyrosine hydroxylase, tyrosinase, and dopamine transporter and increase dopaminergic neurotransmission.
- Restrict acetylcholinesterase activity to reduce tremor.
- Affect redox stress response in the dopaminergic neuron (reduce oxidative stress).
- Inhibit lipid peroxidation in dopaminergic neuron.
- Reduce and prevent dopaminergic neuronal apoptosis.
- Prevent diminution of dopamine transporter and vesicular monoamine transporter 2.
- Prevent downregulation of Bcl-2 and upregulation of Bax in dopaminergic neuron.
- Regulate expression of DAT and VMAT2 in dopaminergic neuron.
- Reduce excitotoxicity in dopaminergic neuron.
- Regulate the p21/GADD45 cell death pathways in dopaminergic neuron.
- PAKalpha, p21-activated kinase 1 (PAK1) survival pathways in dopaminergic neuron.

3.3.5 Restless Leg Syndrome

Restless leg syndrome is a neurological disorder characterized by the compulsion to move the legs with or without paresthesia, worsening symptoms as a result of inactivity, lessening symptoms as a result of activity, and worsening symptoms during the evening and night (Satija and Ondo 2008). In 1685, physician Thomas Willis became the first person to describe its symptoms (Guo et al. 2017). The disorder has a prevalence rate of 10% in Caucasian populations (Satija and Ondo 2008). Brain iron levels are believed to be involved in the pathology of restless leg syndrome. An iron deficiency can be caused by dysregulation in the transportation of iron through the blood-brain barriers, which results from impairments in the microvasculature iron regulatory protein. Iron is necessary for oxidative phosphorylation, oxidative transportation, and myelin production. A deficiency can lead to cellular damage when neuromelanin-containing and dopamine-producing cells interact with inhibited iron uptake by neurons, and the pathophysiology of restless leg syndrome results.

Furthermore, iron deficiency can result in alterations of the brain's dopaminergic system, impairing the medial pain system and leading to the discomfort associated with restless leg syndrome (Guo et al. 2017). Treatment for restless leg syndrome primarily involves dopaminergic agonists, pramipexole, and ropinirole (Guo et al. 2017). Pramipexole can reduce oxidative stress and lessen nigrostriatal pathway damage (Singh and Parmar 2022). Ropinirole impedes adenylyl cyclase and calcium channels and activates potassium channels by stimulating central and peripheral nervous system postsynaptic D_2 dopamine receptors (Rewane and Nagalli 2022). Reishi and *Agaricus* species can increase dopaminergic neurotransmission and reduce the symptoms of restless leg syndrome. There is a flourishing interest in the novel and new iron-rich super-nutrients from various natural bioactives such as algae and mushrooms. In addition, these super-nutrients synergistically combine antioxidant and anti-inflammatory effects in addition to their iron content. Thus, the presence of iron in mycotherapeutics can further aid in the treatment of restless leg syndrome.

3.3.6 Tremor

A tremor is a rhythmic and oscillatory involuntary movement (Louis 2019). In 1874, Pietro Burresi became the first person to use the term "essential tremor" (Louis et al. 2008). Essential tremor is one of the most common neurological disorders, with an estimated 0.9% of the world's population being affected by it (Song et al. 2021). The causation of tremors in patients is thought to be related to dysfunction in the basal ganglia-cerebellum-thalamic circuit and the dentate-olivary circuit. In the basal ganglia-cerebellum-thalamic circuit, cortical motor activity is curbed by increased activity in the global pallidus internus, which sends inhibitory projections to the ventrolateral thalamus that, in turn, connect to the motor cortex. Motor cortical activity is also facilitated by the cerebellar nuclei, which send glutaminergic excitatory projections to the ventrolateral thalamus that, in turn, are sent to the motor cortex. On the other hand, in the dentate-olivary circuit, the inferior olivary nucleus is important to the generation of tremors. Calcium channels usually mediate the regular oscillatory depolarizations demonstrated by the circuit's neurons. Altercations in these oscillations, which play a part in pacing temporal coordination and cerebellar motor learning, can produce tremors (Kamble and Pal 2018). Pharmacological therapies for tremors include muscarinic antagonists such as benztropine, anticonvulsants, beta-adrenergic antagonists, benzodiazepines/GABAergic agents, calcium channel blockers, and atypical neuroleptic agents. Primidone, an anticonvulsant, and propranolol (beta-adrenergic antagonist) have been the most effective therapeutic approach. These synthetic drugs have been shown to decrease the amplitude of tremors by 50% (Rajput and Rajput 2014).

Mycotherapeutics have been shown to both increase and decrease tremors. The mushrooms *Agaricus blazei* (Venkatesh Gobi et al. 2018) and *Amauroderma rugosum* (Li et al. 2021a), used to treat Parkinson's disease, have been shown to decrease tremors associated with the condition. Conversely, acute encephalopathy was observed in patients with renal dysfunction after consuming sugihiratake mushrooms, with one of the symptoms being tremors (Nishizawa 2005). Muscle tremors were also observed in sheep after they were orally administered the *Ramaria flavobrunnescens* mushroom (Sallis et al. 2000). Tremors can also be induced due to increased cholinergic neurotransmission.

3.3.7 Post-Traumatic Stress Disorder

Post-traumatic stress disorder (PTSD) is a chronic condition that results from exposure to combat, natural disaster, or other sources of severe stress. It causes reexperience, avoidance symptoms, and cognitive dysfunctions in patients (Miao et al. 2018). Despite the term emerging relatively recently in 1980, the notion of PTSD has been under longtime scrutiny, often about those who experience the psychological trauma of war (Andreasen 2010). At a range of 5.4–16.8%, the disorder's prevalence rate in veterans is almost double that of its prevalence rate in civilians (Miao et al. 2018). It has been suggested that the interactions between the neuroendocrine and immune systems are associated with the development of PTSD. The abnormal release of glucocorticoids, caused by the activation of the stress response pathways of the hypothalamic-pituitary-adrenal and the sympathetic nervous system, leads to enhanced metabolism, immunosuppression, and hypothalamic-pituitary-adrenal axis negative feedback inhibition due to the binding of the glucocorticoids to the glucocorticoid receptor. Evidence also indicates that compared to healthy controls, individuals with PTSD have higher plasma levels of tumor necrosis factor-alpha, interleukin-1beta, and interleukin-6, all of which are proinflammatory cytokines (Miao et al. 2018). The most common pharmacological treatments for PTSD are antidepressants, such as selective serotonin reuptake inhibitors (SSRIs) (Miao et al. 2018). One SSRI, fluoxetine, is an inhibitor of the reuptake transporter protein in the presynaptic terminal. This prevents serotonin reuptake by presynaptic serotonin neurons, which project to the prefrontal cortex (Sohel et al. 2022). Another approach used to treat PTSD is the anticonvulsant topiramate, which can block voltage-gated sodium channels and reduce the ability of kainate and AMPA receptors to depolarize membranes (Fariba and Saadabadi 2022).

Magic mushrooms and psilocybin have been shown to relieve PTSD by stimulating nerve cell regrowth in parts of the brain responsible for emotion and memory and by stimulating hippocampal neurogenesis (Catlow et al. 2013). Mycotherapeutics also alleviate PTSD by affecting the monoaminergic neurotransmission in the brain.

3.3.8 Smoking Cessation

Nicotine, a nicotine receptor full agonist; varenicline, a partial nicotine agonist; and bupropion, a dopamine reuptake inhibitor, have been used for smoking cessation (Cahill et al. 2016). Similar to PTSD, magic mushrooms and psilocybin have been shown to relieve the symptoms associated with smoking cessation (Johnson et al. 2017). The symptoms associated with smoking cessation include anger, anxiety, depression, difficulty concentrating, frustration, hunger or increased appetite, insomnia, irritability, nicotine cravings, and restlessness. Psilocybin is extensively considered to be significantly effective in smoking cessation currently. Psilocybin administration resulted in a successful 80% abstinence rate from smoking over 6 months compared to a 35% abstinence rate for patients administered with a partial dopamine agonist, varenicline (Nelson 2014). Based on the current pharmacodynamic mechanisms associated with the treatment of smoking cessation, the mycotherapeutics also directly or indirectly modulate dopaminergic neurotransmission and reduce smoking cessation.

3.3.9 Tic Disorder/Tourette Syndrome

Tic disorder is a condition that causes sudden, quick, erratic, repetitive, and nonrhythmic motor activity or sounds that require the movement of air throughout the nose, mouth, or throat. Both types of tics are involuntary and semi-voluntary. Tourette syndrome is the persistence of motor and vocal tics for over a year (Rampello et al. 2006). While transient tics are relatively common, affecting up to 20% of school-aged children, Tourette syndrome has been estimated to affect anywhere between 2.6 and 38 children out of 1000 (Scahill et al. 2014). Although the exact causation and development behind tic disorders remain unclear, evidence suggests that it involves the loss of inhibition in one of the cortico-striato-thalamo-cortical circuits, which play a part in motor, cognitive, and limbic processes. Another implicated system is the interplay of the dorsal striatum and the prefrontal cortex, which is associated with the development of habits. It has been speculated that a disturbance in the connections between the motivational, cognitive, and sensorimotor circuits results in a problem in habit formation, leading to tics. Hyperactivity of the nigrostriatal dopaminergic pathway, such as supersensitive postsynaptic dopamine receptors and elevated intrasynaptic dopamine release, is also suspected to be involved (Rampello et al. 2006). Dopamine receptor antagonists, such as haloperidol, pimozide, and risperidone, are used but cause adverse effects such as anxiety, extrapyramidal movement, and tardive dyskinesia (Rampello et al. 2006). Magic mushrooms have been shown to reduce tics by preventing neuroinflammation and affect the redox stress response, affecting monoaminergic neurotransmission.

3.3.10 Traumatic Brain Injury

Traumatic brain injury (TBI) is an injury to the head through a bump, blow, or jolt that interferes with the brain's proper functioning (Sacks et al. 2018). TBI is considered a high-risk factor for other conditions, such as post-traumatic stress disorder. Around 1.7 million people in the United States have suffered from TBI, with those 15–19 years old and those 65 and older as the most likely patients (Georges and Das 2022). Globally, TBI also affects millions of people, with the total rates of TBI-caused deaths, emergency department visits, and hospitalizations increasing from 2001 to 2010 (Galgano et al. 2017). The pathology of TBI is related to the primary and secondary head injuries associated with the neurological deficits that characterize the disorder. The primary injury has to do with the original impact on the head, which affects the brain and leads to the secondary injury, which involves a cascade of molecular, chemical, and inflammatory responses resulting from cerebral issues arising from TBI. This cascade depolarizes neurons by releasing excitatory neurotransmitters that increase the calcium level of cells. The increased intracellular calcium activates enzyme caspases, caspases, and free radicals, initiating an apoptotic process that directly or indirectly affects the neuronal cells and degrades them. From this degradation, an inflammatory response that further harms the neuronal cells results, prompting a buildup of the fluids and pressure around the brain and a breach in the blood-brain barrier. When not adequately addressed, the chain reaction of traumatic injury to the brain can cause pathological brain compression and lead to death (Galgano et al. 2017).

Pharmacological therapies for TBI focus on neuroprotection, neurovascular regeneration, and neurorestoration to increase brain repair and prevent secondary injury. As increased intracellular calcium levels are a critical component in the cascade of neuronal damage associated with TBI, calcium channel blockers are considered promising treatments for the condition. L-type and N-type calcium channel blockers neutralize intracellular calcium and avert cellular death. For patients with severe cases of TBI, hyperosmolar agents are utilized. An example is the drug mannitol, which has been observed to significantly reduce the increased intracranial pressure caused by the buildup of fluids around the brain (Galgano et al. 2017). *Coriolus versicolor* and *Hericium erinaceus* have been shown to reduce traumatic brain injury (TBI) by preventing neuroinflammation and affecting the redox stress response (D'Amico et al. 2021).

4 Conclusion

There is a drastic escalation in dopaminergic neurological disorders due to the genetic, physiologic, environmental, and iatrogenic (drug-induced) impact. Positively, there is a prosperity of convincing research data and evidence evolving to support the prophylactic and therapeutic use of medicinal mushrooms in treating dopaminergic diseases associated with the brain. Nevertheless, it is extremely critical for healthcare professionals to appropriately choose the proper mushroom species for preventing or treating neuronal dopaminergic pathologies. Healthcare professionals must select safe (minimal adverse effects and allergic manifestations) and effective mycotherapeutics. To reduce adulterations, the growth and harvest of the mushrooms must be regulated and optimized via good manufacturing methods (GMP) and organic farming. The formulations should be free of toxicants and adulterants. Mycotherapeutics have validated their use in dopaminergic neuronal disorder by two major mechanisms:

(a) Decrease dopaminergic neurodegeneration.
(b) Modulate dopaminergic neurotransmission.

Mycotherapeutics reduce dopaminergic neurodegeneration by decreasing oxidative stress, reducing apoptosis, blocking excitotoxicity, inhibiting inflammation, and increasing mitochondrial functions in dopaminergic neurons. Additionally, mycotherapeutics affect dopaminergic neurotransmission by affecting the synthesis, storage, reuptake, degradation, release, and receptor-mediated neurotransmission.

References

Agunloye OM, Oboh G, Falade AO (2021) *Pleurotus ostreatus* and *Lentinus subnudus* supplemented diets restore altered acetylcholinesterase and butyrylcholinesterase activities and improve antioxidant status in transgenic Drosophila melanogaster model. J Diet Suppl 18(4):372–386. https://doi.org/10.1080/19390211.2020.1772441

Andreasen NC (2010) Posttraumatic stress disorder: a history and a critique. Ann N Y Acad Sci 1208:67–71. https://doi.org/10.1111/j.1749-6632.2010.05699.x

Arciniegas DB (2015) Psychosis. Continuum 21. (3 Behavioral Neurology and Neuropsychiatry):715–736. https://doi.org/10.1212/01.CON.0000466662.89908.e7

Ashizawa T, Xia G (2016) Ataxia. Continuum 22:1208–1226. (4 Movement Disorders). https://doi.org/10.1212/con.0000000000000362

Bai Y, Chen L, Chen Y et al (2019) A maitake (*Grifola frondosa*) polysaccharide ameliorates alzheimer's disease-like pathology and cognitive impairments by enhancing microglial amyloid-β clearance. RSC Adv 9(64):37127–37135. https://doi.org/10.1039/c9ra08245j

Bell IR, Caspi O, Schwartz G et al (2002) Integrative medicine and systemic outcomes research: issues in the emergence of a new model for primary health care. Arch Intern Med 162(2): 133–140. https://doi.org/10.1001/archinte.162.2.133

Cahill K, Lindson-Hawley N, Thomas KH et al (2016) Nicotine receptor partial agonists for smoking cessation. Cochrane Database Syst Rev 2016(5):Cd006103. https://doi.org/10.1002/14651858.CD006103.pub7

Catlow BJ, Song S, Paredes DA et al (2013) Effects of psilocybin on hippocampal neurogenesis and extinction of trace fear conditioning. Exp Brain Res 228(4):481–491. https://doi.org/10.1007/s00221-013-3579-0

Chand SP, Arif H (2022) Depression. StatPearls, Florida

Charlson FJ, Ferrari AJ, Santomauro DF et al (2018) Global epidemiology and burden of schizophrenia: findings from the global burden of disease study 2016. Schizophr Bull 44(6): 1195–1203. https://doi.org/10.1093/schbul/sby058

Chong PS, Khairuddin S, Tse ACK et al (2020) *Hericium erinaceus* potentially rescues behavioural motor deficits through ERK-CREB-PSD95 neuroprotective mechanisms in rat model of 3-acetylpyridine-induced cerebellar ataxia. Sci Rep 10(1):14945. https://doi.org/10.1038/s41598-020-71966-z

Clemente AS, Diniz BS, Nicolato R et al (2015) Bipolar disorder prevalence: a systematic review and meta-analysis of the literature. Braz J Psychiatry 37(2):155–161. https://doi.org/10.1590/1516-4446-2012-1693

Contato AG, Brugnari T, Sibin APA et al (2020) Biochemical properties and effects on mitochondrial respiration of aqueous extracts of basidiomycete mushrooms. Cell Biochem Biophys 78(1):111–119. https://doi.org/10.1007/s12013-020-00901-w

Culpepper L (2014) The diagnosis and treatment of bipolar disorder: decision-making in primary care. Prim Care Companion CNS Disord 16(3):PCC.13r01609. https://doi.org/10.4088/PCC.13r01609

D'Amico R, Trovato Salinaro A, Fusco R et al (2021) *Hericium erinaceus* and *Coriolus versicolor* modulate molecular and biochemical changes after traumatic brain injury. Antioxidants (Basel) 10(6):898. https://doi.org/10.3390/antiox10060898

de Silva RN, Vallortigara J, Greenfield J et al (2019) Diagnosis and management of progressive ataxia in adults. Pract Neurol 19(3):196–207. https://doi.org/10.1136/practneurol-2018-002096

Duman RS (2014) Pathophysiology of depression and innovative treatments: remodeling glutamatergic synaptic connections. Dialogues Clin Neurosci 16(1):11–27. https://doi.org/10.31887/DCNS.2014.16.1/rduman

Elhusseiny SM, El-Mahdy TS, Awad MF et al (2021) Proteome analysis and in vitro antiviral, anticancer and antioxidant capacities of the aqueous extracts of *Lentinula edodes* and *Pleurotus ostreatus* edible mushrooms. Molecules 26(15):4623. https://doi.org/10.3390/molecules26154623

Erbiai EH, da Silva LP, Saidi R et al (2021) Chemical composition, bioactive compounds, and antioxidant activity of two wild edible mushrooms *Armillaria mellea* and *Macrolepiota procera* from two countries (Morocco and Portugal). Biomol Ther 11(4):575. https://doi.org/10.3390/biom11040575

Erdem Guzel E, Kaya Tektemur N, Tektemur A et al (2021) The antioxidant and anti-apoptotic potential of *Pleurotus eryngii* extract and its chitosan-loaded nanoparticles against doxorubicin-induced testicular toxicity in male rats. Andrologia 53(11):e14225. https://doi.org/10.1111/and.14225

Fariba KA, Saadabadi A (2022) Topiramate. StatPearls, Florida

Ferreiro E, Pita IR, Mota SI et al (2018) *Coriolus versicolor* biomass increases dendritic arborization of newly-generated neurons in mouse hippocampal dentate gyrus. Oncotarget 9(68):32929–32942. https://doi.org/10.18632/oncotarget.25978

Galgano M, Toshkezi G, Qiu X et al (2017) Traumatic brain injury: current treatment strategies and future endeavors. Cell Transplant 26(7):1118–1130. https://doi.org/10.1177/0963689717714102

Gannotta R, Malik S, Chan AY et al (2018) Integrative medicine as a vital component of patient care. Cureus 10(8):e3098. https://doi.org/10.7759/cureus.3098

Georges A, Das JM (2022) Traumatic brain injury. StatPearls, Florida

Goetz CG (2011) The history of parkinson's disease: early clinical descriptions and neurological therapies. Cold Spring Harb Perspect Med 1(1):a008862. https://doi.org/10.1101/cshperspect.a008862

Guo S, Huang J, Jiang H et al (2017) Restless legs syndrome: from pathophysiology to clinical diagnosis and management. Front Aging Neurosci 9:171. https://doi.org/10.3389/fnagi.2017.00171

Guo C, Guo D, Fang L et al (2021) *Ganoderma lucidum* polysaccharide modulates gut microbiota and immune cell function to inhibit inflammation and tumorigenesis in colon. Carbohydr Polym 267:118231. https://doi.org/10.1016/j.carbpol.2021.118231

Hafiz S, De Jesus O (2022) Ataxia. StatPearls, Florida

Hany M, Rehman B, Azhar Y, Chapman J (2022) Schizophrenia. StatPearls, Florida

Johnson MW, Garcia-Romeu A, Griffiths RR et al (2017) Long-term follow-up of psilocybin-facilitated smoking cessation. Am J Drug Alcohol Abuse 43(1):55–60. https://doi.org/10.3109/00952990.2016.1170135

Juárez OH, Calderón GD, Hernández GE, Barragán MG (2016) The role of dopamine and its dysfunction as a consequence of oxidative stress. Oxididat Med Cell Longev 2016:9730467. https://doi.org/10.1155/2016/9730467

Kamble N, Pal PK (2018) Tremor syndromes: a review. Neurol India 66(Supplement):S36–s47. https://doi.org/10.4103/0028-3886.226440

Klockgether T, Paulson H (2011) Milestones in ataxia. Mov Disord 26(6):1134–1141. https://doi.org/10.1002/mds.23559

Kwei KT, Kuo SH (2020) An overview of the current state and the future of ataxia treatments. Neurol Clin 38(2):449–467. https://doi.org/10.1016/j.ncl.2020.01.008

Lew SY, Lim SH, Lim LW et al (2020) Neuroprotective effects of *Hericium erinaceus* (Bull.: Fr.) Pers. against high-dose corticosterone-induced oxidative stress in PC-12 cells. BMC Complement Med Ther 20(1):340. https://doi.org/10.1186/s12906-020-03132-x

Li L, Rozolsky DM, Walke L et al (2011) Progressive tremor, truncal ataxia, and acute mental status changes after use of bupropion. Consult Pharm 26(9):665–671. https://doi.org/10.4140/TCP.n.2011.665

Li W, Lee SH, Jang HD et al (2017) Antioxidant and anti-osteoporotic activities of aromatic compounds and sterols from *Hericium erinaceum*. Molecules 22(1):108. https://doi.org/10.3390/molecules22010108

Li J, Li R, Wu X et al (2021a) *Amauroderma rugosum* protects PC12 cells against 6-OHDA-induced neurotoxicity through antioxidant and antiapoptotic effects. Oxidat Med Cell Longev 2021:6683270. https://doi.org/10.1155/2021/6683270

Li H, Xiao Y, Han L et al (2021b) *Ganoderma lucidum* polysaccharides ameliorated depression-like behaviors in the chronic social defeat stress depression model via modulation of Dectin-1 and the innate immune system. Brain Res Bull 171:16–24. https://doi.org/10.1016/j.brainresbull.2021.03.002

Liang C, Li H, Zhou H et al (2012) Recombinant Lz-8 from *Ganoderma lucidum* induces endoplasmic reticulum stress-mediated autophagic cell death in SGC-7901 human gastric cancer cells. Oncol Rep 27(4):1079–1089. https://doi.org/10.3892/or.2011.1593

Lin YE, Wang HL, Lu KH et al (2021) Water extract of *Armillaria mellea* (Vahl) P. Kumm. alleviates the depression-like behaviors in acute- and chronic mild stress-induced rodent models via anti-inflammatory action. J Ethnopharmacol 265:113395. https://doi.org/10.1016/j.jep.2020.113395

Littlewood RA, Vanable PA (2008) Complementary and alternative medicine use among HIV-positive people: research synthesis and implications for HIV care. AIDS Care 20(8):1002–1018. https://doi.org/10.1080/09540120701767216

Louis ED (2019) Tremor. Continuum 25(4):959–975. https://doi.org/10.1212/con.0000000000000748

Louis ED, Broussolle E, Goetz CG et al (2008) Historical underpinnings of the term essential tremor in the late 19th century. Neurology 71(11):856–859. https://doi.org/10.1212/01.wnl.0000325564.38165.d1

Ma XR, Zhou CF, Wang SQ et al (2007) Effects of *Ganoderma lucidum* spores on mitochondrial calcium ion and cytochrome C in epididymal cells of type 2 diabetes rats. Zhonghua Nan Ke Xue 13(5):400–402

Mahmood D, Alenezi SK, Anwar MJ et al (2022) New paradigms of old psychedelics in schizophrenia. Pharmaceuticals (Basel) 15(5):640. https://doi.org/10.3390/ph15050640

Mariotti C, Fancellu R, Di Donato S et al (2005) An overview of the patient with ataxia. J Neurol 252(5):511–518. https://doi.org/10.1007/s00415-005-0814-z

Marzio DL, Fenkel JM (2014) Complementary and alternative medications in hepatitis C infection. World J Hepatol 6(1):9–16. https://doi.org/10.4254/wjh.v6.i1.9

Miao XR, Chen QB, Wei K et al (2018) Posttraumatic stress disorder: from diagnosis to prevention. Mil Med Res 5(1):32. https://doi.org/10.1186/s40779-018-0179-0

Miklowitz DJ, Johnson SL (2006) The psychopathology and treatment of bipolar disorder. Annu Rev Clin Psychol 2:199–235. https://doi.org/10.1146/annurev.clinpsy.2.022305.095332

Naeem A, Aslam M, Mühling KH et al (2021) Lithium: perspectives of nutritional beneficence, dietary intake, biogeochemistry, and biofortification of vegetables and mushrooms. Sci Total Environ 798:149249. https://doi.org/10.1016/j.scitotenv.2021.149249

Nelson L (2014) Magic Mushrooms' help longtime smokers quit. The Johns Hopkins University. https://www.hopkinsmedicine.org/news/media/releases/magic_mushrooms_help_longtime_smokers_quit. Accesed on 28 Sep 2022

Nishizawa M (2005) Acute encephalopathy after ingestion of "Sugihiratake" mushroom. Rinsho Shinkeigaku 45(11):818–820

Nkadimeng SM, Nabatanzi A, Steinmann CML (2020) Phytochemical, cytotoxicity, antioxidant and anti-inflammatory effects of *Psilocybe natalensis* magic mushroom. Plants (Basel) 9(9): 1127. https://doi.org/10.3390/plants9091127

Panda AK, Swain KC (2011) Traditional uses and medicinal potential of *Cordyceps sinensis* of Sikkim. J Ayurveda Integr Med 2(1):9–13. https://doi.org/10.4103/0975-9476.78183

Pangrazzi L, Balasco L, Bozzi Y et al (2020) Oxidative stress and immune system dysfunction in autism spectrum disorders. Int J Mol Sci 21(9):3293. https://doi.org/10.3390/ijms21093293

Patel S, Goyal A (2012) Recent developments in mushrooms as anti-cancer therapeutics: a review. 3 Biotech 2(1):1–15. https://doi.org/10.1007/s13205-011-0036-2

Patel SJ, Kemper KJ, Kitzmiller JP et al (2017) Physician perspectives on education, training, and implementation of complementary and alternative medicine. Adv Med Educ Pract 8:499–503. https://doi.org/10.2147/amep.S138572

Rajput AH, Rajput A (2014) Medical treatment of essential tremor. J Cent Nerv Syst Dis 6:29–39. https://doi.org/10.4137/jcnsd.S13570

Rampello L, Alvano A, Battaglia G et al (2006) Tic disorders: from pathophysiology to treatment. J Neurol 253(1):1–15. https://doi.org/10.1007/s00415-005-0008-8

Raote I, Bhattacharya A, Panicker MM (2007) Serotonin 2A (5-HT_{2A}) receptor function: ligand-dependent mechanisms and pathways. In: Chattopadhyay A (ed) Serotonin receptors in neurobiology. CRC Press/Taylor & Francis, Boca Raton, FL

Rasheed N, Alghasham A (2012) Central dopaminergic system and its implications in stress-mediated neurological disorders and gastric ulcers: short review. Adv Pharmacol Sci 2012: 182671. https://doi.org/10.1155/2012/182671

Rewane A, Nagalli S (2022) Ropinirole. StatPearls, Florida

Rosenthal DS, Webster A, Ladas E (2018) Chapter 156—integrative therapies in patients with hematologic diseases. In: Hoffman R, Benz EJ, Silberstein LE et al (eds) Hematology, 7th edn. Elsevier, pp 2253–2261. https://doi.org/10.1016/B978-0-323-35762-3.00156-6

Rossi P, Buonocore D, Altobelli E et al (2014) Improving training condition assessment in endurance cyclists: effects of *Ganoderma lucidum* and *Ophiocordyceps sinensis* dietary supplementation. Evid Based Complem Alternat Med 2014:979613. https://doi.org/10.1155/2014/979613

Roy R (2010) Integrative medicine to tackle the problem of chronic diseases. J Ayurveda Integr Med 1(1):18–21. https://doi.org/10.4103/0975-9476.59822

Sacks D, Baxter B, Campbell BCV et al (2018) Multisociety consensus quality improvement revised consensus statement for endovascular therapy of acute ischemic stroke. Int J Stroke 13(6):612–632. https://doi.org/10.1177/1747493018778713

Sakashita Y, Abe K, Katagiri N et al (2015) Effect of psilocin on extracellular dopamine and serotonin levels in the mesoaccumbens and mesocortical pathway in awake rats. Biol Pharm Bull 38(1):134–138. https://doi.org/10.1248/bpb.b14-00315

Sallis ES, Correa FR, Raffi MB et al (2000) Experimental intoxication by the mushroom Ramaria Flavo-Brunnescens in sheep. Vet Hum Toxicol 42(6):321–324

Salman MS (2018) Epidemiology of cerebellar diseases and therapeutic approaches. Cerebellum 17(1):4–11. https://doi.org/10.1007/s12311-017-0885-2

Sam SE, Sim KS, Rahman S, Tan YS (2022) Neuroprotective properties of wild medicinal mushroom, *Sanguinoderma rugosum* (Agaricomycetes), extracts against glutamate-induced hippocampal cells. Int J Med Mushrooms 24(3):35–50. https://doi.org/10.1615/IntJMedMushrooms.2022042961

Samsam M, Ahangari R, Naser SA et al (2014) Pathophysiology of autism spectrum disorders: revisiting gastrointestinal involvement and immune imbalance. World J Gastroenterol 20(29): 9942–9951. https://doi.org/10.3748/wjg.v20.i29.9942

Satija P, Ondo WG (2008) Restless legs syndrome: pathophysiology, diagnosis and treatment. CNS Drugs 22(6):497–518. https://doi.org/10.2165/00023210-200822060-00004

Scahill L, Specht M, Page C et al (2014) The prevalence of tic disorders and clinical characteristics in children. J Obsessive Compuls Relat Disord 3(4):394–400. https://doi.org/10.1016/j.jocrd.2014.06.002

Seweryn E, Ziała A, Gamian A et al (2021) Health-promoting of polysaccharides extracted from *Ganoderma lucidum*. Nutrients 13(8):2725. https://doi.org/10.3390/nu13082725

Sharma SR, Gonda X, Tarazi FI (2018) Autism spectrum disorder: classification, diagnosis and therapy. Pharmacol Ther 190:91–104. https://doi.org/10.1016/j.pharmthera.2018.05.007

Singh R, Parmar M (2022) Pramipexole. StatPearls, Florida

Snyderman R, Weil AT (2002) Integrative medicine: bringing medicine back to its roots. Arch Intern Med 162(4):395–397. https://doi.org/10.1001/archinte.162.4.395

Sohel AJ, Shutter MC, Molla M et al (2022) Fluoxetine. StatPearls, Florida

Song J, Xing G, Cao J et al (2016) Investigation of the antidepressant effects of exopolysaccharides obtained from *Marasmius androsaceus* fermentation in a mouse model. Mol Med Rep 13(1): 939–946. https://doi.org/10.3892/mmr.2015.4584

Song P, Zhang Y, Zha M et al (2021) The global prevalence of essential tremor, with emphasis on age and sex: a meta-analysis. J Glob Health 11:04028. https://doi.org/10.7189/jogh.11.04028

Stępnicki P, Kondej M, Kaczor AA et al (2018) Current concepts and treatments of schizophrenia. Molecules 23(8):2087. https://doi.org/10.3390/molecules23082087

Swedish Council on Health Technology (2004) A SBU systematic review summaries. In: Treatment of depression: a systematic review. Swedish Council on Health Technology Assessment (SBU). Copyright © 2004 by the Swedish Council on Health Technology Assessment, Stockholm

Tabish SA (2008) Complementary and alternative healthcare: is it evidence-based? Int J Health Sci (Qassim) 2(1):V–ix

Thapar A, Rutter M (2021) Genetic advances in autism. J Autism Dev Disord 51(12):4321–4332. https://doi.org/10.1007/s10803-020-04685-z

Thibaut F (2018) The mind-body cartesian dualism and psychiatry. Dialogues Clin Neurosci 20(1): 3. https://doi.org/10.31887/DCNS.2018.20.1/fthibaut

Thrash B, Thiruchelvan K, Ahuja M et al (2009) Methamphetamine-induced neurotoxicity: the road to Parkinson's disease. Pharmacol Rep 61(6):966–977. https://doi.org/10.1016/s1734-1140(09)70158-6

Umeo SH, Souza GP, Rapachi PM et al (2015) Screening of basidiomycetes in submerged cultivation based on antioxidant activity. Genet Mol Res 14(3):9907–9914. https://doi.org/10.4238/2015.August.19.25

Valverde ME, Hernández-Pérez T, Paredes-López O (2015) Edible mushrooms: improving human health and promoting quality life. Int J Microbiol 2015:376387. https://doi.org/10.1155/2015/376387

Venkatesh Gobi V, Rajasankar S, Ramkumar M et al (2018) *Agaricus blazei* extract attenuates rotenone-induced apoptosis through its mitochondrial protective and antioxidant properties in SH-SY5Y neuroblastoma cells. Nutr Neurosci 21(9):657–666. https://doi.org/10.1080/1028415x.2017.1337290

Venkatesh GV, Rajasankar S, Swaminathan JWM et al (2019) Antiapoptotic role of *Agaricus blazei* extract in rodent model of Parkinson's disease. Front Biosci (Elite Ed) 11(1):12–19. https://doi.org/10.2741/e842

Venturella G, Ferraro V, Cirlincione F, Gargano ML (2021) Medicinal mushrooms: bioactive compounds, use, and clinical trials. Int J Mol Sci 22(2):634. https://doi.org/10.3390/ijms22020634

Vishvakarma R, Mishra A (2019) Protective effect of a protease inhibitor from *Agaricus bisporus* on *Saccharomyces cerevisiae* cells against oxidative stress. Prep Biochem Biotechnol 49(3):244–254. https://doi.org/10.1080/10826068.2018.1536992

Wachtel-Galor S, Yuen J, Buswell JA et al (2011) *Ganoderma lucidum* (Lingzhi or Reishi): a medicinal mushroom. In: Benzie IFF, Wachtel-Galor S (eds) Herbal medicine: biomolecular and clinical aspects. CRC Press/Taylor & Francis, Boca Raton (FL)

Wu P, Tan H, Zhan J (2020) Optimization of bioprocess extraction of *Poria cocos* polysaccharide (PCP) with *Aspergillus niger* β-glucanase and the evaluation of PCP antioxidant property. Molecules 25(24):5930. https://doi.org/10.3390/molecules25245930

Yu Y, Qian L, Du N et al (2017) *Ganoderma lucidum* polysaccharide enhances radiosensitivity of hepatocellular carcinoma cell line HepG2 through Akt signaling pathway. Exp Ther Med 14(6):5903–5907. https://doi.org/10.3892/etm.2017.5340

Zhang X, Zhang X, Gu S et al (2021) Structure analysis and antioxidant activity of polysaccharide-iron (III) from *Cordyceps militaris* mycelia. Int J Biol Macromol 178:170–179. https://doi.org/10.1016/j.ijbiomac.2021.02.163

Zhao H, Li J, Zhang J et al (2017) Purification, in vitro antioxidant and in vivo anti-aging activities of exopolysaccharides by *Agrocybe cylindracea*. Int J Biol Macromol 102:351–357. https://doi.org/10.1016/j.ijbiomac.2017.04.039

Ziaja K, Muszynska B, Gawalska A et al (2018) *Laetiporus sulphureus*—chemical composition and medicinal value. Acta Sci Pol Hortorum Cultus 17:87–96. https://doi.org/10.24326/asphc.2018.1.8

Zou CX, Wang XB, Lv TM et al (2020) Flavan derivative enantiomers and drimane sesquiterpene lactones from the inonotus obliquus with neuroprotective effects. Bioorg Chem 96:103588. https://doi.org/10.1016/j.bioorg.2020.103588

Biological Activities of Some Edible Mushrooms

Didem Şöhretoğlu and Ayşe Kuruüzüm-Uz

Contents

Abstract Since ancient times, mushrooms have been a valuable food source and traditional medicine worldwide. Edible mushrooms possess high nutritional value and medicinal significance due to the presence of a large number of active ingredients that cause the development of therapeutic functions. They are used to treat severe diseases like microbial and viral infections, cancer, tumors, inflammation, and cardiovascular and immune diseases. Mushrooms contain various bioactive compounds, including α- and ß-glucans, proteoglycan, lectin, phenolic compounds (flavonoids, flavonoids, phenolic acids), polysaccharides, triterpenoids, steroids, lentinan, schizophyllan, lovastatin, pleuran, glycopeptides, alkaloids, dietary fiber, and others. The biological activities of some well-known edible mushrooms are discussed in this chapter.

D. Şöhretoğlu · A. Kuruüzüm-Uz (✉)
Department of Pharmacognosy, Faculty of Pharmacy, Hacettepe University, Ankara, Turkey
e-mail: didems@hacettepe.edu.tr; ayseuz@hacettepe.edu.tr

D. C. Agrawal, M. Dhanasekaran (eds.), *Mushrooms with Therapeutic Potentials*,
https://doi.org/10.1007/978-981-19-9550-7_14

Keywords Anticancer · Biological activity · Edible mushrooms · Macrofungus · Polysaccharides

Abbreviations

ABTS	Azino-bis(3-ethylbenzothiazoline-6-sulfonic acid)
AChE	Acetylcholinesterase
AD	Alzheimer's disease
CAT	Catalase
CDK	Cyclin-dependent kinase
COX	Cyclooxygenase
DPP-4	Dipeptidyl peptidase-4
DPPH	2,2-Diphenyl-1-picrylhydrazyl
EtOH	Ethanol
FRAP	Ferric reducing antioxidant power
GLP	Glucagon-like peptide-1
GSH	Glutathione
H_2O	Water
HDL	High-density lipoprotein cholesterol
IFN-γ	Interferon-γ
IL	Interleukin
iNOS	Inducible nitric oxide synthase
JNK	c-Jun N-terminal kinases
LDL	Low-density lipoprotein cholesterol
LPS	Lipopolysaccharide
MDA	Malondialdehyde
MMK	Matrix metalloproteinase
NF-κB	Nuclear factor kappa-light-chain enhancer of activated B cells
NK	Natural killer
NLRP3	Nod-like receptor protein 3
NO	Nitric oxide
Nrf2	Nuclear factor-E2-related factor 2
PGE2	Prostaglandin E2
PSK	Polysaccharide K/Krestin
PSP	Polysaccharopeptide
SOD	Superoxide dismutase
TC	Total cholesterol
TG	Triglyceride
TLR4	Toll-like receptor 4
TNF	Tumor necrosis factor

1 Introduction

Macrofungi are important natural sources that benefit human health and prevent various diseases. Different mushroom species contain bioactive chemicals (mainly phenolic compounds, such as phenolic acids, flavonoids, and terpenoids). Due to their enormous biological activities, polysaccharide-protein complexes in medicinal mushrooms have attracted researchers' attention worldwide. However, their chemical composition mostly depends on their microhabitat's specific environmental conditions, particularly the surrounding mycelia's physicochemical properties (Raseta et al. 2020). The mushrooms' bioactive compounds exhibit immunomodulatory, anticancer, antidiabetic, antihyperlipidemic, microbiota-modulating, anti-inflammatory, antimicrobial, and antioxidative properties. Prepared extracts from individual mushrooms are used to produce dietary supplements as immunomodulatory agents and cosmetics (as regenerating cosmetics (regenerating the skin and treating atopic dermatitis)). They can be used as natural preservatives or probiotics (Golak-Siwulska et al. 2018b). In the following sections, we have selected some important mushroom species and described their biological activities.

2 *Agaricus bisporus* (J.E. Lange) Imbach

Agaricus bisporus (J.E. Lange) Imbach (Fig. 1a) (*Agaricaceae*) is known as button mushroom, common mushroom, cultivated mushroom, etc. It is one of the world's most common and widely consumed mushrooms.

Fig. 1 (**a**) *Agaricus bisporus*; (**b**) *Amanita caesarea*; (**c**) *Armillaria mellea*; (**d**) *Auricularia polytricha*; (**e**) *Boletus edulis*. Attrition: CC-BY-4.0; https://commons.wikimedia.org

A. bisporus exhibits an antioxidant effect higher than a lot of edible mushrooms. This effect is attributed to its phenolic compounds, serotonin, and tocopherol content (Liu et al. 2013; Reis et al. 2012).

Extracts, polysaccharides, lectins, and propionate of *A. bisporus* showed antidiabetic activity in different test models. It (200 mg/kg) reduced plasma glucose levels by 25% in streptozotocin-induced diabetic rats. It also decreased plasma triglyceride levels by 39%. This mushroom possesses antidiabetic and antihyperlipidemic properties. *A. bisporus* contains not only phytosterols, like all mushrooms, but also lovastatin, which lowers the cholesterol levels in the body to reduce the development of cardiovascular diseases (Ekowati et al. 2018; Jeong et al. 2010; Yamaç et al. 2010; Xu et al. 2013).

Water-soluble extract of *A. bisporus* suppressed the progression of liver fibrosis by the antioxidant, anti-inflammatory, and antiapoptotic mechanisms in vitro and in vivo. The anti-inflammatory action of this mushroom is related to a decrease in toll-like receptor 4 (TLR4) expression and a downregulation of Nod-like receptor protein 3 (NLRP3) inflammasome activation (Gallego et al. 2021). *A. bisporus* α-glucan increased nitric oxide and tumor necrosis factor (TNF)-α production by bone marrow-derived macrophages from mice in vitro (Volman et al. 2009). *A. bisporus* β-glucan inhibited the expression of interleukin (IL)-1β and cyclooxygenase (COX)-2, suggesting that different components of this mushroom exhibit anti-inflammatory action (Smiderle et al. 2013).

A. bisporus extracts or different components inhibited proliferation of cancer cells, including PC3, DU145 prostate cancer, HL-60 leukemia cells, MCF-7 breast cancer cell, and sarcoma 180, and induced apoptosis. Moreover, *A. bisporus* extract inhibited aromatase at the estrogen receptor in vitro in MCF-7 cells and also in vivo in rats (Chen et al. 2006; Golak-Siwulska et al. 2018b; Jeong et al. 2012). *A. bisporus* polysaccharides also possess anticancer effect via immunomodulatory effects. They stimulated the production of nitric oxide (NO), IL-6, and TNF-α and activated nuclear factor kappa-light-chain enhancer of activated B cells (NF-κB) pathway in macrophages but had no effect on the proliferation of human colon cancer cells or murine sarcoma cells. However, it reduced tumor growth in sarcoma 180-inoculated mice (Jeong et al. 2012).

Thus, *A. bisporus* is not only a good source of food but also exhibits medicinal values. It exhibited antioxidant, antidiabetic, antihyperlipidemic, anti-inflammatory, and anticancer properties.

3 *Agaricus blazei* Murill

Agaricus blazei Murill (synonym *Agaricus brasiliensis*) belongs to the *Agaricaceae* family. It is commonly known as almond mushroom, mushroom of the sun, or God's mushroom. It is edible, with a sweet taste and a fragrance of almonds. This mushroom gets much attention due to its immunomodulatory effect. Especially there is a commercial mushroom extract, namely, Andosan, containing 82%

A. blazei, 15% *Hericium erinaceus*, and 3% *Grifola frondosa*. Many clinical trials have tested the immunomodulatory, anticancer, and anti-inflammatory effects of Andosan.

A. blazei extracts and polysaccharides showed anti-inflammatory activity in many studies. *A. blazei* extract rich in cerevisterol suppressed IL-6 secretion and expression of the cyclooxygenase (COX)-2 in vitro (Song et al. 2012). *A. blazei* polysaccharides have anti-inflammatory activity interfering in the biosynthesis and secretion of inflammatory mediators. Oral administration suppressed the expression of NF-κB and intercellular adhesion molecule 1 (ICAM-1). It reduced the serum levels of IL-1β, TNF-α, COX-2, and inducible nitric oxide synthase (iNOS) in vivo (Wang et al. 2013b). The anti-inflammatory effect of *A. blazei*-based extract, Andosan, has been reported in a placebo-controlled clinical study in patients with ulcerative colitis or Crohn's disease. The extract showed a partial reduction in pro- and anti-inflammatory cytokines (Therkelsen et al. 2016). Thus, *A. blazei* possesses anti-inflammatory activity in vitro and in vivo, associated with decreased biosynthesis and secretion of inflammatory mediator expression of enzymes, the polysaccharides, the main responsible for this activity.

The antioxidant effect of *A. blazei* is one of the main underlying mechanisms of this mushroom that is well-documented. It scavenged 2,2-diphenyl-1-picrylhydrazyl (DPPH); azino-bis(3-ethylbenzothiazoline-6-sulfonic acid) (ABTS) radicals also exhibited antioxidant action in ferric reducing antioxidant power (FRAP) assay and lipid peroxidation assay in vitro (Bach et al. 2019; Carneiro et al. 2013; Wei et al. 2020). *A. blazei* extracts also exhibited antioxidant effect in vivo by reducing malondialdehyde (MDA) (a marker of lipid peroxidation) and alleviated glutathione (GSH) level, enhancing superoxide dismutase (SOD) activity (Al-Dbass et al. 2012).

Oral administration of *A. blazei* polysaccharides reduced the triglyceride (TG), total cholesterol (TC), and low-density lipoprotein cholesterol (LDL) and increased high-density lipoprotein cholesterol (HDL) plasma levels. Furthermore, it upregulated the expression of CYP7A1. This enzyme limits the conversion of CH to bile acid in the liver and inhibits the transcription factor SREBP-1C, a factor associated with TG metabolism and lipid deposition in cells (Ji et al. 2014; Li et al. 2020b).

Different extracts of *A. blazei* inhibited α-glucosidase in a dose-dependent manner. They also improved the glucose uptake by insulin-resistant HepG2 cells and alleviated postprandial hyperglycemia (Al-Dbass et al. 2012). *A. blazei* extract decreased reduced serum glucose levels as well as HbA1C compared to control in streptozocin-induced diabetes (Ji et al. 2014; Vitak et al. 2015). In addition, *A. blazei* extracts ameliorated diabetic neuropathy by reducing TNF-α and IL-1β levels, suppressing inflammation, and increasing antioxidant status (Ji et al. 2014). It also exhibited antidiabetic effects in human studies. In a randomized, double-blind, placebo-controlled clinical study, *A. blazei* extract improved insulin resistance among subjects with type 2 diabetes and increased adiponectin concentration after 12 weeks of oral treatment compared to control (Hsu et al. 2007).

A. blazei extract exhibited an antimicrobial effect against several common pathogens. It showed a higher antimicrobial effect against gram-positive bacteria than gram-negative ones. The antimicrobial effect of *A. blazei* is attributed to the phenolic content (Bach et al. 2019).

Oral administration of hydroalcoholic extract of *A. blazei* exhibited anxiolytic-like activity in mice, similar to positive control diazepam (Ali et al. 2021). Ethanol extract of *A. blazei* suppressed the rotenone-mediated decrease in dopamine transporter and vesicular monoamine transporter 2 expressions, inhibited apoptosis, and showed neuroprotection (Gobi et al. 2019).

Many studies have mainly focused on cancer-related activities of *A. blazei*. Extracts and different components of this mushroom inhibited the proliferation of various cancer cell lines related to the modulation of cyclin-dependent kinases and their inhibitors, cell cycle proteins. *A. blazei* activated caspases, stimulated release of cytochrome c and radical oxygen species production, modulated proapoptotic and antiapoptotic proteins in turn, and stimulated apoptosis associated with NF-κB and c-Jun N-terminal kinase (JNK) signaling (Jin et al. 2006; Kim et al. 2011; Matsushita et al. 2018; Yu et al. 2009). A. *blazei* also reduced tumor size and volume in animal models (Ito et al. 1997; Yu et al. 2009). *A. blazei* acts as immunostimulants, both in the proliferation and activation of lymphocytes (NK cells and T cells) and macrophages. These stimulated cells can produce cytokines, mainly IL-8 and IL-6 TNF-α (Bernardshaw et al. 2005; Johnson et al. 2009; Lima et al. 2011). Furthermore, randomized controlled trials also proved the immunomodulating effect of *A. blazei* (Tangen et al. 2015; Therkelsen et al. 2016). According to the literature, *A. blazei* contains compounds, including polysaccharides, proteins, steroids, nucleosides, and phenols, that may modulate tumorigenesis and carcinogenesis through different mechanisms. Anticancer activity seems to be associated with polysaccharides chiefly. The effects of different polysaccharides are likely to be mediated by various cell-surface receptors. The combination of such responses mediated from different polysaccharides conceivably provides greater tumor inhibition than if induced by a single polysaccharide.

In conclusion, *A. blazei* possesses anti-inflammatory, antidiabetic, antihyperlipidemic, anxiolytic, neuroprotective, hepatoprotective, anticancer, and immunomodulatory properties. It seems that *A. blazei* exhibited these biological effects mainly related to anti-inflammatory action. Unlike many other edible mushrooms, there is a commercial product of this mushroom. Current clinical trials have shown promising results in using this mushroom as an immunomodulator agent. However, adequately designed, further clinical studies on standardized extracts are needed.

4 *Amanita caesarea* (Scop.) Pers.

Amanita caesarea (Scop.) Pers. (*Amanitaceae*) (Fig. 1b) is a popular edible mushroom with good flavor and taste. *A. caesarea* methanol extract scavenged DPPH radical stronger than positive controls, butylated hydroxytoluene, and butylated hydroxyanisole with an IC_{50} value of 0.7615 mg/mL. It inhibited the peroxidation of linoleic acid in β-carotene-linoleic acid assay. The inhibition values of *A. caesarea* extract were higher than those of Trolox, similarly to butylated hydroxytoluene, and lower than butylated hydroxyanisole. Methanol, chloroform, and acetone extracts of *A. caesarea* possess an antimicrobial effect with MIC values of 312.5–39 mg/mL against several common pathogens. This mushroom exhibited the highest antimicrobial effect against *Candida albicans* and the lowest antimicrobial effect against *Klebsiella pneumoniae* (Doğan and Akbaş 2013).

An *A. caesarea* polysaccharide, APCS, molecular weight 18,620 Da and 33,500 Da, exhibited good in vitro activity in Alzheimer's disease (AD). Treatment with APCS prior to L-glutamic acid (L-Glu) co-exposure reversed the decreased cell viability, inhibited apoptosis, decreased the accumulation of intracellular ROS, and alleviated mitochondrial membrane potential in HT22 cells (Li et al. 2019). APCS and another polysaccharide, ACPS2, with an average molar mass of 16.6 kDa isolated from *A. caesarea* alleviated Alzheimer's disease-like symptoms in APP/PS1 and BALB mice. Both polysaccharides reduced amyloid deposition, tau hyperphosphorylation, neuroinflammation, and brain damage. These effects were mainly related to its suppression of endoplasmic reticulum stress and oxidative stress through regulation of nuclear factor-E2-related factor 2 (Nrf2) signaling and NF-κB activation (Hu et al. 2021; Li et al. 2019).

5 *Armillaria mellea* (Vahl) P. Kumm

Armillaria mellea (Vahl) P. Kumm (*Tricholomataceae*) (Fig. 1c) is known as a honey mushroom. It is an edible mushroom used for its medicinal and health-promoting properties worldwide. *A. mellea* was used to treat palsy, headache, hypertension, insomnia, dizziness, vertigo, neurasthenia, insomnia, and convulsion. It is also a component of traditional Chinese medicine "Tianma" (Sun et al. 2020).

The phenolic contents of the hydromethanolic extract and ethanolic extract of *A. mellea* were 21.68 and 5.70 mg/g. The hydromethanolic extract exhibited DPPH, 2,2'-ABTS free radical scavenging and reducing abilities with EC_{50} values of 452.60, 140.57, and 129.45 g/mL, respectively (Zavastin et al. 2015). The ethanolic extract was more effective in 15-lipoxygenase, α-glucosidase, and ferrous ion chelation assays (EC_{50} = 67.93, 290.93, and 8.54 g/mL, respectively).

Water extract of *A. mellea* induced maturation of human dendritic cells without induction of cytokine expression (Kim et al. 2008). *A. mellea* and some secondary metabolites of this mushroom possess anticancer properties. Armillarikin induced cell death of human leukemia K562, U937, and HL-60 cells and hepatocellular carcinoma HCC, Huh7, HA22T, and HepG2 cells. It also induced apoptosis by activating procaspase-3, procaspase-8, and procaspase-9, inducing cleavage of PARP (Chen et al. 2014, 2016). 5′-Methoxy-armillasin, 5-hydroxyl-armillarivin, armillaridin, armillartin, armillarin, melleolide B, armillarilin, armillasin, armillarigin, and melleolide exhibited highly cytotoxic activity on HepG2 cells (4.95–37.65μg/mL). Melleolide was the most cytotoxic compound, with an IC_{50}

value of 4.95μg/mL. It induced apoptosis by activating different caspases like armillarikin (Li et al. 2016).

Different extracts, as well as metabolites of this mushroom, have an anti-inflammatory effect. Ethanol extract of *A. mellea* protected lipopolysaccharide (LPS)-induced cell death in THP-1 cells. It attenuated LPS-induced nitric oxide (NO) and prostaglandin E2 (PGE2) production and reduced levels of pro-inflammatory cytokines, including TNF-α, IL-4, and IL-8. This extract suppressed the LPS-induced expression of COX-2 and iNOS but did not change COX-1 (Wu et al. 2007). Similarly, a fraction obtained from a sub-ethyl acetate extract of crude ethanol extract of *A. mellea* significantly suppressed the production of inflammation mediator NO and inflammatory cytokines TNF-α, IL-6, and IL-1 β in a dose-dependent manner. Moreover, the same fraction downregulated the phosphorylation levels of NF-κB p65, inhibitory κB-α (IκB-α), and JNK pathways in BV-2 cells. 5-Hydroxymethylfurfural, 2-furoic acid, 4-hydroxybenzoic acid, vanillic acid, syringate, daidzein, and genistein were isolated from this fraction (Geng et al. 2017). In addition, sulfated polysaccharides containing fucose, galactose, glucose, and mannose as major sugars inhibited TNF-α and IL-6 production associated with suppression of NF-κB (Chang et al. 2013). Different compounds isolated from this mushroom, namely, dehydroarmillylorsellinate, arnamial, armillarin, and melleolide D, inhibited 5-lipoxygenase with IC_{50} values of 0.3 ± 0.1, 1.0 ± 0.2, 5.2 ± 1.4, and >10 mM in neutrophils (König et al. 2019).

Protoilludane sesquiterpenoid aromatic esters of *A. mellea* ameliorated depressive behaviors in depressive mice induced by chronic unpredictable mild stress. They reduced hypothalamic-pituitary-adrenal axis hyperactivity by restoring GR negative feedback regulation caused by chronic unpredictable mild stress. These compounds may enhance synaptic plasticity and function by controlling levels of related proteins CREB, brain-derived neurotrophic factor, and synaptic-related protein PSD95. They may have anti-inflammatory effects on the central nervous system by reducing levels of inflammatory cytokines and NLRP3 protein and inhibiting neuronal apoptosis (Sun et al. 2020).

Oral administration of polysaccharide-enriched *A. mellea* fruiting body extract reduced fasting blood glucose and improved glucose intolerance and insulin resistance in type 2 diabetes mellitus. This extract also modulated lipid metabolism by enhancing lipolysis and suppressing lipogenesis (Yang et al. 2019).

Overall, *A. mellea* is a popular delicious mushroom. Besides, it possesses various biological effects, including anti-inflammatory, antioxidant, cytotoxic, antidepressant, and antidiabetic effects.

6 *Auricularia* Species

Auricularia species are known as wood ears or jelly ears due to their ear-shaped gelatinous fruiting bodies. They are growing widely and also cultivated. They have been widely consumed as edible medicinal mushrooms for thousands of years.

Auricularia species exhibited antioxidant effects both in vitro and in vivo. Antioxidant effect of these mushrooms involved mechanism of action of different effects (Agbor et al. 2022; Chellappan et al. 2016; Chiu et al. 2014; Wu et al. 2010). *A. auricula* extract scavenged ABTS ($EC_{50} = 1.2$ mg/mL), DPPH ($EC_{50} = 3.3$ mg/mL), superoxide ($EC_{50} = 0.7$ mg/mL), and hydroxyl radicals ($EC_{50} = 9.0$ mg/mL) and inhibited peroxidation of egg yolk homogenate ($EC_{50} = 0.1$ mg/mL) (Zeng et al. 2012).

Administration of 200 mg/kg of a β-glucan fraction of *A. polytricha* (Fig. 1d) reduced serum oxidative stress markers, glial cell aggregation, and inflammatory infiltrate compared to streptozotocin-induced diabetic rats and exhibited a neuroprotective effect (Agbor et al. 2022). Different polysaccharides of *Auricularia* species were studied. *A. auricula* polysaccharide comprising mannose, rhamnose, glucuronic acid, glucose, galactose, arabinose, and fucose with a molecular weight of 23.51 kDa increased the NO production and the phagocytosis of RAW264.7 macrophages at 0.5 mg/ml. It also induced the secretion of TNF-α and IL-6 from macrophages (Bao et al. 2020). Another *A. auricula* polysaccharide induced endotoxin tolerance by downregulating NF-κB and altering cytokine secretion (Perera et al. 2020). Oral administration of *A. auricula* polysaccharide fraction also increased the weights of the spleen and thymus in aged mice after oral administration, which suggested a possible effect on the activation and differentiation of the thymus lymphocyte and modulating the immune functions of aged mice (Wu et al. 2010). It has been reported that *Auricularia* polysaccharides bind different membrane receptors like toll-like receptor 4 and Dectin-1, activating the expression of pro-inflammatory cytokines and iNOS. It activated the innate immune system and effector cells, including macrophages, B and T1 lymphocytes, and natural killer (NK) cells (Liu et al. 2022).

Treatment with *A. auricula* polysaccharides for 40 days decreased the serum TC, TG, LDL, and MDA levels. It increased serum SOD and GSH-Px activity dose dependently. Also, it ameliorated left ventricle ejection fraction and left ventricular short axis fraction shortening and improved heart function in aged mice (Wu et al. 2010). *A. auricula-judae* polysaccharides also affect plasma TG, TC, and LDL cholesterol in dietary-induced hyperlipidemic rats. Moreover, polysaccharides of this mushroom downregulated liver adipogenic-related gene expressions, suppressed cholesterol synthesis-related gene levels, upregulated fatty acid oxidation-related gene expressions, and induced cholesterol efflux-related gene expressions, thus improving mice hepatic lipid metabolism (Liu et al. 2022; Zeng et al. 2013). The level of matrix metalloproteinase (MMP)-13 expression in atherosclerotic plaques is related to the progression of plaques. *Auricularia* polysaccharides inhibited atherosclerosis development by reducing the expression of MMP-13 in plaques and decreasing the production of collagen fibers (Wang et al. 2007).

A. auricula polysaccharide treatment at 100 and 400 mg/kg decreased blood glucose levels by promoting glucose metabolism. It prevented diabetic nephropathy by regulating blood urea nitrogen, creatinine, uric protein, and inflammatory-related factors. These effects were associated with modulations of the antioxidative system and NF-κB signaling (Hu et al. 2017).

Due to their antioxidant effects, water extract of *A. polytricha* attenuated paracetamol-induced hepatotoxicity and hepatic lipid accumulation in rat models (Chellappan et al. 2016; Chiu et al. 2014).

A water-soluble *A. auricula-judae* polysaccharide extract enhanced wound healing by promoting fibroblast and keratinocyte proliferation, migration, and invasion, increasing collagen synthesis and decreasing E-cadherin expression (Mapoung et al. 2021).

Protein extracts of *A. auricula-judae* exhibited antimicrobial activity with MIC values of 5μg/mL against *Pseudomonas aeruginosa*, *Staphylococcus aureus*, *Klebsiella pneumonia*, *Bacillus subtilis*, and *Candida albicans.* They showed the best antimicrobial effect against *Escherichia coli* with a MIC value of 2.5μg/mL (Oli et al. 2020).

Auricularia species exhibited a cytotoxic effect on different cancer cells, including HCT-15, Huh-7, SK-MEL-5, SNU-213, A-549, and SNU-484. *A. auricula* induced PARP cleavage; activated caspases 3, 7, and 9; downregulated Bcl-xL; upregulated proapoptotic proteins Bak, Bid, and Bik; and caused apoptosis. Also, it downregulated PRDX1 expression, causing inactivation of antioxidant enzymes (Kang et al. 2020). *A. polytricha* suppressed cell proliferation and induced apoptosis. *A. polytricha* polysaccharides caused cell cycle arrest at the G0/G1 phase and upregulated the expression of cyclin-dependent kinase (CDK) inhibitors p53 and p21, whereas downregulated the expression of cyclin A, cyclin D, and CDK2 in A549 cells (Yu et al. 2014).

Auricularia species possess antioxidant, cardioprotective, antihyperlipidemic, antidiabetic, immunomodulatory, anticancer, antimicrobial, hepatoprotective, and neuroprotective effects. Research on this species mainly focused on *A. auricula-judae* and *A. polytricha* polysaccharides.

7 *Boletus edulis* Bull

Boletus edulis Bull. (Boletaceae) (Fig. 1e) is widely distributed across Europe, Asia, and North America in the Northern Hemisphere. It is one of the most popular edible mushrooms due to its flavor and taste.

Antioxidant activity of *B. edulis*, extracts, fractions, or some components has been studied widely. It scavenged DPPH, ABTS, and ·OH radicals. It also exhibited antioxidant activity in β-carotene bleaching and FRAP, reducing power, ferrous ion chelating, cupric reducing antioxidant capacity assay (Garcia et al. 2022; Guo et al. 2012, 2020a; Heleno et al. 2015; Luo et al. 2012). *B. edulis* polysaccharides enhanced the activity of antioxidant enzymes SOD, GSH, and GSH-Px and reduced MDA levels (Xiao et al. 2018, Zheng et al. 2019).

B. edulis polysaccharides ameliorated carbon tetrachloride-induced hepatic damage revealed by smaller hepatocyte lesions, decreased ALT and AST serum levels, and increased antioxidant capacity (Zheng et al. 2019). *B. edulis* polysaccharides reduced fasting blood glucose, ALT and AST levels, and serum lipid profile, similar to metformin in high-fat-induced and streptozocin-induced diabetes. They exhibited anti-hepatotoxic effects associated with decreased sterol regulatory element-binding protein 1, NF-κB and TNF-α expressions and increased CYP7A1 expression (Xiao et al. 2018).

B. edulis polysaccharides suppressed ovalbumin asthma, reduced airway resistance, reduced lesion degree and mucosubstances, reduced pro-inflammatory responses (lower IL-4 and interferon-γ (IFN-γ) level), and increased anti-inflammatory responses (higher proportion of anti-inflammatory CD4+CD25+-FOXP3+ Treg cells) significantly increased in mouse models of asthma (Wu et al. 2016).

B. edulis extracts, different components including polysaccharides, lectins, and biopolymers (composed of polysaccharides and glycoproteins), exhibited anticancer effects by boosting the immune system and exhibiting toxicity to cancer cells. *B. edulis* polysaccharides increased the spleen and thymus indices, enhanced splenocytes proliferation, increased NK cell and CTL activities in the spleen, and stimulated the secretion of the cytokines IL-2 and TNF-α in Renca tumor-inoculated mice (Wang et al. 2014a). A polysaccharide purified from *B. edulis* exhibited an anticancer effect on non-small cell lung cancer A549 in vitro and in vivo. Moreover, it inhibited A549, MDA-MB-231 human breast cancer cell, and Ca761 mouse breast cancer cell proliferation associated with G1 or S phase cell cycle arrest and downregulated CDK4 level. It also arrested apoptosis by activating the expression of PARP, caspase-3, caspase-8, and caspase-9 and upregulated the Bax/Bcl-2 (Meng et al. 2021; Zhang et al. 2021). *B. edulis* lectin inhibited the proliferation of MCF-7 cells, hepatoma Hep G2 cells, and colorectal cancer HT29 cells and reduced the migratory ability of Runt-related transcription factor 2 expressing melanoma cells in a xenotransplanted zebrafish model (Bovi et al. 2013; Valenti et al. 2020). *B. edulis* biopolymer exhibited an antiproliferative effect in colon cancer cells (LS180) associated with a cell cycle arrest in G_0/G_1 phase by altering the p16/cyclin D/CDK4–6/pRb pathway. It also induced apoptosis in a p53-dependent manner (Lemieszek et al. 2013, 2016).

Water extract of *B. edulis* exhibited antimicrobial and anti-biofilm properties against multidrug-resistant pathogens (*Enterococcus faecium*, *Staphylococcus aureus*, *Klebsiella pneumoniae*, *Acinetobacter baumannii*, *Pseudomonas aeruginosa*, and *Enterobacter species*) in different test systems. This extract contains phenolics, including gallic acid, catechin, 2,4-dihydroxybenzoic acid, 2,5-dihydroxyphenylacetic acid, and protocatechuic acid. It is recommended for use in the prevention of wound infection, particularly by multidrug-resistant pathogens (Garcia et al. 2022).

B. edulis is one of the most delicious mushrooms with multiple health benefits. It possesses antioxidant, antidiabetic, anti-inflammatory, antimicrobial, anti-hepatotoxic, and anticancer properties.

8 *Coprinus comatus* (O.F. Müll.) Pers.

Coprinus comatus (O.F. Müll.) Pers. (*Agaricaceae*) (Fig. 2a) is an edible and medicinal fungus widely distributed in most parts of the world.

Antidiabetic effects of *C. comatus* have been shown in many studies. *C. comatus* extracts or some components like comatin or polysaccharides decreased blood sugar levels by increasing plasma insulin levels through enhanced glucagon-like peptide-1 (GLP-1) and suppression of dipeptidyl peptidase-4 (DPP-4). This mushroom protected the pancreas, decreased blood sugar, and raised insulin and GLP-1 levels by inhibiting DPP-4 and reducing ROS (Ding et al. 2010; Husen et al. 2021; Ratnaningtyas et al. 2019, 2022). Furthermore, *C. comatus* polysaccharides inhibited α-amylase (Cao et al. 2019). Additionally, *C. comatus* also suppressed diabetic nephropathy. It ameliorated dysfunction in the kidney and relieved the renal oxidative stress and inflammation by modulating the PTEN/PI3K/Akt and Wnt-1/β-catenin pathways (Gao et al. 2021).

C. comatus was found to possess antioxidant properties in DPPH scavenging, hydroxyl scavenging, β-carotene bleaching and reducing power tests (Cao et al.

Fig. 2 (**a**) *Coprinus comatus*; (**b**) *Coriolus versicolor*; (**c**) *Flammulina velutipes*; (**d**) *Grifola frondosa*; (**e**) *Pleurotus ostreatus*. Attrition: CC-BY-4.0; https://commons.wikimedia.org

2019; Sihoglu Tepe 2021). It significantly enhanced the GSH-Px, SOD, and catalase (CAT) activities (Gao et al. 2021). *C. comatus* also has an anti-hepatotoxic effect. Oral administration of this mushroom alleviated carbon tetrachloride-induced hepatic damage through increasing antioxidant capabilities, decreasing serum aminotransferase levels, and lipid peroxidation intensity (Stilinović et al. 2020).

This mushroom has beneficial effects on lipid metabolism. *C. comatus* polysaccharides inhibited adipocyte differentiation of 3 T3-L1 cells and high-fat diet-given mice. It caused a significant decrease in lipid accumulation through the downregulation of a major transcription factor involved in the adipogenesis pathway, including PPARγ related to the regulation of the Akt pathway (Park et al. 2020). Moreover, *C. comatus* polysaccharides diminished TC, TG, and LDL-C levels and increased HDL-C levels, indicating that oral administration of *C. comatus* polysaccharides alleviated lipid metabolism (Gao et al. 2021).

Like most mushroom polysaccharides, *C. comatus* polysaccharides have prebiotic effects on normal mice and mice with acute alcoholic liver injury by increasing relative abundance of *Firmicutes* and *Lactobacillaceae* and decreasing the abundance of *Rikenellaceae* (Li et al. 2020a).

C. comatus exhibited an anticancer effect in vitro. It inhibited proliferation and induced apoptosis of LNCaP (androgen-sensitive human prostate adenocarcinoma), U87MG and LN-18 glioblastoma cells, and ovarian cancer cells (ES-2). It induced extrinsic and intrinsic apoptotic pathways by reducing procaspases-3, procaspases-8, and procaspases-9 (Dotan et al. 2011; Nowakowski et al. 2021; Rouhana-Toubi et al. 2015). This mushroom reduced the androgen levels and prostate-specific antigen gene expression in LNCaP cells (Dotan et al. 2011).

In conclusion, *C. comatus* is an edible mushroom with health-promoting effects. Especially antidiabetic effect of this mushroom has been reported in several studies. It also regulated lipid metabolism, reduced hyperlipidemia, and inhibited the proliferation of cancer cells. Like most mushroom polysaccharides, it modulates gut microbiota. Thus, it could be regarded as healthy food. Health-promoting products can be produced from this mushroom. However, current research is limited, and more in vitro and in vivo studies are needed.

9 *Coriolus versicolor* (L. ex Fr.) Quel.

Coriolus versicolor (L. ex Fr.) Quel. (Fig. 2b) also named as *Trametes versicolor* (L.) Lloyd and *Polyporus versicolor* L. belongs to the genus *Coriolus* (*Polyporaceae*) (Cruz et al. 2016). *C. versicolor* is a saprotrophic mushroom species that commonly grows on dead logs, bark, stumps, tree trunks, and branches and lives by using lignocellulosic wastes in the temperate zones of Asia, Europe, and North America (Jo et al. 2010). It was named turkey tail mushroom because of its structural similarity. *C. versicolor* has been a well-known traditional medicine in the east for over 2000 years. It has been reported that *C. versicolor* fungus has antiradical, antioxidant, cytotoxic, anticancer, immunostimulant, and neurological effects. In

Japan and China, due to its immunostimulatory properties, it is used in the treatment of gastric cancer patients as an adjuvant. It is approved and has licensed products and food supplements (Dou et al. 2019; Hobbs 2005; Janjusevic et al. 2018). Recently, its neurological and synergistic effects with other drugs have been investigated.

Physiologically active polysaccharopeptide (PSP) and polysaccharide K (PSK, Krestin) from *C. versicolor* are isolated. 4-Isobutoxyphenyl palmitate, 2-hydroxyheptanoic-1-*O*-β-D-glucopyranosyl-9-methyl-4,8-sphinga-dienine (cerebroside), 3β-linoleyloxyergosta-7,22-diene, 3β-linoleyloxyergosta-7-ene, betulinic acid, terpenoids, flavonoids, and other many phenolic compounds derived from cinnamic acid also have been identified in this mushroom (Habtemariam 2020; Wan 2013).

The anticancer effects of *C. versicolor* were investigated in vivo, in vitro, and in clinical studies. PSK and PSP do not have only cytotoxicity; they also have an immunostimulating effect. At 100μg/ml and less, PSK and PSP concentrations were found to have cytotoxic activity. The extracts can inhibit carcinogenesis and tumor cell growth by activating cancer cell apoptosis. The underlying mechanism of this effect is caspase 3 activation, the key enzyme of apoptosis. As expected, markers of apoptosis induction have been stimulated while suppressing cancer cell survival-associated genes and proteins (antiapoptotic Bcl-2 Bcl-xL, etc.). In addition to the metastasis in cancer patients, inhibition of key angiogenic enzymes such as metalloproteases (MMP-9 or MMP-2) suppresses cell migration and invasion. It can reverse myelosuppression, which is a side effect of medications with chemotherapeutics. In myelosuppressive mice, especially in granulocyte colonies with stimulating factor (GCSF) or IL-3 were reversed myelosuppression by PSK1. In myelosuppressive mice, especially in granulocyte colonies with stimulating factor (GCSF) or IL-3 were reversed by PSK. Generally, it has a proliferative effect, including lymphocytes, monocytes, macrophages, and splenocytes on many cells. In addition, human peripheral blood mononuclear cells increased interleukin 1 production. In many more ways, the immune system positively affects the system (Habtemariam 2020).

Sun et al. (2012) reported that PSK possesses clinical benefits with limited side effects in several cancers, e.g., colorectal, gastric, and lung cancer, as an additive to other conventional adjuvant treatments.

The methanol and the aqueous extract have been tested in the DPPH scavenging experiment. They showed a dose-dependent response. IC_{90} values were found as 178.83, 518.06, and 332.98μg/ml, for methanol extract, aqueous extract, and ascorbic acid, respectively. In the extract, catechin (5.91 mg/ml) and quercetin (29.90 mg/ml) were found, and catechin was more effective in biological activity (Hossen et al. 2021; Janjusević et al. 2017). Another study on the H_2O extract expressed better antioxidant scavenging potential than EtOH, showing the highest activity for the *T. versicolor* ($IC_{50} = 5.6$μg/mL, $IC_{50} = 0.6$μg/mL for DPPH and OH radicals, respectively) (Raseta et al. 2020). It was found that the conformation of the polysaccharides was more important for DPPH radical scavenging activity than monosaccharide composition (Kozarski et al. 2012).

In vitro acetylcholinesterase (AChE) inhibitor activity tests showed that the inhibition was concentration-dependent. A strong degree of inhibition was found at the dose of 100μg/ml, compared to donepezil (89.05%) used as an AChE inhibitor (60.53%). The EtOH extract at 500μg/ml had 44.35% inhibition. AChE inhibitory activity of H_2O and EtOH extracts were found $IC_{50} = 78.01$μg/ml and $IC_{25} = 383.96$μg/ml, respectively. This inhibitory effect is due to the major flavonoids in the extracts, baicalein (21.6μg/g) and quercetin (31.2μg/g). It has also been found that this inhibitory effect is due to synergistic effects of the flavonoids, terpenoids, phenolic compounds, and polysaccharides in the aqueous extract, but the strongest effect belongs to baicalein (Fang et al. 2015; Janjusević et al. 2017).

In light of the literature, there seems to be a need for more research on the anticancer and neurological effects of *C. versicolor*. PSK and PSP from *Coriolus* appear safe and effective as long-term adjuvant immunotherapy in conjunction with or after standard chemotherapy and/or radiation. They may increase survival time, immune function, and tumor-associated symptoms in patients with various types of cancer.

10 *Flammulina velutipes* (Curtis) Singer

Flammulina velutipes (Curtis) Singer (*Physalacriaceae*) (Fig. 2c) is an edible mushroom with many biological activities. It is widely consumed and the world's fifth most cultivated mushroom (Karasoy et al. 2019).

According to several reports, the antioxidant effect is one of the main mechanisms underlying the activities of *F. velutipes*. The ethanol extract of *F. velutipes* exhibited strong antioxidant properties in DPPH and ABTS radical scavenging activities and reducing power. It protected PC12 cells from H_2O_2-induced injury by reducing the LDH release, ROS production, and MDA generation. The extract also enhanced the activities of antioxidant enzymes such as GSH level and SOD. Several phenolic and polyphenolic compounds, including arbutin, epicatechin, phillyrin, apigenin, kaempferol, and formononetin, were isolated from this extract (Hu et al. 2016). Likewise, polysaccharide fraction *F. velutipes* boosted glutathione peroxidase (GSH-Px) and SOD activity against oxidative stress in H_2O_2-stimulated L929 cells. Moreover, they inhibited the production of hydroxyl radicals by chelating metal ions such as Fe^{2+} and Cu^{2+} (Wang et al. 2016).

F. velutipes polysaccharides attenuated carbon tetrachloride-induced acute liver injury revealed by histopathological examination of the liver. They also decreased levels of serum aspartate transaminase (AST), alanine aminotransferase (ALT), triglyceride (TG), total cholesterol (TC), total bile acid (TBA) content, hepatic MDA, and protein carbonyl levels. *F. velutipes* polysaccharides enhanced the activities of antioxidant enzymes CAT and SOD. They reduced pro-inflammatory cytokines (including IL-6, IL-1β, and TNF-α). *F. velutipes* polysaccharides changed the gut microbiota according to 16S rRNA, suggesting that their gut microbiota modulating effect is also involved in anti-hepatotoxic action (Xu et al. 2022).

F. velutipes polysaccharides possess anti-inflammatory properties associated with their immunomodulatory and gut microbiota modulatory effects. They regulated

CD4+, CD8+, ICAM-1, and myeloperoxidase in the serum and colon of rats. It also changed NF-α, IFN-γ, IL-1β, IL-6, and iNOS levels in ulcerative colitis mice. It regulated intestinal microorganisms to reduce inflammation by increasing beneficial bacteria (*Lactobacillus*, *Bifidobacterium*) and decreasing harmful bacteria such as *Clostridium* (Zhao et al. 2020). *F. velutipes* polysaccharides prevented scopolamine-induced learning and memory impairment by mediating gut microbiota composition and inhibiting inflammation (Su et al. 2018). Notably, *F. velutipes* polysaccharides in combination with ginsenosides enhanced cognitive functions in Alzheimer's disease model (Zhang et al. 2018).

Several studies showed that *F. velutipes* polysaccharides stimulate the immune system. They stimulated macrophages, T cells, B cells, and NK cells and increased splenocyte proliferation and antibody release to boost humoral immunity. They boosted the secretion of TNF-α, IL-1β, and IL-6 from macrophages and induced cellular nitric oxide formation. Additionally, it showed that there was a considerable increase in the quantity of beneficial flora as well as the concentration of two types of fecal short-chain fatty acids, namely, isobutyric acid and butyric acid. Apart from these, *F. velutipes* polysaccharides stimulate the immune system through mitogen-activated protein kinases (MAPKs), autophagy, and Akt/NF-κB signaling pathways via TLR4 receptor in vitro and in vivo (Meng et al. 2018; Ye et al. 2020; Yin et al. 2010). *F. velutipes* polysaccharides have anticancer action mainly by immunomodulatory effect rather than toxicity to cancer cells. A polysaccharide fraction of *F. velutipes* reduced tumor weight in sarcoma 180-inoculated mice. However, it did not reduce the proliferation of sarcoma 180 cells in vitro. Our literature survey found no other study reporting direct toxicity to cancer cells or tumors of *F. velutipes* or polysaccharides (Leung et al. 1997).

Thus, *F. velutipes* improves learning and memory ability, regulates cognitive function, and protects the liver. But the most notable effect of it is immunomodulation. Compared to other widely cultivated mushrooms, studies on *F. velutipes* are limited and need more research.

11 *Grifola frondosa* (Dicks.)

Grifola frondosa (Dicks.) Gray (Fig. 2d) is an edible and medicinal mushroom belonging to the *Grifolaceae* family. It is known as maitake (Japan) and gray tree flower (China). This mushroom is mainly distributed in northern temperate regions in Japan, Europe, and America, but it also grows at high altitudes in the subtropics with a temperate climate. It is also cultivated due to high demand. It is used as healthy food and food flavoring agent, and some polysaccharide-containing fractions of *G. frondosa* are commercially available as healthcare products like D-fraction. Apart from D-fraction, a proteoglycan, other polysaccharide fractions, such as MD-fraction, X-fraction, grifolan, and MZ-fraction, also possess a wide range of bioactivities. The China Food and Drug Administration approved some patent drugs containing *G. frondosa* polysaccharides, and also there are many clinical trials on these fractions in different countries (He et al. 2017; Wu et al. 2021).

Antidiabetic effects of *G. frondosa* extracts or polysaccharides have been reported in several studies (Chen et al. 2018; Guo et al. 2020b; Jiang et al. 2020; Ma et al. 2014; Shen et al. 2015; Su et al. 2013). Some studies have shown that *G. frondosa* polysaccharides exhibit antidiabetic effects through the insulin signal pathway. They affected insulin receptors, enhanced insulin sensitivity, and ameliorated insulin resistance. They enhanced glucose uptake by cells, activated the insulin receptor protein in the cell membrane, induced phosphorylation of increased Akt, and overcome insulin resistance. In addition to Akt-GSK-3 signaling, upregulation of phosphorylation of insulin receptor and downregulation of phosphorylation of insulin receptor substrate 1 are also involved in fasting serum glucose level decreasing effect of *G. frondosa* polysaccharides (Ma et al. 2014; Xiao et al. 2015). The antidiabetic effect of *G. frondosa* polysaccharides is also related to their α-glucosidase inhibition. Not only polysaccharides but also nonpolar fractions, ergosterol, oleic and linoleic acid, and pyrrole alkaloids of *G. frondosa* possess α-glucosidase inhibitory potential (Chen et al. 2018; Shen et al. 2015; Su et al. 2013). The antidiabetic effect of this mushroom is also associated with its modulatory potential for gut microbiota (Guo et al. 2020b). *G. frondosa* also ameliorated renal function and attenuated renal inflammatory responses in diabetic mice by decreasing IL-6, IL-1β, TGF-β1, and TNF-α levels, suppressing apoptosis and fibrosis. TLR4/NF-κB signaling is associated with the diabetic nephropathy effect of maitake (Jiang et al. 2020).

G. frondosa has an antihypertensive effect and is associated with the renin-angiotensin system (Preuss et al. 2010). Several researchers also reported antihyperlipidemic effect of maitake mushroom. Dried *G. frondosa* powder reduced serum cholesterol, triglyceride, and phospholipid levels 0.3–0.8 times compared to the control group and increased cholesterol excretion approximately 1.8 times by feces. *G. frondosa* also reduced serum total cholesterol concentration and very low-density lipoprotein levels. The total cholesterol-lowering effect was attributed to increased fecal cholesterol excretion. In addition, *G. frondosa* prevented hyperlipidemia in diabetic mice by altering gut microbiota and regulating hepatic glycolipid metabolism-related genes (Kubo and Nanba 1997; Fukushima et al. 2001; Guo et al. 2020b).

G. frondosa also possesses antiviral effects against hepatitis B virus (HBV), enterovirus 71 (EV71), herpes simplex virus type 1 (HSV-1), and human immunodeficiency virus (HIV). A combination of D-fraction of *G. frondosa* with interferon-alpha (IFN-α) synergistically inhibited HBV. The effect was nine times higher than that of IFN-α (Gu and Sivam 2006; Mayell 2001).

In general, mushroom polysaccharides regulate gut microbiota. Gut microbiota-regulating effects of *G. frondosa* involve various biological effects, including antidiabetic, antihyperlipidemic, and against nonalcoholic fatty liver disease (Gangarapu et al. 2014; Guo et al. 2020b; Friedman 2016). *G. frondosa* polysaccharides modulate intestinal microflora by significantly increasing the relative abundance of *Alistipes* and *Bacteroides* and reducing *Enterococcus* and *Firmicutes* to

Bacteroidetes ratio, the latter suggested possessing fat-lowering effects (Guo et al. 2020b; Friedman 2016). Moreover, *G. frondosa* polysaccharides significantly boosted the proportion of *Allobaculum*, *Bacteroides*, *Bifidobacterium*, and other microbial groups in the cecal microbiota, which may improve the immune system of the host and the defense against nonalcoholic fatty liver disease (Liu et al. 2019).

G. frondosa extracts and polysaccharides also exhibited an antioxidant effect. They scavenged DPPH, hydroxyl, and superoxide radicals in vitro and enhanced the antioxidant status and activities of endogenous antioxidant enzymes (He et al. 2017; Wu et al. 2021).

Different extracts and components of *G. frondosa* possess anticancer properties by suppressing tumor growth or modulating the host immune system. *G. frondosa* polysaccharides and ergosterol derivatives exhibited antiproliferative effects on cancer cells (Chen et al. 2018; Cui et al. 2007; Wang et al. 2013a). They also induced apoptosis associated with altering transmembrane potential, activating different caspases, upregulating proapoptotic proteins like Bax, and downregulating antiapoptotic proteins like Bcl-2, Notch1/NF-κB/p65 signaling (Cui et al. 2007; Wang et al. 2013a; Wu et al. 2021). A heteropolysaccharide fraction also exhibited an antiangiogenic effect (Wang et al. 2014b). *G. frondosa* D-fraction reduced the effective dosage of the chemotherapeutic agent by enhancing cisplatin's antitumor and antimetastatic activities. It also reduced the myelosuppression and nephrotoxicity induced by cisplatin (Masuda et al. 2009a).

G. frondosa polysaccharides enhanced phagocytic activity of macrophages, activated T cells, B cells, and NK cells and promoted the release of cytokines IL-1, IL-2, TNF-α, and IFN-γ (Masuda et al. 2009a; Wang et al. 2013c). D-fraction stimulated the differentiation into Th-1 or Th-2 cells of CD4+ T cells by enhancement of IL-12p70 and IFN-γ secretion (Harada et al. 2003; Kodama et al. 2002). D-fraction boosted a Th-1 dominant response, including the cell-mediated immunity related to cytotoxic T cell activation. Furthermore, D-fraction also induced a Th-2 dominant response through macrophage activation, enhancing humoral immunity rather than cell-mediated immunity (Inoue et al. 2002; Kodama et al. 2004). Other than many mushroom polysaccharides, which may become ineffective if administered orally, D-fraction and MD-fraction can be given orally, making them easy to use.

In conclusion, *G. frondosa* can regulate blood lipids and glucose levels and improve fat metabolism and weight loss. It also exhibits anticancer effects by inhibiting tumor growth and metastasis and modulating the immune system. Most of the activities of *G. frondosa* come from polysaccharides. Such as the Gut microbiota regulating effect of *G. frondosa*. Some of these polysaccharides are commercially available as a dietary supplement and undergo clinical trials for the treatment of cancer. However, well-designed randomized controlled clinical trials are needed.

12 *Pleurotus ostreatus* (Jacq. Ex Fr) P. Kumm.

There are about 40 species in the *Pleurotus* genus, including those with high economic significance, such as *P. ostreatus* and *P. pulmonarius*. Many species contain medicinal components, including polysaccharides, proteins, terpenoids, fatty acids, and polyphenols. It is also reported that oyster mushroom contains phenolic compounds with antioxidative effects. Bioactive substances in this genus exhibit immunostimulatory, antineoplastic, antidiabetic, anti-atherosclerotic, anti-inflammatory, hepatoprotective, and antioxidative properties (Golak-Siwulska et al. 2018a; Piska et al. 2017).

Pleurotus ostreatus (Jacq. Ex Fr) P. Kumm. (*Pleurotaceae*) (Fig. 2e), also known as the oyster mushroom, is found on all continents except Antarctica. These mushrooms grow on various lignocellulosic substrates and form shell-shaped fruiting bodies. It has been cultivated commercially on a large scale since World War I. *P. ostreatus* is an important dietary mushroom with many biological activities. It is also rich in proteins, vitamins, minerals, and oleic and linolenic acids. Flavonoids and phenolic acids such as myricetin, naringenin, hesperidin, formononetin, biochanin A, and p-hydroxybenzoic are also found in sinapic, ferulic, p-coumaric acids, etc. (Piska et al. 2017). These species contain high levels of lovastatin, an approved hypolipidemic drug, and pleuran, an immunomodulating polysaccharide (β-glucan) (Golak-Siwulska et al. 2018a; Piska et al. 2017). It was found to possess an anticancer effect due to its polysaccharides and glucans (Jedinak et al. 2010; Jedinak and Sliva 2008; Wu et al. 2011); antiviral and antibacterial properties due to its proteins (laccase, ribonucleases) and *β*-glucans, respectively (Erjavec et al. 2012; Gashaw et al. 2020; Golak-Siwulska et al. 2018a; Iwalokun et al. 2007; Patel et al. 2012; Wang and Ng 2000), anti-inflammatory and antioxidant activities due to its polysaccharides and phenolics (Mitra et al. 2013; Jayakumar et al. 2007, 2011; Venkatakrishnan et al. 2010); and anti-atopic dermatitis effect due to pleuran (Park et al. 2016). The hypoglycemic activity of *P. ostreatus* has been indicated in various in vivo studies on animals (Piska et al. 2017).

Zhang et al. (2012) isolated two polysaccharide fractions from *P. ostreatus* and showed their potent antioxidative effects. Extracts of *P. ostreatus* increased the activities of known potent antioxidant enzymes in aged rats (Jayakumar et al. 2007).

Gu and Sivam (2006) have found water-soluble proteins or polypeptides from *P. ostreatus* cytotoxic effect and induced cytotoxicity on various cell lines. However, it has the most significant cytotoxicity on human androgen-independent prostate cancer PC-3 cells. Both cytotoxicity and apoptosis-inducing effects were dose-dependent. Furthermore, linear insoluble α-glucan *P. ostreatus* exhibited high cytotoxic activity on HeLa cell lines (Wiater et al. 2011). *P. ostreatus* exhibited an antiproliferative effect on the colorectal cancer cell lines COLO-205 and SW 480 and induced apoptosis of monocytic leukemia THP-1 (Piska et al. 2017; Wu et al. 2011). Facchini et al. (2014) proved that polysaccharides extracted from the *P. ostreatus* mycelium successfully inhibited the development of neoplastic cells of Ehrlich tumor (ET) and sarcoma 180 (S-180). *P. ostreatus* extracts have inhibited the growth of HL-60 cells by arresting the cell cycle related to the inducement of apoptosis (Venkatakrishnan et al. 2010).

P. ostreatus, which contains lovastatin, is known to be effective in preventing cardiovascular disorders because it reduces cholesterol. It was reported that oyster mushroom has antioxidant, immunomodulatory, and antitumor activity and affects breast and colon cancer treatment and prevention.

13 Conclusion

Edible mushrooms are important products in the global trade. They have gradually attracted the attention of the food, pharmaceutical, and cosmetic industries regarding their use as food supplements and in searching for new drug candidates.

The biological activities of edible mushrooms are mainly attributed to the presence of polysaccharides, but the contribution of other chemical constituents, such as proteins, triterpenes, steroids, and phenols, are also responsible for these effects. Edible mushrooms possess promising bioactivities, including antitumor and immunomodulation, antioxidant, anti-hyperglycemia, and anti-inflammatory effects (Table 1).

Mushroom polysaccharides regulate gut microbiota, and this regulation involves most of the bioactivity of mushrooms. Arguably, the immunomodulatory effects of mushroom polysaccharides are one of the most studied and well-known bioactivity. Some polysaccharide fractions are commercially available such as *G. frondosa* D-fraction or Andosan, making them easy to use for health promotion.

Although several randomized clinical trials on some edible mushrooms, including *A. blazei*, *G. frondosa*, and *T. versicolor*, have been conducted; larger randomized studies are required to confirm these interesting findings and potential health benefits.

Table 1 Major biological activity of some well-known edible mushrooms

Mushroom name	Biological activity	Application	References
Agaricus bisporus (J.E. Lange) Imbach	– Antioxidant	In vitro, in vivo	Liu et al. (2013)
	– Antidiabetic	In vivo	Ekowati et al. (2018)
	– Antihyperlipidemic	In vivo	Jeong et al. (2010)
	– Anti-inflammatory	In vitro, in vivo	Gallego et al. (2021), Smiderle et al. (2013), Volman et al. (2009)
	– Cytotoxicity against cancer cells	In vitro, in vivo	Chen et al. (2006), Jeong et al. (2012)
Agaricus blazei Murill	– Antioxidant	In vitro, in vivo	Al-Dbass et al. (2012), Bach et al. (2019), Carneiro et al. (2013), Wei et al. (2020)
	– Antidiabetic	In vivo	Al-Dbass et al. (2012), Ji et al. (2014), Vitak et al. (2015)
	– Antihyperlipidemic	In vivo	Ji et al. (2014), Li et al. (2020b)
	– Anti-inflammatory	In vitro, in vivo, clinical study	Song et al. (2012), Therkelsen et al. (2016), Wang et al. (2013b)
	– Antimicrobial	In vitro	Bach et al. (2019)
	– Anxiolytic	In vivo	Ali et al. (2021)
	– Immunostimulant	In vitro, in vivo, clinical study	Bernardshaw et al. (2005), Johnson et al. (2009), Lima et al. (2011), Tangen et al. (2015), Therkelsen et al. (2016)
	– Cytotoxicity against cancer cells	In vitro, in vivo	Ito et al. (1997), Jin et al. (2006), Kim et al. (2011), Matsushita et al. (2018), Yu et al. (2009)
Amanita caesarea (Scop.) Pers.	– Antioxidant	In vitro	Doğan and Akbaş (2013)
	– Antimicrobial	In vitro	Doğan and Akbaş (2013)
	– Anti-Alzheimer	In vitro, in vivo	Hu et al. (2021), Li et al. (2019)
Armillaria mellea (Vahl) P. Kumm	– Antioxidant	In vitro	Zavastin et al. (2015)
	– Antidiabetic	In vivo	Yang et al. (2019)
	– Antidepressant	In vivo	Sun et al. (2020)
	– Anti-inflammatory	In vitro	Chang et al. (2013), Geng et al. (2017), König et al. (2019), Wu et al. (2007)
	– Cytotoxicity against cancer cells	In vitro, in vivo	Kim et al. (2008), Li et al. (2016)
Auricularia species	– Antioxidant	In vitro, in vivo	Agbor et al. (2022), Chellappan et al. (2016), Chiu et al. (2014), Wu et al. (2010)
	– Antidiabetic	In vivo	Hu et al. (2017)
	– Antihyperlipidemic	In vivo	Liu et al. (2022), Zeng et al. (2013)
	– Anti-inflammatory	In vitro, in vivo	Bao et al. (2020), Perera et al. (2020), Wu et al. (2010)
	– Antimicrobial	In vitro	Oli et al. (2020)

(continued)

Table 1 (continued)

Mushroom name	Biological activity	Application	References
	– Cytotoxicity against cancer cells	In vitro	Kang et al. (2020), Yu et al. (2014)
	– Wound healing	In vivo	Mapoung et al. (2021)
Boletus edulis Bull.	– Antioxidant	In vitro	Garcia et al. (2022), Guo et al. (2012, 2020a), Heleno et al. (2015), Luo et al. (2012), Xiao et al. (2018), Zheng et al. (2019)
	– Antidiabetic	In vivo	Xiao et al. (2018)
	– Anti-hepatotoxic	In vivo	Zheng et al. (2019)
	– Anti-inflammatory	In vivo	Wu et al. (2016)
	– Antimicrobial	In vitro	Garcia et al. (2022)
	– Cytotoxicity against cancer cells	In vitro, in vivo	Bovi et al. (2013), Lemieszek et al. (2013, 2016), Meng et al. (2021), Valenti et al. (2020), Wang et al. (2014a), Zhang et al. (2021)
Coprinus comatus (O.F. Müll.) Pers.	– Antioxidant	In vitro, in vivo	Cao et al. (2019), Sihoglu Tepe (2021), Gao et al. (2021)
	– Antidiabetic	In vivo	Cao et al. (2019), Ding et al. (2010), Husen et al. (2021), Ratnaningtyas et al. (2022)
	– Anti-hepatotoxic	In vivo	Stilinović et al. (2020)
	– Antihyperlipidemic	In vivo	Gao et al. (2021)
	– Cytotoxicity against cancer cells	In vitro	Dotan et al. (2011), Nowakowski et al. (2021), Rouhana-Toubi et al. (2015)
Coriolus versicolor (L. ex Fr.) Quel.	– Antioxidant	In vitro	Hossen et al. (2021), Janjusevic et al. (2018), Kozarski et al. (2012), Raseta et al. (2020)
	– Acetylcholinesterase inhibitory	In vitro	Fang et al. (2015), Janjusević et al. (2017)
	– Cytotoxicity against cancer cells	In vitro, in vivo, clinical study	Habtemariam (2020), Sun et al. (2012)
Flammulina velutipes (Curtis) Singer	– Antioxidant	In vitro	Hu et al. (2016), Wang et al. (2016)
	– Anti-inflammatory	In vitro, in vivo	Su et al. (2018), Zhao et al. (2020)
	– Anti-Alzheimer	In vitro, in vivo	Zhang et al. (2018)
	– Cytotoxicity against cancer cells	In vitro, in vivo	Leung et al. (1997)
	– Immunomodulatory	In vitro, in vivo	Meng et al. (2018), Ye et al. (2020), Yin et al. (2010)
Grifola frondosa (Dicks.)	– Antioxidant	In vitro, in vivo	He et al. (2017), Wu et al. (2021)
	– Antidiabetic	In vivo, in vitro	Chen et al.(2018), Guo et al. (2020b), Jiang et al. (2020), Ma et al. (2014), Shen et al. (2015), Su et al. (2013), Xiao et al. (2015)

(continued)

Table 1 (continued)

Mushroom name	Biological activity	Application	References
	– Antihypertensive	In vivo	Preuss et al. (2010)
	– Antihyperlipidemic	In vivo	Kubo and Nanba (1997), Fukushima et al. (2001), Guo et al. (2020b)
	– Antiviral	In vitro	Gu and Sivam (2006), Mayell (2001)
	– Cytotoxicity against cancer cells	In vitro, in vivo	Chen et al. (2018), Cui et al. (2007), Wang et al. (2013a)
	– Gut microbiota regulating	In vivo	Guo et al. (2020b), Friedman (2016), Liu et al. (2019)
	– Immunomodulator	In vitro	Harada et al. (2003), Kodama et al. (2002), Masuda et al. (2009b), Wang et al. (2013c)
Pleurotus ostreatus (Jacq. Ex Fr) P. Kumm.	– Antioxidant	In vitro, in vivo	Golak-Siwulska et al. (2018a), Piska et al. (2017)
	– Antidiabetic	In vivo	Golak-Siwulska et al. (2018a), Piska et al. (2017)
	– Anti-atherosclerotic	In vivo	Golak-Siwulska et al. (2018a), Piska et al. (2017)
	– Anti-inflammatory	In vitro, in vivo	Golak-Siwulska et al. (2018a), Piska et al. (2017)
	– Cytotoxicity against cancer cell lines	In vitro, in vivo	Gu and Sivam (2006), Jedinak et al. (2010), Jedinak and Sliva (2008), Piska et al. (2017), Venkatakrishnan et al. (2010), Wiater et al. (2011), Wu et al. (2011)

References

Agbor CA, Fischer CE, Agaba EA et al (2022) Neuroprotective effect of beta-D-glucan polysaccharide on hyperglycaemia-induced cerebral injury in diabetic animal model. Eur J Anat 26(2): 217–224. https://doi.org/10.26538/tjnpr/v5i12.24

Al-Dbass AM, Al-Daihan SK, Bhat RS (2012) *Agaricus blazei* Murill as an efficient hepatoprotective and antioxidant agent against CCl4-induced liver injury in rats. Saudi J Biol Sci 19:303–309. https://doi.org/10.1016/j.sjbs.2012.03.004

Ali MM, Baig MT, Huma A et al (2021) Effect of *Agaricus blazei* Murill on exploratory behavior of mice-model. Braz Biol 84:e252575. https://doi.org/10.1590/1519-6984.252575

Bach F, Zielinski AAF, Helm CV et al (2019) Bio compounds of edible mushrooms: *in vitro* antioxidant and antimicrobial activities. Food Sci Technol 107:214–220. https://doi.org/10.1016/j.lwt.2019.03.017

Bao Z, Yao L, Zhang X et al (2020) Isolation, purification, characterization, and immunomodulatory effects of polysaccharide from *Auricularia auricula* on RAW264.7 macrophages. J Food Biochem 44:e13516. https://doi.org/10.1111/jfbc.13516

Bernardshaw S, Hetland G, Ellertsen LK et al (2005) An extract of the medicinal mushroom *Agaricus blazei* Murill differentially stimulates production of pro-inflammatory cytokines in human monocytes and human vein endothelial cells *in vitro*. Inflammation 29:147–153. https://doi.org/10.1007/s10753-006-9010-2

Bovi M, Cenci L, Perduca M et al (2013) BEL-trefoil: a novel lectin with antineoplastic properties in king bolete (*Boletus edulis*) mushrooms. Glycobiology 23(5):578–592. https://doi.org/10.1093/glycob/cws164

Cao H, Ma S, Guo H et al (2019) Comparative study on the monosaccharide compositions, antioxidant and hypoglycemic activities in vitro of intracellular and extracellular polysaccharides of liquid fermented *Coprinus comatus*. Int J Biol Macromol 139:543–549. https://doi.org/10.1016/j.ijbiomac.2019.08.017

Carneiro AAJ, Ferreira ICFR, Dueñas M et al (2013) Chemical composition and antioxidant activity of dried powder formulations of *Agaricus blazei* and *Lentinus edodes*. Food Chem 138:2168–2173. https://doi.org/10.1016/j.foodchem.2012.12.036

Chang CW, Lur HS, Lu MK et al (2013) Sulfated polysaccharides of *Armillaria mellea* and their anti-inflammatory activities via NF-κB suppression. Food Res Int 54:239–245. https://doi.org/10.1016/j.foodres.2013.07.005

Chellappan D, Ganasen S, Batumalai S et al (2016) The protective action of the aqueous extract of *Auricularia polytricha* in paracetamol induced hepatotoxicity in rats. Recent Pat Drug Deliv Formul 10:72–76. https://doi.org/10.2174/1872211309666151030110015

Chen S, Oh SR, Phung S et al (2006) Anti-aromatase activity of phytochemicals in white button mushrooms (*Agaricus bisporus*). Cancer Res 66:12026–12034. https://doi.org/10.1158/0008-5472.CAN-06-2206

Chen YJ, Wu SY, Chen CC et al (2014) *Armillaria mellea* component armillarikin induces apoptosis in human leukemia cells. J Funct Foods 6:196–204. https://doi.org/10.1016/j.jff.2013.10.007

Chen YJ, Chen CC, Huang HL (2016) Induction of apoptosis by *Armillaria mellea* constituent armillarikin in human hepatocellular carcinoma. OncoTargets Ther 9:4773–4783. https://doi.org/10.2147/OTT.S103940

Chen S, Yong T, Xiao C et al (2018) Pyrrole alkaloids and ergosterols from *Grifola frondosa* exert anti-α-glucosidase and anti-proliferative activities. J Funct Foods 43:196–205. https://doi.org/10.1016/j.jff.2018.02.007

Chiu WC, Yang HH, Chiang SC et al (2014) *Auricularia polytricha* aqueous extract supplementation decreases hepatic lipid accumulation and improves antioxidative status in animal model of nonalcoholic fatty liver. Biomedicine 4:29–38. https://doi.org/10.7603/s40681-014-0012-3

Cruz A, Pimentel L, Rodríguez-Alcalá LM et al (2016) Health benefits of edible mushrooms focused on *Coriolus versicolor*: a review. J Food Nutr Res 4(12):773–781. https://doi.org/10.12691/jfnr-4-12-2

Cui FJ, Li Y, Xu YY et al (2007) Induction of apoptosis in SGC-7901 cells by polysaccharide-peptide GFPS1b from the cultured mycelia of *Grifola frondosa* GF9801. Toxicol In Vitro 21: 417–427. https://doi.org/10.1016/j.tiv.2006.10.004

Ding Z, Lu Y, Lu Z et al (2010) Hypoglycaemic effect of comatin, an anti-diabetic substance separated from *Coprinus comatus* broth, on alloxan-induced-diabetic rats. Food Chem 121(1): 39–43. https://doi.org/10.1016/j.foodchem.2009.12.001

Doğan HH, Akbaş G (2013) Biological activity and fatty acid composition of Caesar's mushroom. Pharm Biol 51:863–871. https://doi.org/10.3109/13880209.2013.768272

Dotan N, Wasser SP, Mahajna J (2011) The culinary-medicinal mushroom *Coprinus comatus* as a natural antiandrogenic modulator. Integr Cancer Ther 10(2):148–159. https://doi.org/10.1177/1534735410383169

Dou H, Chang Y, Zhang L (2019) *Coriolus versicolor* polysaccharopeptide as an immunotherapeutic in China. Prog Mol Biol Transl Sci 163:361–381. https://doi.org/10.1016/bs.pmbts.2019.03.001

Ekowati N, Yuniati NI, Hernayanti H et al (2018) Anti-diabetic potentials of button mushroom (*Agaricus bisporus*) on Alloxan-induced diabetic rats. Biosaintifika J Biol Biol Edu 10:655–662. https://doi.org/10.15294/biosaintifika.v10i3.17126

Erjavec J, Kos J, Ravnikar M et al (2012) Proteins of higher fungi – from forest to application. Trends Biotechnol 30:259–273. https://doi.org/10.1016/j.tibtech.2012.01.004

Facchini JM, Alves EP, Aguilera C et al (2014) Antitumor activity of *Pleurotus ostreatus* polysaccharide fractions on Ehrlich tumor and Sarcoma 180. Int J Bio Macromol 68:72–77. https://doi.org/10.1016/j.ijbiomac.2014.04.033

Fang X, Jiang Y, Ji H et al (2015) The synergistic beneficial effects of Ginkgo flavonoid and *Coriolus versicolor* polysaccharide for memory improvements in a mouse model of dementia. Evid Based Complement Altern Med 128394(9). https://doi.org/10.1155/2014/128394

Friedman M (2016) Mushroom polysaccharides: chemistry and antiobesity, antidiabetes, anticancer, and antibiotic properties in cells, rodents, and humans. Foods 5:80. https://doi.org/10.3390/foods5040080

Fukushima M, Ohashi T, Fujiwara Y et al (2001) Cholesterol-lowering effects of maitake (*Grifola frondosa*) fiber, shiitake (*Lentinus edodes*) fiber, and enokitake (*Flammulina velutipes*) fiber in rats. Exp Biol Med 226:758–765. https://doi.org/10.1177/153537020222600808

Gallego P, Luque-Sierra A, Falcon G et al (2021) White button mushroom extracts modulate hepatic fibrosis progression, inflammation, and oxidative stress *in vitro* and in LDLR$^{-/-}$ mice. Foods 10:1788. https://doi.org/10.3390/foods10081788

Gangarapu V, Yildiz K, İnce AT et al (2014) Role of gut microbiota: obesity and NAFLD. Turk J Gastroenterol 25:133–140. https://doi.org/10.5152/tjg.2014.7886

Gao Z, Kong D, Cai W et al (2021) Characterization and anti-diabetic nephropathic ability of mycelium polysaccharides from *Coprinus comatus*. Carbohydr Polym 251:117081. https://doi.org/10.1016/j.carbpol.2020.117081

Garcia J, Rodrigues F, Castro F et al (2022) Antimicrobial, Antibiofilm, and antioxidant properties of *Boletus edulis* and *Neoboletus luridiformis* against multidrug-resistant ESKAPE pathogens. Front Nutr 8:773346. https://doi.org/10.3389/fnut.2021.773346

Gashaw G, Fassil A, Redi F (2020) Evaluation of the antibacterial activity of *Pleurotus* spp. cultivated on different agricultural wastes in Chiro, Ethiopia. Int J Microbiol. https://doi.org/10.1155/2020/9312489

Geng Y, Zhu S, Cheng P et al (2017) Bioassay-guided fractionation of ethyl acetate extract from *Armillaria mellea* attenuates inflammatory response in lipopolysaccharide (LPS) stimulated BV-2 microglia. Phytomedicine 26:55–61. https://doi.org/10.1016/j.phymed.2017.01.005

Gobi VV, Rajasankar S, Johnson VMS et al (2019) Anti-apoptotic role of *Agaricus blazei* extract in rodent model of Parkinson's disease. Front Biosci 11:12–19. https://doi.org/10.2741/E842

Golak-Siwulska I, Kałużewicz A, Spiżewski T et al (2018a) Bioactive compounds and medicinal properties of Oyster mushrooms (sp.). Folia Hortic 30(2):191–201. https://doi.org/10.2478/fhort-2018-0012

Golak-Siwulska I, Kałuzewicz A, Wdowienko S et al (2018b) Nutritional value and health-promoting properties of *Agaricus bisporus* (Lange) Imbach. Herba Pol 64:71–81. https://doi.org/10.2478/hepo-2018-0027

Gu Y-H, Sivam G (2006) Cytotoxic effect of oyster mushroom *Pleurotus ostreatus* on human androgen independent prostate cancer PC-3 cells. J Med Food 9:196–204. https://doi.org/10.1089/jmf.2006.9.196

Guo Y, Deng G, Xu X et al (2012) Antioxidant capacities, phenolic compounds and polysaccharide contents of 49 edible macro-fungi. Food Funct 3(11):1195. https://doi.org/10.1039/C2FO30110E

Guo L, Qian X, Dai H et al (2020a) Antioxidant and antibacterial properties of extracts from different polar solvents from *Boletus edulis* Bull. Food Sci Technol Int 45(04):175–181

Guo WL, Deng JC, Pan YY et al (2020b) Hypoglycemic and hypolipidemic activities of *Grifola frondosa* polysaccharides and their relationships with the modulation of intestinal microflora in diabetic mice induced by high-fat diet and streptozotocin. Int J Biol Macromol 153:1231–1240. https://doi.org/10.1016/j.ijbiomac.2019.10.253

Habtemariam S (2020) *Trametes versicolor* (Synn. *Coriolus versicolor*) polysaccharides in cancer therapy: targets and efficacy. Biomedicine 8(5). https://doi.org/10.3390/biomedicines8050135

Harada N, Kodama N, Nanba H (2003) Relationship between dendritic cells and the D-fraction-induced Th-1 dominant response in BALB/c tumor-bearing mice. Cancer Lett 192:181–187. https://doi.org/10.1016/S0304-3835(02)00716-4

He X, Wang X, Fang J et al (2017) Polysaccharides in *Grifola frondosa* mushroom and their health promoting properties: a review. Int J Biol Macromol 101:910–921. https://doi.org/10.1016/j.ijbiomac.2017.03.177

Heleno SA, Ferreira RC, Antonio AL et al (2015) Nutritional value, bioactive compounds and antioxidant properties of three edible mushrooms from Poland. Food Biosci 11:48–55. https://doi.org/10.1016/j.fbio.2015.04.006

Hobbs C (2005) Medicinal value of Turkey tail fungus *Trametes versicolor* (L.: Fr.) Pilát (Aphyllophoromycetideae). Int J Med Mushrooms 7(3):346–347. https://doi.org/10.1615/IntJMedMushr.v6.i3.100

Hossen SM, Tanim MAH, Hossain MS et al (2021) Deciphering the CNS anti-depressant, antioxidant and cytotoxic profiling of methanol and aqueous extracts of *Trametes versicolor* and molecular interactions of its phenolic compounds. Saudi J Biol Sci 28(11):6375–6383. https://doi.org/10.1016/j.sjbs.2021.07.016

Hsu CH, Liao YL, Lin SC et al (2007) The mushroom *Agaricus brasiliensis* in combination with metformin and gliclazide improves insulin resistance in type 2 diabetes: a randomized, double-blind and placebo-controlled clinical trial. J Altern Complement Med 13:97–102. https://doi.org/10.1089/acm.2006.6054

Hu Q, Yu J, Yang W et al (2016) Identification of flavonoids from Flammulina velutipes and its neuroprotective effect on pheochromocytoma-12 cells. Food Chem 204:274–282. https://doi.org/10.1016/j.foodchem.2016.02.138

Hu XY, Liu CG, Wang X et al (2017) Hyperglycemic and antidiabetic nephritis activities of polysaccharides separated from *Auricularia auricular* in diet-streptozotocin-induced diabetic rats. Exp Ther Med 13:352–358. https://doi.org/10.3892/etm.2016.3943

Hu W, Li Z, Wang W et al (2021) Structural characterization of polysaccharide purified from *Amanita caesarea* and its pharmacological basis for application in Alzheimer's disease: endoplasmic reticulum stress. Food Funct 12:11009. https://doi.org/10.1039/d1fo01963e

Husen F, Hernayanti H, Ekowati N et al (2021) Antidiabetic effects and antioxidant properties of the saggy ink cap medicinal mushroom, *Coprinus comatus* (Agaricomycetes), in streptozotocin-induced hyperglycemic rats. Int J Med Mushrooms 23(10):9–21. https://doi.org/10.1615/IntJMedMushrooms.2021040020

Inoue A, Kodama N, Nanba H (2002) Effect of Maitake (*Grifola frondosa*) D-fraction on the control of the T lymph node Th-1/Th-2 proportion. Biol Pharm Bull 25:536–540. https://doi.org/10.1248/bpb.25.536

Ito H, Shimura K, Itoh H et al (1997) Antitumor effects of a new polysaccharide protein complex (ATOM) prepared from *Agaricus blazei* (Iwade strain 101) "Himematsutake" and its mechanism in tumor-bearing mice. Anticancer Res 17:277–284

Iwalokun BA, Usen UA, Otunba AA et al (2007) Comparative phytochemical evaluation, antimicrobial and antioxidant properties of *Pleurotus ostreatus*. Afr J Biotechnol 6(15):1732–1739

Janjusević L, Karaman M, Sibul F et al (2017) The lignicolous fungus *Trametes versicolor* (L.) Lloyd (1920): a promising natural source of antiradical and AChE inhibitory agents. J Enzyme Inhibit Med Chem 32(1):355–362. https://doi.org/10.1080/14756366.2016.1252759

Janjusevic L, Pejin B, Kaisarevic S et al (2018) *Trametes versicolor* ethanol extract, a promising candidate for health–promoting food supplement. Nat Prod Res 32(8):963–967. https://doi.org/10.1080/14786419.2017.1366484

Jayakumar T, Thomas PA, Geraldine P (2007) Protective effect of an extract of the oyster mushroom, *Pleurotus ostreatus*, on antioxidants of major organs of aged rats. Exp Gerontol 42:183–191. https://doi.org/10.1016/j.exger.2006.10.006

Jayakumar T, Thomas PA, Sheu JR et al (2011) *In-vitro* and *in-vivo* antioxidant effects of the oyster mushrooms *P. ostrea*tus. Food Res Int 44:851–861. https://doi.org/10.1016/j.foodres.2011.03.015

Jedinak A, Sliva D (2008) *Pleurotus ostreatus* inhibits proliferation of human breast and colon cancer cells through p53-dependent pathway. Int J Oncol 33:1307–1313. https://doi.org/10.3892/ijo_00000122

Jedinak A, Dudhgaonkar S, Jiang J et al (2010) *Pleurotus ostreatus* inhibits colitis related colon carcinogenesis in mice. Int J Mol Med 26:643–650. https://doi.org/10.3892/ijmm_00000509

Jeong SC, Jeong YT, Yang BK et al (2010) White button mushroom (*Agaricus bisporus*) lowers blood glucose and cholesterol levels in diabetic and hypercholesterolemic rats. Nutr Res 30:49–56. https://doi.org/10.1016/j.nutres.2009.12.003

Jeong SC, Koyyalamudi SR, Jeong YT et al (2012) Macrophage immunomodulating and antitumor activities of polysaccharides isolated from *Agaricus bisporus* white button mushrooms. J Med Food 15:58–65. https://doi.org/10.1089/jmf.2011.1704

Ji H, Huang H, Chao J et al (2014) Protective effect of *Agaricus brasiliensis* on STZ-induced diabetic neuropathic pain in rats. Evid Based Complement Alternat Med 2014:1. https://doi.org/10.1155/2014/679259

Jiang T, Shen S, Wang L et al (2020) *Grifola frondosa* polysaccharide ameliorates early diabetic nephropathy by suppressing the TLR4/NF-κB pathway. Appl Biochem Biotechnol 194:4093. https://doi.org/10.1007/s12010-022-03976-8

Jin C, Choi YH, Moon D et al (2006) Induction of G2/Marrest and apoptosis in human gastric epithelial AGS cells by aqueous extract of *Agaricus blazei*. Oncol Rep 16:1349–1355. https://doi.org/10.3892/or.16.6.1349

Jo WS, Kang MJ, Choi SY et al (2010) Culture conditions for mycelial growth of *Coriolus versicolor*. Mycobiology 38:195–202. https://doi.org/10.4489/MYCO.2010.38.3.195

Johnson E, Forland DT, Saetre L et al (2009) Effect of an extract based on the medicinal mushroom *Agaricus blazei* Murill on release of cytokines, chemokines and leukocyte growth factors in human blood *ex vivo* and *in vivo*. Scand J Immunol 69:242–250. https://doi.org/10.1111/j.1365-3083.2008.02218.x

Kang MA, Jeon YK, Nam MJ (2020) *Auricularia auricula* increases an apoptosis in human hepatocellular carcinoma cells via a regulation of the peroxiredoxin1. J Food Biochem 44: e13373. https://doi.org/10.1111/jfbc.13373

Karasoy AF, Okuyucu H, Pekşen A (2019) *Flammulina velutipes* mantarı. J Fungus 10:152–162

Kim SK, Im J, Yun CH et al (2008) *Armillaria mellea* induces maturation of human dendritic cells without induction of cytokine expression. J Ethnopharmacol 119:153–159. https://doi.org/10.1016/j.jep.2008.06.032

Kim MO, Moon DO, Jung JM et al (2011) *Agaricus blazei* extract induces apoptosis through ROS-dependent JNK activation involving the mitochondrial pathway and suppression of constitutive NF-κB in THP-1 cells. Evid Based Complement Alternat Med. https://doi.org/10.1093/ecam/nep176

Kodama N, Harada N, Nanba H (2002) A polysaccharide extract from Grifola frondosa, induces Th-1 dominant responses in carcinoma-bearing BALB/cmice. Jpn J Pharmacol 90:357–360. https://doi.org/10.1254/jjp.90.357

Kodama N, Murata Y, Nanba H (2004) Administration of a polysaccharide from *Grifola frondosa* stimulates immune function of normal mice. J Med Food 7:141–145. https://doi.org/10.1089/1096620041224012

König S, Romp E, Krauth V et al (2019) Melleolides from honey mushroom inhibit 5-lipoxygenase via Cys159. Cell Chem Biol 26:60–70. https://doi.org/10.1016/j.chembiol.2018.10.010

Kozarski M, Klaus A, Nikšić M et al (2012) Antioxidative activities and chemical characterization of polysaccharide extracts from the widely used mushrooms *Ganoderma applanatum, Ganoderma lucidum, Lentinus edodes* and *Trametes versicolor*. J Food Compost Anal 26(1–2):144–153. https://doi.org/10.1016/j.jfca.2012.02.004

Kubo K, Nanba H (1997) Anti-hyperliposis effect of Maitake fruit body (*Grifola frondosa*). Biol Pharm Bull 20:781–785. https://doi.org/10.1248/bpb.20.781

Lemieszek MK, Cardoso C, Nunes FHFM et al (2013) *Boletus edulis* biological active biopolymers induce cell cycle arrest in human colon adenocarcinoma cells. Food Funct 4:575–585. https://doi.org/10.1039/C2FO30324H

Lemieszek MK, Ribeiro M, Alves HG et al (2016) *Boletus edulis* ribonucleic acid–a potent apoptosis inducer in human colon adenocarcinoma cells. Food Funct 7:3163–3175. https://doi.org/10.1039/C6FO00132G

Leung MYK, Fung KP, Choy YM (1997) The isolation and characterization of an immunomodulatory and anti-tumor polysaccharide preparation from *Flammulina velutipes*. Int Immunopharmacol 35:255–263. https://doi.org/10.1016/S0162-3109(96)00157-9

Li Z, Wang Y, Jiang B et al (2016) Structure, cytotoxic activity and mechanism of protoilludane sesquiterpene aryl esters from the mycelium of *Armillaria mellea*. J Ethnopharmacol 184:119–127. https://doi.org/10.1016/j.jep.2016.02.044

Li Z, Chen X, Zhang Y et al (2019) Protective roles of *Amanita caesarea* polysaccharides against Alzheimer's disease via Nrf2 pathway. Int J Biol Macromol 121:29–37. https://doi.org/10.1016/j.ijbiomac.2018.09.216

Li W, Wang Y, Sun M et al (2020a) The prebiotic-like effects of *Coprinus comatus* polysaccharides on gut microbiota in normal mice and those with acute alcoholic liver injury: a comparative study. Evid Based Complement Altern. https://doi.org/10.1155/2020/2027570

Li Y, Sheng Y, Lu X et al (2020b) Isolation and purification of acidic polysaccharides from *Agaricus blazei* Murill and evaluation of their lipid lowering mechanism. Int J Biol Macromol 157:276–287. https://doi.org/10.1016/j.ijbiomac.2020.04.190

Lima CUJO, Cordova COA, Nobrega OT et al (2011) Does the *Agaricus blazei* Murill mushroom have properties that affect the immune system? An integrative review. J Med Food 14:2–8. https://doi.org/10.1089/jmf.2010.0017

Liu J, Jia L, Kan J et al (2013) *In vitro* and *in vivo* antioxidant activity of ethanolic extract of white button mushroom (*Agaricus bisporus*). Food Chem Toxicol 51:310–316. https://doi.org/10.1016/j.fct.2012.10.014

Liu L, Li M, Yu M et al (2019) Natural polysaccharides exhibit anti-tumor activity by targeting gut microbiota. Int J Biol Macromol 121:743–751. https://doi.org/10.1016/j.ijbiomac.2018.10.083

Liu Q, Ma S, Li S et al (2022) Dietary supplementation of *Auricularia auricula-judae* polysaccharides alleviate nutritional obesity in mice via regulating inflammatory response and lipid metabolism. Foods 11:942. https://doi.org/10.3390/foods11070942

Luo A, Luo A, Huang J et al (2012) Purification, characterization, and antioxidant activities *in vitro* and *in vivo* of the polysaccharides from *Boletus edulis* Bull. Molecules 17(7):8079–8090. https://doi.org/10.3390/molecules17078079

Ma X, Zhou F, Chen Y et al (2014) A polysaccharide from *Grifola frondosa* relieves insulin resistance of HepG2 cell by Akt-GSK-3pathway. Glycoconj J 31:355–363. https://doi.org/10.1007/s10719-014-9526-x

Mapoung S, Umsumarng S, Semmarath W et al (2021) Skin wound-healing potential of polysaccharides from medicinal mushroom *Auricularia auricula-judae* (Bull.). J Fungi 7:247

Masuda Y, Inoue M, Miyata A et al (2009a) Maitake β-glucan enhances therapeutic effect and reduces myelosupression and nephrotoxicity of cisplatin in mice. Int Immunopharmacol 9:620–626. https://doi.org/10.1016/j.intimp.2009.02.005

Masuda Y, Matsumoto A, Toida T et al (2009b) Characterization and antitumor effect of a novel polysaccharide from *Grifola frondosa*. J Agric Food Chem 57:10143–10149. https://doi.org/10.1021/jf9021338

Matsushita Y, Furutani Y, Matsuoka R et al (2018) Hot water extract of *Agaricus blazei* Murrill specifically inhibits growth and induces apoptosis in human pancreatic cancer cells. BMC Complement Alternat Med 18:319. https://doi.org/10.1186/s12906-018-2385-4

Mayell M (2001) Maitake extracts and their therapeutic potential. Altern Med Rev 6:48–60

Meng Y, Yan J, Yang G et al (2018) Structural characterization and macrophage activation of a hetero-galactan isolated from *Flammulina velutipes*. Carbohydr Polym 183:207–218. https://doi.org/10.1016/j.carbpol.2017.12.017

Meng T, Yu S, Ji H et al (2021) A novel acid polysaccharide from *Boletus edulis*: extraction, characteristics, and antitumor activities *in vitro*. Glycoconj J 38:13–24. https://doi.org/10.1007/s10719-021-09972-0

Mitra P, Khatua S, Acharya K (2013) Free radical scavenging and NOS activation properties of water soluble crude polysaccharides from *Pleurotus ostreatus*. Asian J Pharm Clin Res 6(3): 67–70

Nowakowski P, Markiewicz-Zukowska R, Gromkowska-Kępka K et al (2021) Mushrooms as potential therapeutic agents in the treatment of cancer: evaluation of anti-glioma effects of *Coprinus comatus, Cantharellus cibarius, Lycoperdon perlatum* and *Lactarius deliciosus* extracts. Biomed Pharmacother 133:111090. https://doi.org/10.1016/j.biopha.2020.111090

Oli AN, Edeh PA, Al-Mosawi RM et al (2020) Evaluation of the phytoconstituents of *Auricularia auricula-judae* mushroom and antimicrobial activity of its protein extract. Eur J Integr Med 38: 101176. https://doi.org/10.1016/j.eujim.2020.101176

Park KH, Lee ES, Jin YI et al (2016) Inhibitory effect of *Panax ginseng* and *Pleurotus ostreatus* complex on expression of cytokine genes induced by extract of Dermatophagoides pteronyssinus in human monocytic THP-1 and EoL-1 cells. J Mushroom Sci Prod 14(4): 155–161. https://doi.org/10.14480/JM.2016.14.4.155

Park HJ, Yun J, Jang S-H et al (2020) *Coprinus comatus* cap inhibits adipocyte differentiation via regulation of PPARc and Akt signaling pathway. PLoS One 9(9):e105809. https://doi.org/10.1371/journal.pone.0105809

Patel Y, Naraian R, Singh VK (2012) Medicinal properties of *Pleurotus* species (oyster mushroom): a review. WJFPB 3(1):1–12. https://doi.org/10.5829/idosi.wjfpb.2012.3.1.303

Perera N, Yang FH, Chiu HW et al (2020) Phagocytosis enhancement, endotoxin tolerance, and signal mechanisms of immunologically active glucuronoxylomannan from *Auricularia auricula-judae*. Int J Biol Macromol 165:495–505. https://doi.org/10.1016/j.ijbiomac.2020.09.171

Piska K, Sułkowska-Ziaja K, Muszyńska B (2017) Edible mushroom *Pleurotus ostreatus* (oyster mushroom): its dietary significance and biological activity. Acta Sci Pol 16(1)

Preuss HG, Echard B, Bagchi D et al (2010) Maitake mushroom extracts ameliorate progressive hypertension and other chronic metabolic perturbations in aging female rats. Int J Med Sci 7: 169. https://doi.org/10.7150/ijms.7.169

Raseta M, Popović M, Knezevic P et al (2020) Bioactive phenolic compounds of two medicinal mushroom species *Trametes versicolor* and *Stereum subtomentosum* as antioxidant and antiproliferative agents. Chem Biodivers 17(12):e2000683. https://doi.org/10.1002/cbdv.202000683

Ratnaningtyas NI, Hernayanti EN, Sukmawati D et al (2019) Chicken drumstick mushroom (*Coprinus comatus)* ethanol extract exerts a hypoglycaemic effect in the Rattus norvegicus model of diabetes. Biocatal Agric Biotechnol 19:1–9. https://doi.org/10.1016/j.bcab.2019.101050

Ratnaningtyas NI, Hernayanti EN, Ekowati N et al (2022) Ethanol extract of the mushroom *Coprinus comatus* exhibits antidiabetic and antioxidant activities in streptozotocin-induced diabetic rats. Pharm Biol 60:1126–1113. https://doi.org/10.1080/13880209.2022.2074054

Reis FS, Martins A, Barros L et al (2012) Antioxidant properties and phenolic profile of the most widely appreciated cultivated mushrooms: a comparative study between *in vivo* and *in vitro*. Food Chem Toxicol 50:1201–1207. https://doi.org/10.1016/j.fct.2012.02.013

Rouhana-Toubi A, Wasser SP, Fares F (2015) The shaggy ink cap medicinal mushroom, *Coprinus comatus* (higher basidiomycetes) extract induces apoptosis in ovarian cancer cells via extrinsic and intrinsic apoptotic pathways. Int J Med Mushrooms 17:1127–1136. https://doi.org/10.1615/intjmedmushrooms.v17.i12.20

Shen KP, Su CH, Lu TM et al (2015) Effects of *Grifola frondosa* non-polar bioactive components on high-fat diet fed and streptozotocin-induced hyperglycemic mice. Pharm Biol 53:705–709. https://doi.org/10.3109/13880209.2014.939290

Sihoglu Tepe A (2021) Chemical compositions and antioxidant activities of four different mushroom species collected from Turkey. Int J Second Metab 8:214–226. https://doi.org/10.21448/ijsm.953923

Smiderle FR, Alquini G, Tadra-Sfeir MZ et al (2013) *Agaricus bisporus* and *Agaricus brasiliensis* (1→6)-β-d-glucans show immunostimulatory activity on human THP-1 derived macrophages. Carbohydr Polym 94:91–99. https://doi.org/10.1016/j.carbpol.2012.12.073

Song HH, Chae HS, Oh SR et al (2012) Anti-inflammatory and antiallergic effect of *Agaricus blazei* extract in bone marrow-derived mast cells. Am J Chin Med 40:1073–1084. https://doi.org/10.1142/S0192415X12500796

Stilinović N, Čapo I, Vukmirović S et al (2020) Chemical composition, nutritional profile and in vivo antioxidant properties of the cultivated mushroom *Coprinus comatus*. R Soc Open Sci 7: 200900. https://doi.org/10.1098/rsos.200900

Su CH, Lu TM, Lai MN et al (2013) Inhibitory potential of *Grifola frondosa* bioactive fractions on α-amylase and α-glucosidase for management of hyperglycemia. Biotechnol Appl Biochem 60: 446–452. https://doi.org/10.1002/bab.1105

Su A, Yang W, Zhao L et al (2018) Flammulina velutipes polysaccharides improve scopolamine-induced learning and memory impairment in mice by modulating gut microbiota composition. Food Funct 9:1424–1432. https://doi.org/10.1039/C7FO01991B

Sun C, Rosendahl AH, Wang XD et al (2012) Polysaccharide-K (PSK) in cancer - old story, new possibilities? Curr Med Chem 19(5):757–762. https://doi.org/10.2174/092986712798992020

Sun X, Zhang T, Zhao Y et al (2020) Protoilludane sesquiterpenoid aromatic esters from *Armillaria mellea* improve depressive-like behavior induced by chronic unpredictable mild stress in mice. J Funct Food 66:103799. https://doi.org/10.1016/j.jff.2020.103799

Tangen JM, Tierens A, Caers J et al (2015) Immunomodulatory effects of the *Agaricus blazei* Murrill-based mushroom extract AndoSan in patients with multiple myeloma undergoing high dose chemotherapy and autologous stem cell transplantation: a randomized, double blinded clinical study. Biomed Res Int 2015:718539. https://doi.org/10.1155/2015/718539

Therkelsen SP, Hetland G, Lyberg T et al (2016) Cytokine levels after consumption of a medicinal *Agaricus blazei* Murill-based mushroom extract, AndoSanTM, in patients with Crohn's disease and ulcerative colitis in a randomized single-blinded placebo controlled study. Scand J Immunol 84:323–331. https://doi.org/10.1111/sji.12476

Valenti MT, Marchetto G, Perduca M et al (2020) BEL β-trefoil reduces the migration ability of RUNX2 expressing melanoma cells in xenotransplanted zebrafish. Molecules 25:1270. https://doi.org/10.3390/molecules25061270

Venkatakrishnan V, Shenbhagaraman R, Kaviyarasan V et al (2010) Antioxidant and antiproliferative effect of *Pleurotus ostreatus*. J Phytology 2(1):22–28

Vitak TY, Wasser SP, Nevo E et al (2015) The effect of the medicinal mushrooms *Agaricus brasiliensis* and *Ganoderma lucidum* (higher basidiomycetes) on the erythron system in normal and streptozotocin-induced diabetic rats. Int J Med Mushrooms 17:277–286. https://doi.org/10.1615/IntJMedMushrooms.v17.i3.70

Volman JJ, Helsper JPFG, Wei S et al (2009) Effects of mushroom derived beta-glucan-rich polysaccharide extracts on NO production by bone marrow-derived macrophages and NF-kappaB transactivation in Caco-2 reporter cells: can effects be explained by structure? Mol Nutr Food Res 54:268–276. https://doi.org/10.1002/mnfr.200900009

Wan JM-F (2013) Polysaccharide krestin (PSK) and polysaccharopeptide PSP. In: Kastin AJ (ed) Handbook of biologically active peptides, vol 27. Elsevier, New York, pp 180–184. https://doi.org/10.1016/B978-0-12-385095-9.00027-0

Wang H, Ng TB (2000) Isolation of a novel ubiquitin-like protein from *Pleurotus ostreatus* mushroom with ant-human immunodeficiency virus, translation inhibitory and ribonuclease activities. Biochem Biophys Res Commun 276:587–593. https://doi.org/10.1006/bbrc.2000.3540

Wang C, Ju L, Liu M et al (2007) Effects of *Auricularia auricula* polysaccharide on the expression of MMP-13 in rabbit arteriosclerosis plaque. Chin Rem Clin 14:161–168. https://doi.org/10.1021/acs.jafc.0c05934

Wang CL, Meng M, Liu SB et al (2013a) A chemically sulfated polysaccharide from *Grifola frondosa* induces HepG2 cell apoptosis by notch1-NF-κB pathway. Carbohydr Polym 95:282–287. https://doi.org/10.1016/j.carbpol.2013.02.057

Wang P, Li XT, Sun L et al (2013b) Anti-inflammatory activity of water-soluble polysaccharide of *Agaricus blazei* Murill on ovariectomized osteopenic rats. Evid Based Complement Alternat Med 2013:164817. https://doi.org/10.1155/2013/164817

Wang Y, Fang J, Ni X et al (2013c) Inducement of cytokine release by GFPBW2, a novel polysaccharide from fruit bodies of *Grifola frondosa*, through dectin-1 in macrophages. J Agric Food Chem 61:11400–11409. https://doi.org/10.1021/jf4029915

Wang D, Sun SQ, Wu WZ et al (2014a) Characterization of a water-soluble polysaccharide from *Boletus edulis* and its antitumor and immunomodulatory activities on renal cancer in mice. Carbohydr Polym 105:127–134. https://doi.org/10.1016/j.carbpol.2013.12.085

Wang Y, Shen X, Liao W et al (2014b) A heteropolysaccharide, L-fuco-D-manno-1, 6-α-D-galactan extracted from *Grifola frondosa* and antiangiogenic activity of its sulfated derivative. Carbohydr Polym 101:631–641. https://doi.org/10.1016/j.carbpol.2013.09.085

Wang J, Hu S, Nie S et al (2016) Reviews on mechanisms of in vitro antioxidant activity of polysaccharides. Oxid Med Cell Longev 2016:5692852. https://doi.org/10.1155/2016/5692852

Wei Q, Zhan Y, Chen B et al (2020) Assessment of antioxidant and antidiabetic properties of *Agaricus blazei* Murill extracts. Food Sci Nutr 8:332–339. https://doi.org/10.1002/fsn3.1310

Wiater A, Paduch R, Pleszczyńska M et al (2011) α-(1→3)-D-glucans from fruiting bodies of selected macromycetes fungi and the biological activity of their carboxymethylated products. Biotechnol Lett 33:787–795. https://doi.org/10.1007/s10529-010-0502-7

Wu SJ, Tsai JY, Lai MN et al (2007) *Armillaria mellea* shows antiinflammatory activity by inhibiting the expression of NO, iNOS, COX-2 and cytokines in THP-1 cells. Am J Chin Med 35:507–516. https://doi.org/10.1142/S0192415X07005028

Wu Q, Tan ZP, Liu HD et al (2010) Chemical characterization of *Auricularia auricula* polysaccharides and its pharmacological effect on heart antioxidant enzyme activities and left ventricular function in aged mice. Int J Biol Macromol 46:284–288. https://doi.org/10.1016/j.ijbiomac.2010.01.016

Wu J-Y, Chen C-H, Chang W-H et al (2011) Anti-cancer effects of protein extracts from *Calvatia lilacina, Pleurotus ostreatus* and *Volvariella volvacea*. Evid Based Complement Altern Med. https://doi.org/10.1093/ecam/neq057

Wu S, Wang G, Yang R et al (2016) Anti-inflammatory effects of *Boletus edulis* polysaccharide on asthma pathology. Am J Transl Res 8(10):4478–4489

Wu JY, Siu KC, Geng P (2021) Bioactive ingredients and medicinal values of *Grifola frondosa* (Maitake). Foods 10:95. https://doi.org/10.3390/foods10010095

Xiao C, Wu Q, Xie Y et al (2015) Hypoglycemic effects of *Grifola frondosa* (Maitake) polysaccharides F2 and F3 through improvement of insulin resistance in diabetic rats. Food Funct 6:3567–3575. https://doi.org/10.1039/C5FO00497G

Xiao Y, Xu Q, Zhou X et al (2018) The improvement effect of *Boletus edulis* polysaccharides on the liver injury in type 2 diabetic rats. Food Sci Technol 39(11):297–313

Xu H, Yang YJ, Yang T et al (2013) Statins and stem cell modulation. Ageing Res Rev 12:1–7. https://doi.org/10.1016/j.arr.2012.03.006

Xu Y, Zhang Z, Wang B et al (2022) *Flammulina velutipes* polysaccharides modulate gut microbiota and alleviate carbon tetrachloride-induced hepatic oxidative injury in mice. Front Microbiol 13:847653. https://doi.org/10.3389/fmicb.2022.847653

Yamaç M, Kanbak G, Zeytinoğlu M et al (2010) Pancreas protective effect of button mushroom *Agaricus bisporus* (J.E. Lange) Imbach (Agaricomycetidae) extract on rats with Streptozotocin-induced diabetes. Int J Med Mushrooms 12(4):379–389. https://doi.org/10.1615/IntJMedMushr.v12.i4.50

Yang S, Meng Y, Yan J et al (2019) Polysaccharide-enriched fraction from *Armillaria mellea* fruiting body improves insulin resistance. Molecules 24:46. https://doi.org/10.3390/molecules24010046

Ye J, Wang X, Wang K et al (2020) A novel polysaccharide isolated from *Flammulina velutipes*, characterization, macrophage immunomodulatory activities and its impact on gut microbiota in rats. J Anim Physiol Anim Nutr (Berl) 104:735–748. https://doi.org/10.1111/jpn.13290

Yin H, Wang Y, Wang T et al (2010) Purification, characterization and immuno-modulating properties of polysaccharides isolated from *Flammulina velutipes* mycelium. Am J Chin Med 38:191–204. https://doi.org/10.1142/S0192415X10007750

Yu CH, Kan SF, Shu CH et al (2009) Inhibitory mechanisms of *Agaricus blazei* Murill on the growth of prostate cancer *in vitro* and *in vivo*. J Nutr Biochem 20:753–764. https://doi.org/10.1016/j.jnutbio.2008.07.004

Yu J, Sun RL, Zhao ZQ et al (2014) *Auricularia polytricha* polysaccharides induce cell cycle arrest and apoptosis in human lung cancer A549 cells. Int J Biol Macromol 68:67–71. https://doi.org/10.1016/j.ijbiomac.2014.04.018

Zavastin DE, Mircea C, Aprotosoaie AC et al (2015) *Armillaria mellea*: phenolic content, in vitro antioxidant and antihyperglycemic effects. Rev Med Chir Soc Med Nat Iasi 119(1):273–280. https://doi.org/10.3390/ijms12106367

Zeng WC, Zhang Z, Gao H et al (2012) Characterization of antioxidant polysaccharides from *Auricularia auricular* using microwave-assisted extraction. Carbohydr Polym 89:694–700. https://doi.org/10.1016/j.carbpol.2012.03.078

Zeng F, Zhao C, Pang J et al (2013) Chemical properties of a polysaccharide purified from solid-state fermentation of *Auricularia auricular* and its biological activity as a hypolipidemic agent. J Food Sci 78:H1470–H1475. https://doi.org/10.1111/1750-3841.12226

Zhang Y, Dai L, Kong X et al (2012) Characterization and in vitro antioxidant activities of polysaccharides from *Pleurotus ostreatus*. Int J Biol Macromol 51(3):259–265. https://doi.org/10.1016/j.ijbiomac.2012.05.003

Zhang Y, Li H, Yang X et al (2018) Cognitive-enhancing effect of polysaccharides from *Flammulina velutipes* on Alzheimer's disease by compatibilizing with ginsenosides. Int J Biol Macromol 112:788–795. https://doi.org/10.1016/j.ijbiomac.2017.12.149

Zhang Y, Zhou R, Liu F et al (2021) Purification and characterization of a novel protein with activity against non-small-cell lung cancer in vitro and *in vivo* from the edible mushroom *Boletus edulis*. Int J Biol Macromol 174:77–88. https://doi.org/10.1016/j.ijbiomac.2021.01.149

Zhao R, Ji Y, Chen X et al (2020) Effects of a β-type glycosidic polysaccharide from *Flammulina velutipes* on anti-inflammation and gut microbiota modulation in colitis mice. Food Funct 11:4259–4274. https://doi.org/10.1039/C9FO03017D

Zheng Q, Zhang H, Li W et al (2019) Hepatoprotective effect of *Boletus edulis* polysaccharides on mice with acute hepatic damage. Food Mach 35(12):141–145

Nephroprotective Effects of Four *Ganoderma* Species

Rishi M. Nadar, Jack Deruiter, Suhrud Pathak, Surekha Kadannagari, Jahnavi Dhingra, Satyanarayana Pondugula, Timothy Moore, Dinesh Chandra Agrawal, and Muralikrishnan Dhanasekaran

Contents

Abstract Acute kidney injury (AKI) and chronic kidney disease (CKD) processes have been researched extensively regarding their etiopathology. Nevertheless, AKI- and CKD-associated morbidity and mortality have grown in recent years. Natural bioactives are increasingly being investigated as alternative medicines and nutritional supplements for addressing renal pathologies in humans and animals. In addition, natural bioactives have recently been acknowledged as an alternate source for treating renal pathologies due to their traditional practice and multi-target prophylactic and therapeutic properties. *Ganoderma* bioactives have diverse bio-targets and numerous pharmacodynamic actions that can avoid nephron insults and therefore be utilized to treat various kidney pathologies. *Ganoderma lucidum*, an

R. M. Nadar · J. Deruiter · S. Pathak · S. Kadannagari · J. Dhingra · T. Moore · M. Dhanasekaran (✉)
Department of Drug Discovery and Development, Harrison College of Pharmacy, Auburn University, Auburn, AL, USA
e-mail: dhanamu@auburbn.edu

S. Pondugula
Department of Anatomy, Physiology and Pharmacology, Auburn University, Auburn, AL, USA

D. C. Agrawal (✉)
Department of Applied Chemistry, Chaoyang University of Technology, Taichung, Taiwan
e-mail: dcagrawal@gm.cyut.edu.tw

D. C. Agrawal, M. Dhanasekaran (eds.), *Mushrooms with Therapeutic Potentials*,
https://doi.org/10.1007/978-981-19-9550-7_15

efficient medication and valuable food, is recognized as one of the most highly valued traditional Chinese medicines owing to its ability to improve human health. *G. lucidum* has been used as a nutraceutical and alternative medicine for ages to promote health and treat various ailments. Also, *G. lucidum* prevents and treats numerous kidney disorders due to its pharmacokinetic and therapeutic beneficial properties, such as antioxidant, anti-inflammatory, antitumor, and other pharmacodynamic actions. This chapter analyzes several studies and provides insights into *G. lucidum*'s nephroprotective activities.

Keywords Acute kidney injury · Chronic kidney disease · *Ganoderma lucidum* · Mycotherapeutics · Natural bioactives nephroprotection · Nephrotoxicity · Renal damage

Abbreviations

ADPKD	Autosomal dominant polycystic kidney disease
AKI	Acute kidney injury
CCl4	Carbon tetrachloride
CKD	Chronic kidney disease
CNT2	Concentrative nucleoside transporter 2
DNA	Deoxyribonucleic acid
FASDs	Fetal alcohol spectrum disorders
GFR	Glomerular filtration rate
GLUT9	Glucose transporter 9
HIV	Human immunodeficiency virus
HKC-8 cells	Human-derived renal proximal tubular cell line
OAT1 transporter	Organic anion transporter 1
PKD	Polycystic kidney disease
RIRI	Renal ischemia-reperfusion injury
TGF-β	Transforming growth factor-β
URAT1	Urate transporter
UUO	Unilateral ureteral obstruction

1 Introduction

Ganoderma, polypore fungi, is a genus that belongs to the family *Ganodermataceae*. They are large, dark mushrooms with a shiny surface and a woody texture and can be readily distinguished from other polypores due to a double-walled basidiospore (Sun et al. 2022). *Ganoderma* comprises around 80 species, mainly seen in tropical regions (Luangharn et al. 2021). One species, *Ganoderma lucidum*, is so named because the word *lucidum* is derived from the Latin word "lucidus," which refers to

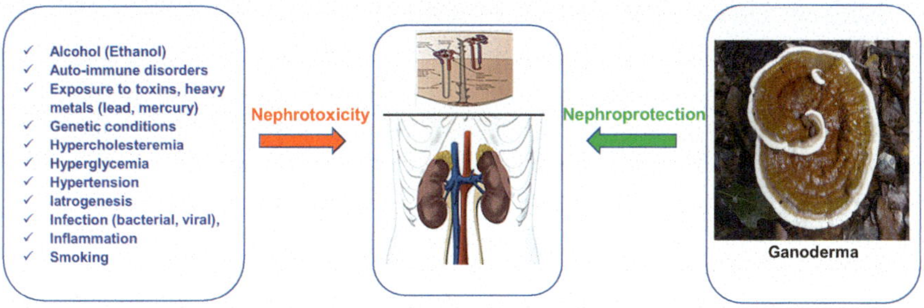

Fig. 1 *Ganoderma* species for treating several renal diseases. OpenStax College, CC BY 3.0 https://creativecommons.org/licenses/by/3.0, via Wikimedia Commons. Artwork by Holly Fischer, CC BY 3.0 https://creativecommons.org/licenses/by/3.0, via Wikimedia Commons. English: NPS Photo, Public domain, via Wikimedia Commons

"shiny" or "brilliant" that applies to the polished appearance of the mushroom exterior. Additionally, *Ganoderma lucidum*, an Asian fungus, has a significant genetic diversity and is used in traditional remedies in Asia, including China and Japan (Wachtel-Galor et al. 2011).

Historically, *Ganoderma* has a lengthy consumption record for fostering well-being and longevity. It is claimed to possess a sequence of divine and transcendent power combined with the fundamental nature of "immortality" (Wachtel-Galor et al. 2011). Therefore, it is deemed the "natural bioactive with sacred vigor/strength," which embodies achievement, happiness, health, celestial power, endurance, and longevity (Wachtel-Galor et al. 2011). Among the nurtured/cultivated fungi, *G. lucidum* is exceptional because it is safe and possesses medicinal value rather than nutritional value (Wachtel-Galor et al. 2011). Currently, numerous health-related commodities, nutraceuticals, and pharmaceutical products incorporate *G. lucidum* as the main ingredient or include it as an additive or supplement (Wong et al. 2020). These marketed commercial products are available in different dosage forms, including powders, dietary supplements, and tea. Commercial health-related commodities, nutraceuticals, and pharmaceutical products are produced from distinct mushroom parts, including mycelia, spores, and fruit bodies (Wachtel-Galor et al. 2011). The specific health-related applications and recognized health benefits include the treatment of cancer, infectious diseases, and inflammatory disorders, control of blood glucose levels, modulation of the immune system, gastrointestinal/hepatoprotection, and nephroprotection (Wachtel-Galor et al. 2011).

Ganoderma species have been used for centuries as nutraceuticals and alternative medicine to improve health and treat numerous renal diseases (Fig. 1). The plethora of beliefs concerning *Ganoderma*'s therapeutic and restorative benefits have been established essentially based on anecdotal evidence, conventional/customary consumption/use, and societal/cultural morals. Kidney diseases and renal pathologies are escalating globally; therefore, novel therapies are required to prevent and treat these lethal disorders. Hence, this book chapter reveals the nephroprotective role of *Ganoderma* (Geng et al. 2020a; Dou et al. 2014).

2 Kidney Diseases/Renal Pathologies

The kidneys are a pair of bean-shaped organs (4–5 inches long) located on either side of the spine, below the ribs in the abdominal cavity. Physiologically, the kidney removes the metabolic waste products (urea), excretes drugs and their metabolites, balances the body's electrolytes and fluids (sodium, potassium, and calcium), releases renin that regulates blood pressure, synthesizes the active form of vitamin D that promotes strong and healthy bones, and controls the production of erythropoietin. Erythropoietin is a hormone needed to produce and mature erythrocytes (red blood cells) (Schoener and Borger 2022). The basic functional units of the kidney are nephrons (tiny filters), and each kidney has millions of nephrons. The pathological conditions or disease states that affect the kidney can be generally classified as autoimmune, congenital, genetic, iatrogenic (drug-induced), infection, inflammation, metabolic, toxin (exogenous and endogenous), tumor, vascular, and others (Table 1). The primary major kidney/renal pathologies are shown in Table 2.

These renal pathologies have unique features and symptoms, such as where the nephron/other parts of the kidney (all or portion) are injured and is the injury reversible or irreversible, which is associated with acute/chronic renal failure. The renal insult or injury can occur due to autoimmune disorders, inflammation, infection (bacterial, viral), alcohol, smoking, hyperglycemia, hypercholesteremia, hypertension, genetic conditions, and exposure to toxins and metals (lead, mercury) which can lead to reversible or progressive, partial loss or permanent, renal function (Table 3).

Renal damage or kidney failure can lead to increased morbidities and mortality. Thus, there is a need for alternative and complementary remedies with or in addition to the current synthetic drugs or natural bioactives. The following section reveals the nephroprotective effects of the *Ganoderma* species.

Table 1 Classification of renal pathologies

Autoimmune (antibody)
Congenital
Genetic
Iatrogenic (drug-induced)
Infection
Inflammation
Metabolic
Toxin (exogenous and endogenous)-induced
Tumor (cancer)
Vascular
Others (cellular insult)

Table 2 The primary major kidney/renal pathologies

• Acid-base disorders – Renal tubular acidosis	• Hydronephrosis
• Acute renal/kidney failure	• Hypertension
• Albuminuria	• Hyperuricemia
• Anemia of chronic kidney disease	• Infections – Cystitis – Pyelonephritis – Renal tuberculosis
• Blockage of renal tubules	• Interstitial nephritis
• Cancer	• Kidney stones
• Chronic renal/kidney failure	• Mineral and bone disorder
• Clots – Renal artery thrombosis	• Neoplasms (renal/kidney)
• Cortical necrosis	• Nephritis – Hereditary nephritis – Proteinuria – Pyelonephritis – Secondary nephritis
• Cystic kidneys	• Nephrolithiasis
• Diabetes insipidus	• Overactive bladder
• Diabetes mellitus	• Papillary necrosis
• Electrolyte disorders	• Prostate disorders – Benign prostatic hypertrophy.
• Fanconi syndrome	• Renal artery obstruction
• Glomerular nephropathies – Primary glomerulonephritis – Rapidly progressive glomerulonephritis	• Vascular renal disease
• Heavy metal-induced renal/ nephrotoxicity	• Uremia
• Hematuria	• Ureterovesical valve malfunction (urine reflux from the bladder into the ureter)
• Hemolytic uremic syndrome	• Vasculitis
• Hepatorenal syndrome	• Zellweger syndrome

3 Nephroprotective Effects of Four *Ganoderma* Species

Ganoderma applanatum, *Ganoderma cochlear*, *Ganoderma lucidum*, and *Ganoderma tsugae* have been found to elicit numerous pharmacodynamic effects. Therefore, mushrooms of this genus have been used to prevent and treat various central nervous systems, peripheral nervous systems, exocrine, endocrine, inflammatory, infectious, and autoimmune disorders, as well as various other diseases. In this chapter, we specifically focus on the nephroprotective effects of a few *Ganoderma* species.

Table 3 Important characteristic features of renal/kidney diseases

Renal/kidney diseases	Important characteristic features
Acute renal failure	Reversible, dehydration
Chronic renal failure	Permanent or partial loss of renal function due to hyperglycemia and hypertension
Diabetic nephropathy	High blood glucose content progressively damages the kidneys
End-stage renal disease	Progressive, complete loss of renal function
Glomerulonephritis	Autoimmune, inflammation
Hypertensive nephropathy	High blood pressure progressively damages the kidneys
Interstitial nephritis	Inflammation of the connective tissue due to allergies/drug, acute renal failure
Nephrogenic diabetes insipidus	Lose the ability to concentrate the urine, thirst, and frequent urination
Nephrolithiasis	Minerals in urine form crystals (stones), pain
Nephrotic syndrome	Substantial amounts of protein in the urine, edema
Nephrotic syndrome	Protein in the urine
Papillary necrosis	All or part of the renal papillae die
Polycystic kidney disease	A genetic condition resulting in large cysts in both kidneys
Pyelonephritis	Bacterial infection (kidney pelvis)
Renal cell carcinoma (kidney cancer)	Due to smoking
Renal cyst	Round or oval fluid-filled pouch with a well-defined outline

3.1 Ganoderma applanatum

Carbon tetrachloride (CCl_4) is present in ambient outdoor and indoor environments (air). However, the significant toxic impacts of inhaled and oral exposure to carbon tetrachloride in humans involve the hepatic, renal, and central nervous systems. The exposure causes headaches, weakness, lethargy, nausea, and vomiting. However, elevated concentrations, chronic inhalation, and oral exposure to carbon tetrachloride induce renal fibrosis and kidney damage in humans (Susilo et al. 2022). Hyperuricemia is a prevalent metabolic disease characterized by higher serum uric levels due to increased uric acid production and/or decreased excretion. Hyperuricemia is associated with renal tubular injury, ensuing tubulointerstitial fibrosis, and gouty nephropathy. Gavaging animals can create models of hyperuricemia with adenine, potassium oxonate, or intraperitoneal administration of inosine 5-monophosphate and guanosine monophosphate.

In animal models, extracts of *Ganoderma* were found to increase renal OAT1 transporter action, decrease renal GLUT9 and URAT1 activity, and decrease gastrointestinal CNT2 function. These actions were proposed to increase renal uric secretion and reduce the absorption of the precursor purines in the gastrointestinal tracts. Also, the polysaccharides of *G. applanatum* exhibit nephroprotective action that avoids the development of renal fibrosis and prevents hyperuricemia-induced nephrotoxicity (Liang et al. 2018).

3.2 Ganoderma cochlear

Renal fibrosis is the ultimate broad lethal pathway associated with various progressive kidney pathologies/renal diseases. Transforming growth factor-β (TGF-β) can stimulate tissue fibrosis by upregulating matrix protein synthesis, preventing matrix degradation, and modifying cell-cell interaction (Isaka 2018). Currently, several approaches focus on TGF-β as a target that inhibits the production, activation, binding to the receptor, and intracellular signaling. Phenolic meroterpenoids (cochlearols) (Fig. 2) isolated from *G. cochlear* have exhibited significant nephroprotective effects. In several studies, the nephroprotection was validated in various renal fibrosis models (TGF-β1-induced fibrosis in rat kidney proximal tubular cells NRK-52e, renal interstitial fibroblast cells NRK-49F, and HKC-8 cells) (Wang et al. 2016, 2019a, b; Dou et al. 2014; Zhang et al. 2020; Meng et al. 2021).

3.3 Ganoderma lucidum *(Reishi)*

In several below-mentioned studies, *G. lucidum* has shown nephroprotective effects.

(i) Due to the consumption and use of mushrooms in beverages, food, health-related commodities, nutraceuticals, and pharmaceutical products, the exposure to fungus is escalating rapidly globally. Consequently, the rate of occurrence of mushroom poisoning and mortality is reported to be escalating drastically. Thus, quite a few new syndromes in mushroom poisoning have been depicted in the last decade, and interestingly "rhabdomyolytic" mushroom toxicity is one of the new syndromes. Few reports on the mushroom, *Russula subnigricans*, show that it induces delayed-onset rhabdomyolysis

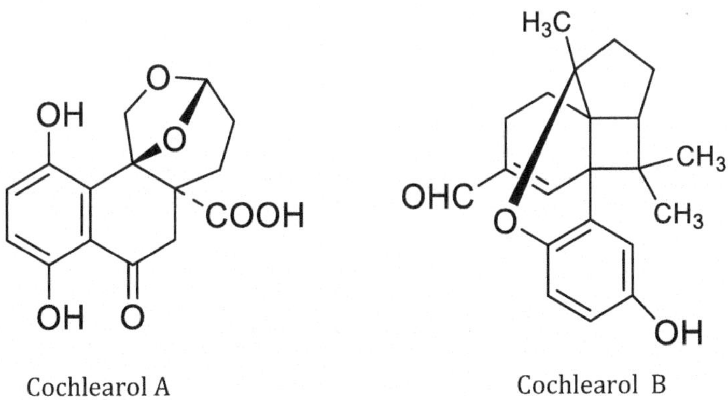

Fig. 2 Structure of the meroterpenoids (cochlearols)

with acute kidney injury, severe hypocalcemia, respiratory failure, ventricular tachycardia, cardiogenic shock, and death in humans (Cho and Han 2016). *G. lucidum* exhibited remedial action in curing *Russula subnigricans* poisoning and could protect against this nephrotoxic fungal-induced nephrotoxicity.

(ii) Anatomically, in humans, the kidneys are the primary excretory organs involved in removing excess fluid and waste from the blood. The blood is filtered in the kidneys through its basic unit, nephrons. An individual nephron encompasses a glomerulus, which is a network of small blood vessels, and they are enclosed in a sac referred to as Bowman's capsule. The filtered waste (urine) flows through tiny tubes and is then transferred from the kidneys to the bladder through ureters for excretion. The glomerular filtration rate (GFR) is the rate of blood flow through the glomerulus, and determination of the GFR helps diagnose renal pathologies early, aiding in treating renal ailments. Furthermore, measurement of the GFR helps to monitor patients with chronic kidney disease (CKD) or other disease states/conditions (hyperglycemia, hypertension, hypercholesteremia) that induce any insult to the nephron/kidneys. Glomerular endothelial dysfunction can result in proteinuria and nephron damage, such as tubulointerstitial fibrosis and glomerulosclerosis, detected in critical nephrosis, such as focal segmental glomerulosclerosis. The glomerular endothelium damage can occur due to oxidative stress and inflammation, leading to the flawed vasodilator release with excessive formation of angiotensin II. This results in hemodynamic imbalance by preferential efferent arteriole constriction resulting in intraglomerular hypertension and glomerulosclerosis. Finally, the efferent constriction diminishes peritubular capillary flow, ultimately resulting in tubulointerstitial fibrosis (thickening, scarring, and damage of the tissue). Additionally, renal fibrosis can be caused by transforming growth factor-β1 (TGF-β1) in kidney proximal tubular cells (Vetchinkina et al. 2016). TGF-β1 is involved in developing fibrosis in various tissues. TGF-β1 has a vital role in tissue fibrosis by upregulating matrix protein synthesis, inhibiting matrix degradation, and altering cell-cell interaction. Similarly, unilateral ureteral obstruction (UUO) renal proximal tubular cell oxidative damage also causes renal fibrosis, and this is considered the pathway of almost all kinds of chronic kidney diseases (CKD) to the end stage of renal diseases (Geng et al. 2020a, b).

G. lucidum has been shown to exert a therapeutic nephroprotective effect due to multifactorial pharmacodynamic action of modulation of immune-circulatory balance, decreasing TGF-β1 hypolipidemic development, vasodilatory mechanism, antiplatelet activity, and improved hemorheology. Combined with ascorbic acid and tocopherol, *G. lucidum* exhibited antioxidant and anti-inflammatory effects, blocking nephrotoxicity and improving renal fibrosis (Futrakul et al. 2003).

(iii) Polycystic kidney disease (PKD) is a common inherited monogenetic disease characterized by the progressive development of clusters of renal cysts primarily within the kidneys, triggering organ enlargement and making the kidney physiologically inactive in due process. These renal cysts in polycystic kidney disease are noncancerous sacs containing water-like fluid which can enlarge significantly, resulting in hypertension, pain (back or side), and a swollen abdomen. Autosomal dominant polycystic kidney disease (ADPKD) is the most prevalent form of PKD, which affects 1 in every 400 to 1000 people. Additionally, ADPKD is a prevalent genetic renal disease globally. ADPKD is usually diagnosed between the ages of 30 and 50; it is referred to as "adult PKD" sometimes. ADPKD patients have defective PKD1 genes (1 out of 6) and PKD2 genes (1 out of 7). Triterpenes of *G. lucidum* impede the formation of renal cyst development by downregulating Ras/MAPK signaling and promoting cell differentiation (Su et al. 2017).

(iv) The amount of albumin in human and animal urine has been used as a diagnostic and prognostic marker to assess the intensity of severity of glomerular function and renal diseases. The urine albumin content usually outperforms serum creatinine and blood urea nitrogen in detecting acute tubular injury. Excess urinary albumin is the intrinsic renal cause of acute kidney injury (ischemia-reperfusion, nephrotoxin, and rhabdomyolysis) and not in either prerenal (secondary to endotoxin) or postrenal (obstructive uropathy) conditions. Increases in urine albumin occur before changes in serum creatinine, which increases only after significant azotemia has developed (Bolisetty and Agarwal 2011). There is a substantial proximal tubular injury during AKI secondary to intrinsic renal triggers. One possible outcome of the tubular damage would be the failure of the proximal tubule to reabsorb albumin, resulting in albuminuria. Likewise, hyperalbuminemia is an increased albumin content in the blood and occurs due to dehydration. Hyperalbuminemia has also been associated with high protein diets. The excess albumin induces oxidative damage, apoptosis, and inflammation due to pro-inflammatory chemokines synthesis in the proximal tubular epithelial cells of humans and animals, resulting in nephrotoxicity. *Ganoderma lucidum* considerably decreases oxidative stress, inflammation, and apoptosis and protects against albumin-induced nephrotoxicity (Min-Chang et al. 2014; Lai et al. 2006).

(v) Hypertension induces vasoconstriction that ultimately damages and weakens the blood vessels throughout the body, including the blood vessels of the kidneys. The damage to the renal blood vessels and vasoconstriction lowers the blood flow. This consequently can prevent the removal of all wastes and extra fluid from the body. The excessive fluid accumulation in the blood vessels can increase the blood pressure, which builds a vicious nephrotoxic progression and triggers severe nephron insult leading to irreversible kidney failure. Interestingly, uncontrolled hypertension can induce the blood vessels (mainly arteries) around the kidneys to narrow, weaken, or harden and decrease blood flow resulting in hypoxia and severe irreversible nephrotoxicity. There are several chemically and surgically induced animal models to investigate the pathophysiology and elucidate the beneficial effects of natural bioactives/synthetic drugs for hypertension. Alerting the sympathetic outflow can be used as an animal model to assess the hypertensive/hypotensive effect. In an animal model of hypertension, the femoral artery and vein were cannulated, one of the kidneys was exposed retroperitoneally, and a branch of the renal nerve was used to integrate renal efferent or afferent nerve activities. In the above animal model of hypertension-induced nephrotoxicity, *Ganoderma lucidum* blocked the hypertensive-mediated insult to the nephrons and protected the kidneys (Lee and Rhee 1990).

(vi) Ingestions and iatrogenic administration of drugs are all too common causes of acute kidney injury. Drugs can induce reversible and irreversible acute and chronic kidney injury based on the nature of the drug, dose, dosage form, and duration of treatment. Several natural bioactives and synthetic drugs trigger acute and chronic kidney injury by altering renal hemodynamics and direct tubular damage, causing renal tubular obstruction, apoptosis, inflammation, and oxidative stress. Cisplatin, a chemotherapeutic administered intravenously (iv), is commonly used to treat bladder, breast, cervical, esophageal, lung, head and neck, ovarian, and testicular cancers, mesothelioma, brain tumors, and neuroblastoma. It is generally administered by intravenous (iv) route of administration. Concerning the mechanism of action, cisplatin (cisplatinum or cis-diamminedichloroplatinum(II)) cross-links with the purine bases on the DNA to form DNA adducts, which can inhibit the repair of the DNA leading to DNA damage and consequently stimulate apoptosis in the cancer cells. The therapeutic challenge or problem with cisplatin is drug resistance, hypersensitivity reactions, and severe adverse effects. The major adverse effects are nephrotoxicity, immunosuppression, increased infections, gastrointestinal disorders, bleeding, and auditory damage. Cisplatin-induced nephrotoxicity was prevented and blocked by *G. lucidum* (Pillai et al. 2011; Mahran and Hassan 2020). Similarly, Adriamycin-induced nephropathy was also blocked by *Ganoderma lucidum.*

(vii) Diabetic nephropathy is one of the most important complications of diabetes mellitus and is considered a principal cause of end-stage renal failure (Hassan et al. 2021). Diabetic mellitus/hyperglycemic nephropathy and renal complication can be induced by the chemical exotoxin (streptozotocin). *G. lucidum* can protect the nephrons and kidneys against hyperglycemia-induced nephrotoxicity in these chemically induced toxic states (He et al. 2006; Hu et al. 2022; Pan et al. 2014).

(viii) Kidney/renal cell cancer (renal cell adenocarcinoma) is a pathology where malignant cancerous cells are found in the lining of tubules in the kidney. In adults, renal cell carcinoma is the most prevalent type of kidney cancer. Smoking and abuse of certain drugs can increase the risk of kidney/renal cancer. The main symptoms include hematuria (blood in the urine) and a lump in the abdomen. *G. lucidum* possesses anticancer action and protects against kidney/renal cell cancer (Vetchinkina et al. 2016; Zhao et al. 2012).

Heavy metal exposure-mediated toxic conditions occur due to extreme exposure to heavy metals such as arsenic, cadmium, lead, mercury, and chromium. Acute or chronic exposure arises from dietary intake, drug-induced, the ecosystem, or dwelling or working environment. Pharmacokinetically, arsenic, cadmium, lead, mercury, and chromium are absorbed and distributed in humans and animals through the skin, inhalation, and ingestion. Heavy metal poisoning occurs due to unexpected, severe, or persistent contact/exposure. Metal toxicity can drastically affect cell/tissue functions, resulting in organ failure and death.

Regarding lead exposure, children are highly vulnerable to lead exposure and toxicity since a child absorbs a significantly higher amount of lead than adults. Lead can cross the placental barrier and can harm the fetus. Lead-based paint and dust found in older buildings, drugs, contaminated air, water, and soil are the major sources of exposure. The major symptoms of acute lead ingestion are nausea, vomiting, diarrhea, and abdominal pain. Lead exposure can also induce anemia, weakness, and kidney and brain damage. However, chronic exposure may induce cellular, tissue, and organ damage and enhance the probability of cancer. Intriguingly, *G. lucidum* protects against lead toxicity-induced nephrotoxicity (Sobowale et al. 2019).

(ix) Renal ischemia-reperfusion injury (RIRI) is one of the leading causes of acute kidney injury (AKI), which can lead to acute renal failure. The development of RIRI is so complicated that it involves many factors, such as inflammatory response, oxidative stress, and cell apoptosis. *G. lucidum* protects due to its antioxidant, anti-inflammatory, and anti-apoptotic activities and can decrease renal fibrosis (Shao et al. 2021; Zhong et al. 2015).

(x) Bismuth treats diarrhea (travelers' diarrhea), sporadic stomach upset, heartburn, and nausea. Furthermore, bismuth possesses an antimicrobial (antibacterial) effect. Bismuth nanoparticles induce severe nephrotoxicity through the AMPK/mTOR pathway. *Ganoderma lucidum* protects against bismuth-induced nephrotoxicity (Yu et al. 2019).

(xi) Excessive ethanol (alcohol) consumption includes binge drinking, heavy drinking, and any drinking by pregnant women or people younger than age 21. Excessive ethanol (alcohol) consumption drastically raises the risk of acute- and chronic-term health risks. Acute ethanol (alcohol) exposure can induce accidents and injuries (burns, motor vehicle crashes, drownings, and falls), enhance violence (assault-physical/sexual, homicide, intimidation, suicide), and unsafe and perilous sexual activities (such as unprotected sex) leading to unintended pregnancy or sexually transmitted diseases (HIV, gonorrhea, syphilis). Ethanol exposure can result in miscarriage, stillbirth, and fetal alcohol spectrum disorders (FASDs) in pregnant women (Bhattacharya et al. 2015, 2020).

In kidneys, ethanol reduces the ability to filter blood and regulate fluid and electrolytes. Moreover, ethanol dehydrates the body, which can affect the structure and physiology of nephrons. Acute and chronic ethanol intake can substantially diminish kidney function, especially in concurrence with prevalent hepatotoxicity. Chronic ethanol intake can decrease the major electrolytes in the blood and severely alter the acid-base balance. Furthermore, ethanol can disrupt the hormonal control mechanisms which regulate renal physiology. Ethanol, by fostering hepatotoxicity, induces an additional harmful impact on the kidneys. *Ganoderma lucidum* significantly reduced ethanol-induced nephrotoxicity (Shieh et al. 2001).

Ganoderma lucidum protects against hyperlipidemic (high-fat diet) and hyperglycemia-induced nephrotoxicity (Huang et al. 2020; Wu et al. 2016), aging-induced nephrotoxicity (Wang et al. 2017), high-cholesterol diet-induced nephrotoxicity (Romero-Córdoba et al. 2020), obesity-induced nephrotoxicity, pesticide (carbofuran)-induced nephrotoxicity, prooxidant-induced renal damage (Lin and Deng 2019), antibiotics (colistin/polymyxin E)-induced nephrotoxicity (Talih et al. 2020), and chronic proteinuric renal diseases. Several human clinical studies have proved the safety to validate further the safety in humans (Zhao et al. 2021; Geng et al. 2019; Joob and Wiwanitkit 2016). Nephroprotection by four species of *Ganoderma* is shown in Table 4.

3.4 Ganoderma tsugae *(Ling Zhi)*

Autoimmune disease is a pathological condition where antibodies target certain tissues/organs, such as nephrons and kidneys, inducing significant renal toxicity. *G. tsugae* was found to improve the survival rate of lupus mice, decrease the amount of proteinuria, decrease serum levels of anti-dsDNA autoantibody, and show evidence of decreased perivascular and parenchyma mononuclear cell infiltration in vital organs, including the kidney (Lai et al. 2001).

4 Conclusion

Nephrotoxicity and kidney diseases may be caused due to autoimmune disorders, fungus/mushroom poisoning, tubulointerstitial fibrosis, renal/kidney cysts, hyperalbuminemia, hypertension, iatrogenesis (chemotherapeutics-cisplatin/Adriamycin and antibiotics-colistin), hyperglycemia, cancer, metal (lead) toxicity, renal ischemic reperfusion, bismuth, alcohol (ethanol), hyperlipidemia (high-fat diet), aging, high-cholesterol diet, obesity, pesticide (carbofuran) exposure, prooxidants, and chronic proteinuric renal diseases. The bioactives of *Ganoderma* have multiple targets and exert numerous pharmacodynamic activities, which can prevent nephron adverse effects and be used to treat various kidney diseases. Natural bioactives have been progressively accepted as alternative remedies and complementary sources for healing kidney diseases in humans and animals. Due to the conventional experience and current knowledge, the multi-target pharmacological characteristics include antioxidation, anti-inflammation, antitumor growth, metastasis, etc.

Table 4 Nephroprotection by *Ganoderma* species

Mushrooms (species)	Nephroprotection	References
Ganoderma applanatum	Protects against carbon tetrachloride-induced renal fibrosis and hyperuricemia	Susilo et al. (2022), Liang et al. (2018)
Ganoderma cochlear	Blocks TGF-β1-induced renal fibrosis	Wang et al. (2016, 2019a, b), Dou et al. (2014), Zhang et al. (2020), Meng et al. (2021)
Ganoderma lucidum	(i) Blocks *Russula subnigricans* poisoning (ii) Protects against Tubulointerstitial fibrosis (iii) Protects against renal/kidney cysts associated with polycystic kidney disease (iv) Protects against hyperalbuminemia (v) Protects against hypertension (vi) Chemotherapeutics (Cisplatin/Adriamycin)-induced iatrogenesis (vii) Diabetic mellitus/hyperglycemic nephropathy (viii) Kidney cancer (ix) Lead toxicity-induced nephrotoxicity (x) Renal ischemia reperfusion injury-induced toxicity (xi) Nephrotoxicity induced by bismuth nanoparticles (xii) Alcohol (ethanol)-induced nephrotoxicity (xiii) Hyperlipidemic (high fat diet) and hyperglycemia-induced nephrotoxicity (xiv) Aging-induced nephrotoxicity (xv) High-cholesterol diet-induced nephrotoxicity (xvi) Obesity-induced nephrotoxicity (xvii) Pesticide (carbofuran)-induced nephrotoxicity (xviii). Prooxidant-induced renal damage (xix) Antibiotics (colistin)-induced nephrotoxicity (xx) Chronic proteinuric renal diseases	Xiao et al. (2003), Futrakul et al. (2003), Su et al. (2017), Lai et al. (2006), Min-Chang et al. (2014), Lee and Rhee (1990), Pillai et al. (2011), Mahran and Hassan (2020), He et al. (2006), Hu et al. (2022), Vetchinkina et al. (2016), Zhao et al. (2012), Sobowale et al. (2019), Shao et al. (2021), Zhong et al. (2015), Yu et al. (2019), Shieh et al. (2001), Huang et al. (2020), Wu et al. (2016), Wang et al. (2017), Hossen et al. (2018), Lin and Deng (2019), Talih et al. (2020), Geng et al. (2020a), Romero-Córdoba et al. (2020)
Ganoderma tsugae	Protect against auto-immune renal disease/Lupus	Lai et al. (2001)

References

Bhattacharya D, Dunaway EP, Bhattacharya S et al (2015) Impaired ILK function is associated with deficits in hippocampal based memory and synaptic plasticity in a FASD rat model. PLoS One 10(8):e0135700. https://doi.org/10.1371/journal.pone.0135700

Bhattacharya D, Fujihashi A, Majrashi M et al (2020) Concurrent nicotine exposure to prenatal alcohol consumption alters the hippocampal and cortical neurotoxicity. Heliyon 6(1):e03045. https://doi.org/10.1016/j.heliyon.2019.e03045

Bolisetty S, Agarwal A (2011) Urine albumin as a biomarker in acute kidney injury. Am J Physiol Renal Physiol 300(3):F626–F627. https://doi.org/10.1152/ajprenal.00004.2011

Cho JT, Han JH (2016) A case of mushroom poisoning with Russula subnigricans: development of rhabdomyolysis, acute kidney injury, cardiogenic shock, and death. J Korean Med Sci 31(7): 1164–1167. https://doi.org/10.3346/jkms.2016.31.7.1164

Dou M, Di L, Zhou LL et al (2014) Cochlearols A and B, polycyclic meroterpenoids from the fungus Ganoderma cochlear that have renoprotective activities. Org Lett 16(23):6064–6067. https://doi.org/10.1021/ol502806j

Futrakul N, Boonyen M, Patumraj S et al (2003) Treatment of glomerular endothelial dysfunction in steroid-resistant nephrosis with Ganoderma lucidum, vitamins C, E and vasodilators. Clin Hemorheol Microcirc 29(3–4):205–210

Geng X, Zhong D, Su L et al (2019) Preventive and therapeutic effect of Ganoderma (Lingzhi) on renal diseases and clinical applications. Adv Exp Med Biol 1182:243–262. https://doi.org/10.1007/978-981-32-9421-9_10

Geng X, Zhong D, Su L et al (2020a) Preventive and therapeutic effect of Ganoderma lucidum on kidney injuries and diseases. Adv Pharmacol 87:257–276. https://doi.org/10.1016/bs.apha.2019.10.003

Geng XQ, Ma A, He JZ et al (2020b) Ganoderic acid hinders renal fibrosis via suppressing the TGF-β/Smad and MAPK signaling pathways. Acta Pharmacol Sin 41(5):670–677. https://doi.org/10.1038/s41401-019-0324-7

Hassan HM, Mahran YF, Ghanim AMH (2021) Ganoderma lucidum ameliorates the diabetic nephropathy via down-regulatory effect on TGFβ-1 and TLR-4/NFκB signalling pathways. J Pharm Pharmacol 73(9):1250–1261. https://doi.org/10.1093/jpp/rgab058

He CY, Li WD, Guo SX et al (2006) Effect of polysaccharides from Ganoderma lucidum on streptozotocin-induced diabetic nephropathy in mice. J Asian Nat Prod Res 8(8):705–711. https://doi.org/10.1080/10286020500289071

Hossen MS, Billah Prince MM, Tanvir EM et al (2018) Ganoderma lucidum and Auricularia polytricha mushrooms protect against carbofuran-induced toxicity in rats. Evid Based Complement Alternat Med 2018:6254929. https://doi.org/10.1155/2018/6254929

Hu Y, Wang SX, Wu FY et al (2022) Effects and mechanism of Ganoderma lucidum polysaccharides in the treatment of diabetic nephropathy in streptozotocin-induced diabetic rats. Biomed Res Int 2022:4314415. https://doi.org/10.1155/2022/4314415

Huang CH, Lin WK, Chang SH et al (2020) Evaluation of the hypoglycaemic and antioxidant effects of submerged Ganoderma lucidum cultures in type 2 diabetic rats. Mycology 12(2): 82–93. https://doi.org/10.1080/21501203.2020.1733119

Isaka Y (2018) Targeting TGF-β signaling in kidney fibrosis. Int J Mol Sci 19(9). https://doi.org/10.3390/ijms19092532

Joob B, Wiwanitkit V (2016) Linzhi (Ganoderma lucidum); evidence of its clinical usefulness in renal diseases. J Nephropharmacol 5(1):9–10

Lai NS, Lin RH, Lai RS et al (2001) Prevention of autoantibody formation and prolonged survival in New Zealand Black/New Zealand white F1 mice with an ancient Chinese herb, Ganoderma tsugae. Lupus 10(7):461–465. https://doi.org/10.1191/096120301678416006

Lai KN, Chan LY, Tang S et al (2006) Ganoderma extract prevents albumin-induced oxidative damage and chemokines synthesis in cultured human proximal tubular epithelial cells. Nephrol Dial Transplant 21(5):1188–1197. https://doi.org/10.1093/ndt/gfk085

Lee SY, Rhee HM (1990) Cardiovascular effects of mycelium extract of Ganoderma lucidum: inhibition of sympathetic outflow as a mechanism of its hypotensive action. Chem Pharm Bull (Tokyo) 38(5):1359–1364. https://doi.org/10.1248/cpb.38.1359

Liang D, Yong T, Chen S et al (2018) Hypouricemic effect of 2,5-dihydroxyacetophenone, a computational screened bioactive compound from Ganoderma applanatum, on hyperuricemic mice. Int J Mol Sci 19(5). https://doi.org/10.3390/ijms19051394

Lin Z, Deng A (2019) Antioxidative and free radical scavenging activity of Ganoderma (Lingzhi). Adv Exp Med Biol 1182:271–297. https://doi.org/10.1007/978-981-32-9421-9_12

Luangharn T, Karunarathna SC, Dutta AK (2021) Ganoderma (Ganodermataceae, Basidiomycota) species from the greater mekong subregion. J Fungi (Basel) 7(10). https://doi.org/10.3390/jof7100819

Mahran YF, Hassan HM (2020) Ganoderma lucidum prevents cisplatin-induced nephrotoxicity through inhibition of epidermal growth factor receptor signaling and autophagy-mediated apoptosis. Oxidative Med Cell Longev 2020:4932587. https://doi.org/10.1155/2020/4932587

Meng XH, Qin FY, Jiang XT et al (2021) (±)-Gancochlearols J - N, renoprotective meroterpenoids from Ganoderma cochlear. Bioorg Chem 112:104950. https://doi.org/10.1016/j.bioorg.2021.104950

Min-Chang G, Wei-Hong T, Zhen X et al (2014) Effects of selenium-enriched protein from Ganoderma lucidum on the levels of IL-1 β and TNF-α, oxidative stress, and NF-κ B activation in ovalbumin-induced asthmatic mice. Evid Based Complement Alternat Med 2014:182817. https://doi.org/10.1155/2014/182817

Pan D, Zhang D, Wu J et al (2014) A novel proteoglycan from Ganoderma lucidum fruiting bodies protects kidney function and ameliorates diabetic nephropathy via its antioxidant activity in C57BL/6 db/db mice. Food Chem Toxicol 63:111–118. https://doi.org/10.1016/j.fct.2013.10.046

Pillai TG, John M, Sara Thomas G (2011) Prevention of cisplatin induced nephrotoxicity by terpenes isolated from Ganoderma lucidum occurring in southern parts of India. Exp Toxicol Pathol 63(1–2):157–160. https://doi.org/10.1016/j.etp.2009.11.003

Romero-Córdoba SL, Salido-Guadarrama I, Meneses ME et al (2020) Mexican Ganoderma lucidum extracts decrease lipogenesis modulating transcriptional metabolic networks and gut microbiota in C57BL/6 mice fed with a high-cholesterol diet. Nutrients 13(1). https://doi.org/10.3390/nu13010038

Schoener B, Borger J (2022) Erythropoietin stimulating agents. In: StatPearls. StatPearls Publishing Copyright © 2022, StatPearls Publishing LLC, Treasure Island, FL

Shao G, He J, Meng J et al (2021) Ganoderic acids prevent renal ischemia reperfusion injury by inhibiting inflammation and apoptosis. Int J Mol Sci 22(19). https://doi.org/10.3390/ijms221910229

Shieh YH, Liu CF, Huang YK et al (2001) Evaluation of the hepatic and renal-protective effects of Ganoderma lucidum in mice. Am J Chin Med 29(3–4):501–507. https://doi.org/10.1142/s0192415x01000526

Sobowale MT, Ozolua RI, Uwaya DO et al (2019) Effects of concurrently administered aqueous extract of lingzhi or reishi medicinal mushroom, Ganoderma lucidum (Agaricomycetes), and lead acetate in rats. Int J Med Mushrooms 21(2):143–154. https://doi.org/10.1615/IntJMedMushrooms.2019029766

Su L, Liu L, Jia Y et al (2017) Ganoderma triterpenes retard renal cyst development by downregulating Ras/MAPK signaling and promoting cell differentiation. Kidney Int 92(6):1404–1418. https://doi.org/10.1016/j.kint.2017.04.013

Sun YF, Xing JH, He XL (2022) Species diversity, systematic revision and molecular phylogeny of Ganodermataceae (polyporales, basidiomycota) with an emphasis on Chinese collections. Stud Mycol 101:287–415. https://doi.org/10.3114/sim.2022.101.05

Susilo RJK, Winarni D, Hayaza S et al (2022) Effect of crude Ganoderma applanatum polysaccharides as a renoprotective agent against carbon tetrachloride-induced early kidney fibrosis in mice. Vet World 15(4):1022–1030. https://doi.org/10.14202/vetworld.2022.1022-1030

Talih G, Kantekin Ç, Göçmen Y et al (2020) Effect of lingzhi or reishi medicinal mushroom, Ganoderma lucidum (Agaricomycetes), capsules on colistin-induced nephrotoxicity. Int J Med Mushrooms 22(5):445–453. https://doi.org/10.1615/IntJMedMushrooms.2020034229

Vetchinkina E, Shirokov A, Bucharskaya A et al (2016) Anti-tumor activity of extracts from medicinal basidiomycetes mushrooms. Int J Med Mushrooms 18(11):955–964. https://doi.org/10.1615/IntJMedMushrooms.v18.i11.10

Wachtel-Galor S, Yuen J, Buswell JA, et al (2011) Ganoderma lucidum (lingzhi or reishi): a medicinal mushroom. In: Benzie IFF, Wachtel-Galor S (eds) Herbal medicine: biomolecular and clinical aspects. CRC Press/Taylor & Francis, Copyright © 2011a by Taylor and Francis Group, LLC, Boca Raton, FL

Wang XL, Zhou FJ, Dou M et al (2016) Cochlearoids F-K: phenolic meroterpenoids from the fungus Ganoderma cochlear and their renoprotective activity. Bioorg Med Chem Lett 26(22): 5507–5512. https://doi.org/10.1016/j.bmcl.2016.10.011

Wang J, Cao B, Zhao H et al (2017) Emerging roles of Ganoderma lucidum in anti-aging. Aging Dis 8(6):691–707. https://doi.org/10.14336/ad.2017.0410

Wang XL, Wu ZH, Di L et al (2019a) Renoprotective meroterpenoids from the fungus Ganoderma cochlear. Fitoterapia 132:88–93. https://doi.org/10.1016/j.fitote.2018.12.002

Wang XL, Wu ZH, Di L et al (2019b) Renoprotective phenolic meroterpenoids from the mushroom Ganoderma cochlear. Phytochemistry 162:199–206. https://doi.org/10.1016/j.phytochem.2019.03.019

Wong JH, Ng TB, Chan HHL et al (2020) Mushroom extracts and compounds with suppressive action on breast cancer: evidence from studies using cultured cancer cells, tumor-bearing animals, and clinical trials. Appl Microbiol Biotechnol 104(11):4675–4703. https://doi.org/10.1007/s00253-020-10476-4

Wu YS, Ho SY, Nan FH et al (2016) Ganoderma lucidum beta 1,3/1,6 glucan as an immunomodulator in inflammation induced by a high-cholesterol diet. BMC Complement Altern Med 16(1): 500. https://doi.org/10.1186/s12906-016-1476-3

Xiao GL, Liu FY, Chen ZH (2003) Clinical observation on treatment of Russula subnigricans poisoning patients by Ganoderma lucidum decoction. Zhongguo Zhong Xi Yi Jie He Za Zhi 23(4):278–280

Yu H, Yang Y, Jiang T et al (2019) Effective radiotherapy in tumor assisted by Ganoderma lucidum polysaccharide-conjugated bismuth sulfide nanoparticles through radiosensitization and dendritic cell activation. ACS Appl Mater Interfaces 11(31):27536–27547. https://doi.org/10.1021/acsami.9b07804

Zhang JJ, Qin FY, Meng XH et al (2020) Renoprotective ganodermaones A and B with rearranged meroterpenoid carbon skeletons from Ganoderma fungi. Bioorg Chem 100:103930. https://doi.org/10.1016/j.bioorg.2020.103930

Zhao H, Zhang Q, Zhao L et al (2012) Spore powder of Ganoderma lucidum improves cancer-related fatigue in breast cancer patients undergoing endocrine therapy: a pilot clinical trial. Evid Based Complement Alternat Med 2012:809614. https://doi.org/10.1155/2012/809614

Zhao A, Yang Y, Pan X et al (2021) Long-term toxicological studies on the Chinese medicine 2036 specialty-Qiangxin recipe in rats. Pharm Biol 59(1):1181–1190. https://doi.org/10.1080/13880209.2021.1967410

Zhong D, Wang H, Liu M et al (2015) Ganoderma lucidum polysaccharide peptide prevents renal ischemia reperfusion injury via counteracting oxidative stress. Sci Rep 5:16910. https://doi.org/10.1038/srep16910

Mechanisms Involved in Edible Mushrooms' Health Beneficial Effects

Diana-Roxana Pelinescu and Ileana Stoica

Contents

Abstract Despite the mushrooms' long application history, only in the last century, more data regarding their nutritional qualities and beneficial effects have been available. The progress in scientific fields such as biology, chemistry, medicine, and pharmacology allowed obtaining data on the mechanisms involved in the mushrooms' beneficial effects. The main mechanisms linked to the presence of different mushrooms' biologically active compounds are antioxidant activity, antimicrobial activity, immunomodulatory effect, and anticancer and antitumor activity.

Although much data has been obtained until now, the study of mushrooms is still far from being completed because many wild mushrooms have not yet been studied, the number of biological compounds is huge, and the mechanisms involved in the beneficial effects of mushrooms are still not fully elucidated.

Keywords Anticancer · Antioxidants · Antimicrobial activity · Edible mushrooms · Immunomodulatory effect

D.-R. Pelinescu (✉) · I. Stoica
Department of Genetics, Faculty of Biology, University of Bucharest, Bucharest, Romania
e-mail: diana.pelinescu@bio.unibuc.ro

D. C. Agrawal, M. Dhanasekaran (eds.), *Mushrooms with Therapeutic Potentials*,
https://doi.org/10.1007/978-981-19-9550-7_16

Abbreviations

CD	Cluster of differentiation
CDC	Centers for Disease Control and Prevention.
CT	Clinical trial
DCs	dendritic cells
DNA	Deoxyribonucleic acid
FIPs	Fungal immunomodulatory proteins
IL	Interleukin
INF	Interferon
iNOS	Inducible nitric oxide synthase
NF-kB	Nuclear factor kappa-light-chain enhancer of activated B cells
NK	Natural killer
NO	Nitric oxide
TLR	Toll-like receptor
TNF	Tumor necrosis factor
USD	United States dollar
WHO	World Health Organization

1 Introduction

The fungi existed 750–850 million years ago (Bonneville et al. 2020), and the mushrooms, which are a part of the Fungi kingdom, were considered by the Romans "Food of the Gods" (Valverde et al. 2015). From ancient times, mushrooms have been known either for their nutritional and culinary characteristics (edible mushrooms) or for their toxic compounds (nonedible mushrooms) (Wasser 2010). The market for mushrooms has been increasing over the years. According to the 2021 Market Research report 2019, the global mushroom edible market size was 28.6 billion USD and is expected to reach 52 billion USD by 2026. Another important part of the mushroom market includes medicinal mushrooms, even if they are not edible. The study of edible mushrooms is far from being completed due to the increased interest in wild mushrooms and the need to determine the safety of their use.

The mushrooms are a group of fungi having distinctive fruiting bodies. Based on molecular biology techniques, including DNA molecule sequencing, Fungi's kingdom is divided into *Zygomycota*, *Chytridiomycota*, *Basidiomycota*, and *Ascomycota* phyla. The mushrooms are considered mainly belong to the *Ascomycota* and *Basidiomycota* phyla. *Ascomycota* phylum is considered the largest phyla of fungi and includes more than 40,000 species distributed worldwide (Zafar et al. 2020). Seven hundred mushroom species are considered safe and beneficial for consumers' health (Li et al. 2021), and 100 mushroom species are cultivated commercially (Zafar et al. 2020). The most common species of edible mushrooms are

Cantharellaceae sp. (chanterelles), *Lycoperdon* sp. and *Calvatia* sp. (puffballs), *Coprinus comatus* (shaggy mane), *Pleurotus ostreatus*, *Pleurotus cystidiosus* (oyster mushrooms), *Boletus* sp. (boletes or porcini mushrooms), *Laetiporus sulphureus* (sulfur shelf), *Agaricus bisporus* (button mushroom), *Morchella esculenta* (morels), *Hericium erinaceus* (bearded tooth), *Volvariella volvacea* (straw mushroom), *Flammulina velutipes* (Eenoki), *Lentinula edodes* (shiitake), *Hypsizygus marmoreus* (beech mushroom), *Pleurotus eryngii* (French horn mushroom), and *Grifola frondosa* (dancing mushroom) (Das et al. 2022).

The wild mushroom species are sometimes specific to different regions and are more or less studied even though some of them are superior to cultivated ones in terms of nutritive value and beneficial effect (Kalač 2013; Nakalembe et al. 2015; Barros et al. 2007; Angelini et al. 2020; Anusiya et al. 2021; Huo et al. 2020).

On an industrial scale, around 20 species are cultivated, of which the most well-known are *Agaricus bisporus*, *Pleurotus* sp., *Lentinula edodes*, *Auricularia* spp., *Volvariella volvacea*, *Flammulina velutipes*, and *Tremella fuciformis.*

Complex studies conducted in recent decades have confirmed not only the nutritional qualities but also the existence of biologically active compounds in mushrooms that improve the health of the consumer (Lu et al. 2020; Sheng et al. 2021; Vamanu et al. 2018; Abdelshafy et al. 2022; Ache et al. 2021; Aida et al. 2009; Zhang et al. 2001; Ko et al. 1995; Końska 2006; Manzi et al. 1999; Ma et al. 2018).

2 Edible Mushrooms' Nutritional Effects

The most important characteristic of edible mushrooms is their nutritional value. Despite the concentration variation of components with nutritional impact, all edible mushrooms contain proteins, glucides, lipids (Table 1), organic acids, enzymes, vitamins, phenols, terpenoids, steroids, lectins, and minerals (Kalač 2013; Assemie and Abaya 2022). The main mushroom vitamins are thiamin (B1), biotin (B7), nicotinamide (B3), folic acid (B9), vitamin D2, vitamin C, and in small amounts, vitamins A and E (Wani et al. 2010). In terms of mineral contents, mushrooms have diverse and high amounts of potassium (K), phosphorus (P), sodium (Na), calcium (Ca), and magnesium (Mg) and lower amounts of copper (Cu), zinc (Zn), iron (Fe), molybdenum (Mo), manganese (Mn), and selenium (Se).

The amount of nutritional components depends mainly on the following:

- The mushroom species.
- The growth conditions.
- Time of harvesting.
- The soil texture.

About 30 years ago, most edible mushrooms were considered to have no medicinal value (Chang 1996). However, numerous scientific studies proved mushroom compounds' direct or indirect beneficial effects in improving the health of consumers (Table 2).

Table 1 The nutritive values of some edible mushroom species and main microelements

Components	Proteins	Glucides	Lipids	Fiber	Ash	Energy (kcal)	Ca	K	P	Na	Mg	References
Mushroom species	g/100 g dry matter						mg/100 g dry matter					Ache et al. (2021)
Termitomyces microcarpus	30.69	44.23	2.17	11.6	11.3	319.27	37.47	1112.76	898.17	12.91	39.03	
Auricularia polytricha	17.44	51.23	2.91	20.69	7.74	301.51	88.62	294	623.96	10.91	83.54	
Polyporus tenuiculus	10.89	58.84	3.22	15.48	11.57	299.49	90.95	428.41	592.25	9.70	94.48	
Volvariella speciosa	19.95	48.44	3.56	4.19	14.13	248.64	12.8	3196.4	612	16.1	7.14	Nakalembe et al. (2015)
Termitomyces tyleranus	21.77	42.99	3.00	3.08	16.87	220.75	15.5	2530.1	794.9	16.8	31.9	
Termitomyces clypeatus	18.00	49.35	3.79	7.69	11.2	250.62	14.8	1869.7	612.3	10.3	10.32	
Agaricus bisporus	33.48	46.17	3.10	20.9	5.7	499	47.0	4015.0	1350.0	5?	4.9	Manikandan (2011), Goyal et al. (2020)
Pleurotus sajor-caju	19.23	63.40	2.70	48.60	6.32	412	73.0	3218.0	1246.0	No data	4.5	
Lentinula edodes	32.93	47.60	3.73	28.8	5.2	387	No data	No data	No data	No data	No data	

Table 2 Example of edible mushrooms and their beneficial effects

Mushrooms	Biologically active compounds	Beneficial effects	References
Cantharellus tubaeformis, *Cantharellus cibarius*, *Boletus edulis*	Vitamin D2	Antioxidant, Ca^{2+} ion homeostasis, immuno-modulatory effect	Cardwell et al. (2008)
Lentinula edodes	b-1,3-D-glucan, Lentinan	Cancer treatment Immunomodulatory effect	Venturella et al. (2021)
Tremella fuciformis	Polysaccharides	Gut microbiota modulation	Wu et al. (2021)
Pleurotus ostreatus	Fractions of proteoglycan	Reduction of tumor and immunomodulatory effect	Sarangi et al. (2006), Jedinak and Sliva (2008)
Grifola frondosa	β-Glucan fraction	Immunomodulatory effect and antitumor effect	Venturella et al. (2021)
Flammulina velutipes	Small molecules enokipodins	Antimicrobial activity	Zeb and Lee (2021)
Albatrellus flettii	Grifolin, neogrifolin, and confluentin	Antimicrobial activity	Zeb and Lee (2021)
Pleurotus eryngii	Polysaccharide	Gut microbiota modulation	Ma et al. (2022)
Auricularia polytricha	Polysaccharide	Anti-inflammatory and gut modulation	Nguepi and Song (2020)
Helvella leucopus	Polysaccharide	Reducing the colonic lesion and gut modulation	Abdureyim et al. (2022)
Lentinus squarrosulus	Carbohydrate and protein fractions	Gut modulation	Ayimbila et al. (2022)
Sparassis crispa	Polysaccharides	Gut modulation, antitumor activity, and immunomodulatory effect	Zhang et al. (2022a, b)
Agaricus bisporus, *Agaricus bitorquis*, *Agaricus campestris*, *Boletus edulis*, *Boletus satanas*, *Flammulina velutipes*, *Hericium erinaceus*	Lectins	Immunomodulatory effect and anticancer effect	Chowdhury et al. (2015)
Calvatia gigantea	2-Pyrrolidinone, 1-dodecene, Ergosterol, Hexadecane, Benzene acetic Acid	Antidiabetic, antioxidant, anti-inflammatory	Ogbole et al. (2019)

The edible mushrooms are a source of valuable biologically active substances, including phenolic and indolic compounds, carotenoids, flavonoids, sesquiterpenoids, glucans, glycoproteins, triterpenoids, sterols, tocopherols, antibiotics, vitamins, and elements, which are present in them in significant quantities and often act synergistically (Kała et al. 2020).

3 Mechanisms Involved in the Beneficial Effects

3.1 Antimicrobial Activity

The studies regarding the beneficial effects of edible mushrooms proved that some compounds found in edible mushrooms have antimicrobial activity against Gram-positive and Gram-negative bacterial strains, yeast, and fungal strains (Table 3). The antimicrobial activity was correlated with the presence of different compounds which are present in different edible mushroom species (Ahmad et al. 2014; De Andrade et al. 2021; Ayodele and Idoko 2011; Contato et al. 2021; Özcan and Ertan 2018; Al-Mazaideh and Al-Swailmi 2021).

The data proved that the edible mushrooms could be a source of a new compound with antimicrobial activity that can be an alternative to antibiotic therapy.

In recent years accumulated data proved the antiviral activity of some biologically active compounds of edible mushrooms like lectins, polysaccharides, and terpenoids (El-Maradny et al. 2021; He et al. 2020; Elhusseiny et al. 2021a, b; Friedman 2016).

3.2 Antioxidant Activity

Molecular biology studies proved the involvement of oxidative stress in some disorders like heart disease, cancer, metabolic diseases, neuronal disorders, and even premature aging (Kozarski et al. 2015). The most frequently involved in oxidative stress are reactive oxygen and nitrogen species generated by endogenous processes or by exogenous factors like pollution, drugs, and radiation. The free radicals could interact with DNA, protein, and lipids and alter them. The free radicals could affect cells and tissues by damaging macromolecules and altering metabolic processes and homeostasis (Mwangi et al. 2022). The human organism possesses a complex antioxidative system to eliminate or minimize reactive species' negative effects and involves enzymatic (catalase, glutathione peroxidases, superoxide dismutase) and nonenzymatic systems (vitamins, carotenoids, polyphenols, uric acid, trace elements, albumin, ceruloplasmin, transferrin, ferritin) (Kozarski et al. 2015). The imbalance between reactive species and the detoxification process leads to oxidative stress. Dietary supplements rich in antioxidants are recommended to help the organism's defense mechanisms and prevent or cure some disorders.

Table 3 Antimicrobial activity of edible mushrooms

Edible mushrooms	Compounds	Species of sensitive strains	References
Agaricus bisporus, Lentinula edodes, Agaricus brasiliensis	Phenolic compounds, Gallic acid, p-hydroxybenzoic	*Staphylococcus aureus, Bacillus cereus, Escherichia coli, Salmonella enteritidis*	Bach et al. (2019)
Armillaria mellea, Calvatia excipuliformis, Clavulina cinerea, Clitocybe gibba, Coprinus micaceus, Craterellus cornucopioides, Laccaria amethystina, Laccaria laccata, Lactarius rufus, Laetiporus sulphureus, Leccinum scabrum, Lycoperdon perlatum, Macrolepiota procera, Marasmius oreades, Pholiota mutabilis, Psilocybe capnoides, Rozites caperata, Sparassis crispa, Xerocomus badius	Phenolic compounds	*Micrococcus luteus, Bacillus subtilis, Escherichia coli, Pseudomonas aeruginosa*	Nowacka et al. (2014)
Termitomyces striatus	Mushroom extract with dichloromethane	*Pseudomonas aeruginosa, Escherichia coli, Staphylococcus aureus, Bacillus subtilis, Candida albicans*	Sitati et al. (2021)
Boletus edulis, Cantharellus cibarius, Craterellus cornucopioides, Hydnum repandum, Agaricus bisporus	Methanol and acetone mushrooms extract	*Pseudomonas aeruginosa, Escherichia coli, Enterococcus faecalis, Staphylococcus aureus, Klebsiella pneumoniae, Serratia marcescens, Candida albicans*	Özcan and Ertan (2018)
Pleurotus eryngii	Mushroom submerged cultivated	*Escherichia coli, Staphylococcus aureus, Pseudomonas aeruginosa, Candida albicans, Candida parapsilosis, Candida tropicalis*	De Andrade et al. (2021)
Volvopluteus gloiocephalus, Clitocybe subconnexa	Phenolic acids	*Escherichia coli, Pseudomonas aeruginosa, Salmonella typhimurium, Enterobacter cloacae, Staphylococcus aureus, Bacillus cereus, Micrococcus flavus, Listeria monocytogenes Aspergillus fumigatus, Aspergillus ochraceus, Aspergillus versicolor, Aspergillus niger,*	Heleno et al. (2015)

(continued)

Table 3 (continued)

Edible mushrooms	Compounds	Species of sensitive strains	References
		Penicillium funiculosum, *Penicillium ochrochloron*, *Trichoderma viride*, *Penicillium verrucosum* var. *cyclopium*	
Termitomyces robustus, *Lentinus squarrosulus*	Aquatic and alcoholic mushroom extract	*Staphylococcus aureus*, *Pseudomonas aeruginosa*, *Salmonella typhi*, *Trichoderma rubrum*, *Aspergillus fumigatus*	Borokini et al. (2016)
Tricholosporum goniospermum	Mushrooms extract – Phenolic compounds	*Pseudomonas aeruginosa*, *Staphylococcus aureus*, *Bacillus cereus*, *Bacillus subtilis*, *Candida albicans*, *Candida tropicalis*, *Candida parapsilosis*	Angelini et al. (2020)
Pleurotus ostreatus, *Lentinula edodes*, *Hypsizygus tessulatus*	Methanolic extract	*Staphylococcus aureus*, *Bacillus subtilis*, *Escherichia coli*, *Klebsiella pneumonia*, *Pseudomonas aeruginosa*, *salmonella typhi*, *Candida albicans*, *Saccharomyces cerevisiae*	Chowdhury et al. (2015)

The edible mushrooms have a high range of natural antioxidant compounds, potential substitutes for synthetic antioxidants, which sometimes have side effects. The antioxidants are found in fruit bodies, mycelium, and even in broth and include phenolic compounds, ascorbic acid, flavonoids, ergothioneine, polysaccharides, glycosides, tocopherols including vitamin E, carotenoids, vitamin A, vitamin D, and minerals (Chun et al. 2021; Angelini et al. 2020; Elhusseiny et al. 2021b; Mahmoud and Abdel-Hadi 2022).

The antioxidants sometimes are specific to the mushroom's family, like in the case of polysaccharides. For example, the chitin-mannan-β-glucans could be found in mushrooms from the *Ascomycete* family, while the *Basidiomycete* family contains chitin-β-glucans (Martinez-Medina et al. 2021). The main phenolic compounds in edible mushrooms are hydroxycinnamic acids (caffeic, ferulic, p-coumaric, and sinapic) and benzoic acids (p-hydroxybenzoic, gallic, gentisic, vanillic, syringic, protocatechuic, s.o.). Trace elements accumulated in edible mushrooms (Zn, Cu, Mn, and Fe) are cofactors of antioxidant enzymes (Kozarski et al. 2015).

The mechanism involved in the edible mushrooms' antioxidant activity are as follows:

- Chain breaking
- Free radical scavengers
- Increasing activity of the liver's oxidative enzymes and glutathione and malondialdehyde levels

- Inhibition or breakdown of lipid peroxides
- Enhancing the synthesis of endogenous antioxidants
- Deactivation of metals
- Hydrogen atom transfer
- Fe^{2+} chelators
- Protectors of mitochondrial components (Chun et al. 2021; Kozarski et al. 2015; Friedman 2016)

In vitro obtained data regarding the edible mushrooms' antioxidant activity support the need to elucidate the mechanisms involved in beneficial effects and the necessity of complex pharmacological and biological studies.

3.3 *Immunomodulatory Effect*

The immune system's function and homeostasis are essential for good health and survival. During the last decades, clinical practice has used immunomodulators, and the market for these products has increased. The immunomodulator products include immunostimulants, immunosuppressors, and immunoadjuvants and are sometimes used to prevent diseases as prophylactic therapy. Like in the case of antioxidants, synthetic compounds used as immunomodulators could be replaced by natural compounds without or with minimal toxicity. The edible and medicinal mushrooms could be a source of immunomodulators due to their bioactive compounds (El Enshasy and Hatti-Kaul 2013; Elhusseiny et al. 2022; Patra et al. 2021).

The studies regarding edible mushroom immunomodulators started more than 25 years ago (Ko et al. 1995; Ohkuma et al. 1983). However, these are far from being completed due to the large number of mushrooms and the broad spectrum of compounds.

The immunomodulatory effect of the edible mushrooms is either direct (Wu et al. 2021; Zhao et al. 2020) or indirect after interaction with gut microbiota (Vlassopoulou et al. 2022; Zhang et al. 2022a, b; Vamanu et al. 2021; Steve and Hui 2020; Ma et al. 2021).

The mechanism involved in immunomodulation depends on the type of mushroom, and the main mechanisms and biologically active compounds are summarized in Fig. 1.

The data obtained from in vitro and in vivo studies have highlighted the ability of either total edible mushroom extract or purified compounds to influence the immune system. Some new evidences of edible mushrooms' immunomodulation effects are summarized in Table 4.

Immunotherapy based on natural bioactive compounds could be a less expensive and efficient alternative to allopathic medicine, including vaccines. The edible and medicinal mushrooms represent a great and promising source of immunomodulatory compounds.

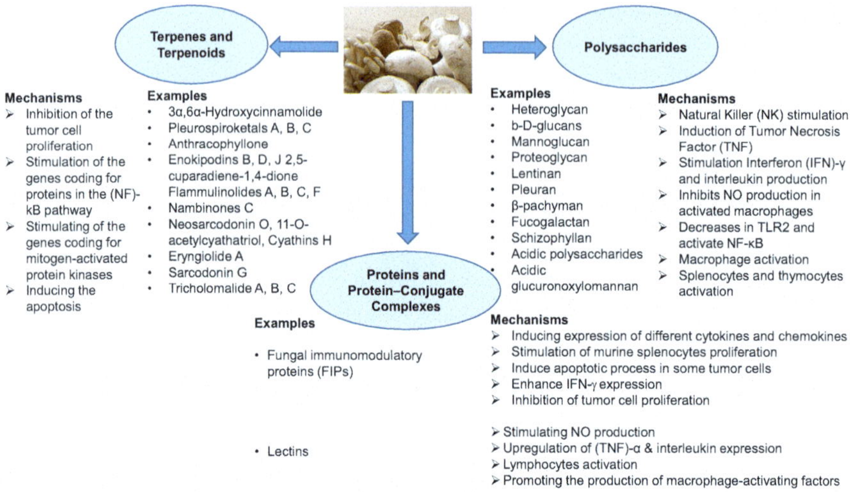

Fig. 1 The mechanisms involved in immunomodulatory effect and main classes of compounds

Table 4 Examples of edible mushrooms' immunomodulatory effects

Mushroom	Model	Effects	Component	References
Lentinula edodes, Agaricus bisporus, Pleurotus ostreatus, Pleurotus columbinus, Pleurotus sajor-caju	Wistar albino rats	Increase the number of white blood cells, increase the level of TNF-α, IFN-γ, and IL-1β, and increase the nitric oxide and the lysozyme concentration	Total extract	Elhusseiny et al. (2021a, b)
Polyporus grammocephalus	In vitro studies	Stimulation of macrophage, splenocytes, and thymocytes	Polysaccharides	Patra et al. (2021)
Lentinula edodes	In vitro studies	Inhibition of CD 3, downregulation of TNF-α, reduction of the NK cytotoxic activity	Selenium-enriched polysaccharide	Kaleta et al. (2021)
Coriolus versicolor	In vitro studies of rat cell line	Increasing the expression of iNOS, activation of macrophages by promoting transcription of TNF-α	Polysaccharides	Zhang et al. (2021)
Volvariella volvacea	Mice C57BL/6 mice	Induces secretion of TNF-α, IL-2, IL-6, and IL-12, p70, promotes the action and function of DCs	Proteins	Li et al. (2021)

3.4 Anticancer and Antitumor Effects

According to the World Health Organization (WHO) and Centers for Disease Control and Prevention (CDC), cancer is the second cause of death in the United States. Despite progress in the medical and pharmaceutical fields, the morbidity rate is still very high. That way, solutions are being sought to prevent and/or treat this disease. There are ongoing clinical trials (phases I and II) in which mushrooms are tested for treating breast, colorectal, and prostate cancer (Panda et al. 2022).

The edible mushrooms' mechanisms involved in anticancer and antitumor effect are exercised in all stages of the appearance, and evolution of the carcinogenic process and includes the following:

- Inactivation of cancerogenic free radicals due to the presence of antioxidants
- DNA repair
- Inducing apoptosis
- Tumor proliferation inhibition due to immune system modulation

The mushrooms' anticancer and antitumor activities include the mechanisms presented before (antioxidant and immunomodulatory activity). Still, some compounds, like lectins, have a direct effect (Końska 2006).

Lectins are carbohydrate-binding proteins with a high degree of selectivity and stereospecificity, which have immunomodulatory, anticancer, and antitumor effects. Edible mushrooms are a natural source of lectins with different chemical structures, molecular masses, and biochemical properties. Lectins have been isolated from different parts of the mushrooms: mycelia, stalk, and caps. The lectin concentration depends on the species, season, location, age, and mycelia growth (Chowdhury et al. 2015).

Other compounds from edible mushrooms with anticancer and antitumor activity are polysaccharides, peptides, and phenols.

In vitro and in vivo animal model studies proved that *Grifola frondosa* (maitake mushroom) polysaccharide (D-fraction) is involved in the anticancer effect by stimulation of apoptosis and blocking of tumor growth in breast cancer (Alonso et al. 2017, 2018). The X – fraction isolated from *Grifola frondosa* fruiting body had antidiabetic activity (Wu et al. 2021).

Some examples of anticancer compounds isolated from the mushrooms are lentinan, krestin, shizophyllan, hispolon, theanine, psilocybin, ganoderic acid, grifolin, cordycepin, antroquinonol, 5-(hydroxymethyl) furan-2-carbaldehyde, 3-isobutyl-1-methoxy-4-(4′-(3-methylbut-2-enyloxy)phenyl)-1H-pyrrole-2, 5-dione, ribonuclease, polysaccharide-peptide complex LB-1b, and polyphenol oxidase (Patel and Goyal 2012; Zhang et al. 2021; Sarangi et al. 2006).

Until now, edible mushrooms' total extracts or purified compounds have been tested to prevent and cure different types of cancer: colorectal, endometrial, gastric, liver, lung, breast, cervical, bladder, miscellaneous tumors, cachexia, pancreatic, prostate, testicular, myeloma, and nasopharyngeal (Panda et al. 2022).

3.5 Other Mechanisms

Some examples of other mechanisms involved in the beneficial effects of edible mushrooms are represented by the following:

- Reduction of total cholesterol, creatinine, aspartate aminotransferase, alanine aminotransferase, plasma glucose, and systolic and diastolic blood pressure – clinical trial (CT) using *Agaricus sylvaticus* (Fortes and Novaes 2011).
- Decreasing the triglyceride and cholesterol levels – CT *Pleurotus ostreatus* (González-Bonilla et al. 2022).
- Modulation of gut microbiota and prevention or curing of obesity by upregulating lipid metabolism, carbohydrate metabolism, bile acid biosynthesis, and downregulation of adipocytokine signaling pathway and steroid hormone biosynthesis – animal model study using *Pleurotus ostreatus* (Hu et al. 2022).
- Improving levels of aminotransferase levels and histopathological features (steatosis, inflammatory foci, pericellular fibrosis) in nonalcoholic steatohepatitis – animal model study using *Ceraceomyces tessulatus* (Suzuki et al. 2022).
- Hypoglycemic and antidiabetic effect – animal model using *Lentinus edodes* polysaccharides (Gong et al. 2022).
- Reduces the neuroinflammation, regulation of neurotrophin synthesis, and modulation of acetylcholinesterase activity – in the animal study using *Agaricus bisporus* (Solano-Aguilar et al. 2021).

4 Conclusion

Mushrooms have a long history of being used as healthy food due to their culinary attributes and beneficial effects. Numerous studies have proved that mushrooms could be considered the "elixir of life," as Chinese culture mentions them, due to their application in the prevention and cure of various diseases like cancer, diabetes, hypertension, metabolic diseases, infectious diseases, etc. The data from the in vitro and in vivo studies, including clinical trials, suggest mushrooms' high applicability potential due to their vast reservoir of biologically active compounds.

References

Abdelshafy AM, Belwal T, Liang Z et al (2022) A comprehensive review on phenolic compounds from edible mushrooms: occurrence, biological activity, application and future prospective. Crit Rev Food Sci Nutr 62(22):6204–6224

Abdureyim Z, Wang L, Tao Q et al (2022) Bachu mushroom polysaccharide alleviates colonic injury by modulating the gut microbiota. Comput Math Methods Med 2022:1–12. https://doi.org/10.1155/2022/1353724

Ache NT, Manju EB, Lawrence MN, Tonjock RK (2021) Nutrient and mineral components of wild edible mushrooms from the Kilum-Ijim forest, Cameroon. Afr J Food Sci 15:152–161. https://doi.org/10.5897/ajfs2021.2089

Ahmad N, Mahmood F, Khalil SA et al (2014) Antioxidant activity via DPPH, gram-positive and gram-negative antimicrobial potential in edible mushrooms. Toxicol Ind Health 30(9):826–834. https://doi.org/10.1177/0748233712463775

Aida FMNA, Shuhaimi M, Yazid M, Maaruf AG (2009) Mushroom as a potential source of prebiotics: a review. Trends Food Sci Technol 20:567–575. https://doi.org/10.1016/j.tifs.2009.07.007

Al-Mazaideh GM, Al-Swailmi FK (2021) Phytonutrients and antimicrobial activities of methanolic extract from Hafr Al-Batin truffles. Int J Pharm Res 33(24B):13–21. https://doi.org/10.9734/jpri/2021/v33i24b31437

Alonso EN, Ferronato MJ, Gandini NA et al (2017) Antitumoral effects of D-fraction from *Grifola frondosa* (Maitake) mushroom in breast cancer. Nutr Cancer 69(1):29–43. https://doi.org/10.1080/01635581.2017.1247891

Alonso EN, Ferronato MJ, Fermento ME et al (2018) Antitumoral and antimetastatic activity of Maitake D-fraction in triple-negative breast cancer cells. Oncotarget 9(34):23396–23412. https://doi.org/10.18632/oncotarget.25174

Angelini P, Venanzoni R, Flores GA et al (2020) Evaluation of antioxidant, antimicrobial and tyrosinase inhibitory activities of extracts from *Tricholosporum goniospermum*, an edible wild mushroom. Antibiotics 9(8):513. https://doi.org/10.3390/antibiotics9080513

Anusiya G, Gowthama Prabu U, Yamini NV et al (2021) A review of the therapeutic and biological effects of edible and wild mushrooms. Bioengineered 12(2):11239–11268. https://doi.org/10.1080/21655979.2021.2001183

Assemie A, Abaya G (2022) The effect of edible mushroom on health and their biochemistry. Int J Microbiol 2022:7p. https://doi.org/10.1155/2022/8744788

Ayimbila F, Siriwong S, Nakphaichit M, Keawsompong S (2022) In vitro gastrointestinal digestion of *Lentinus squarrosulus* powder and impact on human fecal microbiota. Sci Rep 12:2655. https://doi.org/10.1038/s41598-022-06648-z

Ayodele SM, Idoko ME (2011) Antimicrobial activities of four wild edible mushrooms in Nigeria. Int J Sci Nat 2(1):55–59

Bach F, Zielinski AAF, Helm CV et al (2019) Bio compounds of edible mushrooms: in vitro antioxidant and antimicrobial activities. LWT 107:214–220. https://doi.org/10.1016/j.lwt.2019.03.017

Barros L, Calhelha RC, Vaz JA et al (2007) Antimicrobial activity and bioactive compounds of Portuguese wild edible mushrooms methanolic extracts. Eur Food Res Technol 225:151–156. https://doi.org/10.1007/s00217-006-0394-x

Bonneville S, Delpomdor F, Préat A et al (2020) Molecular identification of fungi microfossils in a Neoproterozoic shale rock. Sci Adv 6(4):7599. https://doi.org/10.1126/sciadv.aax7599

Borokini F, Lajide L, Olaleye T et al (2016) Chemical profile and antimicrobial activities of two edible mushrooms (*Termitomyces robustus* and *Lentinus squarrosulus*). J Microbiol Biotechnol Food Sci 5(5):416–423. https://doi.org/10.15414/jmbfs.2016.5.5.416-423

Cardwell G, Bornman JF, James AP, Black LJ (2008) A review of mushrooms as a potential source of dietary vitamin D. Nutrients 10(10):1498. https://doi.org/10.3390/nu10101498

Chang R (1996) Functional properties of edible mushrooms. Nutr Rev 54(11):S91–S93. https://doi.org/10.1111/j.1753-4887.1996.tb03825.x

Chowdhury HMH, Kubra K, Ahmed RR (2015) Screening of antimicrobial, antioxidant properties and bioactive compounds of some edible mushrooms cultivated in Bangladesh. Ann Clin Microbiol Antimicrob 14:8. https://doi.org/10.1186/s12941-015-0067-3

Chun S, Gopal J, Muthu M (2021) Antioxidant activity of mushroom extracts/polysaccharides — their antiviral properties and plausible antiCOVID-19 properties. Antioxidants 10(12):1899. https://doi.org/10.3390/antiox10121899

Contato AG, Aranha GM, de Filho BA et al (2021) Evaluation of the antioxidant and antimicrobial activity of the culinary-medicinal mushroom *Lentinula boryana* (Agaricomycetes) from Brazil. Int J Med Mushrooms 23(10):1–7. https://doi.org/10.1615/INTJMEDMUSHROOMS.2021039966

Das A, Chen CM, Mu SC et al (2022) Medicinal components in edible mushrooms on diabetes mellitus treatment. Pharmaceutics 14(2):436. https://doi.org/10.3390/pharmaceutics14020436

De Andrade CP, de Oliveira AP, Vieira VMC et al (2021) Antimicrobial activity of the edible mushroom *Pleurotus eryngii* (DC.) Quél grown in liquid medium. Acta Sci Biol Sci 2021:43. https://doi.org/10.4025/ACTASCIBIOLSCI.V43I1.58474

El Enshasy HA, Hatti-Kaul R (2013) Mushroom immunomodulators: unique molecules with unlimited applications. Trends Biotechnol 31:668–677

Elhusseiny SM, El-Mahdy TS, Awad MF et al (2021a) Antiviral, cytotoxic, and antioxidant activities of three edible agaricomycetes mushrooms: *Pleurotus columbinus, Pleurotus sajor-caju*, and *Agaricus bisporus*. J Fungi 7(8):645. https://doi.org/10.3390/jof7080645

Elhusseiny SM, El-Mahdy TS, Awad MF et al (2021b) Proteome analysis and in vitro antiviral, anticancer and antioxidant capacities of the aqueous extracts of *Lentinula edodes* and *Pleurotus ostreatus* edible mushrooms. Molecules 26(15):4623. https://doi.org/10.3390/molecules26154623

Elhusseiny SM, El-Mahdy TS, Elleboudy NS et al (2022) Immunomodulatory activity of extracts from five edible basidiomycetes mushrooms in Wistar albino rats. Sci Rep 12:12423. https://doi.org/10.1038/s41598-022-16349-2

El-Maradny YA, El-Fakharany EM, Abu-Serie MM et al (2021) Lectins purified from medicinal and edible mushrooms: insights into their antiviral activity against pathogenic viruses. Int J Biol Macromol 179:239–258. https://doi.org/10.1016/j.ijbiomac.2021.03.015

Fortes RC, Novaes MRCG (2011) The effects of *Agaricus sylvaticus* fungi dietary supplementation on the metabolism and blood pressure of patients with colorectal cancer during post-surgical phase. Nutr Hosp 26:176–186

Friedman M (2016) Mushroom polysaccharides: chemistry and antiobesity, antidiabetes, anticancer, and antibiotic properties in cells, rodents, and humans. Foods 5:1–40. https://doi.org/10.3390/foods5040080

Gong P, Wang X, Liu M et al (2022) Hypoglycemic effect of a novel polysaccharide from *Lentinus edodes* on STZ-induced diabetic mice *via* metabolomics study and Nrf2/HO-1 pathway. Food Funct 13(5):3036–3049. https://doi.org/10.1039/d1fo03487a

González-Bonilla A, Meneses ME, Pérez-Herrera A et al (2022) Dietary supplementation with oyster culinary-medicinal mushroom, *Pleurotus ostreatus* (Agaricomycetes), reduces visceral fat and hyperlipidemia in inhabitants of a rural community in Mexico. Int J Med Mushrooms 24(9):49–61. https://doi.org/10.1615/IntJMedMushrooms.2022044837

Goyal RRK, Grewal RB, Goyal RRK, Grewal RB (2020) Vitamin and mineral content of Agaricus bisporus (white button) and *Pleurotus sajor-caju* (dhingri) mushrooms. Int J Food Sci Nutr 5(3):100–102

He X, Fang J, Guo Q et al (2020) Advances in antiviral polysaccharides derived from edible and medicinal plants and mushrooms. Carbohydr Polym 229:115548. https://doi.org/10.1016/j.carbpol.2019.115548

Heleno SA, Barros L, Martins A et al (2015) Nutritional value, bioactive compounds, antimicrobial activity and bioaccessibility studies with wild edible mushrooms. LWT Food Sci Technol 63(2):799–806. https://doi.org/10.1016/j.lwt.2015.04.028

Hu Y, Xu J, Sheng Y et al (2022) *Pleurotus Ostreatus* ameliorates obesity by modulating the gut microbiota in obese mice induced by high-fat diet. Nutrients 14(9):1868. https://doi.org/10.3390/nu14091868

Huo W, Feng Z, Hu S et al (2020) Effects of polysaccharides from wild morels on immune response and gut microbiota composition in non-treated and cyclophosphamide-treated mice. Food Funct 11(5):4291–4303. https://doi.org/10.1039/d0fo00597e

Jedinak A, Sliva D (2008) *Pleurotus ostreatus* inhibits proliferation of human breast and colon cancer cells through p53-dependent as well as p53-independent pathway. Int J Oncol 33:1307–1313

Kała K, Kryczyk-Poprawa A, Rzewińska A, Zena MB (2020) Fruiting bodies of selected edible mushrooms as a potential source of lovastatin. Eur Food Res Technol 246:713–722. https://doi.org/10.1007/s00217-020-03435-w

Kalač P (2013) A review of chemical composition and nutritional value of wild-growing and cultivated mushrooms. J Sci Food Agric 93(2):209–218. https://doi.org/10.1002/jsfa.5960

Kaleta B, Roszczyk A, Zych M et al (2021) Selective biological effects of selenium-enriched polysaccharide (Se-le-30) isolated from *Lentinula edodes* mycelium on human immune cells. Biomol Ther 11(12):1777. https://doi.org/10.3390/biom11121777

Ko JL, Hsu CI, Lin RH et al (1995) A new fungal immunomodulatory protein, FIP-fve isolated from the edible mushroom, *Flammulina velutipes* and its complete amino acid sequence. Eur J Biochem 228(2):244–249. https://doi.org/10.1111/j.1432-1033.1995.00244.x

Kośka G (2006) Lectins of higher fungi (macromycetes) - their occurrence, physiological role, and biological activity. Int J Med Mushrooms 8(1):19–30. https://doi.org/10.1615/IntJMedMushr.v8.i1.20

Kozarski M, Klaus A, Jakovljevic D et al (2015) Antioxidants of edible mushrooms. Molecules 20:19489–19525

Li JP, Lee YP, Ma JC et al (2021) The enhancing effect of fungal immunomodulatory protein-volvariella volvacea (Fip-vvo) on maturation and function of mouse dendritic cells. Life 11(6):471. https://doi.org/10.3390/life11060471

Lu H, Lou H, Hu J et al (2020) Macrofungi: a review of cultivation strategies, bioactivity, and application of mushrooms. Compr Rev Food Sci Food Saf 19(5):2333–2356. https://doi.org/10.1111/1541-4337.12602

Ma G, Yang W, Zhao L et al (2018) A critical review on the health promoting effects of mushrooms nutraceuticals. Food Sci Hum Wellness 7(2):125–133. https://doi.org/10.1016/j.fshw.2018.05.002

Ma G, Du H, Hu Q et al (2021) Health benefits of edible mushroom polysaccharides and associated gut microbiota regulation. Crit Rev Food Sci Nutr 62(24):6646–6663. https://doi.org/10.1080/10408398.2021.1903385

Ma G, Xu Q, Du H et al (2022) Characterization of polysaccharide from *Pleurotus eryngii* during simulated gastrointestinal digestion and fermentation. Food Chem 370:131303. https://doi.org/10.1016/j.foodchem.2021.131303

Mahmoud OA, Abdel-Hadi SY (2022) Extraction and purification of lovastatin from the edible mushroom *Laetiporus sulphureus* and its antioxidant activity. Egypt J Bot 62(1):169–175. https://doi.org/10.21608/ejbo.2021.82715.1722

Manikandan K (2011) Nutritional and medicinal values of mushrooms. In: Singh M, Vijay B, Kamal S, Wakchaure GC (eds) Mushrooms cultivation, marketing, and consumption. Indian Council of Agricultural Research, New Delhi, India

Manzi P, Gambelli L, Marconi S et al (1999) Nutrients in edible mushrooms: an interspecies comparative study. Food Chem 65(4):477–482. https://doi.org/10.1016/S0308-8146(98)00212-X

Martinez-Medina GA, Chávez-González ML, Verma DK et al (2021) Bio-functional components in mushrooms, a health opportunity: Ergothionine and huitlacoche as recent trends. J Funct Foods 77:104326

Mwangi RW, Macharia JM, Wagara IN, Bence RL (2022) The antioxidant potential of different edible and medicinal mushrooms. Biomed Pharmacother 147:112621. https://doi.org/10.1016/j.biopha.2022.112621

Nakalembe I, Kabasa JD, Olila D (2015) Comparative nutrient composition of selected wild edible mushrooms from two agro-ecological zones, Uganda. SpringerPlus 4:433. https://doi.org/10.1186/s40064-015-1188-z

Nguepi TIS, Song H (2020) Mushrooms bioactive as prebiotics to modulate gut microbiota in relationships with causes and prevention of liver diseases (review). Int J Med Mushrooms 22(6): 509–519. https://doi.org/10.1615/IntJMedMushrooms.2020034706

Nowacka N, Nowak R, Drozd M et al (2014) Analysis of phenolic constituents, antiradical and antimicrobial activity of edible mushrooms growing wild in Poland. LWT Food Sci Technol 59: 689–694. https://doi.org/10.1016/j.lwt.2014.05.041

Ogbole OO, Nkumah AO, Linus AU et al (2019) Molecular identification, in vivo and in vitro activities of *Calvatia gigantea* (macro-fungus) as an antidiabetic agent. Mycology 10:166–173

Ohkuma T, Tanaka S, Ikekawa T (1983) Augmentation of host's immunity by combined cryodestruction of sarcoma 180 and administration of protein-bound polysaccharide, EA6, isolated from *Flammulina velutipes* (Curt. ex Fr.) Sing. in ICR mice. J Pharmacobiodyn 6(2): 88–95. https://doi.org/10.1248/bpb1978.6.88

Özcan Ö, Ertan F (2018) Beta-glucan content, antioxidant and antimicrobial activities of some edible mushroom species. Food Sci Technol 6:47–55. https://doi.org/10.13189/fst.2018.060201

Panda SK, Sahoo G, Swain SS, Luyten W (2022) Anticancer activities of mushrooms: a neglected source for drug discovery. Pharmaceuticals 15(2):176. https://doi.org/10.3390/ph15020176

Patel S, Goyal A (2012) Recent developments in mushrooms as anticancer therapeutics: a review. 3 Biotech 2:1–15. https://doi.org/10.1007/s13205-011-0036-2

Patra S, Maity P, Chakraborty I et al (2021) Structural studies of immunomodulatory (1 → 3)-, (1 → 4)-α glucan from an edible mushroom *Polyporus grammocephalus*. Int J Biol Macromol 168: 649–655. https://doi.org/10.1016/j.ijbiomac.2020.11.121

Sarangi I, Ghosh D, Bhutia SK et al (2006) Anti-tumor and immunomodulating effects of *Pleurotus ostreatus* mycelia-derived proteoglycans. Int Immunopharmacol 6(8):1287–1297. https://doi.org/10.1016/j.intimp.2006.04.002

Sheng K, Wang C, Chen B et al (2021) Recent advances in polysaccharides from *Lentinus edodes* (Berk.): isolation, structures and bioactivities. Food Chem 358:129883. https://doi.org/10.1016/j.foodchem.2021.129883

Sitati CNW, Ogila KO, Waihenya RW, Ochola LA (2021) Phytochemical profile and antimicrobial activities of edible mushroom *Termitomyces striatus*. Evid Based Complement Alternat Med 2021:10p. https://doi.org/10.1155/2021/3025848

Solano-Aguilar GI, Lakshman S, Jang S et al (2021) The effects of consuming white button mushroom *Agaricus bisporus* on the brain and liver metabolome using a targeted metabolomic analysis. Metabolites 11(11):779. https://doi.org/10.3390/metabo11110779

Steve NTI, Hui S (2020) Mushrooms bioactive as prebiotics to modulate gut microbiota in relationships with causes and prevention of liver diseases (review). Int J Med Mushrooms 22(6): 509–519. https://doi.org/10.1615/IntJMedMushrooms.2020034706

Suzuki H, Watanabe K, Arumugam S et al (2022) Meal ingestion of *Ceraceomyces tessulatus* strain BDM-X (Agaricomycetes) protects against nonalcoholic steatohepatitis in mice. Int J Med Mushrooms 24(1):41–52. https://doi.org/10.1615/IntJMedMushrooms.2021041928

Valverde ME, Hernández-Pérez T, Paredes-López O (2015) Edible mushrooms: improving human health and promoting quality life. Int J Microbiol 2015:376387. https://doi.org/10.1155/2015/376387

Vamanu E, Gatea F, Sârbu I (2018) In vitro ecological response of the human gut microbiome to bioactive extracts from edible wild mushrooms. Molecules 23(9):2128. https://doi.org/10.3390/molecules23092128

Vamanu E, Dinu LD, Pelinescu DR, Gatea F (2021) Therapeutic properties of edible mushrooms and herbal teas in gut microbiota modulation. Microorganisms 9(6):1262. https://doi.org/10.3390/microorganisms9061262

Venturella G, Ferraro V, Cirlincione F, Gargano ML (2021) Medicinal mushrooms: bioactive compounds, use, and clinical trials. Int J Mol Sci 22:1–31. https://doi.org/10.3390/ijms22020634

Vlassopoulou M, Paschalidis N, Savvides AL et al (2022) Immunomodulating activity of *Pleurotus eryngii* mushrooms following their in vitro fermentation by human fecal microbiota. J Fungi 8: 329. https://doi.org/10.3390/jof8040329

Wani BA, Bodha RH, Wani AH (2010) Nutritional and medicinal importance of mushrooms. J Med Plant Res 4:2598–2604

Wasser SP (2010) Medicinal mushroom science: history, current status, future trends, and unsolved problems. Int J Med Mushrooms 12(1):1–16

Wu JY, Siu KC, Geng P (2021) Bioactive ingredients and medicinal values of *Grifola frondosa* (Maitake). Foods 10(1):95. https://doi.org/10.3390/foods10010095

Zafar S, Jabeen F, Akram M, Riaz Z, Munir N (2020) Mushroom species and classification. In: Mtewa AG, Egbuna C, Rao GN (eds) Poisonous plants and phytochemicals in drug discovery. Wiley, Hoboken, NJ. https://doi.org/10.1002/9781119650034.ch8

Zeb M, Lee CH (2021) Medicinal properties and bioactive compounds from wild mushrooms native to North America. Molecules 26(2):251. https://doi.org/10.3390/molecules26020251

Zhang M, Cheung PCK, Zhang L (2001) Evaluation of mushroom dietary fiber (nonstarch polysaccharides) from sclerotia of *Pleurotus tuber-regium* (fries) singer as a potential antitumor agent. J Agric Food Chem 49(10):5059–5062. https://doi.org/10.1021/jf010228l

Zhang X, Cai Z, Mao H et al (2021) Isolation and structure elucidation of polysaccharides from fruiting bodies of mushroom *Coriolus versicolor* and evaluation of their immunomodulatory effects. Int J Biol Macromol 166:1387–1395. https://doi.org/10.1016/j.ijbiomac.2020.11.018

Zhang H, Jiang F, Zhang J et al (2022a) Modulatory effects of polysaccharides from plants, marine algae and edible mushrooms on gut microbiota and related health benefits: a review. Int J Biol Macromol 204:169–192. https://doi.org/10.1016/j.ijbiomac.2022.01.166

Zhang W, Hu B, Liu C et al (2022b) Comprehensive analysis of *Sparassis crispa* polysaccharide characteristics during the in vitro digestion and fermentation model. Food Res Int 154:111005. https://doi.org/10.1016/j.foodres.2022.111005

Zhao S, Gao Q, Rong C et al (2020) Immunomodulatory effects of edible and medicinal mushrooms and their bioactive immunoregulatory products. J Fungi 6:1–37. https://doi.org/10.3390/jof6040269

Mycotherapy (Medicinal Mushrooms) as a Potential Treatment for Epilepsy

Rishi M. Nadar, Suhrud Pathak, Jack Deruiter, Fada Alghenaim, Sindhu Ramesh, Graham Lynn, Randall Clark, Timothy Moore, Dinesh Chandra Agrawal, and Muralikrishnan Dhanasekaran

Contents

Abstract Epilepsy is one of the oldest neurological diseases known to humans that affect patients of all ages and sexes worldwide. Consequently, there are various treatment tactics to prevent and treat this recurring neurological condition, such as surgical, pharmacological (synthetic drug), non-pharmacological (diet), and alternative and complementary therapies. However, the current therapies may not be appropriate for all patients, and there are numerous adverse effects and hypersensitivity reactions. Therefore, there is an impending necessity for novel therapies to treat this neurological condition immediately. "Complementary" (therapies used in concert with traditional/conventional remedy) and "alternative" (nontraditional/conventional treatment) therapeutic approaches are currently considered as future

R. M. Nadar · S. Pathak · J. Deruiter · F. Alghenaim · S. Ramesh · G. Lynn · R. Clark · T. Moore · M. Dhanasekaran (✉)
Department of Drug Discovery and Development, Harrison College of Pharmacy, Auburn, University, Auburn, AL, USA
e-mail: dhanamu@auburbn.edu

D. C. Agrawal (✉)
Department of Applied Chemistry, Chaoyang University of Technology, Taichung, Taiwan
e-mail: dcagrawal@gm.cyut.edu.tw

D. C. Agrawal, M. Dhanasekaran (eds.), *Mushrooms with Therapeutic Potentials*,
https://doi.org/10.1007/978-981-19-9550-7_17

healthcare interventions to treat epilepsy prophylactically and therapeutically. Mycotherapy (the use of mushrooms in healthcare) has shown substantial preventative and curative efficacy in treating various human ailments. Hence, in this book chapter, we have identified various beneficial mushrooms with significant anti-epileptic effects. Thus, this may help to curb the pathophysiology, symptoms, and comorbidities associated with neuronal hyperarousal leading to epilepsy.

Keywords Anti-epileptic · Epilepsy · Medicinal mushrooms · Mycotherapy · Neuroprotection seizure

Abbreviations

AD	Alzheimer's disease
AEDs	anti-epileptic drugs
Akt	amino kinase terminal
ALS	amyotrophic lateral sclerosis
ATP depletion	adenosine triphosphate depletion
Aβ	amyloid beta
BBB	blood-brain barrier
BDNF	brain-derived neurotrophic factor
Ca^{2+}	calcium
CAM	complementary alternative medicine
CIM	complementary integrated medicine
CNS	central nervous system
EEG	electroencephalogram
EFSA	European Food Safety Authority
ER	endoplasmic reticulum
ERK	extracellular signal-regulated kinases
FDA	Food and Drug Administration
GLUT1	glucose transporter
HD	Huntington's disease
IL	interleukin
ILAE	International League Against Epilepsy
MEK	mitogen-activated protein kinase/ERK kinase
MMP	mitochondrial membrane potential
MRI	magnetic resonance image
mTOR	mammalian target of rapamycin
NGF	nerve growth factor
PC12	PC12 cell line
PD	Parkinson's disease
PPARγ	peroxisome proliferator activated receptor gamma
PPKAG2	protein kinase AMP-activated non-catalytic subunit gamma 2
ROS	reactive oxygen species

SCN1A mutation	sodium voltage-gated channel alpha subunit 1 mutation
Slc2a4	solute carrier family 2-member 4 gene
TGF	transforming growth factor
TNF	tumor necrosis factor
USA	United States of America
WBC	white blood cell

1 Introduction

Central nervous system (CNS) diseases, or neurological/neurodegenerative disorders, affect all sexes and ages and are escalating globally, presenting a challenge for healthcare worldwide (Kovacs 2017). Additionally, CNS diseases are very expensive to diagnose and treat. The most worldwide prevalent neurological or neurodegenerative disorders are Alzheimer's disease (AD) and other forms of dementia, Parkinson's disease (PD), Huntington's disease (HD), ischemic stroke, epilepsy, depression, psychosis, bipolar disorders, multiple sclerosis, and amyotrophic lateral sclerosis (ALS) (Checkoway et al. 2011). These neurological or neurodegenerative disorders have distinct pathophysiological components triggering the impairment of mood, memory, and movement, thus liable for diminishing an individual's ability to live peacefully in a society (Vaquer-Alicea and Diamond 2019). According to the World Health Organization, neurological and neurodegenerative diseases are exceedingly devastating and impose significant challenges for healthcare globally. Epilepsy is one of the oldest disease states known to humanity (Gross 1992; Vaquer-Alicea and Diamond 2019). Despite the development of a considerable number of anti-epileptic drugs, there remains a need for anti-epileptic medications which are therapeutically effective and safe with minimal adverse effects and without any hypersensitivity reactions (Thijs et al. 2019).

Mycotherapy uses medicinal mushrooms' valuable ingredient*s to treat different illnesses*. Medicinal mushrooms have significant pharmacodynamic potential due to their excellent pharmacokinetic properties and low lipid content and calories (Badalyan et al. 2019); (Agrawal and Dhanasekaran 2021). Regarding pharmacodynamic actions, medicinal mushrooms are a resource of synergistic bioactives that can cross the blood-brain barrier (BBB) (Govindarajulu et al. 2021). They can exert nutraceutical, preventive, curative, and restorative effects in the remedy for several CNS ailments (Lee et al. 2019).

A large volume of scientific literature has established that genetic mutations, oxidative stress (caused by decreased antioxidant and increased prooxidants leading to lipid peroxidation), dysfunction of nuclear and mitochondrial DNA, apoptosis, inflammation, mitochondrial dysfunction (ATP depletion), endoplasmic reticulum (ER) stress, excitotoxicity, overaccumulation of calcium (Ca^{2+}), depletion of growth factors (BDNF, NGF), autophagy, and protein misfolding in the neurons all can contribute to altered neuronal function and various forms of pathology in specific regions of the brain (Thrash et al. 2009; Muralikrishnan et al. 2002; Dhanasekaran et al. 2008; Bhattacharya et al. 2018; Pondugula et al. 2022; Bhattacharya et al. 2015; Daud et al. 2017; Mouli et al. 2015; Ransohoff 2016). Typically, pathology in neuronal pathways correlates with neurological and neurodegenerative disease progression, leading to significant symptoms in affected patients (Thrash et al. 2009). To date, there have been efforts by scientists across the world to enhance the therapy for various neurological disorders. However, most treatments developed to date only lower the progressive, symptomatic impact on neuronal function but do not specifically target the primary cause of the neurological disease (Van Bulck et al. 2019). In many cases, a single drug may not be capable of successfully curing a neurological condition (Gautier 2020). Investigations have demonstrated that herbal and alternative medicine, such as mycotherapy, can be employed to treat neurological disorders (Badalyan et al. 2019; Agrawal and Dhanasekaran 2021; Govindarajulu et al. 2021; Ramesh et al. 2019). Consequently, there has been a current surge of attention in integrative, complementary, and alternative medicine, particularly dietary supplements and functional foods, for preventing and treating various neurological disorders (Yadav et al. 2020).

2 Epilepsy

Epilepsy is a neurological disease in which the brain's electrical activity is disrupted. Epilepsy is an unprovoked seizure separated by more than 1 day (Fisher et al. 2005). The global as well as the national burden of epilepsy is increasing. In the United States, 39 in 1000 people have developed epilepsy in their lifetime, and about three million have been diagnosed with epilepsy. Globally, more than 60 million people are diagnosed with epilepsy (Baulac et al. 2015). Epilepsy affects all age groups and has a similar prevalence among different socioeconomic and racial cohorts. However, its incidence is higher in pediatric and geriatric groups. It has been estimated that 75% of epilepsy cases affect children. It has also been reported that the entire direct cost for each patient affected with epilepsy is from $10,192 to $47,862 (Begley and Durgin 2015). The life expectancy of the epileptic patient is 10 years less than that of normal peers (Gaitatzis et al. 2004).

Epilepsy is classified, according to the affected region, into a "focal" (partial) epilepsy which affects a part of the brain hemisphere, or it could be "generalized" epilepsy which affects the whole brain (Devinsky et al. 2018). Epileptic seizures are subcategorized into generalized onset seizures, which are categorized into the motor and non-motor seizures; focal onset seizure, which is further classified into aware or impaired awareness seizures and motor onset and non-motor onset seizures; and unknown seizures, which are further classified into the motor and non-motor seizures. These categories of seizures are set by the International League Against Epilepsy (ILAE) according to each seizure's different diagnostic characteristics and features. There are distinct stages of seizures, the beginning stage (prodrome and aura), middle stage (ictal), and recovery stage (postictal). The beginning stage phases occur a few seconds before the seizure starts and warn the patient that the seizure is about to occur. The middle stage phase is when the actual seizure starts and the electrical disturbance of the brain continues. The recovery stage is when the seizure is about to end and the body starts to relax (Ali 2018).

Epilepsy is a disease triggered by certain risk factors, classified as external or internal factors. Head trauma, including prenatal injury, brain tumors, inflammation, cerebral hemorrhage, and brain infections, which could be viral or bacterial that alter the patient's neurological functions, are external factors (Aronica and Mühlebner 2017). The internal factors include aging, gene mutation (SCN1A mutation), metabolic disorders (facilitated glucose transporter member 1 (GLUT1) deficiency), and autoimmune diseases like autoimmune encephalitis (Meldrum 1990). However, some seizure cases have no specific known cause. Abnormalities in ion channel conductance, such as blocking of sodium-potassium ATPase, the imbalance between the inhibitory and excitatory neurotransmitters, disruption of the genes that regulate neurotransmitters functions, alterations in neuronal plasticity, electrolyte disturbance, and activation of NMDA receptor all induce epileptogenesis (Huff and Murr 2022; Kapur 2018; Di Giovanni et al. 2016).

Epilepsy has a wide range of symptoms and signs, which depend on the type of epilepsy. The severity is more prominent in the generalized seizure. These range from no alteration in consciousness to memory loss of the events of the attack (Dupont 2010). Furthermore, control of voluntary and involuntary muscles is lost during the seizure. Jerking, collapsing, and body stiffening are common in the generalized seizure (Falco-Walter 2020). Like other neurological disorders, the diagnosis of epilepsy is difficult because most attacks occur outside healthcare institutions and not in the presence of healthcare providers. However, several approaches are used to identify epilepsy, including a thorough evaluation of investigating a patient's medical history, including electroencephalogram (EEG) and neuroimaging (magnetic resonance image (MRI)) findings (Russo 1981). An early and accurate diagnosis is crucial to identify the appropriate treatment modality, which can be reflected as an improvement in the patient's quality of life. Additional diagnostic tests can be performed to determine the epilepsy type and cause, such as positron emission tomography, single-photon emission computed tomography, and magneto encephalogram (Shin et al. 2014).

Several treatment strategies have been developed to manage epilepsy. The cornerstone of the current therapy are medications called "anti-epileptic drugs" (AEDs) (Manford 2017). These drugs control the electrical activity in the brain that causes a seizure. AEDs control more than two-thirds of epilepsy cases, and patients often stop using them because they become seizure-free. More than one drug is typically used, and replacing one with another is a common approach to enhance seizure control (Thijs et al. 2019). The ketogenic, low-carbohydrate, and high-fat diet is an option for children and adults whose epilepsy is not controlled by AEDs. Vagus nerve stimulation, surgery, and deep brain stimulation are other treatment modalities, but their use is limited to specific cases (Ryvlin et al. 2021).

Several hypotheses have been proposed to identify the physiological and the pathological changes associated with epilepsy (Striano and Minassian 2020).

However, most are controversial because epilepsy is a complex disease, and a specific diagnosis is often difficult. Monoamine neurotransmitters, such as serotonin, dopamine, and norepinephrine, are bioactive substances characterized by an amino group connected to an aromatic ring and two carbon chains in their chemical structure (Mastrangelo 2021). They control numerous body functions. Serotonin regulates mood and behavior, and dopamine modulates movement activity, thought, and cognition, while noradrenaline controls cardiovascular and mental functions, including attentiveness and depression (Kurian et al. 2011). It has been reported that monoamines play a key role in regulating epileptic seizures. At high concentrations, they could evoke epilepsy, while within a certain concentration, they have a protective effect against epileptic seizures (Di Giovanni et al. 2016). Association of psychological disorders, such as depression and anxiety with epilepsy, supports the role of monoamine in epilepsy, at least partly in the progression, complication, and response to treatment. These influence patients' compliance with medications, sleep, and quality of life. In some cases, depression can be fatal in patients with epilepsy (Boylan et al. 2004; DiMatteo et al. 2000). Epilepsy, depression, and other neuropsychiatric abnormalities are linked, in part, to monoamine disturbance in the brain.

The conventional prophylactic and therapeutic approaches to treating a disease/pathological condition include surgical, radiological, pharmacological, nutritional, alternative, complementary, psychological, and physical approaches (Schutgens and Clevers 2020; Buchman 2018). Globally, the concept of "integrative healthcare" is currently progressing as a means to better control disease. The "integrative" healthcare approach incorporates a conventional (surgical, radiological, pharmacological) and a complementary approach simultaneously and collectively in an organized manner (Frisch and Rabinowitsch 2019). Integrative healthcare accentuates multimodal interferences, emphasizing curing the patient's entire body instead of focusing on one tissue, organ, or organ system (Boon et al. 2004). Thus, an integrative healthcare approach is well-aligned healthcare between different healthcare professionals and organizations by coordinating traditional and complementary approaches in a synchronized manner to obtain "synergistic healthcare" for the entire body of a patient (Ananth 2009).

Mycotherapy is universally described as the study of using mushrooms (as a whole, or extracts, and bioactive obtained) to treat or prevent diseases (Venturella et al. 2021). Intriguingly, mushrooms have been conventionally and customarily consumed to protect natural well-being and remedy various health ailments throughout the world. Recently, mycotherapy is becoming a popular component within the circle of dietary supplementation (Meade et al. 2022). Dietary supplements are deemed internationally as complementary alternative medicine (CAM) and complementary integrated medicine (Sales-Campos et al. 2021; Massey 2002). Additionally, dietary supplements are among the most frequently employed of all integrative, complementary, and alternative therapies (Burdock 2000). The Food and Drug Administration (FDA) in the United States of America (USA) does not require commercial nutraceutical and pharmaceutical companies to prove safety and efficacy (Denham 2021; Aschenbrenner 2019). Still, the history of the safety of dietary supplements has to be mentioned. Similarly, the European Food Safety Authority (EFSA) has established guidelines for using dietary supplements in health and disease risk reduction claims. It requires toxicological data, and a natural product (botanicals, fungi) can be disclosed exclusively as a food supplement, not as a therapeutic drug (Petkova-Gueorguieva et al. 2019; Silano et al. 2011).

Consequently, for most mushroom-based dietary supplements, safety and efficacy are validated by their customary use, in silico, in vitro, in vivo, and clinical studies (Arroo et al. 2020; Vilariño et al. 2020). Evidence suggests that medicinal mushrooms are effective against allergy, infection (bacteria, fungal, virus), inflammation, cancer, tumor, neurological, hepatological, gastrointestinal, cardiovascular, endocrinological, and nephrological disorders (Badalyan et al. 2019). In this book chapter, we elucidate the effect of mycotherapy on epilepsy.

3 Medicinal Mushrooms to Reduce Seizures and/or Treat Epilepsy

The following sections include brief reports on several mushroom species (Fig. 1a–d) used to reduce seizures or treat epilepsy (Table 1).

3.1 Hericium Erinaceus

H. erinaceus (Fig. 1a), popular as "lion's mane," is a culinary mushroom belonging to the family *Hericiaceae*. The major bioactive compounds of *H. erinaceus* used as mycotherapeutics are erinacines (Fig. 2), aromatic compounds, steroids, alkaloids, and lactones showing anti-inflammatory effects (Jang et al. 2019). *H. erinaceus* can decrease oxidative stress and inflammation and increase immune functions. Also, *H. erinaceus* has been shown to possess antimicrobial, anticancer, antihyperglycemic, anti-fatigue, antihypertensive, antihyperlipidemic, antisenescence, cardiovascular protective, hepatoprotective, neuroprotective, and nephroprotective effects (Chong et al. 2019; Khan et al. 2013; Ghosh et al. 2021; Wang et al. 2019; Chaiyasut et al. 2018; Zhang et al. 2017). The anti-epileptic effect of *H. erinaceus* protected the hippocampal neurons against epileptogenic neurotoxin pilocarpine-induced neurotoxicity (Jang et al. 2019; Chan et al. 2013). *H. erinaceus* inhibited hippocampal cyclooxygenase 2 (COX2) expression and reduced seizures (Jang et al. 2019); (Chan et al. 2013).

Fig. 1 (**a–d**) Mushroom species used to reduce seizures or treat epilepsy. (**a**) *Hericium erinaceus*; (**b**) *Ganoderma lucidum*; (**c**) *Amauroderma rugosum*; (**d**) *Pleurotus ostreatus*. (The pictures of mushrooms are courtesy of Wikimedia Commons; licensed under the Creative Commons Attribution 3.0 and 4.0)

3.2 Ganoderma Lucidum

Commonly known as lingzhi or reishi, *G. lucidum* (Fig. 1b) is a giant, dark mushroom belonging to the family *Ganodermataceae* with a long history of culinary and medicinal usage. The major bioactive compounds of *G. lucidum* are polysaccharides (α/β-D-glucans), alkaloids, triterpenoids (ganoderic acid, ganoderenic acid (Fig. 3), ganoderol, ganoderiol, lucidenic acid), sterols/ergosterol, proteins (LZ-8, LZ-9), nucleosides (adenosine, inosine, uridine), and nucleotides (guanine, adenine) (Martínez-Montemayor et al. 2019). *Ganoderma lucidum* exhibited neuroprotection and anti-epileptic activity against epileptogenic neurotoxin pentylenetetrazole-induced and kainate-induced convulsions and maximal electroshock seizure (Socala et al. 2015). In humans, *G. lucidum* spore powder has been shown to reduce seizure frequency and improve quality of life as measured by the "Quality of Life in Epilepsy Inventory" in patients with epilepsy (Wang et al. 2018; Ahmad et al. 2021; Tello et al. 2013).

Table 1 The pharmacodynamic effects of mushrooms' bioactive compounds to reduce seizure and/or prevent epilepsy

Mushroom species	Bioactive compounds	Pharmacodynamic effect to reduce seizure and/or prevent epilepsy	References
Hericium erinaceus	Erinacines, aromatic compounds, steroids, alkaloids, and lactones	• Anti-inflammatory effect.	Jang et al. (2019)
Ganoderma lucidum	Polysaccharides (α/β-D-glucans), alkaloids, triterpenoids (ganoderic acids, ganoderenic acids, ganoderol, ganoderiol, lucidenic acids), sterols/ergosterol, proteins (LZ-8, LZ-9), nucleosides (adenosine, inosine, uridine), and nucleotides (guanine, adenine)	• Inhibition of intracellular calcium accumulation and stimulation of expression of CaMKII α in epileptic hippocampal neurons. • Anti-inflammatory. • Decreased astrocytic reactivity and reduced the expression of IL-1β and TNF-α.	Martínez-Montemayor et al. (2019) Wang et al. (2018) Wang et al. (2013) Wang et al. (2014) Socala et al. (2015)
Amauroderma rugosum	Linoleic acid, ergosterol, and ethyl linoleate	• Antioxidant. • Anti-inflammatory.	Chan et al. (2013)
Pleurotus ostreatus	Fatty acids, sterols, individual phenolic compounds, terpenic acids, glucans, and chrysin	• Inhibit protease. • Antioxidant.	Sales-Campos et al. (2021)
Armillaria mellea	Ascorbic acid, tannin, total phenolic, total flavonoid, β-carotene, and lycopene)	• Antioxidant, • Anti-inflammatory (decrease IL-1β, TNF-α, and decrease reducing the protein expression of ionized calcium binding adaptor molecule 1). • Repressed nuclear apoptosis (inhibited caspase-3 activation). • Reduced intracellular reactive oxygen species generation and accumulation. • Restored mitochondrial membrane potential (MMP).	Erbiai et al. (2021) Lin et al. (2021) Ojemann et al. (2006)
Ganoderma neo-japonicum	Polysaccharides (terpenes, ganoderic acids, and phenolic compounds	• Affects mitogen-activated protein kinase kinase/extracellular signal-regulated kinase (MEK/ERK1/2) and phosphoinositide-3-kinase/protein kinase B (PI3K/Akt). • Antioxidant DNA repair.	Ling-Sing Seow et al. (2013) Tan et al. (2018) Tan et al. (2015)
Cordyceps militaris	Cordycepin, nucleosides, sterols, cyclic peptides, flavonoids, dihydrobenzofurans, bioxanthracenes, polyketide, terpenes, alkaloids, and phenolics	• Antioxidant. • Anti-inflammatory. • Attenuate ROCK2 protein hyperactivation. • Increase the downregulation of p-Akt/Akt signaling.	Olatunji et al. (2018) Wei et al. (2021)

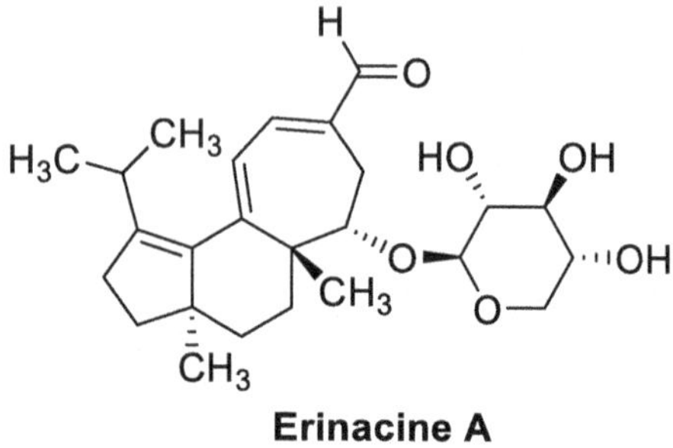

Fig. 2 Chemical structure of erinacine A

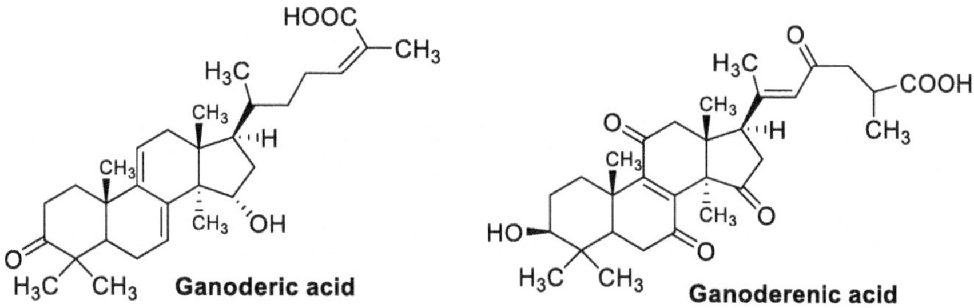

Fig. 3 Chemical structures of ganoderic acid and ganoderenic acid

3.3 Amauroderma Rugosum

Amauroderma rugosum (*Sanguinoderma rugosum*) (Fig. 1c) is a wild medicinal mushroom that belongs to the family *Ganodermataceae* and is conventionally utilized to relieve seizures. *A. rugosum* has been shown to possess ergosterol, linoleic acid, ethyl linoleate (Fig. 4), phenolic compounds, polysaccharides, and triterpenes. *A. rugosum* has been shown to scavenge reactive oxygen species and combat oxidative stress. Additionally, *A. rugosum* can block mitochondrial dysfunction and programmed cell death (apoptosis) induced by the dopaminergic neurotoxin (6-hydroxydopamine (6-OHDA)) in rat dopaminergic (PC12) cells.

Furthermore, dopaminergic neurotoxin 6-OHDA significantly increased the expressions of proapoptotic proteins and decreased the Akt (protein kinase B)/mTOR (mammalian target of rapamycin)- and MEK (mitogen-activated protein kinase kinase)/ERK (extracellular signal-regulated kinases)-dependent signaling pathways. *A. rugosum* exerted significant neuroprotective effects by modulating the above signaling process. This study validates the possible prophylactic and therapeutic use of *A. rugosum* in preventing or treating oxidative, apoptosis, and mitochondrial dysfunction-related neurological and neurodegenerative diseases. The current literature has revealed the antioxidant, anti-inflammatory, neuroprotective, anticancer, antihyperlipidemic, and antibacterial effects of *A. rugosum* (Zheng et al. 2022). Thus the present scientific data imply that *A. rugosum* and its natural

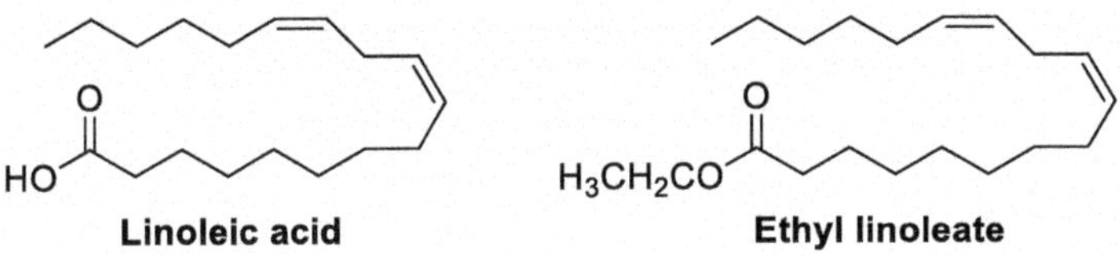

Fig. 4 Chemical structures of linoleic acid and ethyl linoleate

Fig. 5 Chemical structure of chrysin

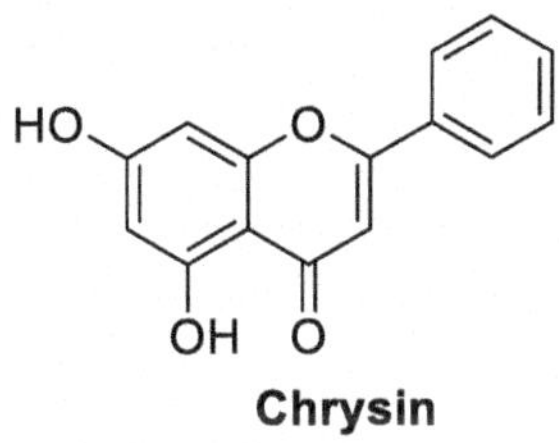

bioactives have significant potent preventative and therapeutic medicinal benefits, particularly for neurological, neurodegenerative, and age-related diseases. *A. rugosum* protects the hippocampal neurons against excitatory neurotoxin and glutamate-induced neurotoxicity (Sam et al. 2022).

3.4 *Pleurotus Ostreatus*

Pleurotus ostreatus (oyster mushroom) (Fig. 1d) is the second most cultivated edible medicinal mushroom belonging to the family *Pleurotaceae*. It contains fatty acids, sterols, individual phenolic compounds, terpenic acids, glucans, and chrysin (Fig. 5). Chrysin (polyphenolic flavonoids) is one of the highly vital bioactive compounds of several fruits, vegetables, and mushrooms. Chrysin in *Oroxylum indicum* has been shown to exert neuroprotection and cardioprotection (Pondugula et al. 2021; Pondugula et al. 2022). Additionally, *Oroxylum* and ethnomedicinal botanicals have been used by indigenous communities in the Indian subcontinent for treating epilepsy (Sharma et al. 2013). Chrysin decreases neuroinflammation considerably and exerts neuroprotection. Extracts of *P. ostreatus* have antitumor, anticancer, antioxidant, nephroprotective, and neuroprotective effects (Sharma et al. 2021; Nabavi et al. 2015).

3.5 *Armillaria Mellea*

Commonly known as honey fungus, *Armillaria* is a basidiomycete fungus in the genus *Armillaria* belonging to the *Physalacriaceae* family. *Armillaria mellea*

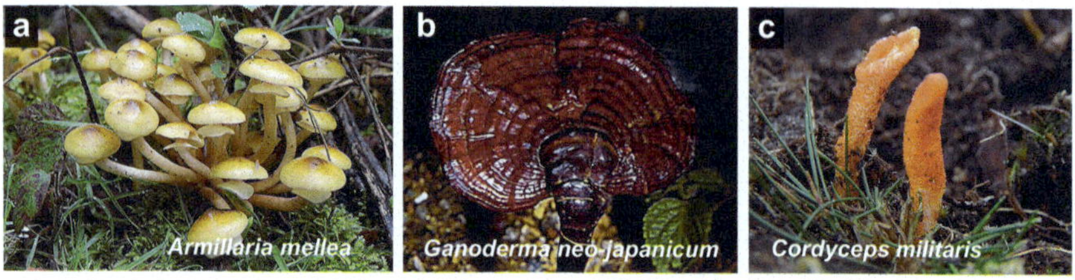

Fig. 6 (**a–c**) Mushroom species used to reduce seizures or treat epilepsy. (**a**) *Armillaria mellea*; (**b**) *Ganoderma neo-japonicum*; (**c**) *Cordyceps militaris.* (The pictures of mushrooms are courtesy of Wikimedia Commons; licensed under the Creative Commons Attribution 3.0 and 4.0)

(Fig. 6a) is an edible medicinal mushroom with several pharmacological activities, such as antioxidant and antiapoptotic effects. Due to potent neuroprotective natural bioactives, *Armillaria mellea* protected against excitotoxins (glutamate) and exogenous neurotoxin (aluminum chloride and D-galactose)-induced neurotoxicity. These endogenous and exogenous neurotoxins increased the generation and accumulation of intracellular free radicals (ROS), stimulated caspase-3 activation, and diminished mitochondrial membrane potential (MMP). However, *A. mellea* exhibited significant neuroprotective effects by blocking the neurotoxic signaling induced by endogenous and exogenous neurotoxins (Delay 1988). In addition, an animal model of Alzheimer's disease blocked the behavioral deficits associated with movement and memory. It inhibited the hippocampal deposition of amyloid beta (Aβ), oxidative damage, cholinergic dysfunction, and p-Tau aggregations (Delay 1988). Ancient Chinese herbal formulations contain ascorbic acid, tannin, vanillin, and the symbiotic fungus *Armillaria*. These formulations have exhibited significant anti-epileptic effects (Ojemann et al. 2006). This natural and alternative anti-epileptic approach is cheap (cost-effective) and has fewer adverse effects than synthetic traditional anti-epileptic drugs. The natural bioactives of *A. mellea* can decrease excitatory neurotransmission and increase inhibitory neurotransmission, leading to decreased neuronal hyperarousal. The decreased neuronal hyperarousal can decrease neuronal excitability and suppress seizures.

3.6 *Ganoderma Neo-Japonicum*

Ganoderma neo-japonicum (Fig. 6b), a polypore fungus in the family *Ganodermataceae*, thrives on *Schizostachyum brachycladum* (decaying bamboo) in Malaysian forests (Sabaratnam et al. 2013). *G. neo-japonicum* is a valuable polypore mushroom that grows on decaying bamboo clumps (*Schizostachyum brachycladum*) in Asia (mainly Malaysia). The Malaysian tribes (Temuans and Temiars) have consumed and utilized basidiocarps of *G. neo-japonicum* for various medicinal benefits. This mushroom has been cultivated to warrant an uninterrupted resource of *G. neo-japonicum* for its therapeutic purpose. The major bioactives of

G. neo-japonicum are sugars/carbohydrates ((1,3)(1,6)-β-D-glucan polysaccharide), proteins, dietary fiber, and micronutrients. The bioactives of *G. neo-japonicum* induce glucose uptake and adiponectin secretion and decrease lipid accumulation in adipocytes. *G. neo-japonicum* increased the expressions of adiponectin, Akt (protein kinase B), PPARγ (peroxisome proliferator activated receptor gamma), PRKAG2 (protein kinase, AMP activated), and Slc2a4 (glucose transporter). This pharmacodynamic action led to the insulin-like effect resulting in glucose uptake and decreasing hyperglycemia. It is indigenous to the Temuan and Temiar tribes that use the basidiocarps to enhance vitality (body strength), cure fever, and reduce seizures. *G. neo-japonicum* possesses antioxidant, genoprotection, and DNA repair properties (Tan et al. 2015; Ling-Sing Seow et al. 2013; Tan et al. 2018). The indigenous people strung and wore *G. neo-japonicum* around their children's necks to prevent and treat epilepsy (Tan et al. 2015).

3.7 Cordyceps Militaris *(Linn)*

Cordyceps militaris (Fig. 6c), belonging to the Cordycipitaceae, is a medicinal mushroom used in tonics in several Asian countries for treating epilepsy, anxiety, and insomnia in patients with depression (Lou et al. 2019; Zhou et al. 2021). It contains cordycepin (Fig. 7), nucleosides, sterols, cyclic peptides, flavonoids, and a variety of alkaloids and phenols (Wu et al. 2021; Ashraf et al. 2020). Its anti-inflammatory and antioxidant actions may benefit its prophylactic and therapeutic management. As a vital factor of the nonspecific or innate immunity of the CNS, neuroinflammation primarily provides neuronal protection by providing immediate repair against endogenous and exogenous neuronal insults, thereby maintaining appropriate neuronal functions (Woodling and Andreasson 2016). Nevertheless, the generation of acute and chronic pro-inflammatory cytokines in the CNS has been linked to weakening of the blood-brain barrier (BBB), resulting in neuronal hyperexcitability in a small region or throughout the brain (Takata et al. 2021; Marchi et al. 2012). Neuroinflammation is principally marked by the increased excessive formation/generation of pro-inflammatory mediators/cytokines in the specific epileptogenic foci (Pracucci et al. 2021). The major pro-inflammatory

Fig. 7 Chemical structure of cordycepin

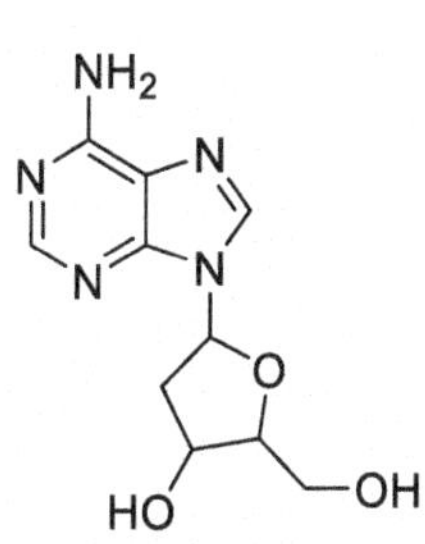

mediators/cytokines and targets are the cyclooxygenase-2/prostaglandin E2, interleukin-1β, transforming growth factor-β, toll-like receptor-4, high-mobility group box 1, and tumor necrosis factor-α (TNF-alpha) (Rana and Musto 2018). Neuroinflammation can cause microglial activation, damage of the BBB endothelial cells, and infiltration of plasma proteins and immune granulocytes (WBC neutrophils) initially and subsequently followed by monocytes within 24 h and lymphocytes within 24–48 h, leading to the upregulation of an array of inflammatory mediators/cytokines (IL-1β, IL-6, TNF-α) and growth factors like TGF-β and BDNF (DiSabato et al. 2016). The neuronal and neurochemical modifications in the CNS are generally detected in several brain regions of epileptic patients and in the brains of an animal model of epilepsy (Saniya et al. 2017). A growing body of research data indicates mutual enablement between inflammation and seizures, implying that impeding the unsought neuroinflammatory signaling within the CNS can deliver innovative therapeutic approaches to prevent and/or treat seizures and epilepsy (Radu et al. 2017).

Interestingly, increased neuronal firing (seizures) enhance the BBB permeability. This can exaggerate and disseminate neuroinflammation via the extravasation of leukocytes and pro-inflammatory mediators/cytokines from blood vessels into the CNS parenchyma (Kaplan et al. 2020). Usually, primary persistent seizures can initiate acute immune and inflammatory responses in the CNS, while the subsequent instinctive recurring seizures trigger and sustain chronic neuroinflammation (Dey et al. 2016). Thus, the communication between the three major active events, immune cell infiltration, BBB interruption, and pro-inflammatory microglial/macrophage polarization, plays a key part in the vicious cycle of neuronal inflammation and seizure. Cordycepin from *Cordyceps militaris* can inhibit neuroinflammation and thus can reduce the increased neuronal firing and can be used to treat epilepsy (Wei et al. 2021).

4 Conclusion

In this chapter, we summarize evidence and knowledge regarding the potential therapeutic benefit of mycotherapy (medicinal mushrooms) in seizure disorders correlated with the bioactive compounds present in culinary (edible) and medicinal mushrooms. Mycotherapeutics can efficiently increase neurogenesis by enhancing various growth factor production, decreasing inflammation and oxidative stress, and affecting neuronal signaling pathways. These neuroprotective mechanisms can decrease neuronal arousal and reduce seizures. Furthermore, prophylactic consumption of selective medicinal mushrooms can diminish or delay age-related neuronal hyperarousal. Even though many medicinal mushrooms are safe to eat, their pharmacokinetic safety profile must be precisely confirmed.

Frequently undervalued in the diet, mushrooms have been consumed and employed as a remedy for centuries throughout the world. Customary and folk remedy practitioners proclaim the restorative and curative value of mushrooms.

Various in vitro, in vivo, and human clinical studies have undoubtedly validated the use of medicinal mushrooms to treat epilepsy. The major bioactive compounds of mycotherapeutics are polysaccharides, indoles, polyphenols, and carotenoids which have been shown to possess potent antioxidant and anti-inflammatory effects, leading to their anti-epileptic potential. To conclude, epidemiological studies, evaluating neuronal and peripheral biomarkers, and randomized clinical trials may assist in shedding light on the potential effect of mycotherapeutics.

References

Agrawal DC, Dhanasekaran M (2021) Medicinal herbs and fungi - neurotoxicity vs. neuroprotection, 1st edn. Springer Nature Singapore Pte Ltd., pp 1–510. https://doi.org/10.1007/978-981-33-4141-8

Ahmad R, Riaz M, Khan A et al (2021) Ganoderma lucidum (Reishi) an edible mushroom; a comprehensive and critical review of its nutritional, cosmeceutical, mycochemical, pharmacological, clinical, and toxicological properties. Phytother Res 35(11):6030–6062. https://doi.org/10.1002/ptr.7215

Ali A (2018) Global health: epilepsy. Semin Neurol 38(2):191–199. https://doi.org/10.1055/s-0038-1646947

Ananth S (2009) Applying integrative healthcare. Explore (NY) 5(2):119–120. https://doi.org/10.1016/j.explore.2008.12.011

Aronica E, Mühlebner A (2017) Neuropathology of epilepsy. Handb Clin Neurol 145:193–216. https://doi.org/10.1016/b978-0-12-802395-2.00015-8

Arroo RRJ, Sari S, Barut B et al (2020) Flavones as tyrosinase inhibitors: kinetic studies in vitro and in silico. Phytochem Anal 31(3):314–321. https://doi.org/10.1002/pca.2897

Aschenbrenner DS (2019) FDA warns of hidden prescription drugs in dietary supplements. Am J Nurs 119(5):22–23. https://doi.org/10.1097/01.NAJ.0000557910.89656.9c

Ashraf SA, Elkhalifa AEO, Siddiqui AJ et al (2020) Cordycepin for health and Well-being: a potent bioactive metabolite of an entomopathogenic medicinal fungus cordyceps with its nutraceutical and therapeutic potential. Molecules 25(12). https://doi.org/10.3390/molecules25122735

Badalyan SM, Barkhudaryan A, Rapior S (2019) Recent progress in research on the pharmacological potential of mushrooms and prospects for their clinical application. In: Agrawal D, Dhanasekaran M (eds) Medicinal mushrooms. Springer, Singapore. https://doi.org/10.1007/978-981-13-6382-5_1

Baulac S, Ishida S, Marsan E et al (2015) Familial focal epilepsy with focal cortical dysplasia due to DEPDC 5 mutations. Ann Neurol 77(4):675–683

Begley CE, Durgin TL (2015) The direct cost of epilepsy in the United States: a systematic review of estimates. Epilepsia 56(9):1376–1387

Bhattacharya D, Dunaway EP, Bhattacharya S et al (2015) Impaired ILK function is associated with deficits in hippocampal based memory and synaptic plasticity in a FASD rat model. PLoS One 10(8):e0135700. https://doi.org/10.1371/journal.pone.0135700

Bhattacharya D, Majrashi M, Ramesh S et al (2018) Assessment of the cerebellar neurotoxic effects of nicotine in prenatal alcohol exposure in rats. Life Sci 194:177–184. https://doi.org/10.1016/j.lfs.2017.12.010

Boon H, Verhoef M, O'Hara D et al (2004) Integrative healthcare: arriving at a working definition. Altern Ther Health Med 10(5):48–56

Boylan L, Flint L, Labovitz D et al (2004) Depression but not seizure frequency predicts quality of life in treatment-resistant epilepsy. Neurology 62(2):258–261

Buchman AL (2018) Nutrition and disease. Gastroenterol Clin N Am 47(1):xiii–xiv. https://doi.org/10.1016/j.gtc.2017.12.002
Burdock GA (2000) Dietary supplements and lessons to be learned from GRAS. Regulat Toxicol Pharmacol 31(1):68–76. https://doi.org/10.1006/rtph.1999.1369
Chaiyasut C, Woraharn S, Sivamaruthi BS et al (2018) Lactobacillus fermentum HP3–mediated fermented Hericium erinaceus juice as a health promoting food supplement to manage diabetes mellitus. J EvidB Based Integrat Med 23:2515690X18765699. https://doi.org/10.1177/2515690X18765699
Chan PM, Kanagasabapathy G, Tan YS et al (2013) *Amauroderma rugosum* (Blume & T. Nees) torrend: nutritional composition and antioxidant and potential anti-inflammatory properties. Evid based Compl Alternat med 2013:304713. 2013. https://doi.org/10.1155/2013/304713
Checkoway H, Lundin JI, Kelada SN (2011) Neurodegenerative diseases. IARC Sci Publ 163:407–419
Chong PS, Fung ML, Wong KH et al (2019) Therapeutic potential of *Hericium erinaceus* for depressive disorder. Int J Mol Sci 21(1). https://doi.org/10.3390/ijms21010163
Daud A, Gill J, Kamra S et al (2017) Indirect treatment comparison of dabrafenib plus trametinib versus vemurafenib plus cobimetinib in previously untreated metastatic melanoma patients. J Hematol Oncol 10(1):3. https://doi.org/10.1186/s13045-016-0369-8
Delay N (1988) AIDS conference in Montreal is a first for Canada. Ont Dent 65(10):31–32
Denham BE (2021) Dietary supplements in the USA: problematic trends. Public Health Nutr 24(9): 2771–2775. https://doi.org/10.1017/s1368980021000665
Devinsky O, Patel AD, Cross JH et al (2018) Effect of cannabidiol on drop seizures in the Lennox–gastaut syndrome. New England J Med 378(20):1888–1897. https://doi.org/10.1056/NEJMoa1714631
Dey A, Kang X, Qiu J et al (2016) Anti-inflammatory small molecules to treat seizures and epilepsy: from bench to bedside. Trends Pharmacol Sci 37(6):463–484. https://doi.org/10.1016/j.tips.2016.03.001
Dhanasekaran M, Tharakan B, Manyam BV (2008) Antiparkinson drug--Mucuna pruriens shows antioxidant and metal chelating activity. Phytother Res 22 (1):6–11. doi:https://doi.org/10.1002/ptr.2109
Di Giovanni G, Svob Strac D, Sole M et al (2016) Monoaminergic and histaminergic strategies and treatments in brain diseases. Front Neurosci 10:541. https://doi.org/10.3389/fnins.2016.00541
DiMatteo MR, Lepper HS, Croghan TW (2000) Depression is a risk factor for noncompliance with medical treatment: meta-analysis of the effects of anxiety and depression on patient adherence. Arch Intern Med 160(14):2101–2107. https://doi.org/10.1001/archinte.160.14.2101
DiSabato DJ, Quan N, Godbout JP (2016) Neuroinflammation: the devil is in the details. J Neurochem 139(Suppl 2):136–153. https://doi.org/10.1111/jnc.13607
Dupont S (2010) Memory and epilepsy Biol Aujourdhui 204(2):181–188. https://doi.org/10.1051/jbio/2010005
Erbiai EH, da Silva LP, Saidi R et al (2021) Chemical composition, bioactive compounds, and antioxidant activity of two wild edible mushrooms Armillaria mellea and Macrolepiota procera from two countries (Morocco and Portugal). Biomolecules 11(4):575. https://doi.org/10.3390/biom11040575
Falco-Walter J (2020) Epilepsy definition, classification, pathophysiology, and epidemiology. Semin Neurol 40(6):617–623. https://doi.org/10.1055/s-0040-1718719
Fisher RS, van Emde BW, Blume W et al (2005) Epileptic seizures and epilepsy: definitions proposed by the international league against epilepsy (ILAE) and the international bureau for epilepsy (IBE). Epilepsia 46(4):470–472. https://doi.org/10.1111/j.0013-9580.2005.66104.x
Frisch NC, Rabinowitsch D (2019) What's in a definition? Holistic nursing, integrative health care, and integrative nursing: report of an integrated literature review. J Holist Nurs 37(3):260–272. https://doi.org/10.1177/0898010119860685
Gaitatzis A, Johnson AL, Chadwick DW et al (2004) Life expectancy in people with newly diagnosed epilepsy. Brain 127(Pt 11):2427–2432. https://doi.org/10.1093/brain/awh267

Gautier L (2020) Welcome to the 10th volume of neurodegenerative disease management. Neurodegenerative Dis Manag 10(1):1–4. https://doi.org/10.2217/nmt-2019-0032

Ghosh S, Nandi S, Banerjee A et al (2021) Prospecting medicinal properties of Lion's mane mushroom. J Food Biochem 45(8):e13833. https://doi.org/10.1111/jfbc.13833

Govindarajulu M, Ramesh S, McKerley G et al (2021) Recent trends in medical mushrooms in a neurodegenerative disorder (Alzheimer's disease). In: Sridhar KR, Deshmukh SK (eds) Advances in macrofungi - pharmaceuticals and cosmeceuticals, 1st edn. CRC Press, Boca Raton, pp 1–328. https://doi.org/10.1201/9781003191278

Gross RA (1992) A brief history of epilepsy and its therapy in the Western hemisphere. Epilepsy Res 12(2):65–74. https://doi.org/10.1016/0920-1211(92)90028-r

Huff JS, Murr N (2022) Seizure. In: StatPearls. StatPearls publishing copyright © 2022. StatPearls Publishing LLC., Treasure Island, FL

Jang HJ, Kim JE, Jeong KH et al (2019) The neuroprotective effect of Hericium erinaceus extracts in mouse hippocampus after pilocarpine-induced status epilepticus. Int J Mol Sci 20(4). https://doi.org/10.3390/ijms20040859

Kaplan L, Chow BW, Gu C (2020) Neuronal regulation of the blood–brain barrier and neurovascular coupling. Nat Rev Neurosci 21(8):416–432. https://doi.org/10.1038/s41583-020-0322-2

Kapur J (2018) Role of NMDA receptors in the pathophysiology and treatment of status epilepticus. Epilepsia Open 3:165–168

Khan MA, Tania M, Liu R et al (2013) *Hericium erinaceus*: an edible mushroom with medicinal values. J Complement Integr Med 10. https://doi.org/10.1515/jcim-2013-0001

Kovacs GG (2017) Concepts and classification of neurodegenerative diseases. Handb Clin Neurol 145:301–307. https://doi.org/10.1016/b978-0-12-802395-2.00021-3

Kurian MA, Gissen P, Smith M et al (2011) The monoamine neurotransmitter disorders: an expanding range of neurological syndromes. Lancet Neurol 10(8):721–733. https://doi.org/10.1016/S1474-4422(11)70141-7

Lee B, Fujihashi A, Govindarajulu M et al (2019) Role of mushrooms in neurodegenerative diseases. In: Agrawal DC, Dhanasekaran M (eds) Medicinal mushrooms - recent progress in research and development. Springer Nature Singapore, pp 223–250. https://doi.org/10.1007/978-981-13-6382-5_8

Lin YE, Wang HL, Lu KH et al (2021) Water extract of Armillaria mellea (Vahl) P. Kumm. Alleviates the depression-like behaviors in acute- and chronic mild stress-induced rodent models via anti-inflammatory action. J Ethnopharmacol 265:113395. https://doi.org/10.1016/j.jep.2020.113395

Ling-Sing Seow S, Naidu M, David P et al (2013) Potentiation of neuritogenic activity of medicinal mushrooms in rat pheochromocytoma cells. BMC Compl Alternat Med 13(1):157. https://doi.org/10.1186/1472-6882-13-157

Lou H, Lin J, Guo L et al (2019) Advances in research on *Cordyceps militaris* degeneration. Appl Microbiol Biotechnol 103(19):7835–7841. https://doi.org/10.1007/s00253-019-10074-z

Manford M (2017) Recent advances in epilepsy. J Neurol 264(8):1811–1824. https://doi.org/10.1007/s00415-017-8394-2

Marchi N, Granata T, Ghosh C et al (2012) Blood–brain barrier dysfunction and epilepsy: pathophysiologic role and therapeutic approaches. Epilepsia 53(11):1877–1886. https://doi.org/10.1111/j.1528-1167.2012.03637.x

Martínez-Montemayor MM, Ling T et al (2019) Identification of biologically active *Ganoderma lucidum* compounds and synthesis of improved derivatives that confer anticancer activities *in vitro*. Front Pharmacol 10. https://doi.org/10.3389/fphar.2019.00115

Massey PB (2002) Dietary supplements. Med Clin North Am 86(1):127–147. https://doi.org/10.1016/s0025-7125(03)00076-2

Mastrangelo M (2021) Epilepsy in inherited neurotransmitter disorders: spotlights on pathophysiology and clinical management. Metabolic Brain Dis 36(1):29–43. https://doi.org/10.1007/s11011-020-00635-x

Meade E, Hehir S, Rowan N et al (2022) Mycotherapy: potential of fungal bioactives for the treatment of mental health disorders and morbidities of chronic pain. J Fungi 8(3). https://doi.org/10.3390/jof8030290

Meldrum BS (1990) Anatomy, physiology, and pathology of epilepsy. Lancet 336(8709):231–234. https://doi.org/10.1016/0140-6736(90)91747-X

Mouli S, Nanayakkara G, AlAlasmari A et al (2015) The role of frataxin in doxorubicin-mediated cardiac hypertrophy. Am J Physiol Heart Circulatory Physiol 309(5):H844–H859. https://doi.org/10.1152/ajpheart.00182.2015

Muralikrishnan D, Ebadi M, Brown-Borg HM (2002) Effect of MPTP on dopamine metabolism in Ames dwarf mice. Neurochem Res 27(6):457–464. https://doi.org/10.1023/A:1019888300957

Nabavi SF, Braidy N, Habtemariam S et al (2015) Neuroprotective effects of chrysin: from chemistry to medicine. Neurochem Int 90:224–231. https://doi.org/10.1016/j.neuint.2015.09.006

Ojemann LM, Nelson WL, Shin DS et al (2006) Tian ma, an ancient Chinese herb, offers new options for the treatment of epilepsy and other conditions. Epilepsy Behav 8(2):376–383. https://doi.org/10.1016/j.yebeh.2005.12.009

Olatunji OJ, Tang J, Tola A et al (2018) The genus Cordyceps: an extensive review of its traditional uses, phytochemistry and pharmacology. Fitoterapia 129:293–316. https://doi.org/10.1016/j.fitote.2018.05.010

Petkova-Gueorguieva ES, Getov IN, Ivanov KV et al (2019) Regulatory requirements for food supplements in the european union and Bulgaria. Folia Med (Plovdiv) 61(1):41–48. https://doi.org/10.2478/folmed-2018-0032

Pondugula SR, Harshan A, Ramesh S et al (2022) Cardioprotective effects of *Oroxylum indicum* extract against doxorubicin and cyclophosphamide-induced cardiotoxicity. Cardiovasc Toxicol 22(1):67–77. https://doi.org/10.1007/s12012-021-09701-x

Pondugula SR, Majrashi M, Almaghrabi M et al (2021) *Oroxylum indicum* ameliorates chemotherapy induced cognitive impairment. PLoS One 16(6):e0252522. https://doi.org/10.1371/journal.pone.0252522

Pracucci E, Pillai V, Lamers D et al (2021) Neuroinflammation: a signature or a cause of epilepsy? Int J Mol Sci 22(13). https://doi.org/10.3390/ijms22136981

Radu BM, Epureanu FB, Radu M et al (2017) Nonsteroidal anti-inflammatory drugs in clinical and experimental epilepsy. Epilepsy Res 131:15–27. https://doi.org/10.1016/j.eplepsyres.2017.02.003

Ramesh S, Majrashi M, Almaghrabi M et al (2019) Overview of therapeutic efficacy of mushrooms. In: Agrawal DC, Dhanasekaran M (eds) Medicinal mushrooms. Springer, Singapore, pp 103–142. https://doi.org/10.1007/978-981-13-6382-5_3

Rana A, Musto AE (2018) The role of inflammation in the development of epilepsy. J Neuroinflam 15(1):144. https://doi.org/10.1186/s12974-018-1192-7

Ransohoff RM (2016) How neuroinflammation contributes to neurodegeneration. Science 353(6301):777–783. https://doi.org/10.1126/science.aag2590

Russo ME (1981) The pathophysiology of epilepsy. Cornell Vet 71(2):221–247

Ryvlin P, Rheims S, Hirsch LJ et al (2021) Neuromodulation in epilepsy: state-of-the-art approved therapies. Lancet Neurol 20(12):1038–1047. https://doi.org/10.1016/s1474-4422(21)00300-8

Sabaratnam V, Kah-Hui W, Naidu M et al (2013) Neuronal health - can culinary and medicinal mushrooms help? J Tradit Complement Med 3(1):62–68. https://doi.org/10.4103/2225-4110.106549

Sales-Campos C, da Silva JF, do Nascimento LBB et al (2021) Nutritional and bioactive properties of an amazon wild oyster culinary-medicinal mushroom, pleurotus ostreatus (Agaricomycetes): contributions to functional food and human health. Int J Med Mushrooms 23(7):79–90. https://doi.org/10.1615/IntJMedMushrooms.2021038780

Sam SE, Sim KS, Rahman S et al (2022) Neuroprotective properties of wild medicinal mushroom, sanguinoderma rugosum (Agaricomycetes), extracts against glutamate-induced hippocampal cells. Int J Med Mushrooms 24(3):35–50. https://doi.org/10.1615/IntJMedMushrooms.2022042961

Saniya K, Patil BG, Chavan MD et al (2017) Neuroanatomical changes in brain structures related to cognition in epilepsy: an update. J Nat Sci Biol Med 8(2):139–143. https://doi.org/10.4103/0976-9668.210016

Schutgens F, Clevers H (2020) Human organoids: tools for understanding biology and treating diseases. Annu Rev Pathol 15:211–234. https://doi.org/10.1146/annurev-pathmechdis-012419-032611

Sharma A, Sharma A, Tripathi A (2021) Biological activities of Pleurotus spp. polysaccharides: a review. J Food Biochem 45(6):e13748. https://doi.org/10.1111/jfbc.13748

Sharma J, Gairola S, Gaur RD et al (2013) Ethnomedicinal plants used for treating epilepsy by indigenous communities of sub-Himalayan region of Uttarakhand. India J Ethnopharmacol 150(1):353–370. https://doi.org/10.1016/j.jep.2013.08.052

Shin HW, Jewells V, Hadar E et al (2014) Review of epilepsy-etiology, diagnostic evaluation and treatment. Int J Neurorehabilitation 1(130):2376–0281

Silano V, Coppens P, Larrañaga-Guetaria A et al (2011) Regulations applicable to plant food supplements and related products in the European Union. Food Function 2(12):710–719. https://doi.org/10.1039/C1FO10105F

Socala K, Nieoczym D, Grzywnowicz K et al (2015) Evaluation of anticonvulsant, antidepressant-, and anxiolytic-like effects of an aqueous extract from cultured mycelia of the lingzhi or reishi medicinal mushroom *Ganoderma lucidum* (higher basidiomycetes) in mice. Int J Med Mushrooms 17(3):209–218. https://doi.org/10.1615/intjmedmushrooms.v17.i3.10

Striano P, Minassian BA (2020) From genetic testing to precision medicine in epilepsy. Neurotherapeutics 17(2):609–615. https://doi.org/10.1007/s13311-020-00835-4

Takata F, Nakagawa S, Matsumoto J et al (2021) Blood-brain barrier dysfunction amplifies the development of neuroinflammation: understanding of cellular events in brain microvascular endothelial cells for prevention and treatment of BBB dysfunction. Front Cell Neurosci 15. https://doi.org/10.3389/fncel.2021.661838

Tan WC, Kuppusamy UR, Phan CW et al (2015) *Ganoderma neo-japonicum* Imazeki revisited: domestication study and antioxidant properties of its basidiocarps and mycelia. Sci Rep 5(1): 12515. https://doi.org/10.1038/srep12515

Tan WC, Kuppusamy UR, Phan CW et al (2018) Cell proliferation and DNA repair ability of *Ganoderma neo-japonicum* (Agaricomycetes): an indigenous medicinal mushroom from Malaysia. Int J Med Mushrooms 20(2):155–163. https://doi.org/10.1615/IntJMedMushrooms.2018025445

Tello I, Campos-Pena V, Montiel E et al (2013) Anticonvulsant and neuroprotective effects of oligosaccharides from Lingzhi or Reishi medicinal mushroom, *Ganoderma lucidum* (higher basidiomycetes). Int J Med Mushrooms 15(6):555–568. https://doi.org/10.1615/intjmedmushr.v15.i6.40

Thijs RD, Surges R, O'Brien TJ et al (2019) Epilepsy in adults. Lancet 393(10172):689–701. https://doi.org/10.1016/s0140-6736(18)32596-0

Thrash B, Thiruchelvan K, Ahuja M et al (2009) Methamphetamine-induced neurotoxicity: the road to Parkinson's disease. Pharmacol Rep 61(6):966–977. https://doi.org/10.1016/S1734-1140(09)70158-6

Van Bulck M, Sierra-Magro A, Alarcon-Gil J et al (2019) Novel approaches for the treatment of Alzheimer's and Parkinson's disease. Int J Mol Sci 20(3). https://doi.org/10.3390/ijms20030719

Vaquer-Alicea J, Diamond MI (2019) Propagation of protein aggregation in neurodegenerative diseases. Annu Rev Biochem 88:785–810. https://doi.org/10.1146/annurev-biochem-061516-045049

Venturella G, Ferraro V, Cirlincione F et al (2021) Medicinal mushrooms: bioactive compounds, use, and clinical trials. Int J Mol Sci 22(2). https://doi.org/10.3390/ijms22020634

Vilariño M, García-Sanmartín J, Ochoa-Callejero L et al (2020) Macrocybin, a natural mushroom triglyceride, reduces tumor growth in vitro and in vivo through caveolin-mediated interference with the actin cytoskeleton. Molecules 25(24). https://doi.org/10.3390/molecules25246010

Wang SQ, Li XJ, Zhou S et al (2013) Intervention effects of Ganoderma lucidum spores on epileptiform discharge hippocampal neurons and expression of neurotrophin-4 and N-cadherin. PLoS One 8(4):e61687. https://doi.org/10.1371/journal.pone.0061687

Wang SQ, Li XJ, Qiu HB et al (2014) Anti-epileptic effect of Ganoderma lucidum polysaccharides by inhibition of intracellular calcium accumulation and stimulation of expression of CaMKII α in epileptic hippocampal neurons. PLoS One 9(7):e102161. https://doi.org/10.1371/journal.pone.0102161. Erratum in: PLoS One 9(10):e111295

Wang GH, Li X, Cao WH, Li J, Wang LH (2018) A retrospective study of Ganoderma Lucidum spore powder for patients with epilepsy. Medicine (Baltimore) 97(23):e10941. https://doi.org/10.1097/md.0000000000010941

Wang X-Y, Zhang D-d, Yin J-Y et al (2019) Recent developments in Hericium erinaceus polysaccharides: extraction, purification, structural characteristics and biological activities. Crit Rev Food Sci Nutri 59(sup1):S96–S115. https://doi.org/10.1080/10408398.2018.1521370

Wei P, Wang K, Luo C et al (2021) Cordycepin confers long-term neuroprotection via inhibiting neutrophil infiltration and neuroinflammation after traumatic brain injury. J Neuroinflam 18(1):137. https://doi.org/10.1186/s12974-021-02188-x

Woodling NS, Andreasson KI (2016) Untangling the web: toxic and protective effects of neuroinflammation and PGE2 signaling in alzheimer's disease. ACS Chem Neurosci 7(4):454–463. https://doi.org/10.1021/acschemneuro.6b00016

Wu X, Wu T, Huang A et al (2021) New insights into the biosynthesis of typical bioactive components in the traditional chinese medicinal fungus Cordyceps militaris. Front Bioeng Biotechnol 9. https://doi.org/10.3389/fbioe.2021.801721

Yadav SK, Ir R, Jeewon R et al (2020) A mechanistic review on medicinal mushrooms-derived bioactive compounds: potential mycotherapy candidates for alleviating neurological disorders. Planta Med 86(16):1161–1175. https://doi.org/10.1055/a-1177-4834

Zhang C, Li J, Hu C et al (2017) Antihyperglycaemic and organic protective effects on pancreas, liver and kidney by polysaccharides from Hericium erinaceus SG-02 in streptozotocin-induced diabetic mice. Sci Rep 7(1):10847. https://doi.org/10.1038/s41598-017-11457-w

Zheng CW, Cheung TMY, Leung GPH (2022) A review of the phytochemical and pharmacological properties of Amauroderma rugosum. Kaohsiung J Med Sci 38(6):509–516. https://doi.org/10.1002/kjm2.12554

Zhou J, Chen X, Xiao L et al (2021) Efficacy and safety of cordyceps militaris as an adjuvant to duloxetine in the treatment of insomnia in patients with depression: a 6-week double- blind, randomized, placebo-controlled trial. Front Psych 12:754921. https://doi.org/10.3389/fpsyt.2021.754921

Natural vs. Synthetic Psilocybin: The Same or Completely Different?

Amza Ali, Mary-Elizabeth Gifford, Henry Lowe, Lorenzo Gordon, and Justin Grant

Contents

Abstract This chapter aims to address the question of whether there is a clinically relevant difference between natural and synthetic psilocybin. Examining the history of when psilocybin was first researched in the late 1950s revealed that both synthetic and natural compounds were explored but lost momentum with the declaration of the

A. Ali
Psyence Group, Toronto, ON, Canada

M.-E. Gifford
Psyence Group, Toronto, ON, Canada

Global Wellness Institute, Miami, FL, USA

H. Lowe
Psyence Group, Toronto, ON, Canada

Biotech R & D Institute, University of the West Indies, Kingston, Jamaica

L. Gordon
Psyence Group, Toronto, ON, Canada

Caribbean School of Medical Sciences, Kingston, Jamaica

J. Grant (✉)
Psyence Group, Toronto, ON, Canada

Department of Pharmacology and Toxicology, Temerty Faculty of Medicine, University of Toronto, Toronto, ON, Canada
e-mail: Justin.grant@utoronto.ca

D. C. Agrawal, M. Dhanasekaran (eds.), *Mushrooms with Therapeutic Potentials*,
https://doi.org/10.1007/978-981-19-9550-7_18

war on drugs in 1971. The past few years have shown a significant rise of research interest in psilocybin in the mental health field. Due to the current regulations and market capitalization, the industry relies on an investment strategy aligned with the traditional pharmaceutical model. Interestingly, there is a long history of popular natural medicines being produced by pharmaceutical companies, and a number of natural products are currently being used in mental health. Research on natural psilocybin, including the phenomenon of the entourage effect, as well as the factors and considerations for future research on natural and synthetic psilocybin, are discussed. Comparing the natural psilocybin molecule to the synthetic psilocybin molecule should be fundamentally similar. Still, the active pharmaceutical ingredient (API) differs from other ingredients in the whole mushroom, and natural extracts may have synergistic effects.

Keywords Entourage effect · Mental health · Natural medicine · Psilocybin · Psilocybin market · Regulations

Abbreviations

5-HT2a	5-Hydroxytryptamine (serotonin) 2a receptor
API	Active pharmaceutical ingredient
ATAI	atai Life Sciences
CAGR	Compound annual growth rate
CMPS	Compass Pathways
CYBN	Cybin
DEA	United States Drug Enforcement Administration
FDA	Food and Drug Administration
FTRP	Field Trip Health
LSD	Lysergic acid diethylamide
MAPS	Multidisciplinary Association for Psychedelic Studies
NASDAQ	National Association of Securities Dealers Automated Quotations
NIH	National Institutes of Health
NUMI	Numinus Wellness
NYU	New York University
PAT	Psychedelic-assisted therapies
PTSD	Post-traumatic stress disorder
SPRAVATO	Esketamine nasal spray
TSXV	Toronto Stock Exchange Venture
UCLA	University of California, Los Angeles
UCSF	University of California, San Francisco
USD	United States dollar
USFDA	United States Food and Drug Administration

1 Introduction

Research on psychedelic psilocybin, commonly known as magic mushrooms, is on the rise and attracting increased interest in academia and industry. Modern psilocybin research owes its start to a 1957 *Life Magazine* article titled "Seeking the Magic Mushroom" by Gordon Wasson. He wrote, "We chewed and swallowed these acrid mushrooms, saw visions, and emerged from the experience awestruck" (Siff 2018).

From the start, the research looked at the synthesized molecule psilocybin and the whole mushroom itself. Wasson introduced the hallucinogenic mushrooms to his friend, the scientist Albert Hofmann, who headed plant medicine research at the Sandoz Laboratories in Switzerland. Dubbed "the father of LSD," Hofmann was the first scientist to identify and synthesize LSD following his research on the rye fungus ergot (Fusar-Poli and Borgwardt 2008). Hofmann holds another first; after his friend Wasson handed over "magic mushrooms" to Hoffman, the Swiss scientist took them into his lab, where he made history. He isolated the psychoactive psychedelic mushroom compounds, which he named psilocin and psilocybin (Hofmann et al. 1958), as shown in Fig. 1.

The following year, 1958, saw the publication of the first five research papers on psilocybin, and a patent on the synthesis of psilocybin was awarded to Hofmann and Troxler (Hofmann and Troxler 1963). Sandoz sold purified 2 mg psilocybin tablets, under the trade name Indocybin, for fundamental and clinical research until 1971 (Geiger et al. 2018). This period is now considered the first golden era in psilocybin research.

From 1958 to 1971, a total of 259 scientific papers on psilocybin mushroom compounds were published, according to NIH's National Library of Medicine (Fig. 2). According to PubMed, scientific publications on psilocybin peaked at 39 in 1968 but rapidly declined after 1971. American President Richard Nixon's declaration of the war on drugs in 1971 brought changes in drug policy and drug laws, and research on psilocybin came to a practical halt as it became a DEA Schedule 1 substance (Augustyn 2016).

In the following years, from 1972 to 2010, psilocybin research slowed by half, from 20 publications per year to ~10. Then in 2011, following activism from

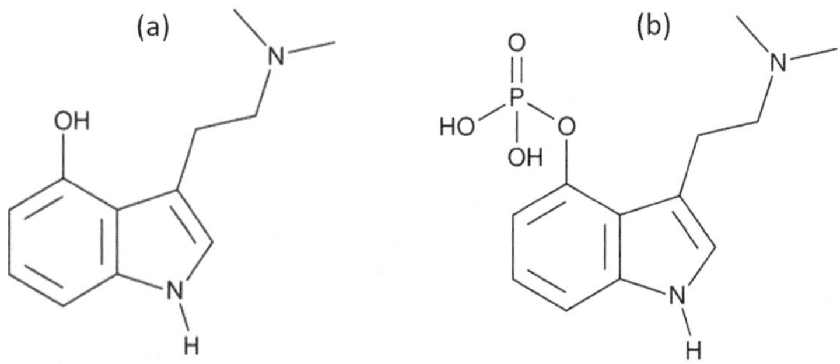

Fig. 1 Chemical structures of (**a**) psilocin and (**b**) psilocybin

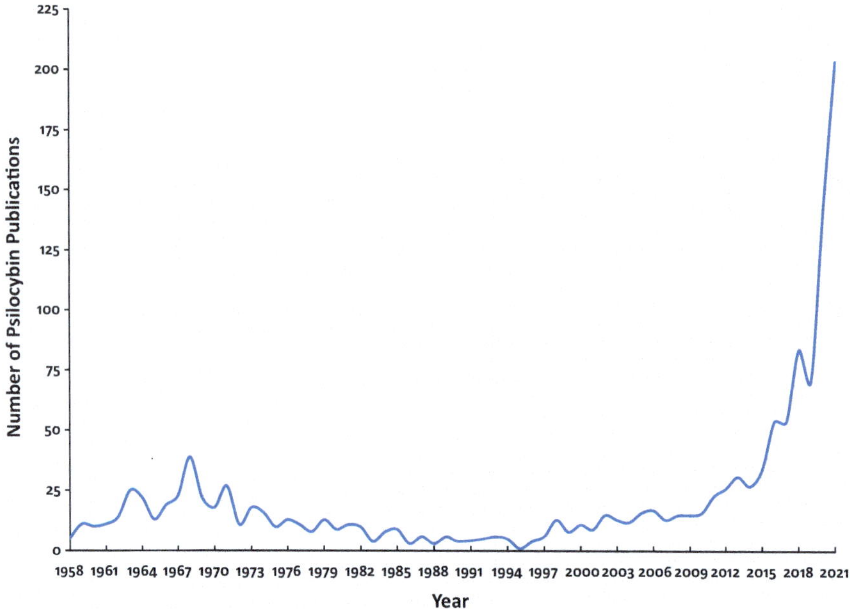

Fig. 2 The number of publications per year on psilocybin from 1958 to 2021. Data was obtained from CSV files from the US National Library of Medicine at the National Institutes of Health (PubMed.gov)

Multidisciplinary Association for Psychedelic Studies (MAPS) and other psychedelic advocates, the number of publications on psilocybin began to increase. In 2020 alone, 140 research publications on psilocybin appeared in print. In 2021, the number of publications on psilocybin increased to 204. This second wave of research publications represents a renaissance of interest in psilocybin that far exceeds the 1960s research efforts (Fig. 2).

As psilocybin research increases, so does awareness and interest. The nonprofit Usona Institute, which advocates greater access to psilocybin, produces pharma-grade synthetic psilocybin for several research investigators and collaborators at prestigious institutions, including the University of Wisconsin-Madison, Johns Hopkins, NYU, Imperial College, UCSF, UCLA, Yale, University of Zurich, Heffter Research Institute, and Swiss Neuromatrix Foundation. Usona provides its agent with supporting documentation for preclinical and clinical studies. Furthermore, Usona is currently enrolling participants in its clinical trial on major depressive disorder and psilocybin. Details of the trial can be found on the Usona website, including of the 100 participants (50% will receive psilocybin, and 50% the placebo niacin). Nearly 14,000 people have expressed interest in enrolling in the study (Usona Institute 2021). To date, only synthetic psilocybin has reached phase 1 and phase 2 clinical trials, according to clinicaltrials.gov. Pharmaceutical trials have not been carried out using naturally derived psilocybin.

Yet, there is an increasing desire for natural medicinal alternatives, as confirmed by the global herbal medicine market and its projected growth. According to GlobalNewsWire.com, in 2019, the global herbal medicine market size was estimated at US$83 billion and is expected to reach US$550 billion by 2030 at a CAGR of 18.9% (InsightSLICE 2021). Plant-based remedies have seen significant growth based on the American Botanical Council's market report (Smith et al. 2021). Future growth continues to look bright, with a 2020 market value of US$405.5 billion and an 18% growth predicted for complementary and alternative medicine products (Intelligence 2020; GrandView Research 2021).

2 Psilocybin as a Natural Medicine

In 1928, the discovery of the antibacterial properties of penicillin by Alexander Fleming became a watershed event in pharmaceutical breakthroughs. Penicillin now stands as a reminder that natural medicine can become replicable and change the course of medicine (Fleming 2001). The pharmacy practice was also revolutionized by introducing the Durham-Humphrey Act of 1951, indirectly contributing to the subsequent rise in prescription drug use (Hoge 1951). Table 1 summarizes natural medicines used in mainstream (western) medicine.

Due to the COVID-19 pandemic, a rise in mental health and substance-use disorders has been reported as having increased rates of anxiety, depression, and

Table 1 Well-known natural medicines

Scientific name of medicinal plant	Drugs derived from the medicinal plant	Indications	Pharmaceutical company
Penicillium chrysogenum L./ *Penicillium rubens* L.	Penicillin	Antibacterial	Abbott Laboratories, Pfizer, Merck, Bristol-Myers Squibb (ACS 1999)
Salix alba L. (willow tree)	Aspirin	Treatment of minor aches, pains, and fevers	Bayer
Papaver somniferum L. (opium poppy)	Morphine	Treatment of severe pain	GlaxoSmithKline
Digitalis purpurea L. (foxglove)	Digitalis (digoxin)	Treatment of heart conditions	GlaxoSmithKline
Taxus brevifolia Nutt. (Pacific/ western yew)	Paclitaxel (Taxol)	Paclitaxel is used for ovarian, breast, lung, bladder, prostate, melanoma, esophageal, and other types of solid tumor cancers, as well as Kaposi's sarcoma (Saville et al. 1995)	Bristol-Myers Squibb

Table 2 Natural plant and fungal medicines for mental health

Scientific name of medicinal plant or fungus	Active ingredients	Suggested clinical use	References
Hypericum perforatum L.	St. John's wort	Depression	(Whiskey et al. 2001)
Salvia hispanica (chia) and *Linum usitatissimum* (flax)	Omega-3	Depression	(Logan 2004; Bloch and Hannestad 2012)
Piper methysticum	Kava	Generalized anxiety	(Sarris et al. 2009)
Griffonia simplicifolia	5-HTP	Depression	(Ribas et al. 2021)
Crocus sativus	Saffron	Depression	(Lopresti and Drum-mond 2014)
Rhodiola rosea L.	SHR-5	Depression	(Darbinyan et al. 2007)
Withania somnifera	Ashwagandha	Anxiety/stress	(Singh et al. 2011)
Matricaria chamomilla	Chamomile	Anxiety/stress	(Srivastava et al. 2010)
Valeriana officinalis	Valerian	Anxiety/stress	(Attele et al. 2000)
Lavandula angustifolia	Lavender	Anxiety/stress	(Kasper 2013)
Passiflora incarnata	Passionflower	Anxiety/stress	(Janda et al. 2020)
Cannabis sativa	Cannabidiol	Anxiety/depression	(García-Gutiérrez et al. 2020)
Panax ginseng	Ginseng	Anxiety/stress	(Kim 2012)
Ocimum tenuiflorum	Tulsi (basil)	Anxiety/stress	(Jamshidi and Cohen 2017)
Psilocybe cubensis[a]	Psilocybin	Anxiety/depression	(Vargas et al. 2020)

[a]The most recent addition to natural medicines for the treatment of anxiety and depression is psilocybin from the mushroom species *Psilocybe cubensis*. Psilocybin has twice received FDA approval for "breakthrough therapy" status for treatment-resistant depression and major depressive disorder (Lowe et al. 2021)

distress. In this context, alternative forms of natural medicines have risen in popularity. According to the American Botanical Council, the market for plant-based remedies has grown substantially (Smith et al. 2021). Table 2 lists some of the popular flora and fungal-based medicines.

Our recent review article discusses the history of natural psilocybin (Lowe et al. 2021). In brief, natural psilocybin has been used for centuries by indigenous groups as part of religious ceremonies. One of the earliest depictions of this was discovered in Spain from prehistoric rock art of *Psilocybe hispanica*, dated 6000 years ago (Akers et al. 2011). The first written record was a sixteenth-century ethnographic research study on the Aztecs, a manuscript referred to as the Florentine Codex. It wasn't until 1799 that hallucinogenic mushrooms were first documented in the *London Medical and Physical Journal*. More formal scientific research studies began in 1958. At the same time, synthetic psilocybin was developed, as described above.

3 The Mushrooming Market

Psychedelic mushrooms are not the only fungal-based mental wellness solution. Mushrooms such as chaga, turkey tail, reishi, and lion's mane have been recognized for their nutraceutical value (Powell 2015). The market for these non-psychedelic mushrooms is booming; retail sales for mushroom-containing food, beverage, and beauty products have significantly increased, with retail sales expected to surpass US $62 billion by 2023 (Baystreet 2021).

As the mental wellness market increases, so does the market for mental health solutions, especially for those intractable conditions such as treatment-resistant depression. The most promising solutions in the pharmaceutical pipeline for mental health are psychedelic-assisted therapies (PAT). The psychedelic drug market size is estimated to reach US$10 billion by 2027, with a CAGR of 12% from 2021 to 2027 (Markets 2021). Although not a classic psychedelic, one notable example is the 2019 US Food and Drug Administration (USFDA) approval of SPRAVATO®, an esketamine nasal spray developed by Johnson and Johnson for use in patients suffering from treatment-resistant depression (Lowe et al. 2021). This demonstrates optimistic support for continued growth in the area of psychedelic pharmaceuticals.

Companies developing pharmaceutical psilocybin, natural or synthetic, have achieved extensive market evaluations, with the top five shown in Table 3. The total aggregate market cap of these five public companies with a focus on psilocybin has reached over US$3.5 billion.

Researchers at distinguished academic institutions are studying whether psilocybin-based medicines can successfully treat mental health issues such as depression, anxiety, alcoholism, substance abuse, and post-traumatic stress disorder (PTSD).

Table 3 List of publicly traded psilocybin companies with over US$100 M market cap

Company	Natural or synthetic psilocybin agent	Market cap[a]	Psilocybin agent	Country based
Atai life sciences	Synthetic	USD 1.867B (NASDAQ: ATAI)	Comp360	Germany
Compass pathways	Synthetic	USD 1.296B (NASDAQ: CMPS)	Comp360	England
Cybin	Synthetic	USD 277.788 M (NASDAQ: CYBN)	CYB003	Canada
Field trip health	Synthetic	USD 253.714 M (NASDAQ: FTRP)	FT104	Canada
Numinus wellness	Natural	CAD 122.655 M (TSXV: NUMI)	PSYBINA RX	Canada

[a]Market caps were recorded from Yahoo Finance on November 19, 2021

4 Natural Vs. Synthetic Medicines

Scientists have a long history of looking to nature to discover novel medicines. One of the earliest documented medical findings was in 1804 when Sertürner purified morphine from opium to produce an analgesic (Lockermann 1951). Synthetic morphine took significantly longer to come to market as morphine was first synthesized in 1956 (Gates and Tschudi 1956).

Natural medicine products were attractive to pharmaceutical companies until the turn of the century in the early 2000s. It is more difficult to intellectually protect novel phytomedicines (through patents) for the simple reason that they occur naturally and because they may contain other compounds. So for those reasons, among others, pharmaceutical companies such as Merck and Bristol-Myers Squibb decreased their investment and closed many in-house natural drug product programs (Beutler 2009).

Synthetic drugs are manufactured with high purity and a defined concentration of active ingredient(s). As a result, the regulatory path is more precise for synthetic products than natural products. A summary of key differences between natural and synthetic medicinal agents is listed in Table 4.

"Natural" does not necessarily mean toxicologically safer or more efficacious. Natural products tend to contain a multiplicity of compounds/secondary metabolites that may not be of particular interest and may even be harmful. The molecular mechanisms of action and the pharmacology of most medicinal plants still have not been fully elucidated due to their complex combination of active molecules, further limiting their application in mainstay medicine. In contrast, during synthetic drug manufacturing/drug development process, lead compound(s) may be identified, isolated, and optimized to treat a particular condition. As a result, undesirable compounds may be removed.

Natural drugs are limited due to a slower production process, sources may be depleted, or natural sources may produce smaller yields. Also, natural sources are sensitive to seasonal variations and biological contaminants. On the other hand, theoretically, synthetic products may be replenished indefinitely during manufacturing and can have more predictable batches.

Table 4 Differences between natural and synthetic compounds

Natural	Synthetic
Complex combination of molecules	Purified compounds
Multidrug effect – "Entourage effect"[a]	Single drug interaction
Variability in production from batch to batch	Predictable batch yields
Slower production	Faster to process

[a]See section below on psilocybin mushroom entourage effect

5 The Psilocybin Mushroom Entourage Effect

Elucidating the interplay between the many molecules in psychoactive mushroom species, which can be described as the *psilocybin mushroom entourage effect*, presents a promising research opportunity. The "entourage effect" is a phenomenon by which compounds within a given plant or fungus work synergistically to enhance its bioactive effects. In the case of *Cannabis sativa L.*, the "entourage effect" is the phenomenon by which cannabinoid and non-cannabinoid secondary metabolites, terpenes, and flavonoids work synergistically to enhance the biological effect of the plant in the treatment of several human conditions, including pain and spasticity (Oreja-Guevara 2012; Giacoppo et al. 2017; Barnes 2006). Though limited and inconclusive, the case may also be the same in *Psilocybe* mushrooms, where the hallucinogenic effects of psilocybin, the main psychoactive compound, may be enhanced by other compounds produced by the mushroom such as psilocin, norpsilocin, baeocystin, norbaeocystin, aeruginascin, and ß-carbolines (Bauer 2019; Gartz 1989; Bauer 2020; Matsushima et al. 2009; Zhuk et al. 2015).

In one particular study, subjects who ingested a species of magic mushrooms known *as Inocybe aeruginascens*, which also produced aeruginascin, experienced euphoric effects as opposed to those who consumed non-aeruginascin species of mushrooms that produced psilocybin and psilocin (Gartz 1989). Therefore, it was hypothesized that aeruginascin synergistically enhanced the psychoactive effects of low levels of psilocybin upon ingestion of *Inocybe aeruginascens* mushrooms (Gartz 1989).

In an animal study, *Psilocybe argentipes* mushroom extract (containing multiple bioactive compounds) was more effective than pure psilocybin at reducing marble-burying behavior in mice (Matsushima et al. 2009). It indicates a synergistic effect of multiple bioactive compounds.

Similar to the ß-carbolines in ayahuasca, the ß-carbolines in *Psilocybe* (psilocybin-containing) mushrooms (cordysinin C, cordysinin D, harmine, harmane, harmol, norharmane, and perlolyrine) may also work synergistically to contribute to an "entourage effect" in mushrooms since, as potent inhibitors of monoamine oxidases, β-carbolines interfere with psilocybin degradation (Blei et al. 2020).

The literature on the possible entourage effect of the various compounds in magic mushrooms is very limited. Therefore, future pharmacological studies are required to elucidate the roles of each metabolite in the overall profile of the biological effects of magic mushrooms.

6 Future Psilocybin Research

From as early as the 1960s to the 1970s, the first extensive scientific studies into psychedelics focused on their effects on brain function and neural connectivity. Classic psychedelic (serotonergic) drugs interact with the serotonin receptors (5-HT/

5-hydroxytryptamine receptors) and their subtypes densely located within the brain (Halberstadt and Geyer 2011; Beliveau et al. 2017; Canal 2018). These receptors mediate emotions and moods such as anxiety and aggression, cognition, sex, learning memory, appetite, and other biological, neurological, and neuropsychiatric processes (Beliveau et al. 2017; Nichols 2020). The mechanisms of action of psilocybin are not completely understood; however, there is research evidence to support that in addition to interaction with 5-HT2a receptors, psilocybin may also have an indirect effect on the mesolimbic dopaminergic pathway, which affects the brain's reward system (de Veen et al. 2017).

Psilocybin has been reported to significantly change brain dynamics and functional connectivity between brain areas (Grimm et al. 2018; Tagliazucchi et al. 2014; Lord et al. 2019; Müller et al. 2018). Positive effects on neurogenesis have also previously been described with another psychedelic, ayahuasca (Morales-García et al. 2017). Psilocybin-induced alteration in brain connectivity involves disintegrating associative networks and integrating sensory function networks (Nichols 2020). The extent to which natural vs. synthetic psilocybin affect or interact with brain receptor occupancy requires further study.

Generally, the use of herbal and synthetic medicines is highly discouraged due to the possibility of adverse drug-drug interactions; however, combinatorial medicine is not completely ruled out. It would require the particular identification of a natural derivate, extensive screening, and further study into interactions with synthetic drugs, which may be costly and time-consuming.

There are several factors and considerations for future research on psilocybin, and these may include the following:

- Cost of extraction vs. cost of synthesis.
- Identifying strains that produce the optimal concentration of active compounds and particular secondary metabolites and comparing them to the yield that can be achieved in the lab via synthesis.
- Continuing studies on molecular mechanisms and the role of other receptors besides 5-HT2a.
- Rigorously evaluating the entourage effect in mushroom species and evaluating combinatorial synergies.
- Determining the ideal dose and dosage form for specific indications.
- Identifying which patients would most benefit and those who are at risk.
- Inclusion of psychedelic-assisted therapy (PAT) – and the establishment of standards and practices for set and setting as well as integration.

7 Summary and Looking Forward

The answer to the question posed in the title of this paper will eventually be known through systematic research, but until then, it will continue to have ardent supporters in each camp. Synthetic and natural psychedelic drugs, including psilocybin, have advantages and disadvantages, and it may be possible to achieve success with either of these compounds in a clinical setting. Regulatory, economic, and educational factors still severely limit both options' market availability in all settings regardless of legal status. Market capitalization for agents such as psilocybin reflects an investment strategy that relies on the traditional pharmaceutical model. However, well-documented cultural awareness and use of mycologically based medicines, rigorous clinical research, and the increased consumer demand for natural and plant-based products will likely incentivize the production and use of natural psilocybin as a therapeutic agent. Ultrahigh dilutions of natural products used in homeopathy have provided intriguing and tantalizing insights into the potential of natural products as healing agents as used over the millennia (Khuda-Bukhsh and Pathak 2008). An exploration that hints at the insights emerging from the rapidly developing field of quantum biology suggests the potential effects on the whole organism from physicochemical phenomena at the subatomic level (Marais et al. 2018). Psilocybin's initial use may be narrowly defined as any new therapeutic agent. Still, its clinical use should steadily expand in the coming years, particularly given psilocybin's favorable safety profile, as supported by many clinical trials.

The distinction between natural and synthetics continues to be blurred as scientists chemically experiment with naturally derived agents, as done for many years. The possibility of the "entourage effect" makes naturally derived psilocybin of particular research interest, driven by the rapid growth in the psychedelic industry and of natural psychedelic products in particular. For the natural product believer, it is tempting to view this molecular complexity inherent in a naturally derived product as more suitable for treating brain disorders, given its known anatomic and neurochemical complexity. It is likely that psilocybin, whether naturally derived or synthetic, will find a secure place in the therapeutic armamentarium for use in the fight against intractable mental health disorders and their associated comorbidities.

References

ACS (1999) Discovery and development of penicillin [Online]. Am Chem Soc. https://www.acs.org/content/acs/en/education/whatischemistry/landmarks/flemingpenicillin.html. Accessed 12 Dec 2021

Akers BP, Ruiz JF, Piper A, Ruck CAP (2011) A prehistoric mural in Spain depicting neurotropic psilocybe mushrooms? Econ Bot 65(2):121–128. https://doi.org/10.1007/s12231-011-9152-5

Attele AS, Xie JT, Yuan CS (2000) Treatment of insomnia: an alternative approach. Altern Med Rev 5(3):249–259

Augustyn A (2016) War on drugs: U.S. history [Online]. https://www.britannica.com/topic/war-on-drugs. Accessed 30 Nov 2021

Barnes MP (2006) Sativex: clinical efficacy and tolerability in the treatment of symptoms of multiple sclerosis and neuropathic pain. Expert Opin Pharmacother 7(5):607–615. https://doi.org/10.1517/14656566.7.5.607

Bauer BE (2019) The entourage effect in magic mushrooms. Psychedelic Sci Rev [Online]. https://psychedelicreview.com/the-entourage-effect-in-magic-mushrooms/. Accessed 02 Dec 2021

Bauer BE (2020) Catching up with the psychedelic entourage effect-part 1: how we got here. Psychedelic Sci Rev [Online]. https://psychedelicreview.com/the-entourage-effect-in-magic-mushrooms/. Accessed 02 Dec 2021

Baystreet (2021) The functional mushroom market could hand you a $62 billion opportunity [Online]. https://www.baystreet.ca/stockstowatch/9666/The-Functional-Mushroom-Market-Could-Hand-You-a-62-Billion-Opportunity. Accessed 01 Dec 2021

Beliveau V, Ganz M, Feng L et al (2017) A high-resolution *in vivo* atlas of the human brain's serotonin system. J Neurosci 37(1):120–128. https://doi.org/10.1523/jneurosci.2830-16.2016

Beutler JA (2009) Natural products as a foundation for drug discovery. Curr Protoc Pharmacol 46: 9.11.11–19.11.21. https://doi.org/10.1002/0471141755.ph0911s46

Blei F, Dörner S, Fricke J et al (2020) Simultaneous production of psilocybin and a cocktail of β-carboline monoamine oxidase inhibitors in "magic" mushrooms. Chemistry 26(3):729–734. https://doi.org/10.1002/chem.201904363

Bloch MH, Hannestad J (2012) Omega-3 fatty acids for the treatment of depression: systematic review and meta-analysis. Mol Psychiatry 17(12):1272–1282. https://doi.org/10.1038/mp.2011.100

Canal CE (2018) Serotonergic psychedelics: experimental approaches for assessing mechanisms of action. Handb Exp Pharmacol 252:227–260. https://doi.org/10.1007/164_2018_107

Darbinyan V, Aslanyan G, Amroyan E et al (2007) Clinical trial of *Rhodiola rosea* L. extract SHR-5 in the treatment of mild to moderate depression. Nord J Psychiatry 61(5):343–348. https://doi.org/10.1080/08039480701643290

de Veen BT, Schellekens AF, Verheij MM, Homberg JR (2017) Psilocybin for treating substance use disorders? Expert Rev Neurother 17(2):203–212. https://doi.org/10.1080/14737175.2016.1220834

Fleming A (2001) On the antibacterial action of cultures of a penicillium, with special reference to their use in the isolation of B. influenzae. 1929. BullWorld Health Organ 79(8):780–790

Fusar-Poli P, Borgwardt S (2008) Albert Hofmann, the father of lsd (1906–2008). Neuropsychobiology 58(1):53–54. https://doi.org/10.1159/000157779

García-Gutiérrez MS, Navarrete F, Gasparyan A et al (2020) Cannabidiol: a potential new alternative for the treatment of anxiety, depression, and psychotic disorders. Biomol Ther 10(11). https://doi.org/10.3390/biom10111575

Gartz J (1989) Biotransformation of tryptamine in fruiting mycelia of *Psilocybe cubensis*. Planta Med 55(3):249–250

Gates M, Tschudi G (1956) The synthesis of morphine. J Am Chem Soc 78(7):1380–1393

Geiger HA, Wurst MG, Daniels RN (2018) DARK classics in chemical neuroscience: psilocybin. ACS Chem Neurosci 9(10):2438–2447. https://doi.org/10.1021/acschemneuro.8b00186

Giacoppo S, Bramanti P, Mazzon E (2017) Sativex in the management of multiple sclerosis-related spasticity: an overview of the last decade of clinical evaluation. Mult Scler Relat Disord 17:22–31. https://doi.org/10.1016/j.msard.2017.06.015

Grimm O, Kraehenmann R, Preller K, Seifritz E, Vollenweider F (2018) Psilocybin modulates functional connectivity of the amygdala during emotional face discrimination. Eur Neuropsychopharmacol 28:691–700

Halberstadt AL, Geyer MA (2011) Multiple receptors contribute to the behavioral effects of indoleamine hallucinogens. Neuropharmacology 61(3):364–381. https://doi.org/10.1016/j.neuropharm.2011.01.017

Hofmann A, Heim R, Brack A, Kobel H (1958) Psilocybin, a psychotropic substance from the Mexican mushroom *Psilocybe mexicana* Heim. Experientia 14(3):107–109. https://doi.org/10.1007/bf02159243

Hofmann A, Troxler F (1963) Esters of indoles. USA Patent 3075992A, 29 January 1963

Hoge JF (1951) The Durham–Humphrey Bill. Food Drug Cosmetic Law J 6 (2):135–141

InsightSLICE (2021) Herbal medicine market global sales are expected to reach US$ 550 Billion by 2030 [Online]. https://www.globenewswire.com/en/news-release/2021/02/16/2176036/0/en/

Herbal-Medicine-Market-Global-Sales-Are-Expected-To-Reach-US-550-Billion-by-2030-as-stated-by-insightSLICE.html. Accessed 30 Nov 2021

Intelligence E (2020) Which trends offer opportunities or pose threats on the European natural ingredients for health products market? CBI [Online]. https://www.cbi.eu/market-information/natural-ingredients-health-products/trends. Accessed 12 Oct 2021

Jamshidi N, Cohen MM (2017) The clinical efficacy and safety of Tulsi in humans: a systematic review of the literature. Evid Based Complement Alternat Med 2017:9217567. https://doi.org/10.1155/2017/9217567

Janda K, Wojtkowska K, Jakubczyk K et al (2020) *Passiflora incarnata* in neuropsychiatric disorders-a systematic review. Nutrients 12(12). https://doi.org/10.3390/nu12123894

Kasper S (2013) An orally administered lavandula oil preparation (Silexan) for anxiety disorder and related conditions: an evidence based review. Int J Psychiatry Clin Pract 17(Suppl 1):15–22. https://doi.org/10.3109/13651501.2013.813555

Khuda-Bukhsh AR, Pathak S (2008) Homeopathic drug discovery: theory update and methodological aspect. Expert Opin Drug Discov 3(8):979–990. https://doi.org/10.1517/17460441.3.8.979

Kim DH (2012) Chemical diversity of Panax ginseng, Panax quinquifolium, and Panax notoginseng. J Ginseng Res 36(1):1–15. https://doi.org/10.5142/jgr.2012.36.1.1

Lockermann G (1951) Friedrich Wilhelm Serturner, the discoverer of morphine. J Chem Educ 28(5):277

Logan AC (2004) Omega-3 fatty acids and major depression: a primer for the mental health professional. Lipids Health Dis 3:25. https://doi.org/10.1186/1476-511x-3-25

Lopresti AL, Drummond PD (2014) Saffron (*Crocus sativus*) for depression: a systematic review of clinical studies and examination of underlying antidepressant mechanisms of action. Hum Psychopharmacol 29(6):517–527. https://doi.org/10.1002/hup.2434

Lord LD, Expert P, Atasoy S et al (2019) Dynamical exploration of the repertoire of brain networks at rest is modulated by psilocybin. NeuroImage 199:127–142. https://doi.org/10.1016/j.neuroimage.2019.05.060

Lowe H, Toyang N, Steele B et al (2021) The therapeutic potential of psilocybin. Molecules 26(10). https://doi.org/10.3390/molecules26102948

Marais A, Adams B, Ringsmuth AK et al (2018) The future of quantum biology. J R Soc Interface 15(148):20180640. https://doi.org/10.1098/rsif.2018.0640

Markets RA (2021) Global psychedelic drugs market report 2020: market size is projected to reach $10.75 billion by 2027. Research and Markets [Online]. https://www.globenewswire.com/en/news-release/2021/03/18/2195240/28124/en/Global-Psychedelic-Drugs-Market-Report-2020-Market-Size-is-Projected-to-Reach-10-75-Billion-by-2027.html. Accessed 02 Dec 2021

Matsushima Y, Shirota O, Kikura-Hanajiri R, Goda Y, Eguchi F (2009) Effects of *Psilocybe argentipes* on marble-burying behavior in mice. Biosci Biotech Biochem 73(8):1866–1868. https://doi.org/10.1271/bbb.90095

Morales-García JA, de la Fuente RM, Alonso-Gil S et al (2017) The alkaloids of *Banisteriopsis caapi*, the plant source of the Amazonian hallucinogen Ayahuasca, stimulate adult neurogenesis *in vitro*. Sci Rep 7(1):5309. https://doi.org/10.1038/s41598-017-05407-9

Müller F, Liechti ME, Lang UE, Borgwardt S (2018) Advances and challenges in neuroimaging studies on the effects of serotonergic hallucinogens: contributions of the resting brain. Prog Brain Res 242:159–177. https://doi.org/10.1016/bs.pbr.2018.08.004

Nichols DE (2020) Psilocybin: from ancient magic to modern medicine. J Antibiot 73(10):679–686. https://doi.org/10.1038/s41429-020-0311-8

Oreja-Guevara C (2012) Clinical efficacy and effectiveness of Sativex, a combined cannabinoid medicine, in multiple sclerosis-related spasticity. Expert Rev Neurother 12(4 Suppl):3–8. https://doi.org/10.1586/ern.12.11

Powell M (2015) Medicinal mushrooms - a clinical guide. Mycology Press, Bamboo Publishing Ltd, Sussex

GrandView Research (2021) Pharmaceutical manufacturing market size, share & trends analysis report (2021–2028) [Online]. https://www.grandviewresearch.com/industry-analysis/pharmaceutical-manufacturing-market. Accessed 12 Jan 2022

Ribas RMG, Oliveira DCL, Silva PC et al (2021) (2021) the different roles of *Griffonia simplicifolia* in the treatment of depression: a narrative review. Int J Complement Alt Med 14(3):167–172. https://doi.org/10.15406/ijcam.2021.14.00553

Sarris J, Kavanagh DJ, Byrne G et al (2009) The kava anxiety depression spectrum study (KADSS): a randomized, placebo-controlled crossover trial using an aqueous extract of Piper methysticum. Psychopharmacol (Berl) 205(3):399–407. https://doi.org/10.1007/s00213-009-1549-9

Saville MW, Lietzau J, Pluda JM et al (1995) Treatment of HIV-associated Kaposi's sarcoma with paclitaxel. Lancet 346(8966):26–28. https://doi.org/10.1016/s0140-6736(95)92654-2

Siff S (2018) R. Gordon Wasson and the publicity campaign to introduce magic mushrooms to mid-century America. Revue française d'études américaines 156(3):91–104

Singh N, Bhalla M, de Jager P, Gilca M (2011) An overview on ashwagandha: a Rasayana (rejuvenator) of Ayurveda. Afr J Tradit Complement Altern Med 8(5 Suppl):208–213. https://doi.org/10.4314/ajtcam.v8i5S.9

Smith T MF, Eckl V, Morton Reynolds C (2021) Herbal supplement sales in u.s. increase by record-breaking 17.3% in 2020 [Online]. https://www.herbalgram.org/news/press-releases/2021/record-breaking-us-herbal-supplement-sales-increase-in-2020/. Accessed 8 Sept 2021

Srivastava JK, Shankar E, Gupta S (2010) Chamomile: a herbal medicine of the past with bright future. Mol Med Rep 3(6):895–901. https://doi.org/10.3892/mmr.2010.377

Tagliazucchi E, Carhart-Harris R, Leech R, Nutt D, Chialvo DR (2014) Enhanced repertoire of brain dynamical states during the psychedelic experience. Hum Brain Mapp 35(11):5442–5456. https://doi.org/10.1002/hbm.22562

Usona Institute (2021) [Online]. https://usonaclinicaltrials.org/major-depressive-disorder-psilocybin-clinical-trial-psil201. Accessed 30 Nov 2021

Vargas AS, Luís Â, Barroso M, Gallardo E, Pereira L (2020) Psilocybin as a new approach to treat depression and anxiety in the context of life-threatening diseases-a systematic review and meta-analysis of clinical trials. Biomedicine 8(9). https://doi.org/10.3390/biomedicines8090331

Whiskey E, Werneke U, Taylor D (2001) A systematic review and meta-analysis of *Hypericum perforatum* in depression: a comprehensive clinical review. Int Clin Psychopharmacol 16(5): 239–252. https://doi.org/10.1097/00004850-200109000-00001

Zhuk O, Jasicka-Misiak I, Poliwoda A et al (2015) Research on acute toxicity and the behavioral effects of methanolic extract from psilocybin mushrooms and psilocin in mice. Toxins (Basel) 7(4):1018–1029. https://doi.org/10.3390/toxins7041018

Preventative and Curative Properties of Reishi and Maitake Mushrooms in Cancer

Suhrud Pathak, Nick Martin, Shannon Kim, Sindhu Ramesh, Rishi M. Nadar, Jack DeRuiter, Timothy Moore, Dinesh Chandra Agrawal, and Muralikrishnan Dhanasekaran

Contents

Abstract Medicinal mushrooms have considerable therapeutic (130 different) properties, including anticancer, antiviral, antibacterial, antiparasitic, antifungal, antioxidant, radical scavenger, cardiovascular, anti-hypercholesterolemic,

S. Pathak · N. Martin · S. Kim · S. Ramesh · R. M. Nadar · J. DeRuiter · T. Moore · M. Dhanasekaran (✉)
Department of Drug Discovery and Development, Harrison School of Pharmacy, Auburn University, Auburn, AL, USA
e-mail: dhanamu@auburbn.edu

D. C. Agrawal (✉)
Department of Applied Chemistry, Chaoyang University of Technology, Taichung, Taiwan
e-mail: dcagrawal@gm.cyut.edu.tw

D. C. Agrawal, M. Dhanasekaran (eds.), *Mushrooms with Therapeutic Potentials*,
https://doi.org/10.1007/978-981-19-9550-7_19

detoxifying, hepatoprotective, and antihyperglycemic actions. In terms of nutrition, mushrooms are considered to be valuable food (nutraceuticals) and a source of therapeutics. Cancer patients worldwide frequently supplement traditional anticancer therapy with complementary alternative drugs and natural bioactive made from organic sources like mushrooms, including the well-known maitake or Reishi medicinal mushroom. *Ganoderma lucidum (G. lucidum* – Reishi*)* and *Grifola frondosa (G. frondosa* – maitake) are species of mushroom with a long history in traditional medicine to lengthen life and improve health. Since the 1970s, they have been intensively grown globally and promoted due to the rising consumption trend. In addition to exerting anticancer characteristics, they are also effective in preventing and treating numerous disorders. The principal bioactive components in *G. lucidum* and *G. frondosa* include terpenoids and polysaccharides, which are responsible for the anticancer effects. This book chapter focuses on the chemotherapeutic properties of *G. lucidum* and *G. frondosa.*

Keywords Anticancer · Maitake · Maitake D-fraction · Medicinal mushroom · Natural bioactives · Nutraceuticals · Reishi

Abbreviations

95-D	High-metastatic human lung cancer cell line
A549	Adenocarcinoma human alveolar basal epithelial cells
AIDS	Acquired immunodeficiency syndrome
AKT(PKB)	Protein kinase B
AP-1	Activator protein 1
BAK1	BCL2 antagonist/killer 1
Bax	Bcl-2-associated X protein
Bcl-2	B-cell leukemia/lymphoma 2
Bcl-xl	B-cell lymphoma-extra large
BW	Body weight
CAV-1	Caveolin 1
Cdc2	Cell division cycle 2
CNS	Central nervous system
COX-2	Cyclooxygenase-2
Cul-3	Cullin 3
EOC cells	Epithelial ovarian carcinomas
FADD	Fas-associated death domain protein
FasL	Fas ligand
Gl-PS	*Ganoderma lucidum* polysaccharides
HCT-116	Human colon cancer cell line
Hep3B	Human hepatoma cell line
HT-29	Human colorectal adenocarcinoma cell line
IBC	Inflammatory breast cancer

IC_{50}	Half maximal inhibitory concentration
ICAM3	Intercellular adhesion molecule 3
IFN-γ	Interferon gamma
IGFBP-7	Insulin growth factor binding protein 7
IL-1	Interleukin 1
IL-6	Interleukin 6
IL-8	Interleukin 8
ITGA2	Integrin subunit alpha 2
JNK	c-Jun N-terminal kinases
LDH	Lactate dehydrogenase
MCF-7	Human breast epithelial cell line
MDA-MB-231	Epithelial human breast cancer cell line
MIP	Molecularly imprinted polymer
MKN-45	Human gastric cancer cell line
MMP-2	Matrix metallopeptidase 2
MMP-9	Matrix metallopeptidase 9
mTOR	Mammalian target of the rapamycin
NF-kappaB	Nuclear factor kappa B
NK	Natural killer cells
NRF2	Nuclear factor erythroid 2-related factor 2
PARP	Poly-(ADP-ribose)-polymerase
PC-3	Human prostate cancer cell
PDEF	Prostate-derived Ets factor
PDGFB	Platelet-derived growth factors
PI3K	Phosphoinositide-3-kinase
qRT-PCR	Real-time quantitative reverse transcription polymerase chain reaction
ROS	Reactive oxygen species
SGC-7901	Human gastric cancer cell line
SMMC-7721	Human hepatocarcinoma cell line
SOD2	Superoxide dismutase 2
SPARC	Secreted protein acidic and rich in cysteine
ST7	Suppression of tumorigenicity 7
STAT-3	Signal transducer and activating transcription 3
SW480	Human isogenic nonmetastatic colorectal cancer cell line
T24	Human bladder carcinoma cell line
TERT	Telomerase reverse transcriptase
TGF-beta	Transforming growth factor
Th1	T helper type 1
TNF-α	Tumor necrosis factor-alpha
uPAR	Urokinase plasminogen activator receptor
VEGF	Vascular endothelial growth factor

1 Introduction

Mushrooms have been used for thousands of years for their potential nutritional and medicinal properties. Hippocrates used and classified the amadou mushroom (Fomes fomentarius) as an anti-inflammatory and used it to cauterize wounds in 450 BC. This Greek physician was among many ancient scientists and physicians who used mushrooms medicinally. Tao Hongjing, an alchemist from fifth-century China, used mushrooms like lingzhi (*Ganoderma lucidum*) and zhu ling (*Dendropolyporus umbellatus*) for a number of ailments. Also, Ötzi, the ice man, reportedly carried a sack of amadou and birch polypore to help him in his travels through the treacherous Alps almost 4500 years ago. Even early Americans used puffballs to help heal their wounds (Stamets and Zwickey 2014). Only recently has this fungus's true medicinal value and power been discovered. Modern research has confirmed many ancient claims associated with the various mushrooms mentioned. Currently, medicinal mushrooms are used as supplements, insecticides, bactericides, herbicides, fungicides, dietary foods, pharmaceuticals, and cosmeceuticals (Gargano et al. 2017). Most of the bioactive properties found in medicinal mushrooms are present not only in the fruiting body but also in the mycelium. A great deal of research is being performed to determine the identity of these mycochemicals, and their potential applications as natural medicines grow more popular.

Cancer is one of the top causes of mortality around the globe. Till now, the most promising treatment options for cancer are based on surgery, radiation, chemotherapy, and immunotherapy. Using synthetic substances, conventional cancer chemotherapy tries to kill or disable tumor cells, while conserving normal cells in the body. However, these medicines often have a small margin of safety, and drug resistance and dose-limiting toxicities are common (Gao and Zhou 2003). Many cancer patients seek complementary and alternative therapy to help battle cancer and lessen the adverse consequences of chemotherapy or radiation. Mycotherapy, a complementary and alternative medicine, has been linked to favorable outcomes in people with cancer in terms of treatment response, side effect reduction, and quality of life enhancement. Studies have highlighted the medicinal mushrooms' broad array of properties and their use in the medical field. However, particular interests focus on medicinal mushrooms' antitumor and immunomodulatory properties, and their anticancer treatments remain a significant challenge.

2 Bioactives in Medicinal Mushrooms

As mentioned previously, bioactive properties within medicinal mushrooms are useful in the treatment and have certain anticancer properties. Polysaccharides are constituents in these medicinal properties (Fig. 1). These polysaccharides carry important biological information about their antitumor, immunomodulatory, antioxidant, anti-inflammatory, antimicrobial, and antidiabetic activity. The structure of the

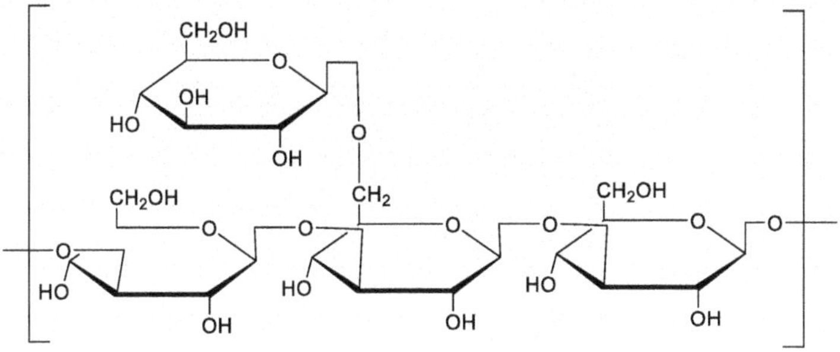

Fig. 1 General partial structure of mushroom polysaccharide

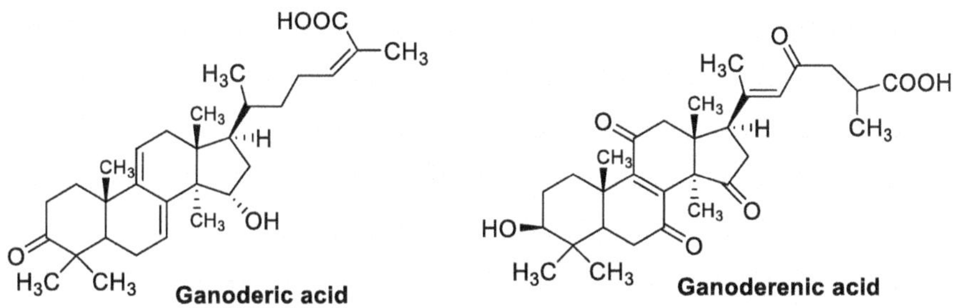

Fig. 2 Chemical structures of ganoderic and ganoderenic acids

molecule determines these specific properties. Weighted degree of branching, side-chain units, backbone linkage, and types of constituent monosaccharides all play a part in influencing the activity of this molecule, but mainly its immunomodulatory properties due to its ability to bind to cell wall receptors and stimulate immune responses. Medicinal mushrooms are used in cancer treatments to reduce negative symptoms of therapy and increase the quality of patient life.

Along with polysaccharides, terpenes are also integral to the mushrooms' bioactive properties. Terpenes are five carbon isoprene units that form the core structure of larger, functionalized terpenoids (Fig. 2). They influence gene expression by stimulating the expression of gene codes for proteins involved in the immune response, producing these anticancer and anti-inflammatory properties. The genus *Ganoderma* includes mushrooms with high terpene concentrations (Venturella et al. 2021).

The anticancer effects of Reishi (*Ganoderma lucidum*) and maitake mushroom (*Grifola frondosa*) are illustrated in Fig. 3.

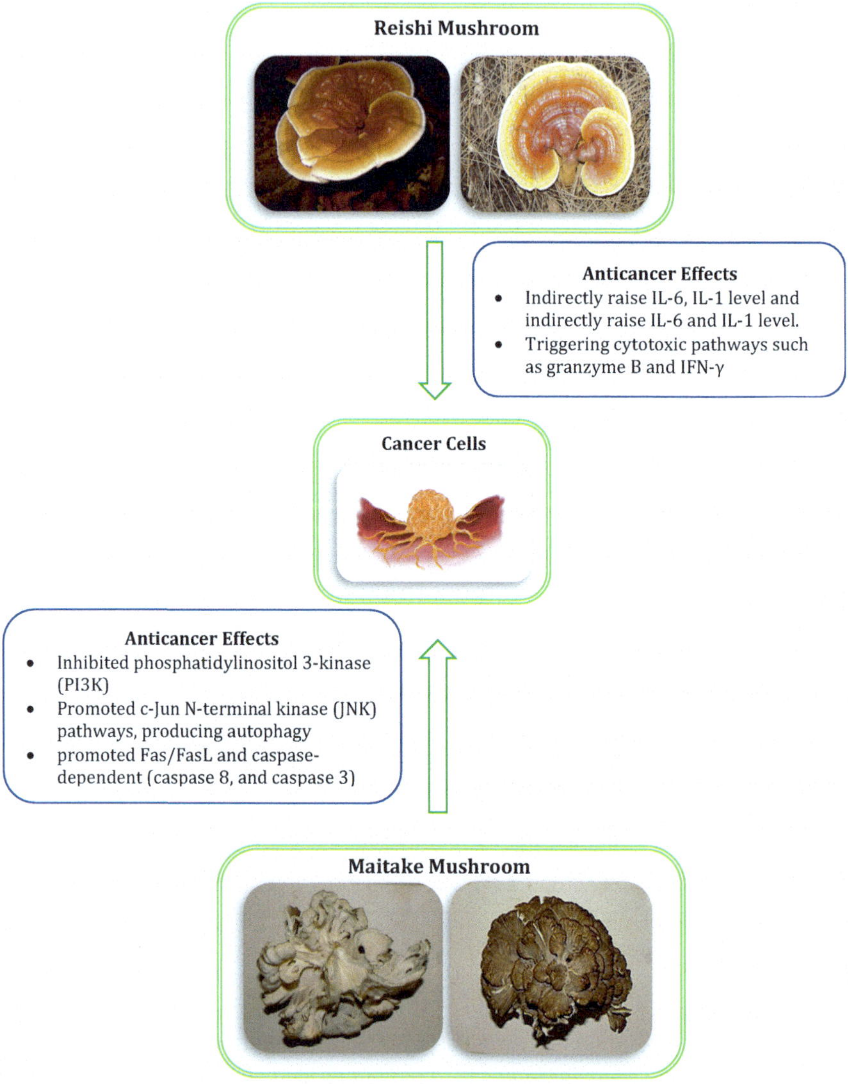

Fig. 3 Preventative and curative properties of Reishi and maitake mushrooms in cancer. Cancer Research UK, CC BY-SA 4.0 <https://creativecommons.org/licenses/by-sa/4.0>, via Wikimedia Commons. Wendell Smith, CC BY-SA 2.0 < https://www.flickr.com/photos/wendellsmith/16805561088> via flickr.com, WolfpackBME, CC BY-SA 4.0 <https://creativecommons.org/licenses/by-sa/4.0>, via Wikimedia Commons, RevDanielElisAxelrod, CC0, via Wikimedia Commons, Sinisa Radic, CC BY-SA 4.0 <https://creativecommons.org/licenses/by-sa/4.0>, via Wikimedia Commons

3 Reishi Mushroom (*Ganoderma Lucidum*)

The Reishi mushroom, scientific name *Ganoderma lucidum*, has long had a place in traditional Asian medicinal techniques, as it is believed to be beneficial to one's health and life span. Some consider it an effective preventative and treatment for various ailments (Unlu et al. 2016). Due to its rarity in nature, the Reishi mushroom was once only available to the members of the nobility. However, as a result of its rising demand, increased attention has been put into growing it, and since 1970, the main source of the mushroom has become cultivation (Unlu et al. 2016). Furthermore, commercial products such as tea and dietary supplements made from various parts, including mycelia, spores, and fruit bodies, have been widely bought and sold (Wachtel-Galor et al. 2011). Most notably, of all its health benefits, the Reishi mushroom is thought to have anticancer properties. This belief has been supported by the results of some laboratory research and a number of preclinical trials. Consequently, the use of the Reishi mushroom as an alternative medicine for cancer patients has been on the rise (Jin et al. 2016).

Although the few clinical results about its ability to treat cancer have been dubious, it is still probable that the Reishi mushroom may be useful to cancer patients. The polysaccharides, dietary fibers, oligosaccharides, triterpenoids, minerals, amino acids, and vitamins that Reishi mushroom contains have been observed to be capable of helping treat hepatopathy, chronic hepatitis, nephritis, hypertension, hyperlipidemia, arthritis, neurasthenia, insomnia, bronchitis, asthma, gastric ulcers, atherosclerosis, leukopenia, diabetes, and anorexia. It also has cytotoxic, cytostatic, anti-metastatic, anti-inflammatory, and immunomodulatory properties. Limited clinical investigations, including case reports and randomized controlled trials, pointed to *G. lucidum* as an alternative adjuvant treatment for boosting the immune system in cancer patients (Cheng and Sliva 2015). One review article points out the diverse mechanism of *G. lucidum*: it can suppress urokinase plasminogen activator and urokinase plasminogen activator receptor (uPAR) expression in cancer cells (Ahmad 2020). However, it should be noted that these mushrooms are utilized primarily as a health supplement and not as a remedy (Batra et al. 2013). For centuries, Reishi mushrooms have been used in East Asian medicine to treat various conditions, including cancer. They exhibit anticancer properties due to their polysaccharide fraction, which helps strengthen the immune system, and their triterpenes, which exhibit a cytotoxic effect against different cancer cells. Several studies have been conducted in vivo, in vitro, and in silico to investigate the anticancer properties of the Reishi mushroom.

3.1 Breast Cancer

Various studies showed that Reishi might be a natural treatment for breast cancer. In one study, extract from Reishi was observed to inhibit the growth of a type of breast cancer cell (Suarez-Arroyo et al. 2013). Cell invasion and the cell spheroids involved in inflammatory breast cancer (IBC) were negatively affected by Reishi. Another study identified the molecular effects of Reishi by concentrating on the phosphoinositide-3-kinase (PI3K)/AKT/mammalian target of the rapamycin (mTOR) pathway, which is a regulator of cell survival and proliferation. They discovered that Reishi inhibits protein synthesis and tumor development by influencing survival and proliferative signaling pathways that affect the translation, implying that Reishi has the potential to be a natural therapy for breast cancer (Suarez-Arroyo et al. 2013). A group of Turkish researchers investigated the time-dependent cytotoxic effects of *G. lucidum* extracts which were tested on MCF-7 cells at 24, 48, and 72 h using five different solvents (ethanol-water, methanol, ethanol, ethyl acetate, and ether). The cytotoxicity findings concluded that *G. lucidum* ether extract was more effective against breast cancer cells ($IC_{50} = 100$ g/mL at 72 h) than the others (Suárez-Arroyo et al. 2017). One clinical study demonstrated that the mean serum levels of TNF-α and IL-8 significantly declined. In contrast, the level of IFN-γ significantly elevated in patients receiving concomitant *G. lucidum* and chemotherapy. The mean serum concentrations of TNF-α, IFN-γ, and IL-8 in individuals who only received chemotherapy were not significantly different before and after treatment (Nidhal et al. 2018).

3.2 Colorectal Cancer

In both mice and rats, the anticarcinogenic effect of a water-soluble extract from the cultivated medium of *G. lucidum* mycelia on the formation of carcinogen-induced colonic aberrant crypt foci (a preneoplastic lesion) was studied, where azoxymethane was used as carcinogens in the rat and N,N′-dimethylhydrazine for a mouse (Lu et al. 2001; Lu et al. 2002). A team directed by Barbara Pence extracted ethanolic extract of *Ganoderma lucidum* and analyzed its activation of apoptosis and the suppression of different cytokine expressions in human colonic carcinoma cell line HT-29. Surprisingly, *Ganoderma lucidum* extract was non-cytotoxic up to a dose of 1000 μg/ml. In addition, the extract effectively reduced the expression of COX-2 and IL-8, MIP, VEGF, and PDGF and reduced nitric oxide production (Ye and Jensen 2017). *Ganoderma lucidum* polysaccharides have also been shown to reduce cell viability in human colon cancer cells (HCT-116 and SW 480) in a concentration-dependent manner. *G. lucidum* polysaccharides can cause apoptosis in HCT-116 cells by increasing the release of lactate dehydrogenase (LDH) and the level of intracellular Ca^{2+}, which activates the Fas-mediated caspase, mitochondrial, and JNK pathways (Sohretoglu and Huang 2018).

3.3 Prostate Cancer

Aside from other cancers, the mushroom can prevent prostate cancer-related angiogenesis by limiting the capillary morphogenesis of human aortic endothelial cells. This is achieved by downregulating PC-3 cells' secretion of VEGF and TGF-beta, thus inhibiting constitutively active AP-1 in prostate cancer cells. All of these effects point to the possibility that Reishi mushrooms have the potential to be used as a treatment option for prostate cancer (Stanley et al. 2005). The in vitro study illustrates that *Ganoderma lucidum* induces apoptosis, ROS accumulation, and cytotoxicity in PC-3 cells. A subsequent carcinogenic transcriptional factor that regulates cell division and apoptosis is the signal transducer and activating transcription (STAT-3). STAT-3 transcription regulation was considered a possible technique to inhibit prostate cell proliferation. In the current study, we demonstrate that *Ganoderma lucidum* inhibits STAT-3 translocation, reducing the production of Bcl-2 and cyclin D1 and declining the expression of Bax, caspase-3, and caspase-9 in PC-3 cells (Wang et al. 2020).

Similarly, one research study on PC-3 cell lines illustrates that *Ganoderma lucidum* reduces cell proliferation by upregulating p21 expression and downregulating cyclin B and Cdc2 expression in a dose- and time-dependent manner. The halt of the cell cycle in the G2/M phase served as another indicator of the suppression of cell proliferation. Furthermore, NF-kappaB-regulated Bcl-2 and Bcl-xl expression was downregulated in PC-3 cells after *Ganoderma lucidum* caused apoptosis (Jiang et al. 2004).

3.4 Leukemia

There is evidence to support the anti-leukemia properties of *Ganoderma lucidum* polysaccharides. *Ganoderma lucidum* polysaccharides potentially indirectly raise IL-6 and IL-1 levels, strengthening in vivo anticancer immunity (Wang et al. 1997). It has been suggested that *Ganoderma lucidum* polysaccharides can enhance the particular cytotoxic T lymphocytes that dendritic cells activate. These lymphocytes targeted the P815 tumor antigens, triggering cytotoxic pathways such as granzyme B and IFN-γ (Cao and Lin 2003).

3.5 Lung Cancer

It has been shown that *G. lucidum* can be used as a treatment for lung cancer (Huang and Cao 2021). To assess the inhibitory effects of triterpenes on cell proliferation and tumor growth, A549 cells and Lewis tumor-bearing mice were employed. Triterpenes from *G. lucidum* may considerably slow tumor development in mice, supporting the mechanism that *G. lucidum* works by boosting IL-6 and TNF-α expression to elicit an immunological response (Feng et al. 2013).

3.6 Melanoma

According to Sun and coworkers, Gl-PS can improve the function of class I molecules of the main histocompatibility complex, increasing the cytotoxicity against melanoma cells (Sun et al. 2012).

3.7 Ovarian Cancer

A study was performed to examine the mushroom's impact on cisplatin's performance against ovarian epithelial carcinoma (EOC) cells. The effects of the cisplatin became significantly pronounced after it was treated with Reishi. Like in prostate cancer cells, cell cycle arrest during phase G2 occurred in the EOC cells after they were treated with Reishi (Zhao et al. 2011).

Reishi mushrooms show some promise as a possible therapeutic for other types of cancer besides inflammatory breast cancer, colon cancer, prostate cancer, leukemia, lung cancer, and melanoma. In addition to epithelial ovarian cancer, there is a possibility that the Reishi mushroom may help in treating lung carcinoma. According to a laboratory study, ganoderic acid Me ((e)-6-(3,15-diacetyloxy-4,4,10,13,14-pentamethyl-2,3,5,6,12,15,16,17-octahydro-1 h-cyclopenta[a]phenanthren-17-yl)-2-methylhept-2-enoic acid), obtained by purifying a Reishi mushroom's fermentation mycelia, has anti-invasion properties. The cell migration of a highly metastatic lung tumor cell line, 95-D, was inhibited dose-dependently and time-dependently by ganoderic acid Me. The acid also had the effect of inhibiting cell-matrix adhesions and aiding in cell homotypic aggregation.

Furthermore, as noted through qRT-PCR and Western blotting, in 95-D cells, ganoderic acid inhibited the expression of matrix metalloproteinases 2 and 9 at both the mRNA and protein levels. All of these various effects indicate that the Reishi mushroom can help to prevent tumor invasion in all cancers and that it has the potential to be incorporated into anti-metastatic cancer treatments as an MMP2/9 inhibitor (Chen et al. 2008). The primary bioactive compounds of the Reishi mushroom (Table 1), which give it its numerous anticancer properties, are triterpenes/triterpenoids and polysaccharides, both of which are major groups of metabolites. Examples of triterpene compounds include ganoderic acid, lucidones, and ganodermic alcohols, all exhibiting strong antitumor, anti-metastatic, and cytotoxic qualities. Examples of major polysaccharides include α-1,3, β-1,3, and β-1,6-D-glucans, all exhibiting pronounced antiangiogenic qualities. Polysaccharides also help strengthen the immune system. Due to these two types of molecules, the Reishi mushroom displays immunostimulating properties and can aid apoptosis,

Table 1 Anticancer bioactives of Reishi mushroom

Bioactives	Anticancer effect	References
• Triterpenes/triterpenoids – Lucidones – Ganodermic alcohols – Ganoderic acid • Polysaccharides – α-1,3, β-1,3 glucans – β-1,6-D-glucans	• Inflammatory breast cancer	Suarez-Arroyo et al. 2013; Suárez-Arroyo et al. 2017; Nidhal et al. 2018)
	• Prostate cancer	Jiang et al. 2004; Stanley et al. 2005; Wang et al. 2020
	• Epithelial ovarian cancer	Zhao et al. 2011
	• Lung carcinoma	Huang and Cao 2021; Feng et al. 2013
	• Colorectal cancer	Lu et al. 2001; Lu et al. 2002; Ye and Jensen 2017; Sohretoglu and Huang 2018
	• Leukemia	Cao and Lin 2003; Wang et al. 1997;
	• Melanoma	Sun et al. 2012

suppressing cell proliferation, metastasis, and invasion (Venturella et al. 2021). As explained when expanding upon its anticancer activity, the Reishi mushroom has displayed properties that give it potential as a therapeutic for inflammatory breast cancer, prostate cancer, epithelial ovarian cancer, and lung carcinoma.

Moreover, clinical and laboratory tests show signs that the Reishi mushroom may be able to help against hepatoma, leukemia, endometrial cancer, and melanoma. In the laboratory study testing its effects against lung carcinoma, fermentation mycelia were purified into ganoderic acid. Given that it has been sold as a dietary supplement, such as tea, it is likely that the mushroom's health benefits can be taken advantage of in these forms. However, the effectiveness of these products as anticancer therapeutics seems not to have been tested.

4 Maitake Mushroom (*Grifola Frondosa*)

The Japanese term for an edible fungus with a huge fruiting body and overlapping caps is maitake (*Grifola frondosa*). It is a top-notch medicinal and culinary mushroom. More and more people are becoming aware of maitake as a rich source of polysaccharide compounds with extraordinary health-promoting potential, specifically having significant antitumor and immunomodulatory activities. *G. frondosa* is a basidiomycetes fungus that belongs to the family of *Grifolaceae* and the order *Polyporales*. When young, *G. frondosa* grows around the stumps of broadleaf trees or trunks and is edible. The northeastern region of Japan's climate is favorable for the growth of *G. frondosa*. It also can grow well in the temperate woods of eastern North America, Eastern Europe, and East Asia. In the United States and Canada, a common fungus is known as sheep's head, king of mushrooms, hen-of-the-woods,

Table 2 Anticancer bioactives of maitake mushroom

Bioactives	Anticancer effect	Reference
• Polysaccharides. • Terpenes. – Sesquiterpenes. • Carbohydrate. – Beta-1,6-D-glucan. – Maitake D-fraction.	• Breast cancer.	Soares et al. 2011; Alonso et al. 2013; Zhang et al. 2017
	• Colorectal cancer.	Kodama et al. 2010; Masuda et al. 2009; Masuda et al. 2017
	• Gastric cancer.	Liu et al. 2020; Cui et al. 2016
	• Hepatocellular carcinoma.	Lin et al. 2016; Zhao et al. 2017
	• Bladder cancer.	Louie et al. 2010; Rajamahanty et al. 2009
	• Pharyngeal and laryngeal cancer.	Kodama et al. 2003; Wong et al. 2020
	• Prostate cancer	Fullerton et al. 2000

and cloud mushrooms. Since it has an excellent source of protein, carbohydrates, dietary fiber, vitamin D2 (ergocalciferol), and minerals (potassium, sodium, calcium, and magnesium) and has a low-fat content and caloric value, *G. frondosa* is edible and is recognized as a nutritious meal. Due to the high levels of trehalose, glutamic, aspartic, and 5′-nucleotide in *G. frondosa*, it has a pleasant, sweet, and umami flavor. Due to its unique and exquisite flavor, *G. frondosa* is utilized as a dry powder food flavoring and a food ingredient (Wu et al. 2021; Mayell 2001). Several studies demonstrated that the maitake mushroom D-fraction, which contains beta-1,6-glucan with beta-1,3 branched chains, has potent anticancer properties (Table 2) by boosting immune-competent cell activity. The immunostimulatory activities of maitake D-fraction on macrophages, natural killer cells, and T cells contribute to the anticancer effects observed in mice with cancer xenografts. In cancer patients receiving concurrent immunotherapy and chemotherapy, D-fraction decreased the growth of breast, hepatic, and pulmonary malignancies.

4.1 Breast Cancer

Soares et al. (2011) showed that in MCF7 cells, maitake D-fraction caused the release of cytochrome c from the mitochondria, which led to mitochondrial malfunction and apoptosis. T lymphocytes, natural killer cells, and macrophages were all stimulated by maitake D-fraction. It modified the expression of several genes, including BCL2 antagonist/killer 1 (BAK1), CAV-1, Cul-3, cyclin E, ICAM3, IGFBP-7, ITGA2, NRF2, ST7, SOD2, and SPARC, which are associated in the initiation of multidrug sensitivity, cell cycle arrest, inhibition of cell growth and proliferation, induced apoptosis, and inhibition of migration and metastasis (Alonso et al. 2013). Zhang et al. (2017) showed that in MCF-7 and MDA-MB-231 breast cancer cells, as well as cancer tissues from mice treated with the *G. frondosa*, polysaccharides increased lactate dehydrogenase release and reactive oxygen species accumulation, prompted mitochondrial dysfunction, upregulated expression of Bax, cleaved caspase-3 and caspase-8, and downregulated expression of B-cell lymphoma 2 (Bcl-2) and Bcl-xl (Zhang et al. 2017). The polysaccharides prevented AKT/glycogen synthase kinase-3β and extracellular signal-regulated kinases, preventing tumor xenograft formation.

4.2 *Colorectal Cancer*

By triggering cell-mediated immunity and the Th1 response, the team at Kobe Pharmaceutical University, Japan, verified that *G. frondosa* β-glucan could inhibit the growth of colon cancer in mice. A low molecular weight protein fraction of *G. frondosa* suppressed the growth of invasive colon cancer in mice, which was considered to be related to elevated IL-1β, TNF-α, and, interestingly, IL-10 levels (Kodama et al. 2010; Masuda et al. 2009; Masuda et al. 2017).

4.3 *Prostate Cancer*

Another in vitro study with human prostate cancer PC-3 cells treated with different concentrations of the pure beta-glucan formulation (a polysaccharide of the maitake mushroom) explored the possible antitumor effect of the maitake mushroom. The results revealed that due to oxidative stress, bioactive beta-glucan, which results in apoptosis, exerted cytotoxic effects and exhibited anticancer activity (Fullerton et al. 2000).

4.4 *Gastric Cancer*

G. frondosa polysaccharide decreased MKN-45 cell proliferation dose-dependently, promoted morphologic alterations, arrested them in the G0/G1 phase, and upregulated the levels of FasL, Fas, FADD, caspase 8, and caspase 3. These alterations demonstrated that *G. frondosa* polysaccharide promoted Fas/FasL and caspase-dependent apoptosis in MKN-45 cells (Liu et al. 2020). *Grifola frondosa* glycoprotein suppressed SGC-7901 proliferation in a dose- and time-dependent manner, elicited typical apoptotic cell interface and nuclear morphology, and suppressed cancer cell proliferation via S phase cell cycle arrest (Cui et al. 2016).

4.5 Hepatocellular Carcinoma

Grifola frondosa inhibited phosphatidylinositol 3-kinase (PI3K) and promoted c-Jun N-terminal kinase (JNK) pathways, producing autophagy. It reduced proliferation, triggered cell cycle arrest, and provoked apoptosis in Hep3B hepatoma cells (Lin et al. 2016). According to one study, D-fraction polysaccharide from *G. frondosa* with vitamin C therapy triggered apoptosis in around 65% of SMMC-7721 cells. After 48 hours of *G. frondosa* therapy, cell cycle analysis revealed cell cycle arrest in the G2/M phase. Furthermore, their analysis revealed that cells treated with the combination of *G. frondosa* showed upregulation of Bax, downregulation of Bcl-2, activation of poly-(ADP-ribose)-polymerase (PARP), and release of cytochrome c, indicating that the mechanism of anticancer activity in the SMMC-7721 hepatocarcinoma cells involved induction of apoptosis (Zhao et al. 2017).

4.6 Bladder Cancer

Research on T24 cells, a cell line established from a human urinary bladder cancer patient, showed that the combination of interferon (IFN)-alpha and maitake mushroom D-fraction decreased proliferation by nearly 75% (Louie et al. 2010). According to one case report, a patient with invasive bladder cancer at high risk of disease recurrence followed a D-fraction regimen (with vitamin C), avoiding all other medicinal measures. The 2-year follow-up revealed no clinical signs of residual illness progression or recurrence, with the possibility of disease remission (Rajamahanty et al. 2009).

Maitake mushrooms have shown some potential as a therapeutic option for cancers other than breast cancer, prostate cancer, colorectal cancer, gastric cancer, hepatocellular carcinoma, and bladder cancer. The findings of one study suggested that the administration of maitake D-fraction during concurrent chemoradiotherapy significantly reduces adverse events and deterioration in the quality of life induced by concurrent chemoradiotherapy in patients with advanced laryngeal and pharyngeal cancer. All cancer patients who received D-fraction alone, without contemporary anticancer medication, experienced increased natural killer cell activity, reduced tumor marker expression, and inhibited metastasis. Maitake mushroom exhibits their anticancer efficacy in tumor-bearing mice by boosting the immune system through the activation of macrophages, T cells, and natural killer (NK) cells (Kodama et al. 2003; Wong et al. 2020). Lastly, one result indicates that D-fraction is a naturally derived therapeutic that might effectively treat canine tumors and other veterinary cancers (Konno 2004).

5 Conclusion

The current book chapter reviewed the reports and scientific articles regarding the possible medicinal approaches of mycotherapy (medicinal mushrooms) as an anticancer treatment. Medicinal mushrooms exhibit a wide range of pharmacological activity and significant nutraceutical properties. The current and future scientific community must make a meaningful contribution to expanding clinical studies and recommending remedies with secure sources and pure genetic makeup to accommodate the increasing interest in mycotherapy. The previous few decades have seen minimal advancement in the search for a cancer cure, owing to evidence-based methods for developing anticancer medications. The enormous structural diversity of naturally occurring bioactives found in mushrooms and the growing understanding of the molecular mechanisms underlying tumor progression and metastasis have created a unique opportunity to develop novel medications that rationally target the abnormal molecular and biochemical signals underlying cancer. Consuming medicinal mushrooms for cancer therapy has been supported by numerous in vitro, in vivo, and human clinical investigations, according to our assessment. *G. lucidum* and *G. frondosa* are the two main mushrooms with anticancer characteristics, specifically polysaccharides, triterpenes, lucidones, and glucans. These main bioactives have been demonstrated to have significant anticancer effects. Intriguingly, Reishi and maitake mushrooms may be an efficient therapeutic strategy with many nutritional values and minimal adverse effects.

References

Ahmad MF (2020) *Ganoderma lucidum*: a rational pharmacological approach to surmount cancer. J Ethnopharmacol 260:113047. https://doi.org/10.1016/j.jep.2020.113047

Alonso EN, Orozco M, Nieto AE et al (2013) Genes related to suppression of malignant phenotype induced by Maitake D-fraction in breast cancer cells. J Med Food 16(7):602–617. https://doi.org/10.1089/jmf.2012.0222

Batra P, Sharma AK, Khajuria R (2013) Probing Lingzhi or Reishi medicinal mushroom Ganoderma lucidum (Higher basidiomycetes): a bitter mushroom with amazing health benefits. Int J Med Mushrooms 15(2):127–143. https://doi.org/10.1615/IntJMedMushr.v15.i2.20

Cao LZ, Lin ZB (2003) Regulatory effect of Ganoderma lucidum polysaccharides on cytotoxic T-lymphocytes induced by dendritic cells *in vitro*. Acta Pharmacol Sin 24(4):321–326

Chen NH, Liu JW, Zhong JJ (2008) Ganoderic acid me inhibits tumor invasion through down-regulating matrix metalloproteinases 2/9 gene expression. J Pharmacol Sci 108(2):212–216. https://doi.org/10.1254/jphs.sc0080019

Cheng S, Sliva D (2015) Ganoderma lucidum for cancer treatment: we are close but still not there. Integrative Cancer Therap 14(3):249–257. https://doi.org/10.1177/1534735414568721

Cui F, Zan X, Li Y et al (2016) Grifola frondosa glycoprotein GFG-3a arrests S phase, alters proteome, and induces apoptosis in human gastric cancer cells. Nutr Cancer 68(2):267–279. https://doi.org/10.1080/01635581.2016.1134599

Feng L, Yuan L, Du M et al (2013) Anti-lung cancer activity through enhancement of immunomodulation and induction of cell apoptosis of total triterpenes extracted from

Ganoderma lucidum (Leyss. Ex Fr.) Karst. Molecules 18(8):9966–9981. https://doi.org/10.3390/molecules18089966

Fullerton SA, Samadi AA, Tortorelis DG et al (2000) Induction of apoptosis in human prostatic cancer cells with beta-glucan (Maitake mushroom polysaccharide). Mol Urol 4(1):7–13

Gao Y, Zhou S (2003) Cancer prevention and treatment by Ganoderma, a mushroom with medicinal properties. Food Rev Int 19(3):275–325. https://doi.org/10.1081/FRI-120023480

Gargano ML, van Griensven LJLD, Isikhuemhen OS et al (2017) Medicinal mushrooms: valuable biological resources of high exploitation potential. Plant Biosystems 151(3):548–565. https://doi.org/10.1080/11263504.2017.1301590

Huang L, Cao Y (2021) Chemical components and cancer immunotherapy of Ganoderma. In: Liu C (ed) The Lingzhi mushroom genome. Springer International Publishing, Cham, pp 177–191. https://doi.org/10.1007/978-3-030-75710-6_10

Jiang J, Slivova V, Valachovicova T et al (2004) Ganoderma lucidum inhibits proliferation and induces apoptosis in human prostate cancer cells PC-3. Int J Oncol 24(5):1093–1099

Jin X, Ruiz Beguerie J, Sze DM et al (2016) Ganoderma lucidum (Reishi mushroom) for cancer treatment. Cochrane Database Syst Rev 4(4):Cd007731. https://doi.org/10.1002/14651858.CD007731.pub3

Kodama N, Komuta K, Nanba H (2003) Effect of Maitake (Grifola frondosa) D-fraction on the activation of nk cells in cancer patients. J Med Food 6(4):371–377. https://doi.org/10.1089/109662003772519949

Kodama N, Mizuno S, Nanba H et al (2010) Potential anti-tumor activity of a low-molecular-weight protein fraction from Grifola frondosa through enhancement of cytokine production. J Med Food 13(1):20–30. https://doi.org/10.1089/jmf.2009.1029

Konno S (2004) Potential growth inhibitory effect of Maitake D-fraction on canine cancer cells. Vet Ther 5(4):263–271

Lin CH, Chang CY, Lee KR et al (2016) Cold-water extracts of Grifola frondosa and its purified active fraction inhibit hepatocellular carcinoma *in vitro* and *in vivo*. Exp Biol Med (Maywood) 241(13):1374–1385. https://doi.org/10.1177/1535370216640149

Liu C, Ji HY, Wu P et al (2020) The preparation of a cold-water soluble polysaccharide from Grifola frondosa and its inhibitory effects on MKN-45 cells. Glycoconj J 37(4):413–422. https://doi.org/10.1007/s10719-020-09932-0

Louie B, Rajamahanty S, Won J et al (2010) Synergistic potentiation of interferon activity with Maitake mushroom D-fraction on bladder cancer cells. BJU Int 105(7):1011–1015. https://doi.org/10.1111/j.1464-410X.2009.08870.x

Lu H, Kyo E, Uesaka T et al (2002) Prevention of development of N,N′-dimethylhydrazine-induced colon tumors by a water-soluble extract from cultured medium of Ganoderma lucidum (Rei-shi) mycelia in male icr mice. Int J Mol Med 9(2):113–117

Lu H, Uesaka T, Katoh O et al (2001) Prevention of the development of preneoplastic lesions, aberrant crypt foci, by a water-soluble extract from cultured medium of Ganoderma lucidum (Rei-shi) mycelia in male F344 rats. Oncol Rep 8(6):1341–1345. https://doi.org/10.3892/or.8.6.1341

Masuda Y, Matsumoto A, Toida T et al (2009) Characterization and anti-tumor effect of a novel polysaccharide from Grifola frondosa. J Agric Food Chem 57(21):10143–10149. https://doi.org/10.1021/jf9021338

Masuda Y, Nakayama Y, Tanaka A et al (2017) Antitumor activity of orally administered Maitake α-glucan by stimulating anti-tumor immune response in murine tumor. PLoS One 12(3): e0173621. https://doi.org/10.1371/journal.pone.0173621

Mayell M (2001) Maitake extracts and their therapeutic potential. Altern Med Rev 6(1):48–60

Nidhal AK, Mohammed A, Hakar AS et al (2018) Immunostimulatory and anti-inflammatory effect of *Ganoderma lucidum* on breast cancer patients. Asian Pac J Cancer Biol 3(2):51–57. https://doi.org/10.31557/APJCB.2018.3.2.51-57

Rajamahanty S, Louie B, O'Neill C et al (2009) Possible disease remission in patient with invasive bladder cancer with D-fraction regimen. Int J Gen Med 2:15–17. https://doi.org/10.2147/ijgm.s4948

Soares R, Meireles M, Rocha A et al (2011) Maitake (D-fraction) mushroom extract induces apoptosis in breast cancer cells by BAK-1 gene activation. J Med Food 14(6):563–572. https://doi.org/10.1089/jmf.2010.0095

Sohretoglu D, Huang S (2018) Ganoderma lucidum polysaccharides as an anti-cancer agent. Anti Cancer Agents Med Chem 18(5):667–674. https://doi.org/10.2174/1871520617666171113121246

Stamets P, Zwickey H (2014) Medicinal mushrooms: ancient remedies meet modern science. Integr Med (Encinitas) 13(1):46–47

Stanley G, Harvey K, Slivova V et al (2005) Ganoderma lucidum suppresses angiogenesis through the inhibition of secretion of VEGF and TGF-beta1 from prostate cancer cells. Biochem Biophys Res Commun 330(1):46–52. https://doi.org/10.1016/j.bbrc.2005.02.116

Suárez-Arroyo IJ, Loperena-Alvarez Y, Rosario-Acevedo R et al (2017) Ganoderma spp.: a promising adjuvant treatment for breast cancer. Medicines 4(1). https://doi.org/10.3390/medicines4010015

Suarez-Arroyo IJ, Rosario-Acevedo R, Aguilar-Perez A et al (2013) Anti-tumor effects of Ganoderma lucidum (Reishi) in inflammatory breast cancer in *in vivo* and *in vitro* models. PLoS One 8(2):e57431. https://doi.org/10.1371/journal.pone.0057431

Sun LX, Lin ZB, Duan XS et al (2012) Enhanced MHC class i and costimulatory molecules on B16F10 cells by Ganoderma lucidum polysaccharides. J Drug Target 20(7):582–592. https://doi.org/10.3109/1061186x.2012.697167

Unlu A, Nayir E, Kirca O et al (2016) Ganoderma lucidum (Reishi mushroom) and cancer. J BUON 21(4):792–798

Venturella G, Ferraro V, Cirlincione F, Gargano ML (2021) Medicinal mushrooms: bioactive compounds, use, and clinical trials. Int J Mol Sci 22(2):634

Wachtel-Galor S, Yuen J, Buswell J et al (2011) Ganoderma lucidum (Lingzhi or Reishi): a medicinal mushroom. In: IFF B, Wachtel-Galor S (eds) Herbal medicine: biomolecular and clinical aspects, 2nd edn. CRC Press/Taylor & Francis, Boca Raton, FL. https://www.ncbi.nlm.nih.gov/books/NBK92757/

Wang SY, Hsu ML, Hsu HC et al (1997) The anti-tumor effect of Ganoderma lucidum is mediated by cytokines released from activated macrophages and T lymphocytes. Int J Cancer 70(6):699–705. https://doi.org/10.1002/(sici)1097-0215(19970317)70:6<699::aid-ijc12>3.0.co;2-5

Wang X, Wang B, Zhou L et al (2020) Ganoderma lucidum put forth anti-tumor activity against PC-3 prostate cancer cells via inhibition of Jak-1/STAT-3 activity. Saudi J Biol Sci 27(10):2632–2637. https://doi.org/10.1016/j.sjbs.2020.05.044

Wong JH, Ng TB, Chan HHL et al (2020) Mushroom extracts and compounds with suppressive action on breast cancer: evidence from studies using cultured cancer cells, tumor-bearing animals, and clinical trials. Appl Microbiol Biotechnol 104(11):4675–4703. https://doi.org/10.1007/s00253-020-10476-4

Wu JY, Siu KC, Geng P (2021) Bioactive ingredients and medicinal values of Grifola frondosa (Maitake). Foods 10(1). https://doi.org/10.3390/foods10010095

Ye H, Jensen AB (2017) *Ganoderma lucidum* on the inhibition of colorectal cancer. Hematol. Med Oncol 2:1–4. https://doi.org/10.15761/HMO.1000126

Zhang P, Li K, Yang G et al (2017) Cytotoxic protein from the mushroom Coprinus comatus possesses a unique mode for glycan binding and specificity. Proc Nat Acad Sci 114(34):8980–8985. https://doi.org/10.1073/pnas.1706894114

Zhao F, Wang YF, Song L et al (2017) Synergistic apoptotic effect of D-fraction from Grifola frondosa and vitamin C on hepatocellular carcinoma SMMC-7721 cells. Integr Cancer Ther 16(2):205–214. https://doi.org/10.1177/1534735416644674

Zhao S, Ye G, Fu G et al (2011) Ganoderma lucidum exerts anti-tumor effects on ovarian cancer cells and enhances their sensitivity to cisplatin. Int J Oncol 38(5):1319–1327. https://doi.org/10.3892/ijo.2011.965

The manufacturer's authorised representative in the EU is Springer Nature Customer Service Centre GmbH, Europaplatz 3, 69115 Heidelberg, Germany. If you have any concerns regarding our products, please contact ProductSafety@springernature.com

Printed and bound by CPI Group (UK) Ltd, Croydon, CR0 4YY
30/09/2026
02225299-0011